Table of Atomic Numbers and Atomic Weights

Name	Symbol	Atomic number	Atomic weight
Actinium	Ac	89	(227)
Aluminum	Al	13	26.98
Americium	Am	95	(243)
Antimony	Sb	51	121.8
Argon	Ar	18	39.9
Arsenic	As	33	74.92
Astatine	At	85	(210)
Barium	Ba	56	137.3
Berkelium	Bk	97	(247)
Beryllium	Be	4	9.012
Bismuth	Bi	83	209.0
Bohrium	Bh	107	(264)
Boron	B	5	10.81
Bromine	Br	35	79.90
Cadmium	Cd	48	112.4
Calcium	Ca	20	40.08
Californium	Cf	98	(251)
Carbon	C	6	12.01
Cerium	Ce	58	140.1
Cesium	Cs	55	132.9
Chlorine	Cl	17	35.45
Chromium	Cr	24	52.00
Cobalt	Co	27	58.93
Copper	Cu	29	63.55
Curium	Cm	96	(247)
Dubnium	Db	105	(262)
Dysprosium	Dy	66	162.5
Einsteinium	Es	99	(252)
Erbium	Er	68	167.3
Europium	Eu	63	152.0
Fermium	Fm	100	(257)
Fluorine	F	9	19.00
Francium	Fr	87	(223)
Gadolinium	Gd	64	157.3
Gallium	Ga	31	69.72
Germanium	Ge	32	72.61
Gold	Au	79	197.0
Hafnium	Hf	72	178.5
Hassium	Hs	108	(265)
Helium	He	2	4.003
Holmium	Ho	67	164.9
Hydrogen	H	1	1.008
Indium	In	49	114.8
Iodine	I	53	126.9
Iridium	Ir	77	192.2
Iron	Fe	26	55.85
Krypton	Kr	36	83.80
Lanthanum	La	57	138.9
Lawrencium	Lr	103	(262)
Lead	Pb	82	207.2
Lithium	Li	3	6.941
Lutetium	Lu	71	175.0
Magnesium	Mg	12	24.31
[illegible]ganese	Mn	25	54.94
[illegible]	Mt	109	(268)
[illegible]	Md	101	(258)
Mercury	Hg	80	200.6
Molybdenum	Mo	42	95.94
Neodymium	Nd	60	144.2
Neon	Ne	10	20.18
Neptunium	Np	93	(237)
Nickel	Ni	28	58.69
Niobium	Nb	41	92.91
Nitrogen	N	7	14.01
Nobelium	No	102	(259)
Osmium	Os	76	190.2
Oxygen	O	8	16.00
Palladium	Pd	46	106.4
Phosphorus	P	15	30.97
Platinum	Pt	78	195.1
Plutonium	Pu	94	(244)
Polonium	Po	84	(209)
Potassium	K	19	39.10
Praseodymium	Pr	59	140.9
Promethium	Pm	61	(145)
Protactinium	Pa	91	(231)
Radium	Ra	88	(226)
Radon	Rn	86	(222)
Rhenium	Re	75	186.2
Rhodium	Rh	45	102.9
Rubidium	Rb	37	85.4
Ruthenium	Ru	44	101.1
Rutherfordium	Rf	104	(261)
Samarium	Sm	62	150.4
Scandium	Sc	21	44.96
Seaborgium	Sg	106	(263)
Selenium	Se	34	78.96
Silicon	Si	14	28.09
Silver	Ag	47	107.9
Sodium	Na	11	22.99
Strontium	Sr	38	87.62
Sulfur	S	16	32.07
Tantalum	Ta	73	180.9
Technetium	Tc	43	(98)
Tellurium	Te	52	127.6
Terbium	Tb	65	158.9
Thallium	Tl	81	204.4
Thorium	Th	90	232.0
Thulium	Tm	69	168.9
Tin	Sn	50	118.7
Titanium	Ti	22	47.88
Tungsten	W	74	183.9
Ununnilium	Uun	110	(272)
Unununium	Uuu	111	(272)
Ununbium	Uub	112	(277)
Uranium	U	92	238.0
Vanadium	V	23	50.94
Xenon	Xe	54	131.3
Ytterbium	Yb	70	173.0
Yttrium	Y	39	88.91
Zinc	Zn	30	65.39
Zirconium	Zr	40	91.22

[illegible]ss number of the isotope of longest half-life.

Available for use with this book:

Two dynamic new learning environments: *Chemistry Interactive 3.0 CD-ROM* and *The Chemistry Place,* an interactive chemistry web site.

CHEMISTRY: INTERACTIVE VERSION 3.0 FOR INTRODUCTORY CHEMISTRY

The first CD-ROM directly tied to an introductory chemistry book, it is designed to deepen students' conceptual understanding and enhance their problem-solving skills. It contains:

- electronically generated problem sets modeled after examples from the text
- tutorials
- animations
- videos
- molecular models

A student can work through a tutorial on ionic formulas, view an animation of an ionic solid dissolving in water, generate problems involving precipitation reactions, or view a hydronium ion from various angles.

THE CHEMISTRY PLACE (www.chemplace.com)

This new subscription-based web site, created by Peregrine Publishers, is available during the academic year 1998–1999 in conjunction with Ebbing/Wentworth: *Introductory Chemistry,* Second Edition. The site offers a rich source of learning activities that range from interactive tutorials and problem-solving exercises to investigative projects. *The Chemistry Place* also contains:

- animations
- a weekly riddle
- a *Scientific American* connection (which allows students and professors to find articles relevant to chemistry)
- a toolkit (comprised of the Periodic Table of the Elements, a scientific calculator, glossary, and equations and constants)

Introductory Chemistry

Second Edition

Darrell D. Ebbing
Wayne State University

R. A. D. Wentworth
Indiana University

Houghton Mifflin Company
Boston New York

Senior Sponsoring Editor *Richard Stratton*
Senior Associate Sponsor *Sue Warne*
Assistant Editor *Marianne Stepanian*
Senior Production/Design Coordinator *Jill Haber*
Senior Manufacturing Coordinator *Priscilla J. Abreu*
Marketing Manager *Penny Hoblyn*

Warning: This book contains text descriptions of chemical reactions and photographs of experiments that are potentially dangerous and harmful if undertaken without proper supervision, equipment, and safety precautions. DO NOT attempt to perform these experiments relying solely on the information presented in this text.

Credits: A list of credits precedes the index.

Cover Design: Stoltze Design
Cover Photograph: Frosted Fallen Fall Leaves, © J. Jamsen/Natural Selection.

Printed in the U.S.A.

Library of Congress Catalog Card Number: 97-72467
ISBN: 0-395-87118-2

3456789-VH-01 00 99

Contents in Brief

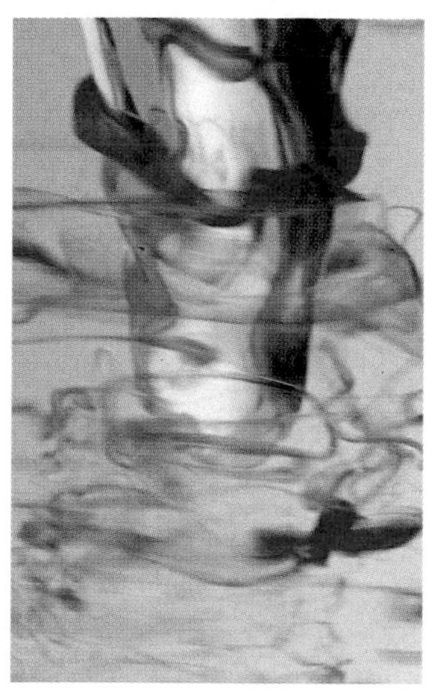

Contents

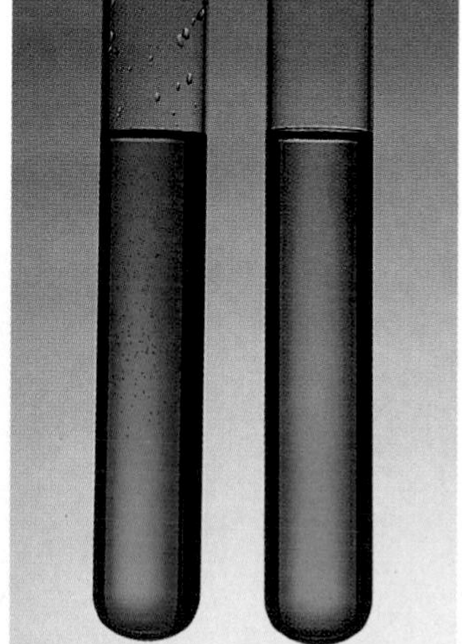

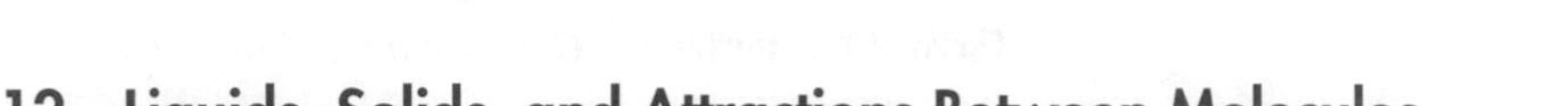

Preface

In writing *Introductory Chemistry,* we have had two types of students in mind. One includes those with no previous background in chemistry, who need an understanding of the basic principles of chemistry to pursue their particular career goals and interests. The second type of student includes those who, while perhaps having had a previous chemistry course, require a thorough review of the basic principles as preparation for taking a general chemistry course.

Though this is a diverse group of students, their requirements seem clear. They want to learn the *skills* needed to advance in their studies. They want to be shown how chemistry is *relevant* to their particular interests and careers. And they are looking for a text that is not only informative but *engaging* and *up to date.* In summarizing what we have done to satisfy these requirements, four areas of the book stand out.

- **Problem-Solving Skills.** Traditionally, textbooks for these courses have relied heavily on problem solving. While we agree with the importance of problem solving to understanding chemistry, we believe it is most important that students acquire *problem-solving skills.* They need explicit help in seeing how to approach a problem. We feel that this requires an approach in which both the text and the worked-out examples help develop these skills. Therefore, every example problem statement is followed by a *problem analysis,* which walks students through the thinking process involved in solving the problem. Only after this problem analysis do we present the full solution to the problem.
- **Conceptual Understanding.** Most instructors agree that problem solving is an important element in chemistry, but the manner in which many textbooks deal with problem solving frequently leads students to see it as a rote process. They memorize specific methods and apply them uncritically to problems. Students need to acquire a vivid feel for the underlying concepts of chemistry so that they can properly place a particular problem in its context. We have directed the text and illustration programs to promote this type of understanding. For instance, the idea of the molecule is central to a chemist's thinking. Early in the book, we introduce molecular models, and we continue to build on the molecular concept. So, before we talk about the gas laws, we describe the kinetic-molecular theory of gases. This gives the student a vivid picture of a gas, which the student can use when approaching gas-law problems.
- **Applications of Chemistry.** We answer students' need to see how chemistry is relevant by introducing everyday applications of chemistry throughout the text, in illustrations, and in boxed essays. We have chosen these applications for student interest and have tried to use an engaging style to capture students' attention. The number of boxed essays *(Chemical Perspectives)* has been nearly doubled with this edition, and the topics are up-to-date and range over many fields, such as nitric oxide as a biochemical messenger (biochemistry, medicine); digital X-ray photography (technology); the greenhouse effect (environment); carbon monoxide and hemoglobin (toxicology); and dinosaurs, human origins, and ancient molecules (archaeology).

■ **Multimedia Package.** We are pleased to be the first introductory chemistry book to offer a CD-ROM directly tied to it. *Chemistry: Interactive 3.0 for Introductory Chemistry* contains electronically generated problem sets modeled after examples from the text, as well as many tools—such as tutorials, animations, videos, and molecular models—to help the student learn problem solving and deepen conceptual understanding. For instance, a student can work through a tutorial on ionic formulas, view an animation of an ionic solid dissolving in water, generate problems involving precipitation reactions, or view the hydronium ion from various angles. The instructors' version of the CD provides many lecture aids, including selected artwork and tables from the text, and an easy-to-use classroom presentation tool. Together, these electronic multimedia supplements provide students and instructors with the most up-to-date learning and teaching environment.

OUTSTANDING FEATURES OF THE BOOK

We have been gratified by the positive comments from users of the first edition, and we have worked hard to retain and strengthen the features of *Introductory Chemistry* that help students develop problem-solving skills, understand fundamental chemical concepts, and appreciate the applications of chemistry to their everyday life. The following features have all been designed to contribute to achieving these goals.

Writing Style

Because many students view chemistry as a challenging subject, we have adopted a friendly, direct writing style that will appeal to students. To aid students in their reading, we have consciously erected verbal signposts that remind them where they have been and where they are going. To show students that chemistry is indeed a subject that is useful and relevant to them, we have worked in as much everyday chemistry as possible. Finally, we have tried to make it clear through organization and illustration that in chemistry, concepts build one on another and each topic introduced has a purpose in developing the subject.

Problem Solving

Problem-solving skills are important because many everyday questions require them; they are also important to an understanding of chemical concepts. Developing these skills requires the student to understand how to approach problems, and it requires practice in solving problems. We have directed much of our effort to helping students gain these problem-solving skills.

When faced with a problem, many beginning students ask, "How do I begin?" To help these students, we included the **Problem Analysis** section with the **Examples** in the text. This problem analysis follows each problem statement, and its purpose is to show the student how to think the problem through, using the information given in the problem and using relevant principles from the text. The **Solution** applies this problem analysis to the specific problem statement. As students work through the problem analysis and solution of many problems, they will begin to see that problem solving is not a matter of grasping thoughts from thin air, but that it involves a process that can be learned.

Learning problem solving involves more than simply reading about how some problems are solved. Students have to solve numerous problems on their own. We have helped by following each worked-out example with a similar **Exercise** and then directing students to similar problems at the end of the chapter (**Try problems . . .).** The end-of-chapter **Practice Problems** are similar to the problems solved in the examples, though some practice problems provide a variation or extension of the problem-solving skill used in the examples. The **Additional Problems** provide more practice and in many cases more difficult problems. All of these problems are in matched pairs; every odd-numbered problem is answered at the back of the book and is followed by a similar even-numbered problem (unanswered).

With this edition, we have added **Practice in Problem Analysis** and **Practice Exam** sections to the end-of-chapter problems. In the Practice in Problem Analysis, we ask students to describe the thinking, or problem analysis, involved in a problem, rather than to solve the problem. Our aim is to get students to focus on this problem analysis and take them away from rote thinking. The purpose of the Practice Exam is to provide students with practice in taking multiple-choice exams.

Study Aids

The job of the textbook is to help students see the connections among ideas and the significance of what they have read. We have done our best to guide students through a chapter by designing a system of *study aids* that they can adapt to their own study program.

Each chapter begins with an **Outline** that shows the structure of the chapter. Note that chapters are divided into *parts,* which are divided into *sections* and *subsections.* This outline can help students, both as they read the chapter for the first time and as they review, to see how topics are related. Each section of a chapter begins with learning **Objectives.** These objectives list concepts and problem-solving skills introduced in the section, and so provide an overview of the significant points in the section.

Throughout the text, **marginal notes** give students additional historical and descriptive information to help engage them in the topic. These notes also provide cross-references to other sections of the text for students to use in reviewing background information.

Chemistry requires that students learn a special vocabulary for expressing ideas. Every time we introduce a **key word,** we note the word by setting it in **boldface type** and following it with an explicit definition of the word in *italic type,* so that students do not have to guess the exact meaning of key words. These key words and some others are gathered in a **Glossary** at the end of the book.

An extensive **Chapter Review** appears at the end of each chapter. **Key Words** are listed, as well as the **Key Equations** appearing in the chapter. Following these lists is a **Summary,** a condensation of the important points covered in the chapter. **Problem-Solving Skills** is a summary list of the different problem-solving skills introduced in the chapter. Each skill is keyed to the worked-out *Examples* in the text that use that problem-solving skill.

Chapter questions follow these review features. Students should be able to answer these **Questions to Test Your Reading** after reading (and perhaps rereading) the chapter. The questions are followed by the end-of-chapter problems described above.

Applications to Chemistry

While our aim is to cover the basic principles of chemistry, we strongly believe that students need to see how these principles relate to the world around them. The principles must appear in the context of everyday chemistry. We have brought applications of chemistry into the book in several ways. First, we have woven applications into the discussion of principles. Where possible, we have used a chemical application as a thread for our discussion, both for interest and to help the students see relationships among principles. Also, we have introduced applications into the examples and problems to give them significance.

The **Chemical Perspectives** are a special feature of the book. These consist of short essays appearing in boxes in the chapter. Each Chemical Perspective deals with a topic of everyday interest—perhaps an environmental concern or chemistry applied to medicine. A Chemical Perspective uses one or more principles in the chapter to illustrate the application of those principles. We have adopted an informal writing style that will serve to engage the students' interest and make for enjoyable reading.

Illustration Program

Chemistry is about three-dimensional objects, and chemical concepts are often best taught with images to reinforce the words. We have gone to great lengths to ensure that the illustration program meets our twofold goal of appropriateness to the topic and of visual interest. The style of many of the figures is, we believe, fresh and modern. For example, zoom-sequence photos are used in some cases to allow students to compare the macroscopic aspects of a piece of matter to its submicroscopic structure. Pictures of a mineral are inset into a photograph of the country where the mineral is mined. Molecular models are generally computer-generated color images. Atoms in these images have been color-coded so that an atom will have the same color throughout the book. In short, much of the photography and artwork is designed to help the student learn from it, as well as enjoy it.

Alternate Edition

This text is also available in a paperback edition, *Fundamentals of Introductory Chemistry,* which contains the first thirteen chapters, covering basic concepts, chemical substances and reactions, atomic structure and bonding, and states of matter.

COMPLETE INSTRUCTIONAL PACKAGE

This text is accompanied by a complete package of instructional materials designed to help instructors and students alike.

For the Student

A **CD-ROM,** *Chemistry: Interactive 3.0 for Introductory Chemistry,* supports the goals of the second edition by helping students visualize molecular behavior and manipulate molecules in three dimensions. Tutorials, animations, videos, molecular models, and problem sets are included. The problem sets will also be available on floppy disk. See your Houghton Mifflin sales representative for more information.

The **Study Guide,** by Susan M. Schelble of the University of Southern Colorado, provides discussion of all key concepts and topics, with worked-out examples incorporated. Important terms and algebra/calculator review are included for each chapter, and additional practice problems are given for all problem-solving skills. Chapters conclude with a practice exam.

The **Lab Manual,** by R. A. D. Wentworth and Karen Pressprich, Indiana University, contains 22 experiments class-tested by hundreds of students. At least one experiment is included for each text chapter, and safety is emphasized throughout.

The **Partial Solutions Manual,** by David Bookin of Mt. San Jacinto College, contains worked-out solutions to all in-chapter exercises and half of the end-of-chapter problems, using strategies emphasized in the text. To ensure the accuracy of the solutions, this supplement and the *Complete Solutions Manual* were checked independently by several instructors.

For the Instructor

A **CD-ROM** product contains the animations, videos, and molecular models that appear on the student version of the CD-ROM. In addition, transparency acetates from the text are included in electronic form, along with a simple-to-use classroom presentation program. The disc is specifically designed to allow instructors to facilitate active learning and to enhance multimedia classroom presentations. The classroom presentation program and electronic versions of the transparencies are also available on floppy disk. See your Houghton Mifflin sales representative for additional information.

The **Complete Solutions Manual,** by David Bookin, contains worked-out solutions to all in-chapter exercises and end-of-chapter problems. Complete answers are also provided for the Questions to Test Your Reading. This is for the convenience of faculty and staff involved in instruction and for instructors who want their students to have solutions for all of the exercises. Departmental approval is required for the sale of the *Complete Solutions Manual* to students.

The **Instructor's Resource Manual with Test Bank,** by Darrell D. Ebbing, R. A. D. Wentworth, Richard S. Perkins, and Leslie N. Kinsland (the latter two of the University of Southwestern Louisiana), is a comprehensive teaching aid that includes chapter descriptions, lecture demonstration sources, and a 1600-item multiple-choice test bank.

The **Computerized Test Bank** is an electronic version of the 1600-item test bank, available in DOS, Windows, and Macintosh versions.

The **Instructor's Resource Manual to Accompany the Lab Manual,** by Karen Pressprich, includes suggested equipment and materials, time requirements, answers to pre-lab and post-lab questions, and sample student data for all experiments.

The **Transparency Package** contains 100 two- and four-color acetates from images and line art used in the text.

Houghton Mifflin Chemistry Videodisc contains video clips of lecture demonstrations and animations of important chemical processes and concepts that can be used in classroom presentations. The disc is available free to adopters of the second edition.

Houghton Mifflin Videotape Series A, B, C, and D provide over 100 lecture demonstrations performed by John Luoma, Cleveland State University; John J. Fortman and Rubin Battino, Wright State University; Patricia L. Samuel, Boston University; and Paul Kelter, University of Nebraska—Lincoln. Series C demonstrations appear on the Houghton Mifflin videodisc as well.

ACKNOWLEDGMENTS

The preparation of an introductory book requires the work of many people. Reviewers have read and commented at various stages of the manuscript, and their work has been indelibly etched in the final text. We owe these people a debt beyond measure. Following is a list of these contributors in alphabetical order:

David W. Ball
Cleveland State University

Hal Bender
Clackamas Community College

Kyle A. Beran
Mesa State College

Michael J. Brabec
Eastern Michigan University

Helen Chichester
College of Alameda

Darwin B. Dahl
Western Kentucky University

Chhabil Dass
The University of Memphis

James E. Hardcastle
Texas Women's University

Kathleen Harter
Community College of Philadelphia

James N. Jacob
University of Rhode Island

Charles Keilin
Larncy College

Leslie N. Kinsland
The University of Southwestern Louisiana

Susanne M. Mathews
Joliet Junior College

Hugh McLean
Florida A&M University

David Nachman
Phoenix College

Barbara Rainard
Community College of Allegheny County

David E. Saleeby
Florence Darlington Technical College

Susan M. Schelble
University of Southern Colorado

Jeffrey A. Schneider
State University of New York at Oswego

Joel Shelton
Bakersfield College

Linda K. Thomas-Glover
Guilford Technical Community College

Vivek Utgikar
The University of Dayton

Robert O. Wiley
Arizona State University

The technical accuracy of the book owes much to Beverly Foote; Catherine Keenan, Chaffey College; John Nash, Purdue University; and David Treichel, Nebraska Wesleyan University, who read all or parts of the book.

Our special thanks go to Richard Stratton, Senior Sponsoring Editor, and Sue Warne, Senior Associate Sponsor, who worked with us from our first planning session in Boston to the final stages of production. Richard and Sue provided us with many creative ideas and the enthusiasm that made this revision possible. Marjorie Anderson was our Developmental Editor for this revision. Her comments and suggestions were most important to us as we produced the manuscript; we are most grateful for her efforts. We also wish to thank Marianne Stepanian, Assistant Editor. Marianne was instrumental in overseeing the development of the ancillary package. When authors sign off on copyedited manuscript, the rigors of production begin. We were very pleased to have Peggy J. Flanagan again as Production Editor. Her attention to the details of turning manuscript to book pages was unflagging, and we are most grateful to her. Douglas Sawyer of Scottsdale Community College did the setup photography for this revision, as he had for the first edition. We thank him for turning our rough ideas into actual photographs. We also thank Jessyca Broekman, Art Editor, Sharon Donahue, Photo Researcher, Jill Haber, Senior Production/Design Coordinator, and Janet Theurer, Designer, for their contributions to the "look" of the finished book.

Finally, we wish to dedicate this book to our wives, Jean Ebbing and Anne Fraker, who by their wit and good humor managed to keep us on an even keel through the inevitable windless days and stormy periods of the writing process.

384 Chapter 12 Liquids, Solids, and Attractions Between Molecules

FIGURE 12.16
The structure of sodium chloride, an ionic solid. Every chloride ion is surrounded by six smaller sodium ions. The chloride ions are approximately close-packed with the much smaller sodium ions in the cavities between the chloride ions.

Wherever possible, artwork links the macroscopic everyday world with microscopic and atomic level illustrations to help students understand what atoms and molecules are and how they behave.

ordered array of molecules together in a molecular solid always include dispersion forces, but dipole-
the melting points
ing points of liqui
often liquids (suc
ture.

Ionic Solids

Ionic solids such a
exist between op
between the part
dipole–dipole for
higher than the m
Unlike molecular
Ionic forces are
can be surrounded
chloride is a good
eral other solids,
However, other ar
The next two e
in them affect thei

234 Chapter 8 Quantities in Chemical Reactions

FIGURE 8.9
Cheese sandwiches, stoichiometric amounts, and limiting reactants. Making cheese sandwiches provides an analogy for reactions with *(A)* stoichiometric amounts and *(B)* a limiting reactant (here, the cheese).

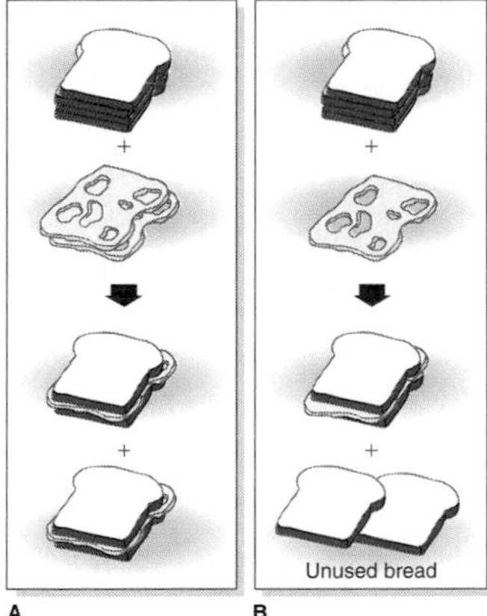

Artwork also helps students get an intuitive feel for the concepts underlying the chemistry.

reactants in stoichiometric amounts, both will be consumed at the completion of the reaction. You can calculate the mass of product obtained from either reactant. On the other hand, if the reactants are not in stoichiometric amounts, there is a limiting reactant. It will be used up and the other reactant will be left over when the reaction is complete. You calculate the amount of any product from the amount of the limiting reactant present at the start of the reaction.

How can you tell if 2.0 mol $Mg(OH)_2$ and 5.0 mol HCl are stoichiometric amounts of reactants? One way is to use the amount of one reactant, say, $Mg(OH)_2$, to calculate the amount of the other reactant (HCl) that would be needed to react with it. If the amount of HCl that you calculate is exactly equal to the amount actually added at the start, the two reactants are present in stoichiometric amounts. Let's do the calculation.

You need to make the following conversion:

2.0 mol $Mg(OH)_2$ —Converts to→ mol HCl

Here is the calculation:

$$2.0\ \cancel{\text{mol Mg(OH)}_2} \times \frac{2\ \text{mol HCl}}{1\ \cancel{\text{mol Mg(OH)}_2}} = 4.0\ \text{mol HCl}$$

In other word… mol $Mg(OH)_2$ reac… with 4.0 mol HCl. Sinc… actually added 5.0 mol HC… dd stoichi… nts of reac…

the gas and the atmosphere. For example, suppose the atmospheric pressure is 751 mmHg and the height of mercury in the arm open to the gas is 63 mm greater than in the end open to the atmosphere. The gas pressure must be 63 mmHg *less* than atmospheric pressure, or

$$P = 751 \text{ mmHg} - 63 \text{ mmHg} = 688 \text{ mmHg}$$

Animations: Kinetic molecular theory and heat transfer; Visualizing molecular motion (single molecule); Visualizing molecular motion (many molecules).

11.3 The Kinetic Molecular Theory of Gases

OBJECTIVE

- Name the theory that explains the behavior of real gases under moderate conditions, and list its four postulates.

The **kinetic molecular theory of gases,** developed during the eighteenth and nineteenth centuries by several scientists, *explains the behavior of gases.* As we will see in Chapter 12, part of this theory can also be used to explain the behavior of liquids and solids.

The kinetic molecular theory applies to *a gas that obeys a simple mathematical relationship between its pressure, volume, temperature, and amount.* ● This gas is called an **ideal gas** and it does not really exist. We can make mathematical predictions about the behavior of an ideal gas, and these predictions will also apply to a real gas as long as moderate conditions of pressure and temperature are maintained. Moderate conditions are pressures that are less than or not much greater than 1 atm and temperatures that are considerably greater than those at which the gases liquefy. We cannot be more specific about these conditions; they vary from gas to gas because the properties of one substance will always differ in many respects from those of another substance. If a real gas is subjected to more extreme conditions, namely, high pressure or low temperature, then deviations from ideal behavior are noticeable. Throughout this chapter, we assume ideal behavior due to moderate conditions. Let's consider the postulates in this theory.

● This relationship, called the *ideal gas law,* is introduced in Section 11.8.

The text promotes an understanding of the underlying concepts of chemistry. For example, the kinetic-molecular theory of gases is discussed before gas law calculations are introduced, to give students a conceptual understanding of gases that will help them understand the gas laws.

Icons in the margin point students to multimedia resources on the CD-ROM that will help deepen conceptual understanding and strengthen problem-solving skills.

he *-ite* ending to *-ous,* and substitute the word *acid* for the word *ion.*

Solution

The formula for the sulfate ion is SO_4^{2-}. Since its charge is −2, the subscript to H in the oxyacid formula is 2. The result is H_2SO_4. To name this compound, change the *-ate* ending in the name of the oxyanion to *-ic* and then substitute the word *acid* for the word *ion.* Thus, the name is sulfuric acid, as shown in Table 5.5.

Exercise 5.10

Give the formula and name of the oxyacid corresponding to the sulfite ion.

(Try Problems 5.45, 5.46, 5.47, and 5.48.)

The number of **Chemical Perspective** boxes has been nearly doubled with this edition, and the topics are up-to-date and cover many fields: for instance nitric oxide as a biochemical messenger (biochemistry, medicine), digital X-ray photography (technology), the greenhouse effect (environment), carbon monoxide and hemoglobin (toxicology), dinosaurs, human origins, and ancient molecules (archaeology).

Chemical Perspective

Beetles, Antiseptics, and Bleaches

The bombardier beetle ensures its survival with a marvelous defense mechanism (Figure 5.10). Unlike the passive defense of the chameleon, who changes its color to match its surroundings, this insect uses chemical warfare against its predators. When it is threatened, it forces a fluid containing hydrogen peroxide and another compound called hydroquinone into an abdominal sac containing enzymes. (Enzymes are "biological catalysts"—that is, biological substances that cause a chemical reaction to speed up without being consumed in the reaction.) The enzymes cause a vigorous and quick reaction between hydrogen peroxide and hydroquinone with the release of a great deal of heat—so much heat that the liquid boils. The beetle discharges the hot liquid and vapor as a fine mist toward the soon-to-be-surprised predator.

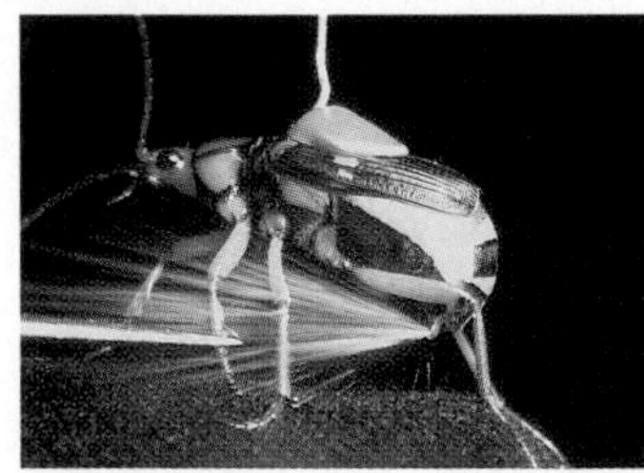

FIGURE 5.10
A bombardier beetle defending itself. The reaction between hydrogen peroxide and hydroquinone in the presence of a catalyst generates enough heat to boil the reaction liquid. The beetle expels this steaming liquid toward a foe.

Hydrogen peroxide, one of the substances used by the beetle, is a very reactive compound. We use it as an antiseptic because it kills harmful microorganisms. We also use it to bleach hair because it reacts with natural pigments and destroys them. Although it is toxic, it is produced in our bodies through biological processes involving oxygen from the air we breathe. Fortunately, we have a biological safeguard that prevents the internal buildup of this substance. Once again, enzymes play an important role. This time they speed up the natural decomposition of hydrogen peroxide, producing water and oxygen gas. One of these enzymes is *catalase.* The effect of catalase from beef liver on hydrogen peroxide is shown in Figure 5.11. The froth you see is caused by bubbles of oxygen from the decomposing hydrogen peroxide.

Water, the other product of the reaction in Figure 5.11, and hydrogen peroxide happen to share one feature: Both are binary molecular compounds of the same elements. They are different compounds, however, because the proportions of these elements differ.

H–O–H	H–O–O–H
H_2O	H_2O_2
Water	Hydrogen peroxide

An extensive **Chapter Review** consists of **Key Words, Key Equations,** and a **Summary. Problem-Solving Skills** is a summary list of the different problem-solving skills introduced in the chapter, each keyed to worked-out *Examples* in the chapter. **Questions to Test Your Reading** provide a basic review of the key concepts of the chapter.

Chapter Review 213

Summary

A *molecular weight* is the sum of the atomic weights of all the atoms in the formula for a molecule of a compound. A *formula weight* is the sum of the atomic weights of all the atoms in the formula unit of a compound. If the formula unit is the formula for a molecule, then the formula weight is identical to the molecular weight.

A *mole* of any element or compound contains *Avogadro's number* (6.022×10^{23}) of atoms, molecules, or formula units. The molar mass of a substance equals the mass of 1 mole of that substance. For an element with a nonmolecular structure, this molar mass equals the atomic weight of that element expressed in grams. If the element occurs as molecules, the molar mass is the molecular weight in grams. For a molecular compound, the molar mass is the compound's molecular weight in grams; for an ionic compound, the molar mass is the compound's formula weight in grams.

The *empirical formula* of a compound is the simplest formula for that substance. It is obtained from the *percentage composition* of the compound, which is expressed as *mass percentages* of the elements that make up the compound. These mass percentages can be determined by chemical analysis in a laboratory, or they can be calculated from the chemical formula if it is known. When you calculate an empirical formula, you convert the mass percentages to ratios of moles that lead you to the subscripts in the formula. A *molecular formula* is a multiple of the empirical formula. An experimental measurement of the molecular weight is required to obtain this multiple.

Problem-Solving Skills

1. **Computing a Molecular Weight from a Formula:** Given a compound's molecular formula, calculate the molecular weight (Example 7.1).
2. **Computing a Formula Weight from a Formula:** Given an ionic compound's formula, calculate the formula weight (Example 7.2).
3. **Determining the Number of Molecules a**[...] **from the Moles of Substance:** Given the a[...] substance in moles, calculate the number of[...] and atoms in the sample (Example 7.3).
4. **Obtaining the Molar Mass of an Element**[...] element, calculate its molar mass (Example 7.[...]
5. **Calculating the Molar Mass of a Compoun**[...] compound, calculate its molar mass (Example[...]
6. **Converting Grams to Moles:** Given a substa[...] in grams, calculate the amount of that substan[...] (Example 7.6).
7. **Converting Moles to Grams:** Given the qu[...] substance in moles, calculate the mass in gram[...] 7.7).
8. **Calculating Mass Percentages from a Formula:** Given the formula of a compound, calculate the mass percentage of each element in the compound (Example 7.8).
9. **Comparing Mass Percentages:** Given two compounds containing a common element, determine which one contains the greater amount of the element per gram of compound (Example 7.9).

Questions to Test Your Reading

7.1 What is the difference between a formula w[...] molecular weight? Define each one.

7.2 Could one substance have both a formula w[...] molecular weight? Explain.

7.3 What is Avogadro's number? What is Avoga[...] ber of chlorine atoms? What is Avogadro's[...] diatomic chlorine molecules?

7.4 What is a mole? What is the molar mass[...]

12.6 The Solid State 385

■ EXAMPLE 12.9 Identifying Types of Solids

Sulfur dioxide (SO_2) is an air pollutant; potassium iodide (KI) is added to common table salt by the distributor to provide iodine in our diets. Identify the type of solid you expect for each substance.

Problem Analysis

Notice that both substances are compounds. Distinguish between ionic and covalent compounds.

Solution

Sulfur dioxide is a molecular substance (both sulfur and oxygen are nonmetals); therefore, it freezes as a molecular solid. Potassium iodide is an ionic substance (potassium is a metal, and iodine is a nonmetal); it exists as an ionic solid.

Exercise 12.9

Classify each of the following solids according to its type: zinc (Zn), sodium bromide (NaBr), and methane (CH_4).

(Try Problems 12.53, 12.54, 12.55, and 12.56.)

■ EXAMPLE 12.10 Determining Relative Melting Points

Does sulfur dioxide or potassium iodide have a higher melting point? Explain your reasoning.

Problem Analysis

Identify the type of solid expected for each substance (see Example 12.9). Determine the types of attractions in each solid and which one has the stronger forces. The compound with stronger attractions between the molecules has a higher melting point.

Solution

As you saw in Example 12.9, sulfur dioxide is a molecular solid, so it is held together by dispersion forces. Potassium iodide is an ionic compound, so it is held together by electrical attractions between oppositely charged ions. Because ionic attractions are stronger than dispersion forces, potassium iodide will have a higher melting point than sulfur dioxide.

Exercise 12.10

Name the type of solid that is formed for each of the following substances: $MgCl_2$, CH_3OH, and Ar. On the basis of the type of solid that you expect, arrange these substances in order of increasing melting point.

(Try Problems 12.57, 12.58, 12.59, and 12.60.)

In-text worked **Examples** guide students through the steps of solving problems

A **Problem Analysis** section shows students how to think the problem through using the information given in the problem and the relevant principles from the text.

The **Solution** applies this problem analysis to the specific problem statement.

Each worked-out Example is followed by a similar **Exercise** and a reference to similar matched-pair problems at the ends of the chapters (**Try Problems . . .**) to give students lots of practice in solving these problems.

The end-of-chapter **Practice Problems** are similar to the problems solved in the Examples, and in some cases provide a variation or extension of the problem-solving skill used in the Examples.

Additional Problems are not keyed to specific topics and are more difficult to solve.

All of these problems are in **matched pairs;** every odd-numbered problem is answered at the back of the book and is followed by a similar even-numbered problem (unanswered).

214 Chapter 7 Chemical Composition

7.7 If hydrogen peroxide has an empirical formula of HO and a molecular weight of 34 amu, what is the molecular formula?

7.8 A compound's molecular formula is $C_6H_{12}O_6$. What is the empirical formula?

Practice Problems

Molecular Weight and Formula Weight (Section 7.1)

7.9 Calculate the molecular weight of each of the following substances.
(a) F_2 (b) PF_5 (c) SO_3
(d) $HC_2H_3O_2$ (e) $C_6H_6O_6$ (f) $C_{12}H_{22}O_{11}$

7.10 Calculate the molecular weight of each of the following substances.
(a) Br_2 (b) SF_6 (c) P_4O_{10}
(d) $H_2C_2O_4$ (e) $C_2H_4Cl_2$ (f) C_3H_7SH

7.11 As grapes ripen, the exceptionally sour taste caused by tartaric acid ($C_4H_6O_2$) disappears as this acid is converted to the sugar glucose ($C_6H_{12}O_6$). Calculate the molecular weight of each of these compounds.

7.12 Indigo ($C_{16}H_{10}N_2O_2$), the dye used to color blue jeans, is derived from a compound known as indoxyl (C_8H_7ON). Calculate the molecular weight of each of these compounds.

7.13 Calculate the formula weight of each of the following compounds.
(a) ZnI_2 (b) $AlBr_3$ (c) C_2H_6
(d) NH_4NO_3 (e) Fe_2O_3 (f) Na_2CrO_4

7.14 Calculate the formula weight of each of the following compounds.
(a) K_2SO_4 (b) $K_2Cr_2O_7$ (c) $Ca_3(PO_4)_2$
(d) $C_{18}H_{36}O_{18}$ (e) Al_2O_3 (f) $CaCO_3$

7.15 Identify the substance that has a formula weight of 125.84 amu.
(a) BCl_3 (b) $MgCl_2$ (c) IF_7
(d) $MnCl_2$ (e) H_3PO_4 (f) $TiCl_4$

7.16 Identify the substance that has a formula weight of 100.46 amu.
(a) $HClO_3$ (b) $BeSO_4$ (c) $PtCl_2$
(d) SF_6 (e) H_2SO_4 (f) $HClO_4$

The Mole and Molar Mass (Sections 7.2 to 7.5)

7.17 How many atoms of each element are in 2.33 mol ethyl mercaptan (C_2H_5SH)?

7.19 How many magnesium ions are in 0.234 mole o nesium nitride?

7.21 What are the molar masses of hydrogen and heli

7.23 What are the molar masses of iron and of sulfur

7.25 What are the molar masses of the following pounds?
(a) CH_3OH (b) PF_5 (c) SO_3
(d) $HC_2H_3O_2$ (e) $C_6H_{12}O_6$ (f) $C_{12}H_{22}O$

7.27 What are the molar masses of the following pounds?
(a) LiH (b) $Mg(ClO_4)_2$ (c) $VOCl$
(d) $Ba_3(PO_4)_2$ (e) C_3H_7Br (f) K_2SO

7.29 How many moles are in 1.00 g of each of the foll elements?
(a) hydrogen (b) lithium (c) sodium
(d) potassium (e) rubidium (f) cesium

ced equ ... s reacti ... the ... How sical states of ach pounds. ... 6 mo ... nds.

Practice in Problem Analysis

For each problem, describe the thinking you would use (the problem analysis) before doing the actual solution, but do not solve the problem.

1. Write a chemical equation that describes how you would prepare barium sulfate from potassium sulfate and another reactant by a precipitation reaction.

2. Write a chemical equation that describes how you would prepare sodium hydroxide from calcium hydroxide and another reactant by a precipitation reaction.

PRACTICE EXAM

1. Consider the chemical equation

$$2H_2O_2(aq) \xrightarrow{Br^-} 2H_2O(l) + O_2(g)$$

Which of the following statements best fits the description given by this equation?
(a) Hydrogen peroxide liquid decomposes to water and oxygen gas.
(b) Hydrogen peroxide solution decomposes with a bromide ion catalyst to give liquid water and oxygen gas.
(c) Hydrogen peroxide liquid decomposes with a bromide ion catalyst to liquid water and oxygen gas.
(d) Hydrogen peroxide solution reacts with bromide ion to give liquid water and oxygen gas.
(e) Hydrogen peroxide decomposes to water and oxygen.

2. Consider the following chemical equation.

$$2Na(s) + 2H_2O(l) \longrightarrow 2NaOH(aq) + H_2(g)$$

Which of the following statements is not true?
(a) Solid sodium reacts with water.
(b) Sodium hydroxide is a product.
(c) The coefficient of hydrogen is one.
(d) As indicated by the (*aq*) on the right, water is a product.
(e) The products are a solution and a gas.

3. Which of the following equations is not balanced?
(a) $NO(g) + O_2(g) \longrightarrow NO_2(g)$
(b) $2Na(s) + Cl_2(g) \longrightarrow 2NaCl(s)$
(c) $HBr + KOH \longrightarrow KBr + H_2O$
(d) $Zn(s) + 2HBr(aq) \longrightarrow ZnBr_2(aq) + H_2(g)$
(e) None of the above

4. Nickel metal reacts with hydrochloric acid to produce hydrogen and a solution of nickel chloride. Which of the following is a balanced equation for this reaction?
(a) $Ni(s) + 2HCl(g) \longrightarrow NiCl_2(aq) + H_2(g)$
(b) $Ni(s) + 2HCl(aq) \longrightarrow NiCl_2(s) + H_2(g)$
(c) $Ni(s) + 2HCl(aq) \longrightarrow NiCl_2(aq) + H_2(g)$
(d) $2Ni(s) + 2HCl(aq) \longrightarrow 2NiCl_2(aq) + H_2(g)$
(e) None of the above

5. Which of the following is a combination reaction?
(a) $Zn(s) + CuBr_2(aq) \longrightarrow ZnBr_2(aq) + Cu(s)$
(b) $2SO_2(g) + O_2(g) \longrightarrow 2SO_3(g)$
(c) $PbO_2(s) \longrightarrow Pb(s) + O_2(g)$
(d) $2HI(aq) + Na_2S(aq) \longrightarrow H_2S(g) + 2NaI(aq)$
(e) None of the above

6. Which of the following is a decomposition reaction?
(a) $Zn(s) + CuBr_2(aq) \longrightarrow ZnBr_2(aq) + Cu(s)$
(b) $SO_2(g) + H_2O(l) \longrightarrow H_2SO_3(aq)$
(c) $PbO_2(s) \longrightarrow Pb(s) + O_2(g)$
(d) $2HI(aq) + Na_2S(aq) \longrightarrow H_2S(g) + 2NaI(aq)$
(e) None of the above

7. Which of the following is a single-replacement reaction?
(a) $Zn(s) + CuBr_2(aq) \longrightarrow ZnBr_2(aq) + Cu(s)$
(b) $SO_2(g) + H_2O(l) \longrightarrow H_2SO_3(aq)$
(c) $PbO_2(s) \longrightarrow Pb(s) + O_2(g)$
(d) $2HI(aq) + Na_2S(aq) \longrightarrow H_2S(g) + 2NaI(aq)$
(e) None of the above

Practice in Problem Analysis asks students to describe the thinking involved in a problem, rather than solving the problem. The goal is to get students to focus on problem analysis and take them away from rote thinking.

Practice Exams appear at the end of every chapter and provide students with practice in taking multiple-choice exams.

Chapter

1 Introduction to Chemistry

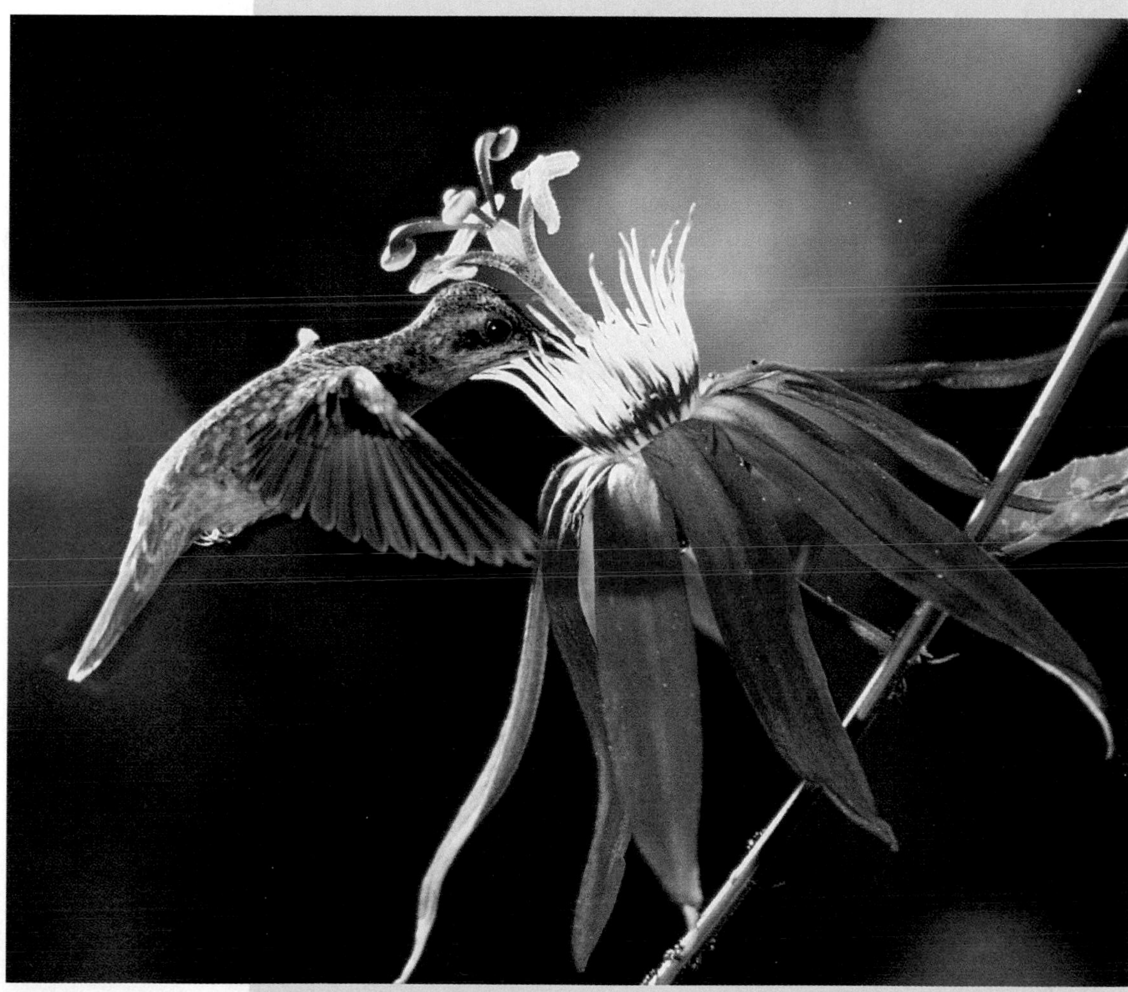

This hummingbird is able to move its wings because of energy provided by chemical reactions in its body.

Chapter Outline

All around you are chemical substances: the air you breathe, the water you drink, the ground you walk on. Anything you can touch or see or smell is composed of chemical substances, including plants and animals and even yourself. Chemistry is the study of these substances. Chemistry is also the study of *chemical reactions* (or chemical changes), in which one set of substances changes into other substances. Chemical reactions may be subtle, as when trees grow in a forest; or they may be more dramatic, as when the forest burns (Figure 1.1).

Chemistry, as you might guess, is an enormous subject, and its influence on the modern world has been equally enormous. Look around you. Perhaps your clothes are made from synthetic fibers of polyester or nylon. These materials are products of chemical processing: Starting with petroleum, air, and water, the chemist or chemical engineer uses various chemical reactions to transform these substances to polyester or nylon. And

FIGURE 1.1
Chemical reactions. *(A)* Subtle chemical reactions are responsible for the growth of the trees in this forest, but the forest fire *(B)* is an example of chemical reactions that cause dramatic changes.

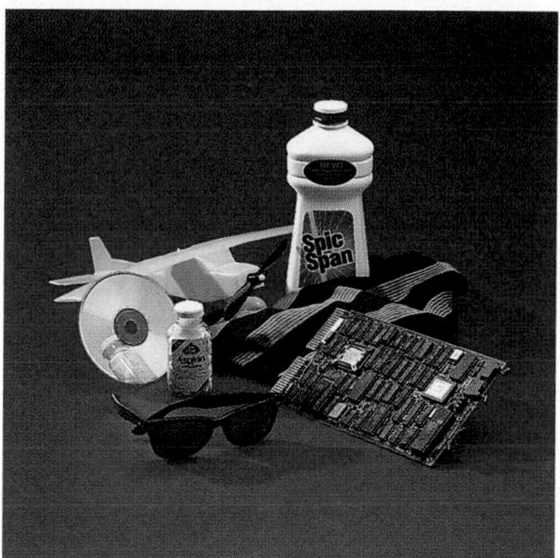

FIGURE 1.2
Products of chemistry. We use chemistry to manufacture many different products.

the dyes used to color these fibers—they, too, are the products of such chemical processing. Almost any manufactured product you can think of—medicines, computer chips, lasers—requires the use of chemistry at some stage of its production (Figure 1.2).

Just as important as these industrial applications, however, are the concepts of chemistry. We explain the world around us through these concepts. For example, what is the origin of the human disease sickle-cell anemia? Why is the presence of ozone in the stratosphere so important to us? Why do scientists think that the dinosaurs died out soon after an asteroid hit the earth about 65 million years ago? These are complex questions, and chemistry has been essential in finding the answers.

In this chapter, we have three objectives. First, we describe in more detail what we mean by chemistry and discuss the role of chemistry in science and technology. Next, we trace the history of chemistry to see how it evolved from the happenstance use of materials to a science based on experiment. Finally, we discuss the interplay of experiment and theory, generally known as the *scientific method.*

1.1 The Science of Chemistry

OBJECTIVES

- Define *matter.*
- Give examples of chemical reactions and describe their characteristics.
- Define *chemistry* and state the major objectives of this science.
- List ways in which chemistry has contributed to other sciences and technology.

Chemistry is a branch of science, which is the systematic study and explanation of natural phenomena. Briefly, chemistry is the study of matter. **Matter** refers to *the material things around you:* this book, the pencil you are writing with, the chair you are sitting in, and the air you are breathing. ● Matter can be a single chemical substance or a complex mixture of substances.

● A **boldface** word is a key word; its definition appears in *italic type.*

Let us define more precisely what we mean by chemistry, starting with the concept of chemical reactions.

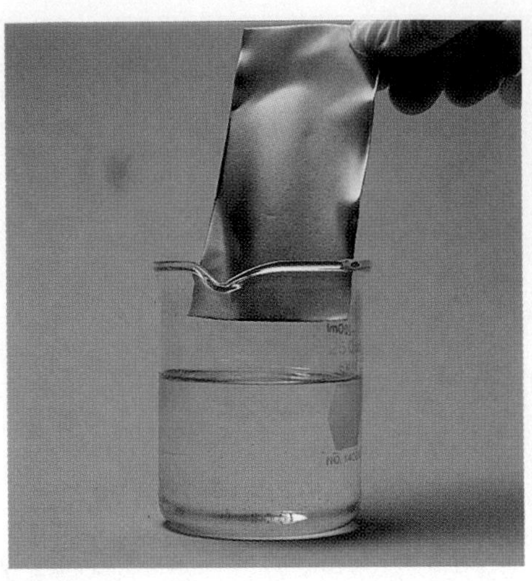

A

B

FIGURE 1.3

Reaction of copper metal and nitric acid. *(A)* Copper metal about to be placed in a beaker of nitric acid. *(B)* After placing the copper in the nitric acid, a red-brown gas called nitrogen dioxide forms. The blue-green color of the solution is due to copper(II) nitrate. Nitrogen dioxide and copper(II) nitrate are entirely new substances formed in this chemical reaction.

FIGURE 1.4

The rusting of iron. The thermometers show that heat evolves when damp iron wool reacts with air to form rust. Rusting causes the temperature to rise.

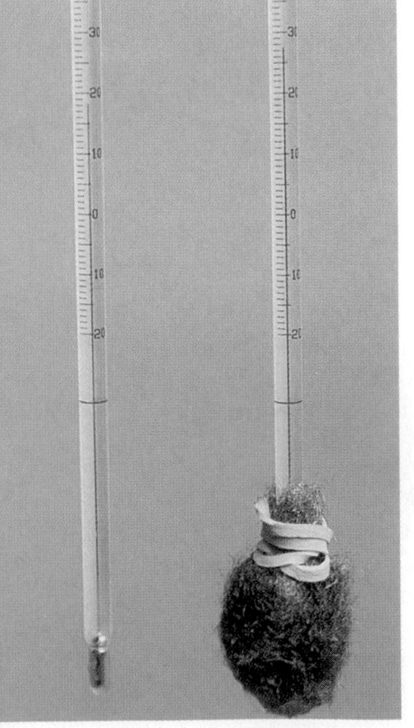

Chemical Reactions and the Definition of Chemistry

An important characteristic of matter is that its form and properties can change completely in a process called a *chemical reaction*. A piece of wood burns in air with a bright flame and yields smoke, gases, and ash. Baking soda (a white, powdery substance) bubbles and fizzes when you add vinegar to it, yielding a clear solution and carbon dioxide gas (the bubbles). Figure 1.3 shows the chemical reaction of copper with nitric acid. Before the reaction, you have a reddish metal and a colorless liquid. After the reaction, you have a blue-green solution and a red-brown gas. You might represent this reaction by the following statement:

$$\text{Copper} + \text{nitric acid} \longrightarrow \text{blue-green solution} + \text{red-brown gas}$$

The arrow means "reacts chemically to give." The substances listed after the arrow are entirely different from those before the arrow—a requirement for a chemical reaction.

Almost all chemical reactions involve energy changes, as well as changes of substances. Heat, light, and sound are forms of energy. You observe energy changes in reactions as the release or absorption of heat and perhaps as the emission of light and sound. When wood burns, it evolves heat and emits light. When iron rusts (a chemical reaction of iron with oxygen in air), it also evolves heat, though this is much less obvious than in the burning of wood (Figure 1.4). The reaction of baking soda with vinegar evolves some heat and sound (fizzing).

Now we can give a more precise definition of chemistry. **Chemistry** is *the science concerned with describing and explaining the different forms of matter and the chemical reactions (and accompanying energy changes) of matter.* We look more closely at the forms of matter and energy in Chapter 3.

Chemistry: The Central Science

Because chemistry is concerned with all matter, it plays a central role in the other sciences and in technology. It has, in fact, been called the central science. For example, when you study introductory biology, you begin by reviewing chemistry, particularly the chemical processes that occur in a cell. And a major area of biological (and chemical) research is the detailed study of cell chemistry. An understanding of many human diseases, such as cancer and AIDS, starts with the study of chemical reactions in the cell.

Video: Magnetic levitation by a superconductor.

Large areas of physics similarly border on chemistry. Consider the study of superconductors. Superconductors are materials that conduct electricity without resistance, so that once an electric current begins to flow in them, it continues to flow indefinitely. Superconductors have been used to build powerful electromagnets, magnets in which a circulating electric current produces a magnetic field. Magnetic resonance imaging (MRI) is an important medical diagnostic tool that uses superconducting magnets (Figure 1.5). New high-speed trains are being developed that use superconducting magnets to levitate, or raise, the trains slightly above the tracks, where they are propelled at high speed (Figure 1.6). At the present time, the materials used in these magnets only become superconducting at very low temperatures (about 460 degrees below zero Fahrenheit). However, physicists and chemists have recently discovered a class of ceramic materials that exhibits superconduction at higher temperatures, closer to room temperature. This discovery is exciting because it may make it possible to design simpler and cheaper superconducting magnets. The development of materials, such as ceramic superconductors, with suitable properties is the province of chemistry.

Geology, the science of the earth, also borders on chemistry. Geochemists are interested in the chemical reactions that occur within the earth or on its surface. For example, they seek the conditions under which minerals and rocks have formed. Natural diamonds are thought to have been formed from carbon deep within the earth, where

FIGURE 1.5
Magnetic resonance imaging. MRI employs a superconducting magnet to obtain a magnetic field. The video monitor in the foreground shows an image of a cross section of the patient's brain. Physicians use images such as this one to diagnose diseases.

FIGURE 1.6
High-speed train. This is a model of a train that will be raised, or levitated, by superconducting magnets. The magnets under the train are attached to the track.

FIGURE 1.7
Synthetic diamonds. These diamonds *(right)* were made by heating graphite *(left)* under high pressure. Pencil "lead" is graphite mixed with clay.

pressure and heat are extreme. Using heat and high pressure, scientists at General Electric Company have prepared industrial-grade diamonds from graphite, a form of carbon used in pencil "lead" (Figure 1.7).

● Can you think of examples in other fields where chemistry has had an impact?

We could cite many more examples, but these should convince you that chemistry is indeed a broad subject, touching on almost every aspect of science and technology. ●

1.2 A Short History of Chemistry

OBJECTIVES

- Describe some early technology that spurred the development of chemistry.
- Describe the view of matter held by the ancient Greek Aristotle.
- State the argument used by the ancient Greeks Leucippus and Democritus to arrive at the notion of atoms.
- Describe the principal goal of the alchemists and their contributions to chemistry.
- Describe Lavoisier's experiments on the combustion of mercury and give his explanation of this combustion.
- State the principal idea of Dalton's atomic theory.

Chemistry has its roots in technology, which is the application of knowledge to practical ends. You can obtain colored pigments, dyes, and perfumes from natural materials using simple techniques, such as mashing leaves and extracting with hot water. Evidence seems to indicate that colored pigments, and possibly dyes and perfumes, were available to Stone Age people. Other materials, such as pottery and metals, however, require complicated chemical processing that no doubt began with chance discoveries.

The discovery of copper metal, for example, occurred about 4500 B.C. when someone must have accidentally heated a blue-green rock (a copper ore) with charcoal obtained from partially burned wood. Later, someone else must have discovered that a mixture of this rock with another rock (an ore of tin) produced bronze when heated with charcoal. The ability to bring about these changes did not come easily to early humans, however; they waited for one significant event, an event described in the following Chemical Perspective.

Chemical Perspective

Primitive Chemists

The discovery of copper and the invention of bronze were important events for humanity. For the first time, our ancestors were able to harness chemical reactions to provide new materials for tools and ornaments. Before these breakthroughs, their tool inventory consisted of pieces of rock and wood, used as they were found. After many millennia, they learned to shape and sharpen them, and even fit them together, but rock remained rock and wood remained wood.

The chance chemical changes that occurred from time to time were unavoidable. Probably the most important change happened when lightning struck a tree and started a fire. Now early humans could cook their food—but no one had the know-how to start a fire whenever desired. The dawn of chemistry and the discoveries of copper and bronze had to wait for the mastery of fire.

When our forebears learned to start a fire, probably as far back as 100,000 B.C. (a date disputed by some archaeologists who believe it happened much earlier), a significant step in our history had occurred. Our ancestors had become chemists in a limited way because they could bring about a few chemical changes whenever they wished. Food could be cooked at any time; bricks and pottery could be hardened; and eventually metals such as copper and iron could be obtained from ores.

Ancient Greek Philosophy

Chemistry, as a science, is not simply a body of facts. It is also the process of explaining those facts. One of the main questions in chemistry is: What is the nature of matter? The oldest known speculations about the nature of matter were those of the ancient Greek philosophers. Aristotle (384–322 B.C.) believed that all material things are made up of four primary or elementary substances: fire, air, water, and earth. He also believed that these substances differed in two properties: hot versus cold and dry versus wet. The primary substances can be represented as the edges of a square, and each one is flanked by their associated properties on the corners of the square (Figure 1.8). For example, cold and dry are the properties associated with the substance earth. In Aristotle's philosophy of matter, a piece of wood must contain a lot of earth because he thought that the wood felt cold and dry.

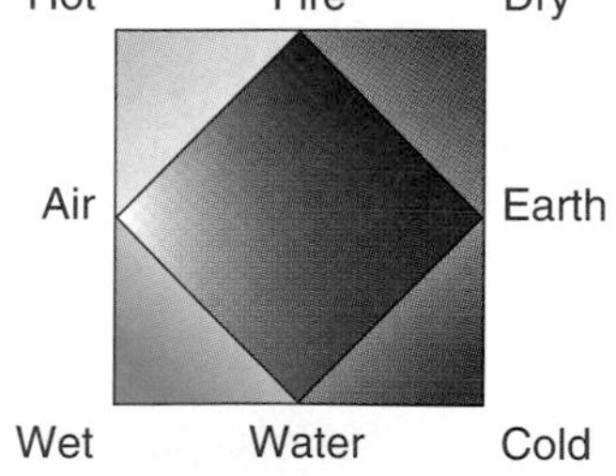

FIGURE 1.8
Aristotle's view of matter. Aristotle believed that all matter is made up of various proportions of the primary substances: fire, air, water, and earth. These primary substances are shown here as edges of a square. Each primary substance was associated with two properties that are shown as corners of the square. Thus, cold and dry were associated with earth.

A different (and minority) view of matter was held by the Greeks Leucippus (fifth century B.C.) and his student Democritus (c. 460–370 B.C.). They believed that matter consists of very small particles, or *atoms,* much too small to be seen by the naked eye, and that these atoms could not be cut into parts. Leucippus and Democritus arrived at this view by posing this question: Could you take a piece of something and cut it into smaller and smaller pieces without limit? Could you, for example, take a piece of copper metal and cut it into two pieces, then cut one of these pieces into two, and one of those pieces into two, and so on, forever? Leucippus and Democritus did not think so. As you continue to subdivide a piece of copper metal, you would come to a single atom of copper that you could not subdivide. Although we currently believe that matter is composed of atoms, many years passed before this idea began to prevail.

Alchemy

Alchemy, which may have originated alongside Greek science, is a subject we often associate with the mystical quest to find a way to change, or transmute, cheap metals such as lead into gold. From the point of view of Aristotle's philosophy, the possibility of transmuting lead to gold did not seem unreasonable. The problem was to find a way to change the proportions of the properties in a material. In their unsuccessful quest to transmute cheap metals to gold, alchemists investigated many materials, heating them for prolonged periods and subjecting them to various treatments. Alchemy was practiced widely until the eighteenth century, when modern chemistry emerged. Its main legacy to chemistry was the development of some basic laboratory techniques and chemical processes.

Modern Chemistry

Modern chemistry relies on experiments to test ideas about matter and chemical reactions. A scientist must be willing to revise his or her conception of the world if it does not fit the results of an experiment. This emphasis on experiment is at odds with the methods of the ancient Greeks, who based their arguments on their philosophical speculations, rather than on experiment as a test of those speculations. Investigators began to base conclusions on experimental results during the sixteenth and seventeenth centuries, and by the eighteenth century the role of experiments was well established.

The work of the French chemist Antoine Lavoisier (1743–1794) is usually cited as the beginning of modern chemistry. Lavoisier insisted on using quantitative methods in his investigations. In particular, he used the balance as a tool to weigh a substance to obtain its *mass,* or *quantity of matter* (Figure 1.9). By weighing substances before and after they were burned, he demonstrated that the phenomenon of burning, or combustion, is a chemical reaction involving a gaseous component of air. Lavoisier also showed that no change in mass occurs during a chemical reaction such as combustion.

When Lavoisier heated the orange substance that we know today as mercury(II) oxide, he found that it decomposed into silvery liquid mercury and a gas. Moreover, the

FIGURE 1.9
A simple chemical balance. You place the material that you wish to weigh on the left pan of the balance. Then you balance this weight by placing known weights on the right pan. Adding up these known weights gives you the mass, or weight, of the material.

combined masses of the mercury and the gas were identical to the original mass of the orange substance (Figure 1.10). ● If he heated mercury with air at lower temperatures, the orange substance was re-formed, and again the total mass was unchanged. Lavoisier proposed that the mercury had combined with a component of the air, and that this component was identical to the gas that was formed when the original orange substance decomposed.

● The Roman numerals and other features of chemical names are explained in Chapter 5.

To test his explanation, Lavoisier heated mercury in a measured volume of air until all of the mercury was converted to the orange substance. He found that a portion of the air had disappeared. This result proved that a component of air had combined with mercury. When he placed a burning candle in the remaining air, it was quickly extinguished, showing that a component of air was required for combustion. He called this component *oxygen*. ●

● Joseph Priestley (1733–1804), a contemporary of Lavoisier and the inventor of carbonated water (soda pop), had already discovered, isolated, and studied oxygen. Although he showed that this gas supports burning, he drew the wrong conclusions about combustion.

In addition to explaining combustion, Lavoisier was the first to devise a list of *chemical elements*, the basic substances from which all other substances can be prepared. Basing his view of an element on an experimental criterion, he defined an element as a substance that cannot be broken down into simpler substances by chemical reaction. He said that experiments would show if a substance was or was not an element.

The use of the balance in quantitative chemical experiments set the stage for modern *atomic theory*. Chemists had found that definite quantities of substances react with each other and produce predictable quantities of new substances. About 1803, the British chemist John Dalton (1766–1844) revived the ancient Greek notion of atoms, but added to it the idea that atoms of a given kind have a definite mass (quantity of matter). With this atomic theory, he was able to explain the quantitative results that chemists had obtained in their experiments. We examine this explanation later. For now, it is sufficient to note that atomic theory has become the central principle of chemistry.

At first many chemists regarded the atom as merely a convenient mental construct and nothing more. Today, however, we have sophisticated instruments that can weigh atoms and even help us visualize them. Figure 1.11 shows an image of iodine atoms on the surface of platinum metal. The image was drawn by computer from signals sent to it

FIGURE 1.10
Decomposing mercury(II) oxide. The orange material is mercury(II) oxide. When heated, it decomposes by a chemical reaction to give mercury metal (the shiny droplets above the orange material) and oxygen gas.

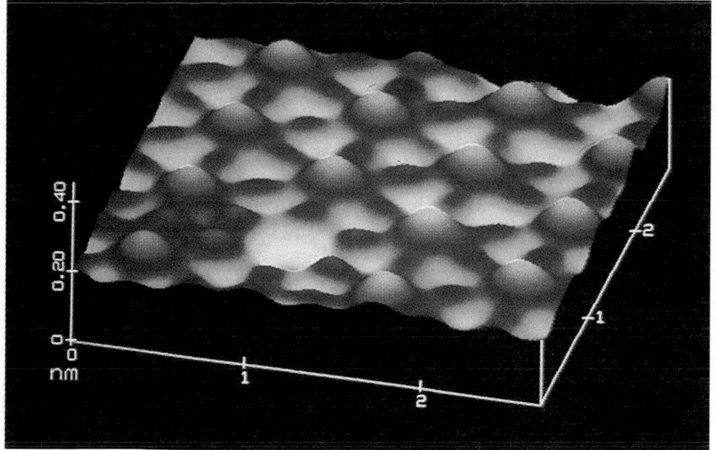

FIGURE 1.11
Image of iodine atoms on a metal surface. This image was drawn by a computer from a scanning tunneling microscope. The iodine atoms are the large peaks with pink tops. (The color is added by the computer.)

● The scanning tunneling microscope was invented in 1981.

by an instrument called a *scanning tunneling microscope.* ● The computer has drawn the iodine atoms as pink-topped peaks.

In this section, we introduced mass, elements, and atoms—three modern chemical concepts. You don't need to learn these concepts yet; we merely wanted to show some of the historical roots of modern chemistry. We will return to each concept in later chapters. We discuss mass in Chapter 2, elements in Chapter 3, and atoms in Chapter 4. We also described the interplay between experiment and explanation in this section. Let's explore that interplay further in the next section.

1.3 The Scientific Method

OBJECTIVES

- Differentiate among the terms *experiment, law, hypothesis,* and *theory.*
- Describe the scientific method.

The intertwined relationship of experiment and explanation that we described in the preceding section can be found in any modern science. The general scheme for carrying out a scientific investigation is called the *scientific method.* Before we describe this procedure in detail, you should know the precise definitions of several terms, some of which we have already used, such as *experiment* and *theory.*

As you have seen, an important requirement of scientific research is the careful design of experiments. An **experiment** is *the observation of some natural phenomenon under controlled circumstances.* Lavoisier, for example, showed that a component of air combined with mercury when both substances were heated in a closed apparatus. Merely heating mercury(II) oxide in the open would not have allowed him to show that a portion of the air had disappeared. Use of a closed vessel was part of the controlled circumstances of the experiment.

After performing many experiments, a scientist may note some pattern in the results. Scientists express such patterns in the form of a **law**, *a simple generalization from experiment.* ● We discuss a number of laws in this book, including the law of conservation of mass. That particular law comes from Lavoisier. He saw a pattern in his experiments: There is no change in mass during a chemical reaction.

● The word *law* is used here in a different sense from its common meaning as a rule established by authority.

Experiment and the statement of experimental results as laws are only one aspect of science; another is explanation. Explanation helps us organize knowledge and predict future events. A **hypothesis** is *a tentative explanation of a law, or regularity of nature.* Dalton explained the law of conservation of mass with his atomic hypothesis (later to become the atomic theory). He said that atoms are not created or destroyed during a chemical reaction—they are merely rearranged. Because the atoms have definite masses, the total mass after the reaction is the same as it was before the reaction.

A hypothesis is the starting point for devising new experiments. Having developed a hypothesis for the combustion of mercury in air, Lavoisier set up quantitative experiments to test this hypothesis. A very important part of any scientific investigation is to devise experiments to test whether or not the hypothesis agrees with experiment. If the hypothesis does not agree with experiment, it must be modified or discarded.

Once a hypothesis successfully passes many experimental tests and can explain a large body of experimental facts, it becomes known as a theory. A **theory** is *a tested explanation of some body of natural phenomena.* For example, the atomic theory explains a chemical reaction as the rearrangement of the atoms of the reacting substances to form new atomic arrangements, the products of the reaction. This theory has been tested by many chemical experiments and in each case has agreed with experimental results. Of course, even if an explanation becomes a theory, it may someday need to be modified if subsequent experiments make this necessary. Although Dalton believed that atoms were indivisible, it is now known that under the right circumstances one atom may break apart to form two different atoms. ●

● The process in which a heavy atom breaks into two lighter atoms is called *fission*. We discuss fission in Chapter 18.

The **scientific method** is *the general process involving experimentation (controlled observation) and explanation (hypothesis and theory) whereby scientific knowledge grows* (Figure 1.12). The scientist devises a series of experiments to test a hypothesis. As a result of these experiments, the scientist may amplify or modify the hypothesis. This scientist or others may then set up additional experiments to test this new hypothesis, which might result in a modified hypothesis, and so forth. In this way, the hypothesis is refined or is replaced by a better one. Eventually, a hypothesis that appears to agree with many experiments will be called a *theory.*

This description of the scientific method is an idealization of how science actually takes place. Science is a human endeavor, and as such it leaves much room for differences of approach and style. Lavoisier was a master at devising the crucial experiment. Dalton, though probably not as good an experimentalist, was able to use other scientists' data effectively, and his ideas were bold. Of course, scientists have all the strengths and failings common to human beings; thus, some have exhibited openness to new ideas and others have had their prejudices.

Notice that scientific investigation (and the scientific method) is not so different from many other kinds of problem solving. You make an observation (your car has sputtered to a stop) and a hypothesis (the car is out of gas). You perform an experiment to check your hypothesis (you fill the gas tank), but you find that the hypothesis is not verified (the car does not start). Now it is time for a new hypothesis (perhaps the fuel line is frozen). And so forth. You can imagine different people approaching similar problems in different ways, but still basically using the scientific method. Of course, some people do not approach problem solving this way. In another example, perhaps the television set blinks, then goes off. Bob kicks the set. This is probably not an example of the scientific method. ●

● Of course, Bob may be doing an experiment. If he thinks there is a loose wire, then kicking the set may be part of Bob's experiment to locate the loose wire. More likely, however, Bob is displaying his frustration—not his scientific method!

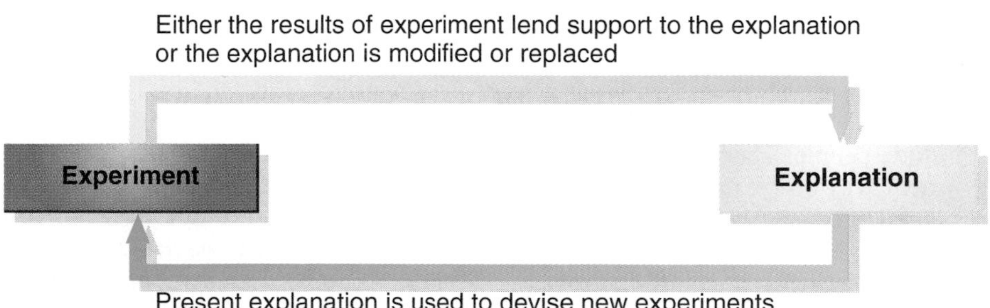
Either the results of experiment lend support to the explanation or the explanation is modified or replaced
Experiment
Explanation
Present explanation is used to devise new experiments

FIGURE 1.12
The scientific method. This diagram shows the interplay between experiment and explanation in science.

Chemical Perspective

The Discovery of Vitamin C: An Application of the Scientific Method

Vitamin C is a nutrient found in fruits and vegetables, such as oranges and broccoli. You also can buy this white, sour-tasting substance in drugstores or supermarkets (Figure 1.13). Its other name is ascorbic acid, which is a shortening of "antiscorbutic (antiscurvy) acid." Scurvy is a disease resulting from a deficiency of vitamin C in the diet. Though few people today have seen anyone with its symptoms (bleeding gums and general weakness), scurvy was common at the beginning of this century. Yet the cure for this disease is simple: just eat fresh fruits and vegetables daily. Why did it take so long to understand this?

Basically, an understanding of the problem required two things: the use of the scientific method and a knowledge of fundamental chemistry. Chemists have used the scientific method only since the late eighteenth century. And it took another 100 years or so before they understood the basic concepts of chemistry.

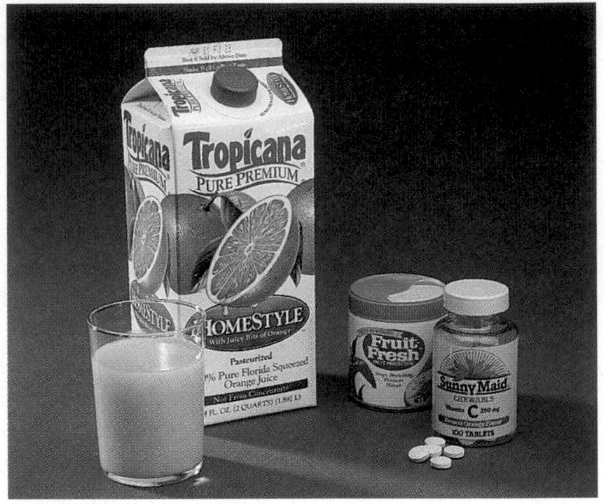

FIGURE 1.13
Some products containing vitamin C. *Left to right:* Orange juice, fruit freshener (which uses vitamin C as an antioxidant), and vitamin C tablets.

Scurvy was a well-known scourge during the age of the seafaring explorers. During a long voyage, the fresh foods that contained vitamin C were eaten in the early weeks, and after months on a diet of salt pork, sailors developed scurvy. An account given of Jacques Cartier's expedition to Newfoundland in 1535 describes scurvy sufferers in terrifying terms: "Their mouths became stinking, their gums so rotten that all the flesh did rot off, even to the roots of their teeth, which almost all fall out."

This account goes on to note that while on land, a crew member was cured of scurvy by drinking sassafras tea. This knowledge might have led to an understanding of the disease with the help of the scientific method. Instead, the cure for scurvy continued to elude people. The idea of isolating effects by experiment was not yet appreciated. So when a sailor was cured of scurvy while on land, it was frequently attributed to "the air of the land" and not to the eating of fresh fruits and vegetables.

The first recorded experiment on scurvy patients was performed about 1746. While at sea, the British surgeon James Lind divided 12 men with scurvy into six groups of two each, feeding those in each group a different diet. Those receiving oranges in their diet quickly recovered from the disease. This knowledge had little immediate effect on sailing practice, however. It was not until 1795, two years after Lind's death, that the British navy endorsed the idea that sailors be given a daily ration of lime juice to combat scurvy. (The slang term *limey* for a British sailor stems from this ration of lime juice.)

Still, until this century people did not understand that scurvy resulted from a lack of some substance in the diet. Without a background in chemistry, they could not grasp the idea that a small quantity of a substance in one's daily diet could alleviate disease. But by the early part of the twentieth century, modern chemistry with its scientific method had become well established. In 1907, scientists in Norway discovered that guinea pigs fed a diet devoid of fresh greens developed the symptoms of scurvy. From this they formed the idea that fresh fruits and vegetables contain a nutritional substance in small amounts whose absence causes scurvy. Within a few years, most scientists understood the concept of vitamins (nutritional substances needed in the diet in only small amounts). The antiscurvy vitamin was soon called vitamin C, although it had yet to be seen.

To prove that there was indeed a vitamin C, scientists needed to isolate it. By the 1920s, the race to isolate vitamin C was on, with chemists in both Europe and the United States rushing toward the finish line. In

1928, the Hungarian chemist Albert Szent-Györgyi (1893–1986) finally isolated vitamin C. Szent-Györgyi (pronounced "Saint Georgie") discovered vitamin C first in the extract of beef adrenal glands, which he obtained from a slaughterhouse. Later, he obtained the vitamin from sweet peppers.

When Szent-Györgyi first tried to publish his discovery, he had not yet understood that the substance he isolated was vitamin C, although he knew that it had sugarlike properties. He called the substance "ignose," using the chemical suffix "-ose" for a sugar and "ign" for his ignorance of the actual nature of the substance. His flippancy angered the editor, who refused to publish his work until he gave the substance a proper name. In a revised manuscript, Szent-Györgyi called it Godnose, but when the editor stood firm, he named the substance hexuronic acid. In 1932, he showed that hexuronic acid was the long-sought vitamin C, which he later renamed ascorbic acid. Szent-Györgyi received the Nobel Prize in medicine for his work in 1937.

CHAPTER REVIEW

Key Words

matter *(p. 3)*
chemistry *(p. 4)*
experiment *(p. 10)*
law *(p. 10)*
hypothesis *(p. 10)*
theory *(p. 10)*
scientific method *(p. 11)*

Summary

Matter is the term applied to the material things around us. An important characteristic of matter is that its form and properties may change completely. These complete changes of matter are called chemical reactions. *Chemistry* is concerned with describing the forms of matter and the chemical reactions they undergo. Because chemistry is concerned with all matter, it plays a central role in the other sciences and in technology.

Chemistry has its roots in early technology, but the earliest ideas about matter came from the ancient Greeks. Aristotle's view of matter (that the different kinds of matter we see are a result of different proportions of four fundamental substances) dominated for nearly 2000 years. In contrast, Leucippus and Democritus believed that matter was composed of atoms. Although it was a minor view at the time, it was revived in the nineteenth century. Alchemy, a mystic quest to find a way to change cheap metals into gold by trying to vary the proportions of properties in the cheap metal, contributed basic laboratory techniques and chemical processes. Modern chemistry developed during the eighteenth century, and one of the most important chemists of this period was the French chemist Antoine Lavoisier (1743–1794). He relied on the use of the balance in chemical experiments. As a result of his experiments, he explained combustion as a chemical reaction involving oxygen in air. He also showed that mass does not change during a chemical reaction.

Scientific research depends on carefully designed *experiments*. After many experiments, a regularity of nature may be observed. Such a regularity can often be stated as a *law*. A *hypothesis*, or tentative explanation of the observed regularity, may also be proposed. If this hypothesis withstands many tests of its validity, it becomes known as a *theory*. In general, an area of science advances by the process of doing experiments based on some hypothesis. As a result of these experiments, the hypothesis may need to be modified. Then additional experiments are carried out to test this new hypothesis. And so on. As a result, the sophistication of the hypothesis (perhaps now called a theory) and the body of scientific knowledge grow. This basic process whereby science advances is known as the *scientific method*.

Questions to Test Your Reading

1.1 Give three examples of matter and state some properties of each.

1.2 Give three examples of chemical reactions. Describe them in words.

1.3 Describe two chemical reactions and the energy changes that accompany each reaction.

1.4 Define chemistry.

1.5 Describe in specific terms how chemistry is important in other sciences.

1.6 Describe some applications of chemistry to everyday living.

1.7 How was copper originally discovered? How was bronze invented?

1.8 Briefly describe Aristotle's view of matter.

1.9 Describe the view of matter held by Leucippus and Democritus.

1.10 What was it that characterized Lavoisier's experiments, especially his experiments on combustion?

1.11 How did Lavoisier explain the combustion of mercury?

1.12 What central theory of chemistry is associated with Dalton? Briefly describe it.

1.13 Define each of the following terms: (a) experiment, (b) law, (c) hypothesis, (d) theory.

1.14 Describe how the scientific method advances a field of science.

PRACTICE EXAM

1. Which statement is true?
(a) Matter refers to the things around you.
(b) Matter does not include mixtures.
(c) Matter does not include gases that you cannot see, such as air.
(d) Matter includes heat and light.
(e) None of the above

2. Which statement is true?
(a) Chemistry is not concerned with describing and explaining the different forms of matter.
(b) Energy is not a concern of chemistry.
(c) Chemistry is concerned with describing and explaining processes such as burning a log or dissolving a piece of copper in an acid.
(d) Chemistry does not play a role in other sciences.
(e) None of the above

3. When Lavoisier heated mercury(II) oxide, he obtained
(a) mercury and water (b) lead and water
(c) only mercury (d) only oxygen
(e) None of the above

4. If you wanted to obtain the mass of a nail, you could do it with a
(a) flask (b) beaker (c) ruler (d) balance
(e) None of the above

5. A simple generalization drawn from the results of experiments is called
(a) a law (b) a theory (c) a hypothesis
(d) the scientific method (e) None of the above

6. A tested explanation of some body of natural phenomena is called
(a) a law (b) a theory (c) a hypothesis
(d) the scientific method (e) None of the above

7. We owe the original idea of atoms to
(a) Leucippus and Democritus
(b) Antoine Lavoisier
(c) John Dalton
(d) Albert Einstein
(e) None of the above

8. We owe the law of conservation of mass to
(a) Leucippus and Democritus
(b) Antoine Lavoisier
(c) John Dalton
(d) Albert Einstein
(e) None of the above

9. We owe the atomic theory to
(a) Leucippus and Democritus
(b) Antoine Lavoisier
(c) John Dalton
(d) Albert Einstein
(e) None of the above

10. Which statement is true?
(a) You can continue dividing a drop of water into smaller drops forever.
(b) When you burn a piece of paper, the combined mass of ashes, smoke, and gases is less than the original mass of the paper.
(c) Antoine Lavoisier's scientific method must have been identical to the scientific method of Albert Szent-Györgyi, the discoverer of vitamin C.
(d) A theory describes an observed regularity in nature.
(e) None of the above

Chapter

Measurement in Chemistry

As you drive a car, the gauges allow you to measure the speed and the amount of gasoline in the tank.

Chapter Outline

Modern chemistry studies matter qualitatively and quantitatively. In Chapter 1, we described a sample of mercury(II) oxide as an orange powder. This is a *qualitative* description of the substance because it does not involve numbers. But when we say that a particular sample of mercury(II) oxide weighs 1.200 ounces, we are describing that sample in a *quantitative* way. Lavoisier was the first to stress the importance of quantitative measurements in chemistry. Only by measuring the masses of substances and volumes of gases involved in a combustion, or burning, was he able to explain the process of combustion. Modern chemistry places even greater reliance on quantitative measurements (Figure 2.1). In this chapter, we introduce the metric system of units. We also describe calculations, present techniques for reporting results, and solve problems using measured quantities.

FIGURE 2.1
A quantitative measurement. The quantity of acid in each liter of this water affects the survival of wildlife.

METRIC UNITS

When you measure the length of an object with a ruler, you are comparing the length of the object with a unit, or standard, of length. If the ruler is divided into inches, the inch is this standard unit. Originally, units of measurement were set up to make it easier for people to trade with one another. Each country developed its own separate system, however, and the relationships between these systems often were not easy to understand. In fact, the relationships between units in a single country were not always simple. You can see this in the units customarily used today in the United States. There are 12 inches to the foot, 3 feet to the yard, 16 ounces to the pound, and so forth. In time, people realized that these systems of units should be simplified and that a single system of units should be adopted.

In 1791, a study committee of the French Academy of Sciences devised the *metric system* of units, in which larger and smaller units of a given quantity are related by multiples of 10. For example, 1 kilometer is exactly 1000 meters. (By comparison, 1 mile equals 5280 feet.) The French Academy also proposed standards for the units that would be internationally acceptable. Over time, these standards have improved. For example, we now define the meter in terms of the wavelength of light. Anyone with the proper equipment can perform the necessary measurements and therefore has access to the metric standard of length. The metric system is now used in most countries of the world and in all scientific work. Even the present system of units in the United States, although not a metric system, is defined in terms of the metric system. For example, the inch is exactly 0.0254 meters.

We begin this part of the chapter by considering the basic ideas involved in making a measurement. Then we explore *scientific notation,* which we need both to define larger and smaller units of the metric system and to do calculations with these units. The remaining sections of this first part are devoted to the study of metric units. In the second part of the chapter, we discuss significant figures. In the last part, we look at calculations and the conversion of one unit to another.

2.1 Measured Numbers and Units

OBJECTIVES

- Identify the two components, or parts, of a measurement.
- Describe the procedure you would use to make a measurement using a standard unit of measure.

FIGURE 2.2
Measuring the volume of gasoline pumped. The measurement consists of a measured number (10.200) and a unit (gallons).

You use measurements almost every day. You look at your watch to determine the time and the speedometer of your car to determine the speed. The numbers you read in these cases are measurements.

The measurements you make normally consist of two parts, or components: a *measured number* and a *unit* of measurement. Suppose you fill your car with gasoline, and the gasoline pump reads 10.200 gallons (Figure 2.2). The 10.200 is the measured number, and gallons is the unit of volume measurement.

The essence of any measurement of a quantity, such as volume, is a comparison with some standard measure for that quantity, which is the unit. Consider this simple example: Suppose you want to measure the volume of liquid in a container, say, milk in the pitcher in Figure 2.3A. You take a measuring vessel whose volume is exactly 1 cup (the cooking measure equal to one-half pint) and fill it with the milk from the pitcher, then pour the milk out of the measuring cup and into another container. You continue this way, as shown in Figure 2.3B, pouring milk from the pitcher into the cup measure and then into the other container, counting the number of cup measures in the pitcher. Of course, when you arrive at the last quantity of milk in the pitcher, after measuring, say, 4 whole cups of liquid, you may not have a full cup. You will have to estimate the fraction of a cup of liquid that you have. Suppose you have 0.2 cup in the final measure. Therefore, you had 4.2 cups of milk in the pitcher (Figure 2.3B).

In measuring the volume of milk in the pitcher, you compared this volume with a standard kitchen measure (the cup), which has an agreed on volume. The cup, then, was the unit of measurement. Of course, you could have used another unit of volume, say, the *liter* (pronounced "lee-ter"), which equals about 1 quart, or you might have used

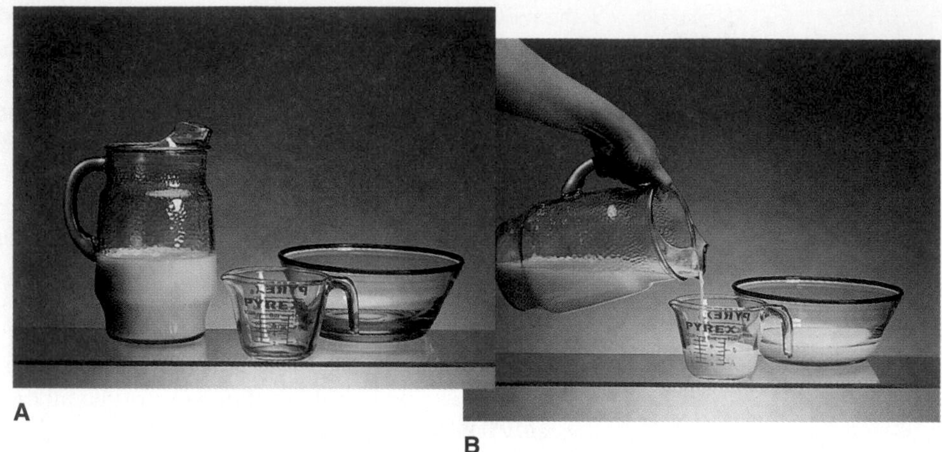

FIGURE 2.3
Measurement of liquid volume. *(A)* A pitcher contains a liquid (milk) whose volume you wish to measure. *(B)* By pouring the milk from the pitcher into the measuring cup, then into the other container, you find that the volume of the liquid is 4.2 cups.

some fraction of a liter. The measured numbers would be different in these cases, although the actual volume of liquid would not change.

All measurements are comparisons with an established standard, which becomes the unit of measurement. The comparison results in a measured number, which gives the multiples or fractions of that unit in the measured quantity. ●

● Think of several measurements and note how in each case the measurement involves a comparison with a standard, or unit.

2.2 Writing Measurements in Scientific Notation

OBJECTIVE

- Convert numbers in decimal form to scientific notation, and vice versa.

If you have used a scientific calculator, you may be familiar with numbers expressed in *scientific notation* (Figure 2.4). Scientific notation is especially useful for representing very large or very small numbers. Such numbers often occur in chemical measurements. For example, the approximate number of oxygen atoms in a liter (about a quart) of air at room temperature and normal barometric pressure is ●

● Barometric pressure varies with weather conditions. "Normal barometric pressure" is the average barometric pressure at sea level.

9,900,000,000,000,000,000,000 atoms

The diameter of an oxygen atom in meters (with 1 meter equal to about a yard) is

0.000,000,000,13 meter

When you try to read these numbers, you find yourself counting the number of zeros to grasp how large or how small the number is. Scientific notation replaces these cumbersome numbers with much more readable numbers that are also easier to use when doing calculations.

Scientific notation is *the representation of a number in the form* $A \times 10^n$, *where* A *is a number with a single nonzero digit to the left of the decimal point, and* n *is a whole number.* For example, you write 210.3 liters in scientific notation as 2.103×10^2 liters. Here, 10^2 means two factors of 10; that is, $10^2 = 10 \times 10$. In other words,

$$2.103 \times 10^2 \text{ liters} = 2.103 \times 10 \times 10 \text{ liters} = 210.3 \text{ liters}$$

In scientific notation, you would write the number of oxygen atoms in a liter of air as 9.9×10^{21}. Note how much easier it is to read and write this number in scientific notation. ●

● An additional treatment of scientific notation is given in Appendix A.

FIGURE 2.4

A scientific calculator. The number on the display is in scientific notation and represents 2.103×10^2.

You call the number n the *exponent* and 10^n the nth *power* of 10. In the quantity 5.003×10^3, the exponent is 3 and the power of 10 is 10^3. The exponent can be a positive number, as it is here, or it can be negative. A negative exponent tells you the number of times 1 is divided by 10. For example, 10 raised to the exponent -3 is equal to 1 divided by three factors of 10.

$$10^{-3} = \frac{1}{10 \times 10 \times 10}$$

To write 7.26×10^{-3} meters in decimal form, you write

$$7.26 \times 10^{-3} \text{ meters} = 7.26 \times \frac{1}{10 \times 10 \times 10} \text{ meter} = 0.00726 \text{ meter}$$

You can easily transform any number written in decimal form to scientific notation by simply moving the decimal point to obtain a number that is between 1 and 9.99. . . . After you do this, you should have one nonzero digit to the left of the decimal point. Count the number of places you moved the decimal point to the left or to the right, and this gives you the exponent in the scientific notation of the number. Let us try this on the number of oxygen atoms in a liter of air, which we gave earlier:

Places moved left

$$9{,}900{,}000{,}000{,}000{,}000{,}000{,}000. = 9.9 \times 10^{21}$$

You move the decimal point to the left until you have only one digit on its left. As you move the decimal point, count the number of digits that you pass (21). In other words, move the decimal point 21 places to the left. This, then, becomes the exponent for the power of 10. The answer is 9.9×10^{21}.

You can write very small numbers in scientific notation just as easily. Consider the diameter of an oxygen atom. In this case, you move the decimal point to the right, counting the number of places moved (10). This number, written with a negative sign, becomes the exponent in the power of 10.

Places moved right

$$0.00000000013 \text{ meter} = 1.3 \times 10^{-10} \text{ meter}$$

The following example illustrates the conversion of numbers in decimal form to scientific notation.

EXAMPLE 2.1 Converting Decimal Numbers to Scientific Notation

Express the following numbers in scientific notation.
(a) 60.0 (b) 0.000456 (c) 6.51

Problem Analysis

Shift the decimal point left or right to obtain a new number between 1 and 9.99 . . . while you count the number of places you move the decimal point. This count becomes the exponent in the power of 10 that you use to multiply by the new number.

(The exponent is positive if the move is to the left, negative if the move is to the right.) The result is the number in scientific notation.

Solution

(a) You shift the decimal point 1 place to the left, giving a number having only one nonzero digit to the left of the decimal point. Now, multiply this number by 10^1.

$$60.0 = 6.00 \times 10^1$$

(b) Shift the decimal point 4 places to the right. Then multiply the result by 10^{-4}.

$$0.000456 = 4.56 \times 10^{-4}$$

(c) The number is already between 1 and 9.99 . . . , so you shift the decimal point 0 places, getting 10^0 as your power of 10. Write the number as 6.51×10^0, or simply leave it as 6.51, because 10^0 equals 1.

Exercise 2.1

Write the following numbers in scientific notation.
(a) 0.0069252 (b) 0.06300 (c) 802.00 (d) 9002 (e) 0.707

(Try Problems 2.21 and 2.22.)

In converting a number written in scientific notation to decimal form, you simply shift the decimal point the number of places given by the exponent in the power of 10. When the exponent is positive, you shift the decimal place to the right. When the exponent is negative, you shift the decimal place to the left.

As an illustration, consider the distance from the earth to the sun, which is 9.3×10^7 miles. To write this in decimal form, you shift the decimal point to the right 7 places. ●

● You shift the decimal point *right* and *reduce* the exponent (from 7 to 0); remember, right and reduce.

$$9.3 \times 10^7 \text{ miles} = 93,000,000. \text{ miles}$$

The result is 93,000,000 miles.

The diameter of a red blood cell is about 8.0×10^{-6} meter. To write this in decimal form, you shift the decimal point 6 places to the left. ●

● You shift the decimal point *left* and make the exponent *larger* (from −6 to 0); remember, left and larger.

$$8.0 \times 10^{-6} \text{ meter} = 0.0000080 \text{ meter}$$

The result is 0.0000080 meter.

The next example further illustrates how to convert a number written in scientific notation to one in decimal form.

■ EXAMPLE 2.2 Converting Scientific Notation to Decimal Form

Convert the following numbers in scientific notation to decimal form.
(a) 8.2×10^{-5} (b) 7.358×10^2

Problem Analysis

Shift the decimal point the number of places indicated by the exponent in the scientific notation. (Shift left if negative; shift right if positive.)

Solution

(a) Shift the decimal point 5 places to the left because the exponent is −5.

$$8.2 \times 10^{-5} = 0.00008\,2$$

The answer is 0.000082.

(b) Shift the decimal point 2 places to the right because the exponent is 2.

$$7.358 \times 10^{2} = 735.8$$

The answer is 735.8.

Exercise 2.2

Convert the following numbers from scientific notation to decimal form.

(a) 4.834×10^{-4} (b) 6.250×10^{3} (c) 4.89×10^{0}

(Try Problems 2.23, 2.24, 2.25, and 2.26.)

2.3 Units of Length, Volume, and Mass

OBJECTIVES

- Describe the roles of base units and prefixes in the modern metric (SI) system.
- Identify the SI base units used to measure length and mass, and the derived SI unit used to measure volume.
- List and define the eight common SI prefixes used in this book.
- Contrast the mass and the weight of an object.

The **metric system** is *a system of units based on the decimal number system*. The modern form of this system of units is called the International System, or SI (from the French *le Système International*). The International System has seven *SI base units*, from which all other units are derived. Smaller and larger units are indicated by writing *base units* with *SI prefixes* to denote multiplication by powers of 10.

SI base units are *those SI units from which all others are derived*. Table 2.1 lists the five base units that we use in this book and includes the symbols, or abbreviations, used to represent them. (Two other base units are the unit of electric current and the unit of luminous intensity.) In this section, we discuss two base units: the meter (a unit of length) and the kilogram (a unit of mass). We also discuss the liter (a derived unit of volume). The kelvin (a unit of temperature) is described later in the chapter, and the mole (a unit of amount of a substance) is described in Chapter 7. The second (a unit of time) is so familiar that it needs no discussion.

One advantage of the metric system over other measurement systems is that it is based on the decimal number system. In SI, a larger or smaller unit than the base unit is

TABLE 2.1
SI Base Units

Quantity	Unit	Symbol
Length	meter	m
Mass	kilogram	kg
Time	second	s
Temperature	kelvin	K
Amount of substance	mole	mol

TABLE 2.2
SI Prefixes

Multiple	Prefix	Symbol
10^6	mega-	M
10^3	kilo	k
10^{-1}	deci	d
10^{-2}	centi	c
10^{-3}	milli	m
10^{-6}	micro	μ*
10^{-9}	nano	n
10^{-12}	pico	p

*Greek letter pronounced "mew."

indicated by an **SI prefix,** which is *a prefix used in SI to indicate a power of 10.* For example, the base unit of length in SI is the meter, which is somewhat longer than a yard. If you want a smaller unit, you could use the centimeter, which is 10^{-2} meter. The prefix *centi-* means 10^{-2}.

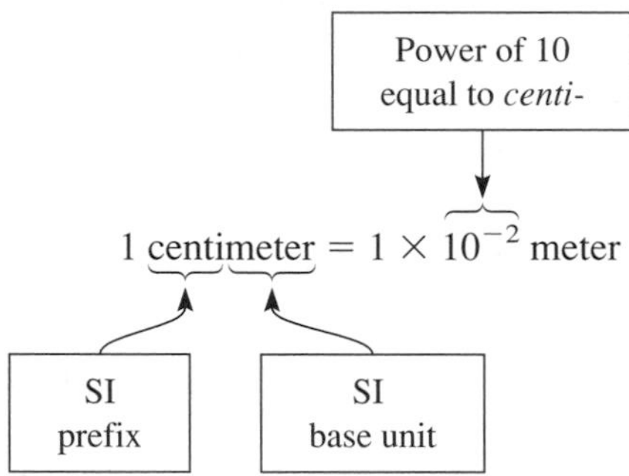

Table 2.2 lists eight common prefixes; these prefixes are the only ones we use in this book. The table also lists the symbols, or abbreviations, adopted for these prefixes.

Length

● Chemists often use the *angstrom,* a unit of length equal to 1×10^{-10} m. The angstrom is a metric unit, but it is not SI, because it is not expressed in terms of an SI prefix and the base unit meter.

The **meter** (m) is *the SI base unit of length.* It equals about 39 inches. To obtain larger and smaller units, combine the meter with appropriate prefixes. If you are interested in distances between cities, you would probably use the kilometer (km), which equals 10^3 m. For the lengths of common objects, you might use the centimeter (cm), which, as you have seen, is 10^{-2} m (Figure 2.5). Table 2.3 lists the sizes of various objects in metric units. ●

TABLE 2.3
Sizes of Various Objects in Metric Units

Object	Size
Oxygen atom (diameter)	0.13 nm, or 130 pm
Red blood cell (diameter)	8 μm
Human little finger (width)	About 1 to 2 cm
Human being (height)	About 1.5 to 2.0 m
City block (average length)	About 0.25 km

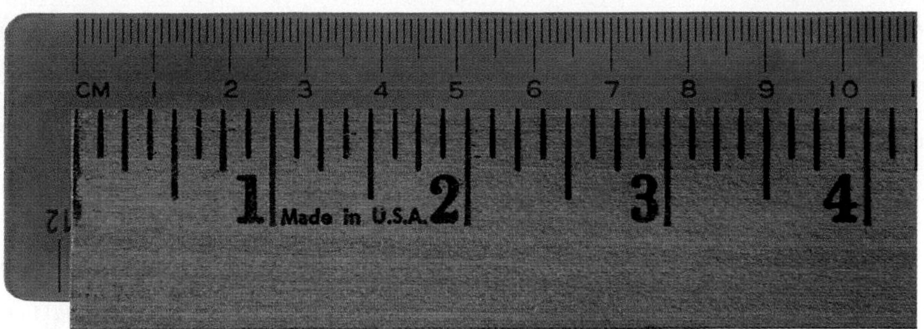

FIGURE 2.5

Comparing centimeters and inches. The top ruler is marked off in centimeters; the bottom one is in inches. The inch is defined as exactly 2.54 cm.

Volume

The **volume** of an object is *the amount of space it occupies.* Volume is a *derived* SI unit; it is derived from the meter, the SI base unit of length. Consider the box shown in Figure 2.6. It measures 43 cm wide, 36 cm deep, and 21 cm high. The volume of the box is

$$\begin{aligned} \text{Volume} &= \text{width} \times \text{depth} \times \text{height} \\ &= 43 \text{ cm} \times 36 \text{ cm} \times 21 \text{ cm} = 3.3 \times 10^4 \times \text{cm} \times \text{cm} \times \text{cm} \\ &= 3.3 \times 10^4 \text{ cm}^3 \end{aligned}$$

● A calculator will give 3.2508×10^4, which rounds to 3.3×10^4. (See Appendix B for use of a scientific calculator.) Section 2.5 presents the rules for rounding.

Note that when we did this calculation, we substituted the measured number and the length unit for each dimension. When we did that, we multiplied the units, as well as the numbers. Thus, the unit of volume comes out of the calculation automatically—in this case by multiplying the three factors of centimeters together. We call this derived unit *cubic* centimeters, abbreviated cm^3. So the volume of the box is 3.3×10^4 cubic centimeters.

Volume is always expressed in units corresponding to length cubed (length × length × length). The SI unit of volume is the base unit of length (meter) cubed, or the cubic meter ($\text{m} \times \text{m} \times \text{m} = \text{m}^3$). The cubic meter, however, is much too large a unit of volume for most laboratory work, so we use a volume unit derived from the decimeter (dm), which is 10^{-1} m (approximately 4 inches). Let's consider an important example: a cube having an edge length of 1 dm. The volume of a cube is 1 dm × 1 dm × 1 dm, or 1 cubic decimeter (dm^3). For most purposes, however, we would say that the volume of this box is 1 liter.

The **liter** (L) is *the derived SI unit of volume equal to 1 cubic decimeter.* Thus, the liter and cubic decimeter are identical units.

$$1 \text{ L} = 1 \text{ dm}^3$$

A liter is somewhat larger than 1 quart. Most laboratory glassware is calibrated in milliliters (mL), or 10^{-3} L (Figure 2.7). One milliliter is equivalent to 1 cubic centimeter.

$$1 \text{ mL} = 1 \text{ cm}^3$$

A milliliter, or cubic centimeter, is equal in volume to about 20 drops of water, and 5 mL is approximately 1 teaspoon.

FIGURE 2.6

Obtaining the volume of a box. You obtain the volume of a box by measuring its width, depth, and height, and then multiplying these dimensions together.

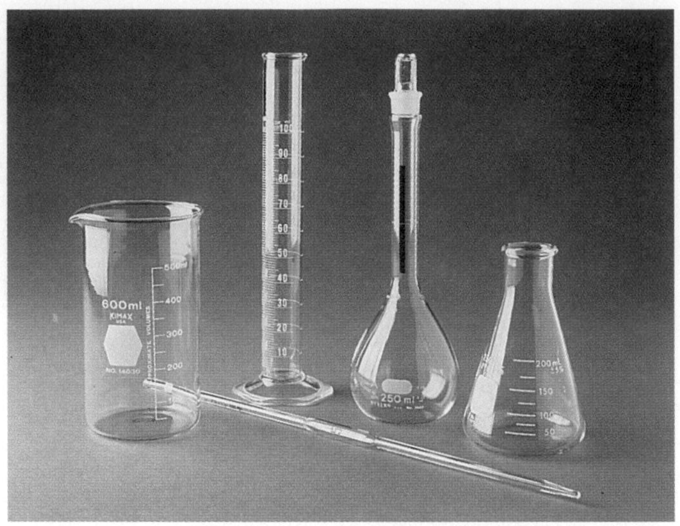

FIGURE 2.7
Laboratory glassware. *Left to right:* A 600-mL beaker, a 100-mL graduated cylinder, a 250-mL volumetric flask, and a 250-mL Erlenmeyer flask. *Front:* A 5-mL pipet (used for dispensing small liquid volumes).

Mass

The mass of an object is the quantity of matter the object contains. In Chapter 1, you saw that the mass of an object can be measured with a simple two-pan balance by comparing the object's mass to the known masses. Although two-pan balances are still used, most laboratories rely on electronic balances (Figure 2.8). Weighing objects on one of these modern balances is fast and simple. You place the object to be weighed on the single pan and note the object's mass on the digital readout. The internal mechanism of this type of balance is similar to that of the two-pan balance, but an electromagnetic device replaces the pan on which you would have put known masses.

The **kilogram** (kg) is *the SI base unit of mass.* This is an unusual base unit because it already has a prefix (*kilo-,* meaning 10^3) in its name. For most purposes, however, we will use the gram (one-thousandth of a kilogram). The kilogram is a little more than 2 pounds, so the gram is a small unit of mass. To visualize how small a gram is, consider that 1 ounce is slightly more than 28 g and a teaspoon of salt weighs about 5 g. In chemistry, we often weigh substances in fractional-gram amounts. The milligram (mg), which is 10^{-3} g, is another frequently used unit of mass.

FIGURE 2.8
A modern balance. The balance has a single pan on which you place objects to be weighed. The mass appears on the digital readout.

Distinction Between Mass and Weight

In common speech, the terms *mass* and *weight* are often used interchangeably. However, the two terms have different meanings, and in scientific work, you must understand the distinction between them. Mass is a *quantity of matter.* As long as an object is unchanged, its mass remains unchanged. Weight, on the other hand, is *the force exerted by gravity on an object* and varies depending on where you are. The force of gravity is the result of the attraction between the object and the earth (or the moon, if you are on the moon). When someone hands you a large book, you feel the downward force of gravity of this book. This downward force is its weight. Although the mass of the book on the earth is identical to its mass on the moon, the book would weigh less on the moon because the moon's force of gravity is less than that of earth.

The weight of an object is proportional to its mass, so two objects of equal mass at the same point on earth have the same weight. You use this equivalence when you weigh an object on a balance. When you balance an object on the left pan against standard masses on the right pan, the weights of both (the forces of gravity on both) are the same. You conclude that the masses are equal. ●

● The process of weighing on a balance compares two weights and requires the force of gravity. The result of the measurement, however, is a mass.

SIGNIFICANT FIGURES

You may have noticed that when you weigh yourself several times in succession on a bathroom scale, the readings of your weight (actually your mass) sometimes differ by a pound or so. This variation in the measurement readings is quite normal, and all measurements exhibit such variations; they are the result of limitations inherent in making any measurement. Of course, you can use a different measuring device to reduce the amount of *uncertainty* in the measurement. Instead of using a spring scale, you could use a physician's balance. Now instead of obtaining variations of a pound or so, the variations may be only a few ounces. Despite your ability to reduce this uncertainty, you cannot eliminate it. In the next sections, we examine uncertainty in measurement and discuss how to use the concept of *significant figures* to report the uncertainty in measured numbers.

2.4 Significant Figures and Uncertainty in Measurement

OBJECTIVES

- Explain why all measurements have some uncertainty in them.
- Count the number of significant figures in a measured number.
- Differentiate between exact numbers and measured numbers.

Let us briefly explore the idea of uncertainty in measurement. Earlier, we described the procedure for measuring the volume of milk in a pitcher. In the laboratory, we often measure liquid volumes using a *graduated cylinder*. To measure the volume of a liquid, you pour the liquid into a graduated cylinder, which is marked off by lines that divide the volume into milliliters (Figure 2.9). You read the volume of liquid by looking at the lower level of its top surface (see the inset in Figure 2.9). Note that the liquid level lies between 32 mL and 33 mL, and is a bit closer to 33 mL than to 32 mL. You estimate the volume to be about 32.6 mL.

You know the first two digits (32) of this measured number with certainty. The last digit, however, is uncertain. Although you estimate the volume to be 32.6 mL, it actually lies somewhere between 32.5 mL and 32.7 mL. If you tell anyone who is familiar with measurements that the volume of liquid is 32.6 mL, that person will understand that the last digit (6) has some uncertainty in it and that the volume of liquid is probably somewhere between 32.5 mL and 32.7 mL.

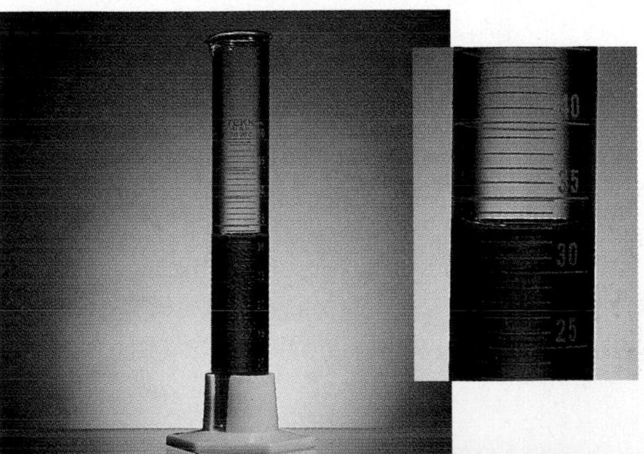

FIGURE 2.9
Measuring volumes with a graduated cylinder. You read the volume by looking at the lowest part of the curved upper surface of the liquid (see inset).

Instead of reporting the volume as 32.6 mL, suppose you report it as 33 mL, which is its volume to the nearest milliliter. Someone reading your report would conclude that the volume lies somewhere between 32 mL and 34 mL. Note that the number of digits, or figures, you use in reporting a measurement conveys important information. We would say that all of the digits in 32.6 are *significant* in the sense that all three are needed to convey your best estimate of the volume.

When you write measured numbers, you give only the significant figures (also called significant digits), neither more nor less. **Significant figures** are *those digits in a measured number (or in the result of calculations with measured numbers) that include all certain digits plus a final one that is somewhat uncertain.* This last digit with some uncertainty in it is called the *least significant digit.* In your volume measurement of 32.6 mL, 6 is the least significant digit.

According to the convention of significant figures, you would be incorrect in reporting your measured volume as 32.60 mL. Writing the volume this way says that the uncertainty in the measured volume occurs in the last digit, which is 0. In other words, the volume lies between about 32.59 mL and 32.61 mL.

Counting Significant Figures

Note the number of significant figures in the measurement 32.6 mL. By reporting the measured number to the appropriate number of significant figures, you are indicating to others the smallest range in which you expect the actual value to lie. If you gave that value as 33 mL, this would say that the range is about 32 mL to 34 mL (which is a larger range than you found in your measurements). If you gave the value as 32.60, this would say that the range is about 32.59 mL to 32.61 mL (which is too small a range).

To write significant figures correctly, it is important to understand how to count the number of significant figures in a measured value. As you will see, this is most important in working with significant figures in calculations. The value 32.6 has three significant figures, which you obtain by simply counting all the digits. However, some numbers contain zeros, and zeros are not always significant digits.

Consider the number 0.0326. You write the first zero simply to draw attention to the decimal point; it is not a significant figure. You could omit this zero (although we prefer to include it). The zero just to the right of the decimal point is also not a significant figure. The easiest way to understand this is to write the number in scientific notation, which is 3.26×10^{-2}. The sole purpose of the zero after the decimal point is to tell you the correct decimal magnitude of the number (the correct power of 10). The number 0.0326 has three significant figures.

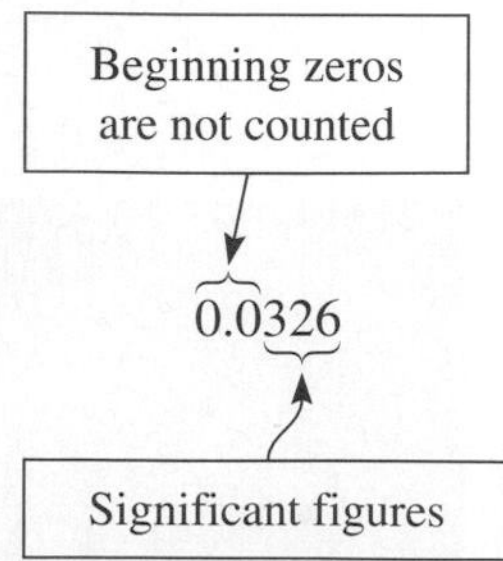

What about the zero at the end of the measured number 0.03260? In this case, the end zero is the least significant figure. Its purpose is to convey the correct range of uncertainty in the measured number. Whenever an end zero is at the right of the decimal point, you can be sure that it is a significant figure. The number 0.03260 has four significant figures.

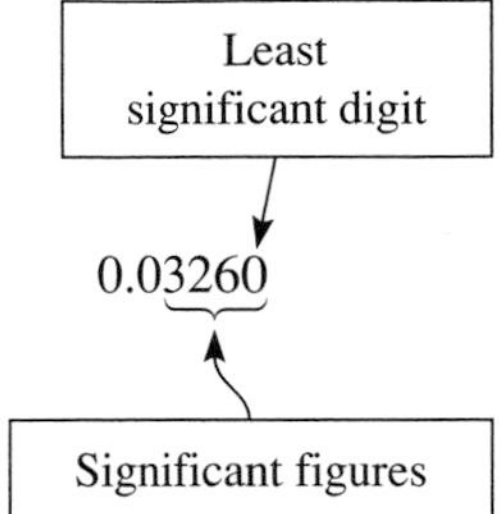

Sometimes the purpose of a zero is not clear. Consider the statement, "The distance between the earth and the sun is 93,000,000 miles." Does this imply 9.3×10^7 miles (in which we have two significant figures because we mean that the sun is between 92 million and 94 million miles away), or does it imply 9.3000000×10^7 miles (in which we have eight significant figures, and we mean that the sun's distance from the earth is between 92,999,999 and 93,000,001 miles)? The answer depends on the purpose of the zeros. Were they written solely to convey the correct magnitude of the number, or were they written to convey the correct uncertainty in the measurement? The person writing the number should clarify the point.

Scientific notation is often the best way to write a number to show which digits are significant. The number 9.3×10^7 has two significant digits in it. In some situations, you may find it useful to add a decimal point as a way of saying that the zeros are significant. For example, if you write 20.°C (note the decimal point), you are saying that the temperature has two significant figures.

The rules for counting significant figures in a number summarize what we have just discussed.

Rules for Counting Significant Figures

1. All digits are significant except zeros at the beginning of a number and possibly zeros at the end of a number. (Each of the numbers 3.56, 3.06, and 0.306 has three significant figures.)
2. End zeros are significant if the number contains a decimal point. (Both 3600. and 3.600 have four significant figures.)
3. End zeros may or may not be significant if the number has no decimal point. Either rewrite the number in scientific notation to clarify the number of significant figures intended, or, if appropriate, add a decimal point.

The following example uses these rules.

■ EXAMPLE 2.3 Counting Significant Figures

Count the number of significant figures in each of the following numbers.

(a) A steel rod is measured and reported to be 10.08 cm long.

(b) A computer reports the time it takes to calculate the class average as 2.150 s.

(c) The volume of solution in a bottle is found to be 280. mL.
(d) You measure the distance you sprinted and write this in your notebook as 0.15 km.
(e) A sample of a shampoo preparation is analyzed; it contains 0.005 g of yellow dye.
(f) The distance between two points is found to be 6800 m.

Problem Analysis

Count all digits, excluding any zeros at the beginning of a number. You should include any zeros at the end of the number in your count if the number has a decimal point in it. If the number does not have a decimal point, but has end zeros, it is not clear how many significant figures are intended.

Solution

(a) Count all digits, including the zeros, since they are neither beginning zeros nor end zeros (rule 1). There are four significant figures.
(b) Count all digits, including the end zero, because the number contains a decimal point (rule 2). There are four significant figures.
(c) Count all digits, including the end zero, because the number contains a decimal point (rule 2). There are three significant figures.
(d) Do not count the beginning zero (rule 1); there are two significant figures.
(e) Do not count the beginning zeros (rule 1); there is one significant figure.
(f) It is not clear how many, if any, of the zeros are significant without more information (rule 3). If the number has three significant figures, you could make this clear by writing it as 6.80×10^3. Then you would count all digits in 6.80 as significant.

Exercise 2.3

How many significant figures are there in each of the following quantities?
(a) 15.07 mL (b) 0.1070 L (c) 58.00 m (d) 4800 km (e) 20.100 g

(Try Problems 2.31, 2.32, 2.33, and 2.34.)

Exact Numbers

So far, the numbers we have discussed have been measured numbers, and in the next section we describe numbers that result from calculations with measured numbers. All of these numbers have some uncertainty associated with them. Here, however, you will encounter *exact numbers,* which arise when you count items or when you define certain units. For example, when you say that there are two other people in the room with you, you mean exactly 2, not 1.9 or 2.1! Also, when you say that there are 12 inches to a foot, you mean exactly 12. ●

● Similarly, there are exactly 2.54 cm in 1 inch.

When you write an exact number, you do not apply the conventions of significant figures to it. When you write that there are 12 inches to the foot, you do not mean that there are only two significant figures in the number 12. In effect, there are an infinite number of significant figures, but it would be impossible to write them. You simply write 12, with the understanding that it is an exact number. There is no uncertainty in an exact number.

2.5 Significant Figures in Arithmetic Results

OBJECTIVES

- Use the rules for obtaining significant figures in an arithmetic result.
- Use the rules for rounding an arithmetic result to a given number of significant figures.

Once you have made some measurements, you may want to do calculations with them. For example, suppose you want to find the floor area of a room. You first measure the length and width of the room; then you multiply the length by the width to find the area. Suppose you find that the length of the room is 6.6 m and the width is 3.8 m. In giving these measurements to two significant figures, you are saying that you know the first digit with certainty, but believe the second digit has some uncertainty in it. To find the area, you multiply the length times the width on your calculator; the calculator result is 25.08 m^2 (meters squared). ●

● Appendix B describes how to use a scientific calculator.

$$\text{Area} = 6.6\text{ m} \times 3.8\text{ m} = 25.08\text{ m}^2 \text{ (calculator result)}$$

You cannot write the answer as this calculator result, however, because it implies less uncertainty than actually exists. Let us look at this further.

Significant Figures in Calculations

If you report the floor area as 25.08 m^2, it means that you know the first three digits (25.0) of the floor area with certainty, but have some uncertainty regarding the last digit. Does this sound correct? Let's try a calculator experiment. The width of the room is 3.8 m. This means that the actual length might be 3.7 m, since there is some uncertainty in the second digit. Suppose you multiply 6.6 m by 3.7 m.

$$6.6\text{ m} \times 3.7\text{ m} = 24.42\text{ m}^2$$

Notice that the result of the calculation has changed from 25.08 m^2 to 24.42 m^2, a change in the second digit. You conclude that the calculated floor area has some uncertainty in the second digit. Instead of leaving the calculated floor area as 25.08 m^2, you write it as 25 m^2 (two significant figures) to reflect this uncertainty. A person reading 25 m^2 would understand that the actual value lies in the approximate range of 24 m^2 to 26 m^2.

In general, because measured numbers have uncertainties in them, the result of a calculation with these numbers has an uncertainty. You express this uncertainty by writing the result to the proper number of significant figures. But how do you choose the proper number of significant figures without doing the type of calculator experiment we just did? There are two rules for deciding the number of significant figures in the result of a calculation: one for the multiplication or division of numbers and one for the addition or subtraction of numbers.

Rules for Significant Figures in Calculations

1. When multiplying or dividing quantities, give as many significant figures in the answer as there are in the quantity with the fewest number of significant figures.
2. When adding or subtracting quantities, give the same number of decimal places in the answer as there are in the quantity with the fewest number of decimal places.

To illustrate rule 1, consider the following calculation:

$$5.7\underline{1}\text{ m} \times 7.\underline{9}\text{ m} = ?$$

The quantities have been written with the proper number of significant figures, and the digit with some uncertainty (least significant digit) is underlined. These uncertainties will affect the answer: according to rule 1, the quantity with the fewer number of significant figures (7.9 m) limits the number of significant figures in the answer (to two). Thus, there should be two significant figures in the answer.

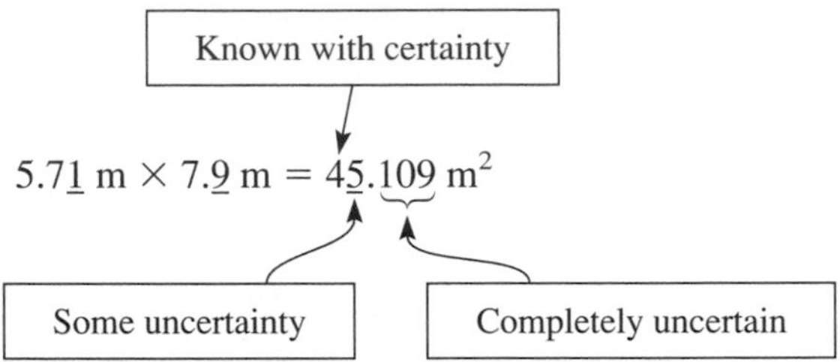

The correct answer is 45 m^2.

Now let's illustrate rule 2. Suppose we add two masses, 296.2 g and 3.246 g.

$$\begin{array}{r} 296.\underline{2}\text{ g} \\ 3.24\underline{6}\text{ g} \\ \hline 299.\underline{4}46\text{ g} \end{array}$$

The two quantities we have added have uncertainties in their last digits, which we have underlined. These uncertainties carry into the answer. Thus, $296.\underline{2}$ is uncertain in the tenths position, and this uncertainty carries into the tenths position of the answer ($299.\underline{4}46$ g). Digits further to the right are then completely uncertain (nonsignificant), so you would drop them and write the answer as 299.4 g. Rule 2, which states that the number of decimal places you write for the answer is the same as in the quantity with the fewer number of decimal places, agrees with this result. Here, 296.2 g has one decimal place, whereas 3.246 g has three decimal places. Therefore, the answer should have only one decimal place. You write the answer as 299.4 g. ●

● Do not confuse the rule for addition and subtraction (which depends on decimal places) with the rule for multiplication and division (which depends on significant figures).

Rounding of Numbers

As you have seen from the previous calculations, the result displayed by a calculator may have more digits than are warranted by the rules for significant figures. In such a case, you need to reduce the calculator result to the correct number of significant figures. The procedure for doing this is called *rounding*. We rounded the answers in the previous calculations by dropping nonsignificant digits. Sometimes, however, you need to add 1 to the last digit that you retain, depending on the digits dropped. Thus, **rounding** is *the procedure of dropping nonsignificant digits in a calculation result and perhaps adjusting the last remaining digit upward.*

Suppose a room is 6.6 m long and 4.83 m wide. The calculator result for the area of the room is

$$6.\underline{6}\text{ m} \times 4.8\underline{3}\text{ m} = 3\underline{1}.878\text{ m}^2$$

Look at the quantities on the left, the first one ($6.\underline{6}$ m) has the fewer number of significant figures (two), so it limits the number of significant figures in the result. In the answer, we have underlined the second digit ($3\underline{1}.878\ m^2$), in which we expect some uncertainty. The digits to the right of this digit are nonsignificant, and we drop them when we round. But we still need to determine if we need to increase the rightmost digit retained. To determine if this digit needs to be increased, we look at the digits that are dropped and apply the following rules. (Some calculators round using these rules.)

Rules for Rounding

1. If the first digit to be dropped is less than 5, leave the preceding digit as it is.
2. If the first digit to be dropped is 5 or greater, increase the preceding digit by 1.

From rule 2, you can see that the number $3\underline{1}.878$ rounds to 32. (The first digit to be dropped is 8, which is greater than 5.) The following example further illustrates the use of significant figures when doing calculations and rounding.

■ EXAMPLE 2.4 Obtaining Significant Figures and Rounding in Calculations

Perform the following arithmetic and give the answers to the correct number of significant figures.

(a) $14.9 - 6.74 =$

(b) $8.924 \div 2.5 =$

Problem Analysis

Carry out each computation on a calculator, rounding the result using the rules for significant figures in calculations and the rules for rounding.

Solution

(a) You use the significant figures rule for the addition and subtraction of numbers. The calculator answer is 8.16. However, the number 14.9 has the fewer number of decimal places (one) and therefore limits the number of decimal places in the answer to one (rule 2 for significant figures in calculations). So you write the result as $8.\underline{1}6$. This number rounds to 8.2 because the first digit dropped (6) is more than 5 (rule 2 for rounding).

$$14.\underline{9} - 6.7\underline{4} = 8.\underline{1}6 \qquad \text{(rounds to 8.2)}$$

(b) You use the significant figures rule for multiplication and division. The quotient of the numbers read from a calculator is 3.5696. The number 2.5 has the fewer number of significant figures and therefore limits the number of significant figures in the answer (rule 1 for significant figures in calculations). The result $3.\underline{5}696$ rounds to 3.6, since the first digit dropped (6) is more than 5 (rule 2 for rounding).

$$8.92\underline{4} \div 2.\underline{5} = 3.\underline{5}696 \qquad \text{(rounds to 3.6)}$$

Exercise 2.4

Carry out the following arithmetic and report the answers to the correct number of significant figures.

(a) $34.1 + 62.98 =$ (b) $112.36 - 84.1 =$ (c) $8.92 \times 3.456 =$
(d) $11.54 \div 3.6 =$

(Try Problems 2.35, 2.36, 2.37, and 2.38.)

If your calculation has two or more steps, it is best to round only the final answer. This ensures that small errors from rounding do not creep into your work. It is also the simplest way to proceed, because you simply enter numbers into your calculator one after the other, retaining the answer from an intermediate step for the next calculation. You will have to keep track of where the least significant figure is in any intermediate result, however, to obtain the correct number of significant figures for the final answer. The next example illustrates this idea.

■ EXAMPLE 2.5 Obtaining Significant Figures in a Multistep Calculation

Carry out the following arithmetic and give the answer to the correct number of significant figures.

$$78.35 \times (5.98 - 5.21) =$$

Problem Analysis

Start the calculation by doing the arithmetic within parentheses first, leaving the intermediate answer in the calculator for the next step. Keep track of significant figures on a sheet of paper by underlining the least significant figure; then, at the end of the calculation, round the final answer.

Solution

First underline the least significant figure in each number in the calculation.

$$78.3\underline{5} \times (5.9\underline{8} - 5.2\underline{1}) =$$

Proceed step by step, starting with the calculation in parentheses first, in which you subtract 5.21 from 5.98.

$$5.9\underline{8} - 5.2\underline{1} = 0.7\underline{7}$$

With 0.77 still in your calculator, the remaining arithmetic is

$$78.3\underline{5} \times 0.7\underline{7} = 6\underline{0}.3295$$

Note that the factor $0.7\underline{7}$ determines the number of significant figures in the multiplication. This is why we have underlined the second figure in the answer. Now you round the calculator answer $6\underline{0}.3295$ to 60., so the final answer has two significant figures. Note that the decimal point after 60 indicates that the zero is one of those significant figures.

Exercise 2.5

Obtain the answers to the following calculations, giving the proper number of significant figures.

(a) $10.5 \times (5.62 + 6.34) =$

(b) $5.84 - \dfrac{1.23}{3.2} =$

(Try Problems 2.39 and 2.40.)

CALCULATIONS

Many of the calculations that you will encounter in introductory chemistry require some simple algebraic manipulations. Others involve a change from one unit to another. We demonstrate both types in the next three sections. We defer calculations with logarithms until Chapter 15.

2.6 Temperature and Changing Temperature Scales

OBJECTIVES

- Define an absolute temperature scale.
- Define the Kelvin temperature scale and give the freezing point and boiling point of water on this scale.
- Define the Celsius and Fahrenheit temperature scales and give the freezing point and boiling point of water on each of these scales.
- Convert temperatures between the Celsius and Kelvin temperature scales.
- Convert temperatures between the Celsius and Fahrenheit temperature scales.

● Do not confuse temperature with *heat,* which is a form of energy.

● We use the Kelvin scale in Chapter 11, where we discuss the properties of gases.

Temperature is measured with a thermometer, and it is a measure of the vigor of motion of atoms in a sample of matter at a given point in the sample. ● At very low temperatures, the atoms move, or vibrate, slowly. When you raise the temperature, the atoms move more violently. You might imagine, then, a temperature at which atomic motion essentially ceases. An *absolute temperature scale* is one in which this temperature is given the value zero. Moreover, it is the lowest possible temperature on this scale. The Kelvin scale (K) is an absolute temperature scale. ● However, the common temperature scale in the United States is the Fahrenheit scale (°F); in Canada and many other places, it is the Celsius (pronounced "sell′-see-us") scale (°C). The Celsius scale is the scale most often used in scientific work, and it is easily related to the Kelvin scale.

Comparison of Temperature Scales

The most common type of thermometer consists of a glass capillary containing a column of liquid, perhaps mercury or alcohol dyed red. Because a liquid expands as the temperature increases, the liquid column rises with increasing temperature. A scale alongside the capillary gives a measure of the temperature. Figure 2.10 compares the Kelvin, Celsius, and Fahrenheit scales.

The **Celsius scale** is *the temperature scale for general use in much of the world and for scientific use worldwide.* On this scale, the freezing point of water is 0°C, and the

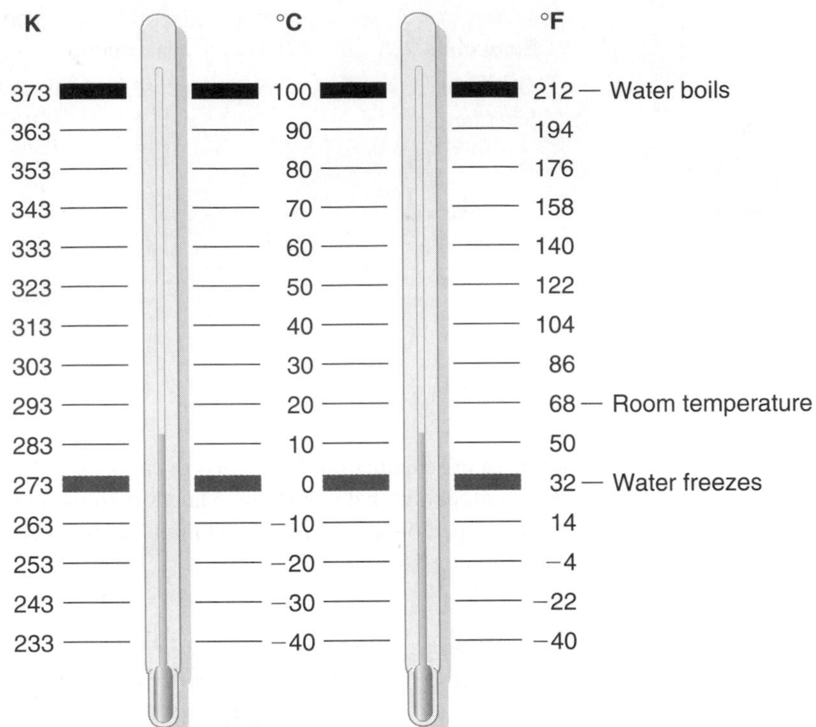

FIGURE 2.10
Comparison of Kelvin, Celsius, and Fahrenheit scales. Water boils at 373 K, 100°C (exactly), and 212°F (exactly); it freezes at 273 K, 0°C (exactly), and 32°F (exactly). Room temperature is about 293 K, 20°C, and 68°F.

boiling point of water at normal barometric pressure is 100°C. On the **Fahrenheit scale,** *the scale in common usage in the United States,* the freezing point of water is 32°F, and the boiling point of water at normal barometric pressure is 212°F. Negative temperatures are possible with both of these scales. For example, liquid nitrogen boils at −321°F and −196°C. Another well-known temperature is discussed in the Chemical Perspective at the end of this section.

In contrast, the **Kelvin scale** is *an absolute temperature scale* on which the lowest possible temperature is zero, or 0 K (read "zero kelvins"). A negative temperature is not possible, unlike the Celsius and Fahrenheit temperature scales. Note that the unit is the **kelvin** (K), which is *the SI base unit of temperature,* and that the degree symbol is not used with the Kelvin scale. The Celsius and Kelvin scales have the same size units, but 0°C is equivalent to exactly 273.15 K. For our purposes, the nearest whole unit will be sufficient. Thus, the freezing point of water on the Kelvin scale is 273 K, and the boiling point is 373 K at normal barometric pressure. Next, we show how to convert temperature from one scale to another.

Relating Kelvin and Celsius Scales

The Kelvin and Celsius scales are related by the following equation:

$$K = {}^{\circ}C + 273$$

Thus, if you have a temperature in Celsius, you change it to Kelvin by adding 273 to it. But what do you do if you have a temperature in Kelvin and want to change it to Celsius? You can obtain the relationship for this problem by algebraic manipulation of the defining equation. We discuss algebraic manipulation of equations in Appendix A, but the essential principle from that discussion is *whatever operation you do to one side of an equation, you must also do to the other side*. Applying this principle, you can solve the defining equation for °C by subtracting 273 from both sides.

$$\text{K} - 273 = {}^\circ\text{C} + 273 - 273$$

Combining the numbers gives

$$\text{K} - 273 = {}^\circ\text{C}$$

or

$$^\circ\text{C} = \text{K} - 273$$

The next example illustrates the use of these equations.

■ EXAMPLE 2.6 Changing Between Celsius and Kelvin Scales

(a) Ethanol (ordinary alcohol) boils at 78°C under normal atmospheric pressure. Express this boiling point on the Kelvin scale.

(b) Liquid oxygen (LOX), used in rocket engines, boils at 90. K under normal atmospheric pressure. Express this boiling point on the Celsius scale.

Problem Analysis

Just remember that K = °C + 273. Use this equation to change from Celsius to Kelvin. When you need to change from Kelvin to Celsius, rearrange this equation to obtain °C = K − 273. Do not clutter your mind with both equations.

Solution

(a) Substitute 78 for °C into the main equation.

$$\text{K} = {}^\circ\text{C} + 273 = 78 + 273 = 351$$

Thus, ethanol boils at 351 K.

(b) Rearrange the main equation to obtain °C in terms of K. Then substitute into this equation. You start with

$$\text{K} = {}^\circ\text{C} + 273$$

Next rearrange the equation to

$$\text{K} - 273 = {}^\circ\text{C} \quad \text{or} \quad {}^\circ\text{C} = \text{K} - 273$$

Now substitute 90. for K in the equation.

$$^\circ\text{C} = \text{K} - 273 = 90. - 273 = -183$$

Thus, liquid oxygen boils at −183°C.

Exercise 2.6

The temperature in Detroit one summer day was 296 K. What was the temperature in degrees Celsius?

(Try Problems 2.41, 2.42, 2.43, and 2.44.)

Relating Celsius and Fahrenheit Scales

We can obtain the relationship between the Celsius and Fahrenheit scales by considering the freezing point and boiling point of water on these two scales (see Figure 2.10). On the Fahrenheit scale, water freezes at 32°F and boils at 212°F, a difference of 180°. On the Celsius scale, water freezes at 0°C and boils at 100°C, a difference of 100°. This means that the Celsius degree is larger than the Fahrenheit degree. The Celsius degree is 180/100, or 1.8, times as big as the Fahrenheit degree. Note that the factor 1.8 is an exact number.

To change from Celsius to Fahrenheit, you need an equation relating the two scales. You can obtain this equation from a simple argument that goes like this. The Celsius degree is larger than the Fahrenheit degree, so you will have fewer Celsius degrees, and more Fahrenheit degrees, between any two temperatures. Therefore, you multiply °C by 1.8. However, remember that the freezing point of water is 0°C, but 32°F. So you also need to add 32 to 1.8°C to obtain degrees Fahrenheit. The equation relating degrees Fahrenheit to degrees Celsius is

$$°F = 1.8°C + 32$$

You can check this by converting 0°C to degrees Fahrenheit (the result is 32°F) and by converting 100°C to degrees Fahrenheit. Here is the calculation:

$$°F = (1.8 \times 100) + 32 = 180 + 32 = 212$$

You can obtain the equation for converting degrees Fahrenheit to degrees Celsius by rearranging the original equation. Subtract 32 from both sides of the equation:

$$°F - 32 = 1.8°C + 32 - 32$$

or $$°F - 32 = 1.8°C$$

or $$1.8°C = °F - 32$$

Now you divide both sides of the equation by 1.8:

$$°C = \frac{°F - 32}{1.8}$$

The conversion between the Celsius and Fahrenheit scales is illustrated in the following example.

■ EXAMPLE 2.7 Changing Between Celsius and Fahrenheit Scales

(a) Normal body temperature is 37.0°C. What is this temperature on the Fahrenheit scale?

(b) A thermometer in your room registers 78°F. What is this temperature on the Celsius scale?

Problem Analysis

Remember that °F = 1.8°C + 32. You can use this equation directly to change from the Celsius scale to the Fahrenheit scale. To change from the Fahrenheit scale to the

Celsius scale, you need to rearrange the equation to solve for degrees Celsius in terms of degrees Fahrenheit. Again, don't try to memorize both equations. Remember the simpler equation (°F = 1.8°C + 32) and obtain the other by rearranging this equation.

Solution

(a) You write the equation relating the Fahrenheit scale to the Celsius scale, then substitute 37.0 for °C:

$$°F = 1.8°C + 32 = (1.8 \times 37.0) + 32 = 66.6 + 32 = 98.6$$

Remember that the factor 1.8 is exact (as is the number 32) and does not affect the number of significant figures in the answer. Normal body temperature is 98.6°F.

(b) You rearrange the main equation, solve for degrees Celsius, and obtain

$$°C = \frac{°F - 32}{1.8}$$

Then you substitute the degrees Fahrenheit into the equation.

$$°C = \frac{7\underline{8} - 32}{1.8} = 2\underline{5}.555556 \qquad \text{(rounds to 26)}$$

Your room temperature is 26°C.

Exercise 2.7

(a) A certain bacterial culture was maintained at 35°C. What is this temperature in degrees Fahrenheit?

(b) A friend runs a fever with a temperature of 102°F. Give this temperature in degrees Celsius.

(Try Problems 2.45, 2.46, 2.47, and 2.48.)

Chemical Perspective

"Normal" Body Temperature

Heat is the inevitable by-product of all chemical reactions within a human body. If that heat were to accumulate, body temperature would rise steadily. However, part of the heat is lost to the surroundings. When the heat that is lost comes into balance with heat production, a person attains a fairly steady body temperature. The "normal" body temperature of an average person is 98.6°F (37.0°C) when the temperature is measured orally. Nevertheless, the oral temperature may vary from as low as 97.0°F (36.1°C) in the morning to as high as 99.0°F (37.2°C) at night. Thus, the body temperature of a healthy person is not really constant; it may vary as much as 2.0°F (about 1°C) during a normal day.

2.7 Problem Solving and Dimensional Analysis

OBJECTIVES

- Define *dimensional analysis.*
- Write an equivalence statement relating a given unit to a desired unit.
- Define *conversion factor.*
- Write a conversion factor, given the relationship (equivalence statement) between two units.
- Convert a quantity from one unit to another, knowing the conversion factor.
- Write a multistep conversion of a quantity from one unit to another.

Many problems that you will encounter later in this book involve a change from one unit to another. However, this type of calculation should be familiar to you already because of everyday situations. For example, suppose that a visitor from Europe reads a road sign in the United States that tells her the distance to a particular city is 213 miles. What is the distance in kilometers (the unit she is accustomed to in Europe)? Here the visitor wishes to change from a distance in which the unit is miles to the equivalent distance in which the unit is kilometers—a change from one unit to another. **Dimensional analysis** will help us do this kind of calculation because it is *a general problem-solving method in which you use the units of quantities to help you decide how to set up the problem.*

How might you convert the 213 miles in the problem to kilometers? To begin, write the conversion that you want to do:

$$213 \text{ mi} \xrightarrow{\text{Convert to}} \text{km}$$

Next, find information that will relate the number of kilometers to the number of miles. The information you need can be found in Table 2.4 in the form of an equivalence:

$$1 \text{ mi} = 1.609 \text{ km}$$

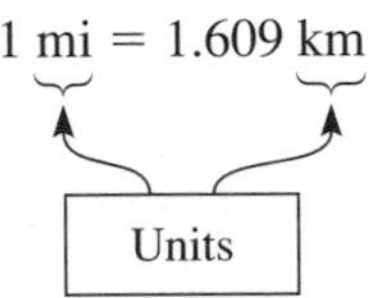

We will treat the labels associated with numbers in such equivalences as units.

Suppose you divide both sides of this equivalence by "1 mi." The unit "mi" cancels on the left side, so that the left side becomes equal to exactly 1.

$$\frac{1 \cancel{\text{mi}}}{1 \cancel{\text{mi}}} = \frac{1.609 \text{ km}}{1 \text{ mi}}$$

$$= 1$$

As a result the right side also equals 1. Moreover, if the factor "1.609 km/1 mi" equals 1, then the reciprocal factor "1 mi/1.609 km" must also equal 1. Thus, the following two factors are each equal to *exactly* 1:

$$\frac{1 \text{ mi}}{1.609 \text{ km}} \quad \text{and} \quad \frac{1.609 \text{ km}}{1 \text{ mi}}$$

Because each factor equals 1, you can multiply any quantity by these factors without changing its value.

These factors can function as *conversion factors.* A **conversion factor** is *a factor equal to 1 that converts a quantity in one unit to an equivalent quantity in another unit.* If you multiply 213 mi by the second of these conversion factors, you will convert 213 miles to an equivalent quantity in kilometers.

TABLE 2.4
Relationships of Some U.S. and Metric Units

Length	Mass	Volume
1 in. = 2.54 cm (exact)	1 lb = 0.4536 kg	1 qt = 0.9464 L
1 yd = 0.9144 m (exact)	1 lb = 16 oz (exact)	4 qt = 1 gal (exact)
1 mi = 1.609 km	1 oz = 28.35 g	
1 mi = 5280 ft (exact)		

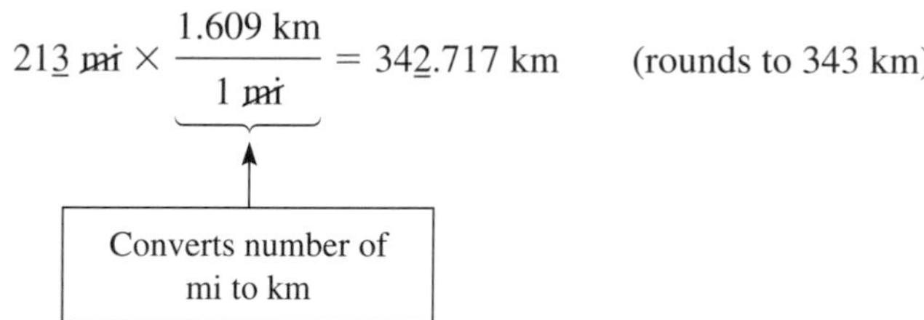

$$21\underline{3}\ \cancel{\text{mi}} \times \frac{1.609\ \text{km}}{1\ \cancel{\text{mi}}} = 34\underline{2}.717\ \text{km} \qquad \text{(rounds to 343 km)}$$

The unit "mi" in the denominator (the bottom of the fraction) of the conversion factor cancels the unit "mi" in the given quantity (213 mi), leaving the desired unit "km" in the numerator (the top of the fraction). You (and the visitor) have accomplished what you set out to do—you know that 213 mi is equivalent to 343 km.

The following steps summarize this approach to problem solving by dimensional analysis.

Steps for Converting Units by Dimensional Analysis

1. Write out the desired conversion. From the problem statement, note the given unit and the desired unit. You want to convert as follows:

$$\text{Quantity expressed in given unit} \xrightarrow{\text{Convert to}} \text{quantity expressed in desired unit}$$

2. Obtain the conversion factor. Write the equivalence statement, then write the conversion factor for the conversion from given unit to desired unit. You write the conversion factor as a ratio with the given unit in the denominator and the desired unit in the numerator. The conversion factor will be a number, or factor, multiplied by a ratio of units:

$$\text{Conversion factor} = \text{factor} \times \frac{\text{desired unit}}{\text{given unit}}$$

3. Perform the conversion calculation. Multiply the quantity in the given unit by the conversion factor:

$$\text{Quantity in given unit} \times \underbrace{\left(\text{factor} \times \frac{\text{desired unit}}{\text{given unit}}\right)}_{\substack{\text{Conversion factor:} \\ \text{converts given unit} \\ \text{to desired unit}}} = \text{quantity in desired unit}$$

● If you have any difficulty doing calculations with powers of 10, see Appendix A for a review of this topic.

One of the most important unit conversions is from a metric unit without a prefix to a corresponding one with a prefix, and vice versa. ● This type of conversion is illustrated in the following example.

■ **EXAMPLE 2.8 Converting from a Metric Unit Without a Prefix to a Metric Unit With a Prefix (and Vice Versa)**

(a) The straight-line distance from Boston to New York is 2.99×10^5 m. How far is this in kilometers?

(b) The volume of water in the world's oceans is 1.35×10^{24} mL. What is this volume in liters?

Problem Analysis

From the problem statement, write down the desired conversion (step 1). Use the meaning of the metric prefix to obtain a conversion factor (step 2). Finally, perform the conversion calculation (step 3).

Solution

(a) The desired conversion is

$$2.99 \times 10^5 \text{ m} \xrightarrow{\text{Convert to}} \text{km}$$

Note that

$$1 \text{ km} = 10^3 \text{ m}$$

Therefore, the conversion factor from meters to kilometers is

$$\frac{1 \text{ km}}{10^3 \text{ m}}$$

because the desired unit is in the numerator and the given unit is in the denominator. The calculation is

$$2.99 \times 10^5 \cancel{\text{m}} \times \underbrace{\frac{1 \text{ km}}{10^3 \cancel{\text{m}}}}_{\text{Converts m to km}} = 2.99 \times 10^2 \text{ km}$$

You get the correct power of 10 by subtracting exponents ($5 - 3 = 2$), giving the power of 10 in the answer as 10^2. Notice that the unit on the bottom of the conversion factor cancels the given unit in the preceding factor.

(b) The conversion that you want is

$$1.35 \times 10^{24} \text{ mL} \xrightarrow{\text{Convert to}} \text{L}$$

Note that the conversion factor from milliliters to liters is

$$\frac{10^{-3} \text{ L}}{1 \text{ mL}}$$

Again, note that the given unit is on the bottom, and the desired unit is on the top. The calculation is

$$1.35 \times 10^{24}\ \cancel{\text{mL}} \times \underbrace{\frac{10^{-3}\ \text{L}}{1\ \cancel{\text{mL}}}}_{\text{Converts mL to L}} = 1.35 \times 10^{21}\ \text{L}$$

As you can see, the unit on the bottom in the conversion factor cancels the given unit in the preceding factor.

Exercise 2.8

Convert the following.

(a) 2.58 g to kg (b) 55.4 cm to m (c) 7.11 L to mL.

(Try Problems 2.49 and 2.50.)

Sometimes you will find it necessary to do multiple conversions to convert a metric unit having a prefix to another metric unit with a prefix. The next example is an illustration of this type of multistep calculation.

EXAMPLE 2.9 Conversion Between Metric Units (Multistep Conversion)

To monitor how quickly a certain prescription drug is absorbed by a patient, a nurse takes a sample of blood from the patient after giving him a prescribed dose. Analysis of the blood shows that it contains 2.9 micrograms (μg) of the drug. What is the mass of this amount of drug expressed in milligrams?

Problem Analysis

Write the conversion from the given metric unit to the SI unit without a prefix, and then from the SI unit without a prefix to the desired metric unit with a prefix (step 1). From these conversions, decide what the conversion factors are (step 2), and then perform the calculation, multiplying the given quantity by the conversion factors (step 3).

Solution

The needed conversions are

$$2.9\ \mu\text{g} \xrightarrow{\text{Convert to}} \text{g} \xrightarrow{\text{Convert to}} \text{mg}$$

The conversion factors from micrograms to grams and from grams to milligrams are derived from the meanings of the prefixes *micro* and *milli* (see Table 2.2):

$$1\ \mu\text{g} = 10^{-6}\ \text{g}$$
$$1\ \text{mg} = 10^{-3}\ \text{g}$$

The conversion factors for this problem are

$$\frac{10^{-6}\ \text{g}}{1\ \mu\text{g}} \quad \text{and} \quad \frac{1\ \text{mg}}{10^{-3}\ \text{g}}$$

To convert from micrograms to grams, you write down the conversion calculation using the first of these factors:

$$2.9\ \cancel{\mu\text{g}} \times \underbrace{\frac{10^{-6}\ \text{g}}{1\ \cancel{\mu\text{g}}}}_{\text{Converts } \mu\text{g to g}} = 2.9 \times 10^{-6}\ \text{g}$$

Then you convert from grams to milligrams using this result and the second conversion factor:

$$2.9 \times 10^{-6}\ \cancel{\text{g}} \times \underbrace{\frac{1\ \text{mg}}{10^{-3}\ \cancel{\text{g}}}}_{\text{Converts g to mg}} = 2.9 \times 10^{-3}\ \text{mg}$$

After you have gained facility with the ideas you need to do multiple conversions, you will see that you can actually string several conversions together in a single line, rather than do them separately. The calculation for this problem becomes

$$2.9\ \cancel{\mu\text{g}} \times \frac{10^{-6}\ \cancel{\text{g}}}{1\ \cancel{\mu\text{g}}} \times \frac{1\ \text{mg}}{10^{-3}\ \cancel{\text{g}}} = 2.9 \times 10^{-3}\ \text{mg}$$

Note how units cancel on the diagonal, leaving you with the desired unit.

Exercise 2.9

Do the following conversions:

(a) 2.45 ns to ms (b) 7.38 kg to mg (c) 5.02 μL to mL

(d) 89.7 cm to km (e) 6.11 ms to μs

(Try Problems 2.51, 2.52, 2.53, and 2.54.)

2.8 Density

OBJECTIVES

- Define *density.*
- Rearrange the equation that defines density, using algebraic manipulations, to obtain equations for volume and mass.
- Calculate one of the quantities—density, mass, or volume—given the other two.

You may hear someone say that iron is heavier than wood. A more precise statement would be that iron has a greater density than wood (or, simply, iron is more dense than wood). This means that a given volume of iron has greater mass than the same volume of wood. The **density** of a substance is *the mass of the substance per unit volume.*

Expressed mathematically, this definition reads

$$\text{Density} = \frac{\text{mass}}{\text{volume}}$$

● The word *per* means "divided by." When you refer to a speed in miles per hour, you mean "miles divided by time in hours."

Thus, the density of a substance is its mass per volume. ● Any substance—solid, liquid, or gas—has a density. The Chemical Perspective at the end of this chapter shows how the density of your body reveals the amount of body fat you have.

Unlike mass or volume, the density of a substance is a property that is independent of the quantity of substance. Whether you have 5.0 g of iron or 50. g of iron, the densities of both samples are the same: 7.87 g/cm^3 (which you read as "7.87 grams per cubic centimeter").

Calculation of Density

Suppose you want to obtain the density of oak wood. You measure the volume of a sample of the wood; then you measure its mass. You find the volume of the sample to be 7.7 cm^3 and its mass to be 5.5 g. Now you substitute these values into the defining equation for the density.

$$\text{Density} = \frac{\text{mass}}{\text{volume}} = \frac{5.5\text{ g}}{7.7\text{ cm}^3} = 0.71\text{ g/cm}^3$$

Note that you obtain the unit of density (g/cm^3) directly from the calculation, as long as you substitute both numbers and units. *It is generally good practice to use units as well as numbers in a calculation.*

Density is an important identifying characteristic of a substance. You are not likely to mistake a gold-plated iron bar for a solid gold bar. Iron metal has a density of 7.87 g/cm^3, whereas gold has a density of 19.3 g/cm^3. So a gold bar would weigh nearly 2.5 times as much as an iron bar of the same dimensions. The next example illustrates the use of density to help identify a substance.

■ EXAMPLE 2.10 Calculating Density from Mass and Volume

A sample of a metal has a volume of 4.05 cm^3 and a mass of 36.2 g. The metal is known to be either iron, nickel, or platinum. The densities of these metals are 7.87 g/cm^3, 8.90 g/cm^3, and 21.45 g/cm^3, respectively. What is the identity of the metal in the sample?

Problem Analysis

Calculate the density of the unknown metal sample by substituting the given data into the defining equation for density. Then compare this value with the densities of the known metals.

Solution

You first calculate the density of the sample.

$$\text{Density} = \frac{\text{mass}}{\text{volume}} = \frac{36.2\text{ g}}{4.05\text{ cm}^3} = 8.9\underline{3}8272\text{ g/cm}^3 \qquad \text{(rounds to 8.94 g/cm}^3\text{)}$$

The calculated density is greater than that of iron and much less than that of platinum. The value 8.94 g/cm^3 nearly equals the density given for nickel (8.90 g/cm^3). The difference is probably the result of experimental error. We conclude that the sample is composed of nickel.

Exercise 2.10

Bromine is a red-brown, poisonous liquid with an unpleasant odor. A chemist first weighed a small graduated cylinder, then poured a sample of bromine into the cylinder and weighed it and the liquid. She also observed the volume of bromine in the graduated cylinder. She reported the following data in her notebook:

Mass of cylinder and bromine	38.53 g
Mass of cylinder	24.59 g
Volume of bromine	4.50 cm^3

What is the density of bromine?

(Try Problems 2.57 and 2.58.)

In certain technical applications, such as medical laboratory tests, you may see specific gravities of samples reported rather than densities. The *specific gravity* of a substance is the ratio of the density of the substance to the density of another substance chosen as a reference (usually water at 4°C).

$$\text{Specific gravity} = \frac{\text{density of substance}}{\text{density of water (4°C)}}$$

Specific gravity has no units because the densities of the substance and water have the same units and therefore cancel. Also, because water at 4°C has a density of 1.000 g/cm^3, the specific gravity of a substance has the same numerical value (to four significant figures) as the density expressed in grams per cubic centimeter. Ethanol, for example, has a density of 0.789 g/cm^3 and a specific gravity of 0.789.

One of the most important uses of density is to convert the volume of a substance to its mass, or vice versa. You can approach these problems in either of two ways: algebraic manipulation of the density equation or dimensional analysis. Let's see how you can use each one of them.

Algebraic Manipulation of the Density Equation

In Section 2.6, we used algebraic manipulation to change the equation relating degrees Fahrenheit in terms of degrees Celsius to give an equation relating degrees Celsius in terms of degrees Fahrenheit. We used a rule from algebra: *Whatever operation you do to one side of an equation, you must also do to the other side.*

You can use this rule to rearrange the defining equation for the density to give two other equations: one that gives the mass of a substance in terms of its density and volume, and another equation that gives the volume in terms of density and mass. Going through the derivations of these equations will give you some practice in the algebraic manipulation of equations.

To get the equation that gives mass in terms of density and volume from

$$\text{Density} = \frac{\text{mass}}{\text{volume}}$$

you multiply *both* sides by volume:

$$\text{Density} \times \text{volume} = \frac{\text{mass}}{\cancel{\text{volume}}} \times \cancel{\text{volume}}$$

After canceling the volume on the right side, you obtain

$$\text{Density} \times \text{volume} = \text{mass}$$

or

$$\text{Mass} = \text{density} \times \text{volume}$$

This equation allows you to obtain the mass of an object if you know its density and volume. For example, suppose you measure out a volume of liquid using a graduated cylinder. You can obtain the mass of this volume, using the previous equation, if you also know the density of the liquid.

Using similar types of manipulations, you can obtain the volume in terms of the mass and density:

$$\text{Volume} = \frac{\text{mass}}{\text{density}}$$

This equation allows you to obtain the volume of an object if you know its mass and density. For example, you can obtain the volume of an irregularly shaped piece of iron by weighing it, then substituting the mass and the density of iron (7.87 g/cm^3) into this equation.

Do not try to memorize either one of the equations we have derived from the defining equation for density. All you need is the defining equation and algebraic manipulation. We postpone looking at examples of these calculations until after we examine how dimensional analysis can be used to solve the same types of problems.

Dimensional Analysis in Density Calculations

You can also think of density as a conversion factor. For example, the density of iron (Fe) is 7.87 g/cm^3 or

$$\text{Density of iron} = \underbrace{\frac{7.87\ \text{g Fe}}{1\ \text{cm}^3\ \text{Fe}}}$$

Converts cm^3 Fe to g Fe

Thus, the density can be used as a conversion factor that converts cm^3 Fe to g Fe.

If you invert the factor from the density equation, you obtain one that converts mass to volume. For example, the conversion factor

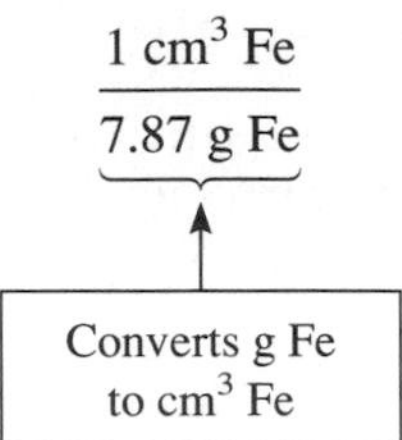

Thus, this conversion factor converts cm^3 Fe to g Fe.

Let's look at some examples in which we solve problems by both algebra and dimensional analysis.

EXAMPLE 2.11 Calculating Mass from Density and Volume

Ethanol (grain alcohol) has a density of 0.789 g/cm^3. If the volume of a sample of alcohol is 39.7 mL, what is the mass?

Problem Analysis

Use either one of the following methods.

1. *Algebraic Method.* Solve the defining equation for density to give the mass in terms of the volume and density. Then substitute the volume and density into this equation and calculate the mass.
2. *Dimensional Analysis.* Use the density of ethanol to obtain a factor that converts from cubic centimeters of ethanol to grams of ethanol.

Solution

1. *Algebraic Method.* Rearrange the defining equation to obtain the mass in terms of density and volume. The equation you obtain is

$$\text{Mass} = \text{density} \times \text{volume}$$

Now substitute the values for density and volume (noting that mL = cm^3).

$$\text{Mass} = 0.789\ g/\cancel{cm^3} \times 39.7\ \cancel{cm^3} = 31.\underline{3}233\ g \qquad \text{(rounds to 31.3 g)}$$

Note that the units cm^3 cancel and give you the unit g, a unit of mass, as expected. If you had inadvertently used the incorrect formula for the mass, you would not have gotten the correct units. By substituting units as well as numbers, you have a check on your work.

2. *Dimensional Analysis.* Use the density to write the conversion factor that converts from cm^3 ethanol to g ethanol. It is

$$\frac{0.789\ g\ ethanol}{1\ cm^3\ ethanol}$$

Use this factor to convert 39.7 mL (or 39.7 cm^3) to g ethanol.

$$39.7\ \cancel{cm^3\ ethanol} \times \frac{0.789\ g\ ethanol}{1\ \cancel{cm^3\ ethanol}} = 31.3\ g\ ethanol$$

Exercise 2.11

An experiment calls for 8.4 cm^3 of nitric acid. You decide to measure out this quantity of nitric acid by weighing the required amount, using the known density of nitric acid (1.50 g/mL). What mass of nitric acid do you need?

(Try Problems 2.59 and 2.60.)

EXAMPLE 2.12 Calculating Volume from the Density and Mass

An experiment requires 35.4 g of chloroform (a liquid). What volume of the liquid would you need to equal this mass of chloroform? The density of chloroform is 1.48 g/cm^3.

Problem Analysis

Use either one of the following methods.

1. *Algebraic Method.* Solve the defining equation for density to give the volume in terms of mass and density. Then substitute the mass and density into this equation and calculate the volume.
2. *Dimensional Analysis.* Use the density of chloroform to obtain a factor that converts from grams of chloroform to cubic centimeters of chloroform.

Solution

1. *Algebraic Method.* Rearrange the defining equation to obtain the volume in terms of density and mass. You obtain

$$\text{Volume} = \frac{\text{mass}}{\text{density}}$$

Now substitute values for the mass and density into this equation:

$$\text{Volume} = \frac{35.4\text{ g}}{1.48\text{ g/cm}^3} = 23.\underline{9}18919\text{ cm}^3 \qquad (\text{rounds to } 23.9\text{ cm}^3)$$

Notice that the units g cancel and you are left with the unit cm^3, which is a unit of volume.

2. *Dimensional Analysis.* Use the density to write the conversion factor that converts from g chloroform to cm^3 chloroform. It is

$$\frac{1\text{ cm}^3\text{ chloroform}}{1.48\text{ g chloroform}}$$

Use this factor to convert 35.4 g chloroform to cm^3 chloroform.

$$35.4\,\cancel{\text{g chloroform}} \times \frac{1\text{ cm}^3\text{ chloroform}}{1.48\,\cancel{\text{g chloroform}}} = 23.9\text{ cm}^3\text{ chloroform}$$

Exercise 2.12

A bar of platinum weighs 3.14×10^4 g. The density of platinum is 21.45 g/cm^3. What is the volume of the platinum bar?

(Try Problems 2.61 and 2.62.)

Chemical Perspective

Your Body Fat from a Density Measurement

Fat, although it has received bad press, is an essential component of the human body. It provides a protective layer around vital organs. And the survival value of fat as an energy reserve is so important that the body converts all excess food energy into fat. The problem, many of us think, is that our bodies are too efficient at doing this.

Body weight (body mass, actually) is frequently used to monitor body fat. It is not a very accurate way to assess body fat, however. You may have heard a story similar to this one. A middle-aged man says, "I weigh the same as I did when I was a student in college." Then a friend says, "But then you had muscles." It is not your weight that you are interested in when you are concerned about body fat. Rather, it is the percent body fat that is important.

The average human body consists of about 20% fat. The rest of the body mass, called lean body mass, is mostly bone and muscle. One of the most accurate ways to measure percent body fat is to take a density measurement of the body. Fat is less dense than water (about 0.92 g/cm^3); and like any substance whose density is less than that of water, it floats. You can test this by dropping a small piece of fat into a glass of water (Figure 2.11). On the other hand, lean body mass has a density equal to about 1.2 g/cm^3. This means that a person's density will be somewhere between 0.92 g/cm^3 and 1.2 g/cm^3, depending on the person's percent body fat. If a person's percent body fat were somewhat greater than 50% (very obese), the body density would be less than 1, and the person would float. To obtain the precise value of percent body fat, you will need to know the precise body density. But how would you obtain someone's body density?

The method of measuring a person's body density is based on the principle of buoyancy, or flotation, first discovered by Archimedes (287?–212 B.C.) of ancient Greece. The story is that Archimedes was asked by the king of Syracuse (part of ancient Greece) to tell him whether the crown that had been made for him was indeed made of pure gold, or whether perhaps the maker of the crown had substituted a cheaper metal for some of the gold he had been given. One day, when Archimedes was taking a bath, it occurred to him how he could tell if the king's crown was pure gold. He became so overjoyed at his insight that he ran out into the streets naked, shouting, "Eureka, eureka. I have found it; I have found it." After Archimedes had refined his insight, it became known as Archimedes' principle.

FIGURE 2.11

Solids floating on liquids. Because fat is less dense than water, fat floats on water. Similarly, a copper penny floats on mercury, a liquid metal, because copper is less dense than mercury. Note, however, that recently minted pennies are copper-clad zinc coins.

Archimedes' principle can be stated this way: *Any object submerged in water will weigh less by the amount of the water displaced by the object.* You can use Archimedes' principle to obtain the density of an object. For example, a piece of iron whose mass is 12.00 g weighs only 10.48 g when weighed underwater. The difference in the weight of the iron out of water and the weight while submerged in water (1.52 g) equals the mass of the water displaced by the iron. From this mass of water, you can obtain the volume of iron. Since the density of water is 1.00 g/cm^3, the volume of the iron is 1.52 cm^3. The density of iron is its mass divided by its volume:

$$\text{Density of iron} = \frac{12.00\ \text{g}}{1.52\ \text{cm}^3} = 7.89\ \text{g/cm}^3$$

In Archimedes' day, gold was the densest substance known. If the king's crown contained another metal (say, silver) in addition to gold, its density would have to be less than that of pure gold (19.3 g/cm^3). In fact, the crown was not pure gold. We are left to imagine what happened to the fellow who made the crown.

The method we described for obtaining the density of iron is basically the method used to obtain body densities. The person is first weighed in air. Then he or she sits in a chair, suspended from a scale, while someone lowers the individual into a tank of water (Figure 2.12). The person must exhale as much air from the lungs as possible while submerged; otherwise, the buoyancy of air will give an erroneous weighing. Not everyone finds this density measurement enjoyable.

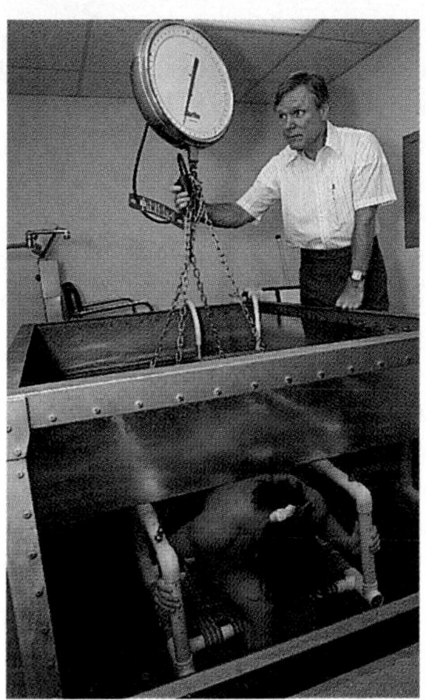

FIGURE 2.12
Determining body density. To obtain a person's density, you weigh the person first in air (usual weighing), then while the person is submerged in water. The person must exhale as much air as possible while submerged for you to obtain a correct density measurement.

CHAPTER REVIEW

Key Words

scientific notation *(p. 18)*
metric system *(p. 21)*
SI base unit *(p. 21)*
SI prefix *(p. 22)*
meter *(p. 22)*
volume *(p. 23)*
liter *(p. 23)*
kilogram *(p. 24)*
significant figures *(p. 26)*
rounding *(p. 30)*
Celsius scale *(p. 33)*
Fahrenheit scale *(p. 34)*
Kelvin scale *(p. 34)*
kelvin *(p. 34)*
dimensional analysis *(p. 38)*
conversion factor *(p. 38)*
density *(p. 42)*

Key Equations

$1\ \text{L} = 1\ \text{dm}^3$

$1\ \text{mL} = 1\ \text{cm}^3$

$\text{K} = {}^\circ\text{C} + 273$

${}^\circ\text{F} = 1.8{}^\circ\text{C} + 32$

$$\text{Density} = \frac{\text{mass}}{\text{volume}}$$

Summary

Measurements consist of two components: a measured number and a unit. Often, measured numbers and results of calculations are expressed in *scientific notation*. It is especially useful for writing very large and very small numbers. *Metric units* are based on the decimal system. The International System (SI) is a particular choice of metric units using seven *SI base units* and a series of *SI prefixes* for the powers of 10. We discussed the following base units: *meter* and *kilogram*. *Volume* is a derived quantity in the International System. The SI unit of volume is the cubic meter, but we more often use the cubic decimeter (dm^3), which is equivalent to the *liter* (L).

A measured number has some uncertainty in it, which you can express by writing the measured number to the correct number of *significant figures*. When using measured numbers in calculations, you must round the answers to the correct number of significant figures. To do this, you must first know the rules for the number of significant figures in a calculation result. There are two rules for this: one for multiplication and division, the other for addition and subtraction. Then you apply the rules for *rounding*.

The SI unit of temperature is the *kelvin,* a unit on the *Kelvin scale*. In scientific use, we often give the temperature on the *Celsius scale,* however. You may have to convert temperature values from one scale to another. To convert degrees Celsius to the Kelvin scale, add 273 (K = °C + 273). To change a temperature reading on the Kelvin scale to Celsius, you will have to rearrange the preceding equation. We also derived a relationship between the *Fahrenheit scale* and the Celsius scale: °F = 1.8°C + 32.

Many problems in introductory chemistry can be solved by *dimensional analysis*. The general approach is this: from the problem statement, you note the unit of the given quantity and note the desired unit. Then, from an equivalence statement, you write down a *conversion factor.* Finally, you multiply the quantity in the given unit by the conversion factor. We illustrated this method of problem solving by showing how to convert from one kind of unit to another.

Density is the mass of a substance per unit volume. You calculate the density of a substance by substituting mass and volume in the defining equation for density. By algebraic manipulation, you can rearrange the defining equation for density to get an equation for mass in terms of volume and density and another equation for volume in terms of mass and density. Instead of using these equations, however, you can view density as a factor that converts volume to mass. By inverting density, you obtain a factor that converts mass to volume.

Problem-Solving Skills

1. **Converting Between Numbers in Decimal Form and Scientific Notation:** Given a number expressed in decimal form, convert the number to one expressed in scientific notation, and vice versa (Examples 2.1 and 2.2).
2. **Counting Significant Figures:** Count the number of significant figures in a given quantity (Example 2.3).
3. **Obtaining Significant Figures and Rounding in Calculations:** Given an arithmetic problem, obtain the answer to the correct number of significant figures (Examples 2.4 and 2.5).
4. **Converting Between Two Temperature Scales:** Given a temperature on the Celsius scale, convert it to kelvins, and vice versa (Example 2.6). Given a temperature on the Celsius scale, convert it to degrees Fahrenheit, and vice versa (Example 2.7).
5. **Converting from One Unit to Another:** Given a quantity expressed in one unit, convert it to another one (Examples 2.8 and 2.9).
6. **Calculating Density from Mass and Volume:** Given the mass and volume of a sample, obtain its density (Example 2.10).
7. **Using Density to Relate Mass and Volume:** Given the volume and density of a sample, calculate the mass (Example 2.11). Given the mass and density of a sample, calculate the volume (Example 2.12).

Questions to Test Your Reading

2.1 Discuss the measurement process, using the example of measuring the length of a rod. In your discussion, show how the measuring process consists of comparing with a standard. Also discuss why we need to note the units as well as the measured numbers in describing a measurement.

2.2 A student entered the number 0.000056 in his calculator and accidentally pressed the equals sign. The calculator display then read 5.6 −5. Can you explain what happened?

2.3 Calculate the following to the correct power of 10:

$$\frac{10^3 \times 10^{-5}}{10^2}$$

If you have any difficulty calculating powers of 10, refer to Appendix A.

2.4 The International System uses seven base units. Only five of these units are needed in this book. What are the three base units discussed in this chapter?

2.5 Suppose you have a given SI base unit. Give the SI prefix for each of the following: (a) a unit that is 1000 times larger, (b) a unit that is 1000 times smaller, (c) a unit that is one-hundredth the original unit, (d) a unit that is one-millionth the original unit.

2.6 The liter is a unit of volume. How many cubic decimeters are there in 1 L? How many cubic centimeters are there in 1 L?

2.7 Suppose you and several friends each measure the length of a room carefully with a meter stick. Later, you and your friends each measure it again using a long tape measure ruled in meters. Discuss the uncertainty in these measurements. Would you expect your results for both measurements to agree exactly with those of your friends? Would you expect your measurements to be more or less uncertain when you use the tape measure rather than the meter stick? Describe the uncertainty in the measured number.

2.8 The mass of a sample was correctly written as 4.830 g. Why would it be incorrect to write 4.8300 g?

2.9 The volume of a liquid was measured several times. The following volumes were obtained: 41.12 mL, 41.18 mL, 41.25 mL, 41.29 mL, 41.05 mL. How many significant figures should be used in reporting the measured volume?

2.10 Which of the following involve examples of exact numbers? (a) You find by counting that there are 23 beans in a jar, (b) you find that the width of a lot is 55 m, (c) there are 3 ft in a yard, (d) a brass weight has a mass of 10 g.

2.11 In doing the subtraction 66.45 − 61.35, a student wrote the answer as 5.100. He reasoned that each of the original numbers has four significant figures, so the answer should have four significant figures. Evaluate his reasoning.

2.12 A student measured the length of a table and found it to be 37.0 in. She calculated the length of the table in feet by dividing by 12. Her arithmetic, including only numbers, is

$$\frac{37.0}{12} = 3.0833333$$

The answer given here is what appeared on her calculator. How would you write the length of the table in feet?

2.13 When you do a calculation involving several steps, should you round after each step or only after the final step? Why?

2.14 Is there a difference between temperature and heat? Explain.

2.15 What is the freezing point of water in degrees Celsius? in degrees Fahrenheit? in kelvins?

2.16 The Kelvin scale is an absolute temperature scale. What does this mean?

2.17 Define density and specific gravity. What common units are used for density? for specific gravity?

2.18 Gypsum is a mineral used to make plaster. The density of gypsum is 2.32 g/cm^3. On which, if either, of the following liquids will gypsum float? (a) carbon tetrachloride (density = 1.60 g/cm^3), (b) methylene iodide (density = 3.33 g/cm^3). Explain.

2.19 A grain is a unit of mass. There are 15.43 grains in 1 g. Write the factor for converting from grains to grams.

2.20 What are the factors for converting from micrograms to milligrams?

Practice Problems

Scientific Notation (Section 2.2)

2.21 Convert the following numbers to scientific notation.
(a) 30,402 (b) 0.00658 (c) 209.00
(d) 0.0000035 (e) 1046

2.22 Give the following numbers in scientific notation.
(a) 0.0000067 (b) 670.5 (c) 78.000
(d) 0.00819 (e) 5901

2.23 Convert the following numbers in scientific notation to decimal form.
(a) 8.9045×10^2 (b) 5.126×10^{-3}
(c) 6.13×10^0 (d) 7.302×10^{-5}
(e) 2.03759×10^4

2.24 Write the following numbers in decimal form.
(a) 6.31×10^{-3} (b) 4.84×10^0
(c) 2.5689×10^2 (d) 3.3×10^{-5}
(e) 8.19783×10^3

2.25 Give the following numbers in decimal form. Some of these numbers are in proper scientific notation, others are not.
(a) 2.68×10^1 (b) 4.598×10^{-3}
(c) 0.783×10^{-2} (d) 85.36×10^2
(e) 3.1891×10^3

2.26 Convert the following numbers to decimal form. Some of these numbers are in proper scientific notation, others are not.
(a) 7.9823×10^{-5} (b) 1.005000×10^4
(c) 0.00872×10^{-2} (d) 595.633×10^3
(e) 6.3×10^{-1}

SI Base Units and Prefixes (Sections 2.3)

2.27 Write each of the following in terms of the correct power of 10 and SI base unit.
(a) 3.69 mm (b) 34.89 km
(c) 92 ps (d) 56.7 ng

2.28 Write each of the following in terms of the correct power of 10 and SI base unit.
(a) 453 mg (b) 51 cm
(c) 103 dm (d) 4.57 pg

2.29 Write each of the following in terms of the SI prefix that corresponds to the given power of 10.
(a) 4.93×10^3 m (b) 2.3×10^{-6} s
(c) 5.68×10^{-3} g (d) 1.568×10^{-2} m
(e) 9.3×10^{-9} g

2.30 Write each of the following in terms of the SI prefix that corresponds to the given power of 10.
(a) 6.15×10^{-9} g (b) 7.48×10^{-2} m
(c) 3.9×10^{-3} g (d) 4.37×10^{-6} s
(e) 8.2×10^3 m

Significant Figures (Sections 2.4 and 2.5)

2.31 Count the number of significant figures in the following quantities.
(a) 0.5006 L (b) 69.0 yd (c) 105.0 s
(d) 590 cm^3 (e) 90. in.

2.32 Give the number of significant figures in each of the following.
(a) 900 mi (b) 900.0 mi (c) 90.00 dm
(d) 0.56 g (e) 0.00830 mL

2.33 Which one of the following numbers has three significant figures?
(a) 4.508 (b) 0.008 (c) 0.00800
(d) 45.09 (e) 0.095

2.34 Which one of the following numbers has four significant figures?
(a) 0.325 (b) 50.0 (c) 56.908
(d) 0.0005 (e) 0.4589

2.35 The following numbers were obtained after arithmetic operations on a calculator. Round each answer to the indicated number of significant figures.
(a) 308.00679 to four significant figures
(b) 45.150873 to four significant figures
(c) 126.55129 to four significant figures
(d) 50.509150 to six significant figures
(e) 1.2937850 to six significant figures

2.36 The following numbers were obtained after arithmetic operations on a calculator. Round each answer to the indicated number of significant figures.
(a) 0.00525500 to three significant figures
(b) 1.2554999 to four significant figures
(c) 49.666666 to two significant figures
(d) 816.05000 to four significant figures
(e) 815.95100 to four significant figures

2.37 Perform the following calculations; then round the answers to the correct number of significant figures.
(a) $561.25 + 68.499 =$ (b) $75.23 - 70.451 =$
(c) $94.2 \times 2.456 =$ (d) $2.1 \times 44.56 =$
(e) $\dfrac{5.897}{7.8} =$

2.38 Do the indicated arithmetic and round the answers to the proper number of significant figures.
(a) $26 - 23.1 =$ (b) $687.2 + 5.64 =$
(c) $989.123 - 898.4981 =$ (d) $39.45 \times 19.8 =$
(e) $\dfrac{8.90}{563.29} =$

2.39 Perform the following calculations; then round the answers to the correct number of significant figures.
(a) $\dfrac{87.12 - 9.236}{5.489} =$
(b) $65.456 \times (23.698 - 20.613) =$
(c) $\dfrac{583.21 - 502.13 + 10.23}{90.135} =$
(d) $\dfrac{73.54}{23.6} - 3.10 =$
(e) $(34.82 \times 6.4 \times 0.1439) - 2.68 =$

2.40 Carry out the following arithmetic, reporting the answers to the proper number of significant figures.
(a) $\dfrac{26.5 \times 56.98}{67.19} =$
(b) $89.23 \times (56.23 - 51.65) =$
(c) $\dfrac{4.981}{9.105 - 8.987} =$
(d) $(89.1 \times 0.1256) - 3.51 =$
(e) $0.9548 + \dfrac{5.16}{43.26} =$

Changing Temperature Scales (Section 2.6)

2.41 Perform the following changes of temperature scale.
(a) 56°C to K (b) 322 K to °C
(c) 35°C to K (d) 458 K to °C
(e) 21°C to K

2.42 Perform the following changes of temperature scale.
(a) 125°C to K (b) 188 K to °C
(c) −77°C to K (d) 395 K to °C
(e) 256°C to K

2.43 Gallium is a metal that melts at 30.°C; the metal melts in the palm of the hand. What is this temperature on the Kelvin scale?

2.44 Liquid nitrogen, which is a gas at room temperature, boils at normal pressure at 77 K. What is this temperature on the Celsius scale?

2.45 Convert each of the following to the indicated temperature scale.
(a) 85°F to °C (b) −40°C to °F
(c) −10°F to °C (d) 55°C to °F
(e) 125°C to °F

2.46 Convert each of the following to the indicated temperature scale.
(a) −15°C to °F (b) 75°F to °C
(c) 25°F to °C (d) 45°C to °F
(e) −75°C to °F

2.47 Dry ice is solid carbon dioxide. Carbon dioxide is normally a gas and is present in the atmosphere. The temperature of dry ice evaporating in the air is −78°C. What is this temperature on the Fahrenheit scale?

2.48 A freezing mixture can be made from table salt (sodium chloride) and ice. This mixture gives a temperature of −6°F. What is this temperature in degrees Celsius?

Conversion of Units (Sections 2.6 and 2.7)

2.49 Do the following conversions.
(a) 165 L to mL (b) 48 ng to g
(c) 4.7 s to ms

2.50 Do the following conversions.
(a) 23.1 g to mg (b) 2.15 ns to s
(c) 750. L to mL

2.51 Do the following conversions.
(a) 7.53 in to cm (b) 4.3 oz to g

2.52 Do the following conversions.
(a) 4.34 L to qt (b) 3.7 km to mi

2.53 Do the following conversions.
(a) 569 cm to km (b) 12.5 cm to dm

2.54 Do the following conversions.
(a) 87 ms to ns (b) 250.0 kg to mg

2.55 A unit of length commonly used by chemists is the angstrom, which equals 100 pm (exactly). The radius of a carbon atom is about 0.77 angstrom. What is this radius in millimeters?

2.56 The erg is a non-SI unit of energy equal to 10^{-7} joules. (The joule is the SI unit of energy.) The energy needed to heat 1 L of water from 0°C to 100°C is 4.18×10^{12} ergs. How many kilojoules of energy is this?

Density (Section 2.8)

2.57 A sample of sucrose (table sugar) weighs 11.2 g. If the volume of the sample is 7.06 cm^3, what is the density of sucrose?

2.58 Glycerol is a viscous, sweet-tasting liquid. In an experiment, a 25.0-mL sample was found to weigh 31.6 g. What is the density of glycerol?

2.59 A gem-quality sample of synthetic emerald weighed 0.975 g. When it was added to a small buret containing water, the top surface of the liquid rose from 5.60 mL to 5.97 mL. What is the density of the emerald sample?

2.60 To identify a metal, an experimenter placed a 32.6-g sample in a graduated cylinder. He then poured in 25.0 mL of water, covering the metal and bringing the liquid level in the cylinder to 28.7 mL. What is the density of the metal?

2.61 Acetone, whose density is 0.7908 g/mL, is the liquid solvent in fingernail polish. What is the mass of 24.48 mL of acetone?

2.62 Pentyl acetate has the odor of bananas and is used in banana flavorings. The density of this substance is 0.875 g/mL. What is the mass of 56.7 mL of pentyl acetate?

2.63 Acetic acid is found in vinegar. The pure substance has a density of 1.049 g/mL. What is the volume of 24.5 g of acetic acid?

2.64 Diethyl ether (commonly known simply as ether) has a density of 0.713 g/mL. What volume of diethyl ether will have a mass of 408 g?

Additional Problems

2.65 A pipet is a glass vessel open at both ends that is used to measure out a definite volume of liquid. In an experiment, the volume of water obtained from a particular pipet on several tries was 10.053 mL, 10.062 mL, and 10.046 mL. How would you write the volume from the pipet using the correct number of significant figures?

2.66 Two students measured the mass of the same quantity of sodium chloride (table salt) several times. The first student obtained 29.56 g, 29.58 g, and 29.58 g. The second student obtained 29.41 g, 29.63 g, and 29.50 g. Obtain the average values of the measurements to the correct number of significant figures for each student.

2.67 What is the number of significant figures in each of the following measured quantities?
(a) The observed speed of light is 2.99792458×10^8 m/s.
(b) The number of atoms in a quantity of carbon was found to be 6.02×10^{23}.
(c) The total volume of blood in a patient was calculated to be 6.0 L.
(d) A sample of blood from a patient contained 205 mg cholesterol per deciliter.
(e) A tablet of prescription drug contained 50.0 mg of the drug.

2.68 Give the number of significant figures in each of the following measured quantities.
(a) The mass of an electron has been measured and found to be $9.1093817 \times 10^{-31}$ kg.
(b) The amount of calcium in a 24-hour sample of urine was 265 mg.
(c) A sample of fish contained 0.11 g of mercury.
(d) A sample of copper ore contained 0.5100 mg of copper.
(e) A sample of blood from a patient contained 80.5 mg triglycerides per deciliter.

2.69 Perform the following arithmetic, rounding the answers to the correct number of significant figures.
(a) $(6.912 - 6.045) \times 7.389 =$
(b) $\frac{7.164}{2.8} \times 5.692 =$
(c) $\frac{4.831 - 0.009}{5.81} =$
(d) $\frac{323.4}{45.91} - \frac{59.21}{61} =$
(e) $(7.12 + 3.45) \times 23.45 =$

2.70 Perform the following arithmetic, rounding the answers to the correct number of significant figures.
(a) $5.93 \times (9.3 + 6.2) =$
(b) $\frac{3.69 \times 51.72 \times 43.26}{9.23} =$
(c) $\frac{613.5}{156.8 - 82.9} =$
(d) $\frac{989.3 + 564.5}{687.4 - 600.2} =$
(e) $(15.00 - 9.67) \times (1.54 + 2.56) =$

2.71 The number of water molecules in one-billionth of a drop of water is about 2,000,000,000,000 (one significant figure). Write this number in scientific notation.

2.72 A particular bacillus, or rod-shaped bacterium, is 3.5×10^{-6} m long. Write the number for this length in decimal form instead of in scientific notation.

2.73 Perform the following calculations and express the answers to the correct number of significant figures using scientific notation.
(a) $(5.21 \times 10^{-5}) - (8.40 \times 10^{-6}) =$
(b) $(7.191 \times 10^3) + (5.648 \times 10^5) =$
(c) $(2.4 \times 10^{-4}) \times (4.32 \times 10^{-2}) =$
(d) $\frac{2.19 \times 10^{-3}}{1.5 \times 10^2} =$
(e) $\frac{(1.543 \times 10^3) - (7.541 \times 10^2)}{4.915 \times 10^4} =$

2.74 Perform the following calculations and express the answers to the correct number of significant figures using either the decimal form or scientific notation.
(a) $(5.42 \times 10^4) + (9.21 \times 10^4) =$
(b) $(8.26 \times 10^{-4}) - (3.2 \times 10^{-3}) =$
(c) $(3.189 \times 10^5) \times (4.15 \times 10^{-6}) =$
(d) $\frac{7.5698 \times 10^{-8}}{4.251 \times 10^{-5}} =$
(e) $[(2.56 \times 10^{-6}) - (5.891 \times 10^{-4})] \times (4.5 \times 10^3) =$

2.75 Express each of the following in terms of the SI prefix and unit that puts the number between 1 and 1000.
(a) 5489 m (b) 7.23×10^{-7} g
(c) 2.164×10^4 m (d) 0.000650 L
(e) 0.0023 s

2.76 Express each of the following in terms of the SI prefix and unit that puts the number between 1 and 1000.
(a) 0.00247 g (b) 3.5×10^{-2} m
(c) 0.000005 s (d) 6815 m
(e) 5.214×10^{-10} g

2.77 Sulfur is a yellow solid that melts at 113°C. What is this temperature in degrees Fahrenheit? in kelvins?

2.78 Bromine is a red-brown liquid that freezes at −7°C. What is this temperature in degrees Fahrenheit? in kelvins?

2.79 Hypothermia is a condition in which the body temperature falls below normal. Temperature control by the body becomes unstable if the body temperature falls below 93°F. What is this temperature on the Celsius scale? on the Kelvin scale?

2.80 Frostbite occurs when living tissue freezes, after continued body heat loss makes it impossible for heat to be maintained in the tissue. Living tissue freezes at about 23°F. What is this temperature on the Celsius scale? on the Kelvin scale?

2.81 In an effort to differentiate a white, crystalline substance from several similar substances, an experimenter measured its density. The density of the substance was found to be 1.231 g/cm^3. What is the volume of a sample weighing 45.6 mg?

2.82 Aspirin (acetylsalicylic acid) is frequently given to control fever. Aspirin has a density of 1.35 g/cm^3. A tablet of the substance contains 0.324 g of aspirin. What is the volume of aspirin in the tablet?

2.83 Lipids, natural fatty substances, are carried in the blood by proteins, called lipoproteins. These substances are classified by density. High-density lipoproteins (HDLs) have a density greater than 1.063 g/cm^3. What is the mass of a sample of lipoprotein having this density whose volume is 2.35 cm^3?

2.84 The density of red blood cells is about 1.093 g/cm^3, somewhat higher than that of blood plasma. Thus, red blood cells tend to settle slowly in a bottle of whole blood. What is the mass of cells whose volume is 5.67 cm^3?

2.85 The specific gravity of the sulfuric acid solution in a fully charged lead storage battery is about 1.28. What is the mass of the solution whose volume is 1.50 L?

2.86 Ethylene glycol is sold as an automobile radiator antifreeze. A water–ethylene glycol solution with a specific gravity of 1.048 freezes at −20°C. What is the mass of the solution whose volume is 2.50 L?

2.87 One species of mite measures about 0.5 mm in length. What is this length in micrometers?

2.88 The smallest species of protozoan is about 2 mm long. What is this length in micrometers?

2.89 An adult rainbow trout weighs about 6 lb. What is this mass in kilograms?

2.90 The highest cloud level at temperate latitudes is about 11 km. What is this height in feet?

2.91 Water droplets in a cloud that are 60. μm or more in diameter grow rapidly until they become raindrops. What is this diameter in inches?

2.92 Mt. Mitchell is the highest peak in the United States east of the Mississippi River. Its elevation is 6684 ft. What is this elevation in meters?

2.93 A certain medication is prescribed on the basis of milligrams of medication per kilogram of body weight (mass). A patient weighs 176 lb. What is this mass in kilograms?

2.94 An informational brochure says that a foot race is run over a 5.0-mi course. A runner from Poland would like to know how far this is in metric units. What is this distance in kilometers?

2.95 The posted speed limit in a Canadian city is given as 40 km/hr. What is this speed limit in miles per hour? Assume that the posted limit is given to two significant figures.

2.96 A bottle of seltzer water contains 750 mL of water. What is this volume in quarts? Assume that the volume

Practice in Problem Analysis

For each problem, describe the thinking you would use (the problem analysis) before doing the actual solution, but do not solve the problem.

1. A woman walked at 2.3 mi/hr until she covered 6.0 km. How many seconds did she walk?

2. A cook measures 2.2 qt of a liquid having a density of 0.82 g/mL. What is the mass of the liquid in pounds?

PRACTICE EXAM

1. Which of the following numbers is exact?
   (a) A man finds that he weighs 168 lb.
   (b) A woman determines that the width of her house is 23 ft.
   (c) There are 12 inches in 1 foot.
   (d) The stratosphere begins 10 miles above the surface of the earth.
   (e) None of the above

2. How many significant figures are in the measurement 0.3010 g?
   (a) 4 (b) 3 (c) 2 (d) 1 (e) 0

3. How many significant figures are in the number 4.020×10^9?
   (a) 7 (b) 6 (c) 4 (d) 3 (e) 2

4. In doing the subtraction $37.45 - 36.95$, a student says that the answer is 0.5000. He reasoned that each of the original numbers has four significant figures, so the answer should also have four significant figures. How many significant figures do you think the answer should have?
   (a) 4 (b) 3 (c) 2 (d) 1 (e) 0

5. Suppose you subtract 7.5×10^{-3} from 1.023×10^{-2} and then divide the result by 4.3241×10^{-3}. How many significant figures do you think the answer should have?
   (a) 5 (b) 4 (c) 3 (d) 2 (e) 1

6. A student measured the length of a room and found it to be 16.0 ft. She calculated the length in yards by dividing this number by 3. The answer her calculator gave was 5.3333333. How would you write the answer?
   (a) 5.3333333 yd (b) 5.33 yd (c) 5.3 yd
   (d) 5 yd (e) None of the above

7. The mass of an object is 21 g. Its mass in milligrams is
   (a) 21 mg (b) 2.1×10^{-4} mg
   (c) 2.1×10^4 mg (d) 2.1×10^2 mg
   (e) None of the above

8. A car travels 133 km. The distance in meters is
   (a) 1.33×10^2 m (b) 1.33×10^3 m
   (c) 1.33×10^4 m (d) 1.33×10^5 m
   (e) None of the above

9. The length of an object is 243 cm. Its length in meters is
   (a) 243 m (b) 2.43×10^2 m
   (c) 2.43×10^{-2} m (d) 2.43×10^{-1} m
   (e) None of the above

10. If the temperature of an object is 0 K, the temperature on the Celsius scale is
    (a) 0°C (b) 100°C
    (c) 32°C (d) 273°C
    (e) None of the above

11. If the temperature of a room is 68°F, what is the temperature on the Celsius scale?
    (a) 341°C (b) 20°C
    (c) 154°C (d) −205°C
    (e) None of the above

12. If the temperature of the water in a glass is 37°C, what is the temperature on the Fahrenheit scale?
    (a) 99°F (b) 37°F
    (c) 20°F (d) 6°F
    (e) None of the above

13. What is the volume of 50.0 g of mercury at 20°C? The density of this metal is 13.6 g/cm^3 at that temperature.
    (a) 3.68 cm^3 (b) 1.00 cm^3
    (c) 0.368 cm^3 (d) 0.272 g
    (e) None of the above

14. What is the mass of a piece of copper that has a volume of 12.0 cm^3? The density of copper is 8.92 g/cm^3.
    (a) 0.107 g (b) 1.07 g
    (c) 10.7 g (d) 107 g
    (e) None of the above

15. Which statement is true?
    (a) A measured number has some uncertainty in it.
    (b) Mass is a derived quantity in the SI system.
    (c) The Celsius and Fahrenheit temperature scales have degrees that are the same size.
    (d) Heat is another word for temperature.
    (e) None of the above

Chapter

3 Matter and Energy

Light energy interacting with matter causes the colors of these leaves.

Chapter Outline

Gazing into the night sky, away from city lights, you see what captivated the ancient astronomers. Figure 3.1 shows one example of an arrangement of stars and interstellar gases familiar to both early and modern stargazers. Looking at this spectacular image, it is not difficult to imagine our forebears wondering about the heavens, speculating about their meaning or origin. Present-day astronomers, with knowledge of modern physics, have developed the "big bang" theory of the origin of the universe. According to this theory, everything that is, in the physical sense, began some 15 billion years ago from a speck that exploded into our present universe. By "everything" we mean the *matter* and *energy* of the universe as we currently know it. In chemistry, we are interested in both matter and energy and their relationship to all the things we see in our surroundings.

FIGURE 3.1
The constellation Orion. Astronomers have known about this conspicuous constellation for centuries.

Matter refers to the material things around us. It has mass and occupies space. And it comes in enormous variety. The water of a mountain brook and the fragrant air of a meadow are very different, yet each is an example of a kind of matter. The glowing gases of the stars (and our local star, the sun) are also matter. Among the topics we discuss in this chapter are the ways in which we can classify matter so that we can begin to understand its great diversity.

What will occupy much of our attention in this book is how matter can change, especially when those changes result in new forms of matter with properties that differ from those of the original substance. A log burns in air and produces gases, smoke, and ashes. How is it possible for a log to change into such different kinds of matter? We begin to answer this question in this chapter; in so doing, we will present the background we need to describe the theory of matter in Chapter 4.

Almost all changes of matter involve energy. Energy is associated with matter in motion or with the potential for motion. To move a car, you need a source of energy (gasoline). To freeze water, you need to slow down the motion of particles of matter by removing heat, a form of energy. When a log burns, it gives off heat energy. The chemical energy released by the burning log results in hotter temperatures, and the particles of matter move faster. Within our sun, matter is being converted to energy, which radiates into space, and some of that energy bathes the earth. In the second part of this chapter, we look at energy. We discuss how energy is associated with matter in motion and with the potential to move matter. We also explain how to treat energy quantitatively.

MATTER

Water, as those of us in wintry climates are continually reminded, may be solid as well as liquid. And when you boil water for tea or coffee, you see evidence that it can also be a gas. If you explore other substances, you will find that many of them can also exist in different *physical states*. We address this physical classification of matter in Section 3.1.

Physical state is only one of the ways of classifying matter. After we compare the physical and chemical properties of matter, we discuss the chemical classification of matter into *elements*, *compounds*, and *mixtures*.

3.1 States of Matter

OBJECTIVES

- List and characterize the three states of matter.
- Explain changes of state.
- Identify the physical state of a substance under given conditions.

Many substances exist in all three physical states—solid, liquid, and gas—and change from one state to another when the temperature changes. Water is in liquid form as it runs from the kitchen tap, but when you place a tray of water in the freezer, which is below 0°C, the liquid changes to ice, or solid water. On the other hand, when you pour water into a pan and heat it to 100°C (at normal atmospheric, or barometric, pressure), the liquid boils and passes into steam, which is a gaseous form of water. You can also reverse these transformations. You can *condense* steam on a cool surface, and in condensing, the steam changes to the liquid (Figure 3.2). And, of course, you can melt ice. In melting, the solid changes to the liquid.

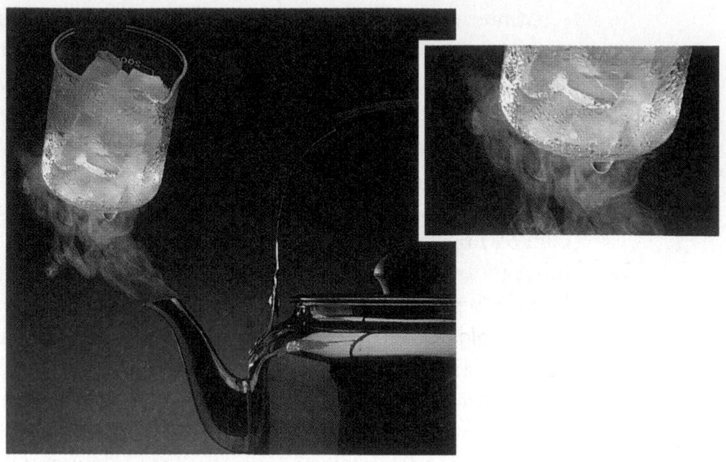

FIGURE 3.2
Water condenses on a cool surface. Water vapor from a teakettle cools on a beaker surface cooled by ice. Note the liquid water dripping from the beaker.

Characterizing the States of Matter

Most kinds of matter can exist in different physical states, depending on the temperature and pressure. Iron, for example, melts to the liquid state when you heat it to 1535°C. Carbon dioxide, which is present in our atmosphere, normally exists as a gas. By compressing it to very high pressure (approximately 57 times that of normal atmospheric pressure), the gas turns to a liquid. The carbon dioxide that is used in some fire extinguishers is in the liquid state. When you release the valve on the cylinder, the pressure drops, causing the liquid to change to the gas that issues from the valve. Many food vendors use dry ice to keep foods cool or frozen. Dry ice is the solid state of carbon dioxide. At normal atmospheric pressure, solid carbon dioxide changes directly to gaseous carbon dioxide without melting to the liquid form. This property accounts for the name "dry ice" (Figure 3.3).

Ice, liquid water, and steam, or water vapor, are chemically the same material; they are just different physical forms of water. Similarly, dry ice, liquid carbon dioxide, and gaseous carbon dioxide are simply different physical forms of the same material. *The three forms of matter—solid, liquid, and gas—*are referred to as the physical states of matter, or simply the **states of matter.**

FIGURE 3.3
Dry ice. The fog spreading from the dry ice consists of fine water droplets formed by condensation of water vapor from air near the very cold dry ice.

Here are the formal definitions of these different states of matter:

Solid: *the form of matter having definite shape and volume*

Liquid: *the form of matter having indefinite shape (taking the shape of its container), but having a definite volume*

Gas: *the form of matter having an indefinite shape and volume (taking the shape and volume of its container)*

You may find it easier to understand these definitions if you have a mental picture of each state of matter. According to the *kinetic molecular theory,* substances such as water and carbon dioxide are made up of extremely small particles of matter, called *molecules.* If matter is composed of atoms, as we explained in Chapter 1, what are molecules? A molecule is an aggregate of either identical or different atoms that are bonded together in a particular order by strong attractive forces. For example, a water molecule consists of two hydrogen atoms bonded to an oxygen atom.

Figure 3.4 gives you a molecular view of a solid, a liquid, and a gas. Each little ball represents a molecule. In a solid, the molecules are closely packed together and in fixed positions (like the particles of sand in a piece of concrete). Because the molecules are in fixed positions, the solid is rigid and so has a definite shape and volume.

In a liquid or gas, the molecules move around quite rapidly. The term *kinetic* in kinetic molecular theory refers to this motion, and the Chemical Perspective at the end of this section provides evidence for it. Because the molecules of a liquid or gas change positions (unlike those in a solid, which have fixed positions), these states have no definite shape. Liquids and gases differ from one another in how closely their molecules are packed together.

You can imagine a liquid as resembling a jar of marbles, which you can pour into a bowl. (A better model would have the jar and bowl sitting on a vibrating table, so that the marbles are continuously moving.) The molecules of a liquid are close together (like the marbles in a jar), and although the molecules are moving around, the liquid always maintains a definite, or fixed, volume. The shape of a liquid varies with the shape of the container, as would the shape of a mass of marbles, first in the jar, later in the bowl.

For a mental picture of a gas, you might think of a swarm of gnats in a jar. The gnats fly around incessantly, occupying the entire container. The gnats in the jar, unlike the marbles, are far apart from one another. If you imagine the jar growing to double its size, you would see the gnats flying around the entire container; the volume they take up would change with the volume of the container. The shape and volume of a mass of gnats would be indefinite, as would the shape and volume of a gas, which similarly consists of a swarm of molecules. And, like the gnats, the molecules in a gas are far apart.

Although many substances can exist in any of the three states of matter (under the proper conditions), some materials exist in only one or two of these states. Paper, for example, is a solid. You cannot melt it to obtain a liquid. If you raise its temperature, it chars rather than melts as a result of chemical decomposition.

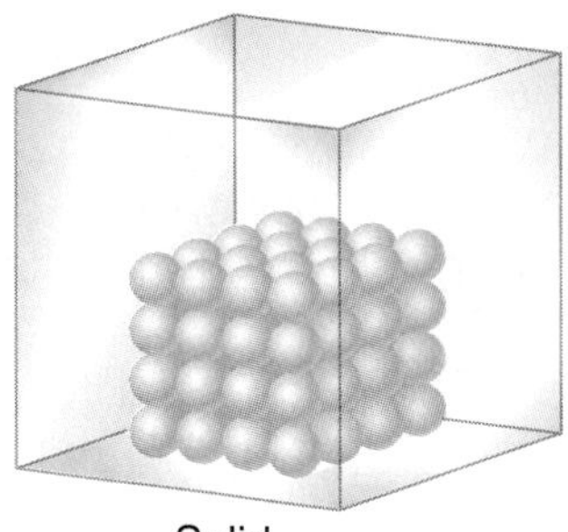

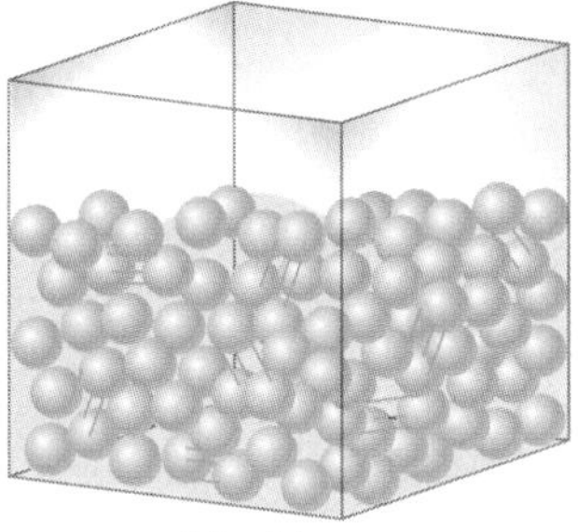

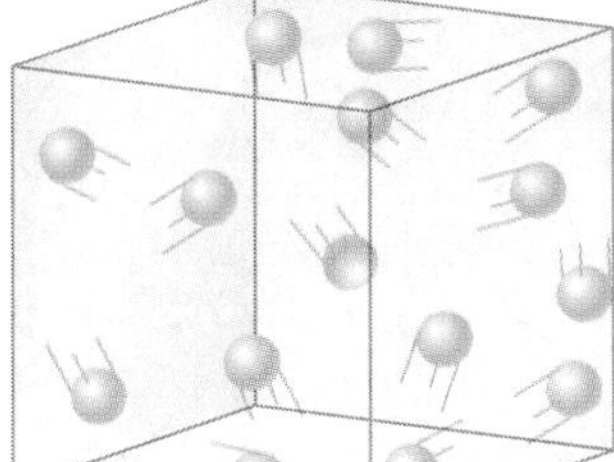

FIGURE 3.4
Molecular models of the states of matter. A molecular solid consists of molecules packed together, with the molecules in fixed positions. A liquid consists of molecules that are close together but moving around. A gas consists of molecules moving around through mostly empty space.

Conditions for Changes of State

Let's use Figure 3.4 to see what happens to water when you raise its temperature. Water is a solid (ice) below 0°C. When you raise the temperature, the state of the water changes from a solid to a liquid at 0°C. We say that it *melts.* When you raise the temperature of the liquid water to 100°C (at normal atmospheric pressure), bubbles of water vapor form in the body of the liquid. (The term *vapor* describes the gaseous state of a substance that normally exists as a liquid or solid.) We say that the water *boils.* Despite these changes of state, *nothing has happened to the water molecules during these changes;* they are still intact. The distance between the water molecules is the only thing that has changed, as shown in Figure 3.4, but the same molecules are still present.

The temperatures at which a substance melts and boils—that is, the *melting point* and *boiling point*, respectively—are properties frequently used to identify a substance. In addition to their use as identifiers, the melting point and boiling point of a substance allow you to determine the state of the substance under given conditions, as the following example illustrates.

■ EXAMPLE 3.1 Identifying the State of a Substance Under Given Conditions

Acetone has a characteristic fragrant odor and is used to make fingernail polish and rubber cement. It melts at −94°C and boils at 56°C. What is the physical state of acetone at 25°C and normal atmospheric pressure?

Problem Analysis

Below the melting point, the substance exists as a solid, although some of it may be in the form of a vapor (gas). Above the melting point, but below the boiling point, the substance exists as a liquid, with some in the form of a vapor. Above the boiling point, the substance exists only as a vapor.

Solution

Acetone would be liquid, because 25°C is below the boiling point (56°C) but above the melting point (−94°C) of acetone (Figure 3.5). Some acetone will be in the form of a vapor above the liquid (which you can smell).

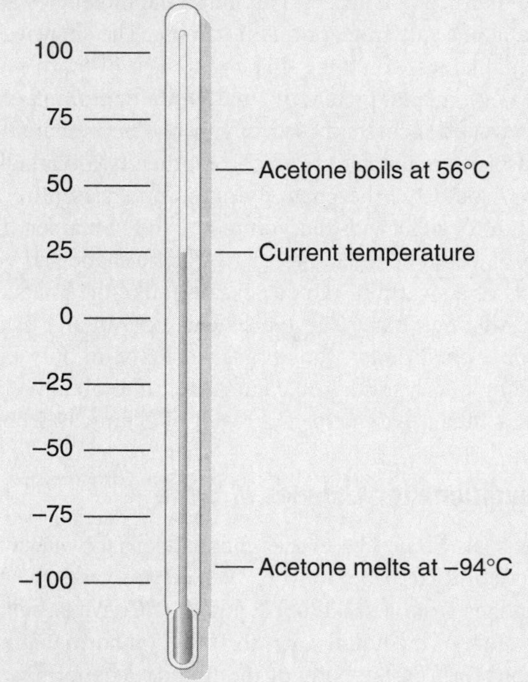

FIGURE 3.5
A comparison of temperatures. The melting point and boiling point of acetone are compared to a temperature of 25°C.

Exercise 3.1

Benzoic acid is used as a food preservative. It melts at 122°C and boils at 249°C. What is the physical state of benzoic acid at 20°C and normal atmospheric pressure?

(Try Problems 3.25, 3.26, 3.27, and 3.28.)

Chemical Perspective

Molecules in Motion

The molecules that make up a liquid or a gas are in constant, random, chaotic motion, according to the kinetic molecular theory of matter. Some of the evidence for this theory is provided by *Brownian motion,* the seemingly spontaneous but erratic movement of particles suspended in liquids or gases.

Brownian motion was discovered in 1827 by Robert Brown, a Scottish botanist, during his study of grains of pollen suspended in water. Using a microscope, Brown discovered that each grain moved erratically as if it were being struck by a series of unseen blows (Figure 3.6). You have seen Brownian motion without

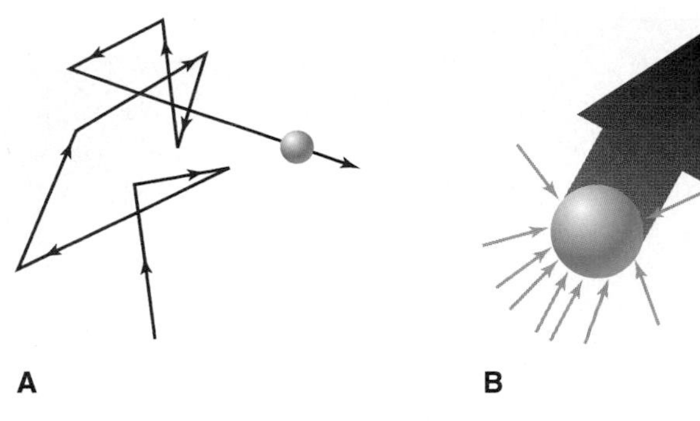

FIGURE 3.6
Brownian motion. *(A)* When it is viewed under a microscope, a pollen particle bounces hither and yon, as if buffeted by many unseen blows. *(B)* The particle moves in a particular direction (shown by the large arrow) because more water molecules are hitting it on one side than on the others (shown by the smaller arrows). The direction changes when the pattern of these collisions changes.

a microscope if you have ever looked closely at a ray of sunlight entering a window. The light lets you see little pieces of dust moving suddenly and chaotically for no apparent reason; they exhibit Brownian motion.

Both types of Brownian motion can be explained by assuming that the suspended particles are being tossed about because constantly moving molecules are bumping into them. Water molecules are colliding with the pollen, and molecules in the air are striking the dust. A suspended particle will move in a particular direction because more molecules are hitting it on one side than the other. The direction will change in another instant, however, because of the constantly varying blows that it receives from these molecules. Thus, the incessant and random motions of the suspended particles reflect the incessant and random motions of the molecules that make up the liquid or the gas.

3.2 Physical and Chemical Changes and Properties

OBJECTIVES

- Differentiate between a physical change and a chemical change.
- Differentiate between a physical property and a chemical property.
- Identify properties as physical or chemical, given a description of a material.

Each different **material,** or *particular kind of matter,* has different properties by which we can characterize it. The melting point and boiling point described in Section 3.1 are two such properties. In general, a property of a material can be a *physical property* or a *chemical property.* To understand the difference between these kinds of properties, we need to distinguish between *physical changes* and *chemical changes.*

A **physical change** is *a change in the form of matter but not in its chemical identity.* The sawing of a log into boards and sawdust is a physical change, because it results in smaller pieces of wood, but it does not alter the chemical identity of each piece. *The same molecules are still present.* The change of liquid water to ice or to steam is a physical change. Water, whether it is in its solid, liquid, or gaseous state, consists of water molecules, so the chemical identity of water is the same in all three states. The process of dissolving salt in water also involves a physical change of the salt. When the salt dissolves, it essentially disperses itself throughout the liquid. The solution appears waterlike, but if it is placed in a dish to evaporate, water molecules escape and salt remains behind (Figure 3.7).

Video: Oxygen, hydrogen, soap bubbles, and balloons.

A **chemical change** (or **chemical reaction**) is *a change in which one or more kinds of matter are transformed into one or more new kinds of matter.* Because a chemical change results in new kinds of matter with new kinds of molecules, you can expect a definite change of characteristics, and often the change is quite dramatic. When a log burns, it forms smoke, gaseous products, and ashes. Burning is a chemical reaction in which the log combines chemically with oxygen in air to form entirely new products with entirely new characteristics. Similarly, when iron rusts, it reacts with oxygen in air to form a red-brown iron oxide (rust). In Lavoisier's experiment described in Chapter 1, mercury metal reacts with oxygen in air to produce orange mercury(II) oxide.

FIGURE 3.7
Evaporation of a salt solution. *(A)* The dish contains salt (sodium chloride) dissolved in water. *(B)* Several days later, the water has evaporated from this solution, leaving the solid salt behind. Dissolving the salt in water is a physical process; it did not change the identity of the salt, as this experiment shows.

You can characterize or identify a material by its various properties. These properties are either physical or chemical. A **physical property** is *a characteristic that can be observed for a material without changing its chemical identity.* Some of these may be associated with physical changes. The melting point and boiling point are examples. Other examples are color and density, where the material is simply observed or a property measured. Using physical properties, you can characterize water as a colorless liquid having a density of about 1.0 g/cm^3 and a freezing point of 0°C. Similarly, you can characterize aspirin in terms of physical properties as a white, crystalline solid having a density of 1.35 g/cm^3 and a melting point of 135°C.

A **chemical property** is *a characteristic of a material that involves a chemical change.* For example, a chemical property of iron is that it reacts with oxygen gas (in air) to produce a red-brown iron oxide (rust). Using both physical and chemical properties, you can characterize iron as a silvery metal that melts at 1535°C and reacts with oxygen to produce a red-brown iron oxide (rust).

FIGURE 3.8
Liquid bromine. Bromine is a red-brown liquid under ordinary conditions.

■ EXAMPLE 3.2 Identifying the Physical and Chemical Properties of a Substance

Bromine is a red-brown liquid that boils at 59°C (Figure 3.8). It reacts readily with many metals, including sodium, with which it yields a white solid. Identify all of the physical properties and all of the chemical properties given in this description.

Problem Analysis

Identify all properties of the substance. Then classify these as physical (involving, at most, physical changes of the material) or chemical (involving chemical changes of the material).

Solution

The physical properties of bromine are as follows: color, red-brown; physical state, liquid; boiling point, 59°C. The chemical properties mentioned are that bromine reacts with many metals, and it reacts with sodium to yield a white solid.

Exercise 3.2

Ethanol is a colorless liquid that freezes at −114°C and boils at 78°C. It burns readily in air, producing carbon dioxide and water. Identify all of the physical properties and all of the chemical properties of ethanol.

(Try Problems 3.29, 3.30, 3.31, and 3.32.)

3.3 Substances and Mixtures

OBJECTIVES

- Define substance and mixture.
- Describe the filtration and distillation methods of separating mixtures.
- Define homogeneous mixture (solution) and heterogeneous mixture, and give examples of each.

The term *material* refers to any particular kind of matter, such as beach sand. Materials are either *substances* or *mixtures* of substances. Beach sand, for example, is a mixture of substances, including calcite (calcium carbonate) from seashells and silica (silicon dioxide) from bits of quartz. If you looked at particles of sand under a microscope, you could separate the seashell fragments from the quartz particles, given the right tools (Figure 3.9).

Until now, we have used the term *substance* in a general way. From now on, however, we will be using the term in a very precise chemical sense. A **substance** is *a material that cannot be separated into different materials by any physical process.* Carbon dioxide and pure water are examples of substances. A physical process is one that involves only physical changes. If you are unable to separate a material into other materials after trying all sorts of physical separation methods, you can probably conclude that the material is a substance. ●

● At one time, it was thought that air was a substance. We now know that it is a mixture of several substances, including nitrogen, oxygen, and argon.

We use the term *mixture* in a special sense also. A **mixture** is *a material that can be separated by physical processes into two or more substances.* Coffee grounds suspended in water and ordinary saltwater are examples of mixtures. To clarify the definitions of substance and mixture, we examine several methods by which you can separate mixtures into their component substances.

FIGURE 3.9
A mixture. Beach sand, depending on the source, may be a mixture of fragments of seashells and quartz particles.

Separating Mixtures into Substances

If the material is obviously a mixture (in which you can see different kinds of particles), you might be able to pick out the particles physically and place them into separate piles. You can even separate a mixture of extremely fine particles this way, if you have a microscope and the proper tools. Although this method would undoubtedly be the easiest way to separate the various materials in beach sand, other mixtures require other methods.

Filtration is a common laboratory and industrial method of physically separating certain mixtures. You can brew coffee by placing ground coffee in a paper filter cup or cone and then pouring hot water over the grounds. The coffee extract passes through the filter paper, leaving the grounds behind. The purpose of the paper cone is to filter the coffee grounds away from the coffee extract. You use filter paper in the laboratory in a similar way to separate a liquid and a suspended precipitate (a fine crystalline solid

FIGURE 3.10
Filtering a precipitate from a solution. Here the experimenter filters a precipitate of lead(II) iodide from the clear solution.

formed in solution by chemical reaction). You pour the solution with the suspended precipitate into a filter cone (Figure 3.10). Clear liquid passes through the filter, and the precipitate remains on the paper. Filtration is a physical process in which a particular mixture (a suspension of a precipitate in a liquid solution) is separated into a solid and a liquid.

Distillation, an embellishment of the physical process of evaporation (see Figure 3.7), is another important physical method of separating a mixture into its components. Let's consider separating the components of ordinary saltwater. Table salt (sodium chloride) is a substance. So, too, is water. When you mix a couple of teaspoons of salt into a cup of water and stir for a while, you obtain a clear liquid. The clear liquid is not a substance, however; it is a mixture. To separate the mixture, you place the salt solution in the distillation flask shown in Figure 3.11 and heat it. When you heat the solution to a high enough temperature, it boils and the water vaporizes. The water vapor passes into the condenser, where it cools and condenses to a liquid. Drops of water run out of the condenser and collect in the receiver flask. At the end of the distillation, pure salt remains in the distillation flask, and pure water has collected in the receiver flask. The mixture has been separated.

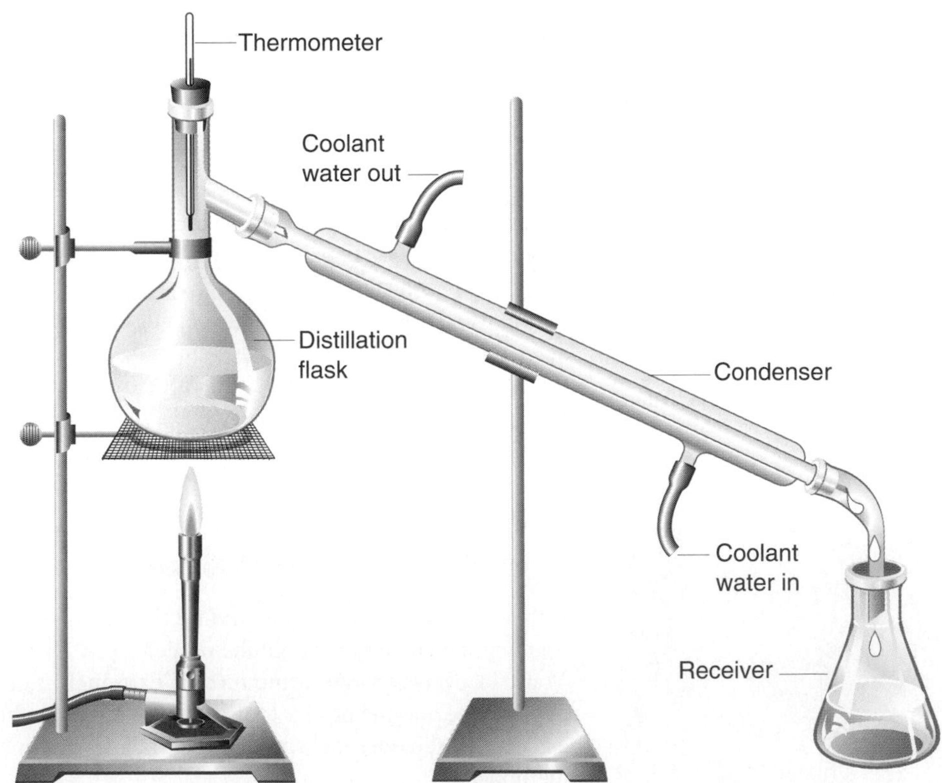

FIGURE 3.11
Distillation of a solution of sodium chloride. Place the solution of sodium chloride in the distillation flask and heat the solution to boiling. Pure water vaporizes, condenses to a liquid inside the cooled condenser, and collects in the receiver flask. Solid sodium chloride remains in the distillation flask at the end of the experiment.

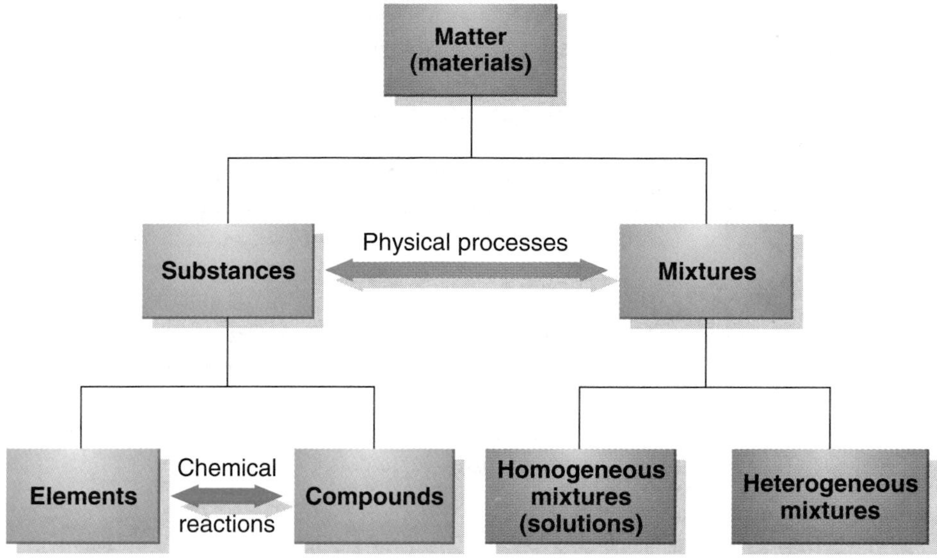

FIGURE 3.12
Classification of matter into substances and mixtures. You can physically mix substances to form a mixture. You can also use physical processes to separate a mixture into substances. Substances can be elements or compounds, as discussed in Section 3.4.

Homogeneous and Heterogeneous Mixtures

Mixtures are of two types. A **homogeneous mixture** (also called a **solution**) is *a mixture that is uniform in its properties throughout.* The liquid mixture you obtain when you thoroughly dissolve sodium chloride in water is a solution, or homogeneous mixture. Density and other physical properties are the same at every point in the liquid. Air is another example of a solution; it is a gaseous mixture of nitrogen, oxygen, and several other substances.

A **heterogeneous mixture** is *a mixture that consists of physically distinct parts with different properties.* If you stir together crystals of sugar and salt, you obtain a heterogeneous mixture. It makes no difference how thoroughly you stir; when you look closely at the mixture, you see individual crystals, each of which is either sugar or salt. Beach sand (see Figure 3.9) is a heterogeneous mixture.

Figure 3.12 summarizes the relationships between materials, substances, mixtures, heterogeneous mixtures, and homogeneous mixtures. Matter is either a substance or a mixture. By physical processes, you can blend together substances to form a mixture. Using other physical processes, you can separate this mixture into its component substances. The figure shows this relationship by a double-headed arrow connecting substances and mixtures. As we show next, you can classify substances themselves as either elements or compounds.

3.4 Elements and Compounds

OBJECTIVES

- Differentiate between an element and a compound, and give examples of each.
- State the law of definite proportions (also called the law of constant composition).

Chemists have prepared and characterized millions of different substances. Interestingly, however, all of these substances consist of only a small number of *elements,* or fundamental substances. Altogether, just over 100 different elements are known. By assembling these elements in different ways, a chemist can make countless new substances called *compounds.*

Definition of an Element

Antoine Lavoisier, who is generally credited with the development of modern chemistry, gave us the first experimentally useful definition of an element. He stated that an **element** is *a substance that cannot be decomposed by any chemical reaction into simpler substances.* ● In Chapter 1, we noted that when you heat mercury(II) oxide, it decomposes into mercury metal and oxygen gas (see Figure 1.10). This means that mercury(II) oxide cannot be an element because it decomposes into other substances (mercury and oxygen). On the other hand, every attempt to decompose mercury or oxygen into other substances has failed. We conclude that mercury and oxygen are elements. By applying this type of reasoning to many different substances, early chemists identified many of the elements. Today, chemists list 112 elements of which Figure 3.13 shows several.

● In the next chapter, where we discuss atoms, we define an element as a substance consisting of only one kind of atom.

Names and Symbols of Elements

You will no doubt recognize the names of some elements, such as carbon, sulfur, iron, and copper. But others, such as gallium, germanium, and cesium, may be new to you, because they are far less common. Where did the names of the elements come from? The names of a few elements have very early origins, because, although the concept of a chemical element is relatively recent, the substances themselves have been known since antiquity. Carbon is an example. The name *carbon* derives from *charcoal,* which is a form of carbon known to anyone who has gathered around a campfire.

The custom today is to allow the discoverer of the element to name it. The names sometimes refer to a peculiarity of the element. William Ramsay and Lord Rayleigh discovered a new gaseous element in air in 1894. They named it *argon* from the Greek *argos,* meaning "inert," because it did not react with other substances. Other names of elements refer to a locality (germanium, from Germany) or to persons (curium, from Marie and Pierre Curie, who discovered the element radium in 1898). ●

● You will see in the next chapter that an element can be assigned a unique number, called the *atomic number.*

Chemists find it convenient to refer to an element by its symbol. The **symbol** of an element is usually *a one- or two-letter abbreviation for the element.* The symbol usually

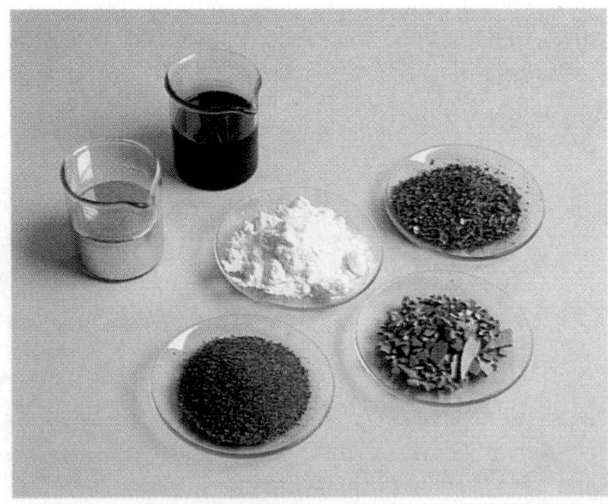

FIGURE 3.13
Some elements. *(Center)* Sulfur. *(From upper left, clockwise)* Mercury, bromine, iodine, chromium, and lead.

comes from the first letter of the name of the element (capitalized), plus perhaps another distinctive letter (in lowercase). For example, the symbol for carbon is C; the symbol for chlorine is Cl. In some cases, however, the name from which the symbol was derived is a foreign name of the element. Because Latin was at one time the language of scholars, many element symbols derive from Latin names. For example, the Latin name for copper is *cuprum,* so the symbol for copper is Cu. Table 3.1 lists some common elements and their symbols, but the inside front cover lists all of the known elements and their symbols.

TABLE 3.1
Some Common Elements

Name of Element	Atomic Symbol	Physical Appearance of Element*
Aluminum	Al	Silvery-white metal
Barium	Ba	Silvery-white metal
Bromine	Br	Reddish-brown liquid
Calcium	Ca	Silvery-white metal
Carbon	C	
Graphite		Soft, black solid
Diamond		Hard, colorless crystal
Chlorine	Cl	Greenish-yellow gas
Chromium	Cr	Silvery-white metal
Cobalt	Co	Silvery-white metal
Copper	Cu (from *cuprum*)	Reddish metal
Fluorine	F	Pale yellow gas
Gold	Au (from *aurum*)	Soft, yellow metal
Helium	He	Colorless gas
Hydrogen	H	Colorless gas
Iodine	I	Bluish-black solid
Iron	Fe (from *ferrum*)	Silvery-white metal
Lead	Pb (from *plumbum*)	Bluish-white metal
Magnesium	Mg	Silvery-white metal
Manganese	Mn	Gray-white metal
Mercury	Hg (from *hydragyrum*)	Silvery-white liquid metal
Neon	Ne	Colorless gas
Nickel	Ni	Silvery-white metal
Nitrogen	N	Colorless gas
Oxygen	O	Colorless gas
Phosphorus (white)	P	Yellowish-white, waxy solid
Potassium	K (from *kalium*)	Soft, silvery-white metal
Silicon	Si	Gray, lustrous solid
Silver	Ag (from *argentum*)	Silvery-white metal
Sodium	Na (from *natrium*)	Soft, silvery-white metal
Sulfur	S	Yellow solid
Tin	Sn (from *stannum*)	Silvery-white metal
Zinc	Zn	Bluish-white metal

*Common form of the element under normal conditions.

FIGURE 3.14
Electrolysis of water. A battery is immersed in a water solution containing a small quantity of salt (added to give a solution that will conduct electricity). Bubbles of hydrogen gas come from the negative terminal, and bubbles of oxygen gas come from the positive terminal.

Compounds

Although you are familiar with a number of elements, most of the substances you are likely to encounter, such as sugar, salt, and aspirin, are compounds. A **compound** is *a substance composed of two or more elements that have been chemically combined.* Thus, mercury(II) oxide is a compound of mercury and oxygen. Water is a compound of the elements hydrogen and oxygen, both of which are gases.

Note that the definition of a compound says that the elements must be "chemically combined." When you put hydrogen gas and oxygen gas together, you obtain a mixture, not a compound. But if you ignite the mixture with a flame, an explosive chemical reaction occurs, forming the compound water. The elements are now chemically combined.

You can separate water into its elements by passing an electric current through it (after adding a little salt to increase the electrical conductivity). The process is a very simple one, called *electrolysis* (Figure 3.14). The electric current provides the energy needed to break the compound into its elements; so it is not a physical separation of the elements, but rather a chemical reaction.

The relationship between substances, elements, and compounds is summarized in Figure 3.12. Elements, as we noted, can react chemically to form compounds, and compounds can be separated by chemical reactions into elements. Figure 3.12 shows this relationship by the double-headed arrow connecting elements and compounds.

By the end of the eighteenth century, chemists had established that a given compound has a definite, constant composition. They summarized this result in the **law of definite proportions** (also known as the **law of constant composition**), which states that *a pure compound, whatever its source, always contains definite, or constant, proportions of the elements by mass.* For example, 1.0000 g of sucrose (table sugar) always contains 0.4210 g of carbon, 0.0648 g of hydrogen, and 0.5142 g of oxygen, chemically combined. It does not matter whether we obtain the sucrose from sugarcane or sugar beets or from any of the many other sources of the substance. By contrast, a mixture of sugar and salt can contain any proportion of the two components of the mixture.

Video: Electrolysis of water.

3.5 Law of Conservation of Mass

OBJECTIVE

- State the law of conservation of mass, and give an example illustrating this law.

Emperor Francis I (1708–1765) of the Holy Roman Empire heated thousands of dollars worth of diamonds in air for a day and found that they "disappeared." Much debate followed on what had happened to the diamonds. Some years later, Lavoisier repeated this experiment, although not on such a grand scale, by placing some diamonds inside a jar of air and exposing them to the sun's heat, using large lenses. What Lavoisier found was that the diamonds burned and produced only carbon dioxide gas. From this experiment in 1772, Lavoisier concluded that diamond is a form of carbon, which chemically combines with oxygen in air, during the process of burning, to yield carbon dioxide. As you saw in Chapter 1, Lavoisier was also able to show that the mass of the substances before burning equals the mass of the substances after burning. Imagine that you burn 12.0 g of diamonds. (This is just a thought experiment!) You would find that the diamonds react with 32.0 g of oxygen and produce 44.0 g of carbon dioxide. We can represent the burning of diamond this way:

$$\underset{12.0\text{ g}}{\text{Diamond (carbon)}} + \underset{32.0\text{ g}}{\text{oxygen}} \longrightarrow \underset{44.0\text{ g}}{\text{carbon dioxide}}$$

In other words, the total mass of the starting substances (12.0 g + 32.0 g) exactly equals the mass of the carbon dioxide (44.0 g).

The **law of conservation of mass,** first stated by Lavoisier, is a fundamental law of modern chemistry. It says that *the total mass remains constant during a chemical reaction.* Thus,

Total mass of substances before reaction = total mass of substances after reaction

ENERGY

Matter is something we can, at least in principle, touch or see. *Energy,* however, is quite different. It is not a material thing, so it is by its nature abstract. Nevertheless, energy is a term used so often in everyday language that it seems quite familiar. A young child jumping up and down has energy, we say. You can see that the child's motion is associated with energy, but energy comes in other forms, too. Heat, light, and electrical energy are also forms of energy. Why? In the next section, we define energy and explore some of the forms it takes.

3.6 Types of Energy

OBJECTIVES

- Define energy.
- Define kinetic energy and potential energy.
- List other forms of energy.
- Define the energy units: joule and calorie.

We can define **energy** as *the potential or capacity to move matter.* The most obvious form of energy is **kinetic energy,** or *the energy of matter in motion.* There is another form of energy, though, where there is no obvious motion, but where there is a potential for motion. *Matter with a potential for motion* has **potential energy.** Thus, the two basic forms of energy are kinetic energy and potential energy. Let's consider an example, so that you can see how these forms of energy are manifested and related. A flashlight battery that you hold in the air has potential energy because it has the potential for motion. If you let it drop, its potential energy is converted to kinetic energy. ●

● A boulder on a hill and water behind a dam also have potential energy because they have the potential for motion: The boulder can roll down the hill and the water can pour down the dam's spillway.

The battery is also a source of another form of energy. It contains chemical substances that can react to produce an electric current. The battery is essentially a store of energy waiting to be tapped; it has *chemical energy.* When you put the battery in a flashlight, this form of energy can be converted to electrical energy first and then to light (Figure 3.15). You can also feel heat, another form of energy, coming from the bulb.

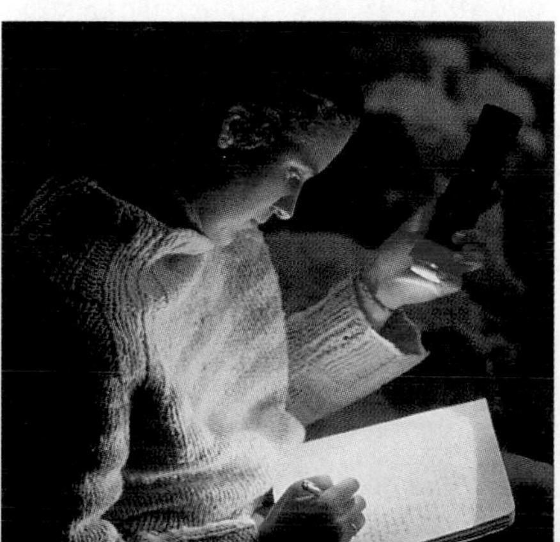

FIGURE 3.15

A flashlight. Chemical substances within this flashlight's batteries have chemical energy that can be converted to electrical energy, light, and heat.

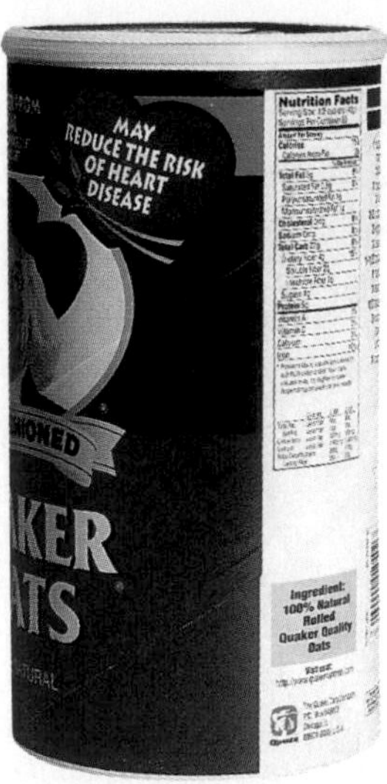

FIGURE 3.16
The caloric content of processed food is on the label. The label shows that a typical serving with a half cup of skim milk will provide 190 Cal (190 kcal or 190,000 cal) of energy.

Energy Units

You have seen that energy can be converted from one type to another, so it is appropriate that all types of energy have the same energy unit. The **joule** is *the SI unit of energy.* It is pronounced "jewl," abbreviated as J, and named after the British physicist who first showed that the different forms of energy are basically the same.

The *calorie* is another unit of energy (but it is not an SI unit). The **calorie** (cal) is *presently defined as exactly 4.184 J, but was originally defined as the quantity of energy needed to raise the temperature of 1 g of water by 1°C.* The new definition in terms of joules is more precise than the older one, but the older definition is a useful one because it relates directly to the quantity of energy needed to heat water.

The "calorie" that is used in many nutrition books and on the labels of processed food is actually a kilocalorie (Figure 3.16). Remember that 1 kcal = 1000 cal. The nutritional unit is often spelled with a capital C (Calorie).

EXAMPLE 3.3 Converting Between Joules and Calories

The daily energy requirements of adults vary by age, body size, and activity level, but a rough average is 2000 Cal, or 2000 kcal, or 2,000,000 cal (2×10^6 cal), enough energy to raise the temperature of 2 million grams of water by 1°C. Express 2×10^6 cal in joules. (Note that 1 cal = 4.184 J.)

Problem Analysis

Use the equivalence statement

$$1 \text{ cal} = 4.184 \text{ J}$$

to obtain a conversion factor from calories to joules. Then do the conversion calculation.

Solution

You wish to convert 2×10^6 cal to joules:

$$2 \times 10^6 \text{ cal} \xrightarrow{\text{Convert to}} \text{joules}$$

From the equivalence statement given in the problem analysis, you can write the following conversion factor:

$$\frac{4.184 \text{ J}}{1 \text{ cal}}$$

The calculation is

$$2 \times 10^6 \cancel{\text{cal}} \times \frac{4.184 \text{ J}}{1 \cancel{\text{cal}}} = \underline{8}.368 \times 10^6 \text{ J} \qquad (\text{rounds to } 8 \times 10^6 \text{ J})$$

Exercise 3.3

A sample of aluminum requires 48.6 J to heat it from 25°C to 48°C. How would you express the quantity 48.6 J in calories?

(Try Problems 3.35, 3.36, 3.37, and 3.38.)

3.7 Heat and Heat Calculations

OBJECTIVES

- State the fundamental property of heat.
- Define specific heat.
- Calculate the heat needed to change the temperature of an object, given the specific heat, mass, and temperature change.
- Calculate the specific heat of a substance, given the heat absorbed, mass, and temperature change.

The fundamental property of heat is this: When you place two objects in contact so that heat can flow between them, heat flows from the hot object to the cold object. The hot object gets cooler, and the cool object gets warmer. In this section, we consider the following types of quantitative questions. You want to warm a beaker of water from room temperature to 50°C; how much heat must you add to the water? Or you want to cool a pan of water to its freezing point; how much heat will you have to remove from the water?

Let's take a look at the following problem. You place a beaker containing 100.0 g of water on a hot plate, and you want to raise the temperature of the water from 20.0°C to 50.0°C. To raise the temperature, you need to add heat. How much heat do you need to add? We will first do this problem in two separate steps, starting with a property of a substance called its *specific heat.* After that, you will see how to do the problem in a single step.

The **specific heat** of a substance is *the quantity of heat required to raise the temperature of 1 g of the substance by 1°C.* Table 3.2 lists the specific heats of some substances. Water has a specific heat of 4.18 J/(g · °C), or 4.18 J per gram per degree Celsius. This means that it takes 4.18 J of heat to raise the temperature of 1 g of water 1°C.

If it takes 4.18 J of heat to raise the temperature of 1 g of water by 1°C (the specific heat of water), how much heat do you need to raise the temperature of 100.0 g of water by 1°C? You simply multiply the specific heat by 100.0 g. Here is the calculation:

$$\underbrace{\frac{4.18\text{ J}}{1\ \cancel{\text{g}} \times 1^\circ\text{C}}}_{\text{Specific heat}} \times \underbrace{100.0\ \cancel{\text{g}}}_{\text{Mass}} = \frac{418\text{ J}}{1^\circ\text{C}}$$

Note that the unit of grams cancels and gives the result 418 J/°C, or 418 J for 1°C.

TABLE 3.2 Specific Heats of Some Substances

Substance	Specific Heat J/(g · °C)	Specific Heat cal/(g · °C)
Aluminum	0.901	0.215
Copper	0.384	0.0918
Ethanol (ethyl alcohol)	2.43	0.581
Iron	0.449	0.107
Lead	0.129	0.0308
Silver	0.234	0.0559
Water	4.18	1.00

Now we are ready to answer the original question: How much heat energy do you need to raise the temperature of 100.0 g of water from 20.0°C to 50.0°C? You already know that this quantity of water requires 418 J to raise the temperature by 1°C. You want to raise the temperature by 50.0°C − 20.0°C (or 30.0°C), so you multiply 418 J/°C by (50.0 − 20.0)°C:

$$\frac{418\ \text{J}}{1\cancel{°\text{C}}} \times \underbrace{(50.0 - 20.0)\cancel{°\text{C}}}_{\text{Temperature change}} = 1.25 \times 10^4\ \text{J}$$

Thus, you must add 1.25×10^4 J of heat to raise the temperature of 100.0 g of water by 30.0°C.

Let's put these two calculations together:

$$\underbrace{\frac{4.18\ \text{J}}{1\ \cancel{\text{g}} \times 1\cancel{°\text{C}}}}_{\text{Specific heat}} \times \underbrace{100.0\ \cancel{\text{g}}}_{\text{Mass}} \times \underbrace{(50.0 - 20.0)\cancel{°\text{C}}}_{\text{Temperature change}} = 1.25 \times 10^4\ \text{J}$$

Note that you multiply specific heat (in units of J per g · °C) by mass (to cancel grams) and by temperature change (to cancel degrees Celsius). We can summarize this result with the following formula:

Heat = specific heat × mass × temperature change

In the preceding calculation, the heat is "heat added" to the water.

You can use a similar calculation to obtain the heat that you must remove from 100.0 g of water to cool it from 20.0°C to 0.0°C (the freezing point):

$$\underbrace{\frac{4.18\ \text{J}}{1\ \cancel{\text{g}} \times 1\cancel{°\text{C}}}}_{\text{Specific heat}} \times \underbrace{100.0\ \cancel{\text{g}}}_{\text{Mass}} \times \underbrace{(20.0 - 0.0)\cancel{°\text{C}}}_{\text{Temperature change}} = 8.36 \times 10^3\ \text{J}$$

You interpret the heat in this calculation as "heat removed," since you know that you are cooling the water. The following example gives more practice in this type of heat calculation.

■ EXAMPLE 3.4 Calculating Heat from Specific Heat, Mass, and Temperature Change

A piece of copper wire has a mass of 35.6 g. How much heat (in joules) would you need to raise its temperature from 25°C to 58°C? The specific heat of copper is listed in Table 3.2 as 0.384 J/(g · °C).

Problem Analysis

Calculate the heat from the following formula:

$$\text{Heat} = \text{specific heat} \times \text{mass} \times \text{temperature change}$$

If you have any difficulty remembering this formula, simply look at the units of specific heat (they are J per g per °C), and note that you will have to multiply specific heat by mass (in g) to cancel g and by temperature change (in °C) to cancel °C.

Solution

The calculation is:

$$\text{Heat} = \underbrace{\frac{0.384\ \text{J}}{1\ \cancel{\text{g}} \times 1\ \cancel{°\text{C}}}}_{\text{Specific heat}} \times \underbrace{35.6\ \cancel{\text{g}}}_{\text{Mass}} \times \underbrace{(58 - 25)\cancel{°\text{C}}}_{\text{Temperature change}}$$

$$= 451.1\ \text{J} \quad (\text{rounds to } 4.5 \times 10^2\ \text{J})$$

Because the copper is being warmed, the heat in this calculation is "heat added." Note the cancellation of units.

Exercise 3.4

A chunk of lead at 26°C and weighing 42 g is placed in a beaker of boiling water at 100°C. How much heat has passed into the lead by the time its temperature becomes equal to that of the boiling water? The specific heat of lead is 0.129 J/(g · °C).

(Try Problems 3.39, 3.40, 3.41, and 3.42.)

The equation relating heat to specific heat, mass, and temperature change can be used to obtain any one of the quantities in terms of the others. Suppose you have a metal, and you wish to determine its specific heat. You take a 58.6-g sample of the metal and heat it from 20.30°C to 26.83°C. You find that this requires 345 J of heat. What is the specific heat? The procedure for solving this problem is explained in the next example.

■ EXAMPLE 3.5 Calculating Specific Heat from Heat, Mass, and Temperature Change

You heat a metal sample weighing 58.6 g by adding 345 J of energy to it. The temperature of the metal rises from 20.30°C to 26.83°C. What is the specific heat of the metal?

Problem Analysis

Start with the equation for heat in terms of the specific heat, mass, and temperature change. Then rearrange the equation to obtain one in which the specific heat is on one side by itself. Substitute values (with units) into this equation.

Solution

Start with the equation

$$\text{Heat} = \text{specific heat} \times \text{mass} \times \text{temperature change}$$

Now rearrange the original equation to get an equation for the specific heat in terms of heat, mass, and temperature change. Let's take that equation and divide both sides by mass times temperature change:

$$\frac{\text{Heat}}{\text{Mass} \times \text{temperature change}} = \frac{\text{specific heat} \times \cancel{\text{mass}} \times \cancel{\text{temperature change}}}{\cancel{\text{mass}} \times \cancel{\text{temperature change}}}$$

The mass and the temperature change cancel on the right-hand side of the equation, leaving the specific heat by itself. Thus,

$$\text{Specific heat} = \frac{\text{heat}}{\text{mass} \times \text{temperature change}}$$

Now you substitute values for the quantities (including units) and do the calculation:

$$\text{Specific heat} = \frac{345 \text{ J}}{58.6 \text{ g} \times (26.83 - 20.30)°\text{C}} = 0.90\underline{1}58836 \text{ J/(g} \cdot °\text{C)}$$

The answer rounds to 0.902 J/(g · °C).

Exercise 3.5

When a sample of platinum weighing 8.75 g has 5.68 J of heat added to it, the temperature rises from 25.47°C to 30.35°C. What is the specific heat of platinum according to these data?

(Try Problems 3.43 and 3.44.)

3.8 Law of Conservation of Energy

OBJECTIVE

- State the law of conservation of energy and give an example.

Energy, as you have seen, comes in various forms, and you can change one form of energy to another. But throughout all of its changes, the total quantity of energy remains fixed. The **law of conservation of energy** states that *energy may be converted from one form to another, but the total quantity of energy remains constant (the total energy is conserved).*

Let's look at an example. Because of its position, the water stored behind a dam has considerable potential energy. The potential energy is converted without loss to kinetic energy as the water flows over the dam and down the spillway (Figure 3.17). Although one form of energy is converted to another, energy is not lost; it is conserved.

FIGURE 3.17
Water over the dam. As water flows down the spillway, the water's potential energy (due to its potential for motion) is converted to kinetic energy (due to its motion).

Now let's look at an example in chemistry. Hydrogen-oxygen fuel cells are being considered for electric cars. A fuel cell is essentially a battery that is fed a supply of fuel (hydrogen, in this case) to produce an electric current. In this cell, hydrogen and oxygen react to produce water and electricity. Such a fuel cell was used in the Apollo spacecraft that went to the moon in 1969 (Figure 3.18), and similar fuel cells are used in the space shuttle orbiters. During operation of the Apollo spacecraft, each gram of hydrogen generated 95 kilojoules (kJ) of electric energy and 48 kJ of heat energy.

Instead of using the hydrogen in a fuel cell, you could burn hydrogen directly in oxygen. How much heat would be produced per gram of hydrogen in that case? The chem-

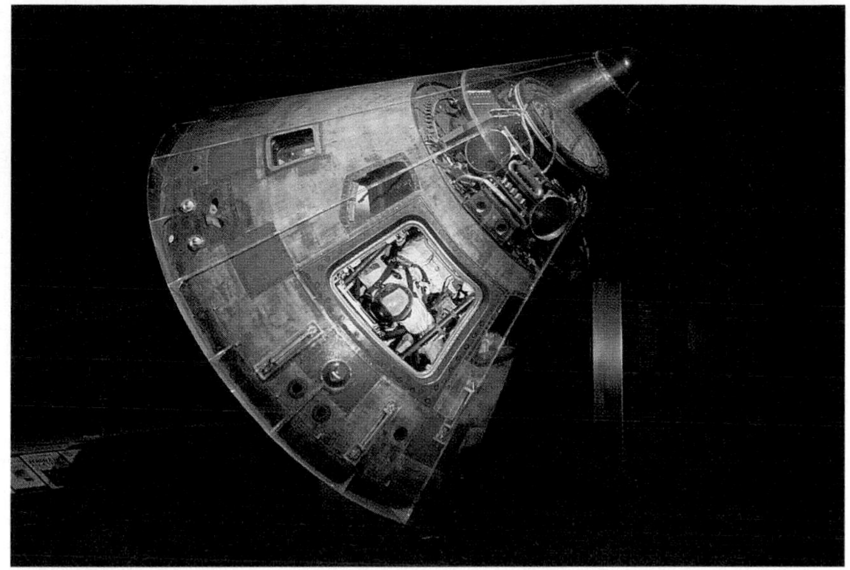

FIGURE 3.18
This spaceship orbited the moon. The electric power required for its lights, computers, and other pieces of apparatus was provided by hydrogen-oxygen fuel cells.

ical change is the same in both cases—hydrogen and oxygen react to produce water—and the total energy released is the same—143 kJ. The fuel cell produces electric energy and heat energy, whereas the direct burning of hydrogen in oxygen releases only heat energy. The heat that is released in the direct burning of 1 g of hydrogen equals the total of the energy released in the fuel cell: 95 kJ + 48 kJ = 143 kJ. Once again, energy is conserved.

Throughout this chapter, we have regarded matter and energy as distinct entities. But we have known since 1905, when Albert Einstein (1879–1955) proposed his theory of relativity, that matter and energy are related entities. Einstein found a mathematical relationship between mass and energy. According to this relationship, if a chemical reaction mixture, or system, releases 143 kJ of heat (such as when 1 g of hydrogen burns in oxygen), the system loses 1.59×10^{-9} g, or 1.59 nanograms (ng). The loss in mass is much too small to measure, however, so we are certainly justified in ignoring mass-energy conversion in a chemical reaction. ●

● The chemical system loses 1.59×10^{-9} g, but its immediate surroundings gain 1.59×10^{-9} g. Conservation of mass does hold exactly for the system plus surroundings.

Chemical Perspective

Energy on Earth

Sixty-five million years ago, the dinosaurs dominated other animals on earth. A short time later, they were extinct. What happened to the dinosaurs? In 1980, a group of scientists found evidence to suggest that an asteroid or comet hit the earth about 65 million years ago, causing a great cloud of dust that covered the earth in darkness for several months. During this lengthy darkness, plants died, and the dinosaurs and many other animals that lived on plants also succumbed. This theory dramatically illustrates the flow of energy on earth.

The ultimate source of energy on earth is the sun. The sun is constantly radiating light (and losing a corresponding quantity of mass, according to Einstein's relationship). A portion of this light energy falls on earth, and some of it is converted by plants to chemical energy in a process called *photosynthesis.* During photosynthesis, carbon dioxide *(continued)* (from air) and water are combined using energy to produce the sugar glucose and oxygen:

Carbon dioxide (from air) + water + energy (sunlight) ⟶ glucose + oxygen

Later, the chemical energy stored in the glucose (and oxygen) is used by the plant for its growth and maintenance. The cells of the plant use glucose and oxygen to produce the energy it needs, effectively reversing photosynthesis:

Glucose + oxygen ⟶ carbon dioxide + water + energy

Much of the energy available on earth has been collected by plants during photosynthesis. All animals ultimately rely on plants for food energy. Grazing animals obtain the energy they need by eating plants. Carnivores are meat-eating animals, many of whom eat grazing animals. Human beings, of course, eat both plants and animals. Ultimately, the plants and therefore the sun provide us with the food energy we need.

Human civilization consumes more than food energy, however. We use energy to heat our homes, to power our cars, and to drive technology. Figure 3.19 shows a chart illustrating the sources of energy consumed this way in the United States. The three largest sources of the energy we consume are petroleum, coal, and natural gas. These are all "fossil fuels." The fossil fuels were formed millions of years ago when aquatic plants and animals were buried and compressed by layers of sediment at the bottoms of swamps and seas. Here again, we can trace the origin of this energy back to plants and therefore to the sun.

Another large source of energy is hydroelectric power. We cannot trace this energy source to plants, but we can trace it back to the sun. Hydroelectric power is electric power generated by river water flowing through a turbine. This energy of the river comes from the sun. The sun's warmth evaporates water, and this water vapor later condenses as rain, which later flows to the rivers.

Only nuclear energy does not originate from the sun. Uranium, which is the source of nuclear energy, has been present on earth since our solar system first formed about 4.5 billion years ago.

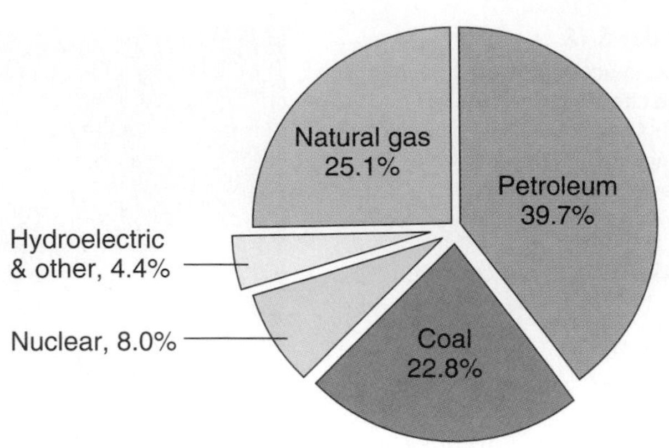

FIGURE 3.19

Sources of the energy consumed in the United States. Data are for 1996. The largest sources—petroleum, coal, and natural gas—are fossil fuels, which originated from plants and animals (and whose energy can therefore be traced to the sun). Hydroelectric power can also be traced to the sun. Nuclear power relies on uranium, which was present when the solar system was formed. *Source:* Energy Information Administration (EIA), from their home page (http://www.eia.doe.gov).

CHAPTER REVIEW

Key Words

states of matter (*p. 60*)
solid (*p. 60*)
liquid (*p. 60*)
gas (*p. 60*)
material (*p. 63*)
physical change (*p. 63*)
chemical change (chemical reaction) (*p. 63*)
physical property (*p. 64*)
chemical property (*p. 64*)
substance (*p. 65*)
mixture (*p. 65*)
homogeneous mixture (solution) (*p. 67*)
heterogeneous mixture (*p. 67*)
element (*p. 68*)
symbol (*p. 68*)
compound (*p. 70*)
law of definite proportions (law of constant composition) (*p. 70*)
law of conservation of mass (*p. 71*)
energy (*p. 71*)
kinetic energy (*p. 71*)
potential energy (*p. 71*)
joule (*p. 72*)
calorie (*p. 72*)
specific heat (*p. 73*)
law of conservation of energy (*p. 76*)

Key Equation

Heat = specific heat × mass × temperature change

Summary

In this chapter, we discussed matter and energy. To begin to make some sense out of the great variety of matter that exists, we introduced two ways of classifying matter. The first way depends on physical state (whether the substance is solid, liquid, or gas). Many substances can exist in any one of the physical states under the right conditions, and each state can be transformed into the other states.

In the second way, we classify matter as an *element, compound,* or *mixture.* To discuss this classification scheme, we first look at the distinction between physical and chemical changes and properties. We define a *substance* as a material that cannot be separated into different materials by any physical process. If the material can be separated by physical processes into more than one substance, we call it a *mixture.* An *element* is a substance that cannot be decomposed by any chemical reaction into simpler substances. Chemists denote each of the elements by a *symbol.* A *compound* is a substance composed of more than one element. According to the *law of definite proportions* (law of constant composition), a compound always contains definite, or constant, proportions of the elements by mass. We also discussed the *law of conservation of mass,* which states that the total mass remains constant during a chemical reaction.

We discussed energy in the second part of the chapter. *Energy* is the capacity to move matter, and it exists in two different forms. *Kinetic energy* is the energy associated with an object by virtue of its motion, but an object with the potential for motion has *potential energy.* Chemical energy, electrical energy, light, and heat are other forms of energy. Any form of energy can be converted to one of the others.

The heat required to raise the temperature of a substance is obtained by multiplying the specific heat of the substance by its mass and by the temperature change. An important fact about energy is that although it can change form, the total quantity of energy remains constant. This result is known as the *law of conservation of energy.*

Problem-Solving Skills

1. **Identifying the State of a Substance Under Given Conditions:** Given the melting point and boiling point of a substance, decide its physical state at a given temperature (Example 3.1).
2. **Identifying the Physical and Chemical Properties of a Substance:** Classify each of the properties given for a substance as either a physical property or a chemical property (Example 3.2).
3. **Converting Between Joules and Calories:** Given the heat added or absorbed in joules, convert the value to calories, and vice versa (Example 3.3).
4. **Calculating Heat from Specific Heat, Mass, and Temperature Change:** Given the specific heat of a substance, its mass, and the temperature change, calculate the heat (Example 3.4).
5. **Calculating Specific Heat from Heat, Mass, and Temperature Change:** Given the heat, mass of a substance, and temperature change, calculate the specific heat of the substance (Example 3.5).

Questions to Test Your Reading

3.1 Give three examples in which a specific substance changes state. What is the substance? What is the name of the change of state it undergoes?

3.2 Name a material that exists as a solid but not in the gas or liquid state.

3.3 What is the name given to the temperature at which a solid melts?

3.4 Define vapor.

3.5 Describe what happens on the molecular level when a substance melts and when it boils.

3.6 Describe what happens on the molecular level when a substance freezes and when its vapor condenses.

3.7 Give three examples of physical changes.

3.8 Give three examples of chemical changes.

3.9 Give three examples of physical properties.

3.10 Give an example of a chemical property for iron and for bromine.

3.11 Define substance and give an example.

3.12 Define mixture and give an example.

3.13 Describe the use of filtration to separate a precipitate from a solution.

3.14 Describe the distillation of a solution of salt in water.

3.15 Give two examples of homogeneous mixtures.

3.16 Give two examples of heterogeneous mixtures.

3.17 Define element. Give three examples.

3.18 Define compound. Give three examples.

3.19 How could we use the law of definite proportions to determine if a material is a compound?

3.20 Define energy.

3.21 Define kinetic energy and potential energy. Give examples of objects with these forms of energy.

3.22 List four other forms of energy.

3.23 Define specific heat. What is the specific heat of liquid water?

3.24 Briefly describe how Einstein's theory of relativity affects Lavoisier's law of conservation of mass.

Practice Problems

States of Matter

3.25 Carbon tetrachloride was once used in dry cleaning, but its use has been discontinued because it is poisonous. Carbon tetrachloride melts at −23°C and boils at 77°C. What is the physical state of carbon tetrachloride at 35°C and normal atmospheric pressure?

3.26 Diethyl ether is commonly called *ether.* The substance is sometimes used as an anesthetic. Diethyl ether has a melting point of −116°C and a boiling point of 35°C. What is the physical state of diethyl ether at 55°C and normal atmospheric pressure?

3.27 Butane is used as the fuel in certain lighters. It melts at −138°C and boils at 0°C. What is the physical state of butane at 20°C and normal atmospheric pressure?

3.28 Sodium chloride is the chemical name of common table salt. It has a melting point of 801°C and a boiling point of 1465°C. What is the physical state of sodium chloride at 950°C and normal atmospheric pressure?

Physical and Chemical Properties

3.29 Hexane is a component of gasoline. It is a colorless liquid that boils at 69°C and freezes at −95°C. Hexane burns in air to produce carbon dioxide and water. Which of these properties of hexane are physical and which are chemical?

3.30 Iodine is a blue-black solid. When heated, it produces a violet vapor. Its melting point is 114°C; its boiling point is 185°C. Iodine reacts with sodium metal to give a white solid. Classify each of the properties given for iodine as either physical or chemical.

3.31 Nitroglycerine is used as a blood vessel dilator to relieve cardiac pain. It is a colorless, oily liquid with a density of 1.59 g/cm^3. Nitroglycerine explodes readily, producing nitrogen, carbon dioxide, water vapor, and oxygen (all gases). State which of the properties of nitroglycerine are physical and which are chemical.

3.32 Lead acetate is a white crystalline substance. It is said to have a sweet taste (but is poisonous), leading to its common name, "sugar of lead." When added to a colorless solution of sodium iodide, lead acetate gives yellow crystals of lead iodide. Which of the properties of lead acetate given here are physical and which are chemical?

Elements, Compounds, and Mixtures

3.33 Classify each of the following as either an element, a compound, or a mixture.
(a) ocean water (b) phosphorus (c) water
(d) milk

3.34 Classify each of the following as either an element, a compound, or a mixture.
(a) syrup (b) aluminum
(c) mercury(II) oxide (d) beach sand

Heat and Specific Heat

3.35 Hydrogen gas burns to produce water. When 1.00 g of hydrogen gas is burned in air, 121 kJ of heat is produced. What is this heat in calories?

3.36 Ammonium nitrate is a white crystalline substance used in medical cold packs. When 1.00 g of ammonium nitrate is dissolved in an excess of water, 321 J of heat is absorbed. What is this heat in calories?

3.37 Methane is the major substance in natural gas. When 1.00 g of methane burns in air, 12.0 kcal of heat is released. What is this heat in joules?

3.38 Magnesium sulfate is a white crystalline substance used in medical hot packs. When 1.00 g of magnesium sulfate is dissolved in an excess of water, 181 cal of heat is released. What is this heat in joules?

3.39 Aluminum is used to make some kitchen pots and utensils. The specific heat of aluminum is 0.901 J/(g · °C). How much heat is required to raise the temperature of a 235-g pan by 15.0°C?

3.40 Copper is sometimes used on the bottom surfaces of kitchen pans. The specific heat of copper is 0.384 J/(g · °C). How much heat is required to raise the temperature of 35.8 g of the metal by 25.0°C?

3.41 Quartz is silicon dioxide. It is used to make certain laboratory vessels. The specific heat of quartz is 0.741 J/(g · °C). How much heat is required to raise the temperature of a 65.4-g vessel from 20.0°C to 38.5°C?

3.42 A ceramic material is used to retain the heat from air warmed by the sun. If its specific heat is 0.649 J/(g · °C), how much heat is absorbed to raise the temperature of 0.500 kg of the material from 20.0°C to 45.0°C?

3.43 An 18.9-g sample of manganese metal is warmed by adding 151.5 J of heat to it. If the temperature increases from 23.2°C to 39.9°C, what is the specific heat of manganese?

3.44 A 28.3-g sample of tin metal is warmed by adding 189.5 J of heat to it. If the temperature of the metal increases from 24.8°C to 55.7°C, what is the specific heat of tin?

3.45 Ethanol (ethyl alcohol) has a specific heat of 2.43 J/(g · °C). If 1071 J of heat is added to 36.0 g of ethanol, what is the increase in temperature of the liquid?

3.46 A sample of 48.6 g of isopropyl alcohol is warmed. If the alcohol, with a specific heat of 2.51 J/(g · °C), absorbs 1628 J of heat, how much does its temperature increase?

3.47 The temperature of a 15.5-g sample of iron is raised by adding 164.8 J of heat to it. If the initial temperature of the iron is 21.5°C, what is its final temperature? The specific heat of iron is 0.449 J/(g · °C).

3.48 A 16.8-g sample of nickel is warmed by adding 104 J of heat to it. If the initial temperature of the nickel is 25.2°C, what is its final temperature? The specific heat of nickel is 0.444 J/(g · °C).

Additional Problems

3.49 Classify each of the following as gas, liquid, or solid under normal room conditions.
(a) carbon dioxide (b) argon (c) chlorine
(d) mercury (e) phosphorus

3.50 Classify each of the following as gas, liquid, or solid under normal room conditions.
(a) copper (b) bromine (c) diamond
(d) oxygen (e) nitrogen

3.51 Which of the following are physical changes and which are chemical changes?
(a) melting of a fat
(b) action of acid on a metal
(c) evaporation of gasoline
(d) solidification of molten iron
(e) burning of a wax

3.52 Which of the following are physical changes and which are chemical changes?
(a) distillation of alcohol
(b) making of soap from fat and lye
(c) filtration of muddy water
(d) grinding of an iron rod
(e) scorching of paper

3.53 Gold has a melting point of 1064°C and a boiling point of 2700°C. What is the physical state of gold at each of the following temperatures, under normal pressures?
(a) −100°C (b) 0°C (c) 500°C
(d) 1000°C (e) 3000°C

3.54 Mercury has a melting point of −39°C and a boiling point of 357°C. What is the physical state of mercury at each of the following temperatures, under normal pressures?
(a) −100°C (b) 0°C (c) 500°C
(d) 1000°C (e) 3000°C

3.55 Bromine has a melting point of $-7°C$ and a boiling point of 59°C. What is the physical state of bromine at each of the following temperatures, under normal pressures?
(a) $-50°C$ (b) 10°C (c) 50°C
(d) 100°C (e) 200°C

3.56 Ethanol (ethyl alcohol) has a melting point of $-114°C$ and a boiling point of 78°C. What is the physical state of ethanol at each of the following temperatures?
(a) $-50°C$ (b) 10°C (c) 50°C
(d) 100°C (e) 200°C

3.57 Identify each of the following as a solution, a heterogeneous mixture, or a substance.
(a) water
(b) water and oil
(c) sodium sulfate dissolved in water
(d) carbon dioxide

3.58 Identify each of the following as a solution, a heterogeneous mixture, or a substance.
(a) ethanol
(b) sodium chloride
(c) tap water
(d) sandy water

3.59 The following is a list of properties of substances. Indicate which properties are physical and which are chemical.
(a) Iodine gives a violet vapor on heating; the vapor condenses back to solid iodine on cooling.
(b) Steel wool burns in air with a bright flash.
(c) Hexane is a liquid that floats on water.
(d) Concentrated sulfuric acid turns sugar into a black char.

3.60 The following is a list of properties of substances. Indicate which properties are physical and which are chemical.
(a) The silvery surface of lead turns a dull gray after exposure to air.
(b) When limestone (calcium carbonate) is heated strongly, it gives a white powder (calcium oxide) and carbon dioxide gas.
(c) Gallium metal melts from the heat of one's hand.
(d) Silver bromide, a yellowish crystalline substance, turns black on exposure to light. The black color is due to silver metal.

3.61 A British thermal unit (Btu) equals 252 calories. How many calories are there in 453 Btu? This is equivalent to how many joules?

3.62 A kilowatt-hour (kWh) is a unit of energy equal to 3.60×10^6 J. An electric bill lists the total energy consumed as 153 kWh. How many joules is this? How many calories?

3.63 A pat of butter provides 35 Cal (35 kcal) on burning. How many grams of water could be warmed from 20.0°C to 35.0°C with the heat from burning this butter?

3.64 A tablespoon of sugar provides 40 Cal (two significant figures) on burning. How many grams of water could be warmed from 25.0°C to 40.0°C with the heat from the burning of this sugar?

3.65 How many joules of heat are required to raise the temperature of 5.42 g of water from 20.8°C to 45.1°C? The specific heat of water is 4.18 J/(g · °C).

3.66 How many joules of heat are required to raise the temperature of 6.87 g of ice from $-15.0°C$ to $-1.5°C$? The specific heat of ice is 2.06 J/(g · °C).

Practice in Problem Analysis

For each problem, describe the thinking you would use (the problem analysis) before doing the actual solution, but do not solve the problem.

1. The temperature of a 2563-g silver bar was 85°C before it was plunged into 966 mL of water. The final temperature of the bar and the water was 36°C. What temperature change did the water experience if the specific heats of silver and water are 0.449 J/(g · °C) and 4.18 J/(g · °C), respectively?

2. A 28-g (1-oz) serving of a breakfast cereal served with 0.5 cup of skim milk has 8 g of protein, 26 g of carbohydrates, and 2 g of fat. If 1 g of protein, 1 g of carbohydrates, and 1 g of fat provide 17 kJ, 17 kJ, and 38 kJ, respectively, what is the caloric content of the cereal?

PRACTICE EXAM

1. Which substance is a liquid under normal room conditions?
(a) oxygen (b) mercury
(c) graphite (d) lead
(e) None of the above

2. Which change is a physical change?
(a) rusting iron
(b) digesting potato chips
(c) evaporating water
(d) burning wood
(e) None of the above

3. Which change is a chemical change?
(a) melting lead (b) cooking an egg
(c) solidifying lava (d) distilling tap water
(e) None of the above

4. Which statement describes a chemical property?
(a) Sodium reacts violently with water.
(b) Sodium is a silvery metal.
(c) Sodium is less dense than lead.
(d) Sodium is a soft metal.
(e) None of the above

5. Which statement describes a physical property?
(a) Silver chloride turns black when it is exposed to light because of the formation of silver metal.
(b) Water decomposes to hydrogen and oxygen when an electric current passes through it.
(c) Iron wool burns in air when it is placed in a flame.
(d) Table salt dissolves in water.
(e) None of the above

6. Which material is an element?
(a) copper (b) carbon monoxide
(c) salt water (d) pure water
(e) None of the above

7. Which material is a compound?
(a) graphite (b) diamond
(c) tap water (d) pure water
(e) None of the above

8. The symbol for copper is
(a) C (b) Co (c) Cp (d) Cu
(e) None of the above

9. Si is the symbol for
(a) silver (b) sodium
(c) silicon (d) selenium
(e) None of the above

10. Which material is a mixture?
(a) graphite (b) diamond
(c) tap water (d) pure water
(e) None of the above

11. Which mixture can be separated by ordinary distillation?
(a) solid sugar and solid salt
(b) water and sand
(c) solid sulfur and iron filings
(d) rubies and diamonds
(e) None of the above

12. Which mixture can be separated by ordinary filtration?
(a) solid sugar and solid salt
(b) water and sand
(c) solid sulfur and iron filings
(d) rubies and diamonds
(e) None of the above

13. When a certain quantity of paper is burned, 234 kcal of heat is produced. What is this heat in kilojoules?
(a) 979 kJ (b) 55.9 kJ (c) 1 kJ (d) 234 kJ
(e) None of the above

14. A metal that may be either aluminum, iron, copper, or lead requires 89.0 J to raise 50.0 g from 26.3°C to 40.1°C. The specific heats of these metals are 0.901 J/(g · °C), 0.449 J/(g · °C), 0.384 J/(g · °C), and 0.129 J/(g · °C), respectively. Identify the metal. Its symbol is
(a) Al (b) Fe (c) Cu (d) Pb
(e) None of the above

15. Which statement is true?
(a) When a ball rolls downhill, the potential energy it loses is one-half of the kinetic energy it gains.
(b) The kinetic energy of an object depends solely on its mass.
(c) The nutritional calorie is identical to the kilojoule.
(d) Einstein showed that matter is fundamentally unrelated to energy.
(e) None of the above

Chapter

4 Atoms, Molecules, and Ions

The Aurora Borealis occurs when molecules in the atmosphere collide with incoming electrons and protons from the sun.

Chapter Outline

According to a recent newspaper account, the owner of a Chevy Malibu wondered what happened when he found a hole through the trunk and gas tank of the old Chevy and a football-size smoking rock lying in a bowl-shaped depression underneath the car. Scientists explained that the car had been hit by a meteoroid, a rock from outer space, noting that the smoking rock was similar to other rocks that have hit the earth. They knew that, moments earlier, a bright streak had been seen in the night sky, presumably the burning trace of this meteoroid through the earth's atmosphere. Like the Chevy owner, we tend to want explanations for our observations, not simply a description of what was observed. In Chapter 3, we described how chemists classify matter as pure *substances* and *mixtures.* Now we want to see how they explain these classifications in terms of a theory of matter.

Substances, you will recall, are either elements or compounds. There are 112 known elements, and every compound is composed of two or more of these elements chemically combined. For example, hydrogen and oxygen are two elements that are gases. When hydrogen gas burns in oxygen gas (present in air), water forms. Thus, water is a compound of hydrogen and oxygen. The elements are held together in some way that makes the compound different from a simple mixture of the elements hydrogen and oxygen. How do we explain this combination? In this chapter, we describe John Dalton's *atomic theory,* which explains the observed differences among the various forms of matter in terms of atoms. The theory states that substances—either elements or compounds—are made up of atoms. Sometimes they occur in collections of identical atoms; sometimes they occur in *molecules;* and sometimes they occur as *ions* (pronounced "eye′-uns"). Together atoms, molecules, and ions constitute the "molecular" basis of matter, one of the themes of this textbook.

ATOMS

Although the idea that matter is composed of atoms is very ancient, the atomic theory of matter in its modern form was developed by the British chemist John Dalton (1766–1844) in about 1803. Dalton could not see atoms but he assumed that they exist in order to account for the quantitative results chemists had found in their experiments and had summarized in two laws: the law of conservation of mass and the law of definite proportions.

Dalton's atomic theory is the basis of modern chemistry. We explain Dalton's theory in the next section. Subsequent sections explain the composition of an atom and how the elements can be classified and arranged according to the properties of their atoms.

4.1 Dalton's Atomic Theory

OBJECTIVE

- State the four main postulates of Dalton's atomic theory.

Although Dalton believed that *atoms* are extremely small particles of matter that retain their identity during chemical reactions, he could not see them. He proposed that they had to exist to explain two chemical laws:

1. The law of conservation of mass, which says that the total mass remains constant during a chemical reaction
2. The law of definite proportions, which says that a pure compound is always composed of the same elements in the same proportions by mass

Recent research with a special instrument known as the scanning tunneling microscope has confirmed Dalton's ideas by allowing chemists to see images of atoms (Figure 4.1). Dalton's **atomic theory,** described in the next paragraphs, is *an explanation of the structure and chemical reactions of matter in terms of atoms.*

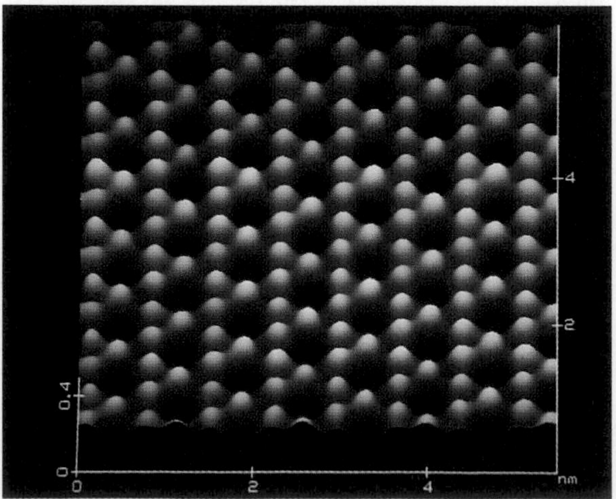

FIGURE 4.1

An image of atoms. This computer-drawn image of a surface was obtained by a scanning tunneling microscope; the image clearly shows individual atoms.

Postulates from Dalton's Atomic Theory

1. All matter is composed of atoms; each **atom** is *an extremely small, chemically indivisible particle.*
2. An **element** is *a type of matter composed of only one kind of atom, which always has certain specific properties.* One of the characteristic properties of the atoms of an element is their average mass, usually called the *atomic weight* of the element.
3. A **compound** is *a type of matter composed of atoms of two or more elements chemically combined in fixed proportions.* The relative numbers of any two kinds of atoms in a compound occur in simple ratios.
4. A **chemical reaction** is *a rearrangement of the atoms present in the reacting substances to give new chemical combinations in the substances formed by the reaction.* Atoms are neither created nor destroyed in a chemical reaction.

Note that the words *element, compound,* and *chemical reaction,* defined in Chapter 3 in simple observational terms, are defined here again but now in terms of the theoretical underlying composition of matter. Let's consider each postulate in more detail.

The first postulate indicates that atoms are indivisible particles. In fact, the word *atom* derives from the Greek, meaning "cannot be cut apart."

The second postulate states that the elements differ because they are composed of different kinds of atoms. The element hydrogen consists only of hydrogen atoms, and the element oxygen consists only of oxygen atoms. Dalton's theory also states that atoms of a given kind have definite properties. For example, the average mass of an atom of a particular element (usually referred to as the element's atomic weight) is a definite value. The average mass of oxygen atoms is about 16 times that of hydrogen atoms; the average mass of carbon atoms is about 12 times that of hydrogen atoms. The concept of atomic weights has played an important role in chemistry, and we discuss it further in Section 4.3.

The third postulate is that compounds are composed of atoms of two or more elements chemically combined in fixed proportions. The relative numbers of any two kinds of atoms in a compound occur in simple ratios. Water, for example, is always composed of hydrogen and oxygen atoms in the ratio 2:1; that is, there are twice as many hydrogen atoms as oxygen atoms in any sample of water. This point explains the law of definite proportions. Because a compound contains atoms in definite, fixed proportions and because atoms have definite masses, compounds must contain elements in definite proportions by mass.

The fourth and final postulate in Dalton's theory is that in a chemical reaction the atoms in the reacting substances are simply rearranged to give new combinations of atoms. New atoms are not created by the chemical reaction, nor are any atoms lost or destroyed. Let's look at an example. Figure 4.2A shows a jar containing hydrogen and oxygen gases before a reaction. You can see that the elements in the jar consist of *diatomic* (two-atom) molecules. ● Figure 4.2B shows the same jar after a chemical reaction. Now it contains molecules of water. Note, however, that the jar contains the same number of hydrogen atoms and the same number of oxygen atoms before and after reaction. Atoms were neither created nor destroyed; the reaction simply caused new combinations of the existing atoms. This explains the law of conservation of mass. Because the number of each kind of atom is the same before and after reaction, the total mass must remain constant.

● We discuss the naturally occurring forms of the elements in Section 4.6.

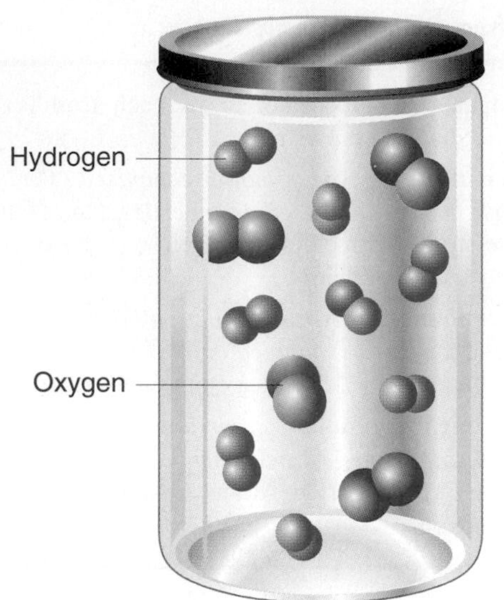

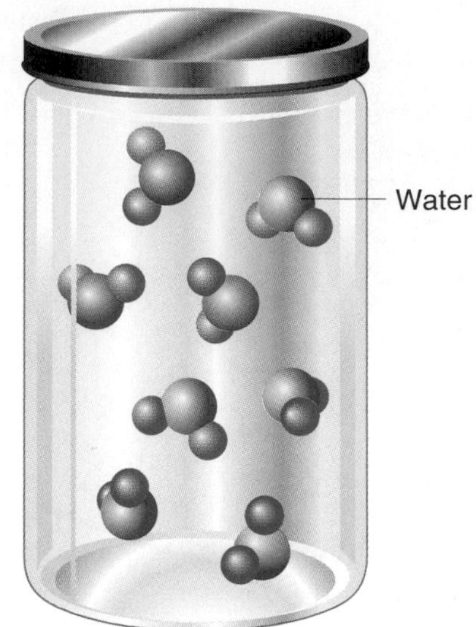

FIGURE 4.2
Reaction of hydrogen and oxygen to give water. *(A)* The jar contains hydrogen gas (light blue) and oxygen gas (red). Each gas contains molecules consisting of two identical atoms; the molecules are diatomic. *(B)* After the reaction, the jar contains water molecules, each composed of one oxygen atom and two hydrogen atoms. Note that the jar contains the same numbers of hydrogen atoms and oxygen atoms before and after the reaction.

You can also see the difference between a mixture and a substance in terms of this atomic picture. Compare the two parts of Figure 4.2. Figure 4.2A shows a *mixture* consisting of two kinds of molecules, hydrogen and oxygen. Figure 4.2B shows a *substance* consisting of only one kind of molecule, water.

Dalton's atomic theory makes it clear that an atom is an extremely small particle of matter that retains its identity during chemical reactions. Nevertheless, an atom is not a fundamental particle because an atom consists of smaller particles, as seen in the next section.

4.2 Particle Structure of the Atom

OBJECTIVES

- Describe the effect of one electric charge on another.
- Briefly describe Thomson's experiment in which he discovered the electron.
- Give Rutherford's explanation of why some alpha particles are scattered nearly backward by metal foils.
- Define the terms *atomic number, mass number,* and *isotope*.
- Write the isotope symbol for an atom of a given number of protons and neutrons.
- Define *percentage abundance* of an isotope.

Although Dalton considered atoms to be indivisible particles, a series of important experiments during the latter part of the nineteenth century and the early part of the twentieth century showed that atoms are themselves composed of particles—we will call them *subatomic particles.* Some of these particles have an electric charge. There are two kinds of electric charge: *positive* and *negative. Two electric charges of the same kind (both negative or both positive) repel one another,* but *two electric charges of opposite kinds (one positive and the other negative) attract one another.* (We will have more to say about the nature of electric charge in the Chemical Perspective that follows this section.) The negatively charged electron was the first subatomic particle to be discovered.

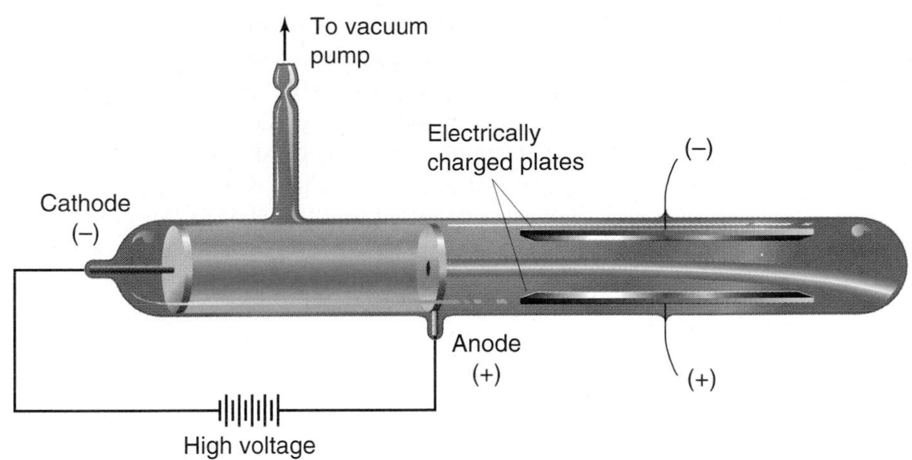

FIGURE 4.3
Electric discharge tube. The tube has two electrodes sealed into it. At the left end is the negative electrode, or cathode. In the middle is a positive electrode, or anode. The cathode emits cathode rays, shown here in green, which are attracted toward the anode. Some of the cathode rays pass through the hole in the anode to produce a beam. The beam of cathode rays continues through two electrically charged plates, where it bends away from the negative plate but toward the positive plate. From this type of result, Thomson concluded that cathode rays are composed of negatively charged particles.

Discovery of the Electron

The fundamental particle of negative electricity was discovered in 1898 by the British physicist J. J. Thomson. He conducted a series of experiments on cathode rays, which are produced in electric discharge tubes. Neon signs, fluorescent lights, and television tubes are all examples of electric discharge tubes. Figure 4.3 diagrams a simple discharge tube. It is constructed by sealing two electrodes, or electrical connections, in a glass tube and evacuating the air from it. When the electrodes are then connected to a high-voltage source, a beam of radiation is emitted from the *cathode,* or negative electrode. Such beams are called *cathode rays.*

Thomson showed that cathode rays travel in straight lines unless bent by electric or magnetic fields (as shown in Figure 4.3). Because cathode rays bend away from a negatively charged plate and toward a positively charged plate, Thomson concluded that these rays consist of negatively charged particles, now called *electrons.* (Later, the mass of the electron was found to be about 1800 times smaller than that of the smallest atom, hydrogen.) ●

● Television tubes use cathode rays to generate the TV image. Cathode rays are directed toward the front surface of the tube, where they strike a material that then phosphoresces a particular color. The cathode rays are bent by the field of an electromagnet, whose strength is varied so that the cathode rays move over the entire front surface of the tube. At the same time, the cathode rays are turned on and off so that an image is generated.

Because Thomson found that cathode rays could be produced using electrodes of various materials, he concluded that electrons are constituents of all atoms, and that cathode rays are produced when the high voltages applied to the electrodes pull electrons from the atoms making up the cathode material. An **electron** is thus *a very light, negatively charged subatomic particle.*

Video: Cathode ray tube.

Nuclear Model of the Atom

Soon after the discovery of the electron, Thomson and others speculated on the nature of the atom. Some of them, including Thomson, argued for the "plum pudding" model. According to this model, the atom consists of a number of electrons (the raisins or "plums" of this model) embedded in a "pudding" of sufficient positive charge to give an electrically neutral atom. This model of the atom proved to be incorrect, however. In 1911 an English physicist named Ernest Rutherford (1871–1937) came up with a different theory.

The experiments in Rutherford's laboratory used *alpha radiation* obtained from a radioactive element, such as radium. The atoms of a radioactive element disintegrate spontaneously over time into fragments consisting of other atoms or subatomic particles. When radium atoms disintegrate, they give off alpha radiation, which consists of alpha particles: fast-moving, positively charged helium atoms minus their electrons. The mass of each alpha particle is about 8000 times greater than that of an electron. When members of Rutherford's laboratory directed alpha radiation at metal foils, they found that most of the alpha particles passed undeflected through the metal foil, but a few were deflected nearly backward. According to the plum pudding model, the alpha particles should have been slowed by the positive charge of the atoms (the "pudding") and deflected only slightly by the very light electrons (the "raisins"). Nothing in the model would explain the backward deflections that were found. The deflections seemed to indicate that the alpha particles were hitting heavy, positively charged objects.

Rutherford concluded that an atom has a **nucleus,** that is, *a positively charged atomic core that takes up very little space in the atom but has most of its mass.* The positively charged nucleus would be expected to attract electrons to it, since electrons have a negative charge. Thus, according to Rutherford's model of the atom, which we now accept as correct, *an atom consists of a nucleus with a number of electrons moving around it.* Normally, an atom does not have an electric charge, so the *number of electrons is just sufficient to neutralize the positive charge of the nucleus.* ●

● Under certain circumstances, neutral atoms can lose or gain electrons to give a charged atom, called an *ion.* We discuss ions in Section 4.8.

Although most of the mass of an atom is in its nucleus (99.95% or more), the nucleus occupies only a very small portion of the atom. Nuclei have diameters of approximately 10^{-15} m, whereas atomic diameters are about 10^{-10} m (0.1 nm), which is a hundred thousand times larger. If the nucleus were the size of a golf ball, the atom would be about 3 miles in diameter. The electrons would move throughout this large space of the atom outside the nucleus.

This nuclear model easily explains the results of alpha particle experiments. As shown in Figure 4.4, when a beam of alpha particles hits a metal foil, most of the alpha

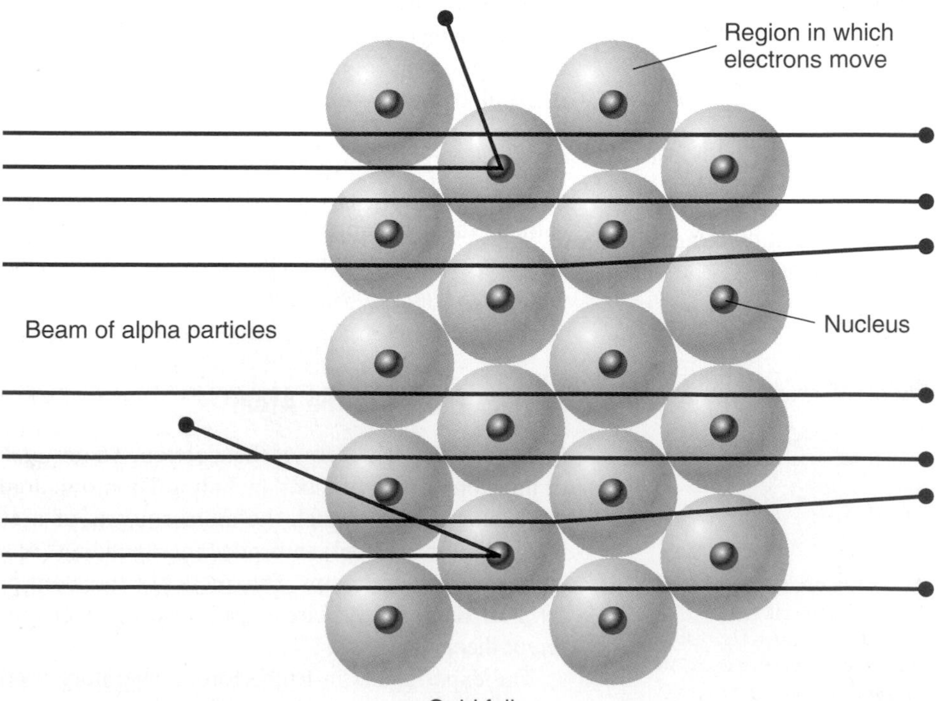

FIGURE 4.4
Scattering of alpha particles by a metal foil. This representation shows that most of the alpha particles passing through the foil are either not deflected or just barely deflected from their original paths. Sometimes, however, an alpha particle hits a metal nucleus and is deflected backward.

TABLE 4.1
Properties of the Electron, Proton, and Neutron

Particle	Mass (kg)	Mass (amu)*	Relative Charge
Electron	9.10939×10^{-31}	0.00055	−1
Proton	1.67262×10^{-27}	1.00728	+1
Neutron	1.67493×10^{-27}	1.00866	0

*The atomic mass unit (amu) equals 1.66×10^{-27} kg; it is defined in Section 4.3.

particles pass through the atoms because atoms are mostly empty space (Figure 4.4). An alpha particle would be only slightly deflected even if it did collide directly with an electron because electrons are about 8000 times lighter than alpha particles. Occasionally, however, an alpha particle collides directly with a heavy, positively charged nucleus, and it is then deflected backward.

Isotopes

The atom, as we have seen, is made up of a nucleus and electrons. Scientists now believe that the nucleus of an atom is itself made up of particles called *protons* and *neutrons.* A **proton** is *a positively charged subatomic particle found in the nucleus.* The charge on the proton is equal to that of the electron, but has the opposite sign (see Table 4.1). As a result, an electron is attracted to a proton. These two particles also differ greatly in mass, the proton being more than 1800 times heavier than the electron. ●

● The nucleus of the most abundant form of the hydrogen atom is simply a proton.

A **neutron** is *an electrically neutral subatomic particle found in the nucleus.* Its mass is almost identical to that of the proton (see Table 4.1). Each different kind of nucleus is made up of different numbers of protons and neutrons. For example, the nucleus of the aluminum atom is composed of 13 protons and 14 neutrons.

We characterize a particular nucleus by two quantities, the *atomic number* and the *mass number.* The **atomic number** is *the number of protons in the nucleus.* Therefore, it is also the number of positive charges on the nucleus. The **mass number** is *the total number of protons and neutrons in the nucleus.* For example, the aluminum nucleus has an atomic number of 13 (the number of protons in the nucleus) and a mass number of 27 (equal to 13 + 14, the total number of protons and neutrons).

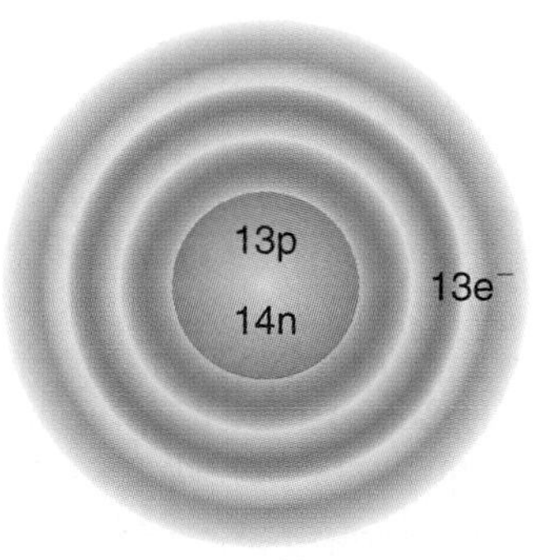

FIGURE 4.5
The aluminum atom. This diagram shows the subatomic particles of the aluminum atom. The nucleus contains 13 protons and 14 neutrons; around the nucleus are 13 electrons.

An atom is normally electrically neutral. This means that *an atom must have as many electrons outside the nucleus as there are protons inside the nucleus.* Because there are 13 protons (positive charges) in the nucleus, there must also be 13 electrons moving about the nucleus so that the atom is electrically neutral. Figure 4.5 diagrams the basic particles (protons, neutrons, and electrons) that make up the aluminum atom.

We have seen that the atomic number determines the number of electrons about the nucleus. Because the electrons in turn determine the chemical properties of the atom, as discussed in Section 4.4, the atomic number is a characteristic of an element. This means that all atoms of an element have the same atomic number. Hydrogen atoms, for example, always have atomic number 1, and oxygen atoms always have atomic number 8.

However, the atoms of an element can have different numbers of neutrons, so they can have different mass numbers. The nuclei of naturally occurring oxygen atoms have

Element	Isotope Mass No.	Percentage Abundance	Element	Isotope Mass No.	Percentage Abundance
Hydrogen	1	99.99	Magnesium	24	78.99
	2	0.02		25	10.00
Carbon	12	98.9		26	11.01
	13	1.1	Aluminum	27	100
Nitrogen	14	99.63	Silicon	28	92.2
	15	0.37		29	4.67
Oxygen	16	99.76		30	3.1
	17	0.04	Phosphorus	31	100
	18	0.2	Sulfur	32	95.02
Fluorine	19	100		33	0.75
Neon	20	90.48		34	4.21
	21	0.27		36	0.02
	22	9.25	Chlorine	35	75.77
Sodium	23	100		37	24.23

TABLE 4.2
Some Isotopes and Their Abundances

mass numbers of 16, 17, and 18 (equivalent to 8, 9, and 10 neutrons, respectively). **Isotopes** are *atoms whose nuclei have the same atomic number but different mass numbers.*

We denote the atomic number and mass number for an atom by putting them on the left side of the atomic symbol, giving the atom's *isotope symbol.* We write the atomic number as a subscript and the mass number as a superscript. For the aluminum atom, we write this isotope symbol as follows.

Mass number ⟶ 27
Atomic number ⟶ 13 Al

$$^{27}_{13}\text{Al}$$

Naturally occurring oxygen consists of the isotopes $^{16}_{8}O$, $^{17}_{8}O$, and $^{18}_{8}O$. These isotopes are also called simply oxygen-16, oxygen-17, and oxygen-18. Naturally occurring aluminum consists of only one isotope, $^{27}_{13}Al$ (aluminum-27).

A sample of oxygen obtained anywhere on earth always contains 99.759% oxygen-16 atoms, 0.037% oxygen-17 atoms, and 0.204% oxygen-18 atoms. In other words, a sample of naturally occurring oxygen containing a total of 100,000 atoms consists on the average of 99,759 $^{16}_{8}O$ atoms, 37 $^{17}_{8}O$ atoms, and 204 $^{18}_{8}O$ atoms. Naturally occurring aluminum, on the other hand, consists of only one isotope; it is always 100% aluminum-27 ($^{27}_{13}Al$).

The percentage of an isotope in a naturally occurring sample of an element is known as the **percentage abundance** of that isotope. For almost all elements, the percentage abundances of the isotopes are the same for all samples of the element, regardless of where on earth the samples were obtained. The reason is that the isotopes of the elements were thoroughly mixed during the formation of the earth, and chemical and most physical processes do not significantly alter this mixture. Table 4.2 lists the isotopes of some elements with their masses and percentage abundances.

EXAMPLE 4.1 Writing Isotope Symbols

The nucleus of an atom contains 26 protons and 27 neutrons. Write the isotope symbol for the atom.

Problem Analysis

The atomic number of the isotope equals the number of protons in the nucleus. The mass number of this isotope equals the sum of the protons and neutrons in the nucleus. Look at the list of elements on the right inside front cover of the book. Find the atomic number of the element, its name, and the corresponding symbol. Write the symbol of the element with the atomic number as the left subscript and the mass number as the left superscript.

Solution

The atomic number of the atom is 26 (the number of protons). The mass number is $26 + 27 = 53$. The list on the right inside front cover of the book shows that the element with an atomic number equal to 26 is iron and that its symbol is Fe. Therefore, the isotope symbol is ${}^{53}_{26}\text{Fe}$.

Exercise 4.1

An atomic nucleus consists of 17 protons and 18 neutrons. What is the isotope symbol of the atom?

(Try Problems 4.23 and 4.24.)

EXAMPLE 4.2 Determining the Number of Protons and Neutrons in a Nucleus

How many protons and neutrons are in the bromine-79 nucleus?

Problem Analysis

Look at the list of elements on the right inside front cover of the book and note the atomic number of the element bromine. The atomic number equals the number of protons. The mass number is given in the name of the isotope (79). Because the mass number is the sum of the number of protons (atomic number) and the number of neutrons, the number of neutrons is given by

$$\text{Number of neutrons} = \text{mass number} - \text{atomic number}$$

Solution

In the list on the right inside front cover, you see that the atomic number of bromine is 35. Therefore, there are 35 protons in the bromine-79 nucleus. The mass number of bromine-79 is 79, so the number of neutrons in the nucleus is derived as follows.

$$\begin{aligned}\text{Number of neutrons} &= \text{mass number} - \text{atomic number}\\ &= 79 - 35 = 44 \text{ neutrons}\end{aligned}$$

Exercise 4.2

Give the number of protons and neutrons in the nucleus of ${}^{19}_{9}\text{F}$.

(Try Problems 4.25 and 4.26.)

Chemical Perspective

Electric Charge and Ben Franklin

Benjamin Franklin (1706–1790) was an early American statesman whose signature is on both the Declaration of Independence and the Constitution of the United States. In addition, Franklin was the first American to gain an international reputation in science. His famous experiment in which he flew a kite with a key attached to the kite string during a storm established that lightning was a form of electricity. This and other work led him to coin words for electric charge such as "positive charge" and "negative charge" and "plus" and "minus" to account for the two types of electric charge that are opposites.

You are already familiar with electric charge if you have ever walked across a nylon carpet in the dry air of a heated home in the winter and received an unexpected shock when you touched a doorknob or someone else. Why did you get the shock? As you walked across the carpet, electrically charged particles rubbed off the carpet onto your shoes. Unless they were conducted away by humid air, the number of these charged particles built up as you walked. They moved over your body, and if you touched a doorknob, they rushed from you to the metal knob. You experienced this rush of so-called static electricity as a small shock.

You can demonstrate the properties of electric charge with some simple experiments. Rub a plastic rod with a piece of fur to give the rod a negative charge. Now bring the plastic rod up to two small corks, each suspended from a nylon string so that they are next to one another (Figure 4.6A). Because the negative charge flows to the corks, they *repel* each other and fly apart (Figure 4.6B). You can give a third cork a positive charge by touching it with a rod that you first rubbed with a piece of silk. If you then place this positively charged cork near one of the negatively charged corks, they will *attract* each other.

A

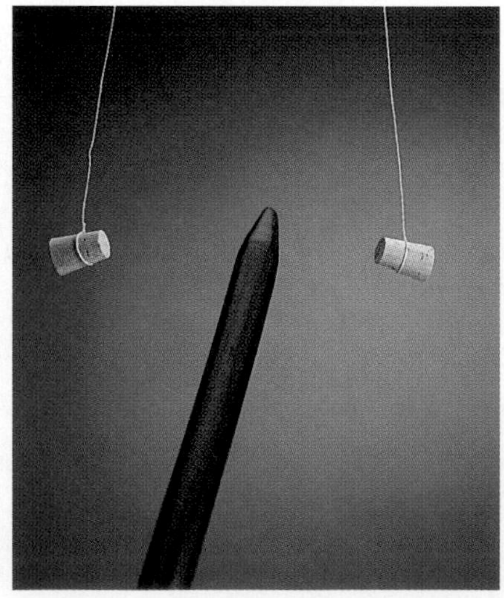

B

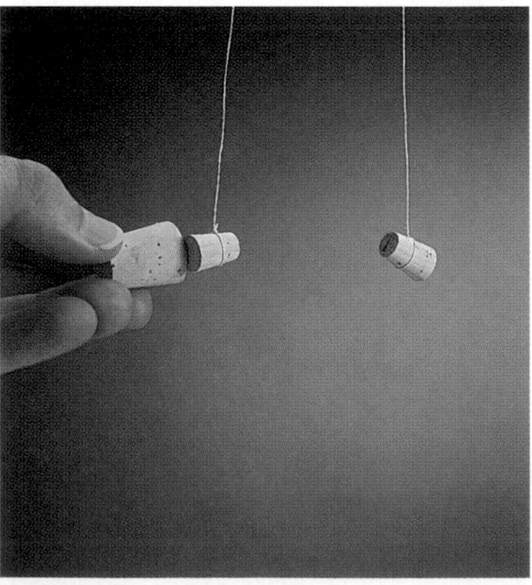

C

FIGURE 4.6
Properties of an electric charge. *(A)* An experimenter brings a negatively charged plastic rod up to two corks. *(B)* The negative charge on the rod flows to the two corks, which now fly apart because they have the same charge. *(C)* The experimenter brings a third cork with a positive charge up to one of the negatively charged corks. The two corks attract because they have opposite charges.

■ **EXAMPLE 4.3 Determining the Number of Electrons in a Neutral Atom**

How many electrons are in the neutral atom of phosphorus-31?

Problem Analysis

Look at the list of elements on the right inside front cover of the book. Note the atomic number corresponding to the element. The atomic number equals the number of positive charges (protons) in the nucleus. The neutral atom has an equal number of negative charges (from electrons). Therefore, in the neutral atom the atomic number also gives the number of electrons.

Solution

Looking at the list of elements on the right inside front cover of the book, you see that phosphorus has an atomic number of 15. This means that there are 15 protons in the phosphorus nucleus, and therefore 15 electrons in the neutral atom.

Exercise 4.3

Give the number of protons, neutrons, and electrons in a neutral atom of boron-11. Where is each of these particles located in the atom?

(Try Problems 4.27, 4.28, 4.29, and 4.30.)

4.3 Atomic Weights

OBJECTIVES

- Define *atomic mass unit.*
- Define *atomic weight.*
- Calculate the atomic weight of an element given the isotope masses and percentage abundances.

Dalton emphasized the relative masses of the atoms of the elements, which he referred to as *atomic weights.* As you will see in succeeding chapters, these atomic weights are indeed one of the most useful properties of the atoms of an element.

Every atom of an element of a given mass number has a definite mass. For example, the mass of any oxygen-16 atom is always 2.6560×10^{-23} g. As you can see, the mass of this atom is incredibly small. If you could count out oxygen atoms into a beaker at a rate of one million a second, it would take you more than 1 billion years to count out enough oxygen atoms to give 1 gram of oxygen!

Clearly, we cannot weigh an oxygen atom, or any other kind of atom, by placing it on a balance. We need special means to obtain the mass of such a small particle. Dalton's solution was to obtain the atomic weights of the atoms relative to one another. For example, hydrogen chloride, which contains equal numbers of hydrogen atoms and chlorine atoms, consists of 35 grams of chlorine for every gram of hydrogen. Thus, chlorine has an atomic weight of about 35 relative to 1 for hydrogen. Atomic weights of the elements continued to be based on the relative masses of substances until recent times. Now more precise values are available through use of an instrument called a *mass spectrometer* (Figure 4.7).

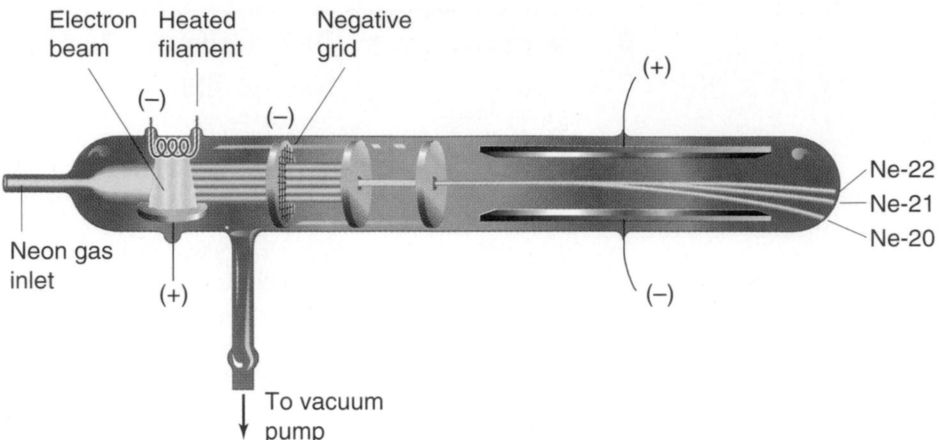

FIGURE 4.7

A mass spectrometer. This sketch shows how a mass spectrometer determines the masses of atoms in a sample of neon. Neon atoms enter the tube at the left, where they are hit by cathode rays. The cathode rays knock electrons from the neon atoms, and the resulting positively charged atoms are pulled toward a negatively charged wire grid. Most of the atoms pass through this grid and form a beam. When the beam passes between two charged plates, it splits into new beams depending on the masses of the isotopes. The amount of deflection of each beam depends on their masses. The mass spectrometer compares the deflection with that of carbon-12 to obtain the mass. (Most mass spectrometers use a magnetic field rather than an electric field because it is easier to obtain accurate magnetic fields.)

Modern Definition of Atomic Weight

Using data from mass spectrometers, modern atomic weights are determined by comparing the mass of an atom to that of the carbon-12 isotope, which is taken as a standard. On this *carbon-12 scale* of atomic masses, the mass of the carbon-12 atom is arbitrarily assigned a value of exactly 12 atomic mass units. An **atomic mass unit (amu)** equals *exactly one-twelfth the mass of a carbon-12 isotope.* On this scale, the aluminum-27 atom has a mass of 26.98 amu; in other words, the mass of the aluminum-27 atom relative to the carbon-12 atom—taken as exactly 12—is 26.98. Notice that the mass of the aluminum-27 isotope (26.98 amu) is approximately equal in magnitude to its mass number (27). This is so because the masses of the protons and neutrons, which make up most of the mass of the atom, are each approximately equal to 1 amu (see Table 4.1). ●

● The mass of an isotope does not exactly equal the sum of its constituent protons and neutrons. When the protons and neutrons come together to form the nucleus, energy is released. According to Einstein, mass is also lost, because mass and energy are related quantities.

The atomic weight of an element is now used to mean the average mass of the natural mixture of isotopes of the element expressed in amu. Aluminum consists of only one isotope, aluminum-27, so its atomic weight equals the mass of aluminum-27, which is 26.98 amu. Since most elements consist of a mixture of isotopes, their atomic weight is an average mass of these isotopes. We define the **atomic weight** of an element to be *the weighted average mass (expressed in atomic mass units) of an atom of the naturally occurring element.*

Calculating Atomic Weights

We can illustrate this definition of atomic weight by calculating the atomic weight of chlorine. First, though, you need to understand what we mean by a weighted average mass. To begin, let us look at a problem in calculating weighted averages that you may find more familiar, the calculation of a course grade, say, in chemistry.

Suppose your chemistry grade is made up of two parts: your le which is 71, and your laboratory grade, which is 82. Suppose also tha grade counts 70% of your final grade and the laboratory grade counts calculate your final grade in the course as follows:

$$
\begin{aligned}
71 \times 0.70 &= 4\underline{9}.7 \\
82 \times 0.30 &= \underline{2\underline{4}.6} \\
&\ 7\underline{4}.3 \qquad \text{(rounds to 74)}
\end{aligned}
$$

The exam grade counts 70% of the total grade, so we multiply the exam grade by 0.70 to give $4\underline{9}.7$. (The underline denotes the least significant digit; we carried an additional digit in the calculations, and we round at the end.) Similarly, since the lab grade counts 30% of the total, we multiply the lab grade by 0.30 to give $2\underline{4}.6$. Thus, we weighted each grade by a certain fraction of the total; that is, we gave a weight of 0.70 to the exam grade and a weight of 0.30 to the lab grade. Then we added these two results to give the total grade, 74, which is the weighted average.

Now let us obtain the atomic weight of chlorine. Using the mass spectrometer, we find that naturally occurring chlorine consists of two isotopes: chlorine-35, with a mass of 34.97 amu, and chlorine-37, with a mass of 36.97 amu. The mass spectrometer also gives us the percentage abundances of these two isotopes. It shows that naturally occurring chlorine consists of 75.77% chlorine-35 atoms and 24.23% chlorine-37 atoms.

Calculating the atomic weight of chlorine from this information is similar to calculating a final course grade. The isotope masses are the "grades," and the percentage abundances are the "grade percentages." To do the calculation, we multiply each isotope mass by its abundance expressed as a decimal fraction and then add these intermediate results to obtain the final result.

$$
\begin{aligned}
34.97\ \text{amu} \times 0.7577 &= 26.5\underline{0}\ \ \text{amu} \\
36.97\ \text{amu} \times 0.2423 &= \underline{\ 8.95\underline{8}\ \text{amu}} \\
&\ 35.4\underline{5}8\ \text{amu}
\end{aligned}
$$

For the data we used, the final result rounds to 35.46 amu. More precise data lead to 35.45 amu as the atomic weight of chlorine.

Atomic weights for other elements have been obtained in a similar way, and a table showing their values is given on the left inside front cover. The next example further illustrates the calculation of atomic weights.

■ EXAMPLE 4.4 Calculating the Atomic Weight of an Element

Gallium is a metallic element used to prepare gallium arsenide for the small lasers employed in compact disc players. The naturally occurring element contains 60.2% gallium-69 atoms and 39.8% gallium-71 atoms. The masses of these isotopes are 68.93 amu and 70.92 amu, respectively. What is the atomic weight of gallium according to these data?

Problem Analysis

Multiply each isotope mass by its abundance expressed as a decimal fraction (the percentage abundance divided by 100). Then add these intermediate results to obtain the atomic weight.

Solution

The calculation is as follows.

$$68.93 \text{ amu} \times 0.602 = 41.\underline{5}0 \text{ amu}$$
$$70.92 \text{ amu} \times 0.398 = 28.\underline{2}3 \text{ amu}$$
$$69.\underline{7}3 \text{ amu}$$

Thus, the atomic weight of gallium is 69.7 amu.

Exercise 4.4

Naturally occurring copper consists of two isotopes, copper-63 and copper-65, with masses of 62.94 amu and 64.93 amu, respectively. The abundances are 69.17% copper-63 and 30.83% copper-65. According to these data, what is the atomic weight of copper?

(Try Problems 4.33, 4.34, 4.35, and 4.36.)

4.4 Periodic Table of the Elements

OBJECTIVES

- Describe how the elements are arranged in a modern periodic table.
- Define *period* and *group* (or family) for a periodic table.
- Locate any element in the periodic table given only its period and group.
- Define *main group* (or representative group) and *transition-metal group.*
- Define *metal, nonmetal,* and *metalloid.*

Soon after Dalton proposed his atomic theory, chemists tried to classify the elements by the properties of the atoms, such as atomic weights. However, until enough elements had been identified and accurate atomic weights obtained for them, these attempts at classification were unsuccessful. In 1860, a conference was held at Karlsruhe, Germany, to resolve a number of fundamental problems in chemistry, including the question of how to obtain accurate atomic weights. Within a few years, chemists had determined relatively accurate atomic weights for most of the 60 or so elements known at that time. This set the stage for the development of the *periodic table,* our present scheme for classifying the chemical elements. This table, which adorns many chemistry classrooms, is so useful that we will refer to it many times throughout this book.

Mendeleev's Periodic Table

The periodic classification of the elements was developed in 1869 by the Russian chemist Dmitri Mendeleev (1834–1907); see Figure 4.8. The periodic table was formulated independently at about the same time by the German chemist Lothar Meyer (1830–1895), but Mendeleev is given the most credit because of the breadth of his work. For example, Mendeleev noted several vacancies in his table and predicted new elements and their properties based on the positions of these vacancies. ● Mendeleev and Meyer both discovered that they could arrange the elements by increasing atomic weight into rows and columns so that the elements in any one column have chemical and physical properties that are either similar or that vary in regular fashion. The name *periodic table* stems from the *periodic,* or recurring, variation of properties as you progress from one element to another in the table.

● Mendeleev predicted the properties of three elements. Later, these elements (germanium, scandium, and gallium) were discovered and shown to have the properties Mendeleev predicted. Chemists were then convinced of the usefulness of the periodic table.

FIGURE 4.8
Dmitri Ivanovich Mendeleev. Mendeleev constructed his periodic table as a part of his effort to systematize chemistry.

The Modern Periodic Table

When Mendeleev tried to arrange all of the elements in his periodic table precisely by their atomic weights, the arrangement did not agree with the elements' properties. Mendeleev thought that the atomic weights might be in error. Later, when the structure of the atom was discovered, scientists noted that the atomic number was a more fundamental characteristic of an element than its atomic weight. When the elements were arranged by atomic number rather than by atomic weight, the difficulties of Mendeleev's arrangement disappeared. Thus, the modern **periodic table** is *a table of the elements ordered by increasing atomic number into rows and columns so that the elements in any one column have similar or regularly changing properties.* A modern form of the periodic table is shown in Figure 4.9. A similar periodic table is also printed on the left inside front cover of this book.

Note that each entry in the table shown in Figure 4.9 lists the atomic number, the atomic symbol, the name of the element, and the atomic weight. The periodic table has been called the most important one-page document in chemistry because it is a convenient way of tabulating such information, in addition to showing the periodic behavior of the elements. Note the basic structure of the table, with the elements divided into rows and columns. A **period** consists of *the elements in any one horizontal row of the periodic table.* A **group** (or **family**) consists of *the elements in any one column of the periodic table.*

The groups, or families, of the periodic table are usually numbered, although no system of numbering has gained the unanimous support of chemists. A numbering scheme frequently seen in North America labels the groups with Roman numerals and A's and B's. However, the International Union of Pure and Applied Chemistry (IUPAC) has recommended that the groups be numbered 1 to 18. The periodic tables in this book show both numbering schemes. When we refer to a group in the periodic table, we will use the traditional North American numbering. *An A group of elements in the periodic table* is called a **main group** (or a **representative group**); *a B group of elements in the periodic table* is called a **transition-metal group.**

As we noted earlier, the elements in any one group have similar or regularly varying chemical and physical properties. This is especially noticeable in some groups. For example, the Group IA elements (which include lithium, sodium, and potassium) are soft metals that react readily with water, and the Group VIIA elements are nonmetals that react readily with most of the metals. ●

● Some of the families have names:

Group IA	the alkali metals
Group IIA	the alkaline earth metals
Group VIIA	the halogens
Group VIIIA	the noble gases

Note that the periods contain varying numbers of elements. The first period consists of only two elements, hydrogen (H) and helium (He). The next two periods each contain 8 main-group elements. The fourth and fifth periods contain 10 transition-metal elements, in addition to 8 main-group elements. The sixth period actually contains 32 elements: 8 main-group elements, 10 transition-metal elements, and 14 additional elements called *inner-transition metals.* Because it is difficult to place a row of 32 elements on a text page if we include all of the usual information, the 14 inner-transition metals are normally placed in separate rows below the main body of the periodic table, with a star in the body of the table to indicate where the rows would be placed.

Metals, Nonmetals, and Metalloids

The elements of the periodic table shown in Figure 4.9 are divided by a heavy "staircase" line into metals on the left (the bulk of the elements) and nonmetals on the right. A **metal** is *a material that has a characteristic luster or shine and that is a relatively good conductor of heat and electricity.* Aluminum, in Group IIIA, is a typical metal. When freshly cut, the surface is quite shiny; after exposure to air this surface becomes

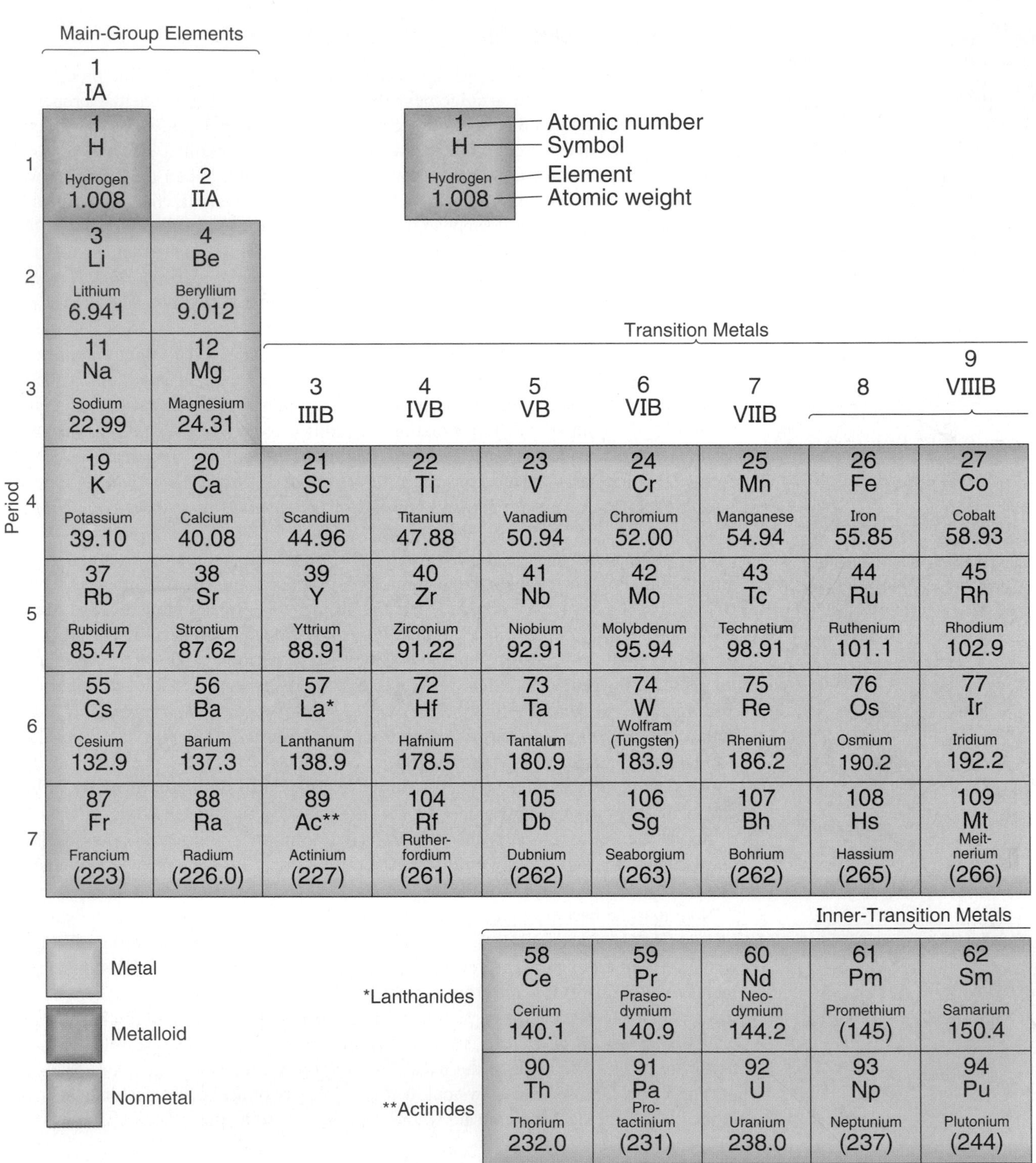

FIGURE 4.9

The modern periodic table. The columns are labeled by numbers from 1 to 18 in the scheme recommended by IUPAC; they are labeled by Roman numerals and the letters A and B in the scheme frequently seen in North America.

Main-Group Elements

10	11 IB	12 IIB	13 IIIA	14 IVA	15 VA	16 VIA	17 VIIA	18 VIIIA
								2 He Helium 4.003
			5 B Boron 10.81	6 C Carbon 12.01	7 N Nitrogen 14.01	8 O Oxygen 16.00	9 F Fluorine 19.00	10 Ne Neon 20.18
			13 Al Aluminum 26.98	14 Si Silicon 28.09	15 P Phosphorus 30.97	16 S Sulfur 32.07	17 Cl Chlorine 35.45	18 Ar Argon 39.95
28 Ni Nickel 58.71	29 Cu Copper 63.55	30 Zn Zinc 65.38	31 Ga Gallium 69.72	32 Ge Germanium 72.59	33 As Arsenic 74.92	34 Se Selenium 78.96	35 Br Bromine 79.90	36 Kr Krypton 83.80
46 Pd Palladium 106.4	47 Ag Silver 107.9	48 Cd Cadmium 112.4	49 In Indium 114.8	50 Sn Tin 118.7	51 Sb Antimony 121.8	52 Te Tellurium 127.6	53 I Iodine 126.9	54 Xe Xenon 131.3
78 Pt Platinum 195.1	79 Au Gold 197.0	80 Hg Mercury 200.6	81 Tl Thallium 204.4	82 Pb Lead 207.2	83 Bi Bismuth 209.0	84 Po Polonium (209)	85 At Astatine (210)	86 Rn Radon (222)
110 Uun Ununnilium (272)	111 Uuu Unununium (272)	112 Uub Ununbium (277)						

63 Eu Europium 152.0	64 Gd Gadolinium 157.3	65 Tb Terbium 158.9	66 Dy Dysprosium 162.5	67 Ho Holmium 164.9	68 Er Erbium 167.3	69 Tm Thulium 168.9	70 Yb Ytterbium 173.0	71 Lu Lutetium 175.0
95 Am Americium (243)	96 Cm Curium (247)	97 Bk Berkelium (247)	98 Cf Californium (251)	99 Es Einsteinium (252)	100 Fm Fermium (257)	101 Md Mende- levium (258)	102 No Nobelium (259)	103 Lr Lawren- cium (260)

dull from the formation of an aluminum oxide coating. Aluminum makes useful kitchen utensils because of its very high heat conductivity. It also has a moderately high electrical conductivity (about one-half that of copper, the metal typically used in electrical-wires). ●

● Most of the known elements are metals.

A **nonmetal** is *an element that does not exhibit the properties of a metal.* Nonmetals are often gases or nonmetallic solids. (Bromine, the only nonmetallic element that is a liquid at ordinary room temperature, is a red-brown liquid with a disagreeable, irritating odor.) The nonmetals oxygen and nitrogen are gaseous components of air. Sulfur is a brittle yellow solid. Carbon occurs in three forms: the two common forms diamond and graphite, and a newly discovered form called buckminsterfullerene. Diamond is a very hard solid that is colorless when pure and perfectly crystalline; graphite is a soft, black solid; and buckminsterfullerene is a dark brown substance.

A few elements along the "staircase" line are classed as metalloids. A **metalloid** is *an element that has properties intermediate between those of a metal and those of a nonmetal.* Silicon is an example. It is a hard, shiny solid that is very brittle, like a hard glass. As with the other metalloids, it is most useful as a semiconductor. A semiconductor is a poor conductor of electricity at room temperature but becomes a good conductor when certain other elements are added to it in very small amounts. Silicon is used to make chips for solid-state electronic devices, including television and computer components.

■ EXAMPLE 4.5 Locating an Element by Group and Period of the Periodic Table

What is the name and symbol of the element in Group VIIB, Period 4?

Problem Analysis

Locate the element in the periodic table (Figure 4.9), which lists the names and symbols of the elements. If you use a periodic table such as the one on the left inside front cover of the book, first locate the atomic symbol. Then refer to the list of elements given on the right inside front cover. Locate the symbol and the corresponding name of the element.

Solution

The element in Group VIIB, Period 4, is manganese, Mn.

Exercise 4.5

Give the name and symbol of the element in Group IIA, Period 3.

(Try Problems 4.39 and 4.40.)

■ EXAMPLE 4.6 Identifying the Metals and Nonmetals in the Periodic Table

What elements in Group IVA are metals?

Problem Analysis

Look at the periodic table given in Figure 4.9 or the one on the left inside front cover of the book. The metallic elements are shown in blue, and the nonmetals are shown in gold.

Solution

The metals in Group IVA are tin (Sn) and lead (Pb).

Exercise 4.6

What elements in Group VA are nonmetals?

(Try Problems 4.41, 4.42, 4.43, and 4.44.)

MOLECULES AND IONS

Substances are either elements or compounds. We begin our study of the atomic composition of elements and compounds in this part of the chapter.

4.5 The "Molecular" Basis of Substances

OBJECTIVES

- Define *molecule.*
- Define *ion.*
- Discuss the meaning of the "molecular" basis of substances.

The left branch of Figure 4.10 shows the atomic structure of the elements. Most of them consist of populations of atoms. Three examples are neon, sodium, and copper. However, a few of the elements, such as hydrogen and oxygen, occur in molecular form. A

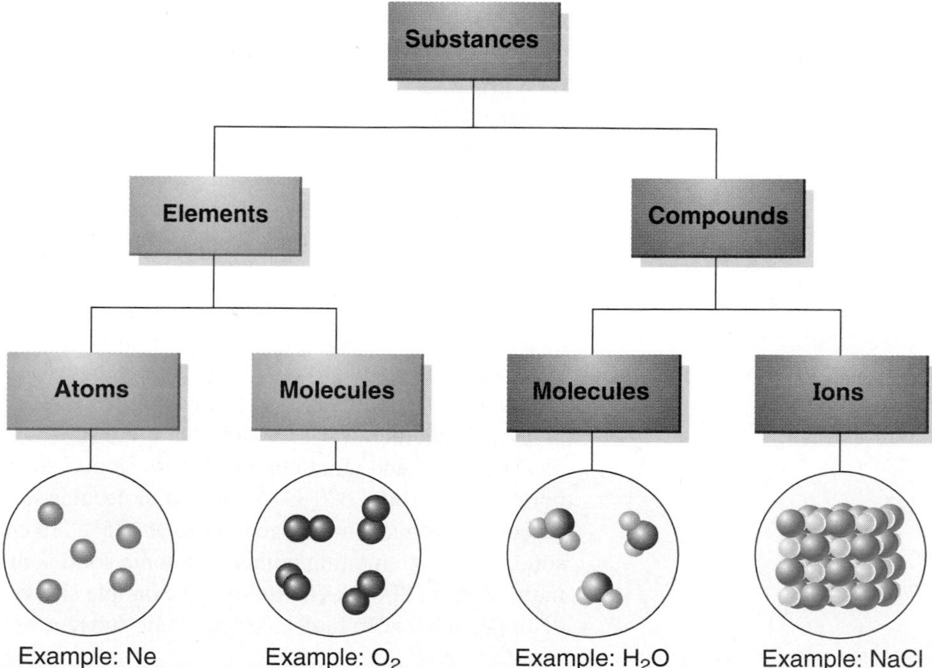

FIGURE 4.10
The "molecular" basis of substances. Although chemists often refer to substances as "molecular," they are using this term loosely. Although a few of the elements are molecular, most of them consist of populations of atoms. Similarly, many compounds consist of molecules, but many others exist as ions.

molecule is *a group of identical or different atoms that are chemically bonded, that is, tightly connected by attractive forces; a molecule is electrically neutral.* A molecule of an element consists of only one kind of atom. For example, the nonmetals, hydrogen and oxygen, occur in nature as *diatomic* (two-atom) molecules.

Compounds (substances with atoms of two or more elements) are made up of either molecules or ions, as shown in the right branch of Figure 4.10. Water is an example of a compound consisting of molecules, with each molecule containing two hydrogen atoms and one oxygen atom. A water molecule (like any other molecule) is a *unit* that can act independently of other molecules—at least under certain conditions. Other compounds, such as sodium chloride (ordinary table salt), exist in the form of ions. An **ion** is *an electrically charged particle derived from an atom or chemically bonded group of atoms by adding or removing electrons.* Sodium chloride and other ionic substances do not contain molecular units; instead, they consist of a vast array of ions, held together by attractions between opposite charges.

Figure 4.10 summarizes the underlying "molecular" basis of substances; it shows the actual building blocks of elements and compounds. The term *molecular* is used loosely because molecules are only one kind of building block; other kinds are atoms and ions. We compare molecules and ions in the next section.

4.6 Comparing Molecular and Ionic Substances

OBJECTIVES

- Predict whether a compound is molecular or ionic given the elements in the compound.
- Compare the melting points of molecular and ionic compounds.

Before we look at the detailed structures of molecular and ionic substances, it is useful to understand the types of elements composing them and the melting-point behaviors of these two types of substances. As a general but important rule, *molecular substances are almost always composed of atoms from nonmetallic elements.* Examples are water (composed of the nonmetals hydrogen and oxygen), ammonia (composed of the nonmetals nitrogen and hydrogen), and sucrose or table sugar (composed of the nonmetals carbon, hydrogen, and oxygen). Another useful and equally important rule is this: *Ionic substances are usually composed of both metallic elements and nonmetallic elements.* (Some molecular substances do contain metal atoms; however, you will not encounter many of these in this book.) Examples of ionic substances are sodium chloride (composed of the metal sodium and the nonmetal chlorine) and sodium hydroxide or lye (composed of the metal sodium and the nonmetals hydrogen and oxygen).

Molecular and ionic substances have different physical properties. In particular, they generally have distinctive melting-point behaviors. Molecular substances usually have relatively low melting points. Many simple molecular substances are gases and liquids at room temperature. (Gases and liquids are substances that have already melted; examples are water and ammonia.) Other molecular substances are solids that have melting points below about 300°C. Aspirin, a molecular substance composed of the nonmetals carbon, hydrogen, and oxygen, melts at 135°C. In contrast, ionic substances are usually solids with high melting points. The ionic solid sodium chloride, or common table salt, melts at about 800°C. The following example shows you how to use these differences to distinguish between molecular and ionic substances.

■ EXAMPLE 4.7 Deciding Whether a Substance Is Molecular or Ionic

For each of the following, decide whether the substance is molecular or ionic; then note whether the melting point of the substance is consistent with this type of substance.

(a) Sulfur dioxide is a gaseous substance at room temperature and is composed of sulfur and oxygen.

(b) Magnesium oxide is a solid substance composed of magnesium and oxygen; it melts at 2800°C.

Problem Analysis

Determine the type of substance, whether molecular or ionic, from the elements making up the substance. Use the following two rules: A substance composed only of nonmetals is generally a molecular substance, and a substance composed of both metals and nonmetals is ionic. If you do not know whether the element is a metal or nonmetal, look at the periodic table on the left inside front cover of the book. (The names and symbols of the elements are listed in a table on the right inside front cover.)

In general, we expect molecular substances to have low melting points (below about 300°C), and we expect ionic substances to have high melting points (above about 300°C).

Solution

(a) Sulfur dioxide is composed of sulfur and oxygen, which are both nonmetals. You conclude that this substance is molecular. Because it is a gas at room temperature (about 25°C), its melting point is below room temperature. This is what you would expect of a molecular substance.

(b) Magnesium oxide is composed of magnesium, a metal, and oxygen, a nonmetal. You conclude that magnesium oxide is an ionic substance. Its melting point is consistent with that of an ionic substance (greater than 300°C).

Exercise 4.7

Methane is a gaseous substance at room temperature and is composed of carbon and hydrogen. Do you expect it to be a molecular or an ionic substance? Does its melting point agree with this conclusion?

(Try Problems 4.45, 4.46, 4.47, and 4.48.)

Some substances, whether molecular or ionic, decompose before they melt and, thus, have no definite melting point. Sucrose is an example; it decomposes giving first caramel and then a carbon char and water vapor. Decomposition is a chemical reaction that changes a substance to other substances.

4.7 Molecular Substances and Their Formulas

OBJECTIVES

- Define *molecular formula* and *structural formula.*
- Give examples of some molecular and structural formulas.
- Give formulas of the elements.

Many important compounds, such as water, sugar, and aspirin, are molecular substances; that is, they are composed of molecules. Molecules are generally so small that even extremely small samples contain tremendous numbers of them. One billionth (10^{-9}) of a drop of water contains about 2 trillion (2×10^{12}) molecules!

Molecules consist of a definite number of identical or different kinds of atoms. Ordinary oxygen, for example, always consists of exactly 2 chemically bonded oxygen atoms. The water molecule consists of just 1 oxygen atom chemically bonded to exactly 2 hydrogen atoms. Molecules of table sugar consist of exactly 12 carbon atoms, 22 hydrogen atoms, and 11 oxygen atoms.

Molecular Formulas

Just as we find it useful to represent atoms by letter symbols (atomic symbols), we find it useful to represent compounds of molecular substances by combinations of atomic symbols to convey both the elements making up the substance and their proportions. *Chemical formulas* are a special notation used to represent chemical substances. An example is a *molecular formula.* A **molecular formula** is *a notation that uses atomic symbols with numeric subscripts to convey the exact number of atoms of the different elements that are in a molecule of substance.* The molecular formula of water is H_2O. The subscript 2 on the H means that the molecule has two hydrogen atoms. No subscript is written when a 1 is meant. Thus O without a subscript means the molecule has one oxygen atom.

■ EXAMPLE 4.8 Writing Molecular Formulas

A molecule of hydrazine contains two nitrogen atoms and four hydrogen atoms. Write the molecular formula of hydrazine, listing the elements in the order given.

Problem Analysis

Write the atomic symbols for the elements with subscripts indicating the number of each kind of atom in the molecule. Write the symbols in the order given in the problem statement. (Do not worry about how to predict this order now; we discuss these conventions later in the book.)

Solution

Hydrazine consists of nitrogen and hydrogen, so the molecular formula will contain the symbols for these elements, N and H. Because the hydrazine molecule contains two atoms of nitrogen and four atoms of hydrazine, the molecular formula of this substance is N_2H_4.

Exercise 4.8

A molecule contains 4 phosphorus atoms and 10 oxygen atoms. Write the molecular formula, listing the elements in the order given.

(Try Problems 4.49 and 4.50.)

Structural Formulas and Molecular Models

Not only does a molecule consist of a definite number of atoms, but the atoms are chemically bonded in definite ways. When we want to use *a chemical formula that shows*

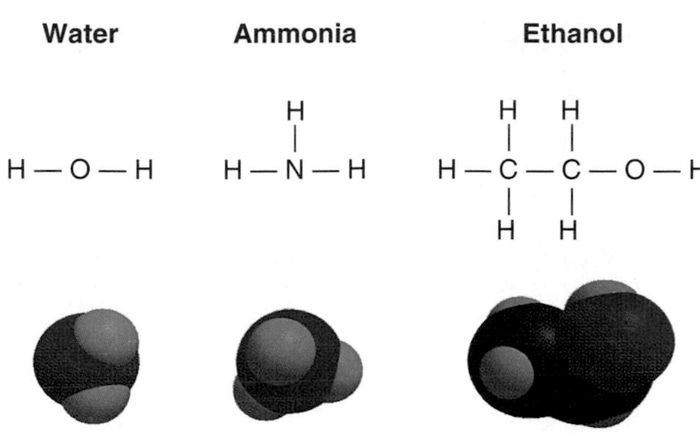

FIGURE 4.11
Structural formulas and molecular models. Structural formulas of water (H_2O), ammonia (NH_3), and ethyl alcohol (C_2H_6O) and computer-drawn models of these molecules. The colors of the atoms in the models are carbon, black; hydrogen, light blue; nitrogen, dark blue; and oxygen, red.

what atoms are bonded to one another, we use a **structural formula.** For example, water has the structural formula H—O—H. This tells us that each hydrogen atom is bonded to the oxygen atom, which is in the center. Each straight line represents a *chemical bond,* which is the strong attractive force that binds atoms together.

The atoms of a molecule are not only bonded to other atoms in a particular order but also arranged in a definite way in three-dimensional space. The atoms in a water molecule, for example, are not in a straight line; they are arranged to form an angular molecule, which we could draw as follows.

$$\overset{\quad O}{H \diagup \quad \diagdown H}$$

Chemists often construct molecular models from plastic or wood to show the shapes of molecules. Computers are also frequently used to generate molecular models that can be rotated on the computer screen. Figure 4.11 shows examples of molecular and structural formulas and models of some simple molecules. ●

● A molecule's shape influences the properties of the substance. One of the reasons medical drugs frequently behave like natural biological substances is that their molecules have similar shapes. Computer programs that can draw and rotate models of large molecules are being increasingly used in chemical research.

Formulas of the Elements

If you refer to the element oxygen in a general way, you use the element symbol, or atomic symbol, O. As a substance, however, oxygen exists either as the O_2 molecule or as the O_3 molecule. The most abundant form of oxygen in air is O_2, which we usually call simply oxygen or, more specifically, dioxygen. The molecular substance O_3 is called ozone. Ozone is present in small amounts in air, where it is produced by ultraviolet rays of the sun. Its characteristic odor is detectable around electrical equipment because electrical discharges can convert O_2 into O_3 molecules. Ozone is also present in the stratosphere, where it filters out the strong ultraviolet rays that are harmful to living cells.

Several other elements also occur naturally as molecules. All of them, including oxygen, are nonmetals, and their formulas are given in the representation of the periodic table shown in Figure 4.12. Notice that most of them consist of *diatomic* (two-atom) molecules. These are H_2, N_2, O_2, and all of the elements in Group VIIA (F_2, Cl_2, Br_2, I_2, and At_2). However, phosphorus forms P_4 molecules, and sulfur forms S_8 molecules. You will need to memorize the molecular forms of these nonmetal elements.

The other elements, both metals and metalloids, are not molecular. They do not have simple molecular structures but consist of a very large, indefinite number of atoms bonded together. These elements are represented simply by their atomic symbols. Diamond, for example, consists of carbon atoms, each bonded to four other carbon atoms

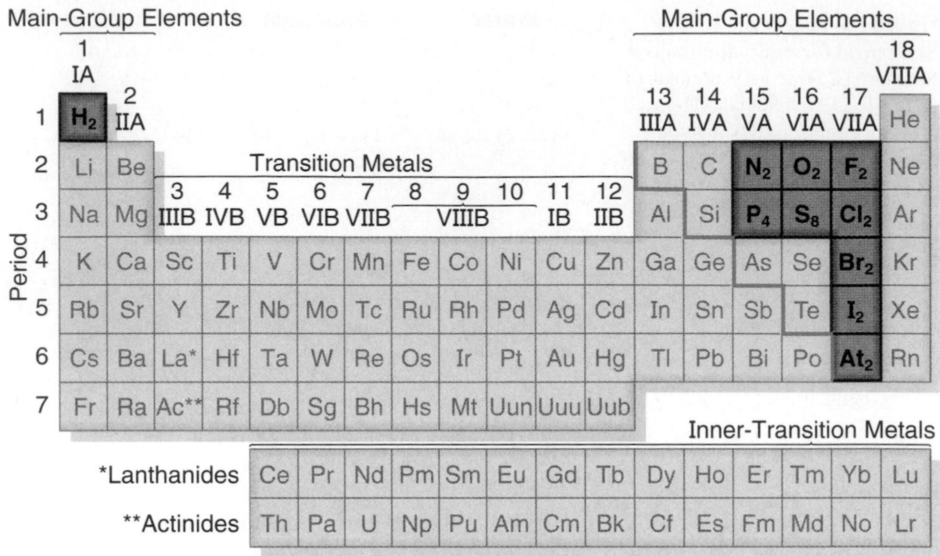

FIGURE 4.12

Molecular form of the elements. Only the elements highlighted in purple occur as molecules; the remainder do not.

to give a three-dimensional structure of an immense but indefinite number of atoms. Graphite also has a three-dimensional structure. We denote both forms of carbon by the symbol C.

4.8 Ionic Substances and Their Formulas

OBJECTIVES

- Define *anion* and *cation.*
- Define *monatomic ion.*
- Define *polyatomic ion.*
- Describe the structure of the sodium chloride crystal, as a simple example of an ionic solid.
- Describe how the relative number of cations and anions in an ionic solid is fixed.

Although atoms are electrically neutral and therefore have equal numbers of positive and negative charges, they can become ions when one or more electrons are transferred from one atom to another. An atom that picks up one or more electrons becomes an **anion** (pronounced "an′-eye′-un"), that is, *a negatively charged ion.* An atom that loses one or more electrons becomes a **cation** (pronounced "cat′-eye′-un"), that is, *a positively charged ion.* An ionic compound contains anions and cations in such numbers that the compound is electrically neutral.

Monatomic Ions

An ion may consist of a single atom or a group of atoms. A **monatomic ion** is *an ion consisting of only one atom.* Usually metal atoms lose electrons, forming monatomic cations. For example, a sodium atom loses one electron when it forms compounds. (The electrons lost are then picked up by the nonmetal atoms to form anions.) We denote the sodium ion as Na^+. We could write the formation of a sodium ion from a sodium atom as follows:

$$Na \longrightarrow Na^+ + e^-$$

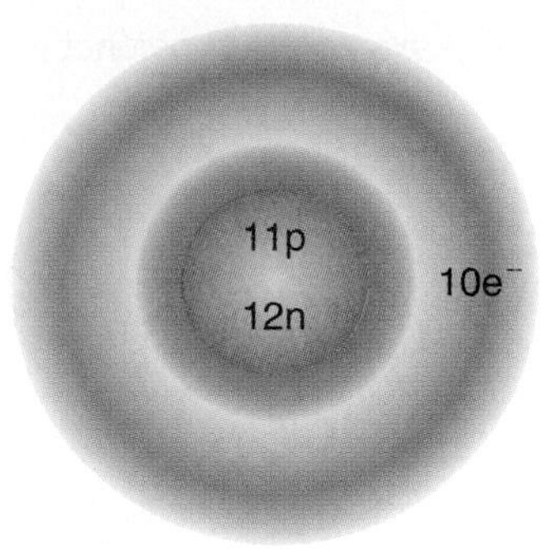

FIGURE 4.13
The sodium-23 ion. Sodium-23 has a nucleus of 11 protons (with a total charge of +11) and 12 neutrons. Around the nucleus, in the ion, are 10 electrons (with a total charge of −10). The charge on the ion is (+11) + (−10) = +1.

Here e^- represents the electron, which will be picked up by a nonmetal atom. Figure 4.13 illustrates the number of protons, neutrons, and electrons in the sodium ion. Note that there is one less electron about the nucleus than there are protons in the nucleus.

A calcium atom usually forms compounds by losing two electrons. The calcium ion therefore has a positive charge of 2 and is denoted Ca^{2+}:

$$Ca \longrightarrow Ca^{2+} + 2e^-$$

The electrons lost by calcium are picked up by the nonmetal atoms to give anions.

Nonmetal atoms often form anions by gaining electrons from a metal atom. For example, a chlorine atom can gain one electron to give the chloride ion, Cl^-. (The monatomic anions have the suffix *-ide* in their name; we explore the naming of ions in Chapter 5.) We can represent the process in which the chlorine atom gains an electron in this way:

$$Cl + e^- \longrightarrow Cl^-$$

Oxygen often forms compounds by gaining two electrons to give the oxide ion, O^{2-}:

$$O + 2e^- \longrightarrow O^{2-}$$

Figure 4.14 shows the number of protons, neutrons, and electrons in the oxide ion.

Polyatomic Ions

An ion can consist of a group of atoms. This group is called a *polyatomic ion.* A **polyatomic ion** is *an ion consisting of two or more atoms chemically bonded but having an excess or a deficiency of electrons so that the entire unit has an electric charge.* The carbonate ion is an example of a polyatomic ion. It consists of a carbon atom to which three oxygen atoms are chemically bonded; this group of atoms has picked up two electrons from somewhere (normally metal atoms), giving the ion two units of negative charge. The carbonate ion is denoted CO_3^{2-}. Note that the formula for a polyatomic ion looks like the formula for a molecule, but it has either a positive charge or a negative charge.

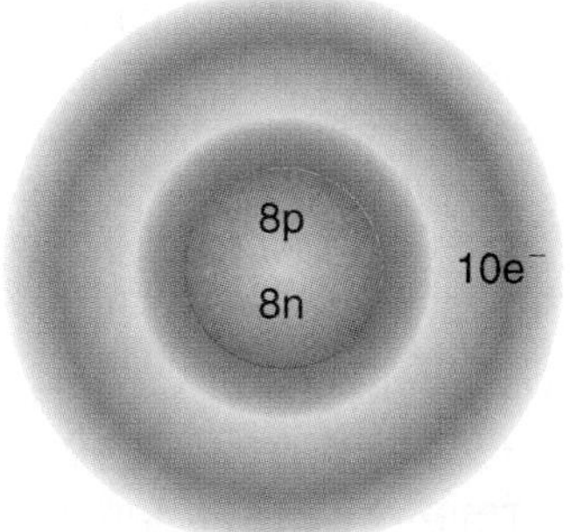

FIGURE 4.14
The oxide ion from oxygen-16. Oxygen-16 has a nucleus of 8 protons (with a total charge of +8) and 8 neutrons. In the ion, there are 10 electrons (with a total charge of −10). The charge on the ion is (+8) + (−10) = −2.

Structure of Ionic Compounds

An ionic compound is a compound composed of cations and anions. Sodium chloride consists of equal numbers of sodium ions, Na^+, and chloride ions, Cl^-. The strong attraction between positive and negative charges holds the ions together to form a crystal of sodium chloride. The number of total ions in the crystal varies, but there is an extremely large number of sodium ions and chloride ions, and because they are in equal numbers the crystal is electrically neutral. Figure 4.15 shows a model of a very small piece of the sodium chloride crystal. Note that each sodium ion is surrounded by six chloride ions and each chloride ion is surrounded by six sodium ions. *Unlike molecular substances, which have molecules as their basic structural units, ionic substances have ions as their basic structural units.*

The relative numbers of cations and anions in an ionic solid are fixed by the requirement that the solid be electrically neutral. Consider two examples. Sodium chloride has

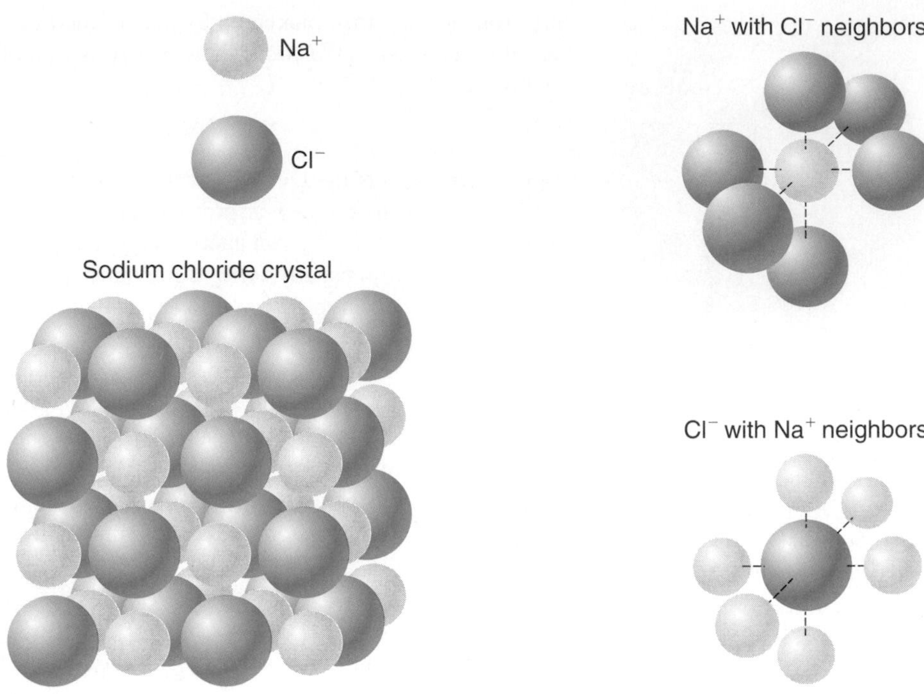

FIGURE 4.15
Model of the sodium chloride crystal. Each sodium ion is surrounded by six chloride ions, and each chloride ion is surrounded by six sodium ions. Because the crystal contains as many chloride ions as sodium ions, it is electrically neutral.

equal numbers of sodium ions and chloride ions, because each kind of ion has the same magnitude of electric charge. Calcium chloride consists of calcium ions and chloride ions in the ratio of 1:2. A Ca^{2+} ion has two positive charges, whereas two Cl^- have a total of two negative charges. The compound is therefore neutral.

You will learn to write formulas for ionic compounds and name them in Chapter 5. For example, you will see that the formula for sodium chloride is NaCl and that the formula for calcium chloride is $CaCl_2$. Although these substances or any other ionic substance do not contain molecules, we do speak of the smallest unit based on the formula of the ionic substance. The **formula unit** is *the group of atoms or ions explicitly symbolized in the formula of a substance.* For example, the formula unit of calcium chloride, $CaCl_2$, consists of one Ca^{2+} ion and two Cl^- ions, as indicated by the formula.

4.9 Electrical Properties of Substances in Solution

OBJECTIVES

- Explain why a water solution of sodium chloride is an electrical conductor, whereas pure water is nonconducting.
- Describe what happens as an ionic solid dissolves in water.

It is commonly appreciated, at least by prudent people, that you do not operate electrical equipment while in the bathtub or while standing in a pool of water. You might

FIGURE 4.16
Electrical conduction by ions. Ions carry electric charge through a solution. When you insert electrodes from a battery into a solution of ions, the cations move to the negative electrode, and the anions move to the positive electrode.

receive a serious electrical shock because water in a bathtub or in a pool is an electrical conductor. It is not so well known that *pure* water is not an electrical conductor. Bath water conducts electricity because it contains ion impurities. Unless a sample of water has been specially purified, it is likely to contain dissolved substances and ions.

Tap water contains various ions from substances that are dissolved in it. Ions are electrically charged and can readily move through a solution, carrying the electrical charge with them. Because an electric current is simply a motion of electric charge, a solution of ions in water will be electrically conducting. Pure water, on the other hand, is composed of molecules, which do not have a net electric charge. Therefore, pure water is a nonconductor.

In general, when an ionic solid dissolves in water, ions that were in fixed positions in the solid move into the water solution. The ions are now free to move about, and when electrodes, or electrical connectors, are placed in the solution the cations move toward the negative electrode and the anions move toward the positive electrode. This motion of ions constitutes the electrical current through the solution. Figure 4.16 shows the motion of ions when electrical connectors from a battery are placed in a solution of sodium chloride.

Figure 4.17 shows a simple experimental apparatus to test the electrical conductivity of a liquid or solution. It consists of two metal strips that serve as electrodes. You connect one electrode directly to one pole of a battery and connect the other electrode to the tip of a flashlight bulb. Then you connect the side of the bulb by a wire to the other pole of the battery. If you were to touch the two electrodes together, you would see the flashlight bulb light up. In this experiment, however, you place the two electrodes in the liquid. If the liquid is an electrical conductor, the bulb lights just as if the electrodes had touched one another directly, because the liquid would carry the current from one electrode to the other. The beaker in Figure 4.17A contains pure water. Note that the bulb does not light, because pure water is not an electrical conductor. The beaker in Figure 4.17B contains a solution of sodium chloride in water. Now the bulb lights.

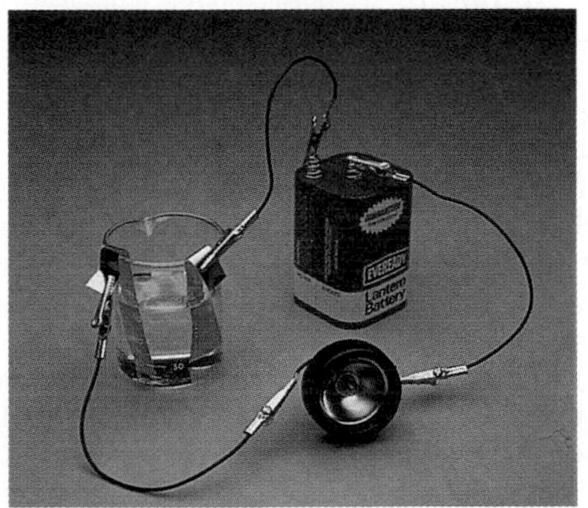

A

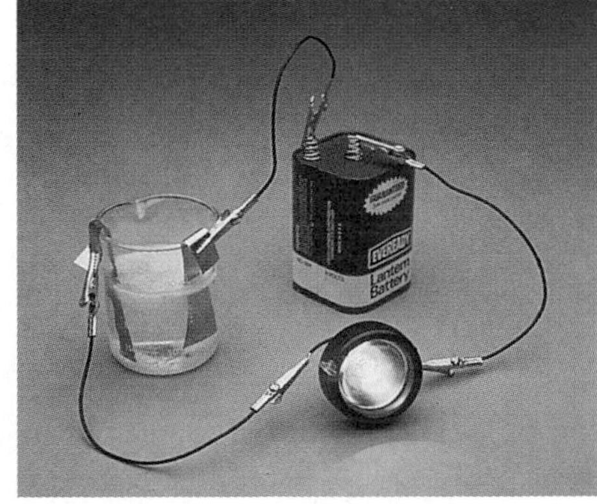

B

FIGURE 4.17
Test of electrical conductivity. *(A)* The beaker contains pure water. The light bulb does not light, showing that pure water is not a conductor of electricity. *(B)* The beaker contains a solution of sodium chloride in water. The light bulb lights up, indicating that the solution is a conductor of electricity.

Chemical Perspective

Seeing Atoms

Do atoms really exist? Today their existence is hardly questioned. Yet as late as the early part of the twentieth century, many prominent scientists refused to believe that atoms were anything more than useful mental constructs. Even if atoms did exist, many felt they would be too small to see. Still, one might establish their existence indirectly by obtaining information about them. If you see a paw print in the woods, you do not need to see the actual bear to know that one lurks nearby. From the print, you might also be able to deduce something about the bear's size. Similarly, perhaps some experiment might establish the size of the atom.

The breakthrough in this problem came from a study of the phenomenon of Brownian motion. Brownian motion is the rapid, erratic movement of very small particles, such as pollen grains, in a liquid at rest (see Chapter 3). The motion of pollen grains is easily visible under an ordinary microscope and is caused by the random movement of molecules of the liquid striking the particles. Although you cannot see the molecules directly, you can see their "paw print." In 1905, Albert Einstein showed that you could relate the Brownian motion of pollen grains to the number of molecules in a given volume of substance. (More precisely, the value obtained is *Avogadro's number,* a very important number in chemistry, as you will see later in the book.) Four years later, the French physicist Jean-Baptiste Perrin performed very exacting experiments in Brownian motion from which he obtained the first value for Avogadro's number and the first estimates of atomic size.

Perrin's estimates of atomic size essentially agree with modern values, showing that atoms are indeed incredibly small. An oxygen atom has a radius of about 6.6×10^{-11} m. It would take more than 600,000 oxygen atoms lined up next to one another to equal the thickness of one page of this text!

Perrin's work established the existence of atoms. Still, the dream of actually seeing atoms remained elusive until 1970. In that year, Albert Crewe at the University of Chicago developed a special microscope, called a *high-resolution scanning electron microscope,* from which he produced the first pictures of thorium atoms in a substance. The thorium atoms, which are relatively large atoms, appear as bright spots in those photographs. Atoms had finally been seen.

More recently, in 1981, Gerd Binnig and Heinrich Rohrer, scientists at the IBM laboratories in Zurich, Switzerland, invented the *scanning tunneling microscope.* This instrument makes it possible to see atoms on surfaces, giving breathtaking views of those atoms. In this instrument, a probe (Figure 4.18) scans a surface, mapping out the position of atoms. A computer gathers the data and draws an image on a screen. The colors are specified by the experimenter and are given merely to show more clearly different regions of the image. Figure 4.1 shows such a computer-drawn image of atoms on a metal surface.

If you can see atoms, maybe you can push them around. In 1990, scientists at IBM in San Jose, California, did just that. Using the probe in a scanning tunneling microscope, they positioned xenon atoms on a nickel metal surface to form the letters IBM (see Figure 4.19).

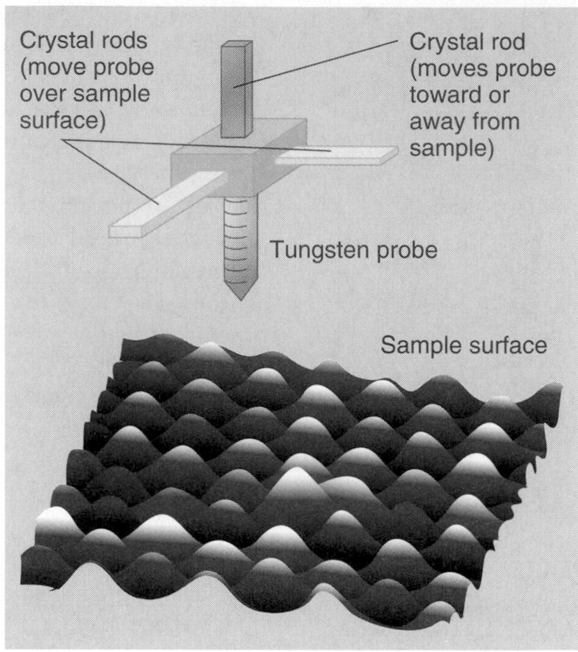

FIGURE 4.18
Probe of a scanning tunneling microscope. A tungsten metal probe, sensitive to small voltage changes between it and the atoms on a surface, is used to scan a sample. The probe is moved automatically across the sample in such a way as to maintain a constant voltage between it and the surface atoms. In this way, the probe maps out the contours of the atoms. Movement of the probe is translated by a computer into an image.

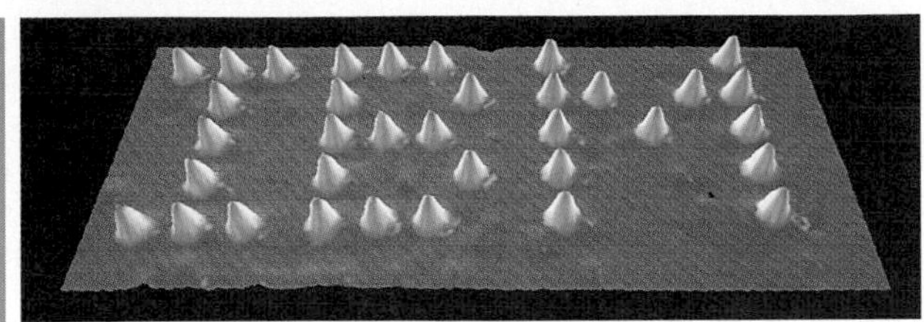

FIGURE 4.19
Moving atoms around. This image obtained by a scanning tunneling microscope shows xenon atoms that have been moved to form the letters IBM.

CHAPTER REVIEW

Key Words

atomic theory *(p. 86)*
atom *(p. 87)*
element *(p. 87)*
compound *(p. 87)*
chemical reaction *(p. 87)*
electron *(p. 89)*
nucleus *(p. 90)*
proton *(p. 91)*
neutron *(p. 91)*
atomic number *(p. 91)*
mass number *(p. 91)*
isotopes *(p. 92)*
percentage abundance *(p. 92)*
atomic mass unit (amu) *(p. 96)*
atomic weight *(p. 96)*
periodic table *(p. 99)*
period *(p. 99)*
group (or **family**) *(p. 99)*
main group (or **representative group**) *(p. 99)*
transition-metal group *(p. 99)*
metal *(p. 99)*
nonmetal *(p. 102)*
metalloid *(p. 102)*
molecule *(p. 104)*
ion *(p. 104)*
molecular formula *(p. 106)*
structural formula *(p. 107)*
anion *(p. 108)*
cation *(p. 108)*
monatomic ion *(p. 108)*
polyatomic ion *(p. 109)*
formula unit *(p. 110)*

Summary

Atomic theory is central to chemistry. According to this theory, all matter is composed of atoms. Elements are composed of one kind of atom, whereas compounds are composed of atoms of two or more elements chemically combined in fixed proportions. A chemical reaction is explained by atomic theory as a rearrangement of the atoms in the reacting substances to give new chemical combinations in the products of the reaction. Dalton's theory readily explained the law of conservation of mass and the law of definite proportions.

Beginning with the latter part of the nineteenth century, chemists and physicists discovered that the atom has a structure. It is composed of a *nucleus* with *electrons* moving around it. The nucleus has a positive charge, whereas the electron has a negative charge; therefore, the nucleus attracts a sufficient number of electrons to it to give a neutral atom. The nucleus is itself composed of *protons* and *neutrons.* Each nucleus is characterized by two numbers, the *atomic number* (the number of protons in the nucleus) and the *mass number* (the number of protons and neutrons in the nucleus). All atoms of a given element have the same atomic number. They may have different mass numbers, however. Atoms with the same atomic number but different mass numbers are called *isotopes.* An atom of aluminum with mass number 27 is designated by the *isotope symbol* $^{27}_{13}Al$. The isotope is also called aluminum-27.

Some elements, such as aluminum, occur naturally on earth as a single isotope. Most elements, however, occur on earth as mixtures of isotopes. These mixtures usually consist of unvarying percentages of the isotopes, called the *percentage abundances.* The *atomic weight* of an element is the weighted average mass (in amu) of an atom of the naturally occurring element. We calculate the atomic weight of an element from the isotope masses and percentage abundances.

The *periodic table* is a table of the elements arranged by increasing atomic number into rows, called *periods,* and columns, called *groups.* The elements in any one group have similar or regularly varying chemical and physical properties. The periodic table is a convenient way to present data for the elements. For example, the periodic table on the inside front cover of this book gives the symbol, atomic number, and atomic weight of each element. It also shows whether an element is a *main-group element* or a *transition element,* and whether it is a *metal,* a *nonmetal,* or a *metalloid.*

Although composed of one kind of atom, elements may occur either as large populations of these atoms or as *molecules* consisting of several of these atoms. Compounds—substances that are composed of atoms of two or more elements—are of two main types: molecular substances and ionic substances. Molecular substances are composed of *molecules,* which are groups of atoms chemically bonded together. We denote them by *molecular formulas,* which give us the number of atoms of each kind in the molecule. If we want to show how the different atoms of a molecule are bonded together, we draw a *structural formula* or construct a molecular model. Ionic substances are composed of *ions,* either *monatomic ions* or *polyatomic ions*. Such substances are composed of *cations* (positively charged ions) and *anions* (negatively charged ions) in such numbers that the overall compound is electrically neutral.

Problem-Solving Skills

1. **Writing Isotope Symbols:** Given the number of protons and neutrons in the nucleus of an atom, write the isotope symbol (Example 4.1).
2. **Determining the Number of Protons, Neutrons, and Electrons in an Atom:** Given the atomic number and mass number of an atom, deduce the number of protons and neutrons in the nucleus (Example 4.2) and the number of electrons outside the nucleus (Example 4.3).
3. **Calculating the Atomic Weight of an Element:** Given the masses (in amu) and percentage abundances of the naturally occurring isotopes of an element, calculate the atomic weight (Example 4.4).
4. **Locating an Element by Group and Period of the Periodic Table:** Given a periodic table like that on the left inside front cover of this book, decide the name and symbol of the element in a given group and period (Example 4.5).
5. **Identifying the Metals and Nonmetals in the Periodic Table:** Given a periodic table like that on the left inside front cover of this book, decide which elements of a group are metals and which are nonmetals (Example 4.6).
6. **Deciding Whether a Substance Is Molecular or Ionic:** Given the elements composing a substance, decide whether the substance is molecular or ionic (Example 4.7).
7. **Writing Molecular Formulas:** Given the number of each kind of atom in a molecule, write the molecular formula (Example 4.8).

Questions to Test Your Reading

4.1 According to Dalton's atomic theory, what is an element? What is a compound? How does Dalton's atomic theory explain a chemical reaction?

4.2 How does atomic theory explain the law of conservation of mass? the law of definite proportions?

4.3 Briefly, how does one produce cathode rays? What is the name of the particle that makes up cathode rays?

4.4 What were the results of the experiments that occurred in Rutherford's laboratory? How did Rutherford explain these results?

4.5 You can describe a particular isotope by its isotope symbol. Describe the information given by the isotope symbol, and the position of each piece of information in this symbol.

4.6 Sodium consists of the single isotope sodium-23, which has 11 protons and 12 neutrons. Describe the sodium-23 atom; that is, identify and give the number of different particles, both inside the nucleus and outside the nucleus.

4.7 Why are the percentage abundances of the isotopes of an element generally the same, regardless of where the element is found?

4.8 What isotope is taken to be the standard for obtaining masses of atoms and atomic weights?

4.9 In Mendeleev's periodic table, the elements were arranged in rows and columns by increasing atomic weights. In what crucial respect does the modern periodic table differ?

4.10 Copper is in Group IB. Is copper a main-group element or a transition-metal element?

4.11 In what ways are the Group IA metals similar to one another?

4.12 Why do modern periodic tables have two rows at the bottom not connected to the main body of the table?

4.13 What is a metalloid? Give an example.

4.14 What is meant by the "molecular" basis of substances?

4.15 Some of the elements consist of molecules. Explain and give an example.

4.16 What are the two main types of compounds? Give an example of each type. How do these two types of compounds tend to differ in their melting points?

4.17 Define molecular formula. Give an example.

4.18 Define structural formula. Give an example.

4.19 What are the molecular formulas of the elements hydrogen, nitrogen, oxygen, chlorine, and sulfur? What would you write for carbon?

4.20 Describe the structure of sodium chloride. What is the ratio of the number of sodium ions to chloride ions in the crystal? How many chloride ions surround each sodium ion?

4.21 Would you expect a water solution of the ionic substance $CaCl_2$ to conduct electricity? Explain.

4.22 Would you expect a water solution of table sugar to conduct electricity? Explain.

Practice Problems

Isotopes

4.23 Write isotope symbols for each of the following.
(a) an atom with 10 protons and 10 neutrons
(b) an atom with 14 protons and 14 neutrons
(c) an atom with 15 protons and 16 neutrons
(d) an atom with 19 protons and 20 neutrons
(e) an atom with 25 protons and 30 neutrons

4.24 Write isotope symbols for each of the following.
(a) an atom with 6 protons and 7 neutrons
(b) an atom with 12 protons and 12 neutrons
(c) an atom with 18 protons and 22 neutrons
(d) an atom with 24 protons and 26 neutrons
(e) an atom with 28 protons and 30 neutrons

4.25 How many protons and neutrons are in each of the following?
(a) nitrogen-14 (b) $^{32}_{15}P$ (c) sulfur-34
(d) $^{35}_{17}Cl$ (e) $^{52}_{24}Cr$

4.26 How many protons and neutrons are in each of the following?
(a) $^{22}_{10}Ne$ (b) $^{29}_{14}Si$ (c) calcium-40 (d) $^{59}_{27}Co$
(e) copper-63

4.27 Give the number of protons, neutrons, and electrons in the zinc-64 atom. Where are these particles located in the atom?

4.28 How many electrons are in the argon-40 atom? How many neutrons are in the nucleus?

4.29 How many electrons are outside the nucleus in the selenium-80 atom?

4.30 How many electrons are in the sulfur-32 atom? in the bromine-79 atom?

4.31 Describe the arsenic-75 atom; that is, what particles and how many are in the nucleus, and what particles and how many are outside the nucleus?

4.32 Describe the silicon-29 atom; that is, what particles and how many are in the nucleus, and what particles and how many are outside the nucleus?

Atomic Weights

4.33 Naturally occurring nitrogen has two isotopes, nitrogen-14 and nitrogen-15. Which isotope is more abundant? The atomic weight is 14.01 amu.

4.34 Naturally occurring vanadium has two isotopes, vanadium-50 and vanadium-51. Which isotope is more abundant? The atomic weight is 50.94 amu.

4.35 Naturally occurring boron consists of two isotopes, boron-10 and boron-11. Boron-10 has a mass of 10.01 amu, and boron-11 has a mass of 11.01 amu. The abundances are 19.7% and 80.3%, respectively. Calculate the atomic weight from these data.

4.36 Natural silver is composed of two isotopes, silver-107 and silver-109. Silver-107 has a mass of 106.90 amu and an abundance of 51.82%. Silver-109 has a mass of 108.90 amu and an abundance of 48.18%. What is the atomic weight of silver calculated from these data?

4.37 An element has three naturally occurring isotopes with the following masses and abundances.

Isotope Mass	Percentage Abundance
19.99 amu	90.92
20.99 amu	0.257
21.99 amu	8.82

What is the atomic weight of the element? What is the name of the element?

4.38 An element has three naturally occurring isotopes with the following masses and abundances.

Isotope Mass	Percentage Abundance
27.98 amu	92.18
28.98 amu	4.71
29.97 amu	3.12

What is the atomic weight of this element? What is the name of the element?

Periodic Table

4.39 Using a periodic table, obtain the symbol corresponding to each of the following elements, and name the element. You may need to refer to the list of elements on the inside front cover of the book.
(a) element in Group IIIA, Period 5
(b) element in Group IVA, Period 3
(c) element in Group VIIB, Period 4
(d) element in Group IVB, Period 4
(e) element in Group VA, Period 4

4.40 Using a periodic table, obtain the symbol corresponding to each of the following elements, and name the element. You may need to refer to the list of elements on the inside front cover of the book.
(a) element in Group IIA, Period 3
(b) element in Group IA, Period 4
(c) element in Group VIA, Period 3
(d) element in Group IB, Period 6
(e) element in Group IVA, Period 5

4.41 What metals are in Group VA? See the periodic table on the inside front cover of this book.

4.42 What metalloids are in Group IVA? See the periodic table on the inside front cover of this book.

4.43 For the element phosphorus, obtain the atomic number, atomic weight, group, and period, and determine whether the element is a metal, nonmetal, or metalloid; use the periodic table on the inside front cover of this book.

4.44 For the element selenium, obtain the atomic number, atomic weight, group, and period, and determine whether the element is a metal, nonmetal, or metalloid; use the periodic table on the inside front cover of this book.

Molecular and Ionic Compounds

4.45 For each of the following pairs of elements, decide whether you expect any compound that forms to be molecular or ionic.
(a) carbon and oxygen
(b) nitrogen and oxygen
(c) lithium and oxygen
(d) calcium and chlorine
(e) sulfur and chlorine

4.46 For each of the following pairs of elements, decide whether you expect any compound that forms to be molecular or ionic.
(a) sodium and nitrogen
(b) phosphorus and sulfur
(c) hydrogen and sulfur
(d) barium and oxygen
(e) magnesium and chlorine

4.47 A compound contains sodium, sulfur, and oxygen. It melts at 884°C. What type of compound should this be, molecular or ionic? Does your prediction agree with the melting point of the compound?

4.48 A compound contains carbon, hydrogen, and iodine. It melts at 120°C. What type of compound should this be, molecular or ionic? Does your prediction agree with the melting point of the compound?

Molecular Formulas

4.49 Write molecular formulas for the following molecules, listing the elements in the order given.
(a) a molecule consisting of two hydrogen atoms, one sulfur atom, and four oxygen atoms
(b) a molecule consisting of two chlorine atoms and seven oxygen atoms
(c) a molecule consisting of two nitrogen atoms and four oxygen atoms
(d) a molecule consisting of three carbon atoms and eight hydrogen atoms
(e) a molecule consisting of two nitrogen atoms and four fluorine atoms

4.50 Write molecular formulas for the following molecules, listing the elements in the order given.
(a) a molecule consisting of two hydrogen atoms and one sulfur atom
(b) a molecule consisting of one phosphorus atom and five chlorine atoms
(c) a molecule consisting of six carbon atoms and six hydrogen atoms
(d) a molecule consisting of one hydrogen atom, one nitrogen atom, and three oxygen atoms
(e) a molecule consisting of one hydrogen atom, one chlorine atom, and four oxygen atoms

Additional Problems

4.51 Sodium-24 is a radioactive isotope used to diagnose restricted blood circulation, for example, in the legs. How many electrons, protons, and neutrons are in the sodium-24 isotope?

4.52 Technetium-99 is a radioactive isotope used to diagnose brain tumors, which preferentially absorb the isotope. How many electrons, protons, and neutrons are in the technetium-99 isotope?

4.53 An isotope of calcium has a mass of 46.95 amu. Write the isotope symbol for this atom.

4.54 An isotope of iron has a mass of 53.94 amu. Write the isotope symbol for this atom.

4.55 Two isotopes of nickel are $^{58}_{28}Ni$ and $^{60}_{28}Ni$. How do these isotopes differ? How are they alike?

4.56 Two isotopes of zinc are $^{64}_{30}Zn$ and $^{66}_{30}Zn$. How do these isotopes differ? How are they alike?

4.57 The mass of an electron is 0.00055 amu; the mass of a proton is 1.00728 amu. How much heavier is the proton than the electron? (What is the ratio of proton mass to electron mass?)

4.58 The mass of a proton is 1.00728 amu; the mass of a neutron is 1.00866 amu. How much heavier is the neutron than the proton? (What is the ratio of neutron mass to proton mass?)

4.59 Chromium has the following isotope masses and percentage abundances.

Isotope Mass	Percentage Abundance
49.95 amu	4.31
51.94 amu	83.76
52.94 amu	9.55
53.94 amu	2.38

Calculate the atomic weight of chromium.

4.60 Iron has the following isotope masses and percentage abundances.

Isotope Mass	Percentage Abundance
53.94 amu	5.84
55.94 amu	91.68
56.94 amu	2.17
57.93 amu	0.31

Calculate the atomic weight of iron.

4.61 In general, the elements in any group of the periodic table tend to increase in metallic character as you proceed down a column. List the elements in Group IVA, and after each element note whether it is a metal, metalloid, or nonmetal. Does your list agree with the general rule stated at the beginning of this problem?

4.62 In general, the elements in any period of the periodic table tend to decrease in metallic character going from left to right. List the elements in the second period, and after each element note whether it is a metal, metalloid, or nonmetal. Does your list agree with the general rule stated at the beginning of this problem?

4.63 Chloroform was formerly used as an anesthetic, but its use has been discontinued because it causes liver damage. The chloroform molecule contains one carbon atom, one hydrogen atom, and three chlorine atoms. Write the molecular formula for chloroform.

4.64 Formaldehyde has been used to preserve biological specimens. Today it is used mainly in the preparation of plastics. The formaldehyde molecule contains one carbon atom, two hydrogen atoms, and one oxygen atom. Write the molecular formula for formaldehyde.

4.65 Aluminum sulfate is used in water purification. The polyatomic sulfate ion is SO_4^{2-}. Explain what the formula of this ion means.

4.66 Sodium hydrogen carbonate is also known as baking soda. The polyatomic hydrogen carbonate ion is HCO_3^-. Explain what the formula of this ion means.

Practice in Problem Analysis

For each problem, describe the thinking you would use (the problem analysis) before doing the actual solution, but do not solve the problem.

1. The formula unit of a compound consists of two aluminum atoms and three sulfur atoms. What is the formula of this substance?

2. A soft element dissolves in water with the release of hydrogen gas. It combines chemically with chlorine to form a compound with a relatively high melting point. The compound will dissolve in water to give a solution that conducts electricity. What can you say about the identity of the element?

PRACTICE EXAM

1. According to Dalton's theory, the fundamental particle of all matter is the
(a) electron (b) proton (c) atom
(d) neutron (e) None of the above

2. Atoms that have the same atomic number but different mass numbers
(a) never occur (b) are called isotopes
(c) are called ions (d) are called neutrons
(e) None of the above

3. Which of the following are isotopes?
(a) nitrogen-13 and carbon-13
(b) hydrogen-3 and helium-3
(c) hydrogen-2 and hydrogen-3
(d) carbon monoxide and carbon dioxide
(e) None of the above

4. How many protons are in an atom of carbon -13?
(a) 6 (b) 7 (c) 13 (d) 19
(e) None of the above

5. How many neutrons are in an atom of carbon -13?
(a) 6 (b) 7 (c) 13 (d) 19
(e) None of the above

6. How many neutrons are in an atom of $^{14}_{6}C$?
(a) 6 (b) 7 (c) 13 (d) 19
(e) None of the above

7. Elements in the same family in the periodic table
(a) have the same number of protons.
(b) have the same number of electrons.
(c) have similar chemical properties.
(d) are in the same horizontal row.
(e) None of the above

8. What elements in Group VIA are metals?
(a) All of them
(b) oxygen (O), sulfur (S), and selenium (Se)
(c) tellurium (Te)
(d) polonium (Po)
(e) None of the above

9. Which pair of elements will combine chemically to form an ionic compound?
(a) magnesium and fluorine
(b) hydrogen and sulfur
(c) carbon and oxygen
(d) chlorine and bromine
(e) None of the above

10. Which pair of elements will combine chemically to form a molecular compound?
(a) mercury and sulfur
(b) hydrogen and sulfur
(c) lead and sulfur
(d) copper and sulfur
(e) None of the above

11. Which statement is true?
(a) Ionic compounds tend to have relatively low melting points.
(b) An anion is a positively charged ion.
(c) Metal atoms often form compounds by gaining electrons from a nonmetal atom.
(d) An atom becomes an ion by gaining protons.
(e) None of the above

12. How many electrons are in Mg^{2+}?
(a) 14 (b) 12 (c) 10 (d) 2
(e) None of the above

13. How many oxygen atoms are indicated by the formula $Al(NO_3)_3$?
(a) 1 (b) 3 (c) 6 (d) 9
(e) None of the above

14. How many nitrogen atoms are indicated by the formula $Al(NO_3)_3$?
(a) 1 (b) 3 (c) 6 (d) 9
(e) None of the above

15. Which statement is true?
(a) An aqueous solution containing a compound derived from sodium and chlorine should conduct electricity.
(b) The solid form of a compound derived from sodium and chlorine should conduct electricity.
(c) An aqueous solution of ordinary table sugar, a substance composed of carbon, hydrogen, and oxygen, should conduct electricity.
(d) Pure water is a good conductor of electricity.
(e) None of the above

Chapter

5 Chemical Formulas and Names

Fumes from hydrochloric acid.

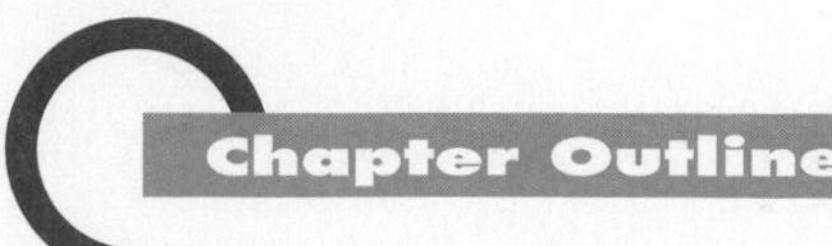

Chapter Outline

Have you ever read the ingredients label on grocery-store products or those on over-the-counter drugs? If you have, you have seen the names of many chemical substances (Figure 5.1). For example, a box of ordinary salt might list salt, calcium silicate, and potassium iodide. Salt is the common name for a substance that chemists call sodium chloride. Calcium silicate and potassium iodide are formal, or *systematic,* names of chemical substances. (The salt producer adds a small quantity of calcium silicate to help prevent the salt from clumping in humid weather. The producer adds potassium iodide as a nutritional source of iodide ion.) Millions of chemical substances exist, and chemists have devised systematic ways of naming them so that the names convey information about their composition and structure. The systematic name of a substance unequivocally identifies it. Thus, the systematic name of salt is sodium chloride. The

FIGURE 5.1

Label on a grocery-store product. If you become a label-reader, you will see the names of many chemical substances. Many of these are of organic compounds, which we do not discuss in this chapter, but others are of inorganic compounds, which we do discuss. All of the ingredients listed on this can of baking powder, except starch, are inorganic compounds. Baking soda, for instance, is the common name for what chemists call sodium hydrogen carbonate.

name, as you will see in this chapter, says that the substance consists of sodium ions and chloride ions. ●

● There are more than 10 million known compounds, and thousands of new ones are discovered each year.

In this chapter, you will learn how to write chemical formulas and names for *inorganic* compounds, which are mostly substances that do not contain the element carbon. (Simple compounds of carbon, such as carbon dioxide, are usually classified as inorganic.) We will not discuss *organic* compounds—generally more complicated compounds containing carbon. Although we identify a few inorganic compounds by common names such as water, ammonia, alum, and gypsum, these names fail to give us any clues about the compositions or structures of the compounds they represent. For instance, what is gypsum? Unless you are familiar with this substance, the name means little. On the other hand, calling it by its chemical name of calcium sulfate immediately tells you that it is composed of calcium ions and sulfate ions (once you understand the rules of naming). Fortunately, simple inorganic compounds can be named systematically by knowing only a few rules. You will learn these rules in this chapter.

IONIC COMPOUNDS

In this first part of the chapter, we look at the formulas and names of ionic compounds. (We introduced ions and ionic compounds in Chapter 4.)

Before systematic naming rules were adopted, chemical substances were often named for some easily observed property (such as sugar of lead from its sweet but poisonous taste) or perhaps for the person who discovered the substance (Glauber's salt). However, once chemists understood that materials had a unique composition and structure, they used this knowledge to give systematic names to substances. In the next three sections, we deal with the formulas and systematic names of binary ionic compounds, those containing only two elements. In Section 5.4 of this first part, we explain the formulas and names of ionic compounds containing more than two elements.

Exploration: Formation of ionic compounds from ions.

5.1 Formulas for Binary Ionic Compounds

OBJECTIVES

- Predict the charge on a main-group monatomic ion.
- Write the formula for a binary ionic compound, given the ions.

An ionic compound consists of cations (positive ions) and anions (negative ions). An example is table salt (sodium chloride), whose formula is NaCl. The compound consists of sodium ions (the cations) and chloride ions (the anions). In writing the formula of an ionic compound, we write the formula for the cation (Na^+) before the formula for the anion (Cl^-). We do not include the positive and negative signs in the formula of the compound.

The familiar ionic substance, sodium chloride, contains only two elements: the metal sodium and the nonmetal chlorine. It is an example of a *binary compound,* which is one composed of only two elements. Specifically, a **binary ionic compound** is *a compound composed of ions from only two elements.* In most cases, these compounds consist of a metal cation and a nonmetal anion. Other examples of binary ionic compounds are CaO, K_2O, and MgF_2. Note that the word *binary* does not refer to the relative number of ions in a formula (conveyed by the subscripts); it means that only two elements are present.

A compound cannot have an electric charge; it must have a net charge of zero. Therefore, an ionic compound must consist of equal numbers of positive and negative charges. You may recall seeing the structure of NaCl in Figure 4.15. The compound consists of equal numbers of sodium ions and chloride ions, giving a neutral compound. We can summarize what we have just discussed in a set of rules for writing ionic compounds.

Rules for Writing Formulas for Ionic Compounds

1. An ionic compound must contain both cations and anions.
2. The formula for the cation is always written before the formula for the anion.
3. The total number of positive charges on the cations must equal the total number of negative charges on the anions so that the net charge is zero. This determines the cation-to-anion ratio, hence the subscripts in the formula of the compound.

In this section and in Section 5.2, we discuss compounds composed of monatomic ions (ions consisting of single atoms). To write the formulas of such compounds, you will need to know the charges on the ions. Let us look at some simple rules for predicting the charges on monatomic ions.

Predicting the Charge on a Monatomic Ion

In most cases, you can use the periodic table to predict the number of electrons lost by an atom of a main-group metal (a metallic element in an A group of the periodic table). The number of electrons lost by main-group metal atoms usually equals the periodic-table group number (the Roman numeral) of the element. For instance, calcium is in Group IIA. The Roman numeral of this group is II, so we expect the atom to lose two electrons. Some of the main-group metallic elements of larger atomic number tend to lose two fewer electrons than what you would predict from the group number. Lead, the element of largest atomic number in Group IVA, forms the cation Pb^{2+} (having a positive charge of $4 - 2 = +2$). Tin, which is just above lead in Group IV, also forms an ion with a charge of $+2$ (Sn^{2+}). See Table 5.1 for a list of common main-group cations.

The transition metals, those in the B columns of the periodic table, often lose a variable number of electrons. For example, iron atoms form compounds by losing either

TABLE 5.1
Some Common Cations and Anions

Cation Symbol	Name	Anion Symbol	Name
Li^+	Lithium ion	F^-	Fluoride ion
Na^+	Sodium ion	Cl^-	Chloride ion
K^+	Potassium ion	Br^-	Bromide ion
Rb^+	Rubidium ion	I^-	Iodide ion
Cs^+	Cesium ion	O^{2-}	Oxide ion
Mg^{2+}	Magnesium ion	S^{2-}	Sulfide ion
Ca^{2+}	Calcium ion	N^{3-}	Nitride ion
Ba^{2+}	Barium ion	P^{3-}	Phosphide ion
Al^{3+}	Aluminium ion		

two or three electrons. The ions are called iron(II) ion and iron(III) ion, respectively, and are denoted Fe^{2+} and Fe^{3+}. There is no simple rule for predicting the number of electrons lost, although most transition metals do form an ion by losing two electrons, as well as ions of other charges.

Atoms of the nonmetals usually form ions having a charge equal to the group number minus 8. Oxygen follows this rule; it is in Group VIA and forms an ion having a charge equal to $6 - 8 = -2$. See Table 5.1 for a list of other main-group anions; note that they, too, follow the rule. A summary of rules for predicting the charge on a monatomic ion follows.

Rules for Predicting the Charge on a Monatomic Ion

1. The charge on a main-group (A group) metal ion equals the group number of the metal. (Some main-group metals of large atomic number have ions equal to the group number minus 2; common examples are Sn^{2+} and Pb^{2+}.)
2. The charge on a transition-metal ion is not easy to predict, because most of them have two or more ions of different charge. Most transition metals have an ion of charge +2, in addition to others. However, silver, zinc, and cadmium have only the following common ions: Ag^{+}, Zn^{2+}, and Cd^{2+}.
3. The charge on a nonmetal ion equals the group number minus 8 (giving a negative number).

■ EXAMPLE 5.1 Predicting the Charge on a Monatomic Ion

What charge is expected for the monatomic ion of sulfur?

Problem Analysis

Use the position of the main-group element in the periodic table to decide on its group number. If the element is a metal, the element will form a monatomic ion having a positive charge equal to the group number. (Some metals of large atomic number, such as tin and lead, tend to form ions with a charge equal to the group number minus 2.) If the element is a nonmetal, it forms a monatomic ion with a negative charge equal to the group number minus 8.

Solution

Sulfur is a nonmetal, so we expect an anion. Because sulfur is in Group VIA, the group number is 6, and the expected anion charge is $6 - 8 = -2$.

Exercise 5.1

What charge is expected for the monatomic ion of barium?

(Try Problems 5.11 and 5.12.)

Determining the Formula of a Binary Ionic Compound

Consider sodium chloride (NaCl) again. Like any other compound, it does not have an overall, or net, charge, even though it is composed of ions, which are charged. Each sodium ion (Na^{+}) has a +1 charge and each chloride ion (Cl^{-}) has a −1 charge, as you would expect from the positions of the elements in the periodic table. Thus, one Na^+ ion must accompany each Cl^- ion so that the net charge is zero:

Na^+	$+1$
Cl^-	-1
	0

This is what the formula NaCl tells you. If you were to write in the charges for the ions (which we do not normally do), the formula would appear as Na^+Cl^-. This formula shows that the compound contains one Na^+ ion for each Cl^- ion.

Let us see how we use the idea that a compound has zero net charge to obtain the formula of an ionic compound given the ions from which it is composed. Consider the compound composed of calcium ions (Ca^{2+}) and chloride ions (Cl^-). Note that every Ca^{2+} ion must be accompanied by two Cl^- ions, in order to give a net charge of zero:

Ca^{2+}	$1 \times (+2) = +2$
Cl^- Cl^-	$2 \times (-1) = -2$
	0

Thus, the formula of this compound (calcium chloride) is $CaCl_2$.

We can use a simple rule to obtain the formula of an ionic compound from the formulas of the ions. *The magnitude of the charge on one ion becomes the subscript of the other ion in the formula of the compound.*

$$Ca^{(2)+} \; Cl^{(1)-} = CaCl_2$$

Note that we write $CaCl_2$ and not Ca_1Cl_2; a 1 is understood when no subscript appears in a formula.

Similarly, the formula of the compound containing Al^{3+} and N^{3-} can be obtained from the charges on the ions.

$$Al^{(3)+} \; N^{(3)-} = Al_3N_3$$

With ionic compounds, the subscripts should be reduced to give the ratio with the smallest whole numbers. The 3:3 ratio in Al_3N_3 should be reduced to 1:1, so that the formula becomes Al_1N_1, or AlN.

■ EXAMPLE 5.2 Determining the Formula for a Binary Ionic Compound

Sapphires and rubies are naturally occurring crystals of aluminum oxide containing traces of other metal ions that impart the characteristic colors to the gemstones (Fig-

FIGURE 5.2
A ruby. Rubies are composed principally of colorless aluminum oxide. The red color is due to traces of Cr^{3+} ions that have replaced the colorless Al^{3+} ions.

ure 5.2). Aluminum oxide contains Al^{3+} and O^{2-} ions. What is the formula for this substance?

Problem Analysis

Use the magnitude of the charge on one ion as the subscript for the other ion. Verify that the net charge is zero. Reduce the ratio of subscripts to the smallest whole numbers.

Solution

Obtain the subscripts for the compound containing Al^{3+} and O^{2-} from the charges on the ions:

$$Al^{③+} O^{②-} = Al_2O_3$$

You can verify that the net charge is zero:

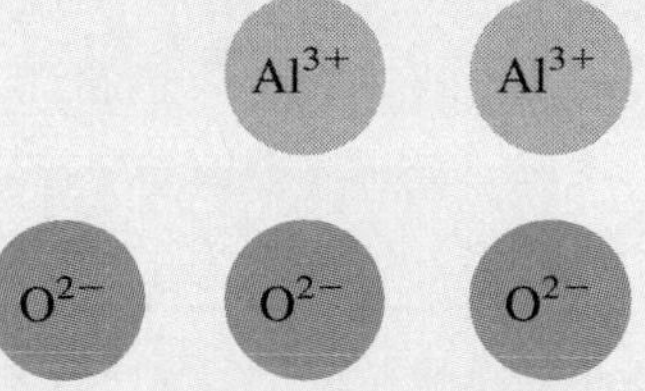

Thus, the formula is Al_2O_3.

Exercise 5.2

Determine the formulas for compounds containing the following ions.
(a) Li^+ and F^- (b) Li^+ and S^{2-} (c) Li^+ and N^{3-}

(Try Problems 5.15, 5.16, 5.17, and 5.18.)

With a little practice, you can write the formulas of binary ionic compounds quickly and easily. In the next two sections, we show you how to name these compounds.

5.2 Naming Binary Ionic Compounds When the Metal Forms a Single Cation

OBJECTIVE

- Assign the systematic name, given the formula of a binary ionic compound, when the metal can form a single kind of cation.

Usually the cation (positive ion) in a binary ionic compound is derived from a metal, and the anion (negative ion) is derived from a nonmetal. In this section, we describe the naming of binary ionic compounds in which the metal forms only one kind of cation. Table 5.1 lists some ions of this type. Sodium, for example, forms compounds with the Na^+ ion, but has no compounds with ions such as Na^{2+}, Na^{3+}, and so forth. Thus, the only binary ionic compound that sodium forms with chlorine is NaCl and never $NaCl_2$ or $NaCl_3$. The metals in the main groups of the periodic table—those in the A columns—generally fall in this category (Figure 5.3), but tin and lead are exceptions to this rule. Compounds containing transition metals (B columns of the periodic table), as well as tin and lead in Group IVA, are named in the next section.

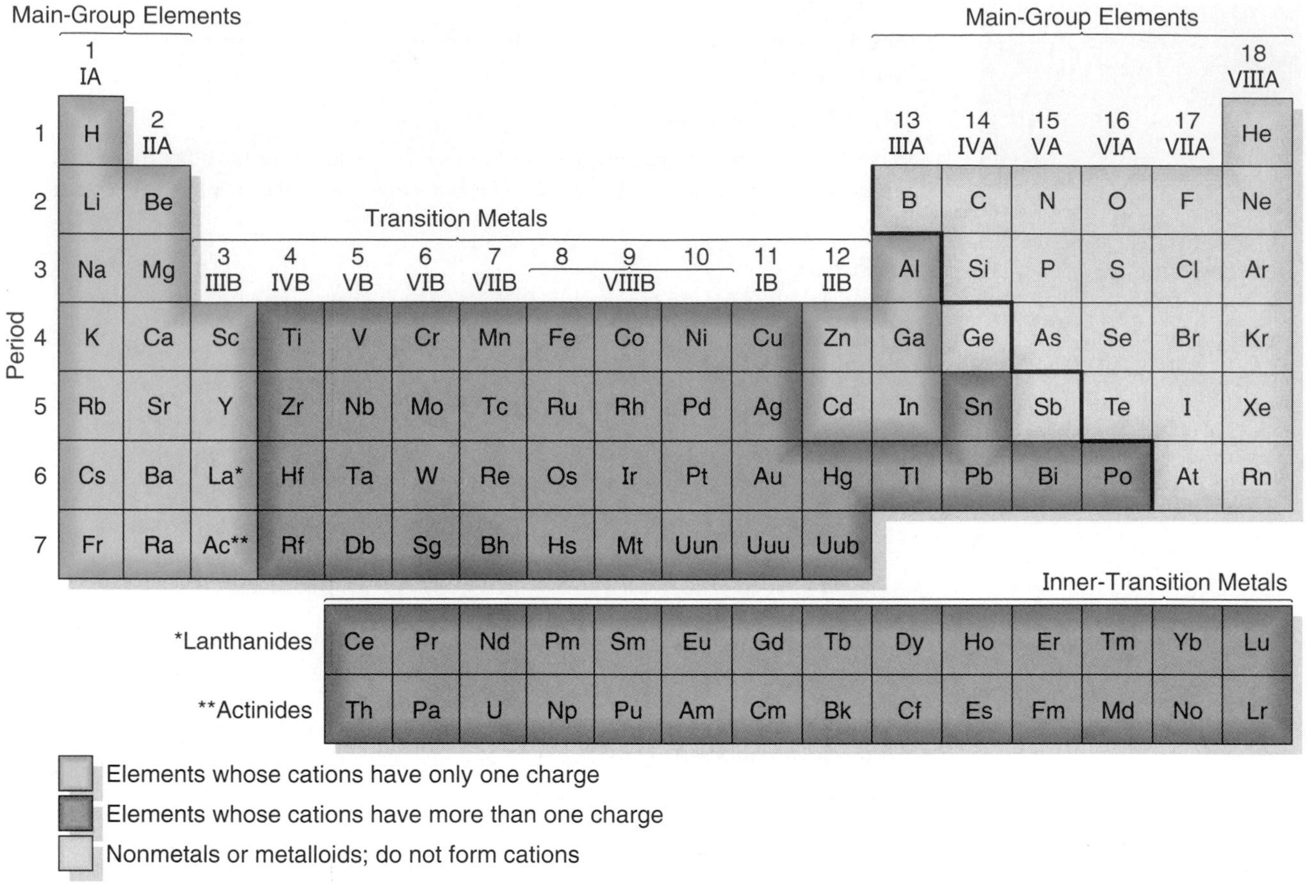

Main-Group Elements | Transition Metals | Main-Group Elements

Period	1 IA	2 IIA	3 IIIB	4 IVB	5 VB	6 VIB	7 VIIB	8 VIIIB	9 VIIIB	10 VIIIB	11 IB	12 IIB	13 IIIA	14 IVA	15 VA	16 VIA	17 VIIA	18 VIIIA
1	H																	He
2	Li	Be											B	C	N	O	F	Ne
3	Na	Mg											Al	Si	P	S	Cl	Ar
4	K	Ca	Sc	Ti	V	Cr	Mn	Fe	Co	Ni	Cu	Zn	Ga	Ge	As	Se	Br	Kr
5	Rb	Sr	Y	Zr	Nb	Mo	Tc	Ru	Rh	Pd	Ag	Cd	In	Sn	Sb	Te	I	Xe
6	Cs	Ba	La*	Hf	Ta	W	Re	Os	Ir	Pt	Au	Hg	Tl	Pb	Bi	Po	At	Rn
7	Fr	Ra	Ac**	Rf	Db	Sg	Bh	Hs	Mt	Uun	Uuu	Uub						

Inner-Transition Metals

*Lanthanides	Ce	Pr	Nd	Pm	Sm	Eu	Gd	Tb	Dy	Ho	Er	Tm	Yb	Lu
**Actinides	Th	Pa	U	Np	Pu	Am	Cm	Bk	Cf	Es	Fm	Md	No	Lr

Elements whose cations have only one charge

Elements whose cations have more than one charge

Nonmetals or metalloids; do not form cations

FIGURE 5.3

Periodic table showing the categories of cation charges. Shown here is the distribution of elements whose cations have only one charge and elements whose cations have more than one charge.

Chemists use specific rules to name binary ionic compounds. Here are the rules when the metallic element forms a single kind of cation.

Rules for Naming a Binary Ionic Compound When the Metal Forms a Single Cation

1. The cation is named for the metallic element from which it is obtained. Table 5.1 lists the names and charges of the cations of some of these metallic elements.
2. The anion name is derived from the element by retaining the root name and adding the suffix *-ide*. Some examples of such anions are given in Table 5.1.
3. The binary ionic compound is named by giving the name of the cation (omitting the word *ion*) followed by the name of the anion (again omitting the word *ion*).

To demonstrate how to apply these rules, let us name the compound whose formula is CaF_2 (Figure 5.4). This is a binary compound because it contains two elements. You would expect it to be ionic because one of the elements is a metal (Ca) and the other is a nonmetal (F). Because calcium is a main-group element (and not one of the exceptions; see Figure 5.3), it forms only one cation. You can use the rules just given to name

FIGURE 5.4
Calcium fluoride mineral. Mexico is rich in traditions and archaeological artifacts such as this Mayan pyramid. Mexico is also the world's leading source of fluorite, the mineral shown here. The principal component of fluorite is calcium fluoride (CaF_2).

this substance. Rule 1 tells you that the cation is named calcium ion, after the metallic element. Rule 2 tells you that the name of the anion is fluoride ion, which is derived from the element's name (fluorine) by retaining the root (*fluor*) and adding the suffix *-ide*. Finally, rule 3 tells you that the name of the compound is calcium fluoride. Note that the name of a binary ionic compound does not convey any information about the subscripts in the compound's formula. ●

● The name of an ionic compound does not need to include the subscripts because you can derive them from the ion formulas.

■ EXAMPLE 5.3 Naming a Binary Ionic Compound

Name each of the following compounds.
(a) KBr (b) Li_2O (c) AlI_3

Problem Analysis

Verify that you are dealing with a binary ionic compound and that it contains a metal that forms only one kind of cation. Then follow the rules for naming such binary ionic compounds. By rule 1, the cation is named for the metallic element from which it is derived. According to rule 2, the name of the anion is the root of the nonmetal name plus the suffix *-ide*. Rule 3 tells you that the name of the compound is the cation name followed by the anion name.

Solution

The compound KBr contains K^+ and Br^-; Li_2O contains Li^+ and O^{2-}; and AlI_3 contains Al^{3+} and I^-. Each of these substances is a binary ionic compound in which the metal forms a single cation (see Figure 5.3). The names of the compounds are (a) potassium bromide, (b) lithium oxide, and (c) aluminum iodide.

Exercise 5.3

Name each of the following compounds.
(a) K_2S (b) Li_3N (c) Al_2O_3

(Try Problems 5.21, 5.22, 5.23, and 5.24.)

So far we have named a compound using its formula. However, when you see the name of a compound in print, you should be able to deduce its formula. The process we use essentially reverses the steps given in Example 5.3. For instance, if you are given the name barium fluoride, you know that it contains barium ions (the cations) and fluoride ions (the anions). Because barium is a member of Group IIA in the periodic table, the charge on the ion is +2, and its formula is Ba^{2+}. The formula for the fluoride ion is F^- because the root (*fluor*) indicates that the element is fluorine (F) from Group VII, and therefore the charge is $7 - 8 = -1$. Once you know the formulas for the ions, you can write the formula of the compound by making the magnitude of the charge on one ion the subscript of the other (as you did in Example 5.2) to achieve a net charge of zero. You find that the formula is BaF_2.

■ EXAMPLE 5.4 Writing the Binary Ionic Formula from a Name

Give the formula for the binary ionic compound named potassium oxide.

Problem Analysis

Obtain the names of the ions from the name of the compound, and then write their corresponding formulas. The charge on the cation equals the group number. The charge on the anion equals the group number minus 8. Use the formulas for the ions to obtain the formula of the compound (as in Example 5.2).

Solution

The ions in potassium oxide are the potassium ion, K^+, and the oxide ion, O^{2-} (see Table 5.1). You obtain the compound's formula by using the magnitude of the charge on one ion as the subscript of the other ion:

$$K^{①+} \; O^{②-} = K_2O_1 \text{ or } K_2O$$

Exercise 5.4

Determine the formulas of barium sulfide and lithium chloride.

(Try Problems 5.25 and 5.26.)

5.3 Naming Binary Ionic Compounds When the Metal Forms Several Cations

OBJECTIVES

- Determine the charge on a metal ion, given the formula of a compound.
- Assign the Stock name to a binary ionic compound, given its formula when the metal can form several kinds of cations.
- Assign the classical name to a binary ionic compound, given its formula when the metal can form several kinds of cations.

Tin, lead, and the transition metals (B group elements) can form ions with two or more different charges (see Figure 5.3). The charge on an ion is important because ions of the same element with different charges have different physical and chemical properties.

Consider chromium. Figure 5.5 shows that Cr^{2+} is blue in aqueous solution, but Cr^{3+} under certain conditions is green. ●

● Cations of the main-group elements thallium, Tl, and bismuth, Bi, also have more than one possible charge.

The multiplicity of possible charges for the cations of these elements has an important consequence: These elements can form two or more binary ionic compounds with another element. For example, the transition-metal iron forms two compounds with oxygen: black FeO and reddish-brown Fe_2O_3, the principal component of rust. Each substance has unique properties. Yet, according to the rules from the preceding section, each would be named iron oxide. Clearly, we need a naming system that distinguishes between these compounds. Two alternative systems are used for naming compounds in which the cation can have one of several possible charges: the modern *Stock system* (preferred) and the *classical system.* They use different names for the cation, but otherwise the rules for naming binary ionic compounds (rules 2 and 3 in Section 5.2) remain the same.

We are going to name a compound given its formula. To obtain the name of the compound, though, you will need a way to obtain the charge on the cation from the compound's formula, because the Stock method and classical method use this charge in naming.

Determining the Charge on a Metal Ion

You obtain the subscripts in the formula of an ionic compound from the ion charges. To obtain the ion charges from the formula, you simply reverse this procedure. The subscripts in the compound's formula give you the smallest cation/anion charge ratio. For example, the formula FeO (with subscripts 1:1) tells you that the ratio of cation charge to anion charge is $+1:-1$. Therefore, the charges are some multiple of this (that is, $+1:-1$, or $+2:-2$, or higher multiple). Since you know that the oxide ion is O^{2-}, the cation must be Fe^{2+}.

Let's try this with Fe_2O_3. The ratio of subscripts is 2:3, so the ratio of charges of cation to anion is $3:-2$. Since you know that the oxide ion is O^{2-}, the cation must be Fe^{3+}.

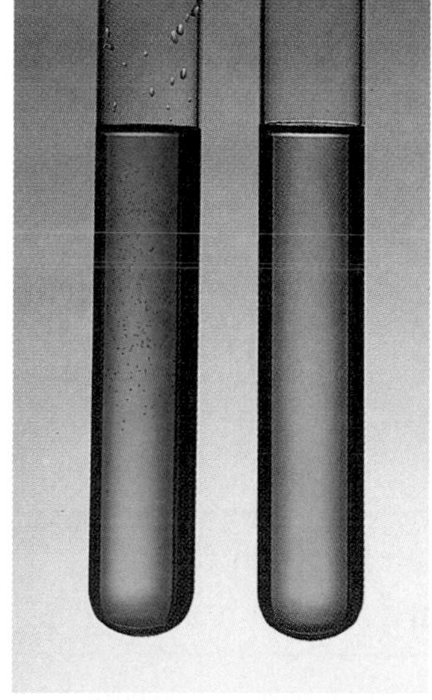

FIGURE 5.5
Colors of two chromium ions. Ions of the same element with different charges are often different colors in aqueous solution or in compounds. The solutions shown here contain Cr^{2+} *(left)* and Cr^{3+} *(right)*. Under other conditions, Cr^{3+} can be red-violet. Chromium is a transition metal.

The Stock System

In the Stock system of naming binary ionic compounds (named for Alfred Stock, who orginated the method), the magnitude of the positive charge on the cation is designated by a Roman numeral in parentheses immediately following the name of the metallic element. (Do not leave a space between the name of the metallic element and the first parenthesis.) The name of Fe^{2+} is iron(II) ion; the name of Fe^{3+} is iron(III) ion. Stock names for some positive ions are given in Table 5.2.

The naming of any compound by the Stock system follows the rules given in Section 5.2, except that the cation name is its Stock name. The name for FeO (containing Fe^{2+}, as we have seen) is iron(II) oxide, and the name for Fe_2O_3 (containing Fe^{3+}) is iron(III) oxide.The Stock method is especially valuable because it can be used for naming compounds in which the metallic element has three or more cations. Here are some manganese oxides: MnO (containing Mn^{2+} ions), Mn_2O_3 (containing Mn^{3+} ions), and MnO_2 (containing Mn^{4+} ions). Their Stock names are manganese(II) oxide, manganese(III) oxide, and manganese(IV) oxide, respectively.

You do not normally use the Stock system of naming when a pair of elements forms only one binary ionic compound. For example, you would name the binary compound

TABLE 5.2
Stock and Classical Names for Some Cations

Cation Symbol	Name According to Stock System	Classical System
Fe^{2+}	Iron(II) ion	Ferrous ion
Fe^{3+}	Iron(III) ion	Ferric ion
Co^{2+}	Cobalt(II) ion	Cobaltous ion
Co^{3+}	Cobalt(III) ion	Cobaltic ion
Cu^{+}	Copper(I) ion	Cuprous ion
Cu^{2+}	Copper(II) ion	Cupric ion
Sn^{2+}	Tin(II) ion	Stannous ion
Sn^{4+}	Tin(IV) ion	Stannic ion
Pb^{2+}	Lead(II) ion	Plumbous ion
Pb^{4+}	Lead(IV) ion	Plumbic ion

of sodium and chlorine as sodium chloride, rather than sodium(I) chloride, because only one compound is possible.

The Classical System

An older system of naming, the classical system, is also used, but is falling into disfavor because of its shortcomings. In this method, the names of the two most common ions derived from a metallic element share a common root but are distinguished by the suffix *-ous* for the cation with the lower charge and *-ic* for the cation with the higher charge. Latin roots are used if the symbol for the element is derived from a Latin word. For example, the symbol for iron is Fe because it is derived from the Latin word *ferrum*. The root that we get from this word is *ferr-*. The classical names for Fe^{2+} and Fe^{3+} are ferrous ion and ferric ion, and the names for FeO and Fe_2O_3 are ferrous oxide and ferric oxide, respectively. Some classical names for ions are shown in Table 5.2.

The main shortcoming of the classical method is its inability to name compounds when the metal has three or more cations. For example, this method cannot be used to name MnO, Mn_2O_3, and MnO_2. Another shortcoming is its inability to give the precise charge on the cation.

■ EXAMPLE 5.5 Naming a Binary Ionic Compound When the Metal Can Have Several Charges

Provide names for CuCl and $CuCl_2$ according to (a) the Stock method and (b) the classical method. (The root name for copper is *cupr-* from the Latin *cuprum*.)

Problem Analysis

Determine the charge on the metal ion in the compound. In the Stock method, convert the magnitude of that charge to a Roman numeral and insert it within parentheses after the cation name. In the classical method, add *-ous* or *-ic* to the root name of the cation, depending on whether the cation charge is the lower or higher one. In both methods, the compound's name is the cation name followed by the anion name.

Solution

You must know the charges on the copper ions regardless of which method you use. For CuCl, the subscript ratio is 1:1, so the ratio of cation charge to anion charge is 1:−1. Because the chloride ion charge is −1, the charge on the copper ion is +1. The ion is Cu^+. Turning to $CuCl_2$, you see that the subscript ratio is 1:2, so the ratio of cation charge to anion charge is 2:−1. Because the chloride ion charge is −1, the charge on the copper ion is +2. The ion is Cu^{2+}.

(a) The names of these cations by the Stock method are

Cu^+: copper(I) ion

Cu^{2+}: copper(II) ion

You can verify these names by looking at Table 5.2. The names of the compounds become

CuCl: copper(I) chloride

$CuCl_2$: copper(II) chloride

(b) According to the classical method, you obtain the name of the ion with the lower charge (Cu^+) by adding *-ous* to the root *cupr-*. Thus, the name is cuprous ion. Similarly, the name of the ion with the higher charge (Cu^{2+}) is cupric ion. These names can be found in Table 5.2. The names of the compounds become

CuCl: cuprous chloride

$CuCl_2$: cupric chloride

Exercise 5.5

Using both the Stock system and the classical system, give names to the compounds whose formulas are SnO and SnO_2. (The root name is *stann-* for *stannum,* the Latin name for tin.)

(Try Problems 5.27, 5.28, 5.29, and 5.30.)

5.4 Compounds with Polyatomic Ions

OBJECTIVES

- State the names and formulas of the polyatomic ions given in Table 5.3.
- Write formulas for compounds containing polyatomic ions.
- Name compounds containing polyatomic ions.

Lithium carbonate, an ionic compound used to treat people suffering from manic depression, is composed of the cation Li^+ and the anion CO_3^{2-}. You can see from its formula, Li_2CO_3, that this is not a binary ionic compound; it contains three elements rather than two. Lithium carbonate is a compound containing a polyatomic ion (CO_3^{2-}). A *polyatomic ion* contains a group of chemically bonded atoms that function as a single unit (see Section 4.8). These atoms have an excess or deficiency of electrons, so that the entire unit has an electric charge. Other examples of polyatomic ions are the hydroxide ion, OH^-, and the ammonium ion, NH_4^+. We discuss the formulas and names of compounds containing polyatomic ions in this section.

Table 5.3 lists some common polyatomic ions. Among the polyatomic ions are the oxyanions. An **oxyanion** is *a negatively charged polyatomic ion that contains an atom*

TABLE 5.3
Names of Some Common Polyatomic Ions

Ion Formula	Name	Ion Formula	Name
NH_4^+	Ammonium ion	HCO_3^-	Hydrogen carbonate ion†
OH^-	Hydroxide ion		
CN^-	Cyanide ion	PO_4^{3-}	Phosphate ion
MnO_4^-	Permanganate ion	HPO_4^{2-}	Hydrogen phosphate ion
NO_2^-	Nitrite ion		
NO_3^-	Nitrate ion	$H_2PO_4^-$	Dihydrogen phosphate ion
SO_3^{2-}	Sulfite ion	ClO^-	Hypochlorite ion
SO_4^{2-}	Sulfate ion	ClO_2^-	Chlorite ion
HSO_4^-	Hydrogen sulfate ion*	ClO_3^-	Chlorate ion
CO_3^{2-}	Carbonate ion	ClO_4^-	Perchlorate ion

* Sometimes called bisulfate ion
† Sometimes called bicarbonate ion

of some element plus one or more oxygen atoms. An example is the carbonate ion, CO_3^{2-}, in the lithium drug. It is an oxyanion because its carbon atom is chemically bonded to three oxygen atoms. ●

● The names of most polyatomic anions end in *-ate* or *-ite*. There are two prominent exceptions, the cyanide ion (CN^-) and the hydroxide ion (OH^-), which are named as if they were monatomic ions.

The table contains several series of oxyanions, each series consisting of a particular element with different numbers of oxygen atoms. The oxyanions of nitrogen, NO_2^- and NO_3^-, are such a series. If there are only two oxyanions in a series, the name of the one with the fewer number of oxygen atoms ends in *-ite,* and the name of the one with the greater number ends in *-ate.* Thus, NO_2^- is the nitrite ion and NO_3^- is the nitrate ion.

If there are more than two oxyanions in a series, the names include the prefix *per-* for the oxyanion with the greatest number of oxygen atoms and the prefix *hypo-* for the oxyanion with the fewest oxygen atoms. When the prefix is *per-,* the suffix is always *-ate,* and when the prefix is *hypo-,* the suffix is always *-ite.* Note the oxyanions of chlorine:

ClO^- : ***hypo***chlor***ite*** ion
ClO_2^-: chlor***ite*** ion
ClO_3^-: chlor***ate*** ion
ClO_4^-: ***per***chlor***ate*** ion

You must memorize the formulas and names of the polyatomic ions listed in Table 5.3 in order to be able to write the formulas and names of compounds containing these ions. The rules we just discussed for naming these ions will help you do this. (Your instructor may add to or subtract from this list.)

Formulas for compounds containing polyatomic ions are obtained in the same manner as those for binary ionic compounds, where we used the fact that a compound cannot have a net charge. Consider the compound consisting of lithium ions, Li^+, and carbonate ions, CO_3^{2-}. The −2 charge of a carbonate ion requires two lithium ions, each bearing a charge of +1, to achieve a net charge of zero:

$$Li^{(1)+} \quad CO_3^{(2)-} = Li_2(CO_3)_1 \text{ or } Li_2CO_3$$

The formula we write is Li_2CO_3. In general, you will need to place parentheses around the formula of the polyatomic ion before adding its subscript. Here, we removed the parentheses when we dropped the subscript 1.

As another example, consider the compound containing Ca^{2+} and HCO_3^- ions. Two HCO_3^- ions are required for each Ca^{2+} ion so that the net charge of the compound is zero:

$$Ca^{2+} \quad HCO_3^{1-} = Ca_1(HCO_3)_2 \text{ or } Ca(HCO_3)_2$$

The formula is $Ca(HCO_3)_2$. Note that parentheses are needed because the subscript 2 applies to the entire polyatomic unit. ●

● Both calcium ion (Ca^{2+}) and hydrogen carbonate ion (HCO_3^-) are found in drinking water. Calcium ion and some other metal ions in water are responsible for its "hardness," in which the metal ion combines with soap to produce a curdy scum.

The naming of compounds containing polyatomic ions is similar to that for binary ionic compounds. *You write the name of the cation followed by the name of the anion.* The substance whose formula is Li_2CO_3 contains lithium ions (the cations) and carbonate ions (the anions). Therefore, the name of the compound is lithium carbonate. Similarly, the compound whose formula is $Ca(HCO_3)_2$ is called calcium hydrogen carbonate because it contains calcium cations and hydrogen carbonate anions, HCO_3^- (see Table 5.3).

The next example further illustrates how to determine the formula and name of an ionic compound containing a polyatomic ion.

EXAMPLE 5.6 Determining the Formula and Name of a Compound Containing a Polyatomic Ion

Give the formula and name of a compound containing Mg^{2+} and OH^- ions. This compound is used as an antacid in milk of magnesia.

Problem Analysis

To obtain the formula, make sure that the net charge on the compound is zero by taking the magnitude of the charge of one ion as the subscript of the other. To obtain the name of the compound, give the name of the cation (minus the word *ion*) first, followed by the name of the polyatomic ion (minus the word *ion*). If you do not know the name of the polyatomic ion, look it up in Table 5.3.

Solution

Because the net charge on this compound must be zero, you can see that every Mg^{2+} ion requires two OH^- ions:

$$Mg^{2+} \quad OH^{1-} = Mg_1(OH)_2 \text{ or } Mg(OH)_2$$

Thus, the formula is $Mg(OH)_2$. Parentheses enclose the entire formula of the polyatomic ion because the subscript 2 applies to the whole ion. To name this compound, you note that the name of the cation is magnesium ion and the name of the anion is hydroxide ion (see Table 5.3). The name of the compound is magnesium hydroxide.

Exercise 5.6

Provide formulas and names for compounds containing the following ions.
(a) K^+ and SO_3^{2-} (b) Ba^{2+} and ClO^- (c) NH_4^+ and CN^-

(Try Problems 5.31, 5.32, 5.33, and 5.34.)

It is also important to be able to write the formula for a compound containing polyatomic ions from its name. For example, if you are given the name potassium nitrate, you know that this compound is composed of potassium ions (K^+) and nitrate ions (NO_3^-). You also know that the formula is KNO_3 because the number of K^+ ions must equal the number of NO_3^- ions so that the net charge is zero. The next example will give you more practice.

FIGURE 5.6

Three names for the same substance. The systematic name of the substance shown in various containers is sodium hydrogen carbonate. In an older naming scheme this substance was called sodium bicarbonate, which is still a very common name. On grocery shelves, it goes by the name baking soda.

■ EXAMPLE 5.7 Writing the Formula from the Name of a Compound Containing a Polyatomic Ion

Give the formula for sodium hydrogen carbonate (Figure 5.6).

Problem Analysis

Identify the ions in the compound. If you have memorized the names and formulas in Table 5.3, this task will be easier. Once you know the ions in the compound, you can use their charges and the fact that the net charge on the compound must be zero to find the subscripts in the formula.

Solution

The ions in sodium hydrogen carbonate are Na^+ and HCO_3^-. Because the charges on these ions are +1 and −1, respectively, each Na^+ ion requires one HCO_3^- ion so that the net charge is zero:

$$Na^{①+} \; HCO_3^{①-} = Na_1(HCO_3)_1 \text{ or } NaHCO_3$$

Thus, the formula is $NaHCO_3$.

Exercise 5.7

Give the formula for barium phosphate.

(Try Problems 5.35 and 5.36.)

MOLECULAR COMPOUNDS

Molecular compounds, discussed briefly in Chapter 4, usually consist of chemically bonded nonmetal atoms in units called *molecules,* rather than being composed of ions. The following section describes how binary molecular compounds are named. A **binary molecular compound** is *a molecular compound composed of only two elements.*

5.5 Binary Molecular Compounds

OBJECTIVE

- Name a binary molecular compound using Greek prefixes.

Water, H_2O, and ammonia, NH_3, are examples of binary molecular compounds (Figure 5.7). Each consists of two nonmetals. Although we have given them common names, there is a system for naming this type of compound. For simplicity, chemists want this system to be similar to the one used for binary ionic compounds, but two problems emerge. First, we need a way to decide which element is to be written first in the formula and in the name. With ionic compounds, the metal is written first, but in a binary mole-

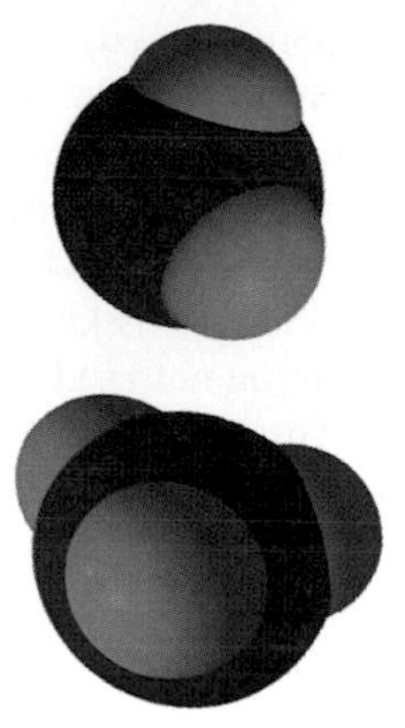

FIGURE 5.7
Water and ammonia molecules. Molecular models of water *(top)* and ammonia *(bottom)*. The hydrogen atoms are shown in light blue, whereas oxygen is red and nitrogen is dark blue.

cular compound both elements are nonmetals. Second, we want the name to suggest the complete formula (the subscripts as well as the elements) because nonmetals often form several compounds between the same two elements. With ionic compounds, you can write the formula as long as you know the charges on the ions; but there are no ions in a binary molecular compound. Let's see how we handle these problems.

We deal with the first problem this way. We write the nonmetal with more metallic character first, paralleling the naming of ionic compounds. Roughly, metallic character increases as the group number of the element in the periodic table decreases, and it increases within a group as we go down the column. The actual sequence we use for the order is as follows:

← Increasing metallic character

Element	Si C	P N	H	Se S	I Br Cl	O	F
Group	IVA	VA		VIA	VIIA		

Decreasing metallic character →

You can reproduce this sequence easily. From a periodic table, write down just the nonmetals starting at the bottom of the leftmost group (Group IVA). (The metalloids act chemically like the nonmetals, so you can include them, too; we have included silicon in our sequence because its compounds are relatively common, but we have omitted the others.) Go upward in the column, then move to the next column on the right until you have gotten through Group VIIA.

Let's see how you use this sequence. Consider a binary compound of nitrogen and oxygen. The symbol for nitrogen appears to the left of the symbol for oxygen in this sequence, so you would write N before O in the formula and in the name of a binary molecular compound of nitrogen and oxygen.

In naming a binary molecular compound, you name the second element by adding *-ide* to its root name, just as you did with binary ionic compounds. In other words, it is named as if it were an anion.Thus, a compound containing nitrogen and oxygen would be named as a nitrogen ox*ide*.

To deal with the second problem, we use Greek prefixes to denote the number of atoms of each element in the molecule (corresponding to the subscripts in the formula). In this Greek-prefix system, *mono-* means "one," *di-* means "two," *tri-* means "three," and so forth. Table 5.4 lists the prefixes up to eight.

Consider a molecule consisting of two atoms of nitrogen and three atoms of oxygen. As you have seen, nitrogen (N) is listed before oxygen (O), so the formula is N_2O_3. The compound is a nitrogen oxide. You specify the number of atoms of each element by adding the appropriate prefix before the name of each element. The complete name is dinitrogen trioxide.

The prefix *mono-* for the first-named element is generally omitted. Thus, NO is named nitrogen monoxide, rather than mononitrogen monoxide. Note also that, for ease

TABLE 5.4
Prefixes for Numbers

Prefix	Number	Prefix	Number
Mono	1	Penta	5
Di	2	Hexa	6
Tri	3	Hepta	7
Tetra	4	Octa	8

of pronunciation, the letter *o* is not repeated: we say monoxide, not mono-oxide. Similarly, N_2O_5 is named dinitrogen pentoxide, rather than dinitrogen pentaoxide. See the Chemical Perspective at the end of this section for recent discoveries about the biological importance of NO (as well as a comment about its common name, nitric oxide).

There are a few exceptions to these rules. HCl is called hydrogen chloride, not hydrogen monochloride. Only one binary compound of hydrogen and chlorine is known, so prefixes are not needed to distinguish compounds and are by tradition not used in such cases. The other binary compounds of hydrogen with halogens (Group VIIA elements) are similarly named without prefixes. The binary compounds of hydrogen with the Group VIA elements H_2S (hydrogen sulfide), H_2Se (hydrogen selenide), and H_2Te (hydrogen telluride) are by tradition named without prefixes. Thus, H_2S is called hydrogen sulfide. Although H_2S was for a long time the only binary compound of hydrogen and sulfur, others have since been discovered. Thus, you would not be wrong in giving the name of H_2S as dihydrogen sulfide.

The following list summarizes the rules for naming binary molecular compounds.

Rules for Naming Binary Molecular Compounds

1. The element with more metallic character is named first, using the full name of the element. More precisely, you use the sequence given in the text to establish the first-named element.
2. The second element is named as if it were an anion.
3. The number of atoms of each element in the molecule is designated by a Greek prefix (Table 5.4). The prefix *mono-* is usually not used for the first element.
4. For ease in pronunciation, the final *a* or *o* of a prefix may be dropped when combined with the name of an element that begins with a vowel.

EXAMPLE 5.8 Naming a Binary Molecular Compound

Give the formula and systematic name for a compound whose molecule contains three atoms of chlorine and one atom of phosphorus.

Problem Analysis

Determine the order of the elements in the formula and name (using the sequence described in the text). Subscripts in the formula denote the number of atoms of each element. In the name, you list the elements in the compound in the same order as in the formula, naming the second element as if it were an anion. Then, you add prefixes to the element names corresponding to the subscripts in the formula.

Solution

Following the sequence of elements, P is given before Cl in the formula and the name. The formula PCl_3 indicates 1 atom of P (no subscript) and 3 atoms of Cl (subscript 3). To name the compound, you name phosphorus first, then chlorine, as in the formula. You name the chlorine as if it were an anion: chloride. Finally, you indicate the number of chlorine atoms by the prefix *tri-*. Note that you do not use the prefix *mono-* with the first-named element (phosphorus). The name is phosphorus trichloride.

Exercise 5.8

Give the formula and systematic name for each compound whose molecule contains:

(a) two atoms of phosphorus and three atoms of oxygen
(b) three atoms of fluorine and one atom of chlorine
(c) four atoms of chlorine and one atom of carbon

(Try Problems 5.37, 5.38, 5.39, and 5.40.)

Chemical Perspective

NO—A Molecular Messenger

Molecules of the human body are generally composed of many hundreds, and often thousands, of atoms. Imagine the surprise of biochemists (biological chemists) a few years ago when they discovered that the molecule NO, composed of just two atoms, has the role of a molecular messenger, carrying a message from one cell to others. For instance, they discovered that NO plays some role in nerve impulse transmission by acting as a messenger between nerve cells. They also discovered that certain immune cells of the body release NO molecules, which then move as a chemical cloud to surround tumor cells, thereby interfering with their ability to produce energy. In this case, NO is more a warrior than a messenger. Since this discovery, researchers have published more than 10,000 papers on the various roles of NO in fruit flies, horseshoe crabs, and other animals.

Let's pause briefly to note the naming of this compound. From the text, you know that the systematic name of NO is nitrogen monoxide. Often, though, this molecule goes by its common name of nitric oxide, from an earlier systematic name. The *-ic* ending in nitric means that it has a greater proportion of oxygen (1 O for each N) than does nitr*ous* oxide, N_2O, another nitrogen oxide (which has ½ O for each N). Today, however, seven nitrogen oxides (N_2O, NO, N_2O_3, NO_2, N_2O_4, N_2O_5, and NO_3) are known! So, the inadequate *-ous*/*-ic* system was supplanted by the prefix system. Try naming the other nitrogen oxides using prefixes.

Nitric oxide has also been found to play a role in the regulation of blood pressure. The cells lining the interior of blood vessels release NO when stimulated by various substances in blood. Each cell releases not a cloud, but a small puff of NO. When the puff of NO reaches the muscle tissue in the blood vessel, the tissue relaxes. The blood vessel expands, or dilates, which lowers the blood pressure. Each puff of NO carries the message *relax, relax*.

This discovery of nitric oxide's role in blood pressure regulation has solved an age-old puzzle. Nitroglycerin, the substance used in dynamite, has been used for more than a century to prevent or treat certain heart attacks (angina pectoris). An angina attack occurs when a blood vessel to the heart is partly blocked, perhaps by a blood clot or by a fatty deposit. But how does nitroglycerin work? Now we know. Nitroglycerin decomposes slowly in the body to yield nitric oxide, which dilates the blood vessels, allowing more blood to flow. Figure 5.8 shows a patient with a nitroglycerin skin patch.

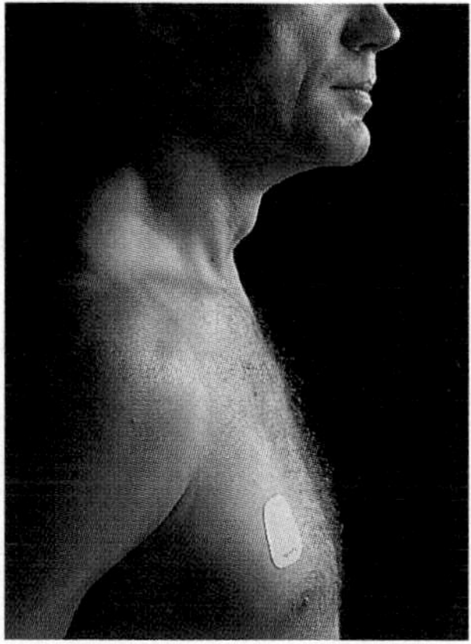

FIGURE 5.8
Nitroglycerin patch. Nitroglycerin has been used for more than a hundred years to treat and prevent angina attacks. Here a patient wears a patch that dispenses nitroglycerin steadily over a period of time.

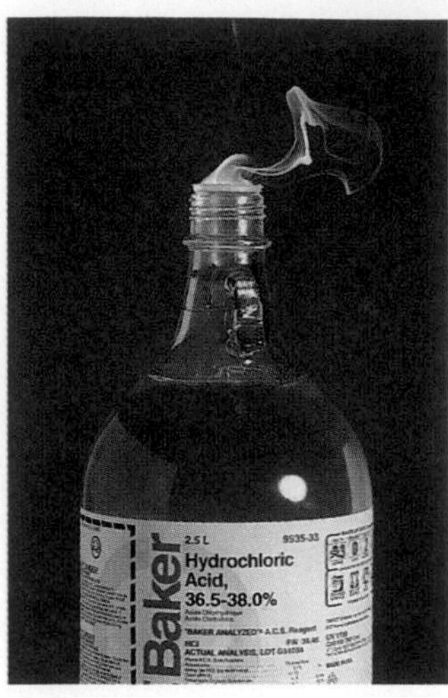

FIGURE 5.9
Hydrochloric acid. This acid is frequently available in chemical laboratories. It is also sold in hardware stores as muriatic acid for cleaning mortar from brick. Note the fumes above the bottle. Hydrogen chloride gas evolves from the concentrated solution, where it combines with moisture in the air to produce this characteristic fuming.

ACIDS

You have certainly heard about stomach acid, acid rain, and battery acid. The word *acid* comes from the Latin *acidus,* meaning "sour," since acids in water have a sour, or tart, taste. Limes, lemons, and grapefruit have a sour taste because of citric acid, and vinegar has its sour taste because of acetic acid. The sour taste is due to hydrogen ion (H^+). Thus, an **acid** is *a compound that produces hydrogen ion, H^+, when the compound is dissolved in water.* Although acids are discussed in depth in Chapter 15, the next two sections deal with naming these important compounds.

5.6 Naming Binary Acids

OBJECTIVE

- Name the acid solution corresponding to a binary compound of hydrogen with a nonmetal.

The substance HCl is a binary molecular compound called hydrogen chloride. When you dissolve hydrogen chloride gas in water, it forms a solution we call hydrochloric acid, which we write as HCl(*aq*) (Figure 5.9). Your digestion is aided by the presence of this acid in your stomach. Hydrochloric acid is an example of a **binary acid,** *an acid solution that forms when you dissolve a binary molecular compound of hydrogen and another nonmetallic element in water.*

You name a binary acid by adding the prefix *hydro-* and the suffix *-ic* to the root name of the nonmetal, followed by the word *acid.* Thus, the name of the acid formed from hydrogen chloride, HCl, is hydro + chlor + ic + acid.

■ EXAMPLE 5.9 Naming a Binary Acid

When you dissolve the binary compound HF in water, it forms an acid used to etch glass. Give the name for this binary acid.

Problem Analysis

Identify the nonmetal other than hydrogen in the binary compound. Add the prefix *hydro-* and the suffix *-ic* to the root name of the nonmetal, and follow it by the word *acid.*

Solution

The nonmetal is fluorine. The name of the binary acid is hydro + fluor + ic + acid, or hydrofluoric acid.

Exercise 5.9

Give the name for HBr(aq).

(Try Problems 5.43 and 5.44.)

5.7 Naming Oxyacids

OBJECTIVE

- Write the name and formula of the oxyacid corresponding to a given oxyanion.

The oxyacids are related to the oxyanions, which we described in Section 5.4. An **oxyacid** is *a molecular substance containing hydrogen, oxygen, and another element that*

TABLE 5.5
Some Oxyacids

Acid Formula	Name	Acid Formula	Name
$HC_2H_3O_2$*	Acetic acid	H_2CO_3	Carbonic acid
HNO_2	Nitrous acid	$HClO$	Hypochlorous acid
HNO_3	Nitric acid	$HClO_2$	Chlorous acid
H_2SO_3	Sulfurous acid	$HClO_3$	Chloric acid
H_2SO_4	Sulfuric acid	$HClO_4$	Perchloric acid
H_3PO_4	Phosphoric acid	H_3BO_3	Boric acid

* Only the first hydrogen atom in the formula is acidic.

when added to water yields hydrogen ion (H^+) and the corresponding oxyanion. For example, nitric acid (whose formula is HNO_3) dissolves in water to give H^+ ion and NO_3^- ion (nitrate ion). Although the pure substance HNO_3 consists of molecules, it yields ions as if it were an ionic compound when it dissolves in water.

You can obtain the formula for an oxyacid from the formula of the corresponding oxyanion. Imagine that you form the acid from H^+ ions and the oxyanion. Because the acid is a compound and therefore electrically neutral, you need to add as many H^+ ions as necessary to counter the negative charge on the oxyanion. Thus, the magnitude of the charge on the oxyanion gives you the subscript for hydrogen in the formula for the oxyacid. (This is the method we used in Example 5.2 to obtain the formula of an ionic substance from its ions.) For example, the formula for the oxyacid corresponding to the oxyanion PO_4^{3-} is H_3PO_4.

$$H^{①+} \; PO_4^{③-} = H_3(PO_4)_1 \text{ or } H_3PO_4$$

(Note that we write the H atoms at the beginning of the formula for the acid.)

You can derive the name of an oxyacid from its corresponding oxyanion name. If the oxyanion name ends in *-ate,* change this to *-ic* and substitute the word *acid* for the word *ion*. If the oxyanion name ends in *-ite,* change this to *-ous* and substitute the word *acid* for the word *ion*. Thus, the acid corresponding to the carbonate ion is carbonic acid. Similarly, the acid corresponding to the nitrite ion is nitrous acid. The formulas and names of some oxyacids are given in Table 5.5.

■ EXAMPLE 5.10 Determining the Formula for an Oxyacid and Naming It

Give the formula and name of the oxyacid derived from the sulfate ion. This acid is found in car batteries.

Problem Analysis

To determine the formula for an oxyacid, take the subscript for hydrogen from the magnitude of the charge on the polyatomic anion. (The formulas and charges of some polyatomic anions are listed in Table 5.3.) To name the oxyacid, change the

-ate ending of an oxyanion to *-ic* or the *-ite* ending to *-ous*, and substitute the word *acid* for the word *ion*.

Solution

The formula for the sulfate ion is SO_4^{2-}. Since its charge is -2, the subscript to H in the oxyacid formula is 2. The result is H_2SO_4. To name this compound, change the *-ate* ending in the name of the oxyanion to *-ic* and then substitute the word *acid* for the word *ion*. Thus, the name is sulfuric acid, as shown in Table 5.5.

Exercise 5.10

Give the formula and name of the oxyacid corresponding to the sulfite ion.

(Try Problems 5.45, 5.46, 5.47, and 5.48.)

Chemical Perspective

Beetles, Antiseptics, and Bleaches

The bombardier beetle ensures its survival with a marvelous defense mechanism (Figure 5.10). Unlike the passive defense of the chameleon, who changes its color to match its surroundings, this insect uses chemical warfare against its predators. When it is threatened, it forces a fluid containing hydrogen peroxide and another compound called hydroquinone into an abdominal sac containing enzymes. (Enzymes are "biological catalysts"—that is, biological substances that cause a chemical reaction to speed up without being consumed in the reaction.) The enzymes cause a vigorous and quick reaction between hydrogen peroxide and hydroquinone with the release of a great deal of heat—so much heat that the liquid boils. The beetle discharges the hot liquid and vapor as a fine mist toward the soon-to-be-surprised predator.

Hydrogen peroxide, one of the substances used by the beetle, is a very reactive compound. We use it as an antiseptic because it kills harmful microorganisms. We also use it to bleach hair because it reacts with natural pigments and destroys them. Although it is toxic, it is produced in our bodies through biological processes involving oxygen from the air we breathe. Fortunately, we have a biological safeguard that prevents the internal buildup of this substance. Once again, enzymes play an important role. This time they speed up the natural decomposition of hydrogen peroxide, producing water and oxygen gas. One of these enzymes is *catalase*. The effect of catalase from beef liver on hydrogen peroxide is shown in Figure 5.11. The froth you see is caused by bubbles of oxygen from the decomposing hydrogen peroxide.

FIGURE 5.10
A bombardier beetle defending itself. The reaction between hydrogen peroxide and hydroquinone in the presence of a catalyst generates enough heat to boil the reaction liquid. The beetle expels this steaming liquid toward a foe.

Water, the other product of the reaction in Figure 5.11, and hydrogen peroxide happen to share one feature: Both are binary molecular compounds of the same elements. They are different compounds, however, because the proportions of these elements differ.

H–O–H (bent) — H_2O — Water

H–O–O–H — H_2O_2 — Hydrogen peroxide

FIGURE 5.11

The decomposition of hydrogen peroxide. When pureed beef liver is poured into a solution of hydrogen peroxide in water, the oxygen evolved causes foam.

Compare the formulas carefully, and you will see that a molecule of hydrogen peroxide has one more oxygen atom than a molecule of water. One extra oxygen atom makes a world of difference! The bombardier beetle cannot use water to begin its defense mechanism, nor do we use it as an antiseptic or bleach. Clearly, the properties of a compound depend strongly on the relative proportions of the elements in it.

CHAPTER REVIEW

Key Words

binary ionic compound *(p. 121)*
oxyanion *(p. 131)*
binary molecular compound *(p.134)*
acid *(p. 138)*
binary acid *(p. 138)*
oxyacid *(p. 138)*

Summary

The formulas and systematic names of chemical substances are based on composition and structure. We write the *formula of an ionic compound* by writing the formula of the cation followed by the formula of the anion (omitting ion charges). The subscripts in the formula of the compound are determined from the fact that a compound can have no net charge. The result is that the magnitude of the charge on one ion becomes the subscript for the other.

To write the formula of a *binary ionic compound,* you need to be able to predict the charge on a monatomic ion. The charge on the monatomic ion of a main-group element usually equals the group number if it is a cation or the group number minus 8 if it is an anion. The charges on transition-metal ions are not easily predictable.

You name a binary ionic compound composed of a metal that forms only one kind of cation as follows. You name the cation after the metal and follow this by the anion name, which consists of the root name of the nonmetal combined with the suffix *-ide*. If the metal can form cations of different charge, you can use the *Stock method* to name the compound.

In this case, the charge on the cation is written as a Roman numeral within parentheses after the metal name. The *classical method* of naming such compounds uses a root name for the metal to which you add a suffix, either *-ous* or *-ic*, depending on whether this is the lower or higher charged cation.

Many ionic compounds contain *polyatomic ions*, many of which are *oxyanions*. These consist of some element combined with oxygen. Most of these elements form a series of two or more oxyanions. If the series consists of two ions, we name the ion containing the fewer number of oxygen atoms from the nonmetal root with the suffix *-ous;* the other ion we name from the nonmetal root with the suffix *-ic*. If the series contains more than two ions, we add prefixes *per-* for the greatest number of oxygens and *hypo-* for the fewest number of oxygens. To write formulas and name compounds of the common polyatomic anions, you will need to memorize the formulas and names of these ions.

A *binary molecular compound* is a compound of two nonmetals and is named very much like a simple binary ionic compound. However, the order of the elements in the formula and name of the compound is the order in the sequence described in the text. Also, you add a Greek prefix to each element name to specify the number of its atoms in the molecule, except that the prefix *mono-* is not given for the first-named element.

An acid is a compound that yields a hydrogen ion in a water solution; it is this ion that gives an acid like the citric acid of lemon juice a sour taste. A *binary acid* is a water solution of the binary compound of hydrogen and another nonmetal. You name such acid solutions by adding the prefix *hydro-* to the root name of the other nonmetal and appending the suffix *-ide*. The *oxyacids* are related to the oxyanions. You name an oxyacid from its corresponding oxyanion. If the anion suffix is *-ate*, you change it to *-ic;* if it is *-ite*, you change it to *-ous*. Then, you add the word *acid*.

Problem-Solving Skills

1. **Predicting the Charge on a Monatomic Ion:** Given a main-group element, predict the charge expected on a monatomic ion (Example 5.1).
2. **Determining the Formula for a Binary Ionic Compound:** Given a monatomic cation and a monatomic anion, write the formula for the binary ionic compound (Example 5.2).
3. **Naming a Binary Ionic Compound:** Given the formula for a binary ionic compound in which the metal ion can have only one charge, deduce its systematic name (Example 5.3).
4. **Writing the Binary Ionic Formula from a Name:** Given the name of a binary ionic compound, write its formula (Example 5.4).
5. **Naming a Binary Ionic Compound When the Metal Can Have Several Charges:** Given a formula for a binary ionic compound in which the metal ion can have several charges, use the Stock method and the classical method to name the compound (Example 5.5).
6. **Determining the Formula and Name of a Compound Containing a Polyatomic Ion:** Given two ions, a cation and a polyatomic anion, determine the formula and name of the compound (Example 5.6).
7. **Writing the Formula from the Name of a Compound Containing a Polyatomic Ion:** Given the name for a compound containing a polyatomic ion, write the formula for the compound (Example 5.7).
8. **Naming a Binary Molecular Compound:** Given the formula of a binary molecular compound, deduce its systematic name (Example 5.8).
9. **Naming a Binary Acid:** Given the formula for a binary acid, deduce its name (Example 5.9).
10. **Determining the Formula for an Oxyacid and Naming It:** Given an oxyanion, deduce the formula and name of the corresponding oxyacid (Example 5.10).

Questions to Test Your Reading

5.1 Define binary ionic compound. Give an example.

5.2 What ion is given first when naming a binary ionic compound?

5.3 What are the names of the ions derived from the elements sodium, potassium, magnesium, and calcium?

5.4 What are the names of the ions derived from the elements fluorine, chlorine, bromine, and iodine?

5.5 What are the names of the Cu^+ and Cu^{2+} ions according to the Stock method? What are their classical names?

5.6 What are the names of the Pb^{2+} and Pb^{4+} ions according to the Stock method? What are their classical names?

5.7 Define *polyatomic ion* and *oxyanion*. Is a polyatomic ion an oxyanion? If your answer is no, give an example of a polyatomic ion that is not an oxyanion and an example of one that is an oxyanion.

5.8 Define *binary molecular compound.* How does a binary molecular compound differ from a binary ionic compound?

5.9 Define *acid* and *binary acid.* Give an example of a binary acid.

5.10 Define *oxyacid.* Give an example.

Practice Problems

Binary Ionic Compounds (Sections 5.1 and 5.2)

5.11 What charge is expected for the monatomic ions of each of the following elements?
(a) lithium (b) bromine (c) nitrogen
(d) magnesium (e) calcium

5.12 What charge is expected for the monatomic ions of each of the following elements?
(a) fluorine (b) radium (c) selenium
(d) gallium (e) potassium

5.13 Sodium carbonate contains sodium ions and carbonate ions (CO_3^{2-}). Its formula is Na_2CO_3. How many of each kind of ion are in each formula unit of the compound?

5.14 Aluminum sulfate contains aluminum ions and sulfate ions (SO_4^{2-}). Its formula is $Al_2(SO_4)_3$. How many of each kind of ion are in each formula unit of the compound?

5.15 A compound consisting of aluminum ions and oxide ions is used to manufacture synthetic rubies. What is the formula for this compound?

5.16 When phosphorus is heated with calcium, the product is a compound that contains calcium ions and phosphide ions. What is the formula for this compound?

5.17 Determine the formulas for the compounds containing the following ions.
(a) Mg^{2+} and O^{2-} (b) Ca^{2+} and Cl^-
(c) Na^+ and O^{2-} (d) Al^{3+} and F^-
(e) Cs^+ and Br^- (f) Zn^{2+} and N^{3-}

5.18 Determine the formulas for the compounds containing the following ions.
(a) Cd^{2+} and S^{2-} (b) K^+ and F^-
(c) Ba^{2+} and I^- (d) Na^+ and N^{3-}
(e) Ag^+ and S^{2-} (f) Mg^{2+} and P^{3-}

5.19 Identify the ions in the following binary ionic compounds.
(a) Li_2O (b) BaO (c) $ZnCl_2$
(d) K_3P (e) NaBr (f) CaI_2

5.20 Identify the ions in the following binary ionic compounds.
(a) SrI_2 (b) Ca_3P_2 (c) Na_2S
(d) MgS (e) LiI (f) MgF_2

5.21 The formula for a substance often added to salt is KI. How would you name this compound?

5.22 The formula for one of the substances often found in fertilizers is KCl. How would you name this compound?

5.23 Name each of the following compounds.
(a) LiBr (b) $CaBr_2$ (c) Rb_2O
(d) Ba_3N_2 (e) RaS (f) AlF_3

5.24 Name each of the following compounds.
(a) NaI (b) Cs_2S (c) BaO
(d) $AlCl_3$ (e) Mg_3P_2 (f) $MgBr_2$

5.25 Write the formulas for the following compounds.
(a) barium oxide (b) strontium chloride
(c) sodium sulfide (d) sodium bromide
(e) aluminum sulfide (f) potassium nitride

5.26 Write the formulas for the following compounds.
(a) calcium iodide (b) barium bromide
(c) silver oxide (d) calcium oxide
(e) sodium phosphide (f) aluminum fluoride

Binary Ionic Compounds When the Metal Can Have More Than One Charge (Section 5.3)

5.27 Write the name of each of the following compounds using the Stock method and the classical method.
(a) Co_2O_3 (b) CoO (c) PbO_2 (d) PbO
(e) CuO (f) Cu_2O

5.28 Write the name of each of the following compounds using the Stock method and the classical method.
(a) SnO_2 (b) SnO (c) Fe_2S_3 (d) FeS
(e) CoF_3 (f) CoF_2

5.29 Is the correct name used in each of the following cases? If not, give the correct name.
(a) $SnCl_4$; tin(II) chloride
(b) PbO_2; lead(IV) oxide
(c) $CuBr_2$; copper dibromide
(d) Fe_2O_3; diiron(II) trioxide
(e) $CoBr_2$; cobalt(II) bromide
(f) $CaCl_2$; calcium dichloride

5.30 Is the correct name used in each of the following cases? If not, give the correct name.
(a) Cu_2O; dicopper(I) oxide
(b) FeS; iron(II) sulfide
(c) $PbCl_4$; lead(IV) chloride
(d) SnI_2; tin(IV) iodide
(e) CoF_3; cobalt(III) fluoride
(f) Co_2S_3; dicobalt trisulfide

Compounds with Polyatomic Ions (Section 5.4)

5.31 Limestone consists of calcium ions and carbonate ions. Write the formula and name of this compound.

5.32 Some antacids contain a substance that consists of aluminum ions and hydroxide ions. Write the formula and name of this compound.

5.33 Write the formulas and names of the compounds containing the following ions.
(a) Ca^{2+} and HCO_3^- (b) NH_4^+ and Cl^-
(c) Mg^{2+} and ClO^- (d) Ag^+ and CN^-
(e) NH_4^+ and NO_3^- (f) Al^{3+} and ClO_4^-
(g) Ba^{2+} and PO_4^{3-} (h) K^+ and NO_2^-
(i) Na^+ and $C_2H_3O_2^-$

5.34 Write the formulas and names of the compounds containing the following ions.
(a) Na^+ and HPO_4^{2-} (b) Na^+ and OH^-
(c) Mg^{2+} and ClO_2^- (d) NH_4^+ and SO_4^{2-}
(e) Ca^{2+} and MnO_4^- (f) Mg^{2+} and ClO_3^-
(g) NH_4^+ and SO_3^{2-} (h) K^+ and CO_3^{2-}
(i) Ba^{2+} and $C_2H_3O_2^-$

5.35 Write formulas for the following compounds.
(a) aluminum hydroxide
(b) barium sulfite
(c) barium sulfate
(d) lithium hydrogen phosphate
(e) lithium dihydrogen phosphate
(f) lithium phosphate

5.36 Write formulas for the following compounds.
(a) calcium hypochlorite
(b) calcium chlorite
(c) calcium chlorate
(d) calcium perchlorate
(e) ammonium acetate
(f) potassium permanganate

Binary Molecular Compounds (Section 5.5)

5.37 Ammonia is the common name for the compound whose formula is NH_3. Give the systematic name for this substance.

5.38 Water is the common name for the compound whose formula is H_2O. Give the systematic name for this substance.

5.39 Each of the following gives the atomic composition of a molecule. What is the formula and systematic name of the corresponding compound?
(a) one atom of oxygen and two atoms of nitrogen
(b) one atom of carbon and four atoms of chlorine
(c) four atoms of fluorine and one atom of silicon
(d) three atoms of fluorine and one atom of chlorine
(e) two atoms of chlorine and seven atoms of oxygen

5.40 Each of the following gives the atomic composition of a molecule. What is the formula and systematic name of the corresponding compound?
(a) three atoms of chlorine and one atom of arsenic
(b) three atoms of iodine and one atom of nitrogen
(c) one atom of bromine and five atoms of fluorine
(d) seven atoms of fluorine and one atom of iodine
(e) three atoms of oxygen and one atom of sulfur

5.41 Write formulas for the following compounds.
(a) nitrogen tribromide (b) xenon tetroxide
(c) oxygen difluoride (d) dichlorine pentoxide
(e) sulfur hexafluoride (f) phosphorus triiodide

5.42 Write formulas for the following compounds.
(a) dinitrogen tetrafluoride (b) nitrogen trichloride
(c) tetraarsenic hexoxide (d) carbon monoxide
(e) carbon dioxide (f) dinitrogen tetroxide

Binary Acids and Oxyacids (Sections 5.6 and 5.7)

5.43 Give the name of the compound and corresponding acid whose formula is H_2S.

5.44 Give the name of the compound and corresponding acid whose formula is HI. (Note: drop the *o* in *hydro-* before the vowel in the nonmetal root.)

5.45 Write names for the following acids.
(a) H_3PO_4 (b) HNO_2 (c) H_2CO_3
(d) HNO_3 (e) $HClO$ (f) $HClO_4$

5.46 Write names for the following acids.
(a) H_2SO_3 (b) H_2SO_4 (c) $HClO_2$
(d) $HClO_3$ (e) H_3BO_3 (f) $HC_2H_3O_2$

5.47 Give formulas for the following acids.
(a) nitrous acid (b) nitric acid
(c) hydroiodic acid (d) sulfuric acid
(e) sulfurous acid (f) hydrobromic acid

5.48 Give formulas for the following acids.
(a) hypochlorous acid (b) phosphoric acid
(c) chloric acid (d) hydrosulfuric acid
(e) hydrochloric acid (f) perchloric acid

Additional Problems

5.49 Baking soda contains sodium ions and hydrogen carbonate ions. What is its formula and chemical name?

5.50 Washing soda contains sodium ions and carbonate ions. What is its formula and chemical name?

5.51 Limestone is a substance whose principal component contains calcium ions and carbonate ions. What is the formula and systematic name of this component?

5.52 Quicklime contains calcium ions and oxide ions. What is its formula and chemical name?

5.53 Write formulas for the following compounds.
(a) ferrous sulfate (b) ferric sulfate
(c) cuprous acetate (d) cupric acetate

5.54 Write formulas for the following compounds.
(a) stannous fluoride (b) stannic fluoride
(c) cobaltous nitrate (d) cobaltic nitrate

5.55 Write formulas for the following compounds. Use the periodic table if necessary.
(a) nickel(II) chloride
(b) titanium(IV) fluoride
(c) manganese(IV) oxide
(d) chromium(III) sulfide

5.56 Write formulas for the following compounds. Use the periodic table if necessary.
(a) vanadium(V) oxide
(b) chromium(II) bromide
(c) manganese(VII) oxide
(d) titanium(II) chloride

5.57 Write formulas for the following compounds. Use the periodic table if necessary.
(a) lead(II) nitrate (b) cobalt(II) hydroxide
(c) copper(II) phosphate (d) copper(I) nitrite
(e) iron(II) chlorite (f) tin(IV) perchlorate

5.58 Write formulas for the following compounds. Use the periodic table if necessary.
(a) cobalt(III) acetate (b) tin(II) hydroxide
(c) copper(II) nitrate (d) iron(III) sulfate
(e) iron(II) sulfate (f) iron(II) sulfite

5.59 Name the following compounds.
(a) $Sn_3(PO_4)_2$ (b) $Fe(NO_3)_2$ (c) $CrSO_4$
(d) $Al(ClO_3)_3$

5.60 Name the following compounds.
(a) $Fe(OH)_3$ (b) $(NH_4)_3PO_4$ (c) $Ca(ClO)_2$
(d) $Cr(NO_3)_3$

5.61 Give the name and formula of the anion corresponding to each of the following oxyacids.
(a) bromic acid; $HBrO_3$
(b) hyponitrous acid; $H_2N_2O_2$
(c) thiosulfuric acid; $H_2S_2O_3$
(d) arsenic acid; H_3AsO_4

5.62 Give the name and formula of the anion corresponding to each of the following oxyacids.
(a) bromous acid; $HBrO_2$
(b) perbromic acid; $HBrO_4$
(c) hypoiodous acid; HIO
(d) diphosphoric acid; $H_4P_2O_7$

Practice in Problem Analysis

For each problem, describe the thinking you would use (the problem analysis) before doing the actual solution, but do not solve the problem.

1. What is the formula of radium chloride?

2. What is the Stock name of $TiCl_4$?

PRACTICE EXAM

1. The formula of the monatomic ion of selenium is
(a) Se^- (b) Se^{4+} (c) Se^{2-} (d) Se^+
(e) None of the above

2. Which one of the following monatomic ions has the incorrect formula?
(a) Ra^{2+} (b) S^{2-} (c) Mg^{2-} (d) P^{3-}
(e) None of the above

3. The formula for the binary compound of Ca^{2+} ion and P^{3-} ion is
(a) CaP (b) CaP_2 (c) Ca_2P (d) Ca_2P_3
(e) None of the above

4. Which of the following has the wrong formula?
(a) KF (b) BaCl (c) Na_2S (d) $AlCl_3$
(e) None of the above

5. The name of the compound AlF_3 is
(a) monoaluminum fluoride
(b) monoaluminum trifluoride
(c) aluminum trifluoride
(d) aluminum fluoride
(e) None of the above

6. Which one of the following compounds is misnamed?
(a) potassium iodide (b) sodium monobromide
(c) sulfur trioxide (d) iron(II) chloride
(e) oxygen difluoride

7. Potassium phosphide has the following formula:
(a) PK_3 (b) KP (c) Po_3Ph (d) K_3P
(e) None of the above

8. Barium sulfide has the following formula:
(a) BaS (b) Ba_2S (c) Ba_2S_2 (d) S_2Ba_2
(e) None of the above

9. What is the systematic name of $CrCl_3$?
(a) chromyl chloride
(b) chromium trichloride
(c) chromium chloride(III)
(d) chromium(III) chloride
(e) None of the above

10. What is the systematic name of $FeCl_3$?
(a) ironic chloride (b) ferrous chloride
(c) ferric chloride (d) ironous chloride
(e) None of the above

11. A compound contains these ions: K^+ and SO_3^{2-}. What is the name of the compound?
(a) potassium sulfide (b) potassium sulfite
(c) potassium sulfate (d) potassium persulfite
(e) None of the above

12. Only one of the following is an ionic compound of a polyatomic anion. Which is it?
(a) sodium phosphide (b) boron trichloride
(c) ferrous chloride (d) aluminum bromide
(e) iron(II) sulfate

13. What is the formula of lithium sulfate?
(a) Li_2SO_4 (b) $Li_2(SO_4)$
(c) Li_2SO_3 (d) $Li(SO_4)_2$
(e) None of the above

14. What is the formula of cobalt(II) phosphate?
(a) $CoPO_4$ (b) $Co(PO_4)_2$
(c) $Co_2(PO_4)_3$ (d) $Co(PO_4)_3$
(e) None of the above

15. A molecule contains one sulfur atom and four chlorine atoms. What is the name of the compound?
(a) monosulfur trichloride
(b) monosulfur tetrachloride
(c) sulfur tetrachloride
(d) sulfur dichloride
(e) None of the above

16. A molecule contains four sulfur atoms and four phosphorus atoms. What is the name of the compound?
(a) sulfur phosphide
(b) phosphorus sulfide
(c) tetraphosphorus tetrasulfide
(d) tetrasulfur tetraphosphide
(e) None of the above

17. You dissolve HBr in water. The name of the solution is
(a) hydrogen bromide (b) hydrogen monobromide
(c) hydrogen bromic acid (d) hydrobromic acid
(e) None of the above

18. Hydrosulfuric acid is made by dissolving what compound in water?
(a) H_2SO_4 (b) HSO_4 (c) HS (d) H_2S
(e) None of the above

19. What is the name of the acid corresponding to the PO_4^{3-} ion?
(a) potassic acid (b) potassous acid
(c) phosphoric acid (d) phosphorous acid
(e) hydrophosphoric acid

20. What is the name of the acid corresponding to the ClO_4^- ion?
(a) hypochlorous acid (b) chlorous acid
(c) chloric acid (d) perchloric acid
(e) None of the above

Chapter

Chemical Reactions and Equations

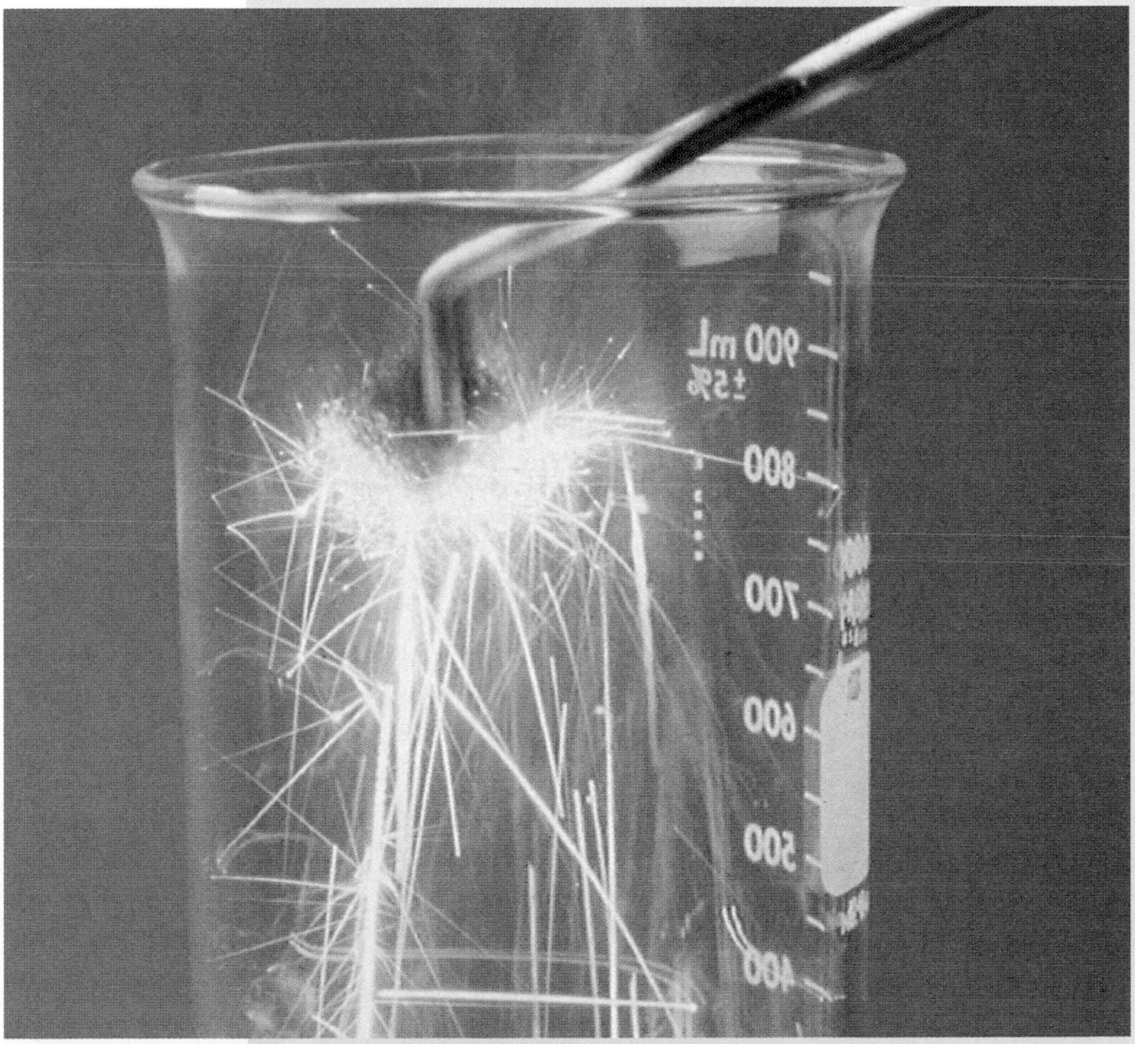

Iron wool will burn in air when it is heated.

What does burning wood or burning natural gas have in common with rusting iron or growing plants? Each is an example of a chemical reaction from our everyday lives (Figure 6.1). A chemical reaction, as you saw in Chapter 3, is a change in which one or more kinds of matter are transformed into one or more new kinds of matter.

Consider the burning of natural gas in the flame on a gas stove (Figure 6.1A). Methane (CH_4), the principal component of natural gas, and oxygen in the air react,

A

B

FIGURE 6.1
Two chemical reactions from everyday life. *(A)* Natural gas (methane) from a gas stove burns because it reacts with oxygen in the air. During the reaction, heat and a flame are produced. *(B)* During its growth, submerged pondweed evolves bubbles of gaseous oxygen, one of the products of a reaction involving carbon dioxide and water.

FIGURE 6.2
Reaction of potassium iodide solution and lead(II) nitrate solution. When the colorless solutions of the reactants are mixed, one of the products is lead(II) iodide, which forms as a yellow solid.

transforming into carbon dioxide and gaseous water (water vapor or steam). Methane and oxygen are reactants. A **reactant** is *a chemical substance involved at the start of a chemical reaction.* Carbon dioxide and water are the products of this reaction. A **product** is *a chemical substance that results from a chemical reaction.* Reactants and products are different kinds of matter, and in every chemical reaction one changes into the other.

Iron rusts because of a chemical reaction: iron and oxygen gas in air are changed by the chemical reaction into iron(III) oxide. Iron and oxygen are the reactants and iron(III) oxide is the product. Plants grow because carbon dioxide and water, the reactants, change into sugar and oxygen gas, the products. A plant needs the sugar as a source of energy. Leaves release the other product, oxygen, through their pores. Although you will not notice the evolution of oxygen if you look at grass or trees (even though it is occurring), you can see it happening with the pondweed in Figure 6.1B.

In this chapter, we look at some of the signals that show that chemical reactions have occurred and see how to express these chemical reactions symbolically in terms of chemical equations. We also examine several types of chemical reactions and some of the reasons they occur.

RECOGNIZING AND SYMBOLIZING CHEMICAL REACTIONS

In most cases, you can recognize a chemical reaction easily. For example, the flame in Figure 6.1A is an undeniable sign of a chemical reaction, whether or not you know what reaction has occurred. What are some other signals for chemical reactions? And how does a chemist explain, perhaps in symbols, what is happening during such reactions? You will find answers to these questions in the next three sections.

6.1 Recognizing Chemical Reactions

OBJECTIVE

- State five telltale signs that a chemical reaction has occurred.

FIGURE 6.3
Zinc metal reacting with hydrochloric acid solution. One of the products of this reaction is hydrogen, which forms gas bubbles.

Many chemical reactions have a way of telling us that they are happening. You might see one or more of the following:

1. The release, or evolution, of heat (Figure 6.1A)
2. The emission of light (Figure 6.1A)
3. The formation of a solid in what was a clear solution (Figure 6.2)
4. The formation of gas bubbles when you add a substance to a solution (Figure 6.3)
5. A change in color (Figure 6.4)

Although these signals tell you when a chemical reaction has occurred, they do not tell you what has happened. A chemist must do experiments to determine the identities of the reactants and products in the reaction, then do further experiments to determine their formulas. Once you know the formulas of the reactants and products, you can write a *chemical equation* that represents what happened in the reaction. The next section tells you about chemical equations.

FIGURE 6.4
Reaction of nickel(II) sulfate solution and ammonia solution. The product of this reaction *(right)* has a distinctly different color from either of the reactants *(left)*.

6.2 Chemical Equations

OBJECTIVES

- Given a chemical equation, identify the reactants and products, as well as the number of molecules or formula units of each substance.
- Given a chemical equation, note the physical states of reactants and products and any reaction conditions, if this information is available in the equation.

A chemical equation has some similarity to a trip sketched on a road map: It tells you where you started—the reactants—and it tells you where you are going—the products—with an arrow pointing the way. A **chemical equation** is *a symbolic way of expressing a chemical reaction.* Consider again the chemical reaction in Figure 6.1A. Methane and oxygen react to form carbon dioxide and water. The chemical equation that describes this reaction is

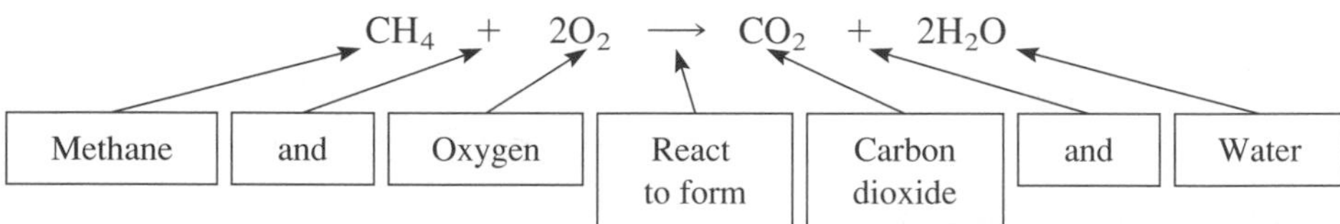

Formulas are used in place of the names of the compounds; + is used in place of "and"; and an arrow means "react to form" or "yield." Table 6.1 lists some symbols commonly used in chemical equations.

The numbers in front of the formulas in the chemical equation are called **coefficients.** When no coefficient appears in front of a formula, a 1 is implied.

$$CH_4 + 2O_2 \longrightarrow CO_2 + 2H_2O$$

Implied 1

Coefficients

For a molecular reaction, as in this example, the coefficients (which should be whole numbers) tell you how many molecules of each reactant are involved and how many molecules of each product are formed. In this example the equation means that one molecule of methane and two molecules of oxygen react to form one molecule of carbon dioxide and two molecules of water. Molecular models that convey the same description

TABLE 6.1
Some Symbols Found in Chemical Equations

Symbol	Meaning
$+$	And
$\longrightarrow$	Reacts to form, or yields
(s)	Solid
(l)	Liquid
(g)	Gas
(aq)	Aqueous (water) solution
$\xrightarrow{\Delta}$	Heat added
$\xrightarrow{Pt}$	Catalyst (in this case, platinum)

are shown in Figure 6.5. Look closely at these models, and count the molecules so you can see that the verbal description of the reaction is captured faithfully by the equation.

If the reaction involves ionic compounds, we must speak of formula units of these compounds rather than molecules. For example, the equation

$$Na_2O + H_2O \longrightarrow 2NaOH$$

means that one formula unit of sodium oxide, an ionic compound, and one molecule of water, a molecular compound, react to form two formula units of sodium hydroxide, another ionic compound.

The physical states of the reactants and products, whether they are gases, liquids, solids, or in solution, are sometimes included in the chemical equation. For example, you use (s) to indicate a solid reactant or product, as in $NaCl(s)$. Similarly, you use (aq) to indicate an aqueous (water) solution of a substance, as in $NaCl(aq)$. Notice that when one of these parenthetical symbols is used, it appears immediately after the formula in the chemical equation. Although you are not required to include the states in the chemical equation, you convey considerably more information when you do. Consider, for example, our equation for the reaction of methane with oxygen:

$$CH_4(g) + 2O_2(g) \longrightarrow CO_2(g) + 2H_2O(g)$$

Note that water is not a liquid in this reaction, but a gas because of the high temperature of the reaction. Here are some other examples using symbols to specify physical state:

$$4Fe(s) + 3O_2(g) \longrightarrow 2Fe_2O_3(s)$$
$$LiOH(s) + CO_2(g) \longrightarrow LiHCO_3(s)$$
$$HCl(aq) + NaOH(aq) \longrightarrow NaCl(aq) + H_2O(l)$$

● The reaction of iron and oxygen is the cause of rust. Rust is essentially iron(III) oxide.

● Lithium hydroxide is used to remove carbon dioxide that astronauts have exhaled in a spacecraft.

Finally, reaction conditions are sometimes written over the arrow, as shown in Table 6.1. For example, if the reactants are heated, a capital Greek delta (Δ) may be placed

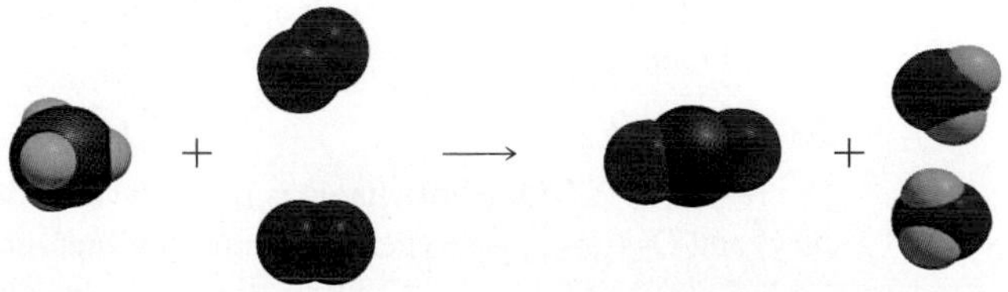

FIGURE 6.5
Representation of the reaction of methane and oxygen. Molecular models represent the balanced equation for the reaction of CH_4 with O_2 to yield CO_2 and H_2O.

over the arrow. Thus, you can write

$$CaCO_3(s) \xrightarrow{\Delta} CaO(s) + CO_2(g)$$

when you want to describe the result of heating a solid sample of calcium carbonate. Similarly, you can indicate any catalyst present. (A **catalyst** is *a substance that causes a chemical reaction to speed up, even though it is not consumed by the reaction.*) You write the formula of the catalyst over the arrow, as in the equation

$$2SO_2(g) + O_2(g) \xrightarrow{Pt} 2SO_3(g)$$

This equation says symbolically that gaseous sulfur dioxide reacts with gaseous oxygen in the presence of a platinum catalyst to yield gaseous sulfur trioxide.

Sometimes it may be necessary to heat the reactants in the presence of a catalyst. If so, both of these reaction conditions can be written in the chemical equation. You write the capital Greek delta (for heat) over the arrow and the formula for the catalyst under the arrow:

$$2KClO_3(s) \xrightarrow[MnO_2]{\Delta} 2KCl(s) + 3O_2(g)$$

This equation describes a reaction in which solid potassium chlorate decomposes when heated in the presence of the catalyst manganese(IV) oxide to form solid potassium chloride and gaseous oxygen.

Recognizing each part of a chemical equation is essential throughout this chapter. Be sure that you can do the following example.

■ EXAMPLE 6.1 Interpreting a Chemical Equation

Joseph Priestley discovered oxygen in 1774 by heating mercury(II) oxide. The reaction is described by the chemical equation

$$2HgO(s) \xrightarrow{\Delta} 2Hg(l) + O_2(g)$$

State in words what the chemical equation means. Your statement should identify the reactants and products, the number of formula units (for ionic compounds) or molecules (for molecular compounds) of every substance, and their physical states. Include any reaction condition specified in your statement.

Problem Analysis

Recognize that the formulas on the left side of the chemical equation represent the reactants and the formulas on the right side are the products. The coefficients (whole numbers) in front of the formulas specify the number of formula units or molecules. The physical state of each substance is given in parentheses. Anything written above or below the arrow indicates reaction conditions. To give a word statement, you need to know the name associated with each formula; see the previous chapter for the rules of naming.

Solution

The reactant is HgO, whose name is mercury(II) oxide. The products are Hg (mercury) and O_2 (gaseous oxygen). The reactant mercury(II) oxide is a solid, whereas one product is a liquid (mercury) and the other is a gas (oxygen). The equation means that two formula units of solid mercury(II) oxide react when heated to yield two atoms of liquid mercury and one molecule of oxygen gas.

Exercise 6.1

For each of the following chemical equations, name the reactants and products, give the number of formula units (for ionic compounds) or molecules (for molecular compounds) of every substance, specify their physical states, and specify the reaction conditions if they are given.

(a) $2H_2(g) + O_2(g) \xrightarrow{Pt} 2H_2O(l)$

(b) $2NaI(aq) + Cl_2(g) \longrightarrow 2NaCl(aq) + I_2(aq)$

(c) $CO(g) + 3H_2(g) \longrightarrow CH_4(g) + H_2O(g)$

(d) $2H_2O_2(aq) \xrightarrow{\Delta} 2H_2O(l) + O_2(g)$

(Try Problems 6.11, 6.12, 6.13, and 6.14.)

Exploration: Conservation of mass and balancing equations.

6.3 Balancing Chemical Equations

OBJECTIVE

- Balance a chemical equation using the smallest possible whole-number coefficients.

A chemical equation is a symbolic summary of experimental information about a chemical reaction. A chemist observes through experiments that methane reacts with oxygen to yield carbon dioxide and water. To begin to write the chemical equation, she must know the formulas of the reactants and products. If she does not know these, she must do more experiments to determine them. The formula of methane is CH_4, the formula for oxygen gas is O_2, whereas that for carbon dioxide is CO_2, and that for water is H_2O. She can now write the following "skeleton" equation:

$$CH_4 + O_2 \longrightarrow CO_2 + H_2O$$

The experiments are now done, but she has one more important step to complete. She must determine the coefficients of each substance in the equation. The process of completing this equation by writing down appropriate whole-number coefficients is called *balancing*. The balanced equation for this reaction is

$$CH_4 + 2O_2 \longrightarrow CO_2 + 2H_2O$$

How do you balance a chemical equation?

The central idea you need was stated in the last point of Dalton's atomic theory, which we discussed in Chapter 4. We can restate the point: A chemical reaction is a rearrangement of atoms in the reactants to give a new arrangement of the same atoms in the products. No atoms are created or destroyed during the reaction. Thus, *the number of atoms of each element must be the same on both sides of the arrow in the chemical equation*. An equation that meets this condition is called a **balanced chemical equation** because it reflects Dalton's theory by requiring a balance, or equality, of the different types of atoms in the equation. Normally, we also ask that the coefficients be as small as possible, so that the balanced equation is the simplest description of the reaction.

An analysis of the preceding equation indicates that it is indeed balanced:

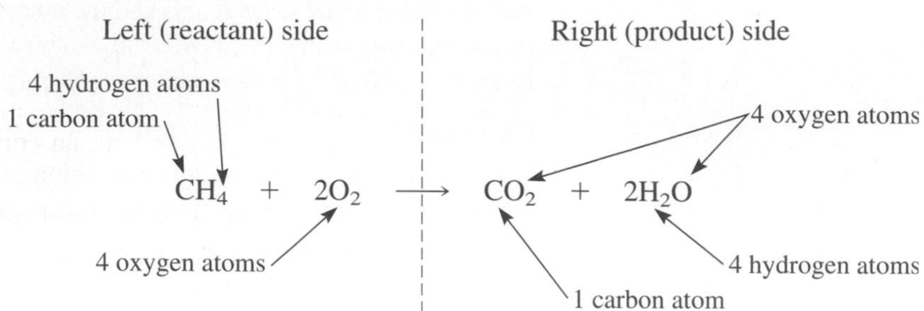

$$
\begin{aligned}
\text{1 carbon atom} &= \text{1 carbon atom} \\
\text{4 hydrogen atoms} &= \text{4 hydrogen atoms} \\
\text{4 oxygen atoms} &= \text{4 oxygen atoms}
\end{aligned}
$$

Remember that the subscripts tell you the number of each type of atom in a formula. Thus, each molecule of oxygen (O_2) contains two atoms of oxygen. Coefficients indicate the number of molecules or formula units. Since the coefficient in front of O_2 is 2, there are two oxygen molecules, or four oxygen atoms, on the left side of the equation. Also, remember that 1 is implied if there is no subscript or coefficient. Thus, the formula for methane (CH_4) on the left side of the equation shows that there is one molecule of this substance, and it contains one carbon atom and four hydrogen atoms. You may understand these ideas better by studying the reaction represented by molecular models in Figure 6.5. Count the atoms in the models to verify that the number of atoms of each element is the same on both sides of the arrow. Notice that no atoms have been created or destroyed. Finally, note that the coefficients are indeed the smallest possible. If you were to double the preceding equation, it would still be balanced, although the coefficients are certainly not the simplest.

$$2CH_4 + 4O_2 \longrightarrow 2CO_2 + 4H_2O$$

The process of balancing a chemical equation is relatively simple, unless the equation itself is quite complicated. We begin with a simplifying rule: *Look for an atom type that occurs in only one formula as a reactant and in only one formula as a product.* Then, choose the coefficients of these two formulas to balance this atom type. You can do this by simply taking the number of atoms on the left as the coefficient on the right, and vice versa. ● Often you can continue to use this rule to balance one or more additional atom types, and as you continue you will determine more and more of the coefficients. Eventually, you may have to balance an atom type that occurs in several formulas, but because most of the coefficients will already have been fixed, this will be easy. Let's see how this works.

● The mathematical idea used here is similar to the one we used to obtain the formula of an ionic compound given its ions.

Consider the burning of propane gas (Figure 6.6). When propane (C_3H_8) burns (reacts with oxygen, O_2), the products are carbon dioxide, CO_2, and water, H_2O, just as they are when methane burns. The skeleton or unbalanced equation is

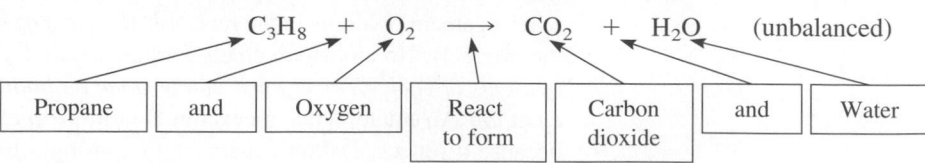

$$C_3H_8 + O_2 \longrightarrow CO_2 + H_2O \quad \text{(unbalanced)}$$

Propane	and	Oxygen	React to form	Carbon dioxide	and	Water

An analysis verifies that the equation is not yet balanced.

FIGURE 6.6

Burning of propane gas. Propane gas from the steel tank burns by reacting with oxygen in the air to give carbon dioxide and water.

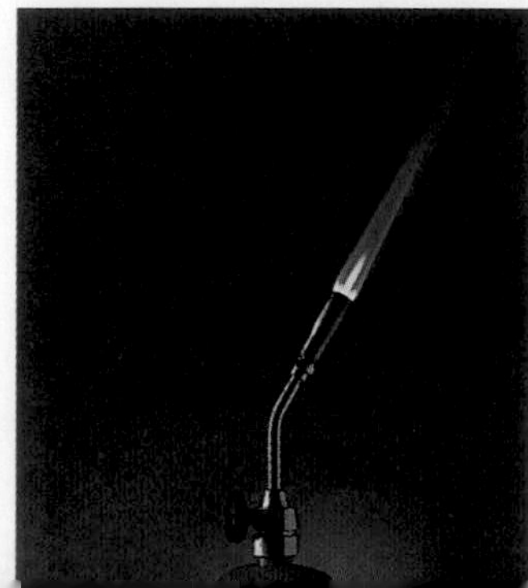

Left side:	3 carbon atoms	Right side:	1 carbon atom
	8 hydrogen atoms		2 hydrogen atoms
	2 oxygen atoms		3 oxygen atoms

The same conclusion can be reached from the molecular models shown in Figure 6.7A.

Now consider our rule for balancing an equation. You look for an atom type that occurs in only one reactant and only one product. Carbon occurs in only C_3H_8 on the left and in only CO_2 on the right. Similarly, hydrogen occurs in only C_3H_8 on the left and in only H_2O on the right. Carbon and hydrogen both satisfy our rule, so you could start with either. On the other hand, while oxygen occurs in only O_2 on the left, it occurs in both CO_2 and H_2O on the right. If you were to start by balancing oxygen atoms, you would have to determine three coefficients at once.

Suppose you start by balancing carbon atoms. Since there are three carbon atoms on the left, you write 3 for the coefficient of CO_2 (on the right). There is one carbon atom on the right, so the coefficient on the left (of C_3H_8) is 1. For the moment, put a 1 in front of C_3H_8 to help you keep track of the coefficients that you have determined.

$$1C_3H_8 + O_2 \longrightarrow 3CO_2 + H_2O \qquad \text{(unbalanced)}$$

Although the equation still needs work, the carbon atoms are now balanced.

Now you proceed to balance the hydrogen atoms. The coefficient of C_3H_8 is already fixed at 1, so you must balance hydrogen atoms by choosing the coefficient of H_2O. There are 8 hydrogen atoms on the left ($1C_3H_8$), so you write 4 in front of H_2O to get 8 hydrogen atoms on the right:

$$1C_3H_8 + O_2 \longrightarrow 3CO_2 + 4H_2O \qquad \text{(unbalanced)}$$

Now only the oxygen atoms need to be balanced. Note that only the coefficient of O_2 remains to be determined. As you can see, there are 10 oxygen atoms on the right side: 6 from the three CO_2 molecules and 4 from the four H_2O molecules. So you can balance the oxygen atoms by writing 5 in front of O_2, on the left side. At the same time, you can omit the unnecessary 1 in front of C_3H_8.

$$C_3H_8 + 5O_2 \longrightarrow 3CO_2 + 4H_2O \qquad \text{(balanced)}$$

Let's check to see that the equation is indeed balanced:

Left side:	3 carbon atoms	Right side:	3 carbon atoms
	8 hydrogen atoms		8 hydrogen atoms
	10 oxygen atoms		10 oxygen atoms

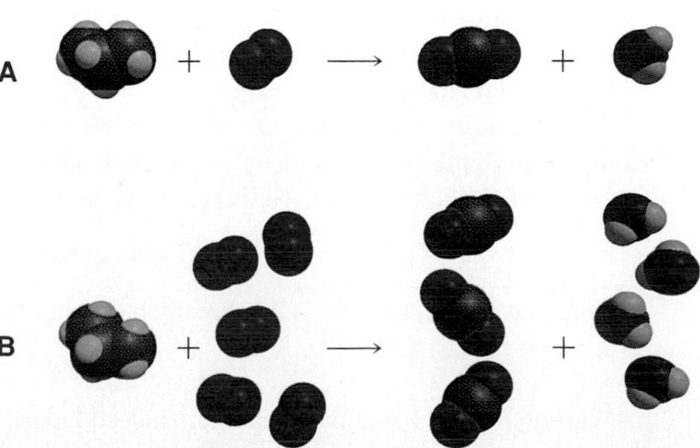

FIGURE 6.7
Representations of the reaction of propane and oxygen. Molecular models represent the *(A)* unbalanced and *(B)* balanced equations for the reaction of C_3H_8 with O_2 to yield CO_2 and H_2O. Count the atoms of each element to verify that *B* is balanced.

You can also see from the molecular models in Figure 6.7B that the equation is balanced. Finally, check to see that the coefficients are the smallest whole numbers possible. (They are.)

The next two examples will give you more practice in balancing chemical equations. In the first one, you are given a skeleton equation and asked to balance it. More is required in the second example because you are given only a word description of the reaction.

■ EXAMPLE 6.2 Balancing an Equation

You can buy hydrogen peroxide, a substance used as an antiseptic and a bleach, in a supermarket as a 3% aqueous solution. It disappears from this solution slowly because of the reaction shown in the following chemical equation:

$$H_2O_2(aq) \longrightarrow H_2O(l) + O_2(g)$$

Balance this equation, if it is not already balanced.

Problem Analysis

Analyze the equation. Is it already balanced? If not, proceed to balance it. *Begin by looking for an atom type that occurs in only one formula as a reactant and in only one formula as a product.* Choose the coefficients of these two formulas to balance this atom type. Simply take the number of atoms on the left as the coefficient on the right, and vice versa. Continue in this way, choosing more coefficients. Eventually, you may have to balance an atom type that occurs in several formulas, but by then most of the coefficients will already have been fixed. After you are done, check to see that the equation is indeed balanced. Also, check that the coefficients are as small as possible.

Solution

An analysis shows that the equation is not balanced.

$$H_2O_2 \longrightarrow H_2O + O_2$$

Left side: 2 hydrogen atoms	Right side: 2 hydrogen atoms
2 oxygen atoms	3 oxygen atoms

Note that hydrogen atoms occur in only the reactant and in only one product. Oxygen, on the other hand, occurs in every formula in the equation. Therefore, you begin by balancing hydrogen atoms. In balancing the equation, you can momentarily omit the parenthetical symbols denoting the physical states of the substances, then add these back at the end.

The number of hydrogen atoms on the left (2) becomes the coefficient of H_2O on the right. Similarly, the number of hydrogen atoms on the right (2) becomes the coefficient of H_2O_2 on the left. You write

$$2H_2O_2 \longrightarrow 2H_2O + O_2$$

Note that the oxygen atoms are already balanced, so the final equation is

$$2H_2O_2(aq) \longrightarrow 2H_2O(l) + O_2(g)$$

You should check that the equation is indeed balanced. Here is the analysis of atoms:

Left side: 2 hydrogen atoms	Right side: 2 hydrogen atoms
2 oxygen atoms	2 oxygen atoms

Note also that the coefficients are the smallest whole numbers possible.

In the first step, you might have noticed that the hydrogen atoms are already balanced. So rather than choosing coefficients of 2 for each compound of hydrogen, you might have chosen 1's.

$$1H_2O_2 \longrightarrow 1H_2O + O_2$$

This certainly balances the hydrogen atoms. Now you proceed to balance the oxygen atoms. Only the coefficient of O_2 (on the right) remains to be fixed. To balance the 2 oxygen atoms on the left, you will need to write ½ for the coefficient of O_2.

$$1H_2O_2 \longrightarrow 1H_2O + \tfrac{1}{2}O_2$$

This fractional coefficient is a bit bothersome, because if you think of this equation in molecular terms, it is difficult to imagine half a molecule. You simply multiply all of the coefficients by 2 to clear the fraction. The final equation is the one we wrote earlier.

Exercise 6.2

Balance the following equations.

(a) $NH_4NO_3(s) \xrightarrow{\Delta} N_2O(g) + H_2O(g)$

(b) $N_2O_5(g) \longrightarrow NO_2(g) + O_2(g)$

(c) $Ca(s) + H_2O(l) \longrightarrow Ca(OH)_2(aq) + H_2(g)$

(d) $As_2S_3(s) + O_2(g) \xrightarrow{\Delta} As_2O_3(s) + SO_2(g)$

(Try Problems 6.15, 6.16, 6.17, and 6.18.)

EXAMPLE 6.3 Going from a Description to a Balanced Equation

Ammonia gas (NH_3) is an important industrial product because it is used extensively by farmers as a fertilizer. It is prepared by the combination of gaseous nitrogen (N_2) and gaseous hydrogen (H_2) in the presence of iron (a catalyst). Write the balanced equation for this reaction. Include symbols for the states of the substances and for the catalyst in the final equation.

Problem Analysis

Write a skeleton equation that conforms to the description of the reaction. Is the equation already balanced? If not, proceed to balance it. Begin *by looking for an atom type that occurs in only one formula as a reactant and in only one formula as a product.*

Solution

Begin by writing a skeleton equation that agrees with the description.

$$N_2 + H_2 \longrightarrow NH_3$$

Analyze the number of atoms of each element:

Left side:	2 nitrogen atoms	Right side:	1 nitrogen atom
	2 hydrogen atoms		3 hydrogen atoms

The analysis shows that the equation is not balanced.

To balance it, note that both nitrogen atoms and hydrogen atoms occur in only one reactant, as well as in the only product. Thus, you could begin by balancing either nitrogen or hydrogen atoms. Suppose you start with nitrogen atoms. There are two nitrogen atoms on the left, so you write 2 for the coefficient of NH_3. Similarly, there is one nitrogen atom on the right, so you write a 1 for the coefficient of N_2:

$$1N_2 + H_2 \longrightarrow 2NH_3 \qquad \text{(unbalanced)}$$

Finally, balance the hydrogen atoms by noting that the six hydrogen atoms on the right side require six on the left. Therefore, place a 3 in front of H_2:

$$1N_2 + 3H_2 \longrightarrow 2NH_3 \qquad \text{(balanced)}$$

After removing the temporary 1 in front of N_2, you obtain

$$N_2 + 3H_2 \longrightarrow 2NH_3 \qquad \text{(balanced)}$$

Check that the equation is indeed balanced. Note also that the coefficients are the smallest whole numbers possible.

Although the equation is balanced, the problem asks you to show the physical states of the substances and any catalyst (iron).

$$N_2(g) + 3H_2(g) \xrightarrow{\text{Fe}} 2NH_3(g)$$

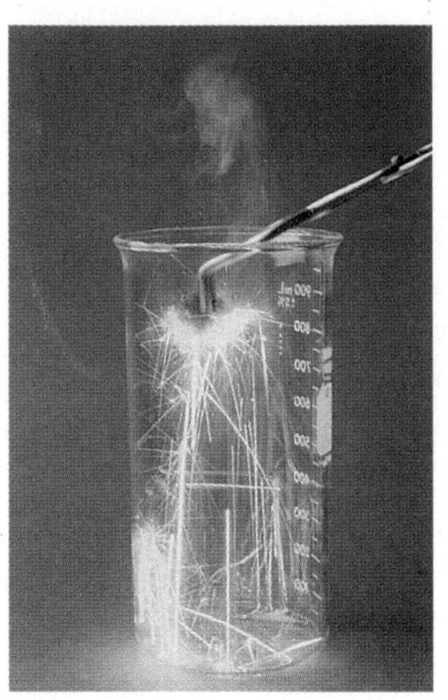

FIGURE 6.8
Iron wool burning. When iron wool is heated, it reacts vigorously with the oxygen in the air to yield iron(III) oxide.

Exercise 6.3

Hot iron wool burns as it reacts with oxygen in the air, as shown in Figure 6.8. The product is iron(III) oxide. Write a balanced chemical equation for this reaction, including a specification of the states of substances.

(Try Problems 6.19, 6.20, 6.21, and 6.22.)

TYPES OF CHEMICAL REACTIONS

We are surrounded by change, almost all of it the result of chemical reactions. Plants grow by combining chemical substances, such as carbon dioxide and water, to give larger, more complicated molecules, including chlorophyll. Chlorophyll is the green substance in leaves that plants require to absorb light for their energy needs. In time, such large molecules tend to break down, or decompose. Plants, and other biological organisms, need to continually repair the results of this natural decomposition through compensating combination reactions that rebuild.

Perhaps you have looked ahead in the chapter. You wonder: How am I going to learn all these different chemical equations? (Don't worry, you will not be asked to do that.) As you might expect, this is a question that chemists have asked themselves. And in answer they have proposed a number of classification schemes. One of these is the *traditional classification scheme,* which we discuss in the remaining sections of this chapter. Almost all of the reactions that we discuss in this book can be understood in terms of this traditional classification scheme.

A large number of reactions, *combination reactions,* occur when chemical substances combine to form larger molecules or units of structure. Many others are *decomposition reactions,* in which substances break down or decompose. Still others can be looked at as though one element has replaced another (a *single-replacement reaction*), or as a reaction in which one piece of a substance replaces another piece of another substance (a *double-replacement reaction*). Other reactions occur when a substance reacts with oxygen (*combustion reactions*). Here then are the different types of chemical reactions that we will explore:

1. Combination reactions
2. Decomposition reactions
3. Single-replacement reactions
4. Double-replacement reactions
5. Combustion reactions

Although we have chosen to discuss the traditional classification scheme in this chapter because of its simplicity and usefulness to us, other classification schemes exist. The traditional scheme looks at a chemical reaction "from a distance." It looks at what chemical substances occur as reactants and what chemical substances result as products. But because chemists like to explain what is happening in terms of the molecules and ions, another important classification scheme has become dominant in recent years. This classification scheme looks "close up" with a theoretical microscope. In one class of reactions, ions in solution combine with others to produce a solid product, which we call a precipitate. So these reactions are called *precipitation reactions.* (In this chapter, we look at these as double-replacement reactions.) In another class of reactions, a proton (a positively charged particle) is transferred from one molecule or ion to another. These are called *acid–base reactions.* And finally, in another large class of reactions, an electron is transferred between molecules or ions. Chemists call these *oxidation–reduction reactions.* We discuss acid–base reactions in Chapter 15 and oxidation–reduction reactions in Chapter 16.

Videos: Ammonium dichromate volcano; Zinc and iodine.

6.4 Combination and Decomposition Reactions

OBJECTIVE

- Identify a chemical equation as a combination reaction or a decomposition reaction if you are given one of these types.

Combination reactions and decomposition reactions are simply the reverse of one another. We consider each in turn.

Combination Reactions

Whenever iron rusts, a **combination reaction**—*a reaction in which two substances chemically combine to form a third*—has occurred. The balanced equation describing the overall formation of rust is

$$4Fe(s) + 3O_2(g) \longrightarrow 2Fe_2O_3(s)$$

This is a combination reaction because iron chemically combines (reacts) with oxygen to form a third substance, iron(III) oxide.

A

B

FIGURE 6.9

Reaction of sodium and chlorine. Sodium metal and chlorine gas *(A)* react violently *(B)*. The product is sodium chloride, common table salt.

Here are three more examples of combination reactions. Sodium metal reacts with chlorine gas (Figure 6.9). The balanced equation for this reaction is

$$2Na(s) + Cl_2(g) \longrightarrow 2NaCl(s)$$

Another combination reaction occurs when liquid bromine is poured on aluminum metal. The product is aluminum bromide:

$$2Al(s) + 3Br_2(l) \longrightarrow 2AlBr_3(s)$$

The vigor of this reaction is shown in Figure 6.10.

The final example is the reaction of solid calcium oxide with sulfur dioxide gas to form solid calcium sulfite:

$$CaO(s) + SO_2(g) \longrightarrow CaSO_3(s)$$

Although the reactants are compounds, you see that two substances have reacted to form a third. ●

● Sulfur dioxide, a gas with a highly irritating odor, is formed when coal and other sulfur-containing substances are burned. It can be removed from air by reactions similar to the one shown here.

A general equation that describes a combination reaction is

$$A + B \longrightarrow AB$$

Compare this equation with the examples we gave.

Decomposition Reactions

Most compounds can be decomposed (broken down) into simpler substances, particularly if energy in the form of heat or light is added to them. For example, silver chloride, AgCl, gradually darkens when exposed to light because it decomposes to silver metal (which looks black when formed as fine particles) and chlorine gas.

$$2AgCl(s) \xrightarrow{\text{light}} 2Ag(s) + Cl_2(g)$$

Some sunglasses take advantage of a similar reaction. These sunglasses darken when

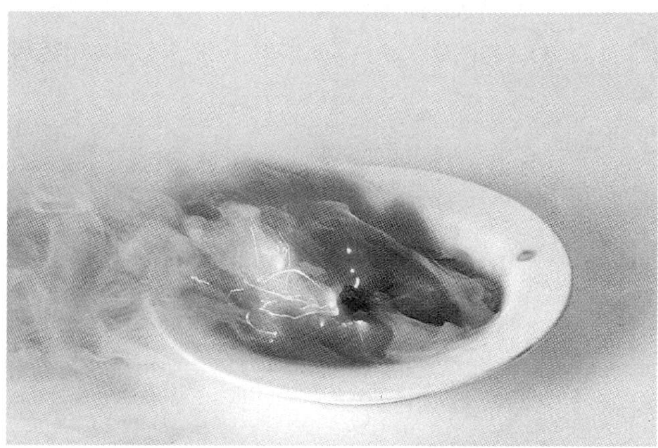

FIGURE 6.10
Reaction of aluminum and bromine. A vigorous reaction occurs when bromine liquid is poured over aluminum foil. The product is aluminum bromide.

they are exposed to strong sunlight. The process reverses, though, when the sunglasses are placed in a shadow.

$$AgCl(\text{in glass}) \xrightarrow{\text{light}} Ag(\text{in glass}) + Cl(\text{in glass})$$

The products in this case are silver atoms and chlorine atoms, rather than the usual forms of the elements. Because the products are held in a rigid glass, they cannot escape, so one chlorine atom cannot combine with another to form Cl_2. Also, because Ag and Cl atoms cannot escape one another, they readily recombine.

The reaction of silver chloride solid to form silver metal and chlorine gas is called a **decomposition reaction**—*a reaction in which a single compound breaks up into two or more other substances*. Let's look at two more examples. Each of these requires heat rather than light to bring about the decomposition; this is the more common way to decompose a substance.

Consider the following reaction of mercury(II) oxide:

$$2HgO(s) \xrightarrow{\Delta} 2Hg(l) + O_2(g)$$

You may recall from Chapter 1 that Antoine Lavoisier examined this reaction (see Figure 1.10). This reaction meets our definition of a decomposition reaction: A single compound has broken up to give two or more substances. In this instance, both products are elements, but that is not a requirement.

Consider the industrial preparation of calcium oxide, which consists of heating limestone, a mineral that is mostly calcium carbonate:

$$CaCO_3(s) \xrightarrow{\Delta} CaO(s) + CO_2(g)$$

In this case, the decomposition leads to two compounds rather than two elements. ●

● Calcium oxide, also known as quicklime, is used extensively in making steel.

A general equation that describes a decomposition reaction is

$$AB \longrightarrow A + B$$

where A and B can be either elements or compounds. Compare this equation with the example equations, to make sure that you understand decomposition reactions. The following example illustrates how to determine whether a reaction is either a combination reaction or a decomposition reaction.

■ EXAMPLE 6.4 Recognizing Combination and Decomposition Reactions

Classify each of the following reactions as either a combination reaction, a decomposition reaction, or neither.

(a) Table sugar (chemically known as sucrose), $C_{12}H_{22}O_{11}$, when heated strongly, turns into carbon and steam:

$$C_{12}H_{22}O_{11}(s) \longrightarrow 12C(s) + 11H_2O(g)$$

(b) Nitrogen monoxide forms during lightning storms from the reaction of gaseous nitrogen and oxygen in the air:

$$N_2(g) + O_2(g) \longrightarrow 2NO(g)$$

(c) The antacid milk of magnesia (a suspension of solid magnesium hydroxide in water) reacts with stomach acid (HCl):

$$Mg(OH)_2(s) + 2HCl(aq) \longrightarrow MgCl_2(aq) + 2H_2O(l)$$

Problem Analysis

Look at the chemical equation. If it looks like $A + B \longrightarrow AB$, it is a combination reaction. If it looks like $AB \longrightarrow A + B$, it is a decomposition reaction. Otherwise, it is neither.

Solution

(a) This is a decomposition reaction, since a compound decomposes into an element and a compound.

(b) This is a combination reaction, because two elements combine to form a third substance.

(c) Since this reaction is not represented by either $A + B \longrightarrow AB$ or $AB \longrightarrow A + B$, it is neither a combination reaction nor a decomposition reaction. You will learn about this type of reaction in Section 6.9.

Exercise 6.4

Classify each of the following reactions according to its type.

(a) $2H_2O_2(aq) \longrightarrow 2H_2O(l) + O_2(g)$

(b) $2Li(s) + Br_2(l) \longrightarrow 2LiBr(s)$

(c) $Mg(s) + 2HBr(aq) \longrightarrow MgBr_2(aq) + H_2(g)$

(Try Problems 6.23, 6.24, 6.25, and 6.26.)

6.5 Single-Replacement Reactions

OBJECTIVES

- Identify a chemical equation as a single-replacement reaction, if you are given one of these types.
- Predict, using the activity series, whether a given equation for a single-replacement reaction will actually occur.

Some metals react with acids or even water. Hydrogen gas forms when the metal dissolves. An example is the reaction of metallic zinc with hydrochloric acid (Figure 6.3):

$$Zn(s) + 2HCl(aq) \longrightarrow ZnCl_2(aq) + H_2(g)$$

Zinc replaces hydrogen in HCl to give $ZnCl_2$ and bubbles of hydrogen gas. Similarly, sodium reacts vigorously with water to give sodium hydroxide and hydrogen gas:

$$2Na(s) + 2H_2O(l) \longrightarrow 2NaOH(aq) + H_2(g)$$

A sodium atom replaces one of the hydrogen atoms in a water molecule. This is clearer if we write the formula of water as HOH.

$$2Na(s) + 2HOH(l) \longrightarrow 2NaOH(aq) + H_2(g)$$

Each of these reactions is a **single-replacement reaction,** *a reaction in which one element reacts by replacing another element in a compound.*

Single-replacement reactions do not have to form hydrogen. For example, when a strip of copper metal is dipped into an aqueous solution of colorless silver nitrate, a single-replacement reaction causes the formation of a blue solution of copper(II) nitrate and solid silver metal (Figure 6.11):

$$Cu(s) + 2AgNO_3(aq) \longrightarrow Cu(NO_3)_2(aq) + 2Ag(s)$$

In this reaction, the element copper has replaced the element silver in its compound.

A general equation that describes a single-replacement reaction is

$$A + BC \longrightarrow AC + B$$

In this case, A and B are elements. Compare this equation with the others in this section to make sure that you understand single-replacement reactions.

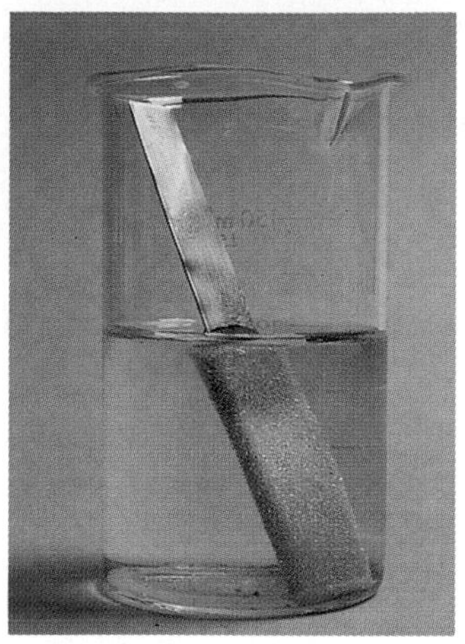

FIGURE 6.11

Reaction of copper metal with silver nitrate solution. When a strip of copper is dipped in a solution of silver nitrate, the single replacement reaction causes the strip to be coated with crystals of silver metal. The blue-green color results from the formation of copper(II) nitrate.

■ EXAMPLE 6.5 Recognizing Single-Replacement Reactions

Which reaction, if any, is a single-replacement reaction?

(a) When table salt, or sodium chloride, is heated with sulfuric acid, the reaction is

$$2NaCl(s) + H_2SO_4(l) \longrightarrow Na_2SO_4(s) + HCl(g)$$

(b) Iron can be prepared from the reaction of aluminum with iron(III) oxide (Figure 6.12):

$$2Al(s) + Fe_2O_3(s) \xrightarrow{\Delta} Al_2O_3(s) + 2Fe(l)$$

(c) When chlorine gas is bubbled through seawater, the bromine (Br_2) can be recovered:

$$Cl_2(g) + 2NaBr(aq) \longrightarrow 2NaCl(aq) + Br_2(aq)$$

Problem Analysis

Examine each equation. If it can be represented by $A + BC \longrightarrow AC + B$, where A and B are elements, it is a single-replacement reaction.

Solution

The reactions in (b) and (c) are single-replacement reactions because they can be represented by $A + BC \longrightarrow AC + B$, where A and B are elements. In these reactions, an element reacts with a compound and displaces another element. The reaction in (a) belongs in a category that is discussed in the next section.

FIGURE 6.12

Reaction of aluminum metal and iron(III) oxide. A mixture of aluminum and iron(III) oxide is ignited. Once started, the reaction mixture burns violently, producing molten iron and an incandescent shower. Molten iron from this reaction is sometimes used for welding.

Exercise 6.5

Does the equation

$$2P(s) + 5Cl_2(g) \longrightarrow 2PCl_5(s)$$

represent a single-replacement reaction? Explain.

(Try Problems 6.27 and 6.28.)

Suppose you were asked if copper would replace zinc from zinc chloride in aqueous solution. In other words, will the following single-replacement reaction occur?

$$Cu(s) + ZnCl_2(aq) \longrightarrow CuCl_2(aq) + Zn$$

You can answer that question by appeal to the **activity series** (see Table 6.2). This is *a list of metallic elements (plus hydrogen) ordered by their relative activities in single-replacement reactions to form a given ion in aqueous solution.* An element A that is above another element B in the activity series will replace that element B from its compound BC. Note that the element zinc, Zn, is above the element copper, Cu, in the activity series. So, you would expect zinc to replace copper from one of its compounds; you would not expect just the opposite, that copper would replace zinc from one of its compounds. The answer to the question at the beginning of this paragraph is no.

$$Cu(s) + ZnCl_2(aq) \longrightarrow \text{no reaction}$$

TABLE 6.2
Activity Series

Metal	Ion Formed
Lithium Li	Li^+
Potassium K	K^+
Barium Ba	Ba^{2+}
Calcium Ca	Ca^{2+}
Sodium Na	Na^+
Magnesium Mg	Mg^{2+}
Aluminum Al	Al^{3+}
Manganese Mn	Mn^{2+}
Zinc Zn	Zn^{2+}
Chromium Cr	Cr^{3+}
Iron Fe	Fe^{2+}
Cadmium Cd	Cd^{2+}
Cobalt Co	Co^{2+}
Nickel Ni	Ni^{2+}
Tin Sn	Sn^{2+}
Lead Pb	Pb^{2+}
Hydrogen H_2	$2H^+$
Copper Cu	Cu^{2+}
Silver Ag	Ag^+
Mercury Hg	Hg^{2+}
Platinum Pt	Pt^{2+}
Gold Au	Au^{3+}

If a reaction does occur in aqueous solution, you will need to know the ion the free metal forms to write the balanced equation. If the metal is a main-group element, you can probably predict this ion. For a transition element, though, there are generally several possible metal ions. Table 6.2 helps you here by listing the ion that a metallic element forms in a single-replacement reaction in aqueous solution.

The next example gives you some practice in using the activity series.

EXAMPLE 6.6 Using the Activity Series

For each of the following, decide whether a reaction occurs. If it does, write a balanced equation, including symbols for the states. If it doesn't occur, write the equation specifying no reaction; that is, write the formulas of the substances, an arrow, followed by the words "no reaction."

(a) Iron metal is placed in an aqueous solution of copper sulfate.
(b) Iron metal is placed in an aqueous solution of magnesium sulfate.

Problem Analysis

Look at the activity series in Table 6.2. This series applies to single-replacement reactions in aqueous solution. If the free metal (iron in this example) is above the metal in the compound, such a reaction will occur. Otherwise, no single-replacement reaction will occur. Now write the appropriate chemical equation. You may need to use Table 6.2 to decide what ion is normally formed in such aqueous reactions.

Solution

(a) Iron is above copper in the activity series. Therefore, iron metal will replace copper in copper sulfate. To write the equation for the single-replacement reaction, you need to know the formula of the iron sulfate that results. Note that Table 6.2 indicates that iron forms Fe^{2+} ion in single-replacement reactions in aqueous solution. Thus, the product compound is of Fe^{2+} and SO_4^{2-}, or $FeSO_4$.

$$Fe(s) + CuSO_4(aq) \longrightarrow FeSO_4(aq) + Cu(s)$$

(b) Iron is below magnesium in the activity series. Therefore, iron will not react with magnesium sulfate in a single-replacement reaction.

$$Fe(s) + MgSO_4(aq) \longrightarrow \text{no reaction}$$

Exercise 6.6

Decide whether or not magnesium will react with silver nitrate in aqueous solution in a single-replacement reaction. If yes, write the balanced equation; otherwise, indicate the equation with the words "no reaction."

(Try Problems 6.29 and 6.30.)

6.6 Double-Replacement Reactions

OBJECTIVES

- Identify a chemical equation as a double-replacement reaction if you are given one of these types.
- Write the equation for a double-replacement reaction that occurs (or may occur) between given reactants.
- Describe in detail what happens when a solution of one ionic compound reacts with the solution of another ionic compound to produce a precipitate.

FIGURE 6.13
A coral reef. The main body of a coral reef is an accumulation of the calcium carbonate skeletons of the marine animals that form colonies on the reef. You can see these coral animals here, as well as the colorful fish that make the reef their home.

Many reactions that you encounter in a first course in chemistry are double-replacement reactions. In this section, we discuss how you can recognize such a reaction, how you would write such a reaction knowing the reactants, and how you can understand why such a reaction might occur. In later sections, we describe specific kinds of double-replacement reactions.

Recognizing Double-Replacement Reactions

Coral reefs, seashells, and limestone have a similar composition (Figure 6.13). They are composed principally of calcium carbonate. Although coral reefs and seashells have a biological origin (as does much limestone), the net chemical reaction for their formation is one involving calcium compounds and carbonate compounds in aqueous solution to produce calcium carbonate. Here is a representative equation showing the formation of calcium carbonate from a calcium compound and a carbonate compound:

$$CaCl_2(aq) + Na_2CO_3(aq) \longrightarrow CaCO_3(s) + 2NaCl(aq)$$

This is a **double-replacement reaction.** It is *a type of reaction in which two compounds exchange parts to form two new compounds*. We can write the following general equation for them:

$$AB + CD \longrightarrow AD + CB$$

Reactions of this type have also been called *metathesis* (me-tath′-e-sis) *reactions*.

When the reactants are ionic, it is the anions that exchange with each other. (Or, if you prefer, you can exchange both cations.) Thus, chloride ion that was associated with calcium ion in calcium chloride becomes associated with sodium ion to form sodium chloride. Similarly, carbonate ion that was associated with sodium ion in sodium carbonate becomes associated with calcium ion in calcium carbonate. When you make this exchange, however, you must remember to write the correct formulas of the products, given their ions (see Examples 5.2 and 5.6).

Writing Double-Replacement Reactions

The reaction shown in Figure 6.2 is a double-replacement reaction. Let's see how you use the idea of an exchange of ions to write the balanced equation for this reaction. The reactants are potassium iodide and lead(II) nitrate, both in aqueous solution. You must first write the formulas of these substances, given their names.

$$KI(aq) + Pb(NO_3)_2(aq) \longrightarrow$$

In a double-replacement reaction, the anions (or cations) are exchanged to give the products. Thus, the nitrate ion becomes associated with the potassium ion to form potassium nitrate, and the iodide ion becomes associated with the lead(II) ion to give lead(II) iodide. The formula of potassium nitrate is KNO_3, and the formula of lead(II) iodide is PbI_2. At the moment, you do not have enough background to know the physical states of these products. We simply note here that they are $KNO_3(aq)$ and $PbI_2(s)$. The equation, not balanced, is

$$KI(aq) + Pb(NO_3)_2(aq) \longrightarrow KNO_3(aq) + PbI_2(s)$$

Now you balance the equation:

$$2KI(aq) + Pb(NO_3)_2(aq) \longrightarrow 2KNO_3(aq) + PbI_2(s)$$

The essential idea, that you write the equation for a double-replacement reaction from the reactants by doing an exchange of ions, is simple enough. But note that you do have to understand how to write formulas from the ions and how to balance a chemical equation.

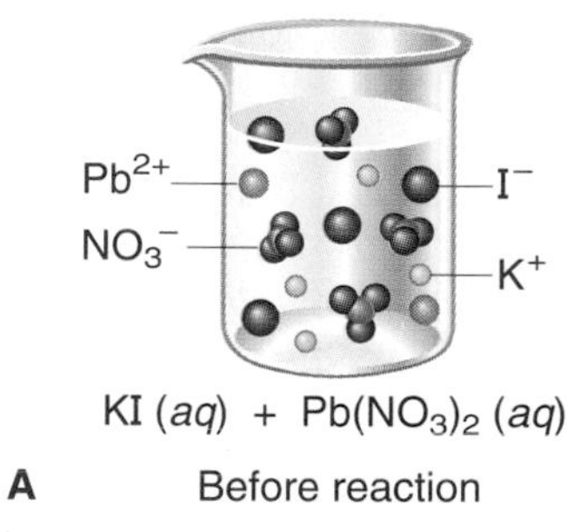

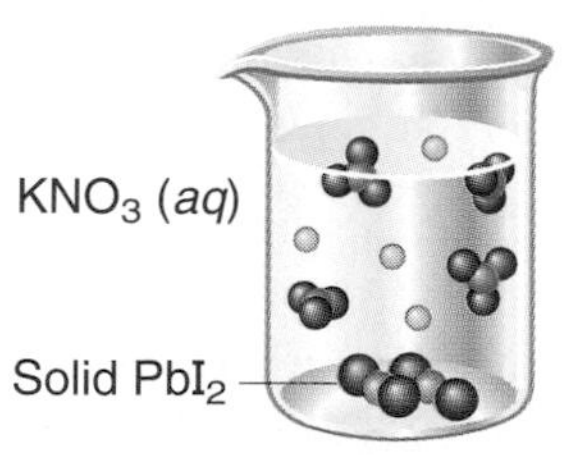

FIGURE 6.14
Reaction of potassium iodide solution with lead(II) nitrate. *(A)* Solution of potassium iodide and lead(II) nitrate. *(B)* When these solutions are mixed, the exchange of ions in the double-replacement reaction leads to the precipitation of lead(II) iodide. K^+ and NO_3^- ions remain in solution.

Understanding Double-Replacement Reactions

Can we explain why a particular double-replacement reaction occurs? Let us look in detail at the reaction of potassium iodide and lead(II) nitrate. When a solid ionic compound such as potassium iodide dissolves in water, the component ions of the solid go into the liquid water, where they are surrounded by water molecules. A solution of potassium iodide, then, consists of potassium ions and iodide ions constantly moving about the solution. In the same way, a solution of lead(II) nitrate consists of lead(II) ions and nitrate ions constantly moving about the solution. When you first mix these two solutions, you have one solution of four different ions: a solution of K^+, I^-, Pb^{2+}, and NO_3^-. See Figure 6.14A. To have a reaction, one of these cations must react with an anion to form a product in such a way that these ions are effectively removed from the solution. Otherwise, all you have is a solution of four different kinds of ions, but no reaction.

In the present example, lead(II) ions react with iodide ions to form a solid product. Figure 6.2 shows the formation of this product as a yellow cloud of solid crystals, which we call a *precipitate.* Each lead(II) iodide crystal consists of Pb^{2+} ions and I^- ions; but these ions are now separated from the K^+ ions and the NO_3^- ions in the solution. You no longer have simply a solution of four different kinds of ions. In effect, the formation of a precipitate removes some ions from the solution and in this sense "drives" the reaction to particular products. See Figure 6.14B.

Ions can be removed from the solution of four different kinds of ions in other ways. One way is for a cation to react with an anion to produce a molecular substance that then leaves the solution as a gas. The formation of a gaseous product is an effective way to drive a reaction to products.

A more subtle way to remove ions from a solution is for a cation to react with an anion to produce an especially stable molecular substance that then remains in the solution. A common reaction of this sort is a neutralization reaction. In this sort of reaction, hydrogen ions from an acid react with hydroxide ions from a compound called a *base* to form water, a very stable molecular substance. Here hydrogen ions and hydroxide ions are removed as water.

Here then are three common double-replacement reactions.

1. Formation of a solid (precipitate)
2. Formation of a gas
3. Formation of water (neutralization)

We consider each in the following three sections.

Chemical Perspective

Carbon Dioxide and the Permian Extinction

Carbon dioxide on earth has a complex relationship with living organisms. For instance, the acidic solution of carbon dioxide produced by the bacterial decay of dead plants is instrumental in the weathering of rocks on earth. We can represent this weathering by the following double-replacement reaction:

$$CaSiO_3(s) + 2H_2CO_3(aq) \longrightarrow Ca(HCO_3)_2(aq) + H_2SiO_3(aq)$$

Here $CaSiO_3$ is an approximate representation of a calcium silicate rock, and H_2SiO_3 is an approximate representation of the silicic acid produced, which yields silicon dioxide, SiO_2, in the form of quartz sand ($H_2SiO_3 \longrightarrow H_2O + SiO_2$). The calcium ion in $Ca(HCO_3)_2(aq)$ eventually washes into the rivers and then to the seas.

Marine organisms require this calcium ion for shell-building. These organisms eat plants, which have used carbon dioxide from the atmosphere, burning them for energy and yielding carbon dioxide. We can write the net result of this shell-building as a double-replacement equation, in which the calcium ion in seawater (represented here as calcium chloride) reacts with carbonic acid (from carbon dioxide and water) to produce the calcium carbonate shell:

$$CaCl_2(aq) + H_2CO_3(aq) \longrightarrow CaCO_3(s) + 2HCl(aq)$$

To form calcium carbonate, hydrogen ions (represented here by HCl) must be pumped away from the shell-producing cells. The shells of many dead organisms settle to the bottom of the oceans, forming sediments. The overall result is that carbon dioxide is slowly removed from the atmosphere by being incorporated in sedimentary calcium carbonate rock.

Recently several biologists and geochemists have implicated carbon dioxide in the mass extinction of animals during the Permian period, 250 million years ago. Carbon dioxide from the atmosphere dissolves in the oceans. It is even more prevalent at great depths, because carbon dioxide dissolves better at the high pressures of these depths. Soda pop is mainly a solution of carbon dioxide, produced by dissolving the gas under pressure. When you snap open a can of pop and release the pressure, carbon dioxide fizzes out (Figure 6.15). According to the hypothesis of these scientists, during this Permian period a colder climate over an interval of years significantly cooled the surface waters of the oceans. The deep ocean waters, containing lots of dissolved carbon dioxide, began to rise, much as hot air rises. When these waters reached the surface, they quickly charged the surface waters with an excess of carbon dioxide, and many marine animals died, unable to cope with these higher concentrations of carbon dioxide. Then these surface waters released their recently acquired load of carbon dioxide into the atmosphere, and other animals died. This release of

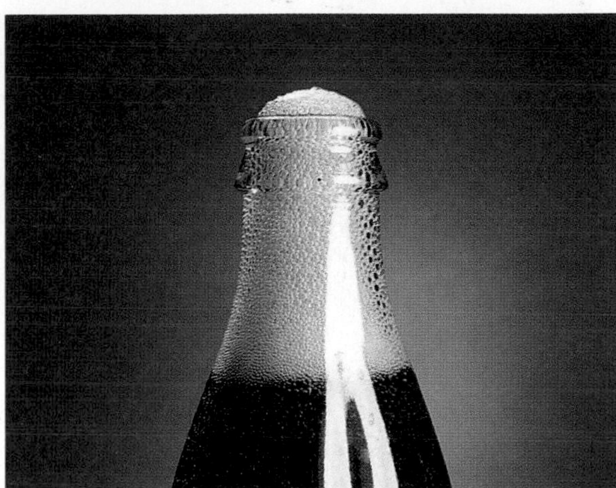

FIGURE 6.15
Carbon dioxide foam. When you open a bottle or can of warm pop, reducing the pressure, carbon dioxide fizzes out of the solution.

carbon dioxide from the oceans was like a giant burp of gas from an enormous can of pop. During this Permian period, about 90% of all animal species suddenly became extinct. A similar event, on a smaller scale, has occurred in recent times; see Figure 6.16.

FIGURE 6.16
Aerial view of Lake Nyos, 1986. More than a thousand people and many cattle died when asphixiated by a blanket of carbon dioxide gas that came from an enormous fountain of water and gas that shot up from the lake on August 21, 1986. Apparently, water at the bottom of the lake that was highly charged with carbon dioxide was somehow forced to the top, perhaps when the lake surface became especially cool.

Exploration: Principle reactions.

6.7 Double Replacement: Solid Forms (Precipitation)

OBJECTIVES

- Identify a chemical equation as a precipitation reaction, if you are given one of these types.
- Predict whether a precipitation reaction occurs, given the potential reactants.

Let's return to the equation we used in the last section:

$$2KI(aq) + Pb(NO_3)_2(aq) \longrightarrow 2KNO_3(aq) + PbI_2(s)$$

As you have seen, this reaction proceeds because the formation of solid lead(II) iodide drives the reaction by removing Pb^{2+} and I^- ions from the solution. *A solid product that forms when two solutions are mixed* is called a **precipitate.** In this reaction, the product lead(II) iodide is a precipitate. *The reaction in which a precipitate forms* is called a **precipitation reaction.** Such reactions depend on the formation of an *insoluble* product.

Insoluble compounds are *compounds that do not dissolve appreciably in a liquid* such as water; they are said to be insoluble in that liquid. (Compounds that dissolve are *soluble*.) Lead(II) iodide, one of the products in the previous reaction, is insoluble in water. ● Potassium nitrate, the other product in this reaction, is soluble and is present in the solution as K^+ and NO_3^- ions. (Solubility is discussed at length in Chapter 13.)

● No substance is completely insoluble in any liquid. For example, lead(II) iodide is not totally insoluble in water, since a small amount will dissolve (about 0.001 g/L) at 25°C. For this reason, some chemists use the expression "sparingly soluble" instead of "insoluble."

You can predict whether a precipitation reaction between two ionic substances occurs by asking whether the potential products are soluble or insoluble. If one of the

TABLE 6.3
The Solubilities of Some Common Ionic Compounds

Soluble Compounds	Exceptions
Lithium, sodium, potassium, and ammonium compounds	
Acetates and nitrates	
Hydrogen carbonates (bicarbonates)	
Fluorides	CaF_2
Chlorides, bromides, and iodides	Lead(II), silver, and mercury(I) compounds
Sulfates	$CaSO_4$, $SrSO_4$, $BaSO_4$, and $PbSO_4$
Oxalates	CaC_2O_4
Insoluble Compounds	**Exceptions**
Carbonates and phosphates	Lithium, sodium, potassium, and ammonium compounds
Hydroxides	LiOH, NaOH, KOH, $Ca(OH)_2$, and $Ba(OH)_2$
Sulfides	Li_2S, Na_2S, K_2S, $(NH_4)_2S$, CaS, and BaS

possible products is insoluble, it precipitates, and a precipitation reaction occurs. Table 6.3 classifies some common ionic compounds as soluble or insoluble. Use this table to help you decide whether a particular compound is soluble or not.

Let's see how this works. Suppose you did not know that KI and $Pb(NO_3)_2$ reacted, and you were asked: Will they give a precipitation reaction? First, you write down the potential double-replacement reaction, writing just the formulas:

$$KI + Pb(NO_3)_2 \longrightarrow KNO_3 + PbI_2 \qquad \text{(unbalanced)}$$

Table 6.3 notes that the chlorides, bromides, and iodides are soluble, except those of lead(II), silver, and mercury(I). Thus, you expect potassium iodide to be soluble and lead(II) iodide to be insoluble in aqueous solution. Table 6.3 also notes that nitrates are soluble. You can now write the previous equation with state labels, writing the label (*aq*) whenever a substance is soluble and (*s*) whenever it is insoluble:

$$KI(aq) + Pb(NO_3)_2(aq) \longrightarrow KNO_3(aq) + PbI_2(s)$$

You expect PbI_2 to precipitate, so a precipitation reaction occurs. In the final step, you balance the equation.

The following example will give you some practice in predicting whether or not two ionic substances will undergo a precipitation reaction.

■ EXAMPLE 6.7 Deciding If a Precipitate Will Form

(a) Oxalic acid ($H_2C_2O_4$) is a poison because it combines with the calcium ion, an ion vital to proper muscle control. Without this ion, muscle tissues go into spasm. If the reactants are represented by

$$H_2C_2O_4(aq) + CaCl_2(aq) \longrightarrow$$

will a precipitate form during the reaction? If so, write a balanced chemical equation that shows the states of the products. The oxalate ion is $C_2O_4^{2-}$.

(b) A chef mixes aqueous solutions of salt substitute (KCl) and baking soda, a substance whose chemical name is sodium hydrogen carbonate ($NaHCO_3$). Will a precipitation reaction occur? If so, write a balanced chemical equation that shows the states of the products.

Problem Analysis

Momentarily assume that a double-replacement reaction occurs between potential reactants. Write the equation for this reaction by exchanging the anions (or the cations) of the reactants to obtain the products. If one of the products is insoluble according to Table 6.3, it will precipitate. Therefore, a precipitation reaction occurs. No reaction occurs if both "products" in the double-replacement equation are soluble ionic compounds.

Solution

(a) Write the equation for the assumed double-replacement reaction:

$$H_2C_2O_4(aq) + CaCl_2(aq) \longrightarrow CaC_2O_4 + HCl$$

Table 6.3 tells you that CaC_2O_4 (calcium oxalate) is insoluble and that HCl is soluble, so add (*s*) after CaC_2O_4 and (*aq*) after HCl. Thus, reaction occurs because CaC_2O_4 precipitates. Here is the balanced equation with appropriate labels for the physical states:

$$H_2C_2O_4(aq) + CaCl_2(aq) \longrightarrow CaC_2O_4(s) + 2HCl(aq)$$

(b) Write the equation for the assumed double-replacement reaction:

$$KCl(aq) + NaHCO_3(aq) \longrightarrow KHCO_3 + NaCl$$

Table 6.3 says that sodium and potassium compounds are soluble. The double-replacement equation with labels for the states is

$$KCl(aq) + NaHCO_3(aq) \longrightarrow KHCO_3(aq) + NaCl(aq)$$

A precipitation reaction does not occur. In fact, no reaction occurs, because both "products" are soluble ionic compounds.

Exercise 6.7

Limestone, a material whose principal component is calcium carbonate, is a sedimentary rock used for statuary, building blocks, and manufacturing purposes. Will calcium carbonate precipitate when solutions of ammonium carbonate and calcium bromide are mixed? If so, write a balanced equation for the chemical reaction.

(Try Problems 6.31, 6.32, 6.33, and 6.34.)

6.8 Double Replacement: Gas Forms

OBJECTIVE

- Write the balanced equation for the reaction of a carbonate, a sulfite, or a sulfide with an acid to form a gas.

The reaction between acetic acid ($HC_2H_3O_2$) in vinegar and baking soda ($NaHCO_3$), two kitchen chemicals, is an excellent example of a reaction that yields a gaseous product (Figure 6.17). The first step is the double-replacement reaction to form carbonic acid:

$$\underset{\text{Acetic acid}}{HC_2H_3O_2(aq)} + \underset{\text{Sodium hydrogen carbonate}}{NaHCO_3(aq)} \longrightarrow \underset{\text{Sodium acetate}}{NaC_2H_3O_2(aq)} + \underset{\text{Carbonic acid}}{H_2CO_3(aq)}$$

However, carbonic acid is unstable. It decomposes to give carbon dioxide and water:

$$H_2CO_3(aq) \longrightarrow CO_2(g) + H_2O(l)$$

The overall result is

$$HC_2H_3O_2(aq) + NaHCO_3(aq) \longrightarrow NaC_2H_3O_2(aq) + \underbrace{CO_2(g) + H_2O(l)}_{H_2CO_3(aq)}$$

The foam that you see in Figure 6.17 consists of gas bubbles of carbon dioxide. When this gaseous substance bubbles out of the solution, it drives the reaction forward. In general, any carbonate compound will react with an acid to evolve carbon dioxide. ●

● Geologists use this reaction of carbonates to identify carbonate minerals in a rock. They treat the rock with hydrochloric acid to see if it fizzes. If it does, and if the gas is odorless, the rock contains a carbonate mineral.

Similarly, compounds containing the sulfite ion (SO_3^{2-}) react with acids to form sulfurous acid (H_2SO_3), which decomposes to give sulfur dioxide, a gas with a strongly irritating odor, and water. An example is the reaction between potassium sulfite and hydrochloric acid:

$$K_2SO_3(aq) + 2HCl(aq) \longrightarrow 2KCl(aq) + \underbrace{SO_2(g) + H_2O(l)}_{H_2SO_3}$$

Table 6.4 lists several kinds of compounds that evolve gases when treated with acids.

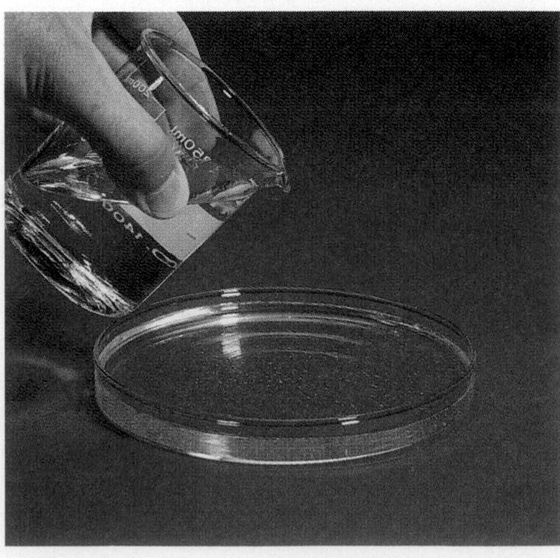

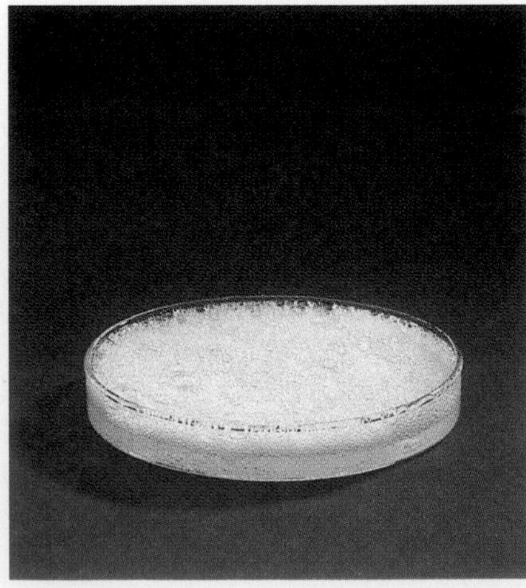

FIGURE 6.17
Reaction of sodium hydrogen carbonate with acetic acid. A solution of baking soda (sodium hydrogen carbonate) reacts with acetic acid in vinegar to yield sodium acetate solution and bubbles of carbon dioxide.

TABLE 6.4
Types of Compounds That Evolve Gases When Treated with Acids

Type of Compound	Gas	Example
Carbonates	CO_2	$Na_2CO_3(aq) + 2HCl(aq) \longrightarrow 2NaCl(aq) + H_2O(l) + CO_2(g)$
Sulfites	SO_2	$Na_2SO_3(aq) + 2HCl(aq) \longrightarrow 2NaCl(aq) + H_2O(l) + SO_2(g)$
Sulfides	H_2S	$Na_2S(aq) + 2HCl(aq) \longrightarrow 2NaCl(aq) + H_2S(g)$

The next example will give you practice in writing double-replacement reactions in which a gas forms.

EXAMPLE 6.8 Writing a Chemical Equation When a Gas Forms

Sphalerite, the principal ore of zinc, consists mainly of zinc sulfide (ZnS). When this compound is treated with hydrochloric acid, a double-replacement reaction occurs. Write a balanced chemical equation for this reaction, identifying the physical states of the substances.

Problem Analysis

Write the formulas of the reactants, then exchange anions (or cations) to complete the equation for the double-replacement reaction. Note whether any product is a gas (H_2S) or forms a gas by decomposition (H_2CO_3 or H_2SO_3). The formation of a gas effectively drives the reaction. If one of the products is H_2CO_3 or H_2SO_3, complete the equation by replacing this substance with its decomposition products. Balance the equation and add the physical states of the substances.

Solution

Begin by writing the reactants:

$$ZnS + HCl \longrightarrow$$

Exchanging the anions, but writing the correct formulas for products, gives the following balanced double-replacement equation:

$$ZnS + 2HCl \longrightarrow ZnCl_2 + H_2S$$

Note that one product is a gas (H_2S), so you can be sure that the reaction occurs as written. Now add labels for the states of the substances. Referring to Table 6.3, you see that zinc sulfide is insoluble, so add (*s*) after ZnS. Also, zinc chloride is soluble, so add (*aq*) after $ZnCl_2$. Hydrochloric acid is an aqueous solution, so add (*aq*) after HCl. Finally, write (*g*) after H_2S. The result is

$$ZnS(s) + 2HCl(aq) \longrightarrow ZnCl_2(aq) + H_2S(g)$$

Exercise 6.8

Limestone is principally calcium carbonate. Write a balanced equation for the reaction of this substance with nitric acid. Identify the physical states of the reactants and products in the equation.

(Try Problems 6.35, 6.36, 6.37, and 6.38.)

6.9 Double Replacement: Water Forms (Neutralization)

OBJECTIVES

- Define *acid, base,* and *salt.*
- Write the chemical equation for the neutralization reaction of a given acid with a given base.

Acids and bases are often used in laboratories, but you also encounter them in everyday life. Recall from Chapter 5 that an **acid** is *a compound that produces hydrogen ions* (H^+) *when it is dissolved in water.* A **base** is *a compound that produces hydroxide ions* (OH^-) *when it is dissolved in water.* Some examples of acids and bases are shown in Figure 6.18. In addition, Table 6.5 lists some common acids and bases.

Acids readily react with bases. *A reaction that occurs between an acid and a base with the formation of an ionic compound and usually water* is called a **neutralization reaction.** *The ionic compound formed in a neutralization reaction* is referred to as a **salt.**

If water is formed, the neutralization reaction can also be classified as a double-replacement reaction. Let's look at an example. A neutralization reaction occurs when aqueous solutions of hydrochloric acid and sodium hydroxide are mixed:

$$\underset{\text{Acid}}{HCl(aq)} + \underset{\text{Base}}{NaOH(aq)} \longrightarrow \underset{\text{Water}}{H_2O(l)} + \underset{\text{Ionic compound}}{NaCl(aq)}$$

If we write water as HOH, it may be easier for you to see that this is a double-replacement reaction:

$$HCl(aq) + NaOH(aq) \longrightarrow HOH(l) + NaCl(aq)$$

The formation of water removes two of the original four ions present and drives the reaction in the direction of the arrow.

Neutralization reactions are commonplace, as shown in the next example.

FIGURE 6.18

Examples of household acids and bases. Shown are vinegar (acetic acid), a lemon (citric acid), muriatic acid (hydrochloric acid), vitamin C (ascorbic acid), a household cleanser (containing a derivative of phenol, or carbolic acid), aspirin (acetylsalicylic acid), oven cleaner (sodium hydroxide), and drain cleaner (sodium hydroxide).

TABLE 6.5
Some Common Acids and Bases

Acid	Formula
Hydrochloric acid	HCl
Hydrobromic acid	HBr
Hydroiodic acid	HI
Nitric acid	HNO_3
Perchloric acid	$HClO_4$
Sulfuric acid	H_2SO_4
Phosphoric acid	H_3PO_4
Acetic acid	$HC_2H_3O_2$
Base	**Formula**
Lithium hydroxide	$LiOH$
Sodium hydroxide	$NaOH$
Potassium hydroxide	KOH
Calcium hydroxide	$Ca(OH)_2$
Barium hydroxide	$Ba(OH)_2$
Aluminum hydroxide	$Al(OH)_3$

EXAMPLE 6.9 Writing a Chemical Equation for Neutralization

A few antacids contain aluminum hydroxide. Note that this is a base because it contains hydroxide ion, some of which dissolves to yield hydroxide ion in water solution. Write a balanced chemical equation that describes the reaction of this compound with stomach acid (hydrochloric acid), showing the physical states of the reactants and products.

Problem Analysis

Write the formulas of the reactants, then exchange anions (or cations) to complete the equation for the double-replacement reaction. Note that this is a reaction of an acid with a base to form a salt and water; it is a neutralization reaction. Balance the equation and deduce the physical states of the reactants and products.

Solution

You obtain the formula of aluminum hydroxide from its ions: Al^{3+} and OH^-. Now write the reactants:

$$Al(OH)_3 + HCl \longrightarrow$$

Exchange the anions to obtain the products:

$$Al(OH)_3 + HCl \longrightarrow AlCl_3 + H_2O$$

In this neutralization reaction, the base aluminum hydroxide reacts with hydrochloric acid to produce the salt aluminum chloride and water. The balanced chemical equation is

$$Al(OH)_3 + 3HCl(aq) \longrightarrow AlCl_3 + 3H_2O(l)$$

The two physical states that you should know have been included. After consulting Table 6.3, you find that $Al(OH)_3$ is insoluble and $AlCl_3$ is soluble. The final equation is

$$Al(OH)_3(s) + 3HCl(aq) \longrightarrow AlCl_3(aq) + 3H_2O(l)$$

Exercise 6.9

The reaction of sulfuric acid with calcium hydroxide (slaked lime) produces a compound used in plaster and wallboard. Write the balanced chemical equation for this reaction and identify the compound.

(Try Problems 6.39, 6.40, 6.41, and 6.42.)

6.10 Combustion Reactions

OBJECTIVE

- Identify a chemical equation as a combustion reaction, if you are given one of these types.

When most substances burn, they are reacting with oxygen gas in the air. A flame results because a lot of energy in the form of heat is produced very quickly. The example presented at the beginning of this chapter occurs when you light a gas stove.

$$CH_4(g) + 2O_2(g) \longrightarrow CO_2(g) + 2H_2O(g)$$

This type of reaction is called a **combustion reaction.** It is *a reaction of a substance with either pure oxygen or oxygen in the air with the rapid release of heat and the appearance of a flame.* If the substance contains carbon, a product is usually carbon dioxide, but it could be carbon monoxide if only a little oxygen is present. If the substance also contains hydrogen, water is a product.

We rely on combustion reactions to produce energy for our homes and cars, and for industry. For example, a lot of electricity is obtained from coal-fueled power plants. The heat from the combustion of coal is used to convert water to steam, which turns a turbine from which electricity is produced. Because coal is essentially carbon, the reaction is

$$C(s) + O_2(g) \longrightarrow CO_2(g)$$

This equation shows that some combustion reactions can also be classified as combination reactions. Remember that a combination reaction occurs when two substances combine to form a third substance. The burning of magnesium is both a combustion reaction and a combination reaction (Figure 6.19):

$$2Mg(s) + O_2(g) \xrightarrow{\Delta} 2MgO(s)$$

Once you understand the essential components of combustion reactions, you will not have any problems in recognizing them. Can you find any combustion reactions in the next example?

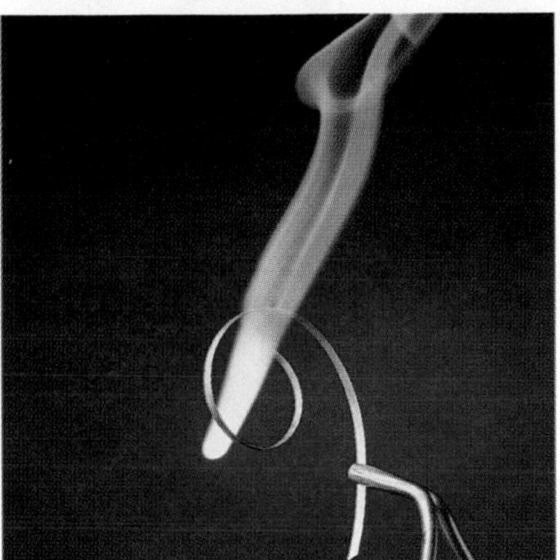

FIGURE 6.19
Magnesium burning. Heated magnesium metal reacts with oxygen in the air to yield magnesium oxide and a bright light. Photographic flashbulbs were based on this reaction. An electric current was used to ignite a magnesium wire in an oxygen-filled bulb.

EXAMPLE 6.10 Recognizing Combustion Reactions

Which of the following reactions is a combustion reaction? Why?

(a) Tums® is an antacid; it contains calcium carbonate that reacts with stomach acid (HCl):

$$CaCO_3(s) + 2HCl(aq) \longrightarrow CaCl_2(aq) + CO_2(g) + H_2O(l)$$

(b) When a cigarette lighter filled with butane is working, the reaction

$$2C_4H_{10}(g) + 13O_2(g) \longrightarrow 8CO_2(g) + 10H_2O(g)$$

provides the heat that you feel and the flame that you see.

(c) The disastrous fire and explosion aboard the hydrogen-filled German zeppelin, the *Hindenburg*, in 1937 was caused by the reaction

$$2H_2(g) + O_2(g) \longrightarrow 2H_2O(g)$$

Problem Analysis

Remember that a combustion reaction is a reaction of a substance with either pure oxygen or oxygen in the air with the rapid release of heat and the appearance of a flame.

Solution

The reaction in (a) is a double-replacement reaction that is driven by the release of a gas. The reactions in (b) and (c) are combustion reactions. They are reactions with oxygen, accompanied by heat and flame.

Exercise 6.10

Which of the following reactions could be combustion reactions? Classify all the reactions and balance all the equations.

(a) $C_3H_8(g) + O_2(g) \longrightarrow CO_2(g) + H_2O(g)$
(b) $Mg(s) + HBr(aq) \longrightarrow MgBr_2(aq) + H_2(g)$
(c) $CS_2(l) + O_2(g) \longrightarrow CO_2(g) + SO_2(g)$

(Try Problems 6.43 and 6.44.)

Chemical Perspective

Hemoglobin and Its Remarkable Reactions

The characteristic color of blood is due to the deep red color of hemoglobin, an important chemical compound in red blood cells. We could not live without hemoglobin, because it carries oxygen from our lungs to every cell in our bodies. Each molecule of hemoglobin consists of four equivalent subunits that are joined together to make a complete molecule. Representations of one of the subunits and the entire molecule are shown in Figure 6.20. How could a molecule that looks like a nest of writhing worms be something that our lives depend on? How does this compound carry oxygen?

Each subunit contains a flat heme group, as shown on the left side of Figure 6.20, and each of these groups can bind chemically to one molecule of oxygen. Since there are four subunits, hemoglobin is capable of carrying four O_2 molecules. Each time a molecule of oxygen is bound, the shape of the hemoglobin molecule changes in a certain way to make it easier to bind the next O_2 molecule. Some chemists use Hb to represent the complete molecule, and we can show hemoglobin's reaction with oxygen as

$$\text{Hb} + 4\text{O}_2 \longrightarrow \text{Hb(O}_2)_4$$

The product of this combination reaction will also lose oxygen during a decomposition reaction:

$$\text{Hb(O}_2)_4 \longrightarrow \text{Hb} + 4\text{O}_2$$

Once again, you can see that combination and decomposition reactions are opposites.

Myoglobin, the oxygen-storage compound of mammalian muscle, resembles one of the subunits of hemoglobin. Since myoglobin (Mb) is equivalent to just one subunit, it binds only one molecule of oxygen:

$$\text{Mb} + \text{O}_2 \longrightarrow \text{Mb(O}_2)$$

but with a higher affinity than hemoglobin. This higher affinity for oxygen allows myoglobin in muscle tissues to take oxygen from oxyhemoglobin in the blood and store it until it is used. This phenomenon resembles a single-replacement reaction and is represented by the equation

$$\text{Hb(O}_2)_4 + 4\text{Mb} \longrightarrow \text{Hb} + 4\text{Mb(O}_2)$$

The highest concentration of myoglobin in humans is found in cardiac muscle because considerable oxygen is required by the heart in times of stress. The myoglobin content of the muscles of diving animals such as whales is greater than that of humans. As a result, these animals are able to remain submerged for long periods of time.

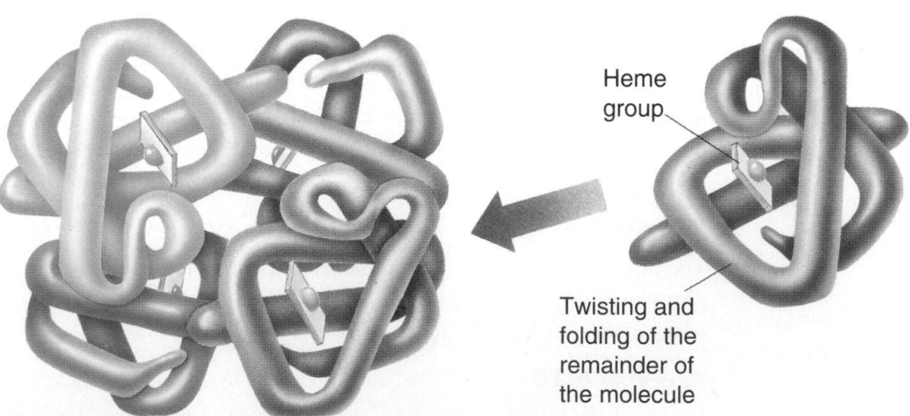

FIGURE 6.20

Hemoglobin. *(Left)* The entire hemoglobin molecule with four subunits. *(Right)* One of the four subunits in hemoglobin. The flat heme group that binds chemically to oxygen is attached to a very long molecule, shown here as a wormlike, twisted cylinder.

CHAPTER REVIEW

Key Words

reactant *(p. 149)*
product *(p. 149)*
chemical equation *(p. 150)*
coefficients *(p. 150)*
catalyst *(p. 152)*
balanced chemical equation *(p.153)*
combination reaction *(p. 159)*
decomposition reaction *(p. 161)*
single-replacement reaction *(p. 163)*
activity series *(p. 164)*
double-replacement reaction *(p. 166)*
precipitate *(p. 169)*
precipitation reaction *(p. 169)*
insoluble *(p. 169)*
acid *(p. 174)*
base *(p. 174)*
neutralization reaction *(p. 174)*
salt *(p. 174)*
combustion reaction *(p. 176)*

Summary

A *chemical reaction* is a change in which one kind of matter is transformed into another kind of matter. In every chemical reaction, *reactants*—chemicals that are present before the reaction—are changed into *products*—chemicals that result from the reaction. Most chemical reactions display certain signals, such as the evolution of heat, the emission of light, the formation of a solid, the formation of gas bubbles, or a change in color.

A chemical reaction is expressed symbolically by a *chemical equation*. In such equations, the formulas for the reactants are placed on the left side of an arrow that points toward the formulas of the products. The equation is *balanced* by placing coefficients (whole numbers) in front of the formulas so that the number of atoms of each element is identical on each side of the arrow.

Chemical reactions can be divided conveniently into at least five different types: *combination reactions* (in which two substances react to form a third), *decomposition reactions* (in which a single compound decomposes into two or more products), *single-replacement reactions* (in which an element reacts with a compound by replacing another element in the compound), *double-replacement reactions* (in which two compounds exchange parts so that two new compounds are formed), and *combustion reactions* (in which a substance reacts with oxygen, producing heat and flame).

Double-replacement reactions are driven in the direction of the arrow by the formation of a solid (precipitation), the formation of a gas, or the formation of water (neutralization). A *neutralization reaction* occurs when an acid reacts with a base. An *acid* is a compound that produces hydrogen ions when it is dissolved in water. A *base* is a compound that produces hydroxide ions when dissolved in water. When acids and bases react, they form an ionic compound (a *salt*) and usually water.

Problem-Solving Skills

1. **Interpreting a Chemical Equation:** Given a chemical equation, identify the reactants and products as well as the number of formula units or molecules and the physical state of each (Example 6.1).
2. **Balancing an Equation:** Given a chemical equation, balance it (Example 6.2).
3. **Going from a Description to a Balanced Equation:** Given a word description of a chemical reaction, deduce the matching chemical equation and balance it (Example 6.3).
4. **Recognizing Combination and Decomposition Reactions:** Given a chemical equation, decide whether it represents a combination reaction, a decomposition reaction, or neither type (Example 6.4).
5. **Recognizing Single-Replacement Reactions:** Given a chemical equation, decide whether it represents a single-replacement reaction (Example 6.5).
6. **Using the Activity Series:** Given a metal and an ionic compound of another metal, decide if a single-replacement reaction will occur (Example 6.6).
7. **Deciding If a Precipitate Will Form:** Given the possible reactants for a chemical reaction, decide if a reaction will occur because a solid (precipitate) will form (Example 6.7).
8. **Writing a Chemical Equation When a Gas Forms:** Given a chemical reaction in which a gas is formed, write a balanced chemical equation (Example 6.8).
9. **Writing a Chemical Equation for Neutralization:** Given an acid and a base, write a balanced chemical equation for neutralization (Example 6.9).
10. **Recognizing Combustion Reactions:** Given a chemical equation, decide whether it represents a combustion reaction (Example 6.10).

Questions to Test Your Reading

6.1 What is a chemical reaction? Define reactant. Define product.

6.2 What is a chemical equation? What is a coefficient? Give an example of a balanced equation.

6.3 What is a combination reaction? Give an example.

6.4 What is a decomposition reaction? Will two or more elements always be the products of this type of reaction? Explain with examples.

6.5 What is a single-replacement reaction? Give an example.

6.6 How do you use the activity series to decide if a particular single-replacement reaction will occur?

6.7 What is a double-replacement reaction? How does it differ from a metathesis reaction?

6.8 How are double-replacement reactions "driven" in the direction of the arrow? Identify three different ways and give an example of each.

6.9 Define acid and base. What is a neutralization reaction? Give an example.

6.10 What is a combustion reaction? Can all combustion reactions be classified as combination reactions? Explain with examples.

Practice Problems

Interpreting Chemical Equations (Section 6.2)

6.11 Identify the reactants and products as well as the number of formula units or molecules and the physical state of each one in the following reactions. If a reaction condition is given, state what it is.
(a) $2K(s) + Br_2(l) \longrightarrow 2KBr(s)$
(b) $P_4(s) + 6Cl_2(g) \longrightarrow 4PCl_3(l)$

6.12 Identify the reactants and products as well as the number of formula units or molecules and the physical state of each one in the following reactions. If a reaction condition is given, state what it is.
(a) $Ca(s) + 2H_2O(l) \longrightarrow Ca(OH)_2(aq) + H_2(g)$
(b) $C_2H_5OH(l) + 3O_2(g) \longrightarrow 2CO_2(g) + 3H_2O(g)$

6.13 Identify the reactants and products as well as the number of formula units or molecules and physical state of each one in the following reactions. If a reaction condition is given, state what it is.
(a) $BaCO_3(s) \xrightarrow{\Delta} BaO(s) + CO_2(g)$
(b) $2H_2O_2(aq) \xrightarrow{KI} 2H_2O(l) + O_2(g)$

6.14 Identify the reactants and products as well as the number of formula units or molecules and physical state of each one in the following reactions. If a reaction condition is given, state what it is.
(a) $2KNO_3(aq) \xrightarrow{\Delta} 2KNO_2(aq) + O_2(g)$
(b) $2CO(g) + O_2(g) \xrightarrow{Pt} 2CO_2(g)$

Balancing Chemical Equations (Section 6.3)

6.15 Balance the following chemical equations.
(a) $CH_3OH + O_2 \longrightarrow CO_2 + H_2O$
(b) $Mg + SiO_2 \longrightarrow MgO + Si$
(c) $Cl_2O_7 + H_2O \longrightarrow HClO_4$
(d) $TiCl_4 + H_2O \longrightarrow TiO_2 + HCl$

6.16 Balance the following chemical equations.
(a) $C_2H_5OH + O_2 \longrightarrow CO_2 + H_2O$
(b) $Fe_2O_3 + H_2 \longrightarrow Fe + H_2O$
(c) $Al_2S_3 + H_2O \longrightarrow Al(OH)_3 + H_2S$
(d) $C_5H_{12} + O_2 \longrightarrow CO_2 + H_2O$

6.17 Balance the following chemical equations.
(a) $CO + H_2 \longrightarrow CH_4 + H_2O$
(b) $Ba + H_2O \longrightarrow Ba(OH)_2 + H_2$
(c) $H_2S + O_2 \longrightarrow H_2O + SO_2$
(d) $H_3PO_3 \longrightarrow H_3PO_4 + PH_3$

6.18 Balance the following chemical equations.
(a) $V_2O_5 + H_2 \longrightarrow V_2O_3 + H_2O$
(b) $C_6H_6 + O_2 \longrightarrow CO_2 + H_2O$
(c) $Al_4C_3 + H_2O \longrightarrow Al(OH)_3 + CH_4$
(d) $NH_3 + O_2 \longrightarrow NO + H_2O$

6.19 Provide a balanced chemical equation from the following description: Nitrogen gas reacts with oxygen gas to give gaseous nitrogen monoxide.

6.20 Provide a balanced chemical equation from the following description: Nitrogen monoxide will react with oxygen gas to give gaseous nitrogen dioxide.

6.21 Provide a balanced chemical equation from the following description: When gaseous nitrogen dioxide reacts with oxygen gas, the product is gaseous dinitrogen pentoxide.

6.22 Provide a balanced chemical equation from the following description: When gaseous dinitrogen pentoxide is heated over a platinum catalyst, nitrogen gas and oxygen gas are formed.

Combination and Decomposition Reactions (Section 6.4)

6.23 Which of the following reactions is either a combination reaction or a decomposition reaction?
(a) $2C_2H_5SH(l) + 9O_2(g) \longrightarrow 4CO_2(g) + 6H_2O(g) + 2SO_2(g)$
(b) $C_3H_6(g) + Br_2(l) \longrightarrow C_3H_6Br_2(l)$
(c) $NH_4Br(s) \xrightarrow{\Delta} NH_3(g) + HBr(g)$
(d) $F_2(g) + Ba(s) \longrightarrow BaF_2(s)$

6.24 Which of the following reactions is either a decomposition reaction or a combination reaction?
(a) $MgCO_3(s) \xrightarrow{\Delta} MgO(s) + CO_2(g)$
(b) $4Li(s) + O_2(g) \longrightarrow 2Li_2O(s)$
(c) $C_2H_4(g) + H_2(g) \xrightarrow{\Delta} C_2H_6(g)$
(d) $2N_2O_5(g) \longrightarrow 4NO_2(g) + O_2(g)$

6.25 Balance each of the following equations. Which of them is a combination reaction or a decomposition reaction?
(a) $CH_4O(l) + O_2(g) \longrightarrow CO_2(g) + H_2O(g)$
(b) $NO_2(g) \longrightarrow NO(g) + O_2(g)$
(c) $N_2(g) + H_2(g) \longrightarrow NH_3(g)$
(d) $Li(s) + O_2(g) \longrightarrow Li_2O(s)$

6.26 Balance each of the following equations. Which of them is a decomposition reaction or a combination reaction?
(a) $C_2H_6O(l) + O_2(g) \longrightarrow CO(g) + H_2O(g)$
(b) $C_2H_6O(l) + O_2(g) \longrightarrow CO_2(g) + H_2O(g)$
(c) $P_4(s) + Cl_2(g) \longrightarrow PCl_3(l)$
(d) $N_2(g) + I_2(g) \longrightarrow NI_3(g)$

Single-Replacement Reactions (Section 6.5)

6.27 Balance each equation and decide which one is a single-replacement reaction.
(a) $Cd(s) + AgNO_3(aq) \longrightarrow Cd(NO_3)_2(aq) + Ag(s)$
(b) $C_3H_6(g) + H_2(g) \longrightarrow C_3H_8(g)$

6.28 Balance each equation and decide which one is a single-replacement reaction.
(a) $O_2(g) + F_2(g) \longrightarrow OF_2(g)$
(b) $Mg(s) + Fe_2(SO_4)_3(aq) \longrightarrow MgSO_4(aq) + Fe(s)$

6.29 Using the activity series (Table 6.2) for each of these, decide whether the reaction occurs as written.
(a) $CdCl_2(aq) + Ni(s) \longrightarrow NiCl_2(aq) + Cd(s)$
(b) $2Al(s) + 3CdCl_2(aq) \longrightarrow 3Cd(s) + 2AlCl_3(aq)$

6.30 Using the activity series (Table 6.2) for each of these, decide whether the reaction occurs as written.
(a) $Mg(s) + SnCl_2(aq) \longrightarrow Sn(s) + MgCl_2(aq)$
(b) $Mn(NO_3)_2(aq) + Ag(s) \longrightarrow AgNO_3(aq) + Mn(s)$

Double-Replacement Reactions (Sections 6.6 to 6.9)

6.31 Will a precipitate form when an aqueous solution of lead nitrate is mixed with an aqueous solution of sodium carbonate? Make your decision based on Table 6.3. If there is a reaction, write the appropriate balanced chemical equation along with the states of the reactants and products.

6.32 Will a precipitate form when an aqueous solution of nickel(II) chloride is mixed with an aqueous solution of sodium sulfide? Make your decision based on Table 6.3. If there is a reaction, write the appropriate balanced chemical equation along with the states of the reactants and products.

6.33 Based on Table 6.3, will a precipitate form during a reaction between the following reactants? If so, write a balanced chemical equation that shows the products and the physical states of all substances.
(a) $MgSO_4 + NaOH \longrightarrow$
(b) $NiCl_2 + NaBr \longrightarrow$
(c) $NaCl + KNO_3 \longrightarrow$
(d) $NH_4NO_3 + NaCl \longrightarrow$

6.34 Based on Table 6.3, will a precipitate form during a reaction between the following reactants? If so, write a balanced chemical equation that shows the products and the physical states of all substances.
(a) $Ba(NO_3)_2 + K_2SO_4 \longrightarrow$
(b) $CaBr_2 + NaNO_3 \longrightarrow$
(c) $AgNO_3 + NaI \longrightarrow$
(d) $Pb(NO_3)_2 + NH_4Cl \longrightarrow$

6.35 Will a gas form when an aqueous solution of sodium carbonate is mixed with an aqueous solution of acetic acid? (Use Table 6.4 to make your decision.) If so, write a balanced equation that shows the products and the physical states of all substances.

6.36 Will a gas form when an aqueous solution of hydrobromic acid is mixed with an aqueous solution of lithium sulfite? (Use Table 6.4 to make your decision.) If so, write a balanced equation that shows the products and the physical states of all substances.

6.37 Will a gas form during a reaction between the following reactants? (Use Table 6.4 to make your decision.) If so, write a balanced chemical equation that shows the products and the physical states of all substances.
(a) $HCl + KBr \longrightarrow$
(b) $Na_2SO_3 + NaOH \longrightarrow$
(c) $(NH_4)_2CO_3 + HCl \longrightarrow$
(d) $Na_2CO_3 + LiI \longrightarrow$

6.38 Will a gas form during a reaction between the following reactants? (Use Table 6.4 to make your decision.) If so, write a balanced chemical equation that shows the products and the physical states of all substances.
(a) $MnS + H_2SO_4 \longrightarrow$
(b) $NaCl + H_3PO_4 \longrightarrow$
(c) $CaCO_3 + HNO_3 \longrightarrow$
(d) $CaCO_3 + NaNO_3 \longrightarrow$

6.39 Which of the following substances are acids or bases?
(a) $NaOH$ (b) HNO_3 (c) $Ca(OH)_2$
(d) HCl (e) KI (f) $NaNO_3$
(g) Li_2SO_4 (h) KOH (i) BaF_2

6.40 Which of the following substances are acids or bases?
(a) H_2SO_4 (b) $Ba(OH)_2$ (c) $NaBr$
(d) HBr (e) $LiNO_3$ (f) $Al(OH)_3$
(g) KCl (h) $HC_2H_3O_2$ (i) $HClO_4$

6.41 Complete and balance the following equations for neutralization reactions.
(a) $HNO_3(aq) + NaOH(aq) \longrightarrow$
(b) $Ba(OH)_2(aq) + HCl(aq) \longrightarrow$
(c) $LiOH(aq) + HBr(aq) \longrightarrow$
(d) $HC_2H_3O_2(aq) + NaOH(aq) \longrightarrow$

6.42 Complete and balance the following equations for neutralization reactions.
(a) $Al(OH)_3(s) + HNO_3(aq) \longrightarrow$
(b) $H_2SO_4(aq) + KOH(aq) \longrightarrow$
(c) $NaOH(aq) + HI(aq) \longrightarrow$
(d) $HClO_4(aq) + Ba(OH)_2(aq) \longrightarrow$

Combustion Reactions (Section 6.10)

6.43 When octane (C_8H_{18}), a component of gasoline, reacts with oxygen, one of the products is always water. The other product can be either carbon dioxide or carbon monoxide, depending on how much oxygen is present. Is this a combustion reaction? Explain and write balanced equations.

6.44 Sometimes butane (C_4H_{10}) in a cigarette lighter burns with a sooty flame when enough oxygen is not present. The soot is carbon. The other product is water. Is this a combustion reaction? Explain and write a balanced equation.

Additional Problems

6.45 Using the following description, write a chemical equation that uses the correct symbols, physical states, and reaction conditions: When one formula unit of solid ammonium chloride (NH_4Cl) is heated, one molecule of gaseous ammonia (NH_3) and one molecule of gaseous hydrogen chloride (HCl) are formed.

6.46 Using the following description, write a chemical equation that uses the correct symbols, physical states, and reaction conditions: Two atoms of solid sodium react with two molecules of liquid water to give two formula units of sodium hydroxide ($NaOH$) in aqueous solution and one molecule of hydrogen gas.

6.47 Carbon dioxide is absorbed from the air by green plants, where it reacts with water under the influence of sunlight to yield glucose and oxygen:

$$CO_2 + H_2O \longrightarrow C_6H_{12}O_6 + O_2$$

This reaction replenishes the oxygen in our atmosphere. Balance the chemical equation.

6.48 In the presence of enzymes from yeast, ordinary table sugar (sucrose) is converted to ethanol and carbon dioxide according to the following unbalanced equation:

$$\underset{\text{Sucrose}}{C_{12}H_{22}O_{11}} + H_2O \longrightarrow \underset{\text{Ethanol}}{C_2H_6O} + CO_2$$

This reaction is called *fermentation,* and it is used to produce alcoholic beverages. Balance the chemical equation.

6.49 Sodium stearate ($NaC_{18}H_{36}O_2$) is often found in soap. It reacts with calcium compounds in hard water to give a curdy, insoluble product, calcium stearate. Write the chemical equation for the reaction of an aqueous solution of sodium stearate with an aqueous solution of calcium chloride. Show the physical states.

6.50 When table sugar ($C_{12}H_{22}O_{11}$) is heated, it turns into a black, charred mass of elemental carbon with the release of steam (water vapor). Write the chemical equation for this reaction, showing the physical states.

6.51 When metallic lithium is heated in nitrogen gas, lithium nitride (Li_3N) is formed. Write a balanced chemical equation for this reaction.

6.52 When metallic magnesium is heated in nitrogen gas, magnesium nitride (Mg_3N_2) is formed. Write a balanced equation for this reaction.

6.53 What drives each of the following reactions in the direction of the arrow? Balance each equation.
(a) $HNO_3(aq) + NaOH(aq) \longrightarrow NaNO_3(aq) + H_2O(l)$
(b) $HNO_3(aq) + NaCN(aq) \longrightarrow NaNO_3(aq) + HCN(g)$
(c) $Pb(NO_3)_2(aq) + NaCl(aq) \longrightarrow NaNO_3(aq) + PbCl_2(s)$

6.54 What drives each of the following reactions in the direction of the arrow? Balance each equation.
(a) $Ni(NO_3)_2(aq) + NaOH(aq) \longrightarrow NaNO_3(aq) + Ni(OH)_2(s)$
(b) $HCl(aq) + Na_2SO_3(aq) \longrightarrow NaCl(aq) + H_2O(l) + SO_2(g)$
(c) $H_3PO_4(aq) + KOH(aq) \longrightarrow K_3PO_4(aq) + H_2O(l)$

6.55 Which substances are insoluble in water?
(a) Na_2SO_4 (b) $LiBr$ (c) $CaCO_3$
(d) $Ba_3(PO_4)_2$ (e) $Cu(OH)_2$ (f) Na_3PO_4
(g) $AgCl$ (h) $AgNO_3$ (i) $PbBr_2$

6.56 Which substances are insoluble in water?
(a) CaF_2 (b) AgF (c) AgI (d) K_2S
(e) $NaOH$ (f) $BaSO_4$ (g) NH_4Br
(h) CuS (i) NH_4NO_3

6.57 Classify each of the following reactions according to its type, and then balance the equation.
(a) $Na(s) + H_2O(l) \longrightarrow NaOH(aq) + H_2(g)$
(b) $CrCl_3(aq) + NaOH(aq) \longrightarrow Cr(OH)_3(s) + NaCl(aq)$
(c) $Zn(s) + CuSO_4(aq) \longrightarrow ZnSO_4(aq) + Cu(s)$
(d) $C_6H_{12}(l) + O_2(g) \longrightarrow CO_2(g) + H_2O(g)$

6.58 Classify each of the following reactions according to its type, and then balance the equation.
(a) $HI(aq) + Al(OH)_3(s) \longrightarrow H_2O(l) + AlI_3(aq)$
(b) $Al(s) + H_3PO_4(aq) \longrightarrow AlPO_4(s) + H_2(g)$
(c) $CaSO_3(s) \longrightarrow CaO(s) + SO_2(g)$
(d) $Al(s) + O_2(g) \longrightarrow Al_2O_3(s)$

6.59 Complete and balance each of the following equations. Show the physical states of the products.
(a) $BaCO_3(s) + HNO_3(aq) \longrightarrow$
(b) $BaCl_2(aq) + H_2SO_4(aq) \longrightarrow$
(c) $Ca(OH)_2(s) + HC_2H_3O_2(aq) \longrightarrow$
(d) $Al(s) + NiSO_4(aq) \longrightarrow$

6.60 Complete and balance each of the following equations. Show the physical states of the products.
(a) $Ni(OH)_2(s) + HCl(aq) \longrightarrow$
(b) $MnCl_2(aq) + NaOH(aq) \longrightarrow$
(c) $KOH(aq) + H_3PO_4(aq) \longrightarrow$
(d) $MnS(s) + H_2SO_4(aq) \longrightarrow$

6.61 Predict the product or products of the following reactions, and write balanced equations that show the physical states of all compounds.
(a) the reaction of lithium and chlorine
(b) the combustion of solid decane ($C_{10}H_{22}$) in the presence of excess oxygen
(c) the combustion of solid decane ($C_{10}H_{22}$) in the presence of a deficiency of oxygen
(d) the decomposition of solid gold(III) oxide into its elements by heating

6.62 Predict the product or products of the following reactions, and write balanced equations that show the physical states of all compounds.
(a) the decomposition of sulfurous acid in an aqueous solution
(b) the reaction between an aqueous solution of sodium hydroxide and perchloric acid
(c) the combination of sodium oxide and sulfur trioxide
(d) the combustion of solid phenol (C_6H_6O) in the presence of excess oxygen

6.63 When lead(II) sulfide is heated in oxygen, the products are solid lead(II) oxide and gaseous sulfur dioxide. Write a balanced equation for this reaction. Show the physical states of all compounds and show that heat was used.

6.64 When gaseous ammonia (NH_3) burns in oxygen over a platinum catalyst, gaseous nitrogen monoxide and water vapor are formed. Write a balanced equation for this reaction. Show the physical states of all compounds and show that a catalyst was used.

6.65 Elemental iron does not occur in nature. Instead, it is prepared in a blast furnace from an iron ore such as hematite (Fe_2O_3). The principal reaction occurs when solid iron(III) oxide reacts with carbon monoxide gas to yield molten (liquid) iron and carbon dioxide. Write a balanced equation for this reaction, showing the physical states of all compounds.

6.66 The battery in a car produces electricity through the reaction of solid lead, solid lead(IV) oxide, and aqueous sulfuric acid. The products of this reaction are solid lead(II) sulfate and water. Write a balanced equation for this reaction and show the physical states of all compounds.

Practice in Problem Analysis

For each problem, describe the thinking you would use (the problem analysis) before doing the actual solution, but do not solve the problem.

1. Write a chemical equation that describes how you would prepare barium sulfate from potassium sulfate and another reactant by a precipitation reaction.

2. Write a chemical equation that describes how you would prepare sodium hydroxide from calcium hydroxide and another reactant by a precipitation reaction.

PRACTICE EXAM

1. Consider the chemical equation

$$2H_2O_2(aq) \xrightarrow{Br^-} 2H_2O(l) + O_2(g)$$

Which of the following statements best fits the description given by this equation?
(a) Hydrogen peroxide liquid decomposes to water and oxygen gas.
(b) Hydrogen peroxide solution decomposes with a bromide ion catalyst to give liquid water and oxygen gas.
(c) Hydrogen peroxide liquid decomposes with a bromide ion catalyst to liquid water and oxygen gas.
(d) Hydrogen peroxide solution reacts with bromide ion to give liquid water and oxygen gas.
(e) Hydrogen peroxide decomposes to water and oxygen.

2. Consider the following chemical equation.

$$2Na(s) + 2H_2O(l) \longrightarrow 2NaOH(aq) + H_2(g)$$

Which of the following statements is not true?
(a) Solid sodium reacts with water.
(b) Sodium hydroxide is a product.
(c) The coefficient of hydrogen is one.
(d) As indicated by the (*aq*) on the right, water is a product.
(e) The products are a solution and a gas.

3. Which of the following equations is not balanced?
(a) $NO(g) + O_2(g) \longrightarrow NO_2(g)$
(b) $2Na(s) + Cl_2(g) \longrightarrow 2NaCl(s)$
(c) $HBr + KOH \longrightarrow KBr + H_2O$
(d) $Zn(s) + 2HBr(aq) \longrightarrow ZnBr_2(aq) + H_2(g)$
(e) None of the above

4. Nickel metal reacts with hydrochloric acid to produce hydrogen and a solution of nickel chloride. Which of the following is a balanced equation for this reaction?
(a) $Ni(s) + 2HCl(g) \longrightarrow NiCl_2(aq) + H_2(g)$
(b) $Ni(s) + 2HCl(aq) \longrightarrow NiCl_2(s) + H_2(g)$
(c) $Ni(s) + 2HCl(aq) \longrightarrow NiCl_2(aq) + H_2(g)$
(d) $2Ni(s) + 2HCl(aq) \longrightarrow 2NiCl_2(aq) + H_2(g)$
(e) None of the above

5. Which of the following is a combination reaction?
(a) $Zn(s) + CuBr_2(aq) \longrightarrow ZnBr_2(aq) + Cu(s)$
(b) $2SO_2(g) + O_2(g) \longrightarrow 2SO_3(g)$
(c) $PbO_2(s) \longrightarrow Pb(s) + O_2(g)$
(d) $2HI(aq) + Na_2S(aq) \longrightarrow H_2S(g) + 2NaI(aq)$
(e) None of the above

6. Which of the following is a decomposition reaction?
(a) $Zn(s) + CuBr_2(aq) \longrightarrow ZnBr_2(aq) + Cu(s)$
(b) $SO_2(g) + H_2O(l) \longrightarrow H_2SO_3(aq)$
(c) $PbO_2(s) \longrightarrow Pb(s) + O_2(g)$
(d) $2HI(aq) + Na_2S(aq) \longrightarrow H_2S(g) + 2NaI(aq)$
(e) None of the above

7. Which of the following is a single-replacement reaction?
(a) $Zn(s) + CuBr_2(aq) \longrightarrow ZnBr_2(aq) + Cu(s)$
(b) $SO_2(g) + H_2O(l) \longrightarrow H_2SO_3(aq)$
(c) $PbO_2(s) \longrightarrow Pb(s) + O_2(g)$
(d) $2HI(aq) + Na_2S(aq) \longrightarrow H_2S(g) + 2NaI(aq)$
(e) None of the above

8. Which of the following is a double-replacement reaction?
(a) $Zn(s) + CuBr_2(aq) \longrightarrow ZnBr_2(aq) + Cu(s)$
(b) $SO_2(g) + H_2O(l) \longrightarrow H_2SO_3(aq)$
(c) $PbO_2(s) \longrightarrow Pb(s) + O_2(g)$
(d) $2HI(aq) + Na_2S(aq) \longrightarrow H_2S(g) + 2NaI(aq)$
(e) None of the above

9. Which of the following reactions do you expect will actually occur? Use Table 6.2.
(a) $Cu(s) + MgCl_2(aq) \longrightarrow Mg(s) + CuCl_2(aq)$
(b) $2Ag(s) + Cu(NO_3)_2(aq) \longrightarrow Cu(s) + 2AgNO_3(aq)$
(c) $Al(s) + 3AgNO_3(aq) \longrightarrow 3Ag(s) + Al(NO_3)_3(aq)$
(d) $3Ag(s) + Al(NO_3)_3(aq) \longrightarrow Al(s) + 3AgNO_3(aq)$
(e) None of the above

10. Which of the following reactions do you expect will actually occur? Use Table 6.3.
(a) $BaCl_2 + Li_2SO_4 \longrightarrow BaSO_4 + 2LiCl$
(b) $BaSO_4 + 2LiCl \longrightarrow BaCl_2 + Li_2SO_4$
(c) $LiNO_3 + KCl \longrightarrow KNO_3 + LiCl$
(d) $KNO_3 + LiCl \longrightarrow LiNO_3 + KCl$
(e) None of the above

11. Which of the following equations correctly describes the reaction of potassium carbonate with hydrochloric acid?
(a) $KHCO_3(aq) + HCl(aq) \longrightarrow KCl(aq) + H_2CO_3(aq)$
(b) $KHCO_3(aq) + HCl(aq) \longrightarrow KCl(aq) + H_2O(l) + CO_2(g)$
(c) $K_2CO_3(aq) + 2HCl(aq) \longrightarrow 2KCl(aq) + H_2O(l) + CO_2(g)$
(d) $K_2CO_3(aq) + HCl(aq) \longrightarrow KCl(aq) + H_2CO_3(aq)$
(e) None of the above

12. Which of the following equations correctly describes the reaction of potassium sulfide with sulfuric acid?
(a) $K_2S(aq) + H_2SO_4(aq) \longrightarrow K_2SO_4(aq) + H_2S(g)$
(b) $K_2S(aq) + H_2SO_3(aq) \longrightarrow K_2SO_3(aq) + H_2S(g)$
(c) $K_2S(aq) + H_2SO_4(aq) \longrightarrow K_2SO_4(aq) + H_2S(aq)$
(d) $K_2SO_3(aq) + H_2SO_4(aq) \longrightarrow K_2SO_4(aq) + H_2SO_3(g)$
(e) None of the above

13. Which of the following equations correctly describes the neutralization of potassium hydroxide by sulfuric acid?
(a) $2KH(aq) + H_2SO_4(aq) \longrightarrow K_2SO_4(aq) + 2H_2(g)$
(b) $2KH(s) + H_2SO_4(aq) \longrightarrow K_2SO_4(aq) + 2H_2(g)$
(c) $KOH(aq) + H_2SO_4(aq) \longrightarrow KHSO_3(aq) + H_2O_2(l)$
(d) $KOH(s) + H_2SO_4(aq) \longrightarrow K_2SO_4(aq) + H_2O(l)$
(e) None of the above

14. Which of the following equations correctly describes the preparation of lithium nitrate by a neutralization reaction?
(a) $LiOH(aq) + HNO_3(aq) \longrightarrow LiNO_3(s) + H_2O(l)$
(b) $Li_2CO_3(aq) + 2HNO_3(aq) \longrightarrow 2LiNO_3(aq) + CO_2(g) + H_2O(l)$
(c) $LiOH(aq) + HNO_3(aq) \longrightarrow LiNO_3(aq) + H_2O(l)$
(d) $LiH(aq) + HNO_3(aq) \longrightarrow LiNO_3(aq) + H_2(g)$
(e) None of the above

15. Which of the following equations correctly describes the combustion reaction of benzene, $C_6H_6(l)$?
(a) $C_6H_6(l) + 3H_2(g) \longrightarrow C_6H_{12}(l)$
(b) $C_6H_6(l) + 12H_2O(l) \longrightarrow 6CO_2(g) + 15H_2(g)$
(c) $C_6H_6(l) + 9O_2(g) \longrightarrow 6CO_2(g) + 3H_2O(g)$
(d) $2C_6H_6(l) + 3O_2(g) \longrightarrow 12C(s) + 6H_2O(g)$
(e) None of the above

Chapter

7 Chemical Composition

Each of these drops of mercury contains about 10^{21} mercury atoms.

Chapter Outline

Chemical composition and chemical formula are related. Once you have determined the composition of a compound (the relative masses of different elements in the substance), you can deduce its formula (the relative number of different kinds of atoms in the substance), and vice versa. This chapter is about this relationship between mass and number.

Let's look at this relationship. Imagine an enormous giant whose world is made up of extremely small nuts and bolts. These are the atoms of his world, and the molecules in his world consist of definite combinations of these nuts and bolts. He has prepared a pure compound that appears to him to be a pile of fine orange sand. To you, this sand pile is enormous, but you can see the individual atoms (which look like the red and yellow pieces of a child's game) and the molecules, each consisting of two nuts screwed onto a bolt (Figure 7.1). You write the formula as RY_2, where R stands for the red bolt and Y stands for the yellow nut. But how can the giant obtain the formula without the ability to see individual atoms and molecules?

The giant adds 2 kilograms of the compound to a shaker that vibrates the molecules to nuts and bolts atoms, and using a sieve, separates them into piles, a red one (of bolts) and a yellow one (of nuts). Now what? He cannot simply compare the sizes of the two piles to obtain the relative number of atoms in the molecule because the size of the pile depends on atom size and how atoms are stacked. But suppose he knows that a red atom has twice the mass of a yellow atom; that is, the giant knows the relative masses of the atoms (their atomic weights). By simply weighing the piles of atoms obtained from the compound, the giant can obtain the formula (well, almost).

In his experiment with the compound, the giant finds that the red pile and yellow pile each weighs 1 kilogram. Because the red atom has twice the mass of a yellow one, the red pile must contain only half as many atoms. Or, equivalently, the yellow pile must contain twice as many atoms as the red one. Therefore, each molecule must contain twice as many yellow atoms as red ones. The formula could be RY_2. (It could also be R_2Y_4, or R_3Y_6, or any molecular formula that has twice as many yellow atoms as red

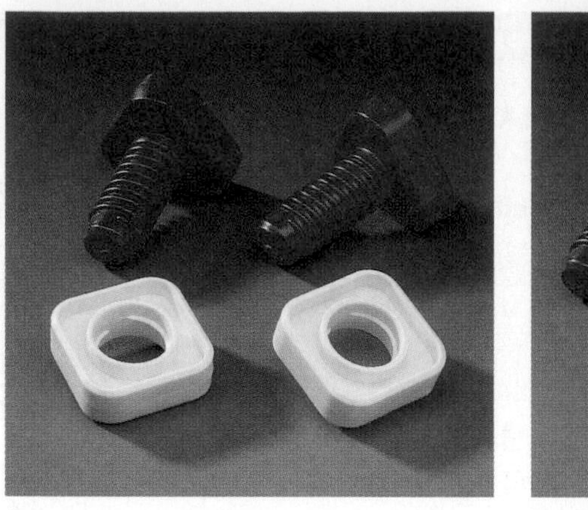
A

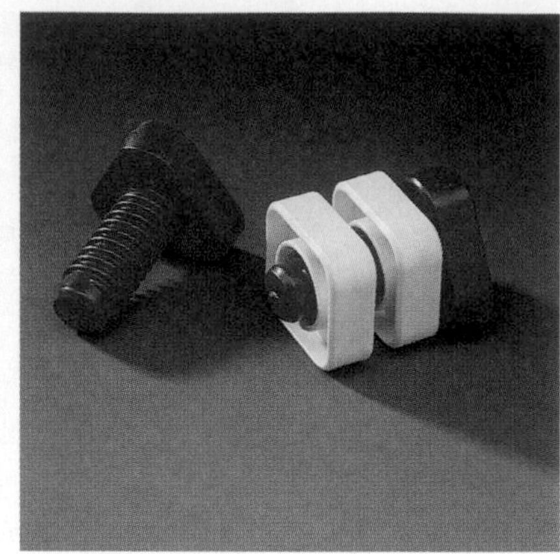
B

FIGURE 7.1
Nuts and bolts represent atoms and molecules in our fanciful world of the giant. *(A)* A yellow nut, Y, represents one type of atom and a red bolt, R, represents another type. *(B)* The assembly of two nuts and a bolt represents a three-atom molecule RY_2.

ones, so the giant will need some information about the mass of the molecule to pin down the formula.)

The problems we face in trying to obtain the formula of a compound are similar to those faced by the giant in our story. And the approach that we use is similar. To be able to tackle a problem of some complexity, however, we will need to develop the mole concept. We do this in the first part of the chapter. Then, we look at how to determine the formula of a compound.

MOLECULAR WEIGHTS, FORMULA WEIGHTS, AND MOLES

A normal sample of a substance consists of a tremendous number of molecules or formula units. In discussing molecules in Chapter 4, we noted that one-billionth of a drop of water contains two trillion molecules. This statement implies that atoms, molecules, and ions are incredibly small. On first thought, this extreme smallness might seem to be an enormous impediment to advancing chemical knowledge. Yet, as you saw from our story in the chapter opening, we can make headway without actually dealing with these large numbers or small sizes. Chemists have devised the mole concept as a convenient way of skirting these issues. (But when we need to talk about numbers of atoms or similar questions, the mole concept can help us then, too.) We begin by defining molecular weight and formula weight. ●

● The scanning tunneling microscope, a recent instrument of research, does allow us to see atoms and molecules and in some cases to manipulate them.

7.1 Molecular Weight and Formula Weight

OBJECTIVE

- Calculate a compound's molecular weight or formula weight, given its compound.

In a candy store, a customer takes jelly beans from a bin and puts them in a bag; a salesperson then weighs the bag to find how much candy the customer has taken. Suppose

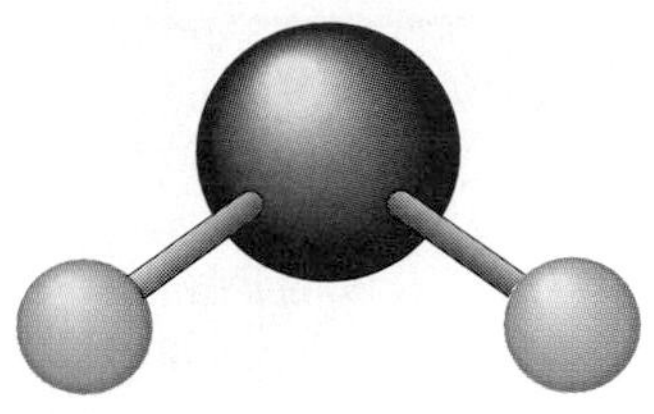

FIGURE 7.2
Molecular model of H_2O. The mass of a molecule, H_2O in this case, equals the sum of the masses of its atoms. (The sticks here represent chemical bonds, which are simply the forces of attraction between atoms.)

the customer bought 113 g, or about 4 ounces, of jelly beans, each jelly bean weighing 1.0 g. How many jelly beans did the customer buy? Clearly, 113. We are able to count large collections of things by weighing them if we know the mass of a single item. This principle also applies to atoms and molecules. In this section, we discuss the masses of atoms and molecules.

The atoms of each element have their own characteristic mass; it is called the atomic weight of the element, and it is expressed in atomic mass units (amu). Similarly, a molecule of a specific compound has a characteristic mass, the molecular weight, which is also expressed in atomic mass units. The **molecular weight** of a compound is *the sum of the atomic weights of the atoms in the molecule.*

Let's obtain the molecular weight of water, H_2O (Figure 7.2). The periodic table on the inside front cover of this book shows that the atomic weights of hydrogen and oxygen are 1.008 and 16.00 amu, respectively. Using AW to mean atomic weight, you write

$$\begin{array}{lcr} 2 \times \text{AW of hydrogen} = 2 \times 1.008 \text{ amu} = & & 2.016 \text{ amu} \\ 1 \times \text{AW of oxygen} = & & \underline{16.00 \text{ amu}} \\ \text{Molecular weight of water} = & & 18.02 \text{ amu} \end{array}$$

● When adding or subtracting quantities, the answer should have the same number of decimal places (digits after the decimal point) as there are in the quantity with the fewest decimal places.

■ EXAMPLE 7.1 Computing a Molecular Weight from a Formula

Molecular compounds known as CFCs (chlorofluorocarbons) have been used as refrigerants in air conditioners and refrigerators. They are being replaced, however, because it is believed that they have destroyed some of the upper-atmosphere ozone, which shields us from high-energy ultraviolet radiation. One of these CFCs is $CClF_3$. Calculate its molecular weight.

Problem Analysis

Obtain the sum of the atomic weights of all atoms in the formula of the compound. Remember that atomic weights and molecular weights are expressed in atomic mass units (amu).

Solution

Obtain the atomic weights of carbon, chlorine, and fluorine from the periodic table. They are 12.01, 35.45, and 19.00 amu, respectively. Because the formula is $CClF_3$, write

$$\begin{array}{lr} 1 \times \text{AW of carbon} = & 12.01 \text{ amu} \\ 1 \times \text{AW of chlorine} = & 35.45 \text{ amu} \\ 3 \times \text{AW of fluorine} = 3 \times 19.00 \text{ amu} = & \underline{57.00 \text{ amu}} \\ \text{Molecular weight of } CClF_3 = & 104.46 \text{ amu} \end{array}$$

Exercise 7.1

Another class of compounds known as HFCs (hydrofluorocarbons) has been proposed as a replacement for CFCs because HFCs appear to be less harmful to our environment. What is the molecular weight of the HFC whose formula is $C_2H_2F_4$?

(Try Problems 7.9, 7.10, 7.11, and 7.12.)

Whereas the term *molecular weight* normally applies to molecular compounds, the term *formula weight* can be used with either molecular compounds or ionic compounds. The **formula weight** of a substance is *the sum of the atomic weights of all of the atoms in a formula unit of the substance.* ●

● Chemists sometimes use the term *molecular weight* loosely when they mean *formula weight*.

■ EXAMPLE 7.2 Computing a Formula Weight from a Formula

Sodium chloride (NaCl), the substance we call salt or table salt, has been important to humans for thousands of years (Figure 7.3). In fact, the word *salary* comes from the Latin word *sal,* meaning "salt," because Roman soldiers were paid in part with a ration of salt. Calculate the formula weight of this ionic compound.

FIGURE 7.3
The Dead Sea. This body of water in Israel is a legendary source of salt (sodium chloride) for seasoning.

Problem Analysis

Calculate the formula weight similar to the way you would calculate a molecular weight. Obtain the sum of the atomic weights of all atoms in the formula unit, expressed in atomic mass units.

Solution

The formula for sodium chloride is NaCl. From the periodic table, obtain the atomic weights of sodium and chlorine. They are 22.99 and 35.45 amu, respectively. Then write

$$\begin{aligned} 1 \times \text{AW of sodium} &= 22.99\ \text{amu} \\ 1 \times \text{AW of chlorine} &= \underline{35.45\ \text{amu}} \\ \text{Formula weight of sodium chloride} &= 58.44\ \text{amu} \end{aligned}$$

Exercise 7.2

Give the formula weights for the ionic substance $Mg(OH)_2$ and the molecular substance PCl_5.

(Try Problems 7.13, 7.14, 7.15, and 7.16.)

You saw in Chapter 4 that atomic masses are relative masses; that is, the mass of any atom is compared with that of carbon-12, which is taken as a standard. You have just learned how to calculate the molecular weight of a molecule and the formula weight of a formula unit in atomic mass units. These masses are also compared with that of

● Remember from Chapter 4 that carbon-12 is the isotope of carbon with six protons and six neutrons.

carbon-12. If you could show a connection between relative masses in atomic mass units and actual masses in grams, the units we use when we weigh things, you would be able to count out atoms, molecules, and formula units by weighing. This relationship is explained in the next section. ●

Chemical Perspective

Using Molecular Weights: Discovery of Buckyball

Until recently, graphite (the principal component of pencil lead) and diamond were thought to be the only major forms of the element carbon. Then in 1985, Harold Kroto and Richard E. Smalley began a collaboration that culminated in the discovery of an exotic molecular form of carbon known informally as "buckyball."

Prior to this work Kroto, a chemist at the University of Sussex in Brighton, England, had succeeded in preparing a number of carbon-containing compounds that were later detected in outer space, products formed in the carbon-rich surface of red giant stars (Figure 7.4). Smalley, a chemist at Rice University in Houston, Texas, and his students had constructed an apparatus to focus the energy of a laser beam on a solid to vaporize it to atoms, which on cooling recombine to form large molecules. These molecules are whisked into a mass spectrometer, a device that measures the molecules' masses, or molecular weights. The molecular weights help identify the products.

FIGURE 7.4
The sun setting over East Africa. Several billion years from now, our sun will evolve into a red giant star. Although cooler than our present sun, and therefore redder in color, it will have become much larger, perhaps engulfing the earth.

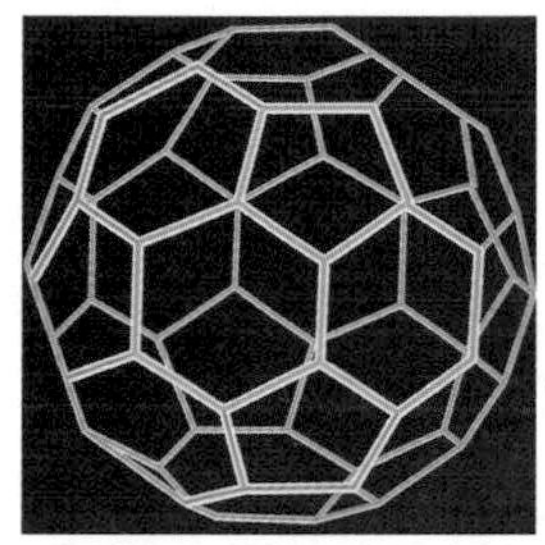

FIGURE 7.5
Buckminsterfullerene (buckyball). *(A)* A frame model of C_{60}. Each corner of the polygons has a carbon atom. *(B)* The frame model, if filled in with solid sides, would look like this soccer ball.

Kroto and Smalley placed graphite in the laser beam in Smalley's apparatus to simulate the carbon-rich conditions at the surface of red giant stars. They found a range of carbon-containing molecules in the products, but were struck by the enormous abundance of one with a molecular weight of 720 amu, corresponding to the molecule C_{60} (60×12 amu $=$ 720 amu). The direction of their research was about to take a sharp turn from red giant stars back to earth!

Why was this molecule so stable? Perhaps it has a special shape. Smalley, using scissors, paper, and tape, began building molecular models to find one that might be particularly stable. Late one night, he succeeded. He constructed a ball of 20 hexagons and 12 pentagons, with each of its 60 corners representing a carbon atom. The model was symmetrical and sturdy enough to bounce on the floor. Indeed, it was simply a paper soccer ball! In 1990, chemists prepared a red-brown substance having formula C_{60} and confirmed that the molecule had a soccer-ball shape. Smalley and Kroto named the molecule buckminsterfullerene (informally shortened to buckyball), after the famous architect R. Buckminster Fuller, who had designed domed buildings from polygons (Figure 7.5).

7.2 The Mole

OBJECTIVES

- Define Avogadro's number and the mole.
- Calculate how many atoms are present in given moles of a substance.

How many atoms are there in a given sample of a substance? What is the mass in grams of an atom of a particular element? We need answers to questions such as these to give us some idea of the magnitudes involved in the world of atoms, molecules, and ions. Many ingenious experiments performed by chemists and physicists have established these quantities experimentally. The essence of all these experiments is capsulized in one quantity: *Avogadro's number.* With this number and a table of atomic weights, you can answer questions such as "How many aluminum atoms are there in one gram of the metal?"

Avogadro's number is *the number of atoms in exactly 12 g of carbon-12*. Carbon-12, you may remember, is the isotope chosen as the reference in setting up an atomic weight scale. On this scale, the mass of the carbon-12 isotope is set equal to exactly 12 amu. The quantity of carbon-12 used in defining Avogadro's number is conveniently chosen to be the numerical value of this atomic mass, but is expressed in grams to give a normal size sample. (Twelve grams of carbon is roughly two teaspoonsful.) Experiments show that Avogadro's number is 6.0221367×10^{23}. We round this to four significant figures. Thus, there are 6.022×10^{23} atoms in 12 g of carbon-12. ●

● Avogadro's number is named in honor of Amedeo Avogadro, an Italian physicist (1776–1856).

The enormity of this number is difficult to grasp. It might help to think in terms of pennies rather than atoms. Imagine everyone on earth counting pennies at the rate of one per second. It would take the world's nearly six billion people more than three million years to count Avogadro's number of pennies.

Most people, including chemists, do not want to deal with large numbers, so they count in groups of items rather than by single items. For example, we prefer to shop for two dozen eggs and three six-packs of soft drinks rather than 24 eggs and 18 cans of soft drinks. Chemists avoid using 6.022×10^{23} atoms, molecules, or formula units by dealing with *moles* of atoms, molecules, or formula units. A mole is a specific number of things, just as a dozen is 12 items. More precisely, a **mole** (abbreviated **mol**) is *the quantity of a substance that contains Avogadro's number (6.022×10^{23}) of atoms, molecules, or formula units of that substance (as expressed by its formula).* For example, 1 mole of carbon, C, contains $6.022\ 3\ 10^{23}$ C atoms, 1 mole of oxygen, O_2, contains $6.022\ 3\ 10^{23}$ O_2 molecules, and 1 mole of sodium chloride, NaCl, contains $6.022\ 3\ 10^{23}$ NaCl formula units. ●

● One mole of things, no matter what they are, contains 6.022×10^{23} things. For instance, 1 mole of nails contains 6.022×10^{23} nails.

There is a precise relationship between moles and mass. Thus, exactly 12 g of carbon-12 is 1 mole of carbon-12. When you take a mole of carbon-12 by weighing out exactly 12 g, you have counted out 6.022×10^{23} atoms in the same way that you count out 12 eggs every time you choose a one-dozen carton of eggs. We explore this relationship between moles and mass in the next section.

Figure 7.6 shows a mole of various substances. Note the widely differing volumes of these substances. Their masses are quite different, too. This is to be expected, since the mass and volume depend on the formula unit. Volume also depends on the arrangement of units in the substance.

Consider the beaker of water in Figure 7.6. Since that beaker contains 1 mole of water, H_2O, you know it contains 6.022×10^{23} H_2O molecules. Moreover, each molecule of water is a chemical combination of two hydrogen atoms and one oxygen atom. Therefore, exactly 1 mole of water consists of $2 \times 6.022 \times 10^{23}$ hydrogen atoms and $1 \times 6.022 \times 10^{23}$ oxygen atoms. In other words, 1 mole of water contains 2 moles of

FIGURE 7.6
One-mole quantities of various substances. *Left to right:* Copper (63.55 g), mercury (200.6 g), naturally occurring carbon (12.01 g), water (18.02 g), and sodium chloride or table salt (58.44 g). Although each quantity has a different mass, each one contains 6.022×10^{23} atoms, molecules, or formula units.

hydrogen atoms and 1 mole of oxygen atoms. By using moles you can keep track of the numbers of atoms in a sample of molecules as easily as you can the numbers of molecules. The next example shows you how to calculate the number of molecules and atoms from the moles of substance.

■ EXAMPLE 7.3 Determining the Number of Molecules and Atoms from the Moles of Substance

The formula for the ordinary oxygen in our air is O_2, but the formula for ozone, the substance in the upper atmosphere that shields us from harmful ultraviolet radiation from the sun, is O_3. How many O_3 molecules are there in 1.37 moles of ozone? How many oxygen atoms are present?

Problem Analysis

Remember that 1 mole of atoms, molecules, or formula units always contains 6.022×10^{23} of these things. The subscripts in a formula tell you how many atoms of each element are in a molecule or formula unit. Be prepared to use dimensional analysis and conversion factors for problems like this one, referring to Section 2.7 if necessary to refresh your memory. You will need to make two conversions.

Moles of O_3 $\xrightarrow{\text{Convert to}}$ Molecules of O_3 $\xrightarrow{\text{Convert to}}$ Atoms of O

Solution

Step 1. Begin the first part of the problem by writing the desired conversion.

$$1.37 \text{ mol } O_3 \xrightarrow{\text{Convert to}} \text{molecules } O_3$$

What is the conversion factor? Remember that 1 mole contains 6.022×10^{23} things, so

$$1 \text{ mol } O_3 = 6.022 \times 10^{23} \text{ molecules } O_3$$

Write the factor that converts moles to molecules from this relationship. Recall that you place the given unit (mol O_3) on the bottom of the conversion factor and the desired unit (molecules O_3) on the top.

$$\frac{\overbrace{6.022 \times 10^{23} \text{ molecules } O_3}^{\boxed{\text{Desired unit}}}}{\underbrace{1 \text{ mol } O_3}_{\boxed{\text{Given unit}}}}$$

Multiply the given quantity (1.37 mol O_3) by the conversion factor.

$$1.37 \cancel{\text{mol } O_3} \times \frac{6.022 \times 10^{23} \text{ molecules } O_3}{1 \cancel{\text{mol } O_3}} = 8.25 \times 10^{23} \text{ molecules } O_3$$

Step 2. The next conversion you need is

$$8.25 \times 10^{23} \text{ molecules } O_3 \xrightarrow{\text{Convert to}} \text{O atoms}$$

The formula O_3 states that each molecule consists of three oxygen atoms. Consequently, you can write

$$1 \text{ molecule } O_3 = 3 \text{ O atoms}$$

The conversion factor will have the given unit (molecules O_3) on the bottom and the desired unit (O atoms) on the top, or

$$\frac{3 \text{ O atoms}}{1 \text{ molecule } O_3}$$

Multiplying the quantity from step 1 (8.25×10^{23} molecules O_3) by this conversion factor gives you

$$8.25 \times 10^{23} \cancel{\text{molecules } O_3} \times \frac{3 \text{ O atoms}}{1 \cancel{\text{molecule } O_3}} = 2.48 \times 10^{24} \text{ O atoms}$$

Look at each conversion factor. Can you see why each one was used and what would have happened if each one had been inverted?

Exercise 7.3

Human blood normally contains a small amount of a sugar called glucose, a compound whose formula is $C_6H_{12}O_6$. How many molecules are present in 0.225 mol of glucose? How many atoms of each element?

(Try Problems 7.17, 7.18, 7.19, and 7.20.)

In this section, we have given you some feeling for Avogadro's number and our counting unit, the mole. We have also shown that you can easily count out atoms of carbon-12 by weighing a sample of carbon-12. In the next section, we show how to count out atoms, molecules, and formula units of other substances by weighing.

7.3 Molar Mass

OBJECTIVES

- Define *molar mass*.
- Calculate the molar mass of a substance, given its formula.

What makes the mole concept so useful is that we can relate a mole of any substance to its mass if we know its formula. Therefore, we can effectively count out Avogadro's number of atoms, molecules, or formula units by weighing a substance on a balance, a relatively simple process. We begin by defining the **molar mass** of a substance as *the mass of 1 mole*. Usually, the molar mass is given in units of grams per mole. Thus, carbon-12 has a molar mass of 12 g (exactly). Let's examine the molar masses of elements and compounds.

Elements

Most of the elements in the periodic table are metals. These are not molecular substances, and we normally represent them by their atomic symbols. Most of the nonmetals, however, are molecular substances and have molecular formulas. Hydrogen, nitrogen, oxygen, and all of the elements of Group VIIA (fluorine, chlorine, bromine, and iodine) normally exist as diatomic (two-atom) molecules: H_2, N_2, O_2, F_2, Cl_2, Br_2, and I_2. White phosphorus (one form of the element phosphorus) exists as P_4 molecules, and sulfur exists as S_8 molecules. Carbon (in the form of graphite or diamond) and silicon exist as nonmolecular solids, and we represent them by their atomic symbols. ●

● Carbon has a molecular form, C_{60}. See the Chemical Perspective at the end of Section 7.1.

The molar mass of a nonmolecular element equals its atomic weight expressed in grams (assuming its formula is the atomic symbol). This important result follows from the definition of the mole. A mole of something always consists of the same number of items (atoms, molecules, or formula units). So the molar mass will be proportional to the relative mass of the item. If the mass of that item is twice that of carbon-12, the molar mass is twice that of carbon-12. Carbon-12 has an atomic mass of 12 amu and a molar mass of 12 g. Copper, Cu, has an atomic weight of 63.55 amu, which is $63.55/12 = 5.296$ times the mass of the carbon-12 atom. Therefore, the molar mass of copper is 5.296 times 12 g or 63.55 g. Note that the molar mass of such an element equals the numerical value of its atomic weight expressed in grams.

The molar mass of a molecular element equals its molecular weight expressed in grams. Nitrogen, N_2, has a molecular weight of 28.0 amu (2×14.0 amu), which is 2.33 times that of the atomic mass of carbon-12. Therefore, the molar mass of N_2 is 2.33 times 12 g, or 28.0 g. To obtain the molar mass of a molecular element, you first calculate its molecular weight. The molar mass has the same numerical value, but in units of grams.

■ EXAMPLE 7.4 Obtaining the Molar Mass of an Element

What are the molar masses of (a) potassium and (b) oxygen?

Problem Analysis

Determine the formula of the element. Thus, if the element is a metal or nonmolecular nonmetal, the formula is the atomic symbol, and its molar mass is the atomic weight expressed in grams. If the element is a molecular nonmetal, the molar mass is the molecular weight expressed in grams.

Solution

(a) Potassium is a metal; its molar mass is its atomic weight expressed in grams, or 39.10 g.

(b) Oxygen occurs as diatomic molecules. The molecular weight is

$$2 \times \text{AW of O} = 2 \times 16.00 \text{ amu} = 32.00 \text{ amu}$$

The molar mass of this element is its molecular weight expressed in grams, or 32.00 g.

Exercise 7.4

What are the molar masses of iodine and radium?

(Try Problems 7.21, 7.22, 7.23, and 7.24.)

Compounds

The molar mass of a substance equals its formula weight (if ionic) or molecular weight (if molecular) expressed in grams. Sodium chloride, NaCl, an ionic compound, has a formula weight of 58.44 amu. Therefore, the molar mass of NaCl is 58.44 g. Water, H_2O, has a molecular weight of 18.02 amu. Its molar mass is 18.02 g. To obtain the molar mass of any compound, you first calculate its formula weight or molecular weight, as appropriate. The molar mass has the same numerical value, but is given in units of grams.

■ EXAMPLE 7.5 Calculating the Molar Mass of a Compound

What is the molar mass of hydrogen peroxide, H_2O_2?

Problem Analysis

The molar mass of an ionic compound is its formula weight expressed in grams. The molar mass of a molecular compound is its molecular weight expressed in grams.

Solution

The molecular weight of H_2O_2 is

$$\begin{aligned} 2 \times \text{AW of H} &= 2 \times 1.008 \text{ amu} = 2.016 \text{ amu} \\ 2 \times \text{AW of O} &= 2 \times 16.00 \text{ amu} = \underline{16.00 \text{ amu}} \\ & \qquad\qquad\qquad\qquad\;\; 34.02 \text{ amu} \end{aligned}$$

The molecular weight of H_2O_2 is 34.02 amu, so the molar mass is 34.02 g.

Exercise 7.5

What is the molar mass of sodium phosphate, Na_3PO_4?

(Try Problems 7.25, 7.26, 7.27, and 7.28.)

7.4 Molar Masses in Calculations: Grams to Moles

OBJECTIVE

- Calculate the moles of substance present, given the mass of that substance and its chemical formula.

Now that you know how to find the mass of 1 mole of any substance, you can relate the mass of any sample of substance to the moles of that substance, and in effect count the number of atoms, molecules, and formula units in the sample. Suppose the graphite (a nonmolecular form of natural carbon) in pencil "lead" has a mass of 0.76 g and you wish to know how many moles of carbon are in this sample. You first calculate the molar mass of carbon, C, from its atomic weight (12.01 amu) in grams. You then obtain the conversion factor to convert grams to moles.Thus,

$$1 \text{ mol C} = 12.01 \text{ g C}$$

Because the desired unit is moles and the given unit is grams, you write

$$\frac{1 \text{ mol C}}{12.01 \text{ g C}}$$

Now you multiply the grams of graphite by this conversion factor:

$$0.76 \text{ g}\cancel{\text{C}} \times \frac{1 \text{ mol C}}{12.01 \text{ g}\cancel{\text{C}}} = 0.063 \text{ mol C}$$ ●

● Remember that the number of significant figures in multiplication and division is governed by the quantity with the fewest number of significant figures.

Note that the units on the bottom of the conversion factor must cancel the units of the mass of the sample (g C). The next example gives you more practice in converting grams to moles.

■ EXAMPLE 7.6 Converting Grams to Moles

How many moles of each of the following substances are present?
(a) A piece of iron whose mass is 42.6 g
(b) A sample of chlorine with a mass of 4.331 g
(c) A quantity of common table sugar, also known as sucrose ($C_{12}H_{22}O_{11}$), with a mass of 2.01 g
(d) A pile of table salt (sodium chloride) with a mass of 13 g

Problem Analysis

Obtain the molar mass of each substance, and use it to convert the mass of the sample to moles. Note that the units on the bottom of the conversion factor must cancel the units of the mass of the sample. If you cannot cancel the units, you probably have the conversion factor written upside down.

Solution

(a) Iron is a metal, so its molar mass is its atomic weight expressed in grams, or 55.85 g. The quantity of iron in moles is

$$42.6 \text{ g}\cancel{\text{Fe}} \times \frac{1 \text{ mol Fe}}{55.85 \text{ g}\cancel{\text{Fe}}} = 0.763 \text{ mol Fe}$$

(b) Elemental chlorine is a nonmetal that occurs as Cl_2. The atomic weight is 35.45 amu and the molecular weight of the molecule is

$$2 \times 35.45 \text{ amu} = 70.90 \text{ amu}$$

The molar mass is 70.90 g. The quantity of chlorine in moles in this sample is

$$4.331 \cancel{\text{g Cl}_2} \times \frac{1 \text{ mol Cl}_2}{70.90 \cancel{\text{g Cl}_2}} = 0.06109 \text{ mol Cl}_2$$

(c) The formula weight of sucrose ($C_{12}H_{22}O_{11}$) is derived as follows.

$$\begin{aligned} 12 \times \text{AW of carbon} &= 12 \times 12.01 \text{ amu} = 144.1 \text{ amu} \\ 22 \times \text{AW of hydrogen} &= 22 \times 1.008 \text{ amu} = 22.176 \text{ amu} \\ 11 \times \text{AW of oxygen} &= 11 \times 16.00 \text{ amu} = \underline{176.0 \text{ amu}} \\ \text{Formula weight of sucrose} &= 342.3 \text{ amu} \end{aligned}$$

Consequently, the molar mass of sucrose is 342.3 g. The amount of sucrose in moles is

$$2.01 \cancel{\text{g sucrose}} \times \frac{1 \text{ mol sucrose}}{342.3 \cancel{\text{g sucrose}}} = 5.87 \times 10^{-3} \text{ mol sucrose}$$

(d) As you saw in Example 7.2, the formula weight of sodium chloride (NaCl) is 58.44 amu, so its molar mass is 58.44 g. The quantity of sodium chloride in moles is

$$13 \cancel{\text{g NaCl}} \times \frac{1 \text{ mol NaCl}}{58.44 \cancel{\text{g NaCl}}} = 0.22 \text{ mol NaCl}$$

Exercise 7.6

Calculate how many moles of magnesium iodide are in 53.8 g.

(Try Problems 7.29, 7.30, 7.31, and 7.32.)

7.5 Molar Masses in Calculations: Moles to Grams

OBJECTIVE

- Calculate a substance's mass in grams, given its moles and chemical formula.

In the preceding section, you learned how to calculate the moles of a substance from its mass in grams. Let's now consider the opposite type of problem: If you know how many moles of a substance are desired, you can calculate the mass you must weigh out. How many grams of chromium must you weigh out if you want 2.22 moles? To do this problem, you need to make one conversion,

2.22 mol Cr —Convert to→ g Cr

Chromium is a metal. Because the atomic weight of this element is 52.00 amu, the molar mass is 52.00 g. Thus,

$$1 \text{ mol Cr} = 52.00 \text{ g Cr}$$

Since the given unit is "mol Cr" and the desired unit is "g Cr," the conversion factor is

$$\frac{52.00 \text{ g Cr}}{1 \text{ mol Cr}}$$

Then the number of grams of chromium in 2.22 moles is

$$2.22\ \cancel{\text{mol Cr}} \times \frac{52.00 \text{ g Cr}}{1\ \cancel{\text{mol Cr}}} = 115 \text{ g Cr}$$

This type of calculation will be useful in Chapter 8 when you wish to determine how much of a substance to weigh out for use in a chemical reaction.

■ EXAMPLE 7.7 Converting Moles to Grams

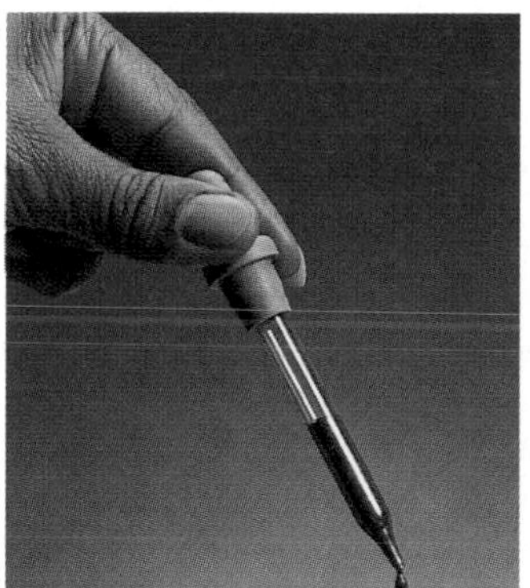

FIGURE 7.7
Mercury. Mercury is the only metal that is a liquid at 25°C.

Consider the drop of mercury in Figure 7.7. A chemist obtained its mass, from which she showed that the drop contained 0.003390 mole. What mass did she find?

Problem Analysis

Obtain the formula for mercury metal so that you can obtain its molar mass. Use the molar mass to convert moles to mass in grams.

0.003390 mol Hg —Convert to→ g Hg

Solution

Mercury is a metal, so its formula is the symbol of the element. Therefore, its molar mass is its atomic weight expressed in grams, or 200.6 g. The mass of the drop is

$$0.003390\ \cancel{\text{mol Hg}} \times \frac{200.6 \text{ g Hg}}{1\ \cancel{\text{mol Hg}}} = 0.6800 \text{ g Hg}$$

Exercise 7.7

Suppose a single drop of water contains 0.00278 mole H_2O. Calculate the mass of this drop.

(Try Problems 7.33, 7.34, 7.35, and 7.36.)

7.6 Percentage Composition

OBJECTIVES

- Define the percentage composition of a compound.
- Obtain the mass percentage of each element in a compound, given its formula.

When we say that the formula of water is H_2O, we are stating that a molecule of water is composed of two atoms of hydrogen and one atom of oxygen. We are giving the numbers of atoms of each element in the formula unit. Another way to state the composition of a compound is by giving its *percentage composition,* which refers to the proportion of each element present by mass, rather than by numbers of atoms.

Let's review briefly what is meant by percentage. When a pie is sliced into two equal sections, each section is one-half of the whole pie. Instead of using a fraction such as one-half, we can say the same thing with percentages. *Percent* means parts per hundred parts. With our pie, the whole pie corresponds to 100 parts, and one-half of the pie is one-half of 100 parts or 50 parts. Thus, each slice represents 50 percent (50%) of the whole pie.

When we speak of the **percentage composition** of a compound, we mean *the mass percentages of each element in the compound.* **Mass percentage** of an element in a compound is equivalent to *the grams of that element in 100 grams of the compound.* For example, water is 11.2% H and 88.8% O, by mass. Of course, these percentages should add up to 100% (within experimental error).

$$
\begin{aligned}
\%\,\text{H} &= 11.2\% \\
\%\,\text{O} &= \underline{88.8\%} \\
 &\ 100.0\%
\end{aligned}
$$

These mass percentages are equivalent to saying that 100.0 g of water consists of 11.2 g H and 88.8 g O.

Throughout the remainder of this chapter, we demonstrate several uses for percentage composition. Here is an example. Hydrogen is an expensive rocket fuel in the U.S. space program. You can obtain hydrogen by decomposing water. How much hydrogen can you obtain from 100 g of water? The percentage composition of water indicates that water is 11.2% hydrogen by mass. That means you can get 11.2 g of hydrogen from 100.0 g of water.

You can obtain the percentage composition of a compound from its formula. It is convenient to assume that you have 1 mole of the compound. Then, you can interpret its formula in molar terms. Each subscript in the formula represents the moles of element in the compound. For example, 1 mole of H_2O contains 2 mol H and 1 mol O. Convert each of these molar quantities to grams of element. The mass percent of an element is the mass of element divided by the mass of compound, multiplied by 100%. Here are the steps.

1. Calculate the molar mass of a compound from its formula.
2. Interpret the formula in terms of moles of elements and convert these molar quantities to grams.
3. Divide the mass of each element (from step 2) by the molar mass of the compound (from step 1) and multiply by 100%:

$$\text{Mass \% of element} = \frac{\text{mass of element in 1 mol of compound}}{\text{molar mass of compound}} \times 100\%$$

■ EXAMPLE 7.8 Calculating Mass Percentages from a Formula

Miners and spelunkers (cave explorers) in the past used carbide lamps fastened onto their caps as a light source (Figure 7.8). Water contained in the top of the lamp dripped onto calcium carbide, CaC_2, to produce acetylene, C_2H_2, a flammable gas,

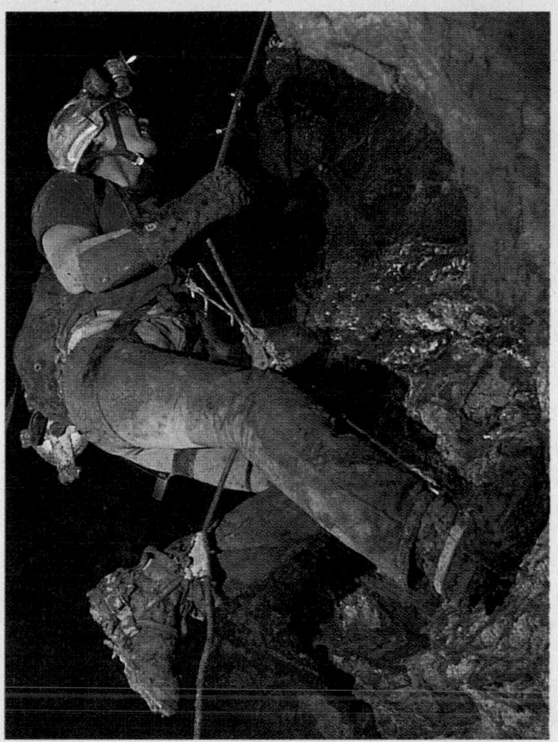

FIGURE 7.8
Acetylene lights the way. Burning acetylene provides light for spelunkers.

which when lit produced a bright light. Calculate the mass percentages of each element in acetylene.

Problem Analysis

Calculate the molar mass from the formula; decide how many grams of each element are in the molar mass of the compound; divide each of these masses by the molar mass of the compound and multiply by 100% to get the mass percentage.

Solution

Step 1. Calculate the molecular weight of C_2H_2.

$$\begin{aligned} 2 \times \text{AW of carbon} &= 2 \times 12.01 \text{ amu} = 24.02 \text{ amu} \\ 2 \times \text{AW of hydrogen} &= 2 \times 1.008 \text{ amu} = \underline{2.016 \text{ amu}} \\ \text{Molecular weight of } C_2H_2 &= 26.04 \text{ amu} \end{aligned}$$

Thus, the molar mass of acetylene is 26.04 g.

Step 2. One mole of C_2H_2 contains 2 moles of carbon and 2 moles of hydrogen. Carbon, C, has a molar mass of 12.01 g. Therefore, the mass of carbon in 2 mol C is

$$2 \cancel{\text{mol C}} \times \frac{12.01 \text{ g C}}{1 \cancel{\text{mol C}}} = 24.02 \text{ g C}$$

Similarly, hydrogen has a molar mass of 1.008 g, so the mass of 2 mol H is

$$2 \cancel{\text{mol H}} \times \frac{1.008 \text{ g H}}{1 \cancel{\text{mol H}}} = 2.016 \text{ g H}$$

You can take a shortcut by picking out 24.02 amu for carbon and 2.016 amu for hydrogen from the addition column of your molecular weight calculation (step 1) and expressing them in grams. You can see that every 26.04 g of acetylene consists of 24.02 g C and 2.02 g H.

Step 3. Calculate the mass percentage of each element.

$$\text{Mass \% of carbon} = \frac{24.02 \text{ g}}{26.04 \text{ g}} \times 100\% = 92.24\%$$

$$\text{Mass \% of hydrogen} = \frac{2.02 \text{ g}}{26.04 \text{ g}} \times 100\% = 7.76\%$$

Note that the sum of the mass percentages is 100.00%.

Exercise 7.8

Verify that the mass percentages of the elements in water are 11.2% H and 88.8% O.

(Try Problems 7.37, 7.38, 7.39, and 7.40.)

You can use mass percentages to show that a given mass of one compound contains more of some particular element than the same mass of another compound. This knowledge might be important if you are trying to produce an element for industrial or other large-scale use. A few pages ago we mentioned the production of hydrogen for rocket fuel from the decomposition of water. Water is 11.2% hydrogen by mass. On a gram-for-gram basis, however, it seems better to produce hydrogen from methane (CH_4), which is 25.14% hydrogen by mass. Other factors may influence the choice of starting material, however. ●

● Hydrogen gas is produced commercially by reacting methane with steam; carbon monoxide is the other product.

■ EXAMPLE 7.9 Comparing Mass Percentages

Hematite (Fe_2O_3) and siderite ($FeCO_3$) are minerals that can be used to manufacture pure iron (Figure 7.9). If you were a miner, which mineral would you prefer to mine to get more iron per gram of mineral?

Problem Analysis

Calculate the percentage of iron in each mineral to see which has the higher mass percentage of iron.

Solution

Step 1. Calculate the molar mass of Fe_2O_3.

$$\begin{aligned} 2 \times \text{AW of iron} &= 2 \times 55.85 \text{ amu} = 111.70 \text{ amu} \\ 3 \times \text{AW of oxygen} &= 3 \times 16.00 \text{ amu} = \underline{48.00 \text{ amu}} \\ \text{Formula weight of } Fe_2O_3 &= \phantom{3 \times 16.00 \text{ amu} =\ } 159.70 \text{ amu} \end{aligned}$$

Thus, the molar mass of Fe_2O_3 is 159.70 g. Next, calculate the molar mass of $FeCO_3$.

FIGURE 7.9
Two minerals containing iron. Both hematite *(left)* and siderite *(right)* contain iron.

$$
\begin{aligned}
1 \times \text{AW of iron} &= & 55.85 \text{ amu} \\
1 \times \text{AW of carbon} &= & 12.01 \text{ amu} \\
3 \times \text{AW of oxygen} &= 3 \times 16.00 \text{ amu} = & \underline{48.00 \text{ amu}} \\
\text{Formula weight of } FeCO_3 &= & 115.86 \text{ amu}
\end{aligned}
$$

Therefore, the molar mass of $FeCO_3$ is 115.86 g.

Step 2. Decide how many grams of iron are in each molar mass. One mole of Fe_2O_3 contains 2 moles of iron. As a result, 1 mole of this substance contains

$$2 \cancel{\text{mol Fe}} \times \frac{55.85 \text{ g Fe}}{1 \cancel{\text{mol Fe}}} = 111.70 \text{ g Fe}$$

(You could have seen from the addition column in step 1 that those 2 moles contribute 111.70 g of iron to the compound's molar mass of 159.70 g.) On the other hand, 1 mole of $FeCO_3$ contains 1 mole of iron. Therefore, it contains 55.85 g Fe in a molar mass of 115.86 g of compound.

Step 3. Calculate the mass percentage of iron in each compound.

$$\text{Mass \% Fe in } Fe_2O_3 = \frac{111.70 \cancel{\text{g}}}{159.70 \cancel{\text{g}}} \times 100\% = 69.94\%$$

$$\text{Mass \% Fe in } FeCO_3 = \frac{55.85 \cancel{\text{g}}}{115.86 \cancel{\text{g}}} \times 100\% = 48.20\%$$

Since Fe_2O_3 has the greater mass percentage of iron (meaning more iron per gram of the mineral), you would prefer to mine hematite.

Exercise 7.9

Ammonium nitrate (NH_4NO_3) and urea (NH_2CONH_2) can be used as fertilizers because each is a source of nitrogen, a key element in plant growth. Which compound contains the most nitrogen per gram?

(Try Problems 7.41 and 7.42.)

We have shown how to calculate the mass percentages of the elements in a compound from the formula of the compound. In the last part of this chapter, we show you how to determine the formula of a compound from the experimentally determined mass percentages of the elements.

DETERMINING CHEMICAL FORMULAS

Every chemical compound has a chemical formula that can be determined by a combination of experiments and calculations. As an example, we consider Albert Szent-Györgyi's determination of the chemical formula of ascorbic acid (vitamin C). You may recall from the Chemical Perspective in Chapter 1 that he obtained this compound as white crystals from an extract of adrenal glands and from sweet peppers. In the remainder of this chapter, we consider what he must have done to identify the chemical formula. He began with a chemical analysis of ascorbic acid, in which he burned a sample of ascorbic acid in oxygen. From the masses of carbon dioxide and water that were produced, he obtained the masses of carbon, hydrogen, and oxygen present in the sample

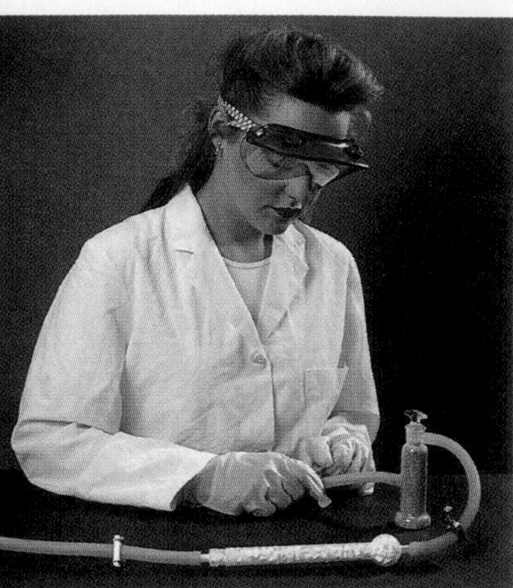

FIGURE 7.10
Chemical analysis. A chemist analyzes a sample to obtain the mass percentages of carbon and hydrogen.

and from them the percentage composition (Figure 7.10). We will not go into the details of those calculations here, but Section 7.7 illustrates a very simple example of a chemical analysis we can use to obtain the percentage composition. Subsequent sections show you how to use the data from a chemical analysis to obtain the formula of ascorbic acid.

7.7 Chemical Analysis and Mass Percentages

OBJECTIVE

- Describe a simple chemical analysis, that of HgO, from which you can obtain the mass percentages of the elements.

The following example illustrates the principles involved in a chemical analysis of a compound to obtain its percentage composition. When a particular compound of mercury and oxygen is heated, it decomposes to give metallic mercury and oxygen gas. If you weighed a sample of this compound, heated it until it decomposed, and then weighed the products separately, you could determine the compound's percentage composition. Suppose you began with a 5.00-g sample of the compound and, after the decomposition, obtained 4.63 g of mercury and 0.37 g of oxygen. You can calculate the mass percentage of any element in a weighed sample of a compound as follows:

$$\text{Mass \% of element in compound} = \frac{\text{mass of element in given mass of compound}}{\text{mass of compound}} \times 100\%$$

(This equation differs only slightly from the one used in Section 7.6; here you are using a weighed quantity, rather than assuming 1 mole.) These are the calculations for the mass percentages of the elements:

$$\text{Mass \% Hg} = \frac{\text{mass Hg in compound}}{\text{mass of compound}} \times 100\%$$

$$= \frac{4.63\text{ g}}{5.00\text{ g}} \times 100\% = 92.6\%$$

$$\text{Mass \% O} = \frac{\text{mass O in compound}}{\text{mass of compound}} \times 100\%$$

$$= \frac{0.37\text{ g}}{5.00\text{ g}} \times 100\% = 7.4\%$$

These percentages account for all of the compound because

$$\begin{aligned} \%\text{ Hg} &= \ \ 92.6\% \\ \%\text{ O} &= \ \ \ \ 7.4\% \\ \hline & \ \ 100.0\% \end{aligned}$$

● Sometimes rounding errors will cause the sum of the mass percentages to differ slightly from 100%.

These results are equivalent to saying that 100.0 g of this compound consists of 92.6 g Hg and 7.4 g O.

Szent-Györgyi used the same basic method to obtain the percentage composition of ascorbic acid. He first showed that the compound contained carbon, hydrogen, and oxygen. Then, using methods that are only a little more complicated than the one described for the compound of mercury and oxygen, he showed that the mass percentages of the elements in ascorbic acid are

$$\begin{aligned} \%\,C &= 40.9\% \\ \%\,H &= 4.5\% \\ \%\,O &= \underline{54.6\%} \\ & 100.0\% \end{aligned}$$

Thus, 100.0 g of ascorbic acid consists of 40.9 g C, 4.5 g H, and 54.6 g O. We continue to work with these results throughout Sections 7.8 and 7.9 to show how he determined the chemical formula of this substance.

7.8 Empirical Formulas

OBJECTIVES

- Define *empirical formula*.
- Calculate the empirical formula for a compound, given the experimental mass percentages of the elements in it.

Once you have determined the percentage composition of a new compound, you can use this information with a table of atomic weights and the mole concept to calculate the empirical, or simplest, formula for the compound. An **empirical formula** for a compound is *the chemical formula with the smallest possible whole-number subscripts.* The empirical formula does not necessarily give the actual number of atoms in a molecule, only their relative numbers. For example, hydrogen peroxide has an empirical formula of HO, meaning that the compound contains equal numbers of H and O atoms. Its molecular formula, however, is H_2O_2, meaning that each molecule consists of two H atoms and two O atoms. The procedure for calculating the empirical formula of a compound follows.

Steps for Calculating an Empirical Formula

1. Write each mass percentage of an element as the mass in grams of the element in 100 g of the compound.
2. Convert these masses to moles using the atomic weights of the elements. These mole quantities are proportional to the number of atoms of each element in the compound and therefore proportional to the subscripts in the formula.
3. Divide each of the mole quantities you obtained in step 2 by the smallest quantity. The smallest quantity will give precisely 1. The resulting values are essentially the subscripts in the empirical formula. If all of these values are whole numbers (or very close to it), they are precisely the subscripts in the empirical formula, and you are finished. Otherwise, you must go to step 4.
4. If one (or more) of the subscripts from the preceding step is not a whole number, multiply all of the subscripts by some small whole number such as 2 or 3 so that all of the subscripts become whole numbers (or very close to it). To find the correct multiplier, begin by multiplying all of the subscripts by 2 and see if the subscripts are now whole numbers. If not, try multiplying by 3. Continue this procedure until you obtain whole numbers for all the subscripts. ●

● Experimental errors can sometimes cause the calculated subscripts to differ slightly from integers.

Let's calculate the empirical formula of ascorbic acid. At the end of the preceding section, we gave the percentage composition of ascorbic acid as follows: % C = 40.9%,

% H = 4.5%, and % O = 54.6%. Step 1 says to write each mass percentage as the mass in 100 g of compound. Thus, 100 g of ascorbic acid contains 40.9 g of carbon, 4.5 g of hydrogen, and 54.6 g of oxygen.

Step 2 tells you to convert the masses of each element to moles. The conversion factor that takes you from g C to mol C comes from the atomic weight of C, which is 12.01 amu. Therefore, the conversion factor is 1 mol C/12.01 g C. The calculation becomes

$$\text{Moles of carbon} = 40.9\ \text{g}\cancel{\text{C}} \times \frac{1\ \text{mol C}}{12.01\ \text{g}\cancel{\text{C}}} = 3.41\ \text{mol C}$$

You use similar procedures for hydrogen and oxygen.

$$\text{Moles of hydrogen} = 4.5\ \text{g}\cancel{\text{H}} \times \frac{1\ \text{mol H}}{1.008\ \text{g}\cancel{\text{H}}} = 4.5\ \text{mol H}$$

$$\text{Moles of oxygen} = 54.6\ \text{g}\cancel{\text{O}} \times \frac{1\ \text{mol O}}{16.00\ \text{g}\cancel{\text{O}}} = 3.41\ \text{mol O}$$

Step 3 instructs you to divide each of these quantities by the smallest one—in this case, 3.41 moles.

$$\text{Subscript for carbon} = \frac{3.41\ \cancel{\text{mol}}}{3.41\ \cancel{\text{mol}}} = 1.00$$

$$\text{Subscript for hydrogen} = \frac{4.5\ \cancel{\text{mol}}}{3.41\ \cancel{\text{mol}}} = 1.3$$

$$\text{Subscript for oxygen} = \frac{3.41\ \cancel{\text{mol}}}{3.41\ \cancel{\text{mol}}} = 1.00$$

Thus, you obtain $C_{1.00}H_{1.3}O_{1.00}$. However, the subscript for H is not a whole number, so you must go on to step 4.

According to step 4, you multiply the subscripts by small whole numbers until the subscript for H is a whole number or close to it. First, multiply by 2.

$$(C_{1.00}H_{1.3}O_{1.00})_2 = C_{2.00}H_{2.6}O_{2.00}$$

Since you do not get a whole number for the H subscript, try multiplying by 3.

$$(C_{1.00}H_{1.3}O_{1.00})_3 = C_{3.00}H_{3.9}O_{3.00}$$

Since the subscript for H is very close to a whole number, you conclude that the empirical formula of ascorbic acid is $C_3H_4O_3$. (Remember from Section 2.4 on significant figures that the last digit in a calculated value is uncertain; the subscript 3.9 for H actually means a number in the range 3.8 to 4.0.) We illustrate this procedure again, for aspirin, in the following example.

■ EXAMPLE 7.10 Calculating an Empirical Formula

Aspirin is a compound consisting of C, H, and O (Figure 7.11). The mass percentages, obtained from chemical analysis, are 60.0% C, 4.5% H, and 35.5% O. Calculate the empirical formula.

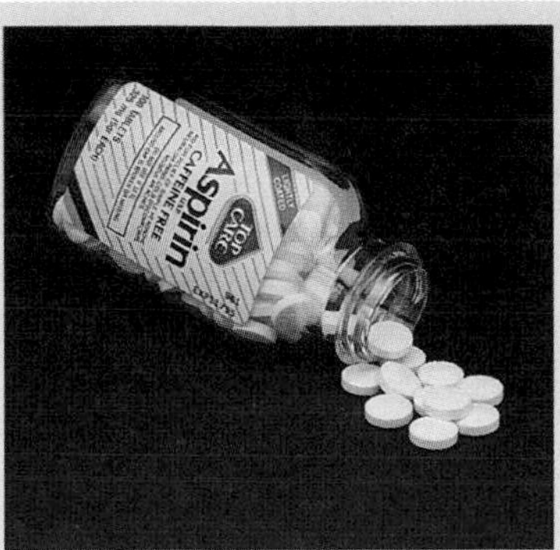

FIGURE 7.11
Aspirin. This drug is a compound containing carbon, hydrogen, and oxygen.

Problem Analysis

Write the grams of each element in 100 g of the compound; convert each of these to moles; and divide each of these mole quantities by the smallest of them. If the results are whole numbers (or very close to it), these are the subscripts. If not, obtain the subscripts by multiplying the results by small whole numbers until you obtain whole numbers for all the subscripts.

Solution

Step 1. Since the percentage composition is 60.0% C, 4.5% H, and 35.5% O, you can see that 100.0 g of aspirin contains

$$\begin{array}{r l} 60.0 & \text{g carbon} \\ 4.5 & \text{g hydrogen} \\ \underline{35.5} & \text{g oxygen} \\ 100.0 & \text{g total} \end{array}$$

Step 2. The quantity (in moles) corresponding to the mass of each element in 100 g of aspirin is as follows:

$$\text{Moles of carbon} = 60.0\ \cancel{\text{g C}} \times \frac{1 \text{ mol C}}{12.01\ \cancel{\text{g C}}} = 5.00 \text{ mol C}$$

$$\text{Moles of hydrogen} = 4.5\ \cancel{\text{g H}} \times \frac{1 \text{ mol H}}{1.008\ \cancel{\text{g H}}} = 4.5 \text{ mol H}$$

$$\text{Moles of oxygen} = 35.5\ \cancel{\text{g O}} \times \frac{1 \text{ mol O}}{16.00\ \cancel{\text{g O}}} = 2.22 \text{ mol O}$$

Step 3. Because 2.22 moles is the smallest of these quantities, you divide the moles of each element by 2.22 moles to obtain the subscripts.

$$\text{Subscript for carbon} = \frac{5.00\ \cancel{\text{mol}}}{2.22\ \cancel{\text{mol}}} = 2.25$$

$$\text{Subscript for hydrogen} = \frac{4.5\ \cancel{\text{mol}}}{2.22\ \cancel{\text{mol}}} = 2.0$$

$$\text{Subscript for oxygen} = \frac{2.22\ \cancel{\text{mol}}}{2.22\ \cancel{\text{mol}}} = 1.00$$

You obtain $C_{2.25}H_{2.0}O_{1.00}$. Because the first subscript is not a whole number, you must go on to step 4.

Step 4. Multiply the entire formula by small whole numbers until every subscript is a whole number.

$$(C_{2.25}H_{2.0}O_{1.00})_2 = C_{4.50}H_{4.0}O_{2.00}$$
$$(C_{2.25}H_{2.0}O_{1.00})_3 = C_{6.75}H_{6.0}O_{3.00}$$
$$(C_{2.25}H_{2.0}O_{1.00})_4 = C_{9.00}H_{8.0}O_{4.00}$$

Because you obtained whole numbers in the last step, you know that the empirical formula for aspirin is $C_9H_8O_4$.

Exercise 7.10

The mass percentages of the elements in methanol, a possible alternative for gasoline, are 37.5% C, 12.6% H, and 49.9% O. What is the empirical formula?

(Try Problems 7.43, 7.44, 7.45, and 7.46.)

7.9 Molecular Formulas

OBJECTIVES

- Define a molecular formula.
- Calculate the molecular formula of a substance, given its empirical formula and molecular weight.

In the last section, you discovered that the empirical formula of ascorbic acid is $C_3H_4O_3$. Let's see how you can obtain its molecular formula. The **molecular formula** is *a multiple of the corresponding empirical formula,* or

$$(\text{Empirical formula})_n = \text{molecular formula}$$

where n is a whole number. The molecular formula of ascorbic acid could be any whole-number multiple of $C_3H_4O_3$. The possibilities are

$$(C_3H_4O_3)_1 = C_3H_4O_3$$
$$(C_3H_4O_3)_2 = C_6H_8O_6$$
$$(C_3H_4O_3)_3 = C_9H_{12}O_9$$

and so forth.

Table 7.1 shows a few examples of compounds, their empirical formulas, values of n, and their molecular formulas. The empirical formula and molecular formula of water are the same, so $n = 1$. For many compounds, however, this is not true. As we men-

TABLE 7.1
Empirical Formulas and Corresponding Molecular Formulas

Compound	Empirical Formula	n	Molecular Formula
Water	H_2O	1	$(H_2O)_1 = H_2O$
Hydrogen peroxide	HO	2	$(HO)_2 = H_2O_2$
Acetylene	CH	2	$(CH)_2 = C_2H_2$
Benzene	CH	6	$(CH)_6 = C_6H_6$

tioned earlier, hydrogen peroxide has an empirical formula HO, but a molecular formula H_2O_2, so $n = 2$. Moreover, two different compounds may have the same empirical formula but different molecular formulas. Examples in Table 7.1 are acetylene (C_2H_2) and benzene (C_6H_6). The empirical formula of both substances is CH.

Again consider the equation

$$(\text{Empirical formula})_n = \text{molecular formula}$$

You must know n if you want to calculate the molecular formula from the empirical formula. To get n, note that the molecular formula weight must be some whole-number multiple of the empirical weight, or

$$n \times \text{empirical formula weight} = \text{molecular weight}$$

When you rearrange this equation, you obtain

$$n = \frac{\text{molecular weight}}{\text{empirical formula weight}}$$

Thus, once you determine the empirical formula of a compound, you can calculate its empirical formula weight. If you are given an experimental measurement of the compound's molecular weight, you can calculate n, and then you can obtain the molecular formula.

Let's return to ascorbic acid, whose empirical formula is $C_3H_4O_3$. Szent-Györgyi found experimentally that its molecular weight is 176 amu. How do you obtain the molecular formula from these data? First, you obtain the empirical formula weight of $C_3H_4O_3$.

$$\begin{aligned}
3 \times \text{AW of carbon} &= 3 \times 12.01 \text{ amu} = 36.03 \text{ amu} \\
4 \times \text{AW of hydrogen} &= 4 \times 1.008 \text{ amu} = 4.032 \text{ amu} \\
3 \times \text{AW of oxygen} &= 3 \times 16.00 \text{ amu} = \underline{48.00 \text{ amu}} \\
\text{Formula weight of } C_3H_4O_3 &= \phantom{3 \times 16.00 \text{ amu} =\ } 88.06 \text{ amu}
\end{aligned}$$

Thus, the empirical formula weight is 88.06 amu.

The value of n for this compound is:

$$n = \frac{\text{molecular weight}}{\text{empirical formula weight}} = \frac{176 \not{\text{amu}}}{88.06 \not{\text{amu}}} = 2.00$$

The molecular formula is equal to the empirical formula multiplied by n, or

$$(C_3H_4O_3)_2 = C_6H_8O_6$$

Thus, the molecular formula of ascorbic acid is $C_6H_8O_6$. We illustrate this procedure again in the following example.

■ EXAMPLE 7.11 Calculating a Molecular Formula from an Empirical Formula and a Molecular Weight

In Example 7.10, you saw that the empirical formula of aspirin is $C_9H_8O_4$. The molecular weight of this substance is 180 amu. What is the molecular formula?

Problem Analysis

Calculate the empirical formula weight. Obtain n from the relationship

$$n = \frac{\text{molecular weight}}{\text{empirical formula weight}}$$

and the molecular formula from

$$\text{Molecular formula} = (\text{empirical formula})_n$$

Solution

Begin by computing the empirical formula weight of $C_9H_8O_4$.

$$\begin{aligned}
9 \times \text{AW of carbon} &= 9 \times 12.01 \text{ amu} = 108.09 \text{ amu} \\
8 \times \text{AW of hydrogen} &= 8 \times 1.008 \text{ amu} = 8.064 \text{ amu} \\
4 \times \text{AW of oxygen} &= 4 \times 16.00 \text{ amu} = \underline{64.00 \text{ amu}} \\
\text{Empirical formula weight of } C_9H_8O_4 &= 180.15 \text{ amu}
\end{aligned}$$

Next, calculate n.

$$n = \frac{\text{molecular weight}}{\text{empirical formula weight}} = \frac{180 \cancel{\text{amu}}}{180.15 \cancel{\text{amu}}} = 0.999$$

Clearly, n must be 1, and the molecular formula of aspirin is identical to its empirical formula, $C_9H_8O_4$.

Exercise 7.11

Ethyl butyrate, the substance responsible for the odor of pineapples, has the empirical formula C_3H_6O. An experiment shows that the molecular weight is 116 amu. What is the molecular formula?

(Try Problems 7.47, 7.48, 7.49, and 7.50.)

Let's put together the entire procedure for obtaining the molecular formula of a compound. Figure 7.12 outlines this procedure. Use this outline as you examine the following example.

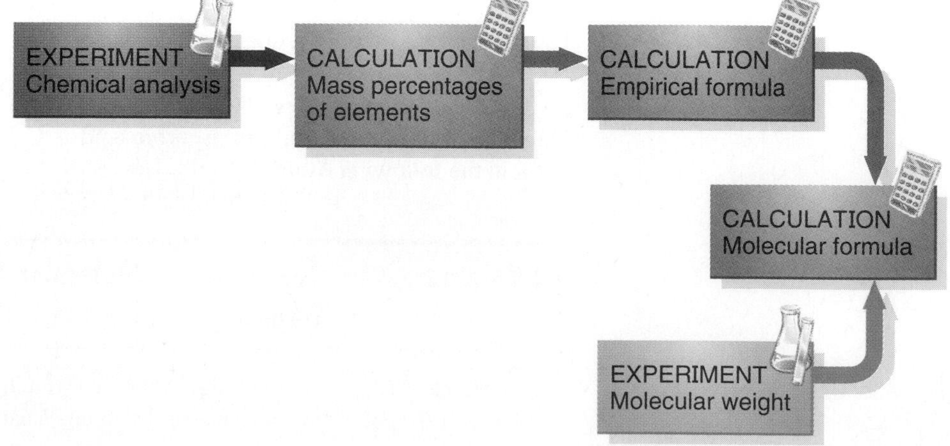

FIGURE 7.12

Flowchart for finding a molecular formula. An outline of the strategy used in obtaining the molecular formula of a compound.

■ EXAMPLE 7.12 Calculating a Molecular Formula from Mass Percentages and a Molecular Weight

The mass percentages of H and O in a compound used for bleaching are 5.9% H and 94.1% O. The molecular weight is 34 amu. Calculate the molecular formula.

Problem Analysis

Use the mass percentages of the elements in the compound to calculate the empirical formula. Then calculate the empirical formula weight. Next, obtain n by dividing the molecular weight by the empirical formula weight. Finally, obtain the molecular formula by multiplying all of the empirical formula's subscripts by n.

Solution

Begin by calculating the empirical formula according to the four-step procedure.

Step 1. The mass percentages tell you that 100.0 g of this compound contains 5.9 g H and 94.1 g O.

Step 2. The moles of each element in 100.0 g of this substance are as follows:

$$\text{Mol H} = 5.9\ \cancel{\text{g H}} \times \frac{1\ \text{mol H}}{1.008\ \cancel{\text{g H}}} = 5.9\ \text{mol H}$$

$$\text{Mol O} = 94.1\ \cancel{\text{g O}} \times \frac{1\ \text{mol O}}{16.00\ \cancel{\text{g O}}} = 5.88\ \text{mol O}$$

Step 3. Dividing by 5.88 mol, you obtain

$$\text{Subscript for hydrogen} = \frac{5.9\ \cancel{\text{mol}}}{5.88\ \cancel{\text{mol}}} = 1.0$$

$$\text{Subscript for oxygen} = \frac{5.88\ \cancel{\text{mol}}}{5.88\ \cancel{\text{mol}}} = 1.00$$

Since the subscripts are whole numbers, step 4 is not needed.

The empirical formula is HO with the following empirical formula weight:

$$\begin{aligned} 1 \times \text{AW of hydrogen} &= 1.008\ \text{amu} \\ 1 \times \text{AW of oxygen} &= 16.00\ \text{amu} \\ \hline \text{Empirical formula weight of HO} &= 17.01\ \text{amu} \end{aligned}$$

Because the molecular weight is 34 amu, the value of n is:

$$n = \frac{\text{molecular weight}}{\text{empirical formula weight}} = \frac{34\ \cancel{\text{amu}}}{17.01\ \cancel{\text{amu}}} = 2.0$$

Thus, the molecular formula is $(HO)_2$ or H_2O_2.

Exercise 7.12

Experiment shows that the percentage composition of a compound is 92.2% C and 7.8% H and that the molecular weight is 78.1 amu. What is the molecular formula?

(Try Problems 7.51, 7.52, 7.53, and 7.54.)

Chemical Perspective

A Cancer Drug from an Unlikely Experiment

Cancer kills thousands of people each year in the United States. Millions of dollars have been spent on research aimed at curing this disease, with only limited success. But *serendipity,* the faculty for making fortunate, unexpected, and accidental discoveries, has helped find a drug to fight cancer.

In 1964, Barnett Rosenberg and his colleagues, then at Michigan State University, were studying the effects of electricity on the growth of bacteria. They passed an electric current through a culture of bacteria by way of platinum electrical connections. After a short time, they noted that cell division in the bacteria had ceased. Why did this happen? Rosenberg and his coworkers began experimenting to find the answer.

Metallic platinum, like other metals, is an excellent electrical conductor. However, unlike most other metals, it is usually chemically inert. Yet a diligent investigation of the culture medium revealed traces of a platinum compound. Apparently, a small quantity of platinum in the electrodes had been converted to this compound, which then migrated into the culture medium. They then showed that this platinum compound did indeed stop cell division. Rosenberg thought that this compound, now called *cisplatin,* might be an effective anticancer drug by stopping runaway cell division, the characteristic of cancer. Their next step was to identify the molecular formula of this substance.

Experiments showed that the percentage composition was 65.1% Pt, 9.3% N, 2.0% H, and 23.6% Cl and that the molecular weight was 300 amu. Calculations from these data indicated that the molecular formula was $PtN_2H_6Cl_2$. Rosenberg's workers then began to search the literature, looking for published information on a compound with this molecular formula. They discovered that the compound was first prepared in 1844! The structural formula had been worked out by the Nobel-Prize-winning chemist Alfred Werner in 1893.

```
Cl      NH3
  \    /
    Pt
  /    \
Cl      NH3
```

Rosenberg was correct in his guess that this compound might be useful as an anticancer drug. Cisplatin (Figure 7.13) has proved to be extremely effective against bladder, ovarian, and testicular cancers.

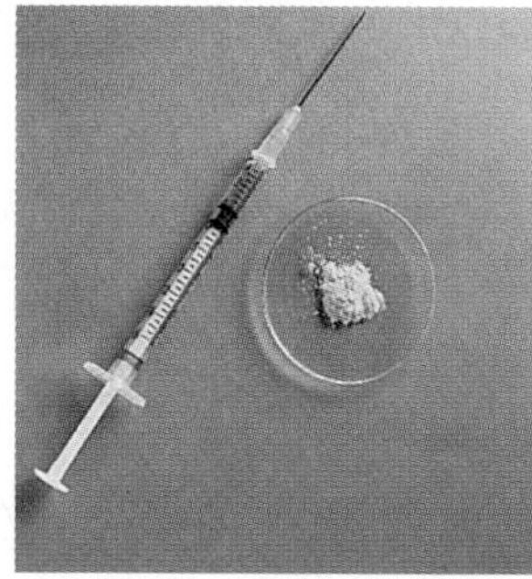

FIGURE 7.13
Cisplatin. Crystals of cisplatin and a hypodermic syringe containing a solution of this substance.

CHAPTER REVIEW

Key Words

molecular weight *(p. 189)*
formula weight *(p. 189)*
Avogadro's number *(p. 192)*
mole *(p. 192)*
molar mass *(p. 195)*
percentage composition *(p. 200)*
mass percentage *(p. 200)*
empirical formula *(p. 205)*
molecular formula *(p. 208)*

Key Equations

$$\text{Mass \% of element} = \frac{\text{mass of element in 1 mole of compound}}{\text{molar mass of compound}} = 100\%$$

$$n = \frac{\text{molecular weight}}{\text{empirical formula weight}}$$

$$(\text{Empirical formula})_n = \text{molecular formula}$$

Summary

A *molecular weight* is the sum of the atomic weights of all the atoms in the formula for a molecule of a compound. A *formula weight* is the sum of the atomic weights of all the atoms in the formula unit of a compound. If the formula unit is the formula for a molecule, then the formula weight is identical to the molecular weight.

A *mole* of any element or compound contains *Avogadro's number* (6.022×10^{23}) of atoms, molecules, or formula units. The molar mass of a substance equals the mass of 1 mole of that substance. For an element with a nonmolecular structure, this molar mass equals the atomic weight of that element expressed in grams. If the element occurs as molecules, the molar mass is the molecular weight in grams. For a molecular compound, the molar mass is the compound's molecular weight in grams; for an ionic compound, the molar mass is the compound's formula weight in grams.

The *empirical formula* of a compound is the simplest formula for that substance. It is obtained from the *percentage composition* of the compound, which is expressed as *mass percentages* of the elements that make up the compound. These mass percentages can be determined by chemical analysis in a laboratory, or they can be calculated from the chemical formula if it is known. When you calculate an empirical formula, you convert the mass percentages to ratios of moles that lead you to the subscripts in the formula. A *molecular formula* is a multiple of the empirical formula. An experimental measurement of the molecular weight is required to obtain this multiple.

Problem-Solving Skills

1. **Computing a Molecular Weight from a Formula:** Given a compound's molecular formula, calculate the molecular weight (Example 7.1).
2. **Computing a Formula Weight from a Formula:** Given an ionic compound's formula, calculate the formula weight (Example 7.2).
3. **Determining the Number of Molecules and Atoms from the Moles of Substance:** Given the amount of a substance in moles, calculate the number of molecules and atoms in the sample (Example 7.3).
4. **Obtaining the Molar Mass of an Element:** Given an element, calculate its molar mass (Example 7.4).
5. **Calculating the Molar Mass of a Compound:** Given a compound, calculate its molar mass (Example 7.5).
6. **Converting Grams to Moles:** Given a substance's mass in grams, calculate the amount of that substance in moles (Example 7.6).
7. **Converting Moles to Grams:** Given the quantity of a substance in moles, calculate the mass in grams (Example 7.7).
8. **Calculating Mass Percentages from a Formula:** Given the formula of a compound, calculate the mass percentage of each element in the compound (Example 7.8).
9. **Comparing Mass Percentages:** Given two compounds containing a common element, determine which one contains the greater amount of the element per gram of compound (Example 7.9).
10. **Calculating an Empirical Formula:** Given the mass percentages of the elements in a compound, calculate the empirical formula (Example 7.10).
11. **Calculating a Molecular Formula from an Empirical Formula and a Molecular Weight:** Given an empirical formula of a compound and its molecular weight, calculate the molecular formula (Example 7.11).
12. **Calculating a Molecular Formula from Mass Percentages and a Molecular Weight:** Given the mass percentages of the elements in a compound and the compound's molecular weight, calculate the molecular formula (Example 7.12).

Questions to Test Your Reading

7.1 What is the difference between a formula weight and a molecular weight? Define each one.

7.2 Could one substance have both a formula weight and a molecular weight? Explain.

7.3 What is Avogadro's number? What is Avogadro's number of chlorine atoms? What is Avogadro's number of diatomic chlorine molecules?

7.4 What is a mole? What is the molar mass of a substance? What is the molar mass of lithium? What is the molar mass of oxygen?

7.5 How many molecules of diatomic oxygen are in 1 mole? How many oxygen atoms are in 1 mole of those molecules?

7.6 Ozone is a triatomic form of oxygen with a formula of O_3. How many molecules of ozone are in 1 mole? How many oxygen atoms are in 1 mole of those molecules?

7.7 If hydrogen peroxide has an empirical formula of HO and a molecular weight of 34 amu, what is the molecular formula?

7.8 A compound's molecular formula is $C_6H_{12}O_6$. What is the empirical formula?

Practice Problems

Molecular Weight and Formula Weight (Section 7.1)

7.9 Calculate the molecular weight of each of the following substances.
(a) F_2 (b) PF_5 (c) SO_3
(d) $HC_2H_3O_2$ (e) $C_6H_6O_6$ (f) $C_{12}H_{22}O_{11}$

7.10 Calculate the molecular weight of each of the following substances.
(a) Br_2 (b) SF_6 (c) P_4O_{10}
(d) $H_2C_2O_4$ (e) $C_2H_4Cl_2$ (f) C_3H_7SH

7.11 As grapes ripen, the exceptionally sour taste caused by tartaric acid ($C_4H_6O_2$) disappears as this acid is converted to the sugar glucose ($C_6H_{12}O_6$). Calculate the molecular weight of each of these compounds.

7.12 Indigo ($C_{16}H_{10}N_2O_2$), the dye used to color blue jeans, is derived from a compound known as indoxyl (C_8H_7ON). Calculate the molecular weight of each of these compounds.

7.13 Calculate the formula weight of each of the following compounds.
(a) ZnI_2 (b) $AlBr_3$ (c) C_2H_6
(d) NH_4NO_3 (e) Fe_2O_3 (f) Na_2CrO_4

7.14 Calculate the formula weight of each of the following compounds.
(a) K_2SO_4 (b) $K_2Cr_2O_7$ (c) $Ca_3(PO_4)_2$
(d) $C_{18}H_{36}O_{18}$ (e) Al_2O_3 (f) $CaCO_3$

7.15 Identify the substance that has a formula weight of 125.84 amu.
(a) BCl_3 (b) $MgCl_2$ (c) IF_7
(d) $MnCl_2$ (e) H_3PO_4 (f) $TiCl_4$

7.16 Identify the substance that has a formula weight of 100.46 amu.
(a) $HClO_3$ (b) $BeSO_4$ (c) $PtCl_2$
(d) SF_6 (e) H_2SO_4 (f) $HClO_4$

The Mole and Molar Mass (Sections 7.2 to 7.5)

7.17 How many atoms of each element are in 2.33 moles of ethyl mercaptan (C_2H_5SH)?

7.18 How many atoms of each element are in 1.56 moles of methyl amine (CH_3NH_2)?

7.19 How many magnesium ions are in 0.234 mole of magnesium nitride?

7.20 How many phosphate ions are in 3.31 moles of calcium phosphate?

7.21 What are the molar masses of hydrogen and helium?

7.22 What are the molar masses of silicon and bromine?

7.23 What are the molar masses of iron and of sulfur?

7.24 What are the molar masses of calcium and nitrogen?

7.25 What are the molar masses of the following compounds?
(a) CH_3OH (b) PF_5 (c) SO_3
(d) $HC_2H_3O_2$ (e) $C_6H_{12}O_6$ (f) $C_{12}H_{22}O_{11}$

7.26 What are the molar masses of the following compounds?
(a) CH_3NH_2 (b) SF_6 (c) P_4O_{10}
(d) $H_2C_2O_4$ (e) $C_2H_4Cl_2$ (f) C_3H_7SH

7.27 What are the molar masses of the following compounds?
(a) LiH (b) $Mg(ClO_4)_2$ (c) $VOCl_3$
(d) $Ba_3(PO_4)_2$ (e) C_3H_7Br (f) K_2SO_3

7.28 What are the molar masses of the following compounds?
(a) $ZnCl_2$ (b) $Pb(ClO_3)_2$ (c) Al_2S_3
(d) $(NH_4)_3PO_4$ (e) C_6H_6 (f) CO_2

7.29 How many moles are in 1.00 g of each of the following elements?
(a) hydrogen (b) lithium (c) sodium
(d) potassium (e) rubidium (f) cesium

7.30 How many moles are in 1.00 g of the following elements?
(a) helium (b) neon (c) fluorine
(d) iodine (e) nitrogen (f) phosphorus

7.31 How many moles are in 5.00 g of each of the following compounds?
(a) CH_3OH (b) PF_5 (c) SO_3
(d) $HC_2H_3O_2$ (e) $C_6H_{12}O_6$ (f) $C_{12}H_{22}O_{11}$

7.32 How many moles are in 5.00 g of the following compounds?
(a) CH_3NH_2 (b) SF_6 (c) P_4O_{10}
(d) $H_2C_2O_4$ (e) $C_2H_4Cl_2$ (f) C_3H_7SH

7.33 What is the mass in grams of 1.11×10^{-2} moles of gaseous dihydrogen sulfide?

7.34 What is the mass in grams of 3.45×10^{-4} moles of dinitrogen monoxide?

7.35 A solution of a certain acid contains 1.3×10^{-3} moles of hydrogen ions. What is their mass in grams?

7.36 If a solution contains 6.2×10^{-2} moles of phosphate ions, what is the mass of those ions in grams?

Mass Percentages (Section 7.6)

7.37 Calculate the mass percentages of the elements in oxalic acid ($H_2C_2O_4$).

7.38 The formula for adipic acid is $C_6H_{10}O_4$. Calculate the mass percentages of the elements in this compound.

7.39 Calculate the mass percentages of the elements in *para*-dichlorobenzene ($C_6H_4Cl_2$), a moth repellent.

7.40 The formula for sorbic acid, a mold inhibitor, is $C_6H_8O_2$. Calculate the mass percentages of the elements in this compound.

7.41 Consider cesium chloride (CsCl) and magnesium chloride ($MgCl_2$). Which compound contains more chlorine per gram of compound?

7.42 Consider sodium carbonate (Na_2CO_3) and sodium chloride (NaCl). Which compound contains more sodium per gram of compound?

Empirical Formulas (Section 7.8)

7.43 A compound contains 43.64% phosphorus and 56.36% oxygen. What is the empirical formula?

7.44 A compound contains 36.84% nitrogen and 63.16% oxygen. What is the empirical formula?

7.45 Hydroquinone, a substance used as a photographic developer, contains 65.4% carbon, 5.5% hydrogen, and 29.1% oxygen. What is the empirical formula?

7.46 Malonic acid, a compound used in the manufacture of certain sleeping pills, consists of 34.6% carbon, 3.9% hydrogen, and 61.5% oxygen. What is the empirical formula?

Molecular Formulas (Section 7.9)

7.47 The empirical formula of an insecticide is CHCl. Its molecular weight is 291 amu. Calculate the molecular formula.

7.48 Putrescine, a compound produced during the decay of animal tissues, has an empirical formula of C_2H_6N and a molecular weight of 88 amu. What is the molecular formula?

7.49 An oxide of phosphorus has an empirical formula of P_2O_5 and a molecular weight of 284 amu. What is the molecular formula?

7.50 The empirical formula of an oxide of nitrogen is NO_2, and its molecular weight is 46 amu. What is the molecular formula?

7.51 Oxalic acid contains 26.7% carbon, 2.2% hydrogen, and 71.1% oxygen. The molecular weight is 90 amu. What is the molecular formula?

7.52 Adipic acid consists of 49.3% carbon, 6.9% hydrogen, and 43.8% oxygen. This compound has a molecular weight of 146 amu. Calculate the molecular formula.

7.53 *Para*-dichlorobenzene is composed of 49.0% carbon, 2.7% hydrogen, and 48.2% chlorine. The molecular weight is 147 amu. What is the molecular formula?

7.54 Sorbic acid consists of 64.3% carbon, 7.2% hydrogen, and 28.5% oxygen. It has a molecular weight of 112 amu. Calculate the molecular formula.

Additional Problems

7.55 How many moles and how many atoms of carbon-12 are in 6.3 g?

7.56 How many moles and how many atoms of carbon-12 are in 0.010 g?

7.57 What is the mass of 2.43×10^{24} atoms of carbon-12?

7.58 What is the mass of 8.32×10^{23} atoms of carbon-12?

7.59 If 1 liter of blood from a human adult contains 3.29×10^{-2} moles of sodium ions, what is its mass in milligrams?

7.60 When a sample of hard water is evaporated, 2.02×10^{-3} mole of calcium carbonate is recovered. How many milligrams are present?

7.61 Calculate the moles of chloride ions in 1.11 kg of magnesium chloride. How many chloride ions are present?

7.62 Calculate the moles of sodium ions in an aqueous solution that contains 2.452 kg of sodium phosphate. How many sodium ions are present?

7.63 How many S_8 molecules would there be if the total number of sulfur atoms was Avogadro's number?

7.64 How many P_4 molecules would there be if the total number of phosphorus atoms was Avogadro's number?

7.65 How many calcium ions are in 3.4 g of calcium carbonate?

7.66 How many chloride ions are in 11.1 g of lithium chloride?

7.67 How many hydrogen atoms are in 2.4 g of water?

7.68 How many fluorine atoms are in 1.43 g of iodine pentafluoride?

7.69 A compound contains 58.8% barium, 13.8% sulfur, and 27.4% oxygen. What is the empirical formula?

7.70 A compound contains 38.7% potassium, 13.8% nitrogen, and 47.5% oxygen. What is the empirical formula?

7.71 Potassium manganate is a dark green, crystalline substance that consists of 39.7% potassium, 27.9% manganese, and 32.5% oxygen. Potassium permanganate is a dark purple, crystalline substance that consists of 24.7% potassium, 34.8% manganese, and 40.5% oxygen. What are the empirical formulas of these compounds?

7.72 Sodium chromate is a yellow, crystalline compound that contains 28.4% sodium, 32.1% chromium, and 39.5% oxygen. Sodium dichromate is an orange, crystalline compound that contains 17.5% sodium, 39.7% chromium, and 42.8% oxygen. Calculate the empirical formulas of these substances.

7.73 A certain compound contains only magnesium and oxygen. Chemical analysis shows that 0.4145 g of the compound contains 0.2501 g of magnesium. What is the empirical formula of the compound?

7.74 A certain compound consists solely of carbon and hydrogen. Chemical analysis shows that 0.3122 g of this substance contains 0.2880 g of carbon. What is this compound's empirical formula?

7.75 A compound contains only carbon and oxygen. If 2.200 g are found to contain 0.600 g of carbon, is this substance carbon monoxide or carbon dioxide?

7.76 A compound contains only hydrogen and oxygen. Chemical analysis indicates that 1.701 g contains 0.101 g of hydrogen. Is this substance water or hydrogen peroxide (H_2O_2)?

Practice in Problem Analysis

For each problem, describe the thinking you would use (the problem analysis) before doing the actual solution, but do not solve the problem.

1. How many oxygen atoms are there in 19.4 g of sulfuric acid, H_2SO_4?

2. A newly prepared compound has a composition of 54.5% C, 9.2% H, and 36.3% O (by mass) and a molecular weight of 88 amu. What is the molecular formula?

PRACTICE EXAM

1. What is the molecular weight of nitric acid, HNO_3?
   (a) 23 amu (b) 31 amu (c) 63 amu
   (d) 42 amu (e) None of the above

2. What is the formula weight of ammonium sulfate, $(NH_4)_2SO_4$?
   (a) 32 amu (b) 114 amu (c) 124 amu
   (d) 132 amu (e) None of the above

3. How many molecules are there in 2.43 mol H_2SO_4?
   (a) 1.46×10^{24} (b) 1.46×10^{23}
   (c) 9.8×10^{23} (d) 2.43 (e) None of the above

4. How many Cl atoms are there in 4.83 mol PCl_5?
   (a) 6.02×10^{23} (b) 2.91×10^{23}
   (c) 2.91×10^{23} (d) 1.45×10^{25}
   (e) None of the above

5. What is the molar mass of nickel?
   (a) 6.02×10^{23} (b) 117 g (c) 117 amu
   (d) 58.7 g (e) None of the above

6. What is the molar mass of calcium chloride?
   (a) 111 amu (b) 75.6 amu (c) 75.6 g
   (d) 116 g (e) None of the above

7. How many moles of O atoms are there in a sample of 6.02×10^{23} O_2 molecules?
   (a) 1 mol (b) 2 mol (c) 0.5 mol
   (d) 4 mol (e) None of the above

8. How many moles of NH_3 are there in 34 g?
   (a) 0.50 mol (b) 1.00 mol (c) 1.50 mol
   (d) 2.0 mol (e) None of the above

9. An experiment calls for 2.4 mol sodium hydroxide, NaOH. How many grams is this?
   (a) 34 g (b) 64 g (c) 82 g (d) 92 g
   (e) None of the above

10. What is the mass percentage of N in N_2O_5?
    (a) 13% (b) 26% (c) 40% (d) 43%
    (e) None of the above

11. Which compound contains the greatest mass percentage of oxygen?
    (a) SO_2 (b) SO_3 (c) CO (d) CO_2
    (e) BeO

12. Choose the one that best defines what is meant by empirical formula.
    (a) The experimentally determined formula
    (b) The formula of an ionic compound
    (c) The formula with the smallest whole-number subscripts
    (d) The formula with the most similar formula
    (e) The formula of the simplest compound

13. A compound contains 29.7% S and 70.3% F, by mass. What is its empirical formula?
    (a) SF (b) SF_2 (c) SF_3 (d) SF_4
    (e) None of the above

14. A compound has the empirical formula SCl and a molecular weight of 135 amu. What is its molecular formula?
    (a) SCl (b) SCl_2 (c) S_2Cl (d) S_2Cl_2
    (e) None of the above

15. A gaseous substance has the following percentage composition: 85.6% C and 14.4% H, by mass. Its molecular weight is between 24 and 32 amu. What is its molecular formula?
    (a) CH_2 (b) CH_4 (c) C_2H_2 (d) C_2H_4
    (e) None of the above

Chapter

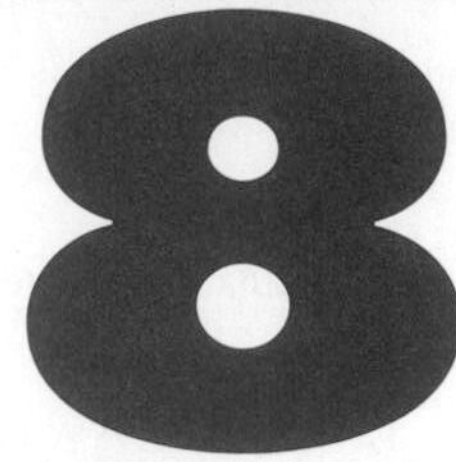

Quantities in Chemical Reactions

Zinc and sulfur react vigorously.

Chapter Outline

Have you ever heard that one antacid is better than another because it will neutralize more stomach acid? Can this claim be true? Before we can answer this question, we need to rephrase it slightly. Could a particular quantity (say, 1 gram) of one of the antacids in Figure 8.1 react with more stomach acid than an identical quantity of one of the others?

Consider another quantitative problem in chemistry. Ammonia (NH_3), an important fertilizer for crops, is produced commercially from the reaction of nitrogen gas with hydrogen gas. Nitrogen is cheap, but hydrogen is expensive. Therefore, it is important to know: How much ammonia can be produced from a given quantity of hydrogen? Without attention to details such as this one, the industrial production of ammonia would not be feasible economically (Figure 8.2). Some other aspects of this crop nutrient are discussed in the Chemical Perspective at the end of this chapter.

The questions we have asked are of the following sort: We want to know how much of a reactant is required to react with a certain amount of another reactant, or how much of a product will be formed. These are questions about **stoichiometry** (stoy′-key-om′-e-tree)—*using a balanced chemical equation to calculate quantities of the reactants and products of a reaction.* You will learn how to work stoichiometry problems in this

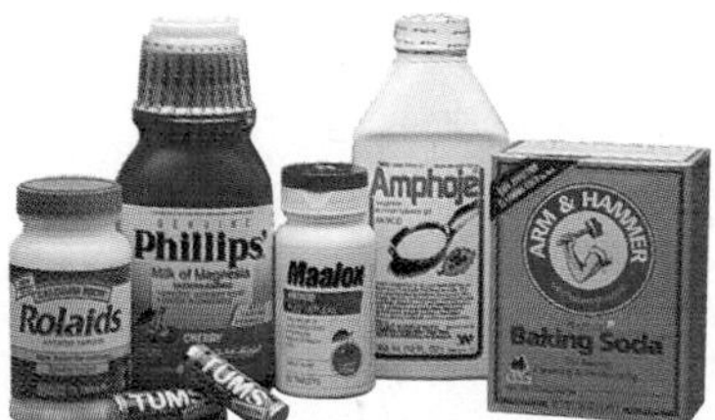

FIGURE 8.1
Antacids. Different antacids containing various chemical substances are available to the consumer.

FIGURE 8.2
An ammonia plant. The production of ammonia in the United States is second only to the production of sulfuric acid. Most ammonia ends up in fertilizers.

chapter. Before you begin, you should be very comfortable with the procedure for balancing chemical equations (Chapter 6), because you require a balanced chemical equation to solve any stoichiometry problem.

MOLE AND MASS CALCULATIONS FROM CHEMICAL EQUATIONS

Stoichiometry calculations are fundamental to practical applications of chemistry. To find out whether one antacid actually neutralizes more stomach acid than another, you need to figure out how much acid is needed to completely use up a fixed quantity (say, 1 gram) of various antacids. To start, you write a balanced chemical equation for each reaction of an antacid with stomach acid (HCl). Each equation tells you how many molecules or formula units of one reactant (here, the antacid) you need for one molecule or formula unit of another reactant (HCl). You can then use concepts from Chapter 7 to relate the mass of one reactant to the corresponding mass required for another reactant. We begin by showing that a balanced chemical equation can be interpreted in terms of moles of reactants and products.

8.1 Interpreting a Balanced Chemical Equation

OBJECTIVE

- Show that you can interpret a balanced chemical equation in terms of formula units and molecules or in terms of moles.

As you learned in Chapter 4, Dalton's atomic theory states that atoms are neither created nor destroyed during a chemical reaction; they are merely rearranged. This requirement is met faithfully when you balance a chemical equation because you make sure that every atom on the left side of the equation also appears on the right side.

For example, when magnesium hydroxide, the principal component of the antacid called milk of magnesia, reacts with hydrochloric acid (stomach acid), the reaction is described by the following balanced equation:

$$Mg(OH)_2(s) + 2HCl(aq) \longrightarrow MgCl_2(aq) + 2H_2O(l)$$

(Remember that hydrochloric acid is a solution of HCl in water.) You can see that atoms have been rearranged, and that one magnesium atom, two oxygen atoms, four hydrogen atoms, and two chlorine atoms appear on each side of the balanced equation. As you saw in Chapter 6, you can interpret this equation to mean that ●

● Remember that we speak of formula units with ionic substances and molecules with molecular substances.

$$1 \text{ formula unit } Mg(OH)_2 + 2 \text{ molecules } HCl \longrightarrow 1 \text{ formula unit } MgCl_2 + 2 \text{ molecules } H_2O$$

The coefficients of the balanced equation specify only the relative numbers of formula units and molecules that enter into the reaction; they do not give the actual numbers of molecules or formula units involved. You can multiply the coefficients in the balanced equation by any number and still have a balanced equation. For example, you can multiply the equation by 8 as follows:

$$8Mg(OH)_2 + 8 \times 2HCl \longrightarrow 8MgCl_2 + 8 \times 2H_2O$$

The equation is still balanced because 8 magnesium atoms, 16 oxygen atoms, 32 hydrogen atoms, and 16 chlorine atoms appear on each side. ● You can multiply by a very large number, such as Avogadro's number (6.022×10^{23}), without changing the relative

● The labels denoting states of matter have been dropped from the equation for simplicity.

numbers of the molecules. If you designate Avogadro's number by N_A, you obtain

$$N_A\ Mg(OH)_2 + N_A \times 2HCl \longrightarrow N_A\ MgCl_2 + N_A \times 2H_2O$$

Avogadro's number of atoms, molecules, or formula units is also called 1 mole, as you saw in Chapter 7. Therefore, this equation can also be written

$$1\ mol\ Mg(OH)_2 + 2\ mol\ HCl \longrightarrow 1\ mol\ MgCl_2 + 2\ mol\ H_2O$$

It says that 1 mole of magnesium hydroxide reacts with 2 moles of hydrogen chloride to produce 1 mole of magnesium chloride and 2 moles of water.

The previous equation is simply an alternative interpretation of the original balanced equation for the reaction.

$$Mg(OH)_2 + 2HCl \longrightarrow MgCl_2 + 2H_2O$$

What you notice is that you can interpret the balanced chemical equation in terms of molecules and formula units of reactants and products,

$$1\ \text{formula unit}\ Mg(OH)_2 + 2\ \text{molecules}\ HCl \longrightarrow 1\ \text{formula unit}\ MgCl_2 + 2\ \text{molecules}\ H_2O$$

or as moles of reactants and products,

$$1\ mol\ Mg(OH)_2 + 2\ mol\ HCl \longrightarrow 1\ mol\ MgCl_2 + 2\ mol\ H_2O$$

depending on your needs.

8.2 Mole Calculations from Chemical Equations

OBJECTIVE

- Calculate the moles of any reactant or product, given a balanced chemical equation and the moles of any other reactant or product.

Stoichiometry involves relating the moles of one reactant or product to the moles of another reactant or product. You will encounter three basic types of stoichiometry problems:

1. How many moles of a reactant do you need to react completely with given moles of another reactant?
2. How many moles of a product can you obtain from given moles of a reactant? (Assume that the reactant is completely consumed at the end of the reaction.)
3. How many moles of a reactant do you need to produce given moles of a product?

Each of these types of problems can be solved identically. You are always given the moles of one substance and are asked to calculate the moles of another substance in the reaction. The solution to any of these problems requires a conversion factor that is a mole ratio derived from the balanced chemical equation. ●

● Conversion factors were discussed in Chapter 2.

Consider an example. How many moles of HCl (in hydrochloric acid) are required to react with 0.334 mol of magnesium hydroxide? The chemical equation

$$Mg(OH)_2 + 2HCl \longrightarrow MgCl_2 + 2H_2O$$

means that 1 mole of $Mg(OH)_2$ reacts with 2 moles of HCl. The conversion factors, or mole ratios, that come from this statement are

$$\frac{2\ mol\ HCl}{1\ mol\ Mg(OH)_2} \quad \text{or} \quad \frac{1\ mol\ Mg(OH)_2}{2\ mol\ HCl}$$

Notice that the coefficients from the balanced chemical equation appear in the mole ratios. In addition, remember that a conversion factor always has the desired unit on the top and the given unit on the bottom. Thus, the ratio on the left side converts moles of $Mg(OH)_2$ to moles of HCl, whereas the ratio on the right converts moles of HCl to moles of $Mg(OH)_2$.

Since the conversion in this problem is

$$0.334 \text{ mol } Mg(OH)_2 \xrightarrow{\text{Converts to}} \text{mol HCl}$$

you use

$$\frac{2 \text{ mol HCl}}{1 \text{ mol } Mg(OH)_2}$$

as the conversion factor. Here is the calculation:

$$0.334 \text{ } \cancel{\text{mol } Mg(OH)_2} \times \frac{2 \text{ mol HCl}}{1 \text{ } \cancel{\text{mol } Mg(OH)_2}} = 0.668 \text{ mol HCl}$$

Thus, 0.334 mol $Mg(OH)_2$ reacts with 0.668 mol HCl. If you started with these quantities of reactants, at the end of the reaction, both would be used up. Consider another example.

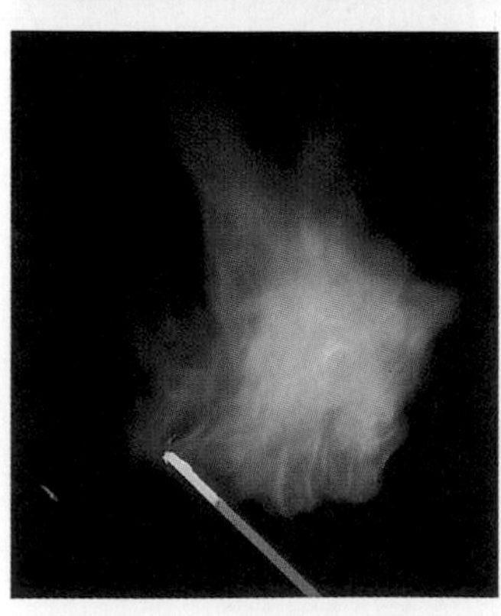

FIGURE 8.3
Reaction of hydrogen and oxygen. Hydrogen in a balloon reacts explosively with oxygen in the air when they are ignited.

■ EXAMPLE 8.1 Using Mole Ratios to Calculate Moles of a Reactant from Moles of Another Reactant

When hydrogen contained in a balloon is touched with a spark, it reacts with the oxygen in the air (Figure 8.3). The equation for this explosive reaction is

$$2H_2(g) + O_2(g) \longrightarrow 2H_2O(g)$$

How many moles of oxygen are needed to react with 2.16 mol hydrogen?

Problem Analysis

Write down the desired conversion: moles given substance $\longrightarrow$ moles desired substance. Use the balanced chemical equation to obtain the proper mole ratio that converts from "moles given" to "moles desired." Multiply the moles of the given substance by the mole ratio. Check your work by noting the cancellation of "moles given," leaving the unit "moles desired."

Solution

The conversion you want is

$$2.16 \text{ mol } H_2 \xrightarrow{\text{Converts to}} \text{mol } O_2$$

Because the balanced chemical equation means that 2 mol H_2 reacts with 1 mol O_2, the correct mole ratio for the desired conversion is

$$\frac{1 \text{ mol } O_2}{2 \text{ mol } H_2}$$

where the desired units are on the top and the given units are on the bottom. The calculation is

$$2.16 \text{ mol H}_2 \times \frac{1 \text{ mol O}_2}{2 \text{ mol H}_2} = 1.08 \text{ mol O}_2$$

The calculation tells you that you need 1.08 mol of oxygen to react completely with 2.16 mol of hydrogen.

Exercise 8.1

When iron rusts, Fe_2O_3 forms. The chemical equation for the reaction is

$$4\text{Fe}(s) + 3\text{O}_2(g) \longrightarrow 2\text{Fe}_2\text{O}_3(s)$$

How many moles of iron are necessary if 5.82 mol O_2 is used?

(Try Problems 8.13, 8.14, 8.15, and 8.16.)

In the next two examples, you will see how to calculate moles of a product from moles of a reactant and moles of a reactant from moles of a product.

EXAMPLE 8.2 Using Mole Ratios to Calculate Moles of a Product from Moles of a Reactant

Calculate the quantity of ammonia in moles that can be obtained from 4.1 mol H_2. The balanced chemical equation is

$$\text{N}_2(g) + 3\text{H}_2(g) \longrightarrow 2\text{NH}_3(g)$$

Assume that sufficient nitrogen is available, so that all of the hydrogen is consumed in the reaction.

Problem Analysis

Use the same strategy that you used in Example 8.1: Write down the desired conversion of moles given substance $\longrightarrow$ moles desired substance. Use the balanced chemical equation to obtain the proper mole ratio that converts from "moles given" to "moles desired." Multiply the moles of the given substance by the mole ratio. Check your work by noting the cancellation of "moles given," leaving the unit "moles desired."

Solution

The conversion is

4.1 mol H_2 $\xrightarrow{\text{Converts to}}$ mol NH_3

The balanced chemical equation means that 3 mol H_2 produce 2 mol NH_3. Because of the desired conversion, you want a mole ratio with mol NH_3 on the top and mol H_2 on the bottom, or

$$\frac{2 \text{ mol NH}_3}{3 \text{ mol H}_2}$$

Next, do the calculation.

$$4.1 \, \cancel{\text{mol H}_2} \times \frac{2 \text{ mol NH}_3}{3 \, \cancel{\text{mol H}_2}} = 2.7 \text{ mol NH}_3$$

Thus, you obtain 2.7 moles of ammonia from 4.1 moles of hydrogen (providing enough nitrogen is available).

Exercise 8.2

How many moles of carbon dioxide will be obtained when 8.4 moles of propane (C_3H_8) is burned? The equation for the reaction is

$$C_3H_8(g) + 5O_2(g) \longrightarrow 3CO_2(g) + 4H_2O(g)$$

(Try Problems 8.17, 8.18, 8.19, and 8.20.)

■ EXAMPLE 8.3 Using Mole Ratios to Calculate Moles of a Reactant from Moles of a Product

FIGURE 8.4
One of nature's ways of producing ozone. The energy from an electrical storm converts ordinary oxygen (O_2) to ozone (O_3).

Oxygen is converted to ozone during electrical storms (Figure 8.4). The chemical equation is

$$3O_2(g) \longrightarrow 2O_3(g)$$

The same reaction is used commercially to produce ozone. How many moles of oxygen are required to prepare 8.22 mol of ozone?

Problem Analysis

Use the same strategy you used in Examples 8.1 and 8.2: Write down the desired conversion of moles given substance ⟶ moles desired substance. Use the balanced chemical equation to obtain the proper mole ratio that converts from "moles given" to "moles desired." Multiply the moles of the given substance by the mole ratio. Check your work by noting the cancellation of "moles given," leaving the unit "moles desired."

Solution

The conversion you want is

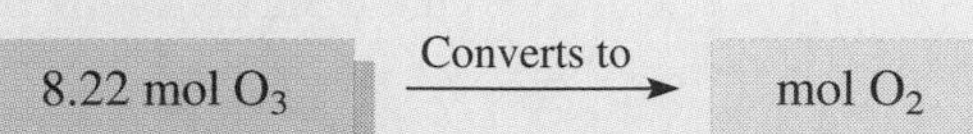

The chemical equation means that 2 mol O_3 comes from 3 mol O_2. The proper mole ratio has mol O_2 on the top and mol O_3 on the bottom, or

$$\frac{3 \text{ mol O}_2}{2 \text{ mol O}_3}$$

The calculation is

$$8.22 \, \cancel{\text{mol O}_3} \times \frac{3 \text{ mol O}_2}{2 \, \cancel{\text{mol O}_3}} = 12.3 \text{ mol O}_2$$

The answer tells you that you need 12.3 moles of oxygen to produce 8.22 moles of ozone.

Exercise 8.3

Octane (C_8H_{18}) is a component of gasoline. When octane burns, the chemical equation for the reaction is

$$2C_8H_{18}(l) + 25O_2(g) \longrightarrow 16CO_2(g) + 18H_2O(g)$$

How many moles of octane are required to give 2.22 mol of water?

(Try Problems 8.21, 8.22, 8.23, and 8.24.)

Review Examples 8.1, 8.2, and 8.3 again and notice their similarities. Each one requires you to have the balanced chemical equation for the reaction, from which you derive the proper mole ratio. The proper mole ratio always has the desired unit on the top and the given unit on the bottom.

At the beginning of this chapter, we asked about the quantity of stomach acid that a given quantity of antacid could neutralize, and about the quantity of ammonia that can be obtained from a given quantity of hydrogen. This section has shown you how to calculate those quantities if they are expressed in moles. The next section shows you how to calculate those quantities in grams.

8.3 Mass Calculations from Chemical Equations

OBJECTIVE

- Calculate the mass of any reactant or product, given a balanced chemical equation and the mass of any other reactant or product.

In Section 8.2, you learned how moles of reactants and products are related by mole ratios obtained from the balanced chemical equation. These relationships are not immediately useful, however, because you have no direct method of measuring moles. However, you can use the concept of molar mass from Section 7.3 to relate moles to mass and so in effect count out the moles of any substance by weighing it. In this section, we calculate the masses of reactants and products involved in a chemical reaction.

In most stoichiometry problems, you begin with the mass of one of the reactants or products (substance A). From there, you do three conversions to obtain the mass of another substance (substance B) in the reaction:

1. Convert the mass of substance A to moles of substance A.
2. Convert moles of substance A to moles of substance B.
3. Convert moles of substance B to mass of substance B.

You have already done these conversions in previous, but separate, calculations. Here we show you how to put these conversions together to do the usual stoichiometry problem.

Let's look again at the reaction of the antacid magnesium hydroxide with hydrochloric acid.

$$Mg(OH)_2(s) + 2HCl(aq) \longrightarrow MgCl_2(aq) + 2H_2O(l)$$

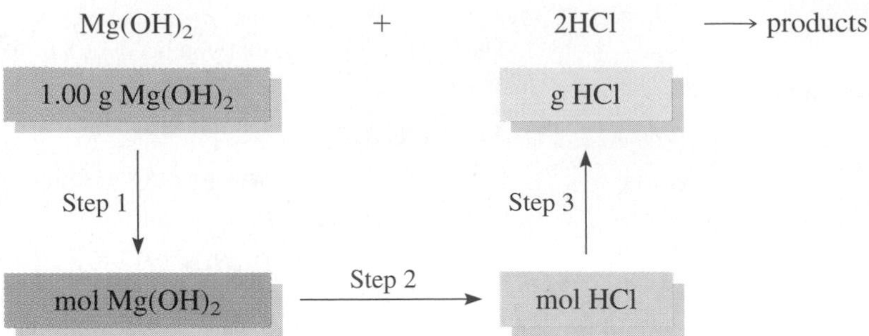

FIGURE 8.5
Conversions involved in a typical stoichiometry calculation. You first convert 1.00 g $Mg(OH)_2$ to mol $Mg(OH)_2$, which you then convert to mol HCl. Finally, you convert mol HCl to g HCl.

What mass of HCl will react completely with 1.00 g of magnesium hydroxide? The calculation involves the three conversion steps shown in Figure 8.5.

In step 1, you convert 1.00 g $Mg(OH)_2$ to moles $Mg(OH)_2$. To do that, you need the molar mass of $Mg(OH)_2$, from which you get the factor to convert mass to moles. The formula weight of $Mg(OH)_2$ is 58.33 amu, so its molar mass is 58.33 g. (You should do the calculation of formula weight; if you have difficulty, see Examples 7.1 and 7.2.) The conversion factor to change g $Mg(OH)_2$ to moles $Mg(OH)_2$ is:

$$\frac{1 \text{ mol } Mg(OH)_2}{58.33 \text{ g } Mg(OH)_2}$$

Note that the given unit, g $Mg(OH)_2$, is on the bottom, whereas the desired unit, mol $Mg(OH)_2$, is on the top. Here is the calculation that converts 1.00 g $Mg(OH)_2$ to moles. (If you do not understand this calculation, see Example 7.6.)

$$1.00 \cancel{\text{ g } Mg(OH)_2} \times \frac{1 \text{ mol } Mg(OH)_2}{58.33 \cancel{\text{ g } Mg(OH)_2}} = 0.017\underline{1}4 \text{ mol } Mg(OH)_2$$

In step 2, you use the balanced chemical equation to convert 0.01714 mol $Mg(OH)_2$ to moles HCl, following the approach developed in Example 8.1. According to the chemical equation, 1 mol $Mg(OH)_2$ reacts with 2 mol HCl. So, the conversion factor is

$$\frac{2 \text{ mol HCl}}{1 \text{ mol } Mg(OH)_2}$$

Note that the given unit, mol $Mg(OH)_2$, is on the bottom, whereas the desired unit, mol HCl, is on the top. Here is the calculation:

$$0.017\underline{1}4 \cancel{\text{ mol } Mg(OH)_2} \times \frac{2 \text{ mol HCl}}{1 \cancel{\text{ mol } Mg(OH)_2}} = 0.034\underline{2}8 \text{ mol HCl}$$

Finally, in step 3, you convert moles HCl to grams HCl. To do that, you need the formula weight, which is 36.46 amu, from which you obtain a molar mass of 36.46 g. (You should do the calculation to obtain the formula weight of HCl.) Step 3 is similar to step 1, except that the desired unit in the conversion factor has grams HCl on top and moles HCl on the bottom. Here is the calculation:

$$0.034\underline{2}8 \cancel{\text{ mol HCl}} \times \frac{36.46 \text{ g HCl}}{1 \cancel{\text{ mol HCl}}} = 1.25 \text{ g HCl}$$

TABLE 8.1
Mass of HCl Neutralized by 1.00 Gram of Antacid

Antacid Ingredient	Reaction	Mass of HCl (g)	Commercial Products*
$NaHCO_3$	$NaHCO_3 + HCl \longrightarrow NaCl + CO_2 + H_2O$	0.434	Baking soda, Alka-Seltzer tablet
$CaCO_3$	$CaCO_3 + 2HCl \longrightarrow CaCl_2 + CO_2 + H_2O$	0.729	Tums
$Mg(OH)_2$	$Mg(OH)_2 + 2HCl \longrightarrow MgCl_2 + 2H_2O$	1.25	Milk of Magnesia
$Al(OH)_3$	$Al(OH)_3 + 3HCl \longrightarrow AlCl_3 + 3H_2O$	1.40	Amphojel

* The commercial product contains other ingredients besides the antacid ingredient.

You can put all three conversion steps into one calculation, saving yourself time. (See Section 2.8 for a discussion of this.) Here is the setup:

$$1.00\ \cancel{g\ Mg(OH)_2} \times \underset{\text{(Step 1)}}{\frac{1\ \cancel{mol\ Mg(OH)_2}}{58.33\ \cancel{g\ Mg(OH)_2}}} \times \underset{\text{(Step 2)}}{\frac{2\ \cancel{mol\ HCl}}{1\ \cancel{mol\ Mg(OH)_2}}} \times \underset{\text{(Step 3)}}{\frac{36.46\ g\ HCl}{1\ \cancel{mol\ HCl}}} = 1.25\ g\ HCl$$

Note how the unit on the bottom of a conversion factor cancels the unit on top (or on the line) in the previous factor, finally leaving the unit that you desire.

At the beginning of this chapter we asked about the quantities of stomach acid that can be neutralized by identical quantities of various antacids. This example shows that 1.00 g $Mg(OH)_2$ can neutralize 1.25 g HCl. The amounts of this acid neutralized by 1.00-g samples of the active ingredient in other antacids are shown in Table 8.1. Bear in mind that a commercial antacid may contain other ingredients besides the active ingredient responsible for the acid neutralization. Antacid tablets, for instance, contain flavorings, in addition to binding agents. And although commercials for antacids may compare the amounts of each needed to neutralize the same quantity of acid, you might want to consider other factors, including cost and taste.

The next three examples provide additional practice in going from the mass of one substance in a chemical reaction to the mass of another. The Chemical Perspective at the end of this section gives a practical example of why this type of calculation is important.

■ EXAMPLE 8.4 Calculating the Mass of a Reactant from the Mass of Another Reactant

Acid rain (containing sulfuric acid, for example) has increased the acidities of many lakes, affecting their ability to sustain fish life (Figure 8.6). Pulverized limestone, which is principally calcium carbonate, has been added to some lakes to reduce their acidity by the following reaction:

$$H_2SO_4(aq) + CaCO_3(s) \longrightarrow CaSO_4(aq) + H_2O(l) + CO_2(g)$$

How many grams of calcium carbonate must be added to reduce the acidity of a lake containing 1.0×10^6 g (1.0×10^3 kg, or about 1 ton) of sulfuric acid to react completely by the previous reaction?

FIGURE 8.6
Effects of acid rain. Acid rain, a mixture of sulfuric acid and nitric acid in rainwater resulting from industrial pollution, has caused severe damage to forests and lakes. Although the lakes retain their beauty, they can no longer sustain life.

Problem Analysis

Write down the required conversions: mass given substance ⟶ moles given substance ⟶ moles desired substance ⟶ mass desired substance. Calculate the molar mass of any substance needed to convert between mass and moles of that substance. Use the balanced chemical equation to obtain the mole ratio that converts from "moles given" to "moles desired." Set up the calculation using the three conversion factors obtained. Check your work by noting the correct cancellation of units.

Solution

You need the following conversions:

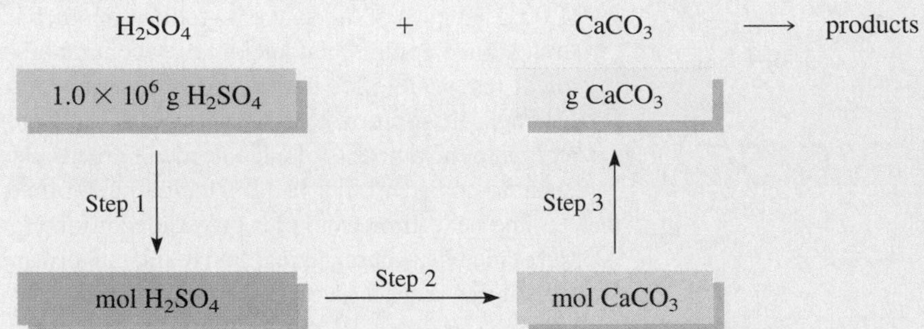

The molar masses of sulfuric acid and calcium carbonate are 98.09 g and 100.09 g, respectively. You need these in steps 1 and 3. Note that 1 mol H_2SO_4 reacts with 1 mol $CaCO_3$. You need this in step 2. Here is the calculation, with the appropriate conversion factors:

$$1.0 \times 10^6 \text{ g } H_2SO_4 \times \underset{(1)}{\frac{1 \text{ mol } H_2SO_4}{98.09 \text{ g } H_2SO_4}} \times \underset{(2)}{\frac{1 \text{ mol } CaCO_3}{1 \text{ mol } H_2SO_4}} \times \underset{(3)}{\frac{100.09 \text{ g } CaCO_3}{1 \text{ mol } CaCO_3}}$$

$$= 1.0 \times 10^6 \text{ g } CaCO_3 \qquad \text{(required)}$$

(The quantity of calcium carbonate is identical to the quantity of sulfuric acid because their molar masses happen to be very close.)

Exercise 8.4

How many grams of iron react with 3.22 g of oxygen if they react according to the following equation?

$$4Fe(s) + 3O_2(g) \longrightarrow 2Fe_2O_3(s)$$

(Try Problems 8.25, 8.26, 8.27, and 8.28.)

EXAMPLE 8.5 Calculating the Mass of a Product from the Mass of a Reactant

How many grams of ammonia can you prepare from 25.0 g of hydrogen? The equation is

$$N_2(g) + 3H_2(g) \longrightarrow 2NH_3(g)$$

Assume that sufficient nitrogen is available.

Problem Analysis

The problem analysis is the same as in the previous example. Write the conversions: mass given substance ⟶ moles given substance ⟶ moles desired substance ⟶ mass desired substance. Calculate the molar masses of "given" and "desired" substances and obtain the mole ratio that converts from "moles given" to "moles desired." Set up the calculation using the three conversion factors. Check your work by noting the correct cancellation of units.

Solution

Here are the conversions you need:

$$N_2 + 3H_2 \longrightarrow 2NH_3$$

The molar masses of hydrogen and ammonia are 2.016 g and 17.03 g, respectively. Since 3 mol H_2 produce 2 mol NH_3, the calculation is

$$25.0\ \cancel{g\ H_2} \times \underset{(1)}{\frac{1\ \cancel{mol\ H_2}}{2.016\ \cancel{g\ H_2}}} \times \underset{(2)}{\frac{2\ \cancel{mol\ NH_3}}{3\ \cancel{mol\ H_2}}} \times \underset{(3)}{\frac{17.03\ g\ NH_3}{1\ \cancel{mol\ NH_3}}} = 141\ g\ NH_3$$

Therefore, you can prepare 141 g of ammonia from 25.0 g of hydrogen.

Exercise 8.5

Phosphorus reacts violently with chlorine according to the equation

$$P_4(s) + 10Cl_2(g) \longrightarrow 4PCl_5(s)$$

Find the mass of phosphorus pentachloride that you can produce from 5.00 g of phosphorus. Assume that sufficient chlorine is present.

(Try Problems 8.29, 8.30, 8.31, and 8.32.)

EXAMPLE 8.6 Calculating the Mass of a Reactant from the Mass of a Product

A sample of hydrogen gas reacts with oxygen gas to produce 13.1 g of gaseous water. How many grams of oxygen have reacted?

Problem Analysis

First, write the balanced chemical equation for the reaction. Then, as in the previous examples, write the conversions: mass given substance $\longrightarrow$ moles given substance $\longrightarrow$ moles desired substance $\longrightarrow$ mass desired substance. Calculate the molar masses of "given" and "desired" substances and obtain the mole ratio that converts from "moles given" to "moles desired." Set up the calculation using the three conversion factors. Check your work by noting the correct cancellation of units.

Solution

The balanced equation is

$$2H_2(g) + O_2(g) \longrightarrow 2H_2O(g)$$

and the conversions are

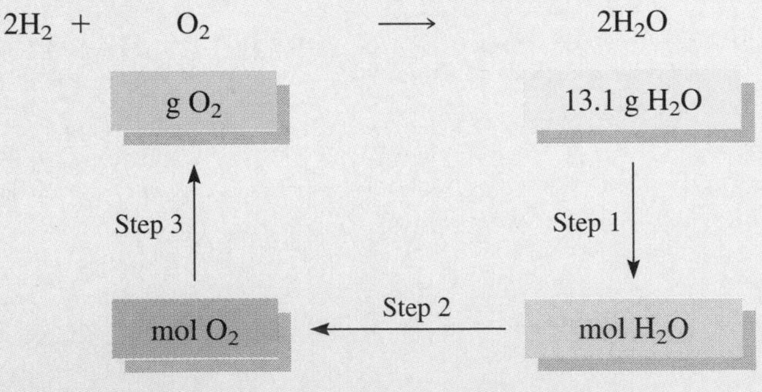

$$2H_2 + O_2 \longrightarrow 2H_2O$$

g O_2 ← Step 3 ← mol O_2 ← Step 2 ← mol H_2O ← Step 1 ← 13.1 g H_2O

The molar mass of H_2O is 18.02 g; that of O_2 is 32.00 g. Since 2 mol H_2O requires 1 mol O_2, the calculation is

$$13.1\ \text{g H}_2\text{O} \times \underset{(1)}{\frac{1\ \text{mol H}_2\text{O}}{18.02\ \text{g H}_2\text{O}}} \times \underset{(2)}{\frac{1\ \text{mol O}_2}{2\ \text{mol H}_2\text{O}}} \times \underset{(3)}{\frac{32.00\ \text{g O}_2}{1\ \text{mol O}_2}} = 11.6\ \text{g O}_2$$

This calculation shows that you need 11.6 g of oxygen to produce 13.1 g of water.

Exercise 8.6

Nitrogen monoxide reacts with oxygen to yield nitrogen dioxide.

$$2NO(g) + O_2(g) \longrightarrow 2NO_2(g)$$

This reaction has been implicated in the formation of photochemical smog. How many grams of oxygen react to form 141 g of nitrogen dioxide?

(Try Problems 8.33, 8.34, 8.35, and 8.36.)

So far, in calculating the amount of desired substance from the amount of a given substance, we have assumed that sufficient amounts of all other reactants are present. But sometimes this is not true, so we must consider what happens when the limited availability of one reactant determines the amount of product that can form. We do that in the remainder of the chapter.

Chemical Perspective

Digital X-Ray Photography

In this age of computers, the most useful form of information, including reading matter and illustrations, is digital, a form in which the information occurs as a series of 0's and 1's in computer memory or on CDs or similar storage media. You can analyze digital information by computer, send it over telephone lines, and store it for easy recovery. You may have seen digital cameras, which use magnetic disks, rather than film, to record a picture. Medical scientists are now working on digital x-ray photography.

The oldest human x-ray photograph was one taken by Wilhelm Roentgen, the discoverer of x rays. In 1895, Roentgen, a German physicist, had discovered that x rays, like light, blacken a photographic plate. After this discovery, he had his wife place her hand over a photographic plate, and he then exposed both to his x rays. The developed negative clearly showed the bones of her hand, as well as the ring on her finger (Figure 8.7).

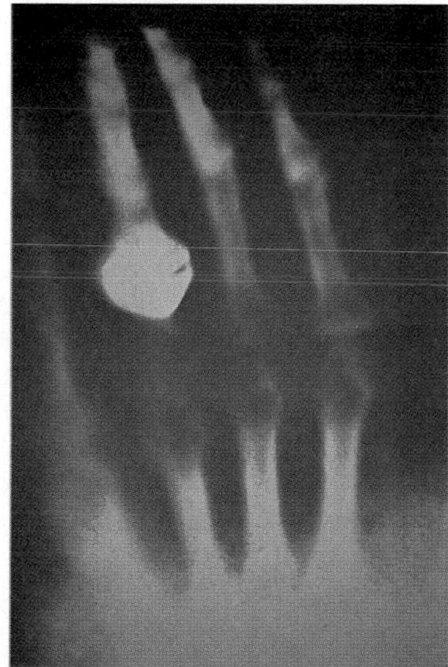

FIGURE 8.7
The first human x-ray photograph. This is a photograph taken by Wilhelm Roentgen of his wife's hand.

Today, x-ray photography, or radiography, is a widely used medical diagnostic tool, and it still uses photographic film. How might scientists make a digital x-ray camera that circumvents photographic film altogether? One approach uses a process similar to that used in a photocopier.

The heart of a photocopier is a drum coated with the element selenium, a *photoconductor.* A photoconductor is a material that releases electrons where light strikes it. Suppose you want to copy a page consisting of black type on white paper. When you place the page on a copier and push the copy button, the drum begins to revolve while an electrode applies

(continued)

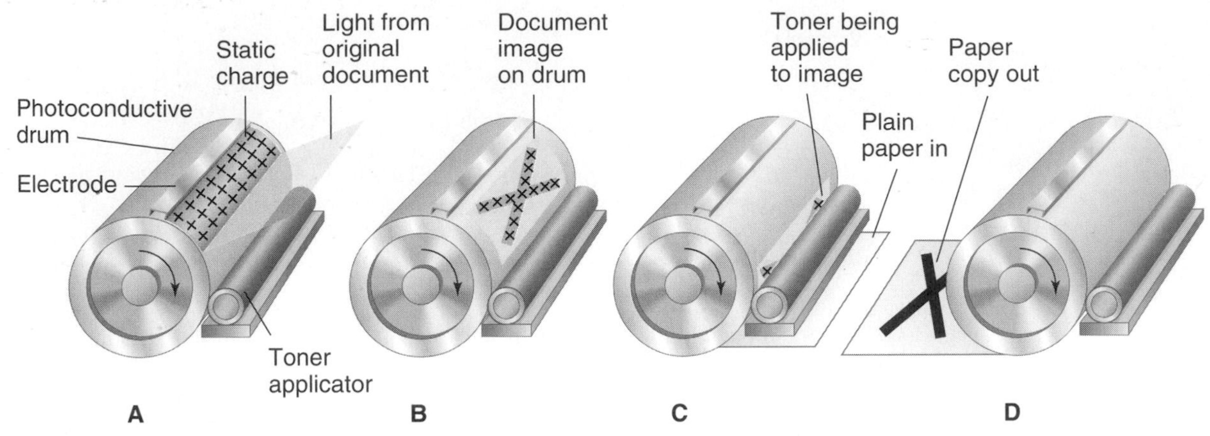

FIGURE 8.8
How a photocopier works. *(A)* An electrode applies a static charge to the rotating drum surface. *(B)* Wherever light from the original document falls on the drum, emitted electrons combine with the static charge. This leaves static charge on the drum only where the original document was black. *(C)* The resulting image moves under the toner applicator. Toner adheres to the static charge, giving a positive image on the drum. Plain paper rolls under the drum, where the toner image is transferred to it. *(D)* The paper copy emerges.

a positive charge of static electricity to its surface (Figure 8.8). Light reflected from the white part of the printed page passes through a lens system and strikes the photoconductor surface of the drum, which releases electrons. These electrons combine with positive charges on the drum surface. Positive charge remains on the drum surface only where the original printed page is black, which does not reflect light. The result is an image on the drum of the original printed page that is composed of static electricity. Here the digital x-ray camera would diverge from the photocopier. A photocopier develops the static electric image with a black powder, called the *toner.* The toner sticks to the drum wherever there is static electricity (much like pieces of lint stick to static on your clothes). The toner image on the drum is then pressed onto paper, which becomes the copy of the printed page. A digital x-ray camera might use a series of transistors in a line passing over the photoconductor surface to interpret the location of electric charges. Signals from the transistors would give a digital image that can be stored electronically.

For the chemist, the challenge is to find a good x-ray photoconductor. Heavy elements, such as bismuth and lead, are the best x-ray absorbers and are candidates for study as x-ray photoconductors. Scientists at the Dupont Experimental Station in Wilmington, Delaware, have recently studied a material composed of nylon and 50% by mass of bismuth(III) iodide. The material has shown good x-ray photoconductor properties and can be easily made into thin sheets. Chemists produce bismuth(III) iodide by a direct combination reaction from the elements. A question that bears on the cost of the material is “How many grams of bismuth metal are needed to produce 1.00 g of the bismuth(III) iodide/nylon photoconductor material?” By now, you should be able to answer this question. (The answer is 0.177 g.)

LIMITING REACTANTS AND PERCENTAGE YIELDS

When a fuel substance burns in air, it generally burns until the fuel is gone (unless someone puts the fire out). At the end of this combustion, one of the reactants (oxygen in air) remains in nearly unlimited amount, whereas the other reactant (the fuel) has been entirely consumed. Consider the reaction shown in Figure 8.3. Hydrogen in the balloon reacts with oxygen in the air according to the chemical equation

$$2H_2(g) + O_2(g) \longrightarrow 2H_2O(g)$$

The balloon contains a limited quantity of hydrogen, but the air contains an almost limitless amount of oxygen. When the reaction is over, all of the hydrogen is consumed but plenty of oxygen remains. Hydrogen is the limiting reactant. A **limiting reactant** is *the reactant that is used up when a reaction goes to completion even though other reactants are not completely consumed.* In the next two sections you will learn to identify the limiting reactant in a chemical reaction and to predict the maximum amount of product that can be formed from it, as well as how much of the other reactants are left over.

The last section of this chapter discusses percentage yield. For various reasons, you usually recover less of a product of a reaction than you calculate from theory. Percentage yield is a way to express this. It is the percentage of the theoretical mass of product that you actually obtain from a reaction.

Exploration: Limiting reactants.

8.4 Identifying Limiting Reactants

OBJECTIVES

- Define stoichiometric amounts of reactants in a chemical reaction.
- Compare a reaction beginning with stoichiometric amounts of reactants with one beginning with a limiting reactant.
- Identify the limiting reactant (if any) in a reaction, given the balanced chemical equation and the amount of each reactant.

If you put reactants into a container in amounts such that all of them are consumed when the reaction is complete, you have placed stoichiometric amounts of reactants into the container. The **stoichiometric amounts** of reactants are *the amounts of reactants that will be entirely consumed if the reaction goes to completion. You obtain the amounts by doing a stoichiometric calculation.* Suppose you have 4.0 g H_2. You need 32.0 g O_2, according to the stoichiometric calculation, to completely react with this H_2. (Can you do this calculation?) So if you put 4.0 g H_2 and 32.0 g O_2 into a reaction container, you have added the stoichiometric amounts of reactants. When the reaction is complete, both reactants will be used up, or consumed.

Suppose, however, that you add less O_2 than the stoichiometric amount. Say, you still add 4.0 g H_2, but only 30.0 g of O_2. In this case, when all 30.0 g O_2 have reacted, you will have some H_2 left over. Oxygen, O_2, is the limiting reactant.

Here is a simple analogy that may help you to grasp this concept of limiting reactant. You decide to make some cheese sandwiches, each sandwich consisting of one piece of cheese between two slices of bread. How many sandwiches can you make from two slices of cheese and four pieces of bread? Clearly, you can make two sandwiches (Figure 8.9A). You have exactly the right amounts of cheese and bread so that nothing will be left over. These exact amounts of sandwich ingredients are analogous to the stoichiometric amounts in a chemical reaction.

Now imagine that you have one slice of cheese and four pieces of bread. Although you have four pieces of bread, you can make only one sandwich because the quantity of cheese limits the number of sandwiches you can make. As shown in Figure 8.9B, you have two pieces of bread left over. The cheese, however, has been used up; it is the "limiting reactant" and determines the amount of product (the number of cheese sandwiches you can make).

Again consider the reaction of magnesium hydroxide with hydrochloric acid.

$$Mg(OH)_2(s) + 2HCl(aq) \longrightarrow MgCl_2(aq) + 2H_2O(l)$$

Suppose you add 2.0 mol $Mg(OH)_2$ and 5.0 mol HCl (as hydrochloric acid) to a beaker. Either these are the stoichiometric amounts of reactants or they are not. If you added the

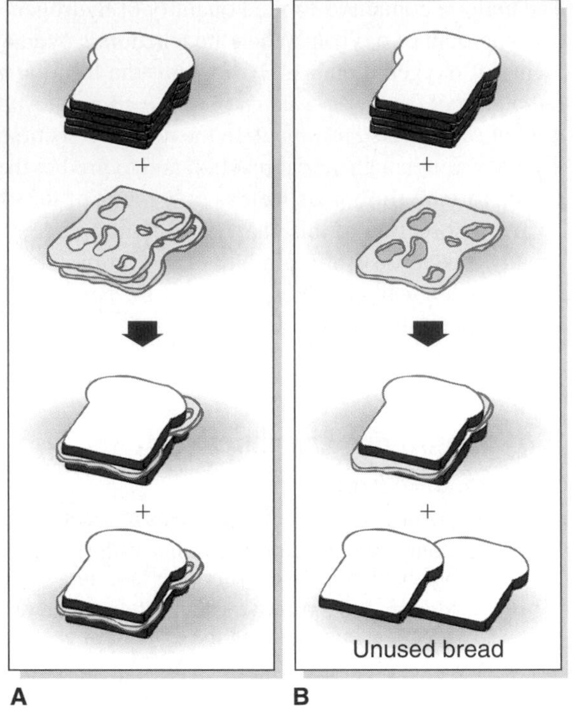

FIGURE 8.9
Cheese sandwiches, stoichiometric amounts, and limiting reactants. Making cheese sandwiches provides an analogy for reactions with *(A)* stoichiometric amounts and *(B)* a limiting reactant (here, the cheese).

reactants in stoichiometric amounts, both will be consumed at the completion of the reaction. You can calculate the mass of product obtained from either reactant. On the other hand, if the reactants are not in stoichiometric amounts, there is a limiting reactant. It will be used up and the other reactant will be left over when the reaction is complete. You calculate the amount of any product from the amount of the limiting reactant present at the start of the reaction.

How can you tell if 2.0 mol $Mg(OH)_2$ and 5.0 mol HCl are stoichiometric amounts of reactants? One way is to use the amount of one reactant, say, $Mg(OH)_2$, to calculate the amount of the other reactant (HCl) that would be needed to react with it. If the amount of HCl that you calculate is exactly equal to the amount actually added at the start, the two reactants are present in stoichiometric amounts. Let's do the calculation.

You need to make the following conversion:

$$2.0 \text{ mol } Mg(OH)_2 \xrightarrow{\text{Converts to}} \text{mol HCl}$$

Here is the calculation:

$$2.0 \cancel{\text{mol } Mg(OH)_2} \times \frac{2 \text{ mol HCl}}{1 \cancel{\text{mol } Mg(OH)_2}} = 4.0 \text{ mol HCl}$$

In other words, 2.0 mol $Mg(OH)_2$ reacts with 4.0 mol HCl. Since you actually added 5.0 mol HCl, you did not add stoichiometric amounts of reactants.

One reactant must be the limiting reactant. Which is it? Since 2.0 mol $Mg(OH)_2$ reacts with 4.0 mol HCl, and you actually start with 5.0 moles, HCl must be in excess. Then, the other reactant, $Mg(OH)_2$, must be used up. It is the limiting reactant.

Perhaps you are a bit bothered because we arbitrarily chose to start by converting 2.0 mol $Mg(OH)_2$ to the amount of HCl that would react with it. Would you get the same result if you had started with 5.0 mol HCl instead? Let's see. You want to know how much $Mg(OH)_2$ would be needed to react with 5.0 mol HCl.

$$5 \cancel{\text{mol HCl}} \times \frac{1 \text{ mol Mg(OH)}_2}{2 \cancel{\text{mol HCl}}} = 2.5 \text{ mol Mg(OH)}_2$$

You need 2.5 mol $Mg(OH)_2$ but have only 2.0 moles. The reactants are not in stoichiometric amounts, and $Mg(OH)_2$ is clearly the limiting reactant (since it was used up before all of the HCl was consumed). It makes no difference which reactant you begin the calculation with. The following summarizes the method for determining the limiting reactant.

Steps for Determining the Limiting Reactant

1. Use the amount of one reactant (call it reactant A) to calculate the amount of the other reactant (call it reactant B) that will react completely with reactant A.
2. Compare the actual amount of B with the amount calculated (the amount needed to react completely with reactant A). There are three possibilities:
   a. The actual amount of B equals the calculated amount. A and B are in stoichiometric amounts. There is no limiting reactant.
   b. The actual amount of B is greater than the calculated amount. B is in excess, so A is the limiting reactant.
   c. The actual amount of B is less than the calculated amount. B is the limiting reactant.

The next example gives you some practice in identifying a limiting reactant. As you proceed, think about the reasoning used in each step.

FIGURE 8.10
Bleach. Many household bleaches contain sodium hypochlorite (NaClO).

EXAMPLE 8.7 Identifying a Limiting Reactant

Solutions of sodium hypochlorite (NaClO) are sold as bleach (Figure 8.10). They are prepared by bubbling chlorine gas into an aqueous solution of sodium hydroxide.

$$2\text{NaOH}(aq) + \text{Cl}_2(g) \longrightarrow \text{NaClO}(aq) + \text{NaCl}(aq) + \text{H}_2\text{O}(l)$$

Is there a limiting reactant when the reaction begins with 3.4 mol NaOH and 2.0 mol Cl_2? If so, what is it?

Problem Analysis

Use the amount of one reactant (A) to calculate the amount of the other (B) that will react completely with A. If the actual amount of B equals the calculated amount, A and B are present in stoichiometric amounts; there is no limiting reactant.

If the actual amount of B is greater than the calculated amount, B is in excess, and A is the limiting reactant. If the actual amount of B is less than the calculated amount, B is the limiting reactant; A is in excess.

Solution

Step 1. What amount of Cl_2 is needed to react completely with the given amount of NaOH (3.4 mol)?

$$3.4\ \cancel{\text{mol NaOH}} \times \frac{1\ \text{mol Cl}_2}{2\ \cancel{\text{mol NaOH}}} = 1.7\ \text{mol Cl}_2$$

Step 2. Compare the actual amount of Cl_2 with the calculated amount. Because the actual amount of Cl_2 (2.0 mol) is greater than the calculated amount (1.7 mol), Cl_2 is in excess and NaOH is the limiting reactant.

Exercise 8.7

The reaction of nitrogen and hydrogen to produce ammonia is

$$N_2(g) + 3H_2(g) \longrightarrow 2NH_3(g)$$

Identify the limiting reactant, if any, in each of the following mixtures of the reactants.

(a) 27 mol N_2 and 81 mol H_2
(b) 2.5 mol N_2 and 6.2 mol H_2
(c) 1.7 mol N_2 and 9.0 mol H_2

(Try Problems 8.37, 8.38, 8.39, and 8.40.)

8.5 Calculations with Limiting Reactants

OBJECTIVE

- Calculate the mass of product formed and the mass of any reactant that may be left over, given the balanced chemical equation and the masses of reactants.

As you saw from making cheese sandwiches, the limiting sandwich ingredient determines the total number of sandwiches you can make. Similarly, *in a limiting reactant problem, you calculate the amount of any product from the amount of the limiting reactant.* (If stoichiometric amounts of reactants are present, however, you can base the calculation of product on any reactant.)

In the previous section, we looked at the reaction of 2.0 mol $Mg(OH)_2$ with 5.0 mol HCl (as hydrochloric acid).

$$Mg(OH)_2(s) + 2HCl(aq) \longrightarrow MgCl_2(aq) + 2H_2O(l)$$

We discovered that 2.0 mol $Mg(OH)_2$ reacts completely with 4.0 mol HCl, so $Mg(OH)_2$ is the limiting reactant and HCl is in excess. Suppose we now ask how much magnesium chloride was produced in this reaction?

You obtain the amount of product from the amount of limiting reactant, which in this case is 2.0 mol $Mg(OH)_2$. You convert 2.0 mol $Mg(OH)_2$ to moles $MgCl_2$:

$$2.0\ \cancel{\text{mol Mg(OH)}_2} \times \frac{1\ \text{mol MgCl}_2}{1\ \cancel{\text{mol Mg(OH)}_2}} = 2.0\ \text{mol MgCl}_2$$

You obtain 2.0 mol $MgCl_2$.

You might also want to know how much of the HCl remains when the reaction is complete. Previously, we found that 2.0 mol $Mg(OH)_2$ reacts with 4.0 mol HCl. Since we started with 5.0 mol HCl, the amount remaining is

$$5.0 \text{ mol HCl} - 4.0 \text{ mol HCl} = 1.0 \text{ mol HCl}$$

The following example shows the entire calculation for another reaction.

■ EXAMPLE 8.8 Calculating the Quantity of a Product in Moles When a Limiting Reactant Occurs

Hydrofluoric acid will etch glass and quartz (SiO_2), as shown in Figure 8.11. The chemical equation for the reaction with quartz is

$$SiO_2(s) + 4HF(aq) \longrightarrow SiF_4(g) + 2H_2O(l)$$

If a reaction begins with 2.00 mol SiO_2 and 2.00 mol HF, how many moles of each substance will be present when the reaction is complete?

FIGURE 8.11
Etched glass. The etching is done by coating the glass object with paraffin wax, scratching a pattern into this wax, then dipping the coated glass object into hydrofluoric acid. The acid reacts wherever it can reach the glass, giving it a frosted appearance.

Problem Analysis

First, determine which reactant is the limiting reactant. Then, calculate the amount of products formed from the amount of the limiting reactant, using the same strategy as in Example 8.2. To find how much of the excess reactant remains, first calculate how much of it was used up by multiplying the amount of limiting reactant available by the appropriate mole ratio. Subtract the amount of excess reactant used from the starting amount.

Solution

To find the limiting reactant, you might ask how much HF is needed to react with the given quantity of SiO_2 (2.00 mol). The calculation is

$$2.00 \cancel{\text{mol SiO}_2} \times \frac{4 \text{ mol HF}}{1 \cancel{\text{mol SiO}_2}} = 8.00 \text{ mol HF}$$

Because only 2.00 mol HF is available, you can see that HF is the limiting reactant. Use the available quantity of this substance to calculate the quantity of each product in moles.

$$2.00\ \cancel{\text{mol HF}} \times \frac{1\ \text{mol SiF}_4}{4\ \cancel{\text{mol HF}}} = 0.500\ \text{mol SiF}_4$$

$$2.00\ \cancel{\text{mol HF}} \times \frac{2\ \text{mol H}_2\text{O}}{4\ \cancel{\text{mol HF}}} = 1.00\ \text{mol H}_2\text{O}$$

Thus, the reaction between 2.00 mol SiO_2 and 2.00 mol HF results in the formation of 0.500 mol SiF_4 and 1.00 mol H_2O. All of the HF, the limiting reactant, will be used up (zero moles left). The quantity of SiO_2 used in the reaction is

$$2.00\ \cancel{\text{mol HF}} \times \frac{1\ \text{mol SiO}_2}{4\ \cancel{\text{mol HF}}} = 0.500\ \text{mol SiO}_2 \qquad (\text{SiO}_2\ \text{used up})$$

The quantity of SiO_2 remaining after the reaction is equal to the original quantity minus the amount that was used up, or

$$2.00\ \text{mol SiO}_2 - 0.500\ \text{mol SiO}_2 = 1.50\ \text{mol SiO}_2 \qquad (\text{SiO}_2\ \text{remaining})$$

Exercise 8.8

When $CaCO_3$, the principal ingredient in some antacids, reacts with hydrochloric acid, the products are calcium chloride, carbon dioxide, and water.

$$\text{CaCO}_3(s) + 2\text{HCl}(s) \longrightarrow \text{CaCl}_2(aq) + \text{CO}_2(g) + \text{H}_2\text{O}(l)$$

If the reaction begins with 1.0 mole of each reactant, how many moles of calcium chloride are formed at the completion of the reaction? How many moles of the excess reactant are left?

(Try Problems 8.41, 8.42, 8.43, and 8.44.)

In the previous example, we were given moles of reactants from which we obtained the limiting reactant. Usually, though, we measure the quantities of reactants by weighing them. In that case, we must first convert masses of reactants to moles to obtain the limiting reactant. The reason is simple: The coefficients in a chemical equation relate numbers of atoms, molecules, formula units, or moles; they do not directly relate masses.

For example, suppose that 20.0 g of hydrogen and 60.0 g of oxygen are mixed and allowed to react. The chemical equation is

$$2\text{H}_2(g) + \text{O}_2(g) \longrightarrow 2\text{H}_2\text{O}(g)$$

Is there a limiting reactant? What mass of water will be formed?

First, you convert the masses of H_2 and O_2 to moles using their molar masses, which are 2.016 g and 32.00 g, respectively.

$$20.0\ \cancel{\text{g H}_2} \times \frac{1\ \text{mol H}_2}{2.016\ \cancel{\text{g H}_2}} = 9.92\ \text{mol H}_2$$

$$60.0\ \cancel{\text{g O}_2} \times \frac{1\ \text{mol O}_2}{32.00\ \cancel{\text{g O}_2}} = 1.88\ \text{mol O}_2$$

Next, you find the limiting reactant. You ask what amount of O_2 is needed to react with the given quantity of H_2 (9.92 mol).

$$9.92 \text{ mol H}_2 \times \frac{1 \text{ mol O}_2}{2 \text{ mol H}_2} = 4.96 \text{ mol O}_2$$

The available quantity of O_2 (1.88 mol) is less than the quantity needed (4.96 mol). Therefore, O_2 is the limiting reactant and H_2 is present in excess.

You use the available quantity of the limiting reactant to calculate the moles of water formed in the reaction.

$$1.88 \text{ mol O}_2 \times \frac{2 \text{ mol H}_2\text{O}}{1 \text{ mol O}_2} = 3.76 \text{ mol H}_2\text{O}$$

Finally, you need to calculate the mass of 3.76 mol H_2O by using water's molar mass (18.02 g).

$$3.76 \text{ mol H}_2\text{O} \times \frac{18.02 \text{ g H}_2\text{O}}{1 \text{ mol H}_2\text{O}} = 67.8 \text{ g H}_2\text{O}$$

Therefore, 67.8 g of water can be formed from a reaction mixture containing 20.0 g of hydrogen and 60.0 g of oxygen.

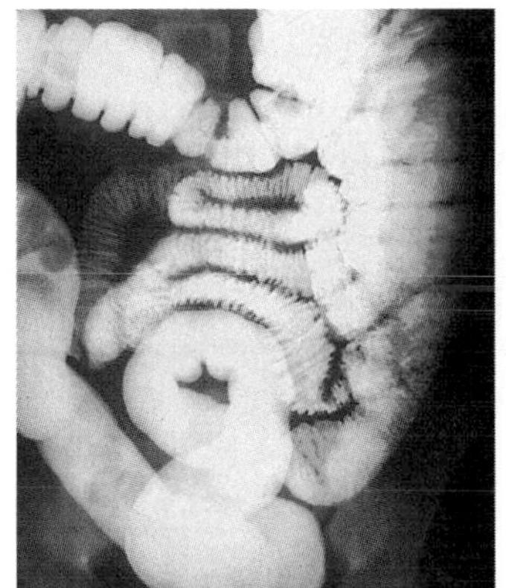

A

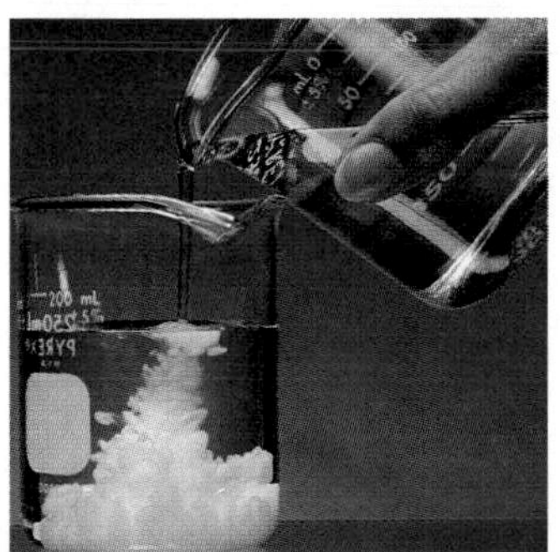

B

FIGURE 8.12

Barium sulfate. *(A)* When it is blended with water and swallowed, barium sulfate makes the digestive system opaque to x rays and therefore visible on x-ray film. *(B)* Barium sulfate can be prepared as a white precipitate by mixing aqueous solutions of barium chloride and sodium sulfate.

■ EXAMPLE 8.9 Calculating the Mass of a Product from the Mass of a Limiting Reactant

Before a person has an x ray taken of his stomach or intestines, he must swallow a suspension of barium sulfate in water. Barium sulfate absorbs x rays, so the stomach and intestines show up on the photograph (Figure 8.12A). You can prepare barium sulfate by mixing solutions of barium chloride and sodium sulfate (Figure 8.12B).

$$\text{BaCl}_2(aq) + \text{Na}_2\text{SO}_4(aq) \longrightarrow \text{BaSO}_4(s) + 2\text{NaCl}(aq)$$

How many grams of barium sulfate form if a solution containing 5.0 g $BaCl_2$ is mixed with a solution containing 5.0 g Na_2SO_4? The molar masses of $BaCl_2$, Na_2SO_4, and $BaSO_4$ are 208.2 g, 142.0 g, and 233.3 g, respectively.

Problem Analysis

Convert the masses of the reactants to moles. Find the limiting reactant (if one is present) using the same strategy as in Example 8.7. Calculate the moles of barium sulfate from the available amount of the limiting reactant. Convert the amount of barium sulfate to grams.

Solution

Convert the mass of each reactant to moles using the reactant's molar mass.

$$5.0 \text{ g BaCl}_2 \times \frac{1 \text{ mol BaCl}_2}{208.2 \text{ g BaCl}_2} = 0.024 \text{ mol BaCl}_2$$

$$5.0 \text{ g Na}_2\text{SO}_4 \times \frac{1 \text{ mol Na}_2\text{SO}_4}{142.0 \text{ g Na}_2\text{SO}_4} = 0.035 \text{ mol Na}_2\text{SO}_4$$

Next, find the limiting reactant. What amount of $BaCl_2$ is needed to react with the given quantity of Na_2SO_4 (0.035 mol)?

$$0.035 \text{ mol } Na_2SO_4 \times \frac{1 \text{ mol } BaCl_2}{1 \text{ mol } Na_2SO_4} = 0.035 \text{ mol } BaCl_2$$

Because the available amount of $BaCl_2$ (0.024 mol) is less than the quantity required by the Na_2SO_4, $BaCl_2$ is the limiting reactant.

Using the available amount of limiting reactant, you calculate the amount of $BaSO_4$ produced:

$$0.024 \text{ mol } BaCl_2 \times \frac{1 \text{ mol } BaSO_4}{1 \text{ mol } BaCl_2} = 0.024 \text{ mol } BaSO_4$$

Finally, you calculate the mass of the $BaSO_4$:

$$0.024 \text{ mol } BaSO_4 \times \frac{233.3 \text{ g } BaSO_4}{1 \text{ mol } BaSO_4} = 5.6 \text{ g } BaSO_4$$

Therefore, the reaction of 5.0 g $BaCl_2$ and 5.0 g Na_2SO_4 results in the formation of 5.6 g $BaSO_4$.

Exercise 8.9

Baking soda, $NaHCO_3$, is often used as an antacid (Table 8.1). It reacts with hydrochloric acid according to the equation

$$NaHCO_3(s) + HCl(aq) \longrightarrow NaCl(aq) + H_2O(g) + CO_2(g)$$

Determine the limiting reactant, if any, when 1.00 g of baking soda is mixed with hydrochloric acid containing 2.00 g of HCl. How many grams of carbon dioxide are produced?

(Try Problems 8.45, 8.46, 8.47, and 8.48.)

8.6 Percentage Yields

OBJECTIVE

- Calculate the percentage yield in a reaction, given the mass of a limiting reactant and the mass of product recovered.

In Sections 8.3 and 8.5, you learned how to calculate the amount of a product from stoichiometric amounts of reactants and from the available quantity of a limiting reactant. The amount of product you calculate is called the **theoretical yield,** *the maximum mass of a product that you can obtain from given amounts of reactants*. For example, the theoretical yield of $BaSO_4$ in Example 8.9 was 5.6 g. This is the maximum mass of $BaSO_4$ that you can obtain from 5.0 g each of $BaCl_2$ and Na_2SO_4.

When reactions occur, however, you rarely obtain the theoretical yield, for at least three reasons. First, when you attempt to recover a product from a reaction, small amounts may be lost even when you are very careful. This is a common problem in the laboratory as well as in an industrial chemical plant. Second, reactions other than the desired reaction may use up some of the reactants. Finally, a reaction may appear to stop before a limiting reactant is used up so that a mixture of reactants and products is obtained; in other words, a particular reaction may not actually go to completion. ● For these reasons, the actual yield is usually less than the theoretical yield. The **actual yield** is *the mass of a product that you actually recover*. We usually use the percentage yield to compare the actual yield with the theoretical yield. The **percentage yield** of a product is *the actual yield (from experiment) expressed as a percentage of the theoretical*

● This problem is due to an unfavorable chemical equilibrium, the subject of Chapter 14.

FIGURE 8.13
Reaction of zinc and sulfur. When a mixture of zinc and sulfur is heated, a fierce reaction causes the formation of zinc sulfide.

yield (from calculations), or

$$\text{Percentage yield} = \frac{\text{actual yield (g)}}{\text{theoretical yield (g)}} \times 100\%$$

Let's consider a representative calculation. Suppose you prepare zinc sulfide by heating a mixture of the elements zinc and sulfur (Figure 8.13).

$$8Zn(s) + S_8(s) \longrightarrow 8ZnS(s)$$

Suppose, further, that you calculate the theoretical yield of zinc sulfide to be 53.7 g from the amounts of the reactants, but you find that the actual yield obtained during an experiment is only 35.2 g. The percentage yield of zinc sulfide is derived as follows:

$$\text{Percentage yield of ZnS} = \frac{35.2\ \cancel{g}}{53.7\ \cancel{g}} \times 100\% = 65.5\%$$

EXAMPLE 8.10 Calculating Theoretical and Percentage Yields

A common type of iron ore contains the mineral hematite (Fe_2O_3). Industrially, metallic iron is obtained from such ore in a blast furnace (Figure 8.14) in a reaction with carbon monoxide.

$$Fe_2O_3(s) + 3CO(g) \longrightarrow 2Fe(l) + 3CO_2$$

Because this process is designed to recover as much metallic iron from as little ore as possible, it maintains an excess of carbon monoxide so that Fe_2O_3 is the limiting reactant.

In a laboratory demonstration of this reaction, you start with 8.65 g of hematite and recover 5.22 g of iron. What is the theoretical yield and the percentage yield?

Problem Analysis

First calculate the molar mass of Fe_2O_3 so that you can convert the given mass of Fe_2O_3 to moles. Since you already know that Fe_2O_3 is the limiting reactant, use this quantity to calculate the theoretical yield of Fe in moles. Convert the theoretical yield to grams, using the molar mass of Fe. Divide the actual yield (5.22 g) by the theoretical yield and multiply by 100% to obtain the percentage yield.

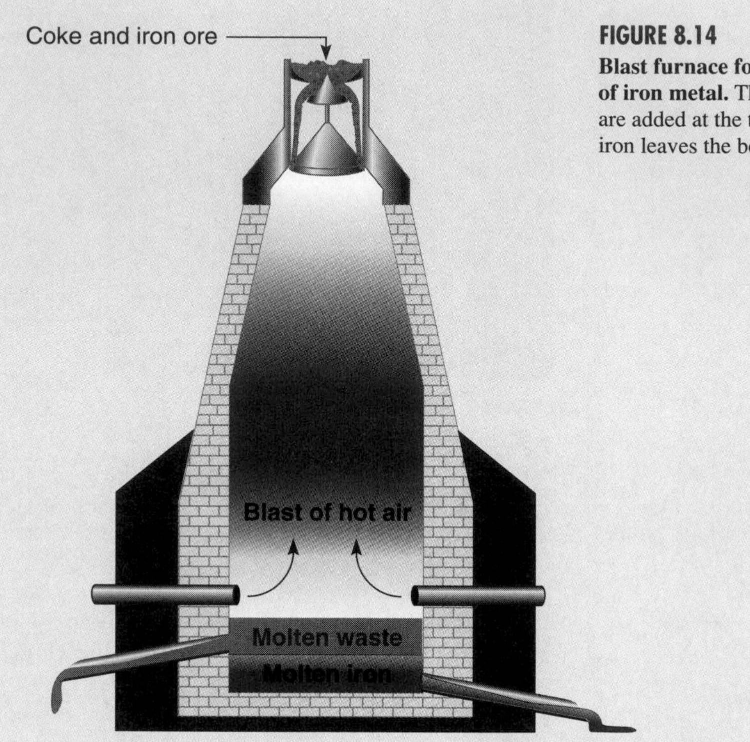

FIGURE 8.14
Blast furnace for production of iron metal. The reactants are added at the top and molten iron leaves the bottom.

Solution

The molar mass of Fe_2O_3 is 159.7 g, so the number of moles of Fe_2O_3 is

$$8.65\ \cancel{g\ Fe_2O_3} \times \frac{1\ mol\ Fe_2O_3}{159.7\ \cancel{g\ Fe_2O_3}} = 0.0542\ mol\ Fe_2O_3$$

You calculate the maximum moles of iron from this quantity of the oxide.

$$0.0542\ \cancel{mol\ Fe_2O_3} \times \frac{2\ mol\ Fe}{1\ \cancel{mol\ Fe_2O_3}} = 0.108\ mol\ Fe \qquad \text{(theoretical)}$$

Now you convert this amount of iron to grams, using the molar mass of Fe, which is 55.85 g.

$$0.108\ \cancel{mol\ Fe} \times \frac{55.85\ g\ Fe}{1\ \cancel{mol\ Fe}} = 6.03\ g\ Fe \qquad \text{(theoretical)}$$

This is the theoretical yield of iron.

Because the actual yield is 5.22 g, the percentage yield is

$$\text{Percentage yield of Fe} = \frac{5.22\ \cancel{g}}{6.03\ \cancel{g}} \times 100\% = 86.6\%$$

Exercise 8.10

Calculate the theoretical yield and percentage yield if 5.00 g of hydrogen was used in the explosion shown in Figure 8.3 and 0.32 g of water was recovered.

(Try Problems 8.49, 8.50, 8.51, and 8.52.)

Chemical Perspective

Nitrogen, the Limiting Reactant for Plant Growth

Although plants require many different nutrients for their growth, nitrogen—more specifically, usable nitrogen—is often the one in the shortest supply. As a result, it is not only an essential nutrient but also the limiting reactant in the processes of plant growth.

Every living thing needs nitrogen because it is a chemical component of proteins and nucleic acids. We get our nitrogen by eating plants and by eating other animals that have eaten plants. Plants get their nitrogen indirectly from the air, the ultimate source of nitrogen, where it occurs as colorless and odorless N_2. In fact, the greatest percentage of the gases in air is N_2—more than 75% by mass.

However, despite its abundance in air, nitrogen is very unreactive, so somehow it must be incorporated into chemical compounds before plants can use it. Several natural processes "fix" nitrogen, that is, convert it to compounds that plants can use. Some nitrogen is fixed by lightning with the formation of nitrogen monoxide (NO), which is then converted to other nitrogen compounds. Nitrogen is also fixed by bacteria in the root nodules of legumes such as peas and beans. This type of fixation results in the formation of ammonia (NH_3). Pure ammonia is a gas under ordinary conditions. The liquid called "ammonia" sold in supermarkets as a cleaning solution is really a solution of ammonia in water.

Up to about 100 years ago, farmers supplemented natural sources of fixed nitrogen with fertilizers such as manure from farm animals, guano (bird droppings) from Peru, and nitrates mined from Chile. As the world population grew, the demand for crops increased enormously and other nitrogen fertilizers were required. Chemists then began to look for ways to use the potentially rich pool of nitrogen in the air, but they were thwarted by nitrogen gas's lack of reactivity.

Ironically, military needs provided the impetus to overcome this difficulty. Shortly before World War I, the German chemist Fritz Haber (1868–1934) showed that nitrogen from the air could be combined with hydrogen gas under certain conditions to form ammonia.

$$N_2(g) + 3H_2(g) \longrightarrow 2NH_3(g)$$

German industrialists supported this research because they wanted to convert the ammonia to nitric acid, a starting material for the production of explosives.

This synthesis reaction is so slow under ordinary conditions that the percentage yield from the elements is virtually zero. Haber needed to find a catalyst to make the reaction occur at a reasonable rate so that he could obtain a significant yield of ammonia. By 1913, he had found the catalyst (iron mixed with metal oxides), as well as the other conditions required to make the process work. Haber received the Nobel Prize in chemistry in 1918 for this successful research. Commercial production of ammonia by his process began in 1921.

About 2×10^7 tons of ammonia are produced annually in the United States by the *Haber process*. A large fraction of this is liquefied and applied directly to farmland (Figure 8.15). Most of the rest is converted to the solid fertilizers ammonium nitrate, NH_4NO_3, ammonium sulfate, $(NH_4)_2SO_4$, and urea, $(NH_2)_2CO$. The extraordinary achievements of modern agriculture are due in large measure to Fritz Haber's discovery of how to fix atmospheric nitrogen. While famine has always been a factor in human history, it is unquestionably true that nitrogen fertilizers have relieved many people from its peril.

It is a bit sobering to realize, however, that the amount of nitrogen fixed worldwide *(continued)*

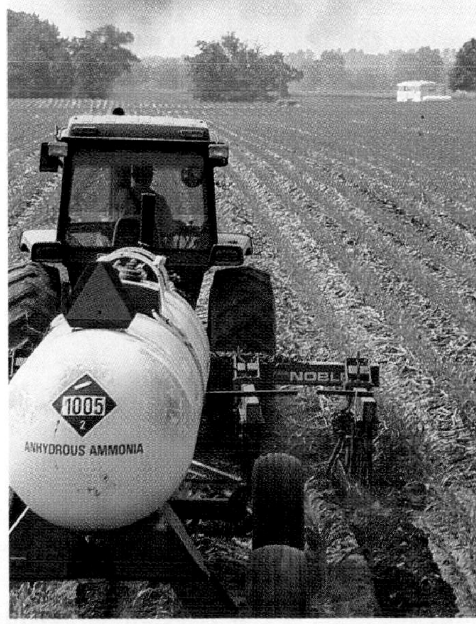

FIGURE 8.15
Liquid ammonia as a fertilizer. Anhydrous (water-free) liquid ammonia is injected into the ground, where it dissolves in moisture in the soil.

by the Haber process today (80 million metric tons; one metric ton equals 1000 kg) is commensurate with the quantity of nitrogen fixed by natural processes (90 to 150 million metric tons). The obvious question is "What happens eventually to all this fixed nitrogen?" Environmental scientists are beginning to suspect that our bountiful supply of fertilizer nitrogen is not entirely benign. For example, some of it leaches from soils, flows into our rivers, and then into coastal estuaries, where it affects the delicate balance of nature. No doubt we must be more creative in solving our fertilizer problems than we have been. But this realization should not detract from Haber's great achievement.

CHAPTER REVIEW

Key Words

stoichiometry *(p. 219)*
limiting reactant *(p. 233)*
stoichiometric amount *(p. 233)*
theoretical yield *(p. 240)*
actual yield *(p. 240)*
percentage yield *(p. 240)*

Key Equation

$$\text{Percentage yield} = \frac{\text{actual yield}}{\text{theoretical yield}} \times 100\%$$

Summary

In practical applications, chemists must calculate the quantities of reactants and products that are involved in a chemical reaction. These problems in *stoichiometry* always begin with a balanced chemical equation for the reaction. The coefficients in a balanced chemical equation specify only the relative numbers of molecules or formula units involved, not the absolute numbers. Therefore, those coefficients can be interpreted in terms of moles of reactants and products. From the coefficients, you can obtain mole ratios that enable you to convert the amount of one reactant or product (in moles) to the amount of any other substance involved in the reaction. Then, using the molar mass of each substance, you can convert the mass of any reactant or product into the mass of any other reactant or product.

Reactants can be mixed in *stoichiometric amounts,* the exact proportions dictated by the balanced chemical equation so that every reactant is used up when the reaction is completed. Sometimes, however, there is a *limiting reactant,* one that is completely used up while the other reactants are in excess. The available quantity of the limiting reactant limits the amount of products that can be formed.

The maximum amount of a product that can be formed from a given quantity of reactants is the *theoretical yield.* It is usually greater than the *actual yield,* the amount of a product that is actually recovered. The *percentage yield* is the actual yield expressed as a percentage of the theoretical yield.

Problem-Solving Skills

1. **Using Mole Ratios to Calculate Moles of One Reactant or Product from Moles of Any Other Reactant or Product:** Given a balanced chemical equation and the quantity of any reactant or product in moles, calculate the amount of any other reactant or product in moles (Examples 8.1, 8.2, and 8.3).
2. **Calculating the Mass of One Reactant or Product from the Mass of Any Other Reactant or Product:** Given a balanced chemical equation and the mass of any reactant or product, calculate the required mass of another reactant or the expected mass of a product (Examples 8.4, 8.5, and 8.6).
3. **Identifying a Limiting Reactant:** Given a balanced chemical equation and the quantities of the reactants in moles, identify the limiting reactant if one is present (Example 8.7).
4. **Calculating the Quantity of a Product in Moles When a Limiting Reactant Occurs:** Given a balanced chemical equation and the quantities of the reactants in moles, identify the limiting reactant if one is present and calculate the required quantities of other reactants in moles and the expected quantities of the products in moles (Example 8.8).

5. **Calculating the Mass of a Product from the Mass of a Limiting Reactant:** Given a balanced chemical equation and the masses of the reactants, identify the limiting reactant if one is present and calculate the masses of the products (Example 8.9).

6. **Calculating Theoretical and Percentage Yields:** Given a balanced chemical equation, the mass of one reactant, and the actual yield of the reaction, calculate the theoretical yield and the percentage yield (Example 8.10).

Questions to Test Your Reading

8.1 What does Dalton's atomic theory tell you about a chemical equation?

8.2 Using both molecules and moles, describe the meaning of the chemical equation

$$CO(g) + 3H_2(g) \longrightarrow CH_4(g) + H_2O(g)$$

8.3 Ethylene, C_2H_4, burns in oxygen to give carbon dioxide and water. Write the chemical equation for this reaction. Give the interpretations for this equation based on molecules and based on moles.

8.4 Hydrogen sulfide, H_2S, burns in oxygen to yield sulfur dioxide and water. Write the chemical equation for this reaction. Is it true that 5.0 mol of hydrogen sulfide will combine with 15 mol of oxygen? Explain.

8.5 What is a mole ratio in the context of this chapter?

8.6 Consider the reaction in Question 8.2 again. Which mole ratio,

$$\frac{1 \text{ mol CO}}{3 \text{ mol } H_2} \quad \text{or} \quad \frac{3 \text{ mol } H_2}{1 \text{ mol CO}}$$

will allow you to calculate the number of moles of carbon monoxide required to react with a known quantity of hydrogen? Why?

8.7 Balance each of the following equations and give the mole ratio that will allow you to convert moles of the reactant indicated in boldface to moles of the product indicated in boldface.

(a) $CO + \mathbf{O_2} \longrightarrow \mathbf{CO_2}$
(b) $CH_4 + \mathbf{O_2} \longrightarrow \mathbf{CO_2} + H_2O$
(c) $C_2H_6 + \mathbf{O_2} \longrightarrow \mathbf{CO_2} + H_2O$
(d) $C_3H_8 + \mathbf{O_2} \longrightarrow \mathbf{CO_2} + H_2O$

8.8 Balance each of the following equations and give the mole ratio that will allow you to convert moles of the reactant indicated in boldface to moles of the product indicated in boldface.

(a) $\mathbf{SO_2} + O_2 \longrightarrow \mathbf{SO_3}$
(b) $CS_2 + \mathbf{Cl_2} \longrightarrow \mathbf{CCl_4} + S_2Cl_2$
(c) $\mathbf{WO_3} + H_2 \longrightarrow W + \mathbf{H_2O}$
(d) $\mathbf{Li} + O_2 \longrightarrow \mathbf{Li_2O}$

8.9 Explain why we know that 1 g of carbon monoxide will not react exactly with 3 g of hydrogen according to the chemical equation in Question 8.2.

8.10 What is a limiting reactant?

8.11 What is the difference between a reaction that uses a limiting reactant and one that uses stoichiometric amounts of the reactants?

8.12 What are theoretical yield, actual yield, and percentage yield?

Practice Problems

Mole Calculations from Chemical Equations (Section 8.2)

8.13 Aluminum and elemental oxygen will react to form aluminum oxide, Al_2O_3. The chemical equation is

$$4Al(s) + 3O_2(g) \longrightarrow 2Al_2O_3(s)$$

How many moles of oxygen are required to react with 1.86 mol of aluminum?

8.14 Elemental nitrogen and hydrogen will react under appropriate conditions to form ammonia, NH_3. This reaction is described by the chemical equation:

$$N_2(g) + 3H_2(g) \longrightarrow 2NH_3(g)$$

How many moles of nitrogen are required to react with 2.54 mol of hydrogen?

8.15 Ethanol, C_2H_5OH, is a component of the fuel called gasohol. The unbalanced chemical equation for the combustion of ethanol is

$$C_2H_5OH(g) + O_2(g) \longrightarrow CO_2(g) + H_2O(g)$$

How many moles of oxygen will be required to react with 8.24 moles of ethanol?

8.16 Octane, C_8H_{18}, is a component of gasoline. The unbalanced chemical equation for the combustion of octane is

$$C_8H_{18}(g) + O_2(g) \longrightarrow CO_2(g) + H_2O(g)$$

How many moles of octane will be required to react with 4.44 mol of oxygen?

8.17 Aluminum will react with iodine to form aluminum iodide. The chemical equation is

$$2Al(s) + 3I_2(s) \longrightarrow 2AlI_3(s)$$

How many moles of aluminum iodide can be obtained from 5.6 mol of iodine and the required amount of aluminum?

8.18 Aluminum will dissolve in sulfuric acid to form aluminum sulfate and gaseous hydrogen according to the chemical equation

$$2Al(s) + 3H_2SO_4(aq) \longrightarrow Al_2(SO_4)_3(aq) + 3H_2(g)$$

How many moles of aluminum sulfate will be formed from 24.3 mol of sulfuric acid and the required amount of aluminum?

8.19 Sodium hydroxide reacts with sulfuric acid to form sodium sulfate and water. The chemical equation is

$$2NaOH(aq) + H_2SO_4(aq) \longrightarrow Na_2SO_4(aq) + 2H_2O(l)$$

How many moles of sodium sulfate can be obtained from 3.00 mol of sodium hydroxide and the required quantity of sulfuric acid?

8.20 Aluminum reacts with oxygen to form aluminum oxide. The chemical equation is

$$4Al(s) + 3O_2(g) \longrightarrow 2Al_2O_3(s)$$

How many moles of aluminum oxide can be obtained from 5.22 mol of aluminum and the required amount of oxygen?

8.21 Nitric acid, HNO_3, is manufactured by a process that allows nitrogen dioxide to react with water. The chemical equation is

$$3NO_2(g) + H_2O(l) \longrightarrow 2HNO_3(aq) + NO(g)$$

How many moles of nitrogen dioxide are required to produce 3.56 mol of nitric acid?

8.22 Elemental phosphorus can be prepared from calcium phosphate, $Ca_3(PO_4)_2$, sand, SiO_2, and carbon in an electric furnace. The chemical equation is

$$2Ca_3(PO_4)_2 + 6SiO_2 + 10C \longrightarrow P_4 + 6CaSiO_3 + 10CO$$

How many moles of carbon are required to produce 134 mol of phosphorus?

8.23 Sodium hydrogen carbonate is the chemical name for baking soda. This substance can be prepared by dissolving gaseous carbon dioxide in an aqueous solution of sodium carbonate. The chemical equation is

$$Na_2CO_3(aq) + CO_2(g) + H_2O(l) \longrightarrow 2NaHCO_3(aq)$$

How many moles of sodium carbonate and carbon dioxide will be required in the preparation of 1.60 mol of sodium hydrogen carbonate?

8.24 A solution of cesium hydroxide can be prepared by mixing aqueous solutions of cesium sulfate and barium hydroxide to react and then removing barium sulfate by filtration. The chemical equation is

$$Cs_2SO_4(aq) + Ba(OH)_2(aq) \longrightarrow 2CsOH(aq) + BaSO_4(s)$$

How many moles of each reactant will be needed if a student wishes to prepare 1.76 mol of cesium hydroxide?

Mass Calculations from Chemical Equations (Section 8.3)

8.25 Sodium will react readily with chlorine according to the chemical equation

$$2Na(s) + Cl_2(g) \longrightarrow 2NaCl(s)$$

Calculate the mass of sodium that is needed to react with 1.00 g of chlorine.

8.26 When iron rusts, the reaction is described by

$$4Fe(s) + 3O_2(g) \longrightarrow 2Fe_2O_3(s)$$

Calculate the mass of oxygen required to react with 1.00 g of iron.

8.27 Chromium will dissolve in hydrochloric acid to form chromium(II) chloride and hydrogen. The chemical equation is

$$Cr(s) + 2HCl(aq) \longrightarrow CrCl_2(aq) + H_2(g)$$

How many grams of hydrochloric acid are needed to react with 1.00 g of chromium?

8.28 Magnesium will burn in air when it is heated. The chemical equation is

$$2Mg(s) + O_2(g) \longrightarrow 2MgO(s)$$

How many grams of oxygen are required for reaction with 1.00 g of magnesium?

8.29 If 1.62 g of calcium carbonate is heated, how many grams of calcium oxide will be obtained when the reaction is finished? The chemical equation is

$$CaCO_3(s) \longrightarrow CaO(s) + CO_2(g)$$

8.30 Copper(II) oxide will decompose when it is heated strongly. The chemical equation is

$$4CuO(s) \longrightarrow 2Cu_2O(s) + O_2(g)$$

How many grams of oxygen can be obtained from 2.64 g CuO?

8.31 Mercury(II) oxide decomposes when it is heated. The chemical equation is

$$2HgO(s) \longrightarrow 2Hg(l) + O_2(g)$$

How many grams of mercury can be obtained from 10.3 g HgO?

8.32 Water will decompose when an electric current is passed through it. The chemical equation is

$$2H_2O(l) \longrightarrow 2H_2(g) + O_2(g)$$

How many grams of oxygen can be obtained from 1.32 g of water?

8.33 Ammonia will react with hydrochloric acid to yield ammonium chloride according to the chemical equation

$$NH_3(aq) + HCl(aq) \longrightarrow NH_4Cl(aq)$$

How many grams of HCl are consumed if 2.36 g NH_4Cl is formed?

8.34 Sodium hydroxide reacts with hydrochloric acid to give sodium chloride and water. The chemical equation is

$$NaOH(aq) + HCl(aq) \longrightarrow NaCl(aq) + H_2O(l)$$

If 23.1 g of sodium chloride is produced in this reaction, how many grams of sodium hydroxide are used up?

8.35 Sodium will react violently with water according to the chemical equation

$$2Na(s) + 2H_2O(l) \longrightarrow 2NaOH(aq) + H_2(g)$$

What mass of sodium takes part in this reaction if 3.5 g of hydrogen is formed?

8.36 Magnesium will react with steam to form magnesium hydroxide and to liberate hydrogen. The chemical equation is

$$Mg(s) + 2H_2O(g) \longrightarrow Mg(OH)_2(s) + H_2(g)$$

If 13.8 g of magnesium hydroxide is formed, how many grams of water took part in the reaction?

Limiting Reactants and Percentage Yield (Sections 8.4 to 8.6)

8.37 Potassium superoxide, KO_2, is used as a source of oxygen in rebreathing masks. The chemical equation for the reaction is

$$4KO_2(s) + 2H_2O(l) \longrightarrow 4KOH(s) + 3O_2(g)$$

Identify the limiting reactant, if any, in each of the following mixtures of reactants.
(a) 6.4 mol KO_2 and 2.1 mol H_2O
(b) 8.4 mol KO_2 and 1.5 mol H_2O
(c) 8.4 mol KO_2 and 2.1 mol H_2O

8.38 When ammonia is burned in the presence of a catalyst, nitrogen monoxide and water are formed as shown in the equation

$$4NH_3(g) + 5O_2(g) \longrightarrow 4NO(g) + 6H_2O(g)$$

Identify the limiting reactant, if any, in each of the following mixtures of reactants.
(a) 4.8 mol NH_3 and 5.8 mol O_2
(b) 12 mol NH_3 and 15 mol O_2
(c) 32 mol NH_3 and 44 mol O_2

8.39 Consider the following unbalanced chemical equations. If 2.0 mol of each reactant is used, which reactant, if any, is the limiting reactant?
(a) $P_4(s) + Cl_2(g) \longrightarrow PCl_3(s)$
(b) $Al(s) + Cl_2(g) \longrightarrow AlCl_3(s)$
(c) $C(s) + Cl_2(g) \longrightarrow CCl_4(l)$

8.40 Consider the following unbalanced chemical equations. If 2.0 mol of each reactant are used, which reactant, if any, is the limiting reactant?
(a) $C(s) + O_2(g) \longrightarrow CO(g)$
(b) $Fe(s) + Cl_2(g) \longrightarrow FeCl_3(s)$
(c) $N_2(g) + H_2(g) \longrightarrow NH_3(g)$

8.41 Sodium will react with chlorine to yield sodium chloride. The chemical equation for this reaction is

$$2Na(s) + Cl_2(g) \longrightarrow 2NaCl(s)$$

If the reaction begins with 3.0 mol of each reactant, how many moles of each substance will be present when the reaction is complete?

8.42 Iron reacts with oxygen to give iron(III) oxide. The chemical equation is

$$4Fe(s) + 3O_2(g) \longrightarrow 2Fe_2O_3(s)$$

If the reaction begins with 1.00 mol of each reactant, how many moles of each substance will be present when the reaction is complete?

8.43 The reaction of aluminum with oxygen is described by the chemical equation

$$4Al(s) + 3O_2(g) \longrightarrow 2Al_2O_3(s)$$

If 2.5 mol of each of the reactants are mixed, how many moles of the reactants and product will be present when the reaction is complete?

8.44 When copper(I) oxide is heated in oxygen, copper(II) oxide is formed. The chemical equation is

$$2Cu_2O(s) + O_2(g) \longrightarrow 4CuO(s)$$

If 3.00 mol of each of the reactants are heated together, how many moles of the reactants and product will be present when the reaction is complete?

8.45 Ammonia will react with hydrochloric acid to yield ammonium chloride according to the chemical equation

$$NH_3(aq) + HCl(aq) \longrightarrow NH_4Cl(aq)$$

If the reaction begins with 1.00 g of each reactant, how many grams of ammonia will be present at the end of the reaction?

8.46 Sodium hydroxide reacts with hydrochloric acid to give sodium chloride and water. The chemical equation is

$$NaOH(aq) + HCl(aq) \longrightarrow NaCl(aq) + H_2O(l)$$

If the reaction starts with 1.00 g each of NaOH and HCl, how many grams of NaCl will be present when the reaction is complete?

8.47 Methanol, CH_3OH, is prepared industrially by the reaction shown in the chemical equation

$$CO(g) + 2H_2(g) \longrightarrow CH_3OH(g)$$

In a laboratory test, 30.0 g of each reactant is added to a reaction vessel. Which reactant, if any, is not completely consumed at the end of the reaction? How many grams of this reactant will be left, and how many grams of methanol will be formed?

8.48 The complete combustion of carbon disulfide occurs by the reaction shown in the chemical equation

$$CS_2(g) + 3O_2(g) \longrightarrow CO_2(g) + 2SO_2(g)$$

If 254 g of each reactant are used, which reactant, if any, is not completely consumed at the end of the reaction? How many grams of this reactant will be left, and how many grams of carbon dioxide will be formed?

8.49 The theoretical yield of a product in a reaction is 4.87 g, but only 4.72 g is recovered. What is the percentage yield?

8.50 The theoretical yield of a product in a reaction is 112 g, but only 4.02 g is recovered. What is the percentage yield?

8.51 The last step in making nitric acid is the reaction of nitrogen dioxide with water. The chemical equation is

$$3NO_2(g) + H_2O(l) \longrightarrow 2HNO_3(aq) + NO(g)$$

If the actual yield of the acid is 44.2 g when 60.0 g of nitrogen dioxide reacts with excess water, calculate the theoretical yield and the percentage yield.

8.52 Nitrobenzene, $C_6H_5NO_2$, is the substance that gives shoe polish its odor. It is made from the reaction of benzene (C_6H_6) and nitric acid according to the chemical equation

$$C_6H_6(l) + HNO_3(aq) \longrightarrow C_6H_5NO_2(l) + H_2O(l)$$

If the actual yield of nitrobenzene is 28.7 g when 20.3 g of benzene reacts with excess nitric acid, calculate the theoretical yield and the percentage yield.

Additional Problems

8.53 Calcium carbonate reacts with hydrochloric acid to yield calcium chloride, carbon dioxide, and water. Write a balanced equation for this reaction. If 22.4 mol of calcium carbonate is used, how many moles of hydrochloric acid are required and how many moles of each product are formed?

8.54 Lithium sulfite, Li_2SO_3, reacts with hydrobromic acid to yield lithium bromide, sulfur dioxide, and water. Write a balanced equation for this reaction. If 2.46 mol of lithium sulfite is used, how many moles of hydrobromic acid are required and how many moles of each product are formed?

8.55 Rust, Fe_2O_3, can be dissolved with hydrochloric acid. The products are iron(III) chloride and water. Write a balanced chemical equation for this reaction. If 1.23 mol of rust is to be dissolved, how many moles of hydrochloric acid are required and how many moles of each product are formed?

8.56 Sulfuric acid reacts with aluminum oxide to form aluminum sulfate and water. Write a balanced chemical equation for this reaction. If you wish to dissolve 3.42 mol of the oxide, how many moles of sulfuric acid are required and how many moles of each product are formed?

8.57 Sodium phosphate, Na_3PO_4, is formed when sodium hydroxide is neutralized with phosphoric acid. Water is the other product. Write a balanced equation for this reaction. If 5.00 g of sodium hydroxide and 7.00 g of phosphoric acid are used, which reactant, if any, is not completely consumed at the end of the reaction? Calculate the mass of this reactant at the end of the reaction and the mass of sodium phosphate that forms.

8.58 Sulfuric acid is formed when sulfur trioxide reacts with water. Write a balanced equation for this reaction. If 4.56 g of sulfur trioxide and 10.0 g of water are used, which reactant, if any, is not completely consumed at the end of the reaction? Calculate the mass of this reactant at the end of the reaction and the mass of sulfuric acid that forms.

8.59 Calcium hydrogen carbonate, $Ca(HCO_3)_2$, is a component of hard water. It can be removed by adding calcium hydroxide, $Ca(OH)_2$, because insoluble calcium carbonate, $CaCO_3$, precipitates. The other product is water. Write a balanced chemical equation for this reaction. Calculate the mass of calcium carbonate that forms from 16.3 g of calcium hydrogen carbonate and the required quantity of calcium hydroxide.

8.60 Sulfuric acid is neutralized by sodium hydroxide. Write a balanced chemical equation for this reaction. How many grams of sodium sulfate are formed when 2.08 g of sodium hydroxide reacts with the required amount of sulfuric acid?

8.61 Barium hydroxide reacts with phosphoric acid to yield barium phosphate and water. Write a balanced chemical equation for this reaction. How many grams of barium hydroxide and phosphoric acid are required to form 3.00 g of barium phosphate?

8.62 Cobalt(II) chloride will react with silver nitrate in water to give cobalt(II) nitrate and solid silver chloride. Write a balanced chemical equation for this reaction. Calculate the mass of each reactant needed to produce 1.50 g of silver chloride.

8.63 Metallic zinc can be obtained from zinc oxide and carbon monoxide at high temperature. The other product is carbon dioxide. The carbon monoxide can be obtained from the combustion of carbon. Write balanced equations for both reactions. What is the maximum amount of zinc that can be obtained from 75.0 g of zinc oxide, 50.0 g of carbon, and excess oxygen?

8.64 Hydrogen cyanide, HCN, can be prepared by a two-step process. First, ammonia is allowed to react with oxygen. The products are nitrogen monoxide, NO, and water. Then, nitrogen monoxide is allowed to react with methane. The products of this reaction are hydrogen cyanide, water, and H_2. Write balanced equations for both reactions. What is the maximum mass of hydrogen cyanide that can be obtained from 24.2 g of ammonia, 25.1 g of methane, and excess oxygen?

8.65 The chemical name for aspirin is acetylsalicylic acid. The formula for this substance is $C_9H_8O_4$. It is prepared by heating salicylic acid, $C_7H_6O_3$, with acetic anhydride, $C_4H_6O_3$. Along with aspirin, acetic acid also forms. Calculate the maximum mass of aspirin that can be formed from 2.00 g of salicylic acid and 4.00 g of acetic anhydride.

8.66 Methyl salicylate, also known as oil of wintergreen, is related to aspirin. Its formula is $C_8H_8O_3$. It can be prepared by heating salicylic acid, $C_7H_6O_3$, with methanol, CH_3OH. Water is also a product. What is the maximum mass of methyl salicylate that can be obtained from 1.50 g of salicylic acid and 11.20 g of methanol?

Practice in Problem Analysis

For each problem, describe the thinking you would use (the problem analysis) before doing the actual solution, but do not solve the problem.

1. When 5.0 g iron wool burns in oxygen, how many grams of iron(III) oxide form according to the following equation?

$$4Fe(s) + 3O_2(g) \longrightarrow 2Fe_2O_3(s)$$

2. You add 5.0 g sodium hydroxide to a solution containing 5.0 g sulfuric acid. How many grams of sodium sulfate form at the completion of the reaction?

$$2NaOH(aq) + H_2SO_4(aq) \rightarrow Na_2SO_4(aq) + 2H_2O(l)$$

PRACTICE EXAM

Problems 1 to 15 refer to the following commercial reaction in which tungsten, W, is produced from tungsten(VI) oxide, WO_3:

$$WO_3(s) + 3H_2(g) \longrightarrow W(s) + 3H_2O(l)$$

1. How many moles of hydrogen do you need to react with 0.24 mol tungsten(VI) oxide?
(a) 0.08 mol (b) 0.12 mol (c) 0.24 mol
(d) 0.48 mol (e) None of the above

2. How many moles of tungsten(VI) oxide would you need to react with 0.24 mol hydrogen?
(a) 0.08 mol (b) 0.12 mol (c) 0.24 mol
(d) 0.48 mol (e) None of the above

3. How many moles of tungsten metal could you produce from 0.24 mol tungsten(VI) oxide?
(a) 0.08 mol (b) 0.12 mol (c) 0.24 mol
(d) 0.48 mol (e) None of the above

4. How much hydrogen would you need to produce 0.24 mol tungsten?
(a) 0.08 mol (b) 0.12 mol (c) 0.24 mol
(d) 0.48 mol (e) None of the above

5. If the reaction produces 0.24 mol tungsten, how many moles of water is also produced?
(a) 0.08 mol (b) 0.12 mol (c) 0.24 mol
(d) 0.48 mol (e) None of the above

6. How many grams of hydrogen do you need to react with 36 g tungsten(VI) oxide?
(a) 0.94 g (b) 1.6 g (c) 14 g (d) 29 g
(e) None of the above

7. How many grams of tungsten(VI) oxide would you need to react with 2.4 g hydrogen?
(a) 0.94 g (b) 1.6 g (c) 14 g (d) 29 g
(e) None of the above

8. How many grams of tungsten metal could you produce from 36 g tungsten(VI) oxide?
(a) 0.94 g (b) 1.6 g (c) 14 g (d) 29 g
(e) None of the above

9. How much hydrogen would you need to produce 48 g tungsten?
(a) 0.94 g (b) 1.6 g (c) 14 g (d) 29 g
(e) None of the above

10. If the reaction produces 48 g tungsten, how many grams of water is also produced?
(a) 0.94 g (b) 1.6 g (c) 14 g (d) 29 g
(e) None of the above

11. You place 24 g tungsten(VI) oxide and 10 g hydrogen in a closed vessel and allow them to react. What is the limiting reactant?
(a) WO_3 (b) H_2 (c) W (d) H_2O
(e) No limiting reactant

12. You place 2.1 mol each of tungsten(VI) oxide and hydrogen in a closed vessel and allow them to react. How many moles of tungsten are produced?
(a) 0.70 mol (b) 1.4 mol (c) 2.1 mol
(d) 4.2 mol (e) None of the above

13. You place 2.1 mol each of tungsten(VI) oxide and hydrogen in a closed vessel and allow them to react. How many moles of water are produced?
(a) 0.70 mol (b) 1.4 mol (c) 2.1 mol
(d) 4.2 mol (e) None of the above

14. You place 24 g of tungsten(VI) oxide and 1.2 g of hydrogen in a closed vessel and allow them to react. How many grams of tungsten are produced?
(a) 6.4 g (b) 8.0 g (c) 19 g (d) 24 g
(e) None of the above

15. You place 24 g of tungsten(VI) oxide and 1.2 g of hydrogen in a closed vessel and allow them to react. If you obtain 6.4 g of tungsten at the completion of the reaction, what is the percentage yield of tungsten?
(a) 25% (b) 34% (c) 54% (d) 80%
(e) None of the above

Chapter

Electron Structure of Atoms

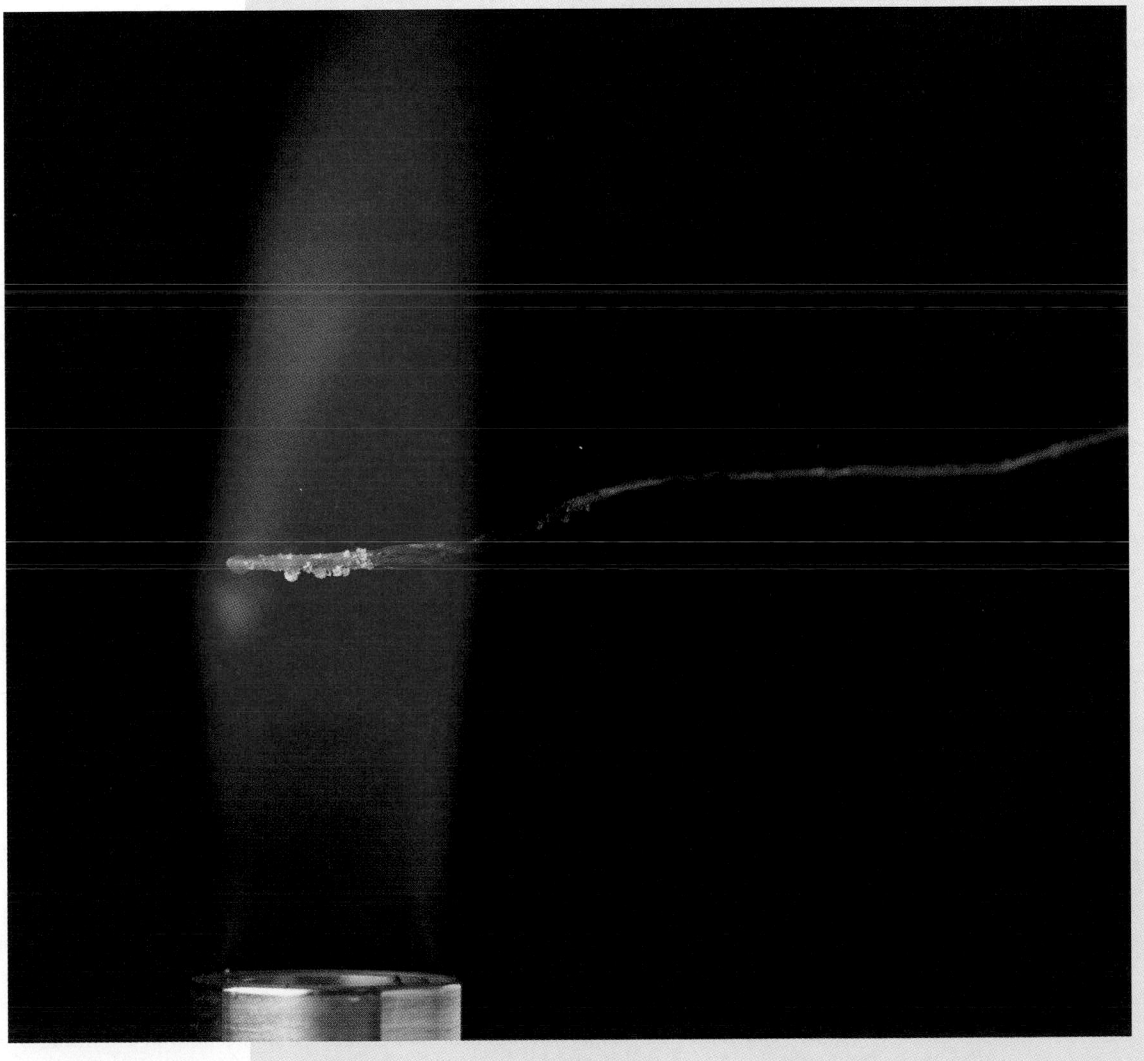

Strontium burns with a red flame.

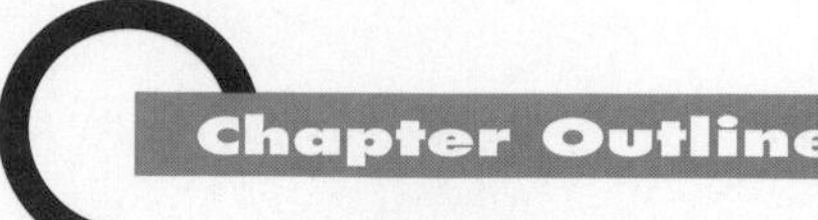

Chapter Outline

Video: Flame tests.

After hydrogen, helium is the second most abundant element in the universe. But how do astronomers know that? It might seem that the philosopher Auguste Comte was on solid ground when he stated in 1844 that, because of the remoteness of the stars, we could never know their chemical composition. Yet only a few years later chemists discovered that under the proper circumstances every element emits a certain combination of colors of light that can identify it, just as your fingerprints can identify you. By analyzing the light coming from stars, astronomers have been able to identify the different chemical elements in the stars and in cosmic space, as well as their relative quantities. It is an interesting historical note that helium was actually found on the sun, in 1868, before it was found on earth, in the 1890s. (The name *helium* comes from the Greek word *helios,* meaning "sun.")

The central principle of chemistry is that all matter is composed of atoms. These atoms emit the colors of light that characterize each element. As you saw in Chapter 4, each atom consists of a nucleus, or positively charged core, and around this nucleus are electrons, which are negatively charged particles. In this chapter, we see how the characteristic light emitted by an atom gives us information about the arrangement of electrons about the atomic nucleus. This arrangement determines the chemical properties of the atom.

You may recall that the atomic number of an element is the number of protons in the nucleus of one of its atoms. The atomic number is a fundamental property, because it also equals the number of electrons around the nucleus in the normal (electrically neutral) atom, and these electrons determine the atom's chemical properties. The periodic table of the elements (that is, their arrangement into the rows and columns based on their chemical and physical properties) is a result of the way the electrons are arranged in

FIGURE 9.1
Fireworks. The brilliant colors of fireworks are caused by heated metal atoms emitting their characteristic wavelengths of visible light.

atoms, and this arrangement or structure depends on the number of electrons and thus on the atomic number.

In this chapter, we explore this electron structure of atoms and the periodic properties of the elements. We answer questions such as these: How are the electrons arranged in different atoms? How do these arrangements explain the periodic table? How do properties such as atomic size vary throughout the periodic table?

ENERGY LEVELS AND ATOMIC ORBITALS

Fundamental discoveries have often been found in phenomena that did not seem fundamental at the time. The clue to the electron structure of atoms lay in the phenomena of colored flames and fireworks.

Fireworks displays (Figure 9.1) were discovered very early by the Chinese, who found that certain substances impart colors to the flames of burning materials. By the eighteenth century, chemists began to use flame tests to identify some elements in compounds. To perform a flame test, you place a sample of a compound in a flame; the color of the flame identifies the element (Figure 9.2). For example, copper compounds impart a blue-green color to flames; strontium and lithium compounds impart a deep red color.

FIGURE 9.2
Flame tests for some elements. A platinum wire loop is dipped into the compound of a metal and inserted into the flame to produce the color associated with the emission from that metal atom. The flames are those of lithium (red), strontium (red), copper (blue-green), and calcium (orange).

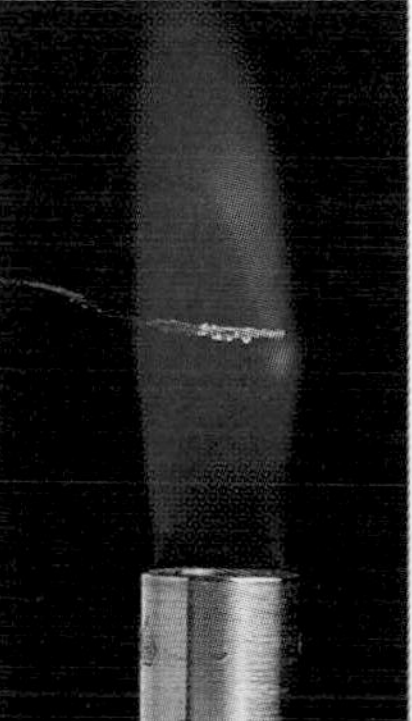
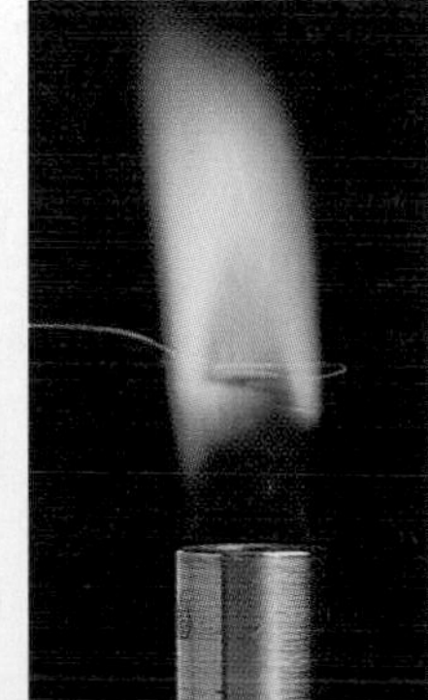
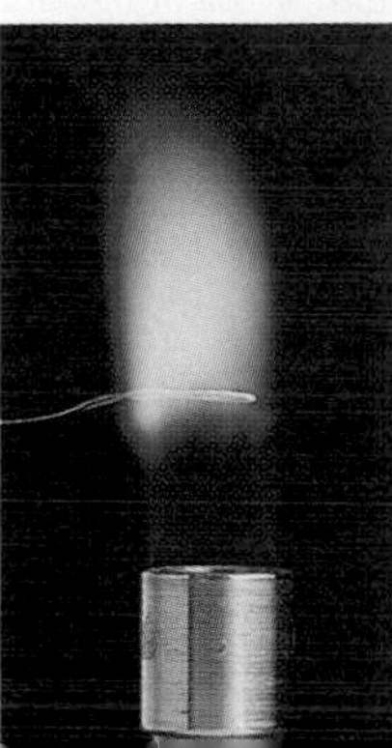

● By colored light, we mean light in the visible region of the spectrum. You see the visible spectrum when you look at a rainbow. Infrared radiation lies beyond the red end of the visible spectrum, and ultraviolet light lies beyond the violet end.

We now know that the heat in a flame breaks a substance into atoms, and in this breakup many of these atoms gain extra energy. The atoms quickly emit this extra energy as light, which in many cases is colored. ● Stars shine (emit visible light) because their atoms are at high temperatures.

At the beginning of the twentieth century, the Danish physicist Niels Bohr (1885–1962) showed that the color of light emitted by atoms depends on the energy emitted, and this in turn depends on the electron structure of the atoms. We describe Bohr's theory of the atom in Section 9.2 and its modern interpretation in Section 9.3. First, we need to say a few words about light and other forms of electromagnetic radiation.

Animations: Electromagnetic wave; Refraction of white light.

9.1 Light and Other Forms of Electromagnetic Radiation

OBJECTIVES

- List various types of electromagnetic radiation.
- Describe the characteristic properties of electromagnetic radiation.
- Relate the color of light to its energy.

Light exhibits wave properties. You can demonstrate this by a simple experiment. Look toward a bright (but not blinding) light, such as a window during the day or a fluorescent light. Now hold your two thumbs in front of your eyes while looking toward the light, and bring their edges together without touching (Figure 9.3). If you are careful, you will see a series of black lines in the area between your thumbs. Light waves flow around each thumb, and where the waves from the two thumbs come together they interact with one another. The black lines are places where the two waves come together and cancel one another, so there is no light wave in these places; you see black (no light). The effect, called *wave interference,* can also be observed with water waves.

Water waves, a disturbance of the water surface, are the most familiar kind of wave. Light, of course, is much less tangible. It is wave disturbance of electric and magnetic fields that move through space. We say that light is a form of *electromagnetic radiation.* Visible light is only one kind of electromagnetic radiation. Other examples are radio and TV waves, the microwaves used in microwave ovens, and the infrared radiation emitted by anything hot, such as a hot toaster coil or a burning log.

White light is a mixture of different light waves, each of a different color. The physicist Isaac Newton first demonstrated this using a glass prism, or glass wedge. When white light falls on a prism, it spreads out to form the rainbow of colors called the visible spectrum (Figure 9.4). A rainbow, in fact, is simply the visible spectrum obtained when white light is spread out by water droplets in the air, which together act like a glass

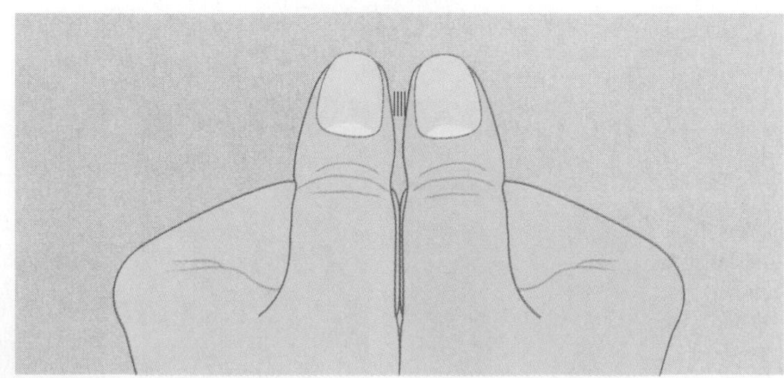

FIGURE 9.3
Wave interference pattern of light. If you bring your thumbs together close to one eye and look toward a bright light, you will see a series of black lines. This is an interference pattern. For steadiness, brace the fingers of one hand on those of the other.

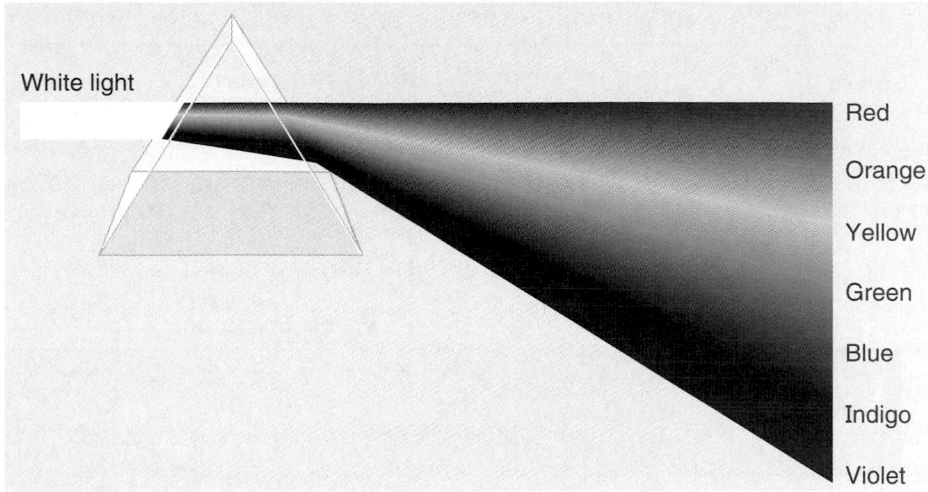

FIGURE 9.4
Visible spectrum of light. When passed through a glass prism, white light can be separated into its component colors, yielding the visible spectrum.

prism. Newton also showed that you could combine the colors of the rainbow to obtain white light.

All of the different kinds of electromagnetic radiation, as well as the different colors of the visible spectrum, differ in their wavelengths. The **wavelength** is *the distance between two peaks (or the distance between two troughs) of a wave* (see Figure 9.5). Violet light, for example, has wavelengths of about 400 nm (4×10^{-7} m), whereas red light has wavelengths of about 700 nm (7×10^{-7} m). Infrared radiation has longer wavelengths than red light, and radio waves are even longer, with wavelengths equal to several centimeters or longer. Ultraviolet light has shorter wavelengths than violet light, and x rays have even shorter wavelengths. The range of wavelengths of electromagnetic radiation, shown in Figure 9.6, is called the *electromagnetic spectrum.* The different kinds of electromagnetic radiation (x rays, visible light, and so on) are said to belong to different regions of the electromagnetic spectrum.

Waves are characterized by *frequency* as well as by wavelength. The frequency of a wave is the number of wavelengths passing a given point in one second. The frequency and wavelength of a wave are related. Waves with short wavelengths have high frequencies, and waves with long wavelengths have low frequencies.

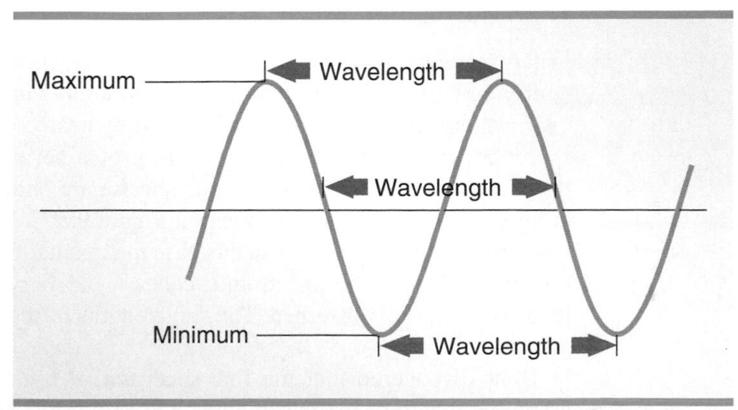

FIGURE 9.5
Wavelengths of light waves. The wavelength of a light wave is the distance between two equivalent points on the wave, such as the maxima.

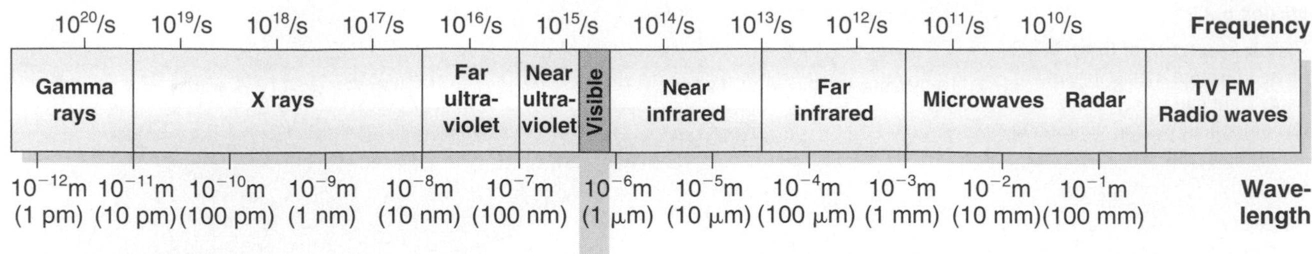

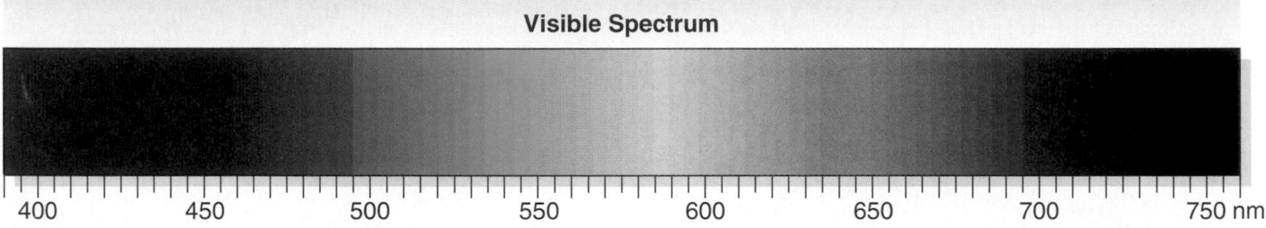

FIGURE 9.6

The electromagnetic spectrum. Different types of light have different wavelengths and frequencies.

Although the wave picture of light is appropriate in describing many of its properties, light also has particle properties. A **photon** is *a particle of light consisting of an extremely small packet of energy.* The quantity of energy in a photon depends on the region of the electromagnetic spectrum or color of the light. A photon of red light has less energy than a photon of blue light. In general, the shorter the wavelength (and the higher the frequency) of light, the greater the energy in a photon of that radiation.

Animation: H_2 line spectrum.

9.2 Bohr's Theory of the Atom

OBJECTIVES

- Compare the line spectrum of an element and the continuous spectrum.
- Define *energy level.*
- Describe how Bohr's theory explains the different colors in the line spectrum of an atom.
- Describe the two concepts of Bohr's atomic theory that are retained in modern atomic theory (quantum mechanics).

When you separate ordinary white light into its components with a prism, you obtain a *continuous spectrum* consisting of a continuous rainbow of colors (Figure 9.4). However, when you separate the light emitted by heated atoms of an element, you see a *line spectrum,* that is, a spectrum consisting of a series of colored lines against a black background (Figure 9.7). These line spectra are characteristic of the particular kind of atom you are looking at, as shown in Figure 9.8. A line spectrum can also be observed when light shines on atoms; in this case the resultant spectrum has black lines on a rainbow background. This spectrum is called an *absorption spectrum* because some wavelengths of light are absorbed. The wavelengths of the lines are the same in both types of spectra.

Video: Pickle light.

Bohr discovered that the line spectrum of hydrogen atoms can be explained by assuming that the electrons in atoms move in orbits about their nuclei with only certain

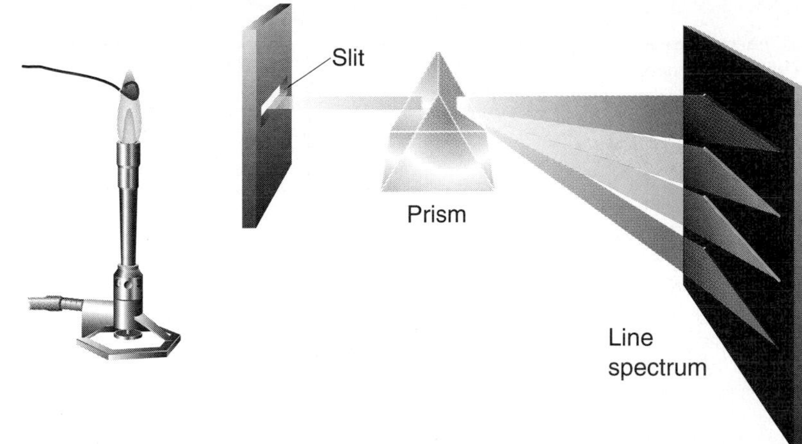

FIGURE 9.7
Measurement of line spectra. A line spectrum is generated when light from heated atoms passes through a prism. The slit produces a sharp beam of light, so that the lines of color coming from the prism are also sharp.

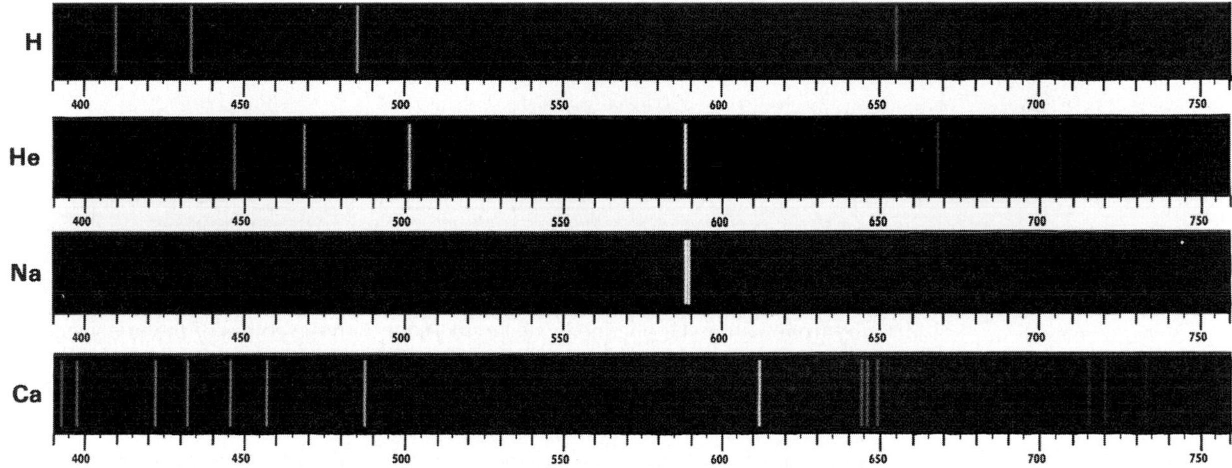

FIGURE 9.8
Line spectra of atoms. The lines are those in the visible region of the spectrum. Shown are spectra of hydrogen (H), helium (He), sodium (Na), and calcium (Ca). Note that the spectrum for each element is unique and so can be used to identify that element.

allowed energies (Figure 9.9). Each electron in an atom can have only one of these allowed energies; it cannot have an intermediate value. An **energy level** of an atom is *one of the allowed energy values that an electron can have.* The innermost orbit of the atom has the lowest energy, and the outer orbits have higher energy.

Each energy level in this diagram is designated by a *quantum number, n,* which is a positive integer, or whole number. Normally, the hydrogen atom exists in the lowest energy level, $n = 1$; but when heated, the atom gains energy and its electron jumps to a higher energy level, with $n = 2, 3, 4$, or higher. Some time later, the electron jumps back (or, as we say, undergoes a *transition*) to a lower level. When it does this, the excess energy is lost as a photon, or particle of light, whose wavelength corresponds to the energy change. Because only certain energy levels are allowed, only certain energy changes can occur in these transitions. The quantity of energy in each photon is associated with its particular wavelength (which determines its color), as already mentioned, so that a collection of similar atoms emits only particular colors of light, corresponding to the line spectrum of the element. Because energy levels of an atom depend on the element, the line spectrum is characteristic of the element.

Bohr had assumed that the electron moves in a specific circular orbit about the nucleus, much like the earth moves about the sun. When he applied his theory to the

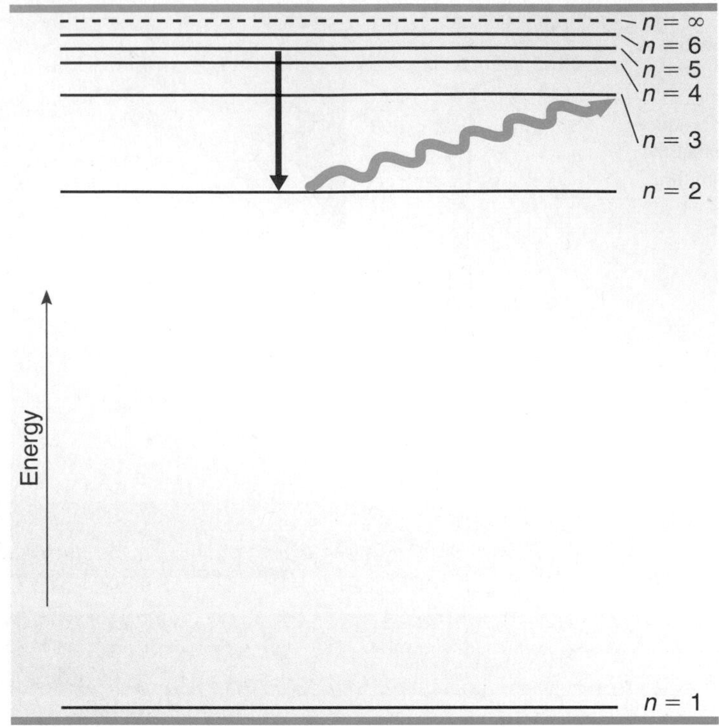

FIGURE 9.9
Allowed energies (energy levels) of the hydrogen atom. The energy of the electron is plotted along the vertical axis, with the lowest energy corresponding to $n = 1$. When the electron is in a level other than $n = 1$, it eventually undergoes a jump, or transition, to lower energy. The diagram shows a transition from $n = 4$ to $n = 2$. When a hydrogen atom undergoes this transition, it emits blue-green light. The energy of the blue-green light matches the energy change of the electron.

hydrogen atom, he obtained exact agreement with the colors of the line spectrum. However, he was unable to obtain similar agreement with the line spectra of other atoms, and his theory has now been supplanted by *quantum mechanics,* a mathematical theory developed in the 1920s. Despite these problems with Bohr's theory, two of its central concepts remain in the new theory: the concept of *energy levels* and the concept of *transitions* between energy levels. In the next section, we discuss how the newer theory views the electron structure of atoms.

Animations: 1*s* orbital; 2 *px* orbital; 2 *py* orbital; and 2 *pz* orbital.

9.3 Orbitals, Electron Shells, and Subshells

OBJECTIVES

- Define *orbital, electron shell,* and *subshell.*
- Describe the values that are possible for the shell quantum number *n,* which denotes the different electron shells.
- Describe the two features that the orbitals in a shell have in common.
- Calculate the number of orbitals in a given shell from the shell quantum number *n*.
- Describe the meaning of any given subshell notation.
- List the different subshells in a given shell using subshell notation.
- State the number of orbitals in a given subshell.
- List the five lowest energy subshells, in order from lowest to highest orbital energy.

The difficulty scientists had in extending Bohr's theory beyond the hydrogen atom led them to look for an entirely new way to explain the nature of the atom. Bohr's theory seemed to say that it is the electrons in an atom that determine its line spectrum and

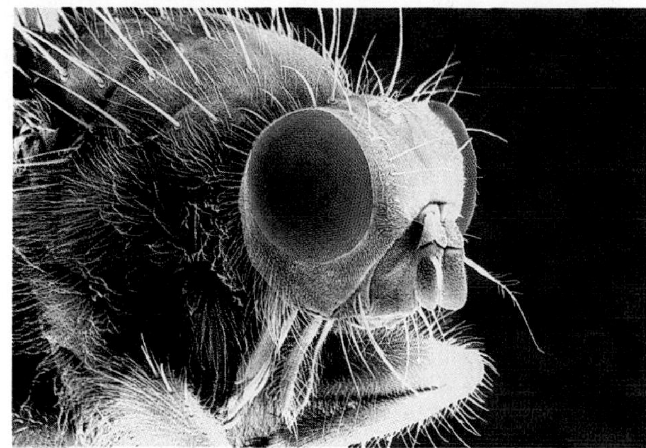

FIGURE 9.10
Image of a fly taken with a scanning electron microscope. As the electron beam scans the specimen, the microscope stores the image data in a computer, which then builds up a composite picture. The colors you see are "false" colors, added to improve the clarity of the image.

maybe its chemistry. But is there a better way to proceed? How should one think about the electrons in an atom?

You may recall from our discussion in Section 9.1 that light has both wave and particle properties. In 1923, the French physicist Louis de Broglie argued that perhaps electrons are similar in having both wave and particle properties. A few years later, it was shown experimentally that electrons moving at high enough energies do in fact have measurable wave properties and can be diffracted like light. Electron microscopes employ electron waves to obtain high magnifications of biological specimens and surfaces of materials (Figure 9.10). De Broglie's idea proved to be the key to solving the question of how electrons behave in atoms.

Starting from de Broglie's wave-particle idea, the German physicist Erwin Schrödinger devised an equation in 1926 that could in principle be solved to obtain the energy levels of any atom. The theory as it subsequently developed became known as *quantum mechanics* and is presently used to describe atoms and molecules. ●

● Quantum mechanics was developed independently by the German physicists Erwin Schrödinger (1887–1961) and Werner Heisenberg (1901–1976). Bohr contributed to its basic philosophy.

Each of the different quantum-mechanical "states" of an atom has an associated mathematical function called its *wave function,* as well as an energy level. The square of the wave function tells you how an electron is distributed in space. That is, it tells you how likely you are to find the electron at a given point in space. Where the square of the wave function is large, the probability of finding the electron is high; where it is small, the probability is low. The square of the wave function then is a *probability distribution.*

To understand what is meant by a probability distribution, picture a fruit fly hovering around a ripe grape hung at the end of a string. Suppose you take a time exposure of this scene. Figure 9.11 shows what you might see on the developed photograph. If somehow you had been presented with this photograph at the beginning of your observation of the fruit fly, you could have used it as a probability distribution to predict where you might expect to find the fruit fly. Though you could not predict precisely where the fruit fly would be at any given moment, you could say where the fruit fly is most likely to be found (at the center of the distribution, near the grape) and where it is much less likely to be (well away from the grape). Although the fruit fly is most likely to be found near the grape, you might find it at very large distances from the fruit, however unlikely.

FIGURE 9.11
Distribution of a fruit fly about a grape. The fruit fly is most often near the fruit, but can be found less often at places away from the fruit.

Orbitals

In quantum mechanics we have to abandon Bohr's idea that electrons move in specific circular orbits about the nucleus. Instead, we describe an electron as occupying an **orbital,** which is *the wave function for an electron in an atom.* Roughly speaking, it is the region of space around the nucleus where the electron is likely to be found. This orbital is much less well defined than an orbit; instead of moving over a definite path, the electron occupies a region of space. The electron's exact position at any given time within that region of space is not precisely predictable. We can only say how likely it is that we will find the electron at a particular spot if we were to look for it there. Although we often draw an orbital as if the electron occupies a definite region of space, this is not quite true. An orbital, unlike an orange or an apple, does not have a definite boundary surface. It is possible, although unlikely, to find an electron a long distance away from the nucleus. In drawings of orbitals, we are depicting the region where the electron is most likely to be found. The Chemical Perspective at the end of this section describes how the scanning tunneling microscope allows us to see atoms and electron distributions.

Electron Shells

The orbitals of an atom are grouped in *electron shells,* with different shells corresponding to regions of space that are different distances from the nucleus. Each electron shell is designated by a shell quantum number, n. This is essentially the quantum number used in Bohr's theory, but its meaning is somewhat different. As in Bohr's theory, it can have any positive, whole-number value from 1 to infinity. The electron shell closest to the nucleus has a quantum number $n = 1$, whereas shells farther from the nucleus have higher quantum numbers, such as $n = 2$ or $n = 3$, and so forth.

The shell quantum number, denoting the electron shell, tells you not only the approximate size of an orbital in a shell (that is, how far the main part of the orbital extends from the nucleus) but also the approximate energy of an orbital in the shell. An orbital of low shell quantum number n has a low energy, whereas an orbital of higher shell quantum number has a higher energy. Thus, we can define an **electron shell** as *a set of orbitals of approximately the same size and energy.*

The number of possible orbitals within a shell equals $n \times n$ or n^2. Thus, the $n = 1$ shell contains 1^2 or one orbital, the $n = 2$ shell contains 2^2 or four orbitals, and the $n = 3$ shell contains 3^2 or nine orbitals. Each orbital can hold only two electrons, so the more orbitals a shell has, the more electrons it can contain.

In summary, an electron shell consists of a set of orbitals of approximately the same size and energy. Each shell contains n^2 number of orbitals, where n is the shell quantum number.

■ EXAMPLE 9.1 Calculating the Number of Orbitals in a Shell

How many orbitals are in the $n = 4$ shell?

Problem Analysis

The number of orbitals in a shell equals the square of the shell quantum number, n^2.

Solution

There are $4^2 = 4 \times 4 = 16$ orbitals in the $n = 4$ shell.

Exercise 9.1

How many orbitals are in the $n = 3$ shell?

(Try Problems 9.35 and 9.36.)

Subshells

The orbitals in a particular shell are grouped into subshells. A **subshell** is *a subset of orbitals of an electron shell, all of which have the same energy and similar shape.* The number of subshells in a shell equals the value of n. For example, the $n = 1$ shell has only one subshell, called an s subshell; the $n = 2$ shell has two subshells, called s and p subshells; and the $n = 3$ shell has three subshells, called $s, p,$ and d subshells. The number of possible subshells for the first 5 shells can be summarized as follows.

$n = 1$ 1 subshell (s)
$n = 2$ 2 subshells (s, p)
$n = 3$ 3 subshells (s, p, d)
$n = 4$ 4 subshells (s, p, d, f)
$n = 5$ 5 subshells (s, p, d, f, g)

It may seem odd that the different subshell types are denoted $s, p, d, f, g,$ rather than a, b, c, d, e. The first four letters, $s, p, d, f,$ originally denoted characteristics of line spectra (s for sharp, for example); today, the letters are simply labels for the different subshells. After f, the letter labels are in alphabetic order: $g, h, i,$ and so forth.

A subshell in a particular shell is denoted by writing the shell quantum number followed by the letter designation of the subshell type. For example, a p-type subshell in the $n = 2$ shell is denoted $2p$. The subshell notation for a d-type subshell in the $n = 3$ shell is $3d$. This same notation is used to denote an orbital in a given subshell and an electron in an orbital. Thus, we refer to the orbitals in the $3d$ subshell as $3d$ orbitals and to the electrons in the $3d$ orbitals as $3d$ electrons.

Each subshell consists of a specific number of orbitals, depending on the type of subshell. An s-type subshell consists of a single orbital. A p-type subshell consists of three orbitals. The number of orbitals in a subshell increases by two in going from one subshell to the next through the sequence $s, p, d, f,$ and so on. See Table 9.1 for the number of orbitals in each shell and subshell.

All the orbitals in each subshell, as we noted in the definition, have a specific shape. All s orbitals have a spherical shape, and p orbitals have a dumbbell shape. Figure 9.12 shows the $1s$ and $2s$ orbitals. Note that the $2s$ orbital is larger than the $1s$ orbital. As mentioned earlier, the $n = 2$ shell has a higher energy and its electrons are located farther from the nucleus than the $n = 1$ shell. The three orbitals in a p subshell differ by their orientation in space. One of the orbitals lies along the x axis, another one lies along the y axis, and the third one lies along the z axis (see Figure 9.13). The orbitals in other types of subshells have more complicated shapes.

It is important not to misinterpret the shapes we draw for the orbitals and the language that we use to refer to electrons and orbitals. The shapes we draw, as in Figure 9.13, are the volumes of the space around an atomic nucleus in which the electron is

TABLE 9.1
Allowable Subshells and Number of Orbitals for Given Shells

Shell	Subshells	Number of Orbitals
$n = 1$	$1s$	1
$n = 2$	$2s, 2p$	$1 + 3 = 4$
$n = 3$	$3s, 3p, 3d$	$1 + 3 + 5 = 9$
$n = 4$	$4s, 4p, 4d, 4f$	$1 + 3 + 5 + 7 = 16$

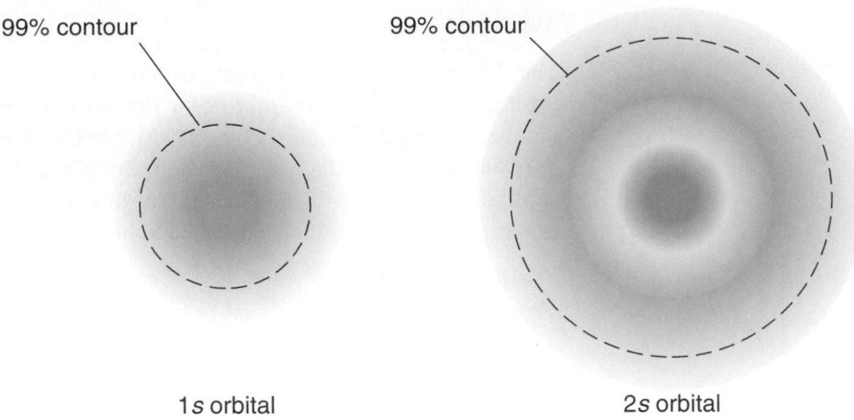

FIGURE 9.12
The shapes and sizes of *s* orbitals. All *s* orbitals have a spherical shape. The larger the value of *n*, the larger the size of the *s* orbital in that shell. The region where there is a high probability of finding a 2*s* electron extends farther from the nucleus than the region where there is a high probability of finding a 1*s* electron. The intensity of color in these drawings indicates the probability of finding an electron. The contours represented by the dashed circles show the size of the orbital in which the electron is likely to be found 99% of the time.

most likely to be found. It is possible, though not likely, to find an electron outside the volume of space that we draw for the orbital. So these shapes do not represent containers in which the electrons are held, even though we often say that the electron is "in" an orbital or that an electron is "contained" in an orbital.

We noted earlier that the shell quantum number was related approximately to the energy of the orbitals. The energy can be defined more precisely by also considering the identity of the subshell. Although the energy is determined approximately by the shell,

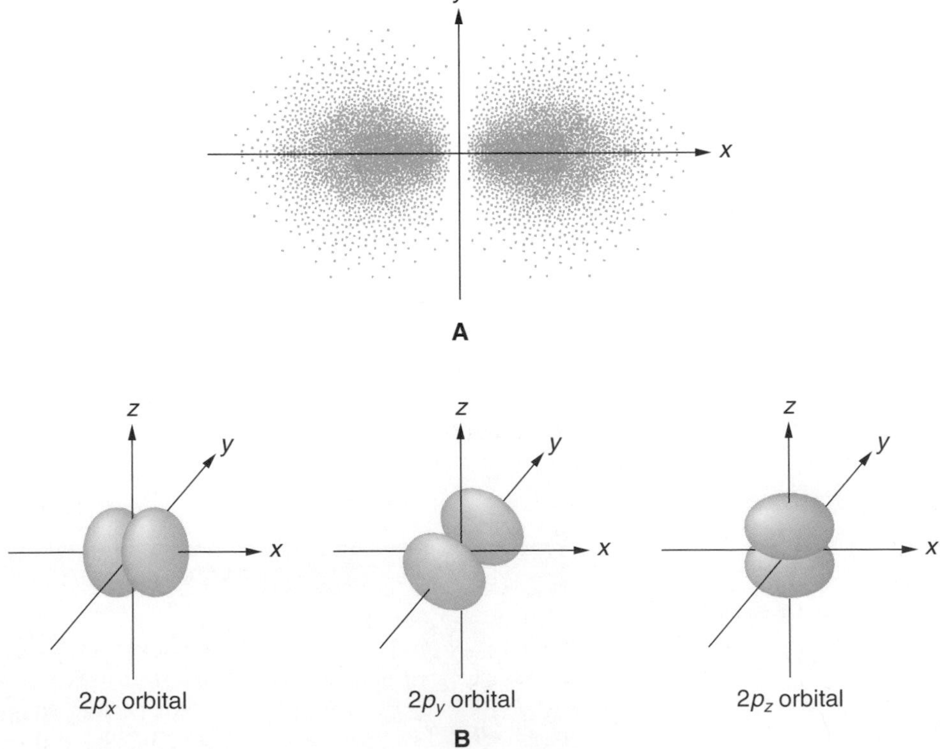

FIGURE 9.13
The shapes of the *p* orbitals. The general shapes are illustrated by the 2*p* orbitals; the orbital $2p_x$ lies along the *x* axis, and so forth.

the energy of an orbital also depends on the subshell within that shell. We discuss the energy of orbitals in detail momentarily.

■ EXAMPLE 9.2 Interpreting Subshell Notation

What is the meaning of the notation $6f$?

Problem Analysis

Recall that the leading number refers to the shell quantum number and the letter refers to the subshell type.

Solution

The notation $6f$ refers to the $n = 6$ shell and an f-type subshell.

Exercise 9.2

What is the meaning of the notation $5d$?

(Try Problems 9.37 and 9.38.)

■ EXAMPLE 9.3 Specifying the Subshells in a Shell

How many subshells are in the $n = 4$ shell? Give the notations for these subshells. How many orbitals are in each subshell?

Problem Analysis

The number of subshells in a shell equals the shell quantum number, n. The notations for each of these begin with n. You list the corresponding subshell letters, starting with s and going through the sequence s, p, d, f, and so forth, until you have listed n letters. You obtain the number of orbitals in these subshells by writing down the odd numbers 1, 3, 5, and so forth, one number for each subshell.

Solution

Because $n = 4$, there are four subshells. Each subshell notation begins with 4. The subshells are $4s$, $4p$, $4d$, and $4f$. The numbers of orbitals in each subshell are 1, 3, 5, and 7, respectively.

Exercise 9.3

Give the notations for all of the subshells in the $n = 3$ shell. How many orbitals are in each subshell?

(Try Problems 9.39 and 9.40.)

Orbital Energies

In the hydrogen atom, all subshells within a shell and all orbitals within a subshell have the same energy, as shown by Figure 9.14. For instance, a $2s$ electron has the same energy as a $2p$ electron. This means that the energy levels in a hydrogen atom can be designated by the shell quantum number n alone, as was done in Bohr's theory.

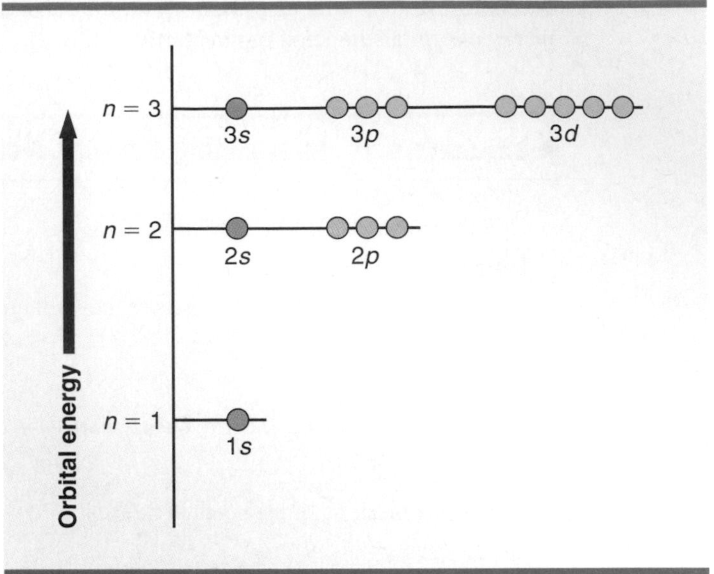

FIGURE 9.14
Orbital energy levels for the hydrogen atom. The energies of subshells and orbitals in the hydrogen atom are determined only by the shell quantum number, *n*. All orbitals in a given shell have the same energy. In this diagram, the energy of shells (and subshells and orbitals) is related to the vertical position of the horizontal line. The orbitals are represented as circles and are grouped into subshells.

However, for atoms with more than one electron, all orbitals of a given subshell have the same energy, and these orbitals have slightly different energy than orbitals in another subshell (in that same shell). The *s* orbital has the lowest energy in a given shell, and after that the energies of the orbitals increase in the order *s, p, d, f,* and so on. For example, the orbitals of the $n = 3$ shell increase in energy in the order 3*s*, 3*p*, 3*d*.

The relative energies of the orbitals are conveniently depicted in an energy-level diagram, similar to the diagram used for the energy levels of the hydrogen atom. The relative energies of the first few orbitals are shown in Figure 9.15. Because a *p*-type subshell contains three orbitals, all at the same energy, these energy levels are shown as three circles at the same level. Note that the energies of the orbitals increase in the order 1*s*, 2*s*, 2*p*, 3*s*, 3*p*.

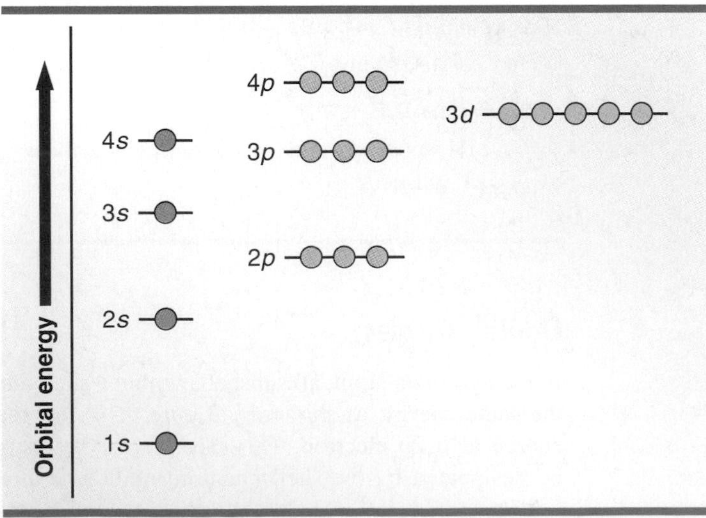

FIGURE 9.15
Orbital energy levels for multielectron atoms. An *s* subshell consists of a single orbital. Each *p* subshell consists of three orbitals; the energy levels are shown as three circles at the same energy. Similarly, the 3*d* subshell has five orbitals and is depicted as five circles at the same energy. Note that the subshells for a given shell (say, 3*s*, 3*p*, and 3*d*) occur at slightly different energies.

Chemical Perspective

Seeing and Moving Atoms

If matter is made of atoms, then show me one. Early chemists would have scoffed at that challenge. Atoms were real enough—they explained what chemists knew about matter—but the possibility of seeing an atom appeared remote. In the best microscope using ordinary light, a barely visible dot might be thousands of atoms wide. How could you hope to see an individual atom? But then the discovery of the wave properties of electrons changed that pessimistic view. The electron microscope, which operates on the same principle as a light microscope but uses electron waves in place of light, gives us glimpses of atoms as dark spots on a photographic plate. Now, the scanning tunneling microscope (STM), invented in the early 1980s and operating on a different principle, shows us atoms and molecules on surfaces like mounds of rock against a flat plain.

To understand the scanning tunneling microscope, imagine that you want to draw a map of a portion of a wilderness area showing the height of its hills and the depths of its valleys. To do that, you walk over the area, carrying an accurate altimeter to measure heights, and as you walk up and down the terrain, the altimeter readings change, and you write down the height values at various points. Back at your desk, you draw the map from your data.

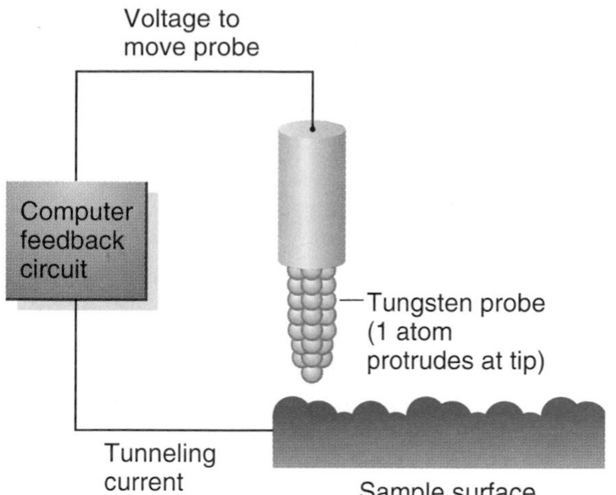

FIGURE 9.16
An STM probe scanning a surface. Note that the probe ends in a single atom. Electrons tunnel from the atom at the tip of the probe to an atom on the surface. The tunneling current from the sample is fed to a computer feedback circuit, which sends a voltage to the probe. The voltage moves the probe up or down so that the tunneling current is kept constant. A mechanism moves the probe so that it scans the sample surface.

With a scanning tunneling microscope, you use a mechanism to scan the point of a needle-like probe over a surface (Figure 9.16). While scanning, the probe traces the hills and valleys of the electron distribution of the atoms on the surface. The probe, so sharp that its tip ends in a single atom, has the function of the altimeter in our map making; its vertical position measures height above the surface. As it follows the contour of the electron distribution on the surface, its movement records the height of the distribution. The scanning of the probe is done under computer control, and the computer records the height data at various points, and can display these data on a computer screen.

How is it possible for the probe to follow the contour of the electron distribution without crashing into the side of an atom? When the single atom at the probe tip comes close enough to an atom on the surface, the electron distributions of the two atoms merge or overlap. An electron in this overlapping region does not know which atom it belongs to. An electron that was on the probe might later find itself on the metal surface. The electron is said to have "tunneled" from the probe to the surface. The movement of electrons from the probe to the surface is nothing more than an electric current. A voltage between the probe and the surface magnifies this tunneling current. As the probe moves downward toward the surface, the tunneling current increases rapidly when the electron distributions overlap significantly. A computer mechanism moves the probe back up to reduce the tunneling current to its former value. So, the probe moves to follow the contour of the electron distribution on the surface by maintaining a constant tunneling current.

Researchers have used the STM probe to nudge or even pick up atoms. Figure 9.17 shows a "quantum corral" constructed by moving 48 iron atoms into a circle on a copper metal surface. The iron atoms look like sharp mountain peaks rising up from a desert floor. Inside the quantum corral, you can see the wavelike distribution of electrons trapped within.

Scientists have become intrigued lately with the possibilities of manufacturing devices at the atomic or molecular level. Recently, IBM scientists in Zurich, Switzerland, have made the world's smallest abacus. An abacus is a simple device that has *(continued)*

FIGURE 9.17
Quantum corral. IBM scientists in San Jose, California, used an STM to arrange 48 iron atoms in a circle. The iron atoms are the sharp peaks; inside the quantum corral is the wavelike ripple of electron distribution within the corral. [IBM Research Division, Almaden Research Center; research done by Donald M. Eigler and coworkers.]

FIGURE 9.18
An abacus. A teacher in China uses the abacus (attached to the blackboard) to solve arithmetic problems by moving the green beads back and forth in a prescribed way.

beads strung on parallel rods; one uses it to do arithmetic by moving the beads back and forth on the rods (Figure 9.18). The "beads" in the miniature abacus are buckminsterfullerene molecules (buckyballs), which are soccerball-shaped carbon molecules. (See the Chemical Perspective at the end of Section 7.1.) A scanning probe moves the buckyball beads of the abacus.

ELECTRON CONFIGURATIONS

In the first part of this chapter, we explored the basis of the electron structure of atoms. We noted that electrons occupy orbitals; that is, they exist in particular regions of space. We saw that the energy of an orbital depends primarily on its electron shell but also depends somewhat on the subshell. Finally, we saw that all orbitals of a given subshell have the same energy. In the next four sections of the chapter, we look at which orbitals of an atom the electrons actually occupy. The lowest energy arrangements of electrons in orbitals, or lowest energy electron configurations, are the ones most responsible for the chemical properties of the elements. Because electron configurations follow a pattern that repeats periodically, we can explain the periodic behavior of the elements that we see in the periodic table.

9.4 Electron Configurations of the First Eighteen Elements

OBJECTIVES

- Define *electron configuration.*
- Interpret the notation for a particular electron configuration.
- Describe the Pauli exclusion principle.
- Write down the electron configurations of the elements from hydrogen to argon, knowing the order of filling of the first five subshells.
- Draw orbital diagrams that correspond to the electron configurations of the elements from hydrogen to argon.

An **electron configuration** is *a particular distribution of electrons among the different subshells of an atom.* We describe an electron configuration by listing the symbols for the occupied subshells one after another, adding a superscript to each symbol to show the number of electrons in the subshell. The oxygen atom in its lowest energy state, for example, has two electrons in its 1*s* subshell, two electrons in its 2*s* subshell, and four electrons in its 2*p* subshell. The notation for this configuration is $1s^22s^22p^4$.

■ EXAMPLE 9.4 Interpreting the Notation for an Electron Configuration

What is the meaning of the notation $1s^22s^22p^63s^23p^3$? What is the total number of electrons in the atom?

Problem Analysis

The shell quantum numbers and the subshell labels indicate which subshells contain electrons. The superscripts give the number of electrons in each of the subshells. The total number of electrons in the atom is the sum of the superscripts.

Solution

The superscripts show that there are two electrons in the 1*s* subshell, two electrons in the 2*s* subshell, six electrons in the 2*p* subshell, two electrons in the 3*s* subshell, and three electrons in the 3*p* subshell. The total number of electrons is $2 + 2 + 6 + 2 + 3 = 15$.

Exercise 9.4

Explain the meaning of the notation $1s^22s^22p^63s^23p^5$.

(Try Problems 9.43 and 9.44.)

Ground-State Electron Configuration

An atom has many different possible electron configurations, but it normally exists in its ground-state electron configuration, or lowest energy electron configuration. This is the most important configuration in determining the chemical properties of the element. From now on, when we use the term *electron configuration* without qualification, we mean the ground-state electron configuration, the configuration of lowest energy.

How do we determine which electron configuration has the lowest energy? It seems reasonable that the electrons would occupy orbitals of the lowest energy. For example, the hydrogen atom has one electron (its atomic number), and this electron goes into the 1*s* orbital because this is the orbital of lowest energy. So the electron configuration of hydrogen is $1s^1$. Helium (He) has atomic number 2. The two electrons of the helium atom go into the 1*s* orbital, giving the configuration $1s^2$.

Pauli Exclusion Principle

Lithium (Li) has atomic number 3, and if the three electrons were to go into the 1*s* orbital, the atom would have the configuration $1s^3$. However, this is not found experimentally—the actual configuration is $1s^22s^1$. How is this discrepancy explained?

In 1925, the German physicist Wolfgang Pauli (1900–1958) discovered that to explain the periodic table, certain electron configurations had to be excluded from consideration (because some configurations that you might write do not represent real possibilities). He found that all discrepancies from experiment could be resolved by adopting a simple idea. According to the **Pauli exclusion principle,** *an orbital can hold no more than two electrons.* Thus, an orbital can hold one or two electrons (or be empty), but cannot hold three electrons. The 1*s* subshell has only one orbital. The configuration $1s^3$ is therefore impossible and must be excluded from consideration. The third electron in lithium cannot go into the 1*s* orbital, because the orbital is filled; it must occupy the orbital of next higher energy, which is 2*s*.

Electron Configurations: Hydrogen to Argon

Because the energies of the orbitals increase in the order 1*s*, 2*s*, 2*p*, 3*s*, 3*p* (as indicated in Figure 9.15), the first five subshells fill in that order. Using this information and the Pauli exclusion principle, we can easily obtain the electron configurations of the first 18 elements of the periodic table. Beryllium, the next one after lithium, has an atomic number of 4, so it has four electrons. Two of these electrons go into the 1*s* orbital (1*s* subshell), filling it. The other two electrons must go into the next higher subshell, the 2*s* subshell (with a single 2*s* orbital). The electron configuration of beryllium is $1s^22s^2$.

What if we have an atom with more than four electrons? Once the 2*s* subshell of the atom is filled, the 2*p* subshell begins to fill. This subshell has three orbitals, each holding two electrons. Therefore, the 2*p* subshell can hold a maximum of six electrons. Elements from atomic number 5 to atomic number 10 have the following configurations, representing the filling of the 2*p* subshell:

atomic number = 5	boron, B	$1s^22s^22p^1$
atomic number = 6	carbon, C	$1s^22s^22p^2$
atomic number = 7	nitrogen, N	$1s^22s^22p^3$
atomic number = 8	oxygen, O	$1s^22s^22p^4$
atomic number = 9	fluorine, F	$1s^22s^22p^5$
atomic number = 10	neon, Ne	$1s^22s^22p^6$

Once the 2*p* subshell has filled, any additional electrons must go into the 3*s* orbital, which is next higher in energy. The configurations of the next two elements are

atomic number = 11	sodium, Na	$1s^22s^22p^63s^1$
atomic number = 12	magnesium, Mg	$1s^22s^22p^63s^2$

The 3*s* subshell has filled; now the 3*p* subshell begins to fill. The next six electron configurations are

atomic number = 13	aluminum, Al	$1s^22s^22p^63s^23p^1$
atomic number = 14	silicon, Si	$1s^22s^22p^63s^23p^2$
atomic number = 15	phosphorus, P	$1s^22s^22p^63s^23p^3$
atomic number = 16	sulfur, S	$1s^22s^22p^63s^23p^4$
atomic number = 17	chlorine, Cl	$1s^22s^22p^63s^23p^5$
atomic number = 18	argon, Ar	$1s^22s^22p^63s^23p^6$

With argon, the 3*p* subshell is filled.

Orbital Diagrams

Sometimes it is useful to describe how the electrons are arranged in orbitals within a subshell. The electron configurations only tell you how many electrons are in a subshell, but not in which orbitals they are located. Consider a 2*p* subshell, which contains three orbitals. There are two ways to fill these orbitals with two electrons. Both electrons might go into the same $2p$ orbital, or each of the electrons might go into different $2p$ orbitals. These situations represent different arrangements of electrons within a given subshell.

Recall that all the orbitals in a subshell have the same energy. Therefore, the three orbitals of a $2p$ subshell have the same energy, and you might expect the different arrangements of two electrons within a $2p$ subshell to have the same energy. Actually, a slight difference in energy occurs as a result of the interaction of the two electrons. Electrons are negatively charged, and two like charges repel. Because of this, the electron arrangement in which the electrons occupy different $2p$ orbitals has lower repulsion and slightly lower energy.

In general, the arrangement of electrons representing the ground state, or lowest energy state, of an atom is the one following **Hund's rule:** *Electrons do not pair up in an orbital unless all orbitals in a subshell already contain one electron.* Let's see how this affects the arrangements of electrons in the ground states of atoms.

The arrangement of electrons can be illustrated by *orbital diagrams* such as those shown in Figure 9.19. In these diagrams, the orbitals are represented by circles and the electrons by arrows. Some arrows point up and others point down. These arrows reflect the magnetic properties of electrons that result when an electron spins about its axis. The *electron spin* can be either clockwise or counterclockwise. Experiments show that when

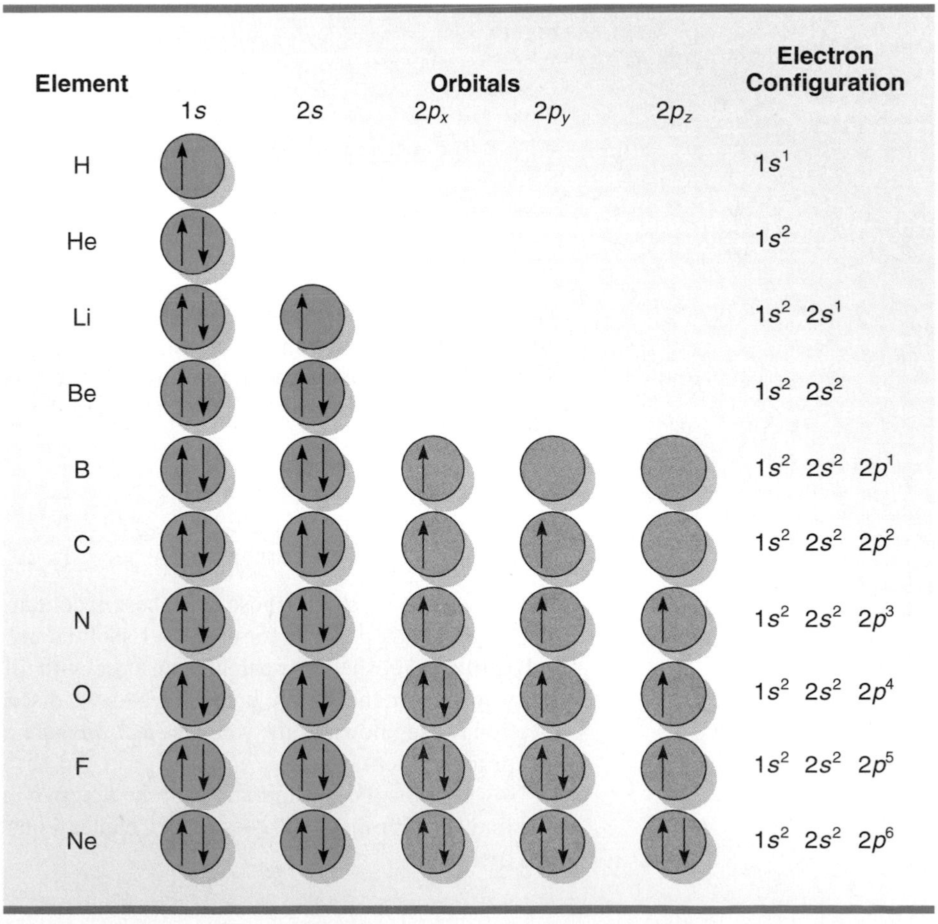

FIGURE 9.19
Orbital diagrams of the first ten elements. The circles represent orbitals grouped by subshell. The arrows represent electrons.

Element	1s	2s	2p_x	2p_y	2p_z	Electron Configuration
H	↑					$1s^1$
He	↑↓					$1s^2$
Li	↑↓	↑				$1s^2\ 2s^1$
Be	↑↓	↑↓				$1s^2\ 2s^2$
B	↑↓	↑↓	↑	○	○	$1s^2\ 2s^2\ 2p^1$
C	↑↓	↑↓	↑	↑	○	$1s^2\ 2s^2\ 2p^2$
N	↑↓	↑↓	↑	↑	↑	$1s^2\ 2s^2\ 2p^3$
O	↑↓	↑↓	↑↓	↑	↑	$1s^2\ 2s^2\ 2p^4$
F	↑↓	↑↓	↑↓	↑↓	↑	$1s^2\ 2s^2\ 2p^5$
Ne	↑↓	↑↓	↑↓	↑↓	↑↓	$1s^2\ 2s^2\ 2p^6$

two electrons occupy a single orbital, they have opposite spins and their magnetic properties cancel one another.

Consider the orbital diagrams for the elements from boron through neon, where electrons are being added to the subshell of $2p$ orbitals. Boron, with five electrons and an electron configuration of $1s^22s^22p^1$, begins to occupy the $2p$ subshell.

$1s$ (↑↓) $2s$ (↑↓) $2p$ (↑)( )( )

Carbon, with six electrons, adds another electron to the $2p$ subshell, this one going into another orbital of the $2p$ subshell.

$1s$ (↑↓) $2s$ (↑↓) $2p$ (↑)(↑)( )

(The unpaired electrons in a subshell point in the same direction, which we usually write as up arrows. We will not go into the reason for this.)

Nitrogen has seven electrons, of which three occupy the $2p$ subshell, each in a different orbital.

$1s$ (↑↓) $2s$ (↑↓) $2p$ (↑)(↑)(↑)

Oxygen has eight electrons, the last of which must pair up in one of the $2p$ orbitals and must have opposite spin from the other electron in that orbital. The orbital diagram is

$1s$ (↑↓) $2s$ (↑↓) $2p$ (↑↓)(↑)(↑)

The elements fluorine and neon successively fill each remaining p orbital, as shown in Figure 9.19.

9.5 Periodicity of Electron Configurations

OBJECTIVE

- Demonstrate the similarity in the outer-shell electron configurations of the elements in any given group of the periodic table.

We now have enough electron configurations of atoms to compare elements in the first periods of any main group. Consider the Group VIIIA elements. All of these elements are relatively unreactive gases, forming only a few compounds. ● For this reason, these elements are called the *noble gases*. Here are the configurations for the first three elements in Group VIIIA.

● No compounds of the noble gases were known until 1962, when several compounds of xenon with fluorine were prepared. Since then, compounds of krypton and radon have been prepared.

atomic number = 2	helium, He	$1s^2$
atomic number = 10	neon, Ne	$1s^22s^22p^6$
atomic number = 18	argon, Ar	$1s^22s^22p^63s^23p^6$

Note that the outer shells (those of highest n) contain completely filled s and p subshells. (No p subshell is possible for the $n = 1$ shell in any atom, including helium, which has only a filled s subshell.) Apparently, an atom with filled s and p subshells has special stability, making it chemically unreactive. As we discuss in detail in Chapter 10, the outer electrons (also known as the *valence electrons*) are responsible for the chemical properties of the element.

Now let's list the configurations of the first two metals of Group IA. These elements, you may recall from the discussion in Chapter 4 on the periodic table, are very reactive, soft metals.

atomic number = 3	lithium, Li	$1s^22s^1$
atomic number = 11	sodium, Na	$1s^22s^22p^63s^1$

Note that each configuration consists of a noble-gas core of electrons plus an outer shell containing an *s* orbital occupied by one electron. To emphasize this, the configurations are frequently abbreviated as follows.

atomic number = 3	lithium, Li	$[He]2s^1$
atomic number = 11	sodium, Na	$[Ne]3s^1$

Each abbreviation consists of square brackets enclosing the symbol of the noble-gas core followed by the remainder of the notation for the configuration. Note that the number of electrons in the outer shell (1) equals the group number (the Roman numeral). This result is also true for other main groups in the periodic table.

The configurations for the first two elements of Group IIA are as follows.

atomic number = 4	beryllium, Be	$1s^22s^2$
atomic number = 12	magnesium, Mg	$1s^22s^22p^63s^2$

Each configuration consists of a noble-gas core plus a doubly occupied *s* orbital. Again note that the number of electrons in the outer shell (2) equals the group number (II).

The configurations for the first two elements of Group VIIA are as follows.

atomic number = 9	fluorine, F	$1s^22s^22p^5$
atomic number = 17	chlorine, Cl	$1s^22s^22p^63s^23p^5$

Each configuration consists of a noble-gas core plus an *s* subshell with two electrons and a *p* subshell with five electrons, for a total of seven outer-shell electrons, equal to the group number (VII).

To summarize, as you go through the elements by increasing atomic number, the general form of the outer-shell electron configuration repeats, and you find that the elements in any given column (group) of the periodic table have an outer-shell configuration with the same general form. The number of outer-shell electrons equals the group number, and the outer shell of electrons primarily determines the chemical properties of the atom. Because all of the elements in a group have similar outer-shell electron configurations, they have similar chemical properties. This pattern explains why the periodic table has the form that it does.

9.6 Using the Periodic Table to Obtain Electron Configurations

OBJECTIVE

- Use the general pattern of the periodic table to obtain the order of filling of the subshells and hence the complete electron configuration for an atom.

By observing the general pattern of the periodic table, you can deduce which subshell is filling, and you can use this information to obtain the electron configuration of any atom. First, notice that the length of a period, or row, in the periodic table depends on the types of subshells that fill in that period. In the first period, the 1*s* subshell fills. Because this subshell has only one orbital, it can hold only two electrons, and therefore the first period has only two elements (hydrogen and helium). In each of the next two periods, an *s* and a *p* subshell fill. These subshells have a total of four orbitals, which hold a total of eight electrons, and therefore the second and third periods each have eight elements. Figure 9.20 is a sketch of a periodic table showing the subshell that is filling

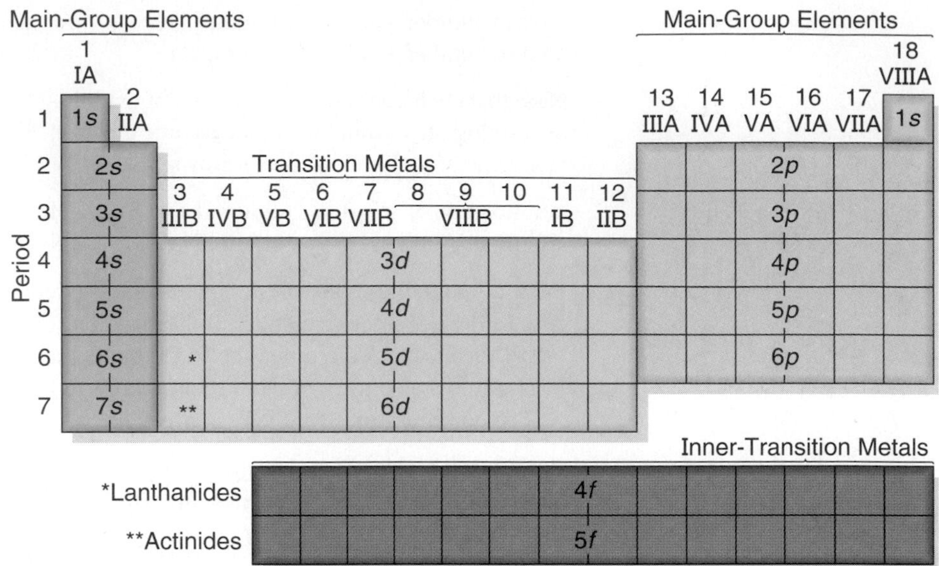

FIGURE 9.20

Periodic table showing the subshells that are filling. The colored areas show different types of subshells that are filling. For example, the orange area at the right corresponds to elements in which a *p* subshell is filling.

for each of the elements. The order of filling in electrons goes from left to right across the periodic table; when a row is filled, the next row fills from left to right, and so on.

With the completion of the third period of elements, the 3*s* and 3*p* subshells have filled. You might now expect the 3*d* subshell to begin to fill. However, if you look at the fourth period of elements, you will see that there are first two elements in which the 4*s* subshell fills. This is then followed by a series of transition elements, in which the 3*d* subshell fills. Apparently, the total energies of the atoms are lower if the 4*s* subshell fills before the 3*d*.

We are now ready to use the periodic table to obtain the electron configuration of an atom. Look again at Figure 9.20. Note that the first two columns on the left side correspond to atoms in which an *s* subshell fills, and the six columns on the right correspond to atoms in which a *p* subshell fills. The series of elements that are interpolated between these columns, the transition metals, correspond to atoms in which a *d* subshell fills. Finally, the two rows of elements below the main body of the periodic table, the inner-transition metals, correspond to atoms in which an *f* subshell fills. The next two examples illustrate how, by knowing this pattern, you can obtain the electron configuration of any atom.

■ EXAMPLE 9.5 Writing the Electron Configuration of an Element in Periods 1 to 3

The element phosphorus (P) is in the third period of Group VA. What is the complete electron configuration of the phosphorus atom? Use the periodic table to obtain the order of filling of the subshells.

Problem Analysis

You can use the pattern of the periodic table to write down the subshells in order of their filling. (At first, you may want to refer to Figure 9.20; but after some practice, you should be able to use any periodic table to do this.) Also, you can use the periodic table to obtain the atomic number, and therefore the total number of electrons in phosphorus. Distribute these electrons among the subshells in the order in which they fill. Remember that an *s* subshell can hold only two electrons, whereas a *p* subshell can hold six electrons.

Solution

Starting at the beginning of the periodic table and going to phosphorus, write down the subshells in the order in which they fill. They are

$$1s, 2s, 2p, 3s, 3p$$

Phosphorus has an atomic number of 15 and therefore 15 electrons. Distribute these electrons to the subshells. After you have filled all the subshells through 3*s*, you will have distributed 12 electrons. Put the remaining three electrons into the 3*p* subshell. You obtain

$$1s^2 2s^2 2p^6 3s^2 3p^3$$

Exercise 9.5

Obtain the complete electron configuration of the chlorine (Cl) atom using the periodic table.

(Try Problems 9.47 and 9.48.)

■ EXAMPLE 9.6 Writing the Electron Configuration of an Element in Periods 4 to 7

Selenium (Se) is in the fourth period of Group VIA (the same group as sulfur). What is its complete electron configuration? Use the periodic table to obtain the order in which the subshells are filled.

Problem Analysis

Looking at the pattern of a periodic table, write down the subshells in the order in which they fill. Use the periodic table to note the atomic number of selenium and therefore the total number of electrons in the atom. Then distribute these electrons to the subshells in the order in which they fill. Remember that an *s* subshell can hold only two electrons, whereas a *p* subshell can hold six electrons, and a *d* subshell can hold ten electrons.

Solution

The subshells, through selenium, are

$$1s, 2s, 2p, 3s, 3p, 4s, 3d, 4p$$

Selenium has an atomic number of 34 and therefore 34 electrons. Distribute these electrons to the subshells, remembering that you can put a maximum of ten electrons into a *d* subshell. After you have filled all of the subshells through 3*d*, you will have distributed 30 electrons. Put the remaining four electrons into the 4*p* subshell. You obtain

$$1s^2 2s^2 2p^6 3s^2 3p^6 4s^2 3d^{10} 4p^4$$

It is convenient to rearrange the subshells in the previous notation so that they are in order by shells. This way, the subshells in the outer shell, or valence shell ($4s^2 4p^4$), will be located at the far right.

$$1s^2 2s^2 2p^6 3s^2 3p^6 3d^{10} 4s^2 4p^4$$

Note that the valence shell has the general form $ns^2 np^4$ (where *n* is the shell quantum number), which is expected for a Group VIA element.

Exercise 9.6

Obtain the complete electron configuration of the germanium (Ge) atom using the periodic table.

(Try Problems 9.49 and 9.50.)

9.7 Valence-Shell Configuration of a Main-Group Element

OBJECTIVES

- Define *valence electron.*
- Use the position of an element in the periodic table to obtain the valence-shell configuration for the atom.

● The valence electrons of atoms interact to form chemical bonds. We discuss this in detail in Chapter 10.

We have said previously that the **valence electrons,** or *outer electrons (those electrons corresponding to largest n)*, are most important in determining the chemical properties of an element. ● Thus, in many cases, you are primarily interested in the configuration of these valence electrons. For example, phosphorus has the complete electron configuration $1s^2 2s^2 2p^6 3s^2 3p^3$. The valence-shell configuration is $3s^2 3p^3$.

You can determine the valence-shell configuration of an atom from the position of the element in the periodic table. The valence-shell configuration of an atom of a main-group element has the form $ns^a np^b$, where *n* is the shell number (also equal to the row number in the periodic table). The letters *a* and *b* indicate the number of electrons in each subshell, and therefore $a + b$ equals the total number of electrons in the valence shell (also equal to the group number). Therefore, from the row, you determine *n*; from the group, you determine $a + b$. You obtain the valence-shell configuration by simply distributing the electrons, first to the *ns* subshell, then to the *np* subshell.

Let us use the periodic table to obtain the valence-shell configuration of the phosphorus atom, which we obtained earlier from its complete electron configuration. Looking at the periodic table on the inside front cover of the book, you see that phosphorus (P) is in Group VA and Period 3. Because phosphorus is in an A group (VA), it is a main-group element. Such an element has a valence shell with electrons in an *s* subshell and possibly a *p* subshell. The general form of the valence-shell configuration is $ns^a np^b$, where *n* is the shell number, or period number. For phosphorus, this equals 3. The total number of valence electrons, $a + b$, equals the group number (Roman numeral V = 5). Two of these electrons will go into the s subshell and the remaining electrons (3) will go into the p subshell. Therefore, the valence-shell configuration is $3s^2 3p^3$. The next example further illustrates this method of obtaining the valence-shell configuration.

■ EXAMPLE 9.7 Obtaining the Valence-Shell Configuration of a Main-Group Element

Obtain the valence-shell configuration of the selenium atom using the periodic table.

Problem Analysis

From the position of the element in the periodic table, you know that the row number equals n, and the group number equals $a + b$. Now, you distribute the valence electrons to the subshells, filling in the ns subshell first. If you have electrons left over, you put these into the np subshell.

Solution

The row number of selenium is 4, so $n = 4$, and the general form of the valence-shell configuration is $4s^a 4p^b$. The group number is 6 (the Roman numeral VI), so $a + b = 6$. You put two of these electrons into the $4s$ subshell and the remaining electrons (four) into the $4p$ subshell. The valence-shell configuration is $4s^2 4p^4$. Note that this is the outer shell of the complete configuration that we obtained for selenium in the previous example.

Exercise 9.7

Obtain the valence-shell configuration of germanium (Ge) using the position of the element in the periodic table.

(Try Problems 9.51 and 9.52.)

PERIODIC PROPERTIES

The periodic table that Mendeleev constructed in the 1850s now forms the organizational basis for the chemistry of the elements. In the previous sections, you saw how you can explain the arrangement of this table in terms of the electron structure of atoms. When you arrange the elements by increasing atomic number, the general form of the valence-shell configurations repeats periodically (Figure 9.21). Because the chemistry of the elements depends on the valence (or outer) electrons of the atoms, the chemical properties of the elements have a periodicity as well. The physical properties of the elements also generally depend on the electron configurations of the atoms, so even these properties behave periodically. We can state these facts as the **periodic law:** *When you arrange the elements by atomic number, their chemical and physical properties vary periodically.*

In the next two sections, we examine the periodic character of two physical properties: atomic and ionic radii (atomic sizes) and ionization energies (ease of removing, or ionizing, electrons from an atom). Both of these properties affect the chemical behavior of the elements.

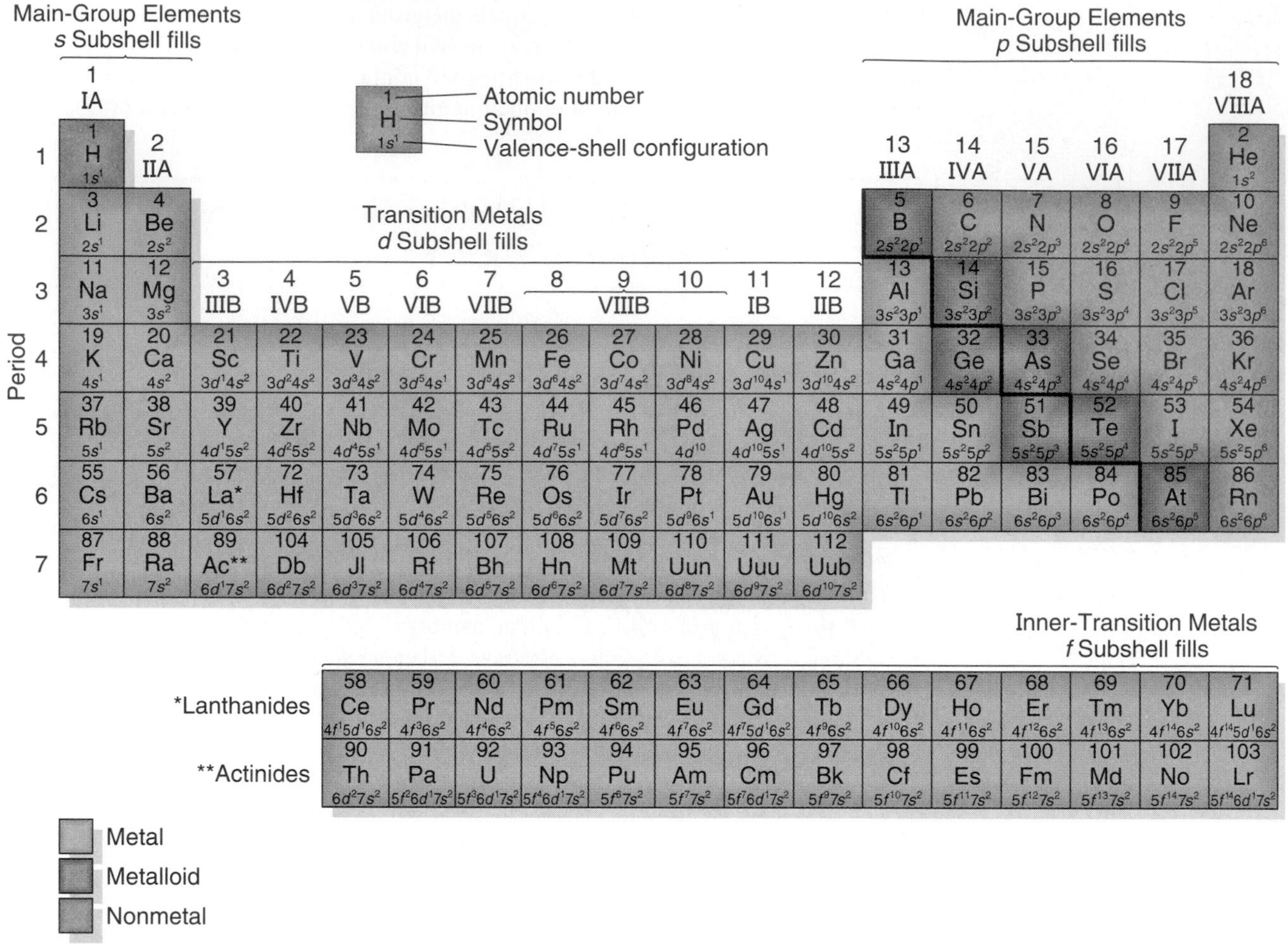

Main-Group Elements *s* Subshell fills

Transition Metals *d* Subshell fills

Main-Group Elements *p* Subshell fills

Key: 1 — Atomic number; H — Symbol; $1s^1$ — Valence-shell configuration

Period	1 IA	2 IIA	3 IIIB	4 IVB	5 VB	6 VIB	7 VIIB	8 VIIIB	9 VIIIB	10 VIIIB	11 IB	12 IIB	13 IIIA	14 IVA	15 VA	16 VIA	17 VIIA	18 VIIIA
1	1 H $1s^1$																	2 He $1s^2$
2	3 Li $2s^1$	4 Be $2s^2$											5 B $2s^22p^1$	6 C $2s^22p^2$	7 N $2s^22p^3$	8 O $2s^22p^4$	9 F $2s^22p^5$	10 Ne $2s^22p^6$
3	11 Na $3s^1$	12 Mg $3s^2$											13 Al $3s^23p^1$	14 Si $3s^23p^2$	15 P $3s^23p^3$	16 S $3s^23p^4$	17 Cl $3s^23p^5$	18 Ar $3s^23p^6$
4	19 K $4s^1$	20 Ca $4s^2$	21 Sc $3d^14s^2$	22 Ti $3d^24s^2$	23 V $3d^34s^2$	24 Cr $3d^54s^1$	25 Mn $3d^54s^2$	26 Fe $3d^64s^2$	27 Co $3d^74s^2$	28 Ni $3d^84s^2$	29 Cu $3d^{10}4s^1$	30 Zn $3d^{10}4s^2$	31 Ga $4s^24p^1$	32 Ge $4s^24p^2$	33 As $4s^24p^3$	34 Se $4s^24p^4$	35 Br $4s^24p^5$	36 Kr $4s^24p^6$
5	37 Rb $5s^1$	38 Sr $5s^2$	39 Y $4d^15s^2$	40 Zr $4d^25s^2$	41 Nb $4d^45s^1$	42 Mo $4d^55s^1$	43 Tc $4d^55s^2$	44 Ru $4d^75s^1$	45 Rh $4d^85s^1$	46 Pd $4d^{10}$	47 Ag $4d^{10}5s^1$	48 Cd $4d^{10}5s^2$	49 In $5s^25p^1$	50 Sn $5s^25p^2$	51 Sb $5s^25p^3$	52 Te $5s^25p^4$	53 I $5s^25p^5$	54 Xe $5s^25p^6$
6	55 Cs $6s^1$	56 Ba $6s^2$	57 La* $5d^16s^2$	72 Hf $5d^26s^2$	73 Ta $5d^36s^2$	74 W $5d^46s^2$	75 Re $5d^56s^2$	76 Os $5d^66s^2$	77 Ir $5d^76s^2$	78 Pt $5d^96s^1$	79 Au $5d^{10}6s^1$	80 Hg $5d^{10}6s^2$	81 Tl $6s^26p^1$	82 Pb $6s^26p^2$	83 Bi $6s^26p^3$	84 Po $6s^26p^4$	85 At $6s^26p^5$	86 Rn $6s^26p^6$
7	87 Fr $7s^1$	88 Ra $7s^2$	89 Ac** $6d^17s^2$	104 Db $6d^27s^2$	105 Jl $6d^37s^2$	106 Rf $6d^47s^2$	107 Bh $6d^57s^2$	108 Hn $6d^67s^2$	109 Mt $6d^77s^2$	110 Uun $6d^87s^2$	111 Uuu $6d^97s^2$	112 Uub $6d^{10}7s^2$						

Inner-Transition Metals *f* Subshell fills

*Lanthanides	58 Ce $4f^15d^16s^2$	59 Pr $4f^36s^2$	60 Nd $4f^46s^2$	61 Pm $4f^56s^2$	62 Sm $4f^66s^2$	63 Eu $4f^76s^2$	64 Gd $4f^75d^16s^2$	65 Tb $4f^96s^2$	66 Dy $4f^{10}6s^2$	67 Ho $4f^{11}6s^2$	68 Er $4f^{12}6s^2$	69 Tm $4f^{13}6s^2$	70 Yb $4f^{14}6s^2$	71 Lu $4f^{14}5d^16s^2$
**Actinides	90 Th $6d^27s^2$	91 Pa $5f^26d^17s^2$	92 U $5f^36d^17s^2$	93 Np $5f^46d^17s^2$	94 Pu $5f^67s^2$	95 Am $5f^77s^2$	96 Cm $5f^76d^17s^2$	97 Bk $5f^97s^2$	98 Cf $5f^{10}7s^2$	99 Es $5f^{11}7s^2$	100 Fm $5f^{12}7s^2$	101 Md $5f^{13}7s^2$	102 No $5f^{14}7s^2$	103 Lr $5f^{14}6d^17s^2$

Metal

Metalloid

Nonmetal

FIGURE 9.21

Periodic table. Metallic elements are shown in blue, nonmetals in orange, and metalloids in purple. The table also shows valence-shell configurations.

9.8 Periodicity of Atomic and Ionic Radii

OBJECTIVES

- Use the trends shown by the atomic radii of main-group elements to predict the relative sizes of two atoms.
- Use the trends shown by the ionic radii of main-group elements to predict the relative sizes of two ions.

The size of an atom depends strongly on its electron configuration and, like the electron configuration, exhibits periodicity among the elements. The size of an atom is generally referred to by its *atomic radius*. The relative atomic radii, or sizes of atoms, for the main-group elements are depicted in Figure 9.22. The figure shows two general trends. First, *the atomic radius increases in any given main group as you move down the group of elements.* This makes sense because the relative size of an atom is determined by the size of its valence shell. Each time you proceed to the next lower element in a column, the atom has an additional shell of electrons and therefore a larger atomic radius. For example, in going from sodium, whose electron configuration is $1s^22s^22p^63s^1$, to potassium, whose electron configuration is $1s^22s^22p^63s^23p^64s^1$, a new electron shell has been added, and there is an increase in atomic radius.

Second, *the atomic radius decreases in any given period of main-group elements as you move across the period.* The reason for this trend is less obvious. As you go from

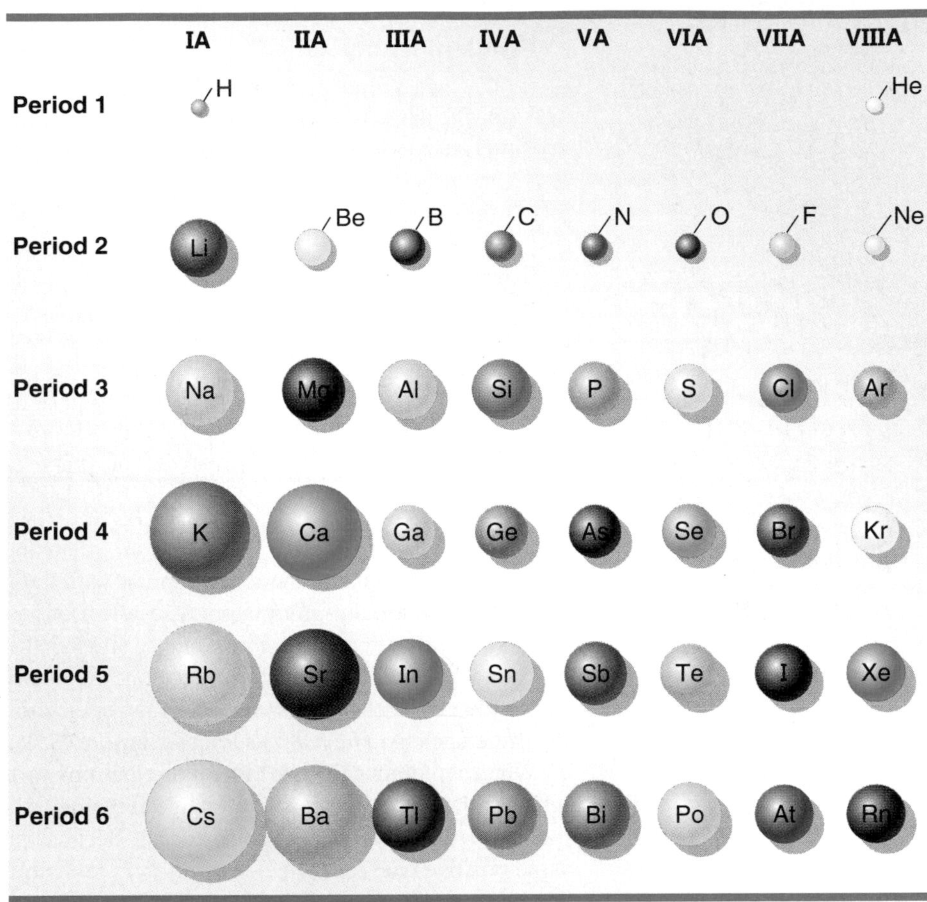

FIGURE 9.22
Atomic radii of the main-group elements. The drawings depict the approximate relative sizes of the main-group atoms. Note the trends in sizes moving down a column and across a row.

one element to the next one on its right, another electron is added to the same valence shell. However, the size of that shell is determined in part by the attractive force of the nucleus on the electrons. In going toward the right from one element to the next within a period, the nuclear charge increases by 1, while *n* remains fixed. For example, in going from sodium (valence-shell configuration $3s^1$) to magnesium (valence-shell configuration $3s^2$), there is no change in the shell quantum number *n,* but the atomic number (positive charge on the nucleus) increases from 11 to 12. The attractive force of the nucleus for the valence-shell electrons increases, and the shell size and atomic radius decrease.

■ EXAMPLE 9.8 Choosing the Element in a Pair of Elements in Any Period or Any Group Whose Atom Has the Greater Atomic Radius

Using a periodic table but without looking at Figure 9.22, choose the element whose atom you expect to have the larger atomic radius in each of the following pairs: (a) Mg, Na; (b) Al, B.

Problem Analysis

(a) The larger atom in any period of elements will correspond to the element farther to the left in the periodic table. Check a periodic table to choose the element.
(b) The larger atom in any group of elements will correspond to the element closer to the bottom of the group in the periodic table. Check a periodic table to choose the element.

Solution

(a) The larger atom is Na. (b) The larger atom is Al.

Exercise 9.8

Which has the smaller atomic radius, F or O?

(Try Problems 9.55 and 9.56.)

The size of an ion, called its *ionic radius,* also depends on its electron configuration. We can readily compare the size of an ion with the size of the atom from which it was formed. When a main-group atom loses electrons to form a cation, it usually loses its outer shell, so the cation is considerably smaller than the atom. More generally, when any atom loses electrons to form a cation, the nuclear charge remains the same, while the number of electrons decreases. The fewer electrons of the cation are pulled tighter to the same nucleus. *The cation radius is smaller than that of the corresponding atom.*

When an atom gains one or more electrons to form an anion, the nuclear charge remains the same (the positive charge remains fixed), while the number of electrons increases (the negative charge increases). Because more electrons are being held by the same positive charge, each electron is held less strongly in the anion than in the corresponding atom, so the orbitals enlarge. *The anion radius is larger than that of the corresponding atom.*

We can also compare the relative sizes of certain ions, either those with the same valence electron configuration in a group or those with the same overall electron configuration. *Ionic size increases in a given group as atomic number increases.* Consider the alkali metals, where we see a regular increase in cation radius as we go down the group. (The symbol *pm* stands for picometers, or 10^{-12} m.)

$$Li^+, 60 \text{ pm}; \quad Na^+, 95 \text{ pm}; \quad K^+, 133 \text{ pm}; \quad Rb^+, 148 \text{ pm}; \quad Cs^+, 169 \text{ pm}$$

We see a similar trend among the *halide ions*—the ions formed by adding an electron to the halogen atoms.

$$F^-, 136 \text{ pm}; \quad Cl^-, 181 \text{ pm}; \quad Br^-, 195 \text{ pm}; \quad I^-, 216 \text{ pm}$$

Just as for atoms, the outer-shell quantum number n increases as we go down a group. Thus, the outer electron shell, and therefore the ion, becomes larger in going down a group.

We can compare the sizes of ions that have the same number of electrons. Such a series of ions is called an *isoelectronic series.* In general, *the ionic radius decreases as the atomic number increases in an isoelectronic series.* Consider the sequence S^{2-}, Cl^-, K^+, Ca^{2+}. You should be able to verify that each of these ions has a total of 18 electrons. This sequence then is an isoelectronic series. The electron configuration of each ion is the same as the atom with 18 electrons (argon), which is $1s^22s^22p^63s^23p^6$. Note how the ionic radius decreases with nuclear charge, or atomic number.

Ion	Radius	Electron Configuration	Nuclear Charge
S^{2-}	184 pm	$1s^22s^22p^63s^23p^6$	16+
Cl^-	181 pm	$1s^22s^22p^63s^23p^6$	17+
K^+	133 pm	$1s^22s^22p^63s^23p^6$	19+
Ca^{2+}	99 pm	$1s^22s^22p^63s^23p^6$	20+

The number of electrons and the electron configuration are constant, but the nuclear charge increases as you move down this series. The ionic radius decreases with increasing atomic number in this series because the electron cloud is being pulled closer to the nucleus by the increasing nuclear charge.

■ EXAMPLE 9.9 Choosing the Element in a Pair of Elements in Any Period or Any Group Whose Ion Has the Greater Ionic Radius

Using a periodic table, choose the element whose ion you expect to have the larger ionic radius in each of the following pairs: (a) Mg^{2+}, Ca^{2+}; (b) Mg^{2+}, Al^{3+}.

Problem Analysis

(a) Magnesium and calcium are in the same group of the periodic table (IIA), and the larger ion in any group of elements will correspond to the element closer to the bottom of the table. Check a periodic table to choose the element.

(b) The ions Mg^{2+} and Al^{3+} are isoelectronic since both have the electron configuration $1s^22s^22p^6$. (They are in Period 3, so their cations have an electron configuration like that of neon.) The larger isoelectronic ion in any period of elements corresponds to the element with the smaller nuclear charge (the smaller atomic number). Check a periodic table to choose the element.

Solution

(a) Calcium is closer to the bottom of the periodic table, so Ca^{2+} is the larger ion.

(b) Aluminum has a nuclear charge of +13, whereas magnesium has a nuclear charge of +12. The larger ion is Mg^{2+}.

Exercise 9.9

Which has the smaller ionic radius, F^- or Cl^-?

(Try Problems 9.57 and 9.58.)

9.9 Periodicity of Ionization Energies

OBJECTIVES

- Define *ionization energy.*
- State the two periodic trends shown by ionization energies for atoms of main-group elements.
- Explain how periodic trends in the ionization energy of atoms relate to periodic trends in their metallic character.
- For reactive nonmetals, compare the relative values of the ionization energy and the electron affinity of an atom.

● This ionization energy is also called the first ionization energy because it is the energy required to remove one electron from an atom. There are also second, third, and higher ionization energies, corresponding to the energy required to remove more electrons, one at a time, from the atom.

The chemical properties of an element depend on the tendency of its atoms to gain or lose electrons. One way this tendency can be described is by the **ionization energy,** *the energy required to remove an electron from an atom.* ● The term *ionization* generally refers to the process of producing ions. You produce a positive ion by removing an electron from an atom. For example, the ionization of the sodium atom (removing an electron) produces a sodium ion and an electron.

$$\text{Na} + \text{energy} \longrightarrow \text{Na}^+ + \text{e}^-$$

A positively charged ion and a negatively charged electron attract one another, so it is reasonable that it requires energy (the ionization energy) to pull an electron away from a positive atomic ion. Ionization energies are usually given in terms of the energy required to remove one electron from each atom in *a mole of atoms*. The ionization energy of sodium is 496 kJ/mol.

Figure 9.23 lists the ionization energies of the main-group elements. There are two main trends in these ionization energies, and their explanation parallels those we gave for atomic radii. First, note that *the ionization energy of an atom decreases in any given group as you move down the column of elements*. As you move down a column, the atom gets larger because it has an additional shell of electrons. As a result, the outer electrons are further from the nucleus and the attraction between the nucleus and the outer electrons is smaller, so the outer electrons are more easily removed. When the electron is easily removed, the ionization energy is small.

Second, note that *the ionization energy of an atom increases in any given period as you move across the row of elements*. The explanation is similar to the one we gave for the decrease in atomic radius as you move across a period. The nuclear attraction for electrons increases as the nuclear charge increases. This means that the electrons are held more strongly, and therefore the ionization energy is greater.

■ EXAMPLE 9.10 Choosing the Element in a Pair of Elements in Any Period or Any Group Whose Atom Has the Smaller Ionization Energy

Using a periodic table but without looking at Figure 9.23, choose the element whose atom you expect to have the smaller ionization energy in each of the following pairs: (a) Mg, Na; (b) Al, B.

Problem Analysis

(a) The atom in any period of elements that has the smaller ionization energy will correspond to the element farther to the left in the periodic table. Check a periodic table to choose the element.

(b) The atom in any group of elements that has the smaller ionization energy will correspond to the element closer to the bottom of the group in the periodic table. Check a periodic table to choose the element.

Solution

(a) The atom with the smaller ionization energy is Na.

(b) The atom with the smaller ionization energy is Al.

Period	IA	IIA	IIIA	IVA	VA	VIA	VIIA	VIIIA
1	H 1312							He 2372
2	Li 520	Be 899	B 801	C 1086	N 1402	O 1314	F 1681	Ne 2081
3	Na 496	Mg 738	Al 578	Si 786	P 1012	S 1000	Cl 1251	Ar 1521
4	K 419	Ca 590	Ga 579	Ge 762	As 947	Se 941	Br 1140	Kr 1351
5	Rb 403	Sr 549	In 558	Sn 709	Sb 834	Te 869	I 1008	Xe 1170
6	Cs 376	Ba 503	Tl 589	Pb 716	Bi 703	Po 812	At 926	Rn 1037

FIGURE 9.23
Ionization energies of the main-group elements. Values are given in units of kJ/mol.

Exercise 9.10
Which atom has the smaller ionization energy, F or O?
(Try Problems 9.59 and 9.60.)

The metallic character of the elements, an important chemical property, tends to parallel ionization energy. A reactive metal is one that loses its electrons easily to form the metal cation. The lowest ionization energies are for the elements at the lower left corner of the periodic table, and these are the elements that have the greatest metallic character. The highest ionization energies are for the elements at the upper right corner of the periodic table, and these are the elements that have the smallest metallic character, the nonmetals.

The most reactive nonmetals are those that have large ionization energy (that is, they tend to hold on to their electrons) and large electron affinity. The *electron affinity* of an atom refers to how strongly an atom tends to gain electrons from other atoms. The elements fluorine, oxygen, and chlorine have the strongest electron affinities and also large ionization energies. Because of these properties, these elements are the most reactive nonmetals. Although the noble gases have the largest ionization energies, they have very small electron affinities. Thus, they are relatively unreactive nonmetals. ●

● The noble-gas atoms have little tendency to lose electrons (high ionization energy) and little tendency to gain electrons (small electron affinity).

Chemical Perspective

The Greenhouse Effect

Carbon dioxide is a minor component of the atmosphere, but a major player in the climate and biology of earth. Much of this carbon dioxide arises from natural processes, such as the decay of plant material. But human activities, especially the burning of fossil fuels (coal, oil, and natural gas), are increasing the amount of carbon dioxide in the air. Climatologists have monitored carbon dioxide in the atmosphere since the late 1950s and have found that the amount of CO_2 in the atmosphere has increased steadily from about 315 ppm (parts per million) in 1958 to about 362 ppm in 1996 (Figure 9.24). Other studies indicate that the atmospheric carbon dioxide content was only about 290 ppm around 1850, when burning of fossil fuels became an important source of energy. Climatologists are concerned about the rising concentration of carbon dioxide in the atmosphere because molecules can absorb and emit radiation, similar to what we observed for atoms, and the selective absorption and reemission by CO_2 of portions of the electromagnetic spectrum (Section 9.1) leads to the *greenhouse effect.*

The principal gases in the atmosphere, O_2 and N_2, are transparent to visible light from the sun and absorb only a portion of its ultraviolet radiation. When the sunlight reaches the surface of the earth, it is absorbed there and heats the earth's surface. The heated surface then radiates this energy back into the atmosphere as infrared radiation, which neither O_2 nor N_2 absorbs. If these molecules were the only constituents of the atmosphere, this radiation would escape into outer space. But other atmospheric molecules, especially CO_2, do absorb infrared radiation, becoming more energetic and resulting in a warmer atmosphere. When these molecules reemit their energy as radiation, they do so in all directions. In this way a significant portion of the radiation is directed back to the earth, so that the sun's warmth is effectively trapped between these molecules and the earth (Figure 9.25). You see this

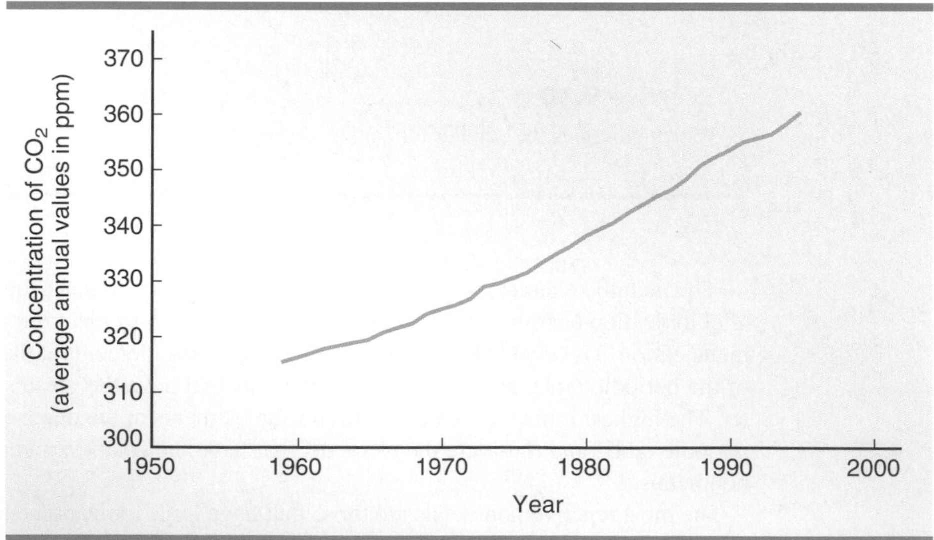

FIGURE 9.24

Concentrations of carbon dioxide in the atmosphere (average annual values at Mauna Loa, Hawaii). Concentrations of carbon dioxide in the atmosphere have been increasing steadily. In 1959, the atmospheric CO_2 concentration was 315.8 ppm (parts per million); in 1995, this had increased to 360.9 ppm. [Data are from Keeling, C. D., and T. P. Whorf, 1996. Atmospheric CO_2 records from sites in the SIO air sampling network. In *Trends: A Compendium of Data on Global Change.* Carbon Dioxide Information Analysis Center, Oak Ridge National Laboratory, Oak Ridge, Tenn.]

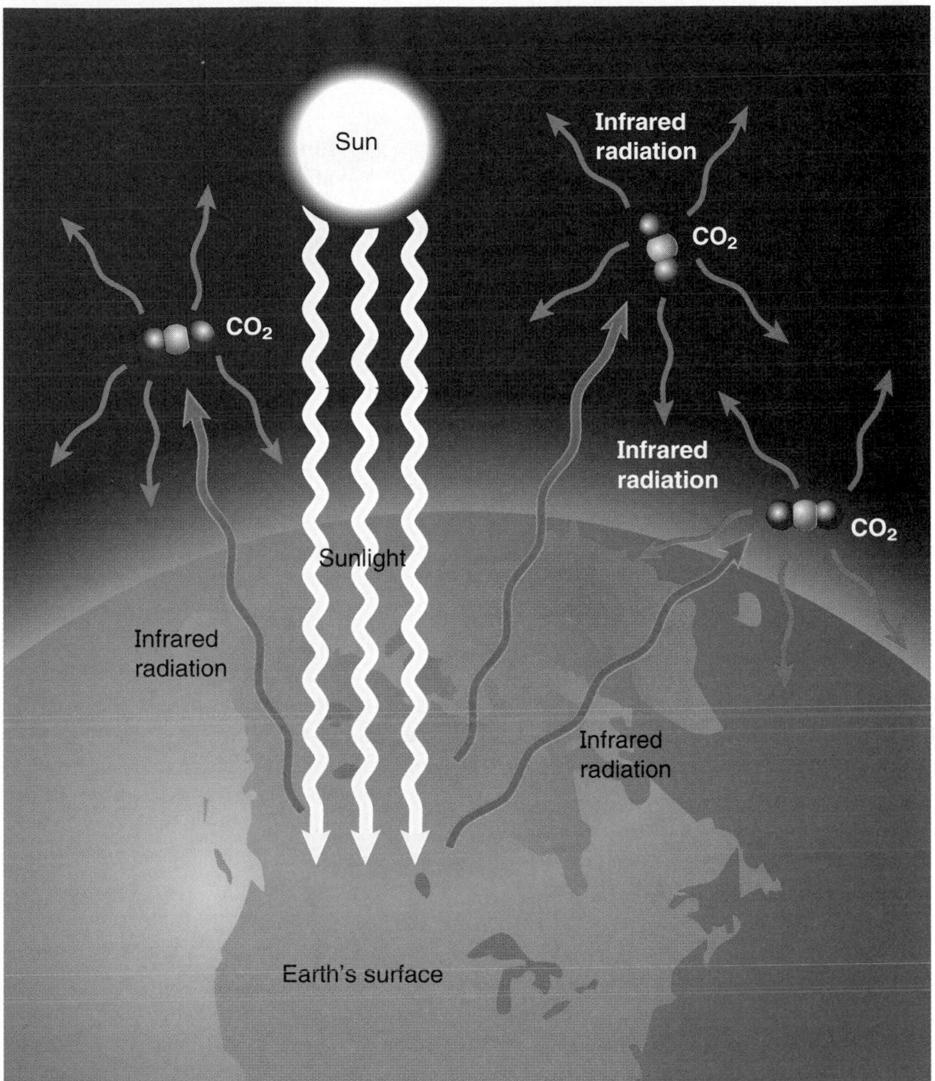

FIGURE 9.25

Greenhouse effect of some gases in the atmosphere. When sunlight passes through the atmosphere, it is absorbed by the earth's surface. This heated surface then radiates this energy back into the atmosphere as infrared radiation. Some gases, such as CO_2, absorb these infrared rays and then reradiate the radiation. A significant fraction of the radiation returns to the surface of the earth. These gases essentially trap the radiation and so act like the glass on a greenhouse.

greenhouse effect when you park your car in the sun. Glass, like the atmosphere, is transparent to visible light but absorbs infrared radiation. It therefore traps the energy of the sun, and the interior of the car warms up.

The greenhouse effect is important on earth both in providing livable temperatures and in moderating the temperature fluctuations that otherwise would occur. The concern of climatologists is that the additional greenhouse effect from the increase of carbon dioxide in the atmosphere might drastically alter our present climate. The prediction of climate, however, is enormously complex and the uncertainties at present are large. A report of the U.S. Global Change Research Program in 1995 tried to assess climate predictions. They rated as "very probable" that the average temperature of the earth's surface would increase between 0.5°C and 2.0°C by 2050. Although this may seem to be a small temperature increase, such small changes can have large effects on local climates. For comparison, it has been estimated that the average global temperature has increased by 3°C to 4°C from the last major ice age, about 18,000 years ago, to the present. The climate predictions for the next century indicate a similar magnitude of global warming, but within a much shorter time span. The question is, can we (and other living things) adjust to such a relatively rapid change?

CHAPTER REVIEW

Key Words

wavelength *(p. 255)*
photon *(p. 256)*
energy level *(p. 257)*
orbital *(p. 259)*
electron shell *(p. 260)*
subshell *(p. 261)*
electron configuration *(p. 267)*
Pauli exclusion principle *(p. 268)*
Hund's rule *(p. 269)*
valence electrons *(p. 274)*
periodic law *(p. 275)*
ionization energy *(p. 280)*

Summary

When you separate the light emitted by the atoms of an element, you obtain a line spectrum characteristic of that element. According to *Bohr's theory of the atom,* a line in a spectrum is produced when an electron in an atom undergoes a *transition* from one *energy level* to a lower one. Energy is lost by the atom as a *photon* of light. The *wavelength* and therefore the color of that light depend on the energy of the photon.

In the modern quantum-mechanical theory of the electron structure of atoms, an electron occupies an *orbital,* or region of space about the atom. Atomic orbitals are grouped in electron *shells,* with the orbitals in a shell having approximately the same size and energy. Each shell has an associated shell quantum number *n,* with $n = 1$ being the shell closest to the nucleus. The orbitals within a shell are further subdivided into *subshells,* each subshell having orbitals of a specific shape and energy. The number of subshells in a shell equals the quantum number *n*. Thus, the $n = 1$ shell has one subshell (*s*); the $n = 2$ shell has two subshells (*s* and *p*); and the $n = 3$ shell has three subshells (*s, p,* and *d*). Each subshell contains a specific number of orbitals; the *s* subshell has one orbital, the *p* subshell has three orbitals, and the *d* subshell has five orbitals.

An *electron configuration* is a particular distribution of electrons among the different subshells of an atom. The electron configuration of lowest energy for the lithium atom is $1s^22s^1$, not $1s^3$, because each orbital can hold no more than two electrons according to the *Pauli exclusion principle.* The subshells fill in a particular order. Using this order and the Pauli exclusion principle, you can reproduce the electron configurations of the atoms. According to *Hund's rule,* one electron goes into each orbital in a subshell before any orbital is occupied by two electrons.

The configurations of the *valence electrons* have a periodic behavior, which explains the periodic chemical properties of the elements displayed in the periodic table. The general form of the outer shells of electrons repeats, so that the outer shell of every element in a given column (group) has the same arrangement of electrons. The number of outer-shell electrons equals the group number. By noting that different subshells are filling in different regions of the periodic table, you can reproduce the order of filling of the subshells, and therefore the complete electron configuration of any element. If you know the position of an element in the periodic table, you can deduce the *valence-shell configuration* of that element. This is the configuration of outer electrons, which determines the chemical properties of the atom.

Because chemical and physical properties depend on the valence-shell configurations of atoms, these properties exhibit the periodicity that you see in such configurations. For instance, *atomic radii* show two general trends: The atomic radius increases in any given group as you move down the column of elements, and the atomic radius decreases in any period as you move across the row of elements. *Ionic radii* show similar trends: The ionic radius increases down a group and decreases with increasing atomic number for a series of *isoelectronic ions. Ionization energy* (the energy required to remove an electron from an atom) decreases in any given group as you move down the column of elements, and ionization energy increases in any given period as you move across the row of elements. Metallic character tends to parallel ionization energy. The elements of the greatest metallic character are those of lowest ionization energy. The most reactive nonmetals have large ionization energy, but also have strong *electron affinity* (strong tendency to gain electrons from other atoms).

Problem-Solving Skills

1. **Calculating the Number of Orbitals in a Shell:** Given the value of *n,* determine the number of orbitals in the shell (Example 9.1).
2. **Interpreting Subshell Notation:** Given a subshell notation, describe its meaning (Example 9.2).
3. **Specifying the Subshells in a Shell:** Given the value of *n* for a shell, determine the number of subshells, the notations for these subshells, and the number of orbitals in each subshell (Example 9.3).

4. **Interpreting the Notation for an Electron Configuration:** Given the notation for an electron configuration, interpret its meaning and obtain the total number of electrons (Example 9.4).
5. **Writing Electron Configurations:** Given an element, use a periodic table to obtain the order of filling of subshells, and from that the complete electron configuration of the element (Examples 9.5 and 9.6).
6. **Obtaining the Valence-Shell Configuration of a Main-Group Element:** Given the position of a main-group element in the periodic table, obtain the valence-shell configuration of the atom (Example 9.7).
7. **Choosing the Element in a Pair of Elements in Any Period or Any Group Whose Atom Has the Greater Atomic Radius:** Given the relative position of two elements in the periodic table, choose which has larger atoms (Example 9.8).
8. **Choosing the Element in a Pair of Elements in Any Period or Any Group Whose Ion Has the Greater Ionic Radius:** Given the relative position of two elements in the periodic table, choose which of their ions is larger (Example 9.9).
9. **Choosing the Element in a Pair of Elements in Any Period or Any Group Whose Atom Has the Smaller Ionization Energy:** Given the relative position of two elements in the periodic table, choose which element has the smaller ionization energy (Example 9.10).

Questions to Test Your Reading

9.1 Describe the light emitted by a heated atom. How does it differ from ordinary white light?

9.2 What are the two principal features of Bohr's theory of the atom?

9.3 What is meant by the term *energy level*?

9.4 Bohr's theory applies quantitatively to only the hydrogen atom. What is the name of the modern theory used to describe all atoms?

9.5 How does an orbital differ from the orbit of an electron in Bohr's theory?

9.6 What is the difference between an electron shell and a subshell? Define the two terms.

9.7 The different electron shells in an atom are designated by the quantum number n. What are the possible values of n?

9.8 How many subshells are there in a shell whose quantum number is n? How do we denote the different subshells?

9.9 What is the general expression for the number of orbitals in the nth shell of an atom? Using the notation for subshells, list the subshells for the $n = 5$ shell. For each subshell, list the number of orbitals it contains. Does the total number of orbitals in the $n = 5$ shell agree with the number obtained from the general expression?

9.10 What is the shape of an s orbital? of a p orbital?

9.11 What is the order (by energy) of the first four subshells of a given shell?

9.12 What is an electron configuration of an atom?

9.13 How does the Pauli exclusion principle affect the number of electrons that can be put into a given subshell? Use the Pauli exclusion principle to obtain the numbers of electrons that can be put into the $n = 1$ and $n = 2$ shells.

9.14 Use the numbers that you obtained in the previous question to explain the lengths of the first two rows of the periodic table.

9.15 Give the electron configurations of the first three noble gases (the elements in Group VIIIA).

9.16 Use the periodic table to write down the first nine electron subshells in the order that they are filled.

9.17 What are the valence electrons of an atom? Why are they of special interest?

9.18 The general form of the valence-shell configuration of a main-group element is ns^anp^b, where $a + b$ equals the number of valence electrons in the atom. Carbon has four electrons in its valence shell, whose quantum number n is 2. What is the valence-shell configuration of the carbon atom?

9.19 State the trends in atomic radius that you see as you progress through the periodic table.

9.20 Explain why you expect the atomic radius to increase as you progress from strontium (Sr) to barium (Ba) in Group IIA.

9.21 State the trends you would see for an isoelectronic series of ions.

9.22 Write an equation that describes the ionization of the potassium atom.

9.23 Explain why you expect the ionization energy to decrease as you progress from strontium (Sr) to barium (Ba) in Group IIA.

9.24 Explain what is meant by electron affinity.

9.25 Compare the ionization energies of metals and nonmetals.

9.26 Which elements have the strongest electron affinities?

Practice Problems

Electromagnetic Radiation (Section 9.1)

9.27 Does red light, with a wavelength of 700 nm, or violet light, with a wavelength of 400 nm, have the higher frequency?

9.28 Does red light, with a wavelength of 700 nm, or infrared light, with a wavelength of 2000 nm, have the higher frequency?

9.29 Which has the higher energy, green light, with a wavelength of 500 nm, or yellow light, with a wavelength of 575 nm?

9.30 Which has the higher energy, microwaves, with a wavelength of 1×10^6 nm, or x rays, with a wavelength of 1 nm?

Bohr's Theory (Section 9.2)

9.31 Which of the energy levels $n = 4$ and $n = 3$ has the greater energy?

9.32 Which of the energy levels $n = 6$ and $n = 2$ has the greater energy?

9.33 An electron in the $n = 5$ level of the hydrogen atom undergoes a transition to the $n = 3$ level and emits a photon. In a similar hydrogen atom, the electron in the $n = 5$ level undergoes a transition to the $n = 2$ level and emits a photon. Which photon has the greater energy?

9.34 An electron in the $n = 5$ level of the hydrogen atom undergoes a transition to the $n = 4$ level and emits a photon. In a similar hydrogen atom, the electron in the $n = 5$ level undergoes a transition to the $n = 3$ level and emits a photon. Which photon has the lesser energy?

Electron Shells and Subshells (Section 9.3)

9.35 How many orbitals are in each of the following shells?
(a) $n = 2$ (b) $n = 5$

9.36 How many orbitals are in each of the following shells?
(a) $n = 1$ (b) $n = 6$

9.37 Describe what is meant by the following notations.
(a) $3p$ (b) $4d$ (c) $2s$

9.38 Describe what is meant by the following notations.
(a) $4f$ (b) $3s$ (c) $5p$

9.39 Write the notations for all of the subshells in the $n = 2$ shell. How many orbitals are in each subshell?

9.40 Write the notations for all of the subshells in the $n = 5$ shell. How many orbitals are in each subshell?

9.41 List the subshells $3d$, $3s$, $3p$ in order, left to right, by increasing energy.

9.42 List the subshells $4d$, $4s$, $4p$, $4f$ in order, left to right, by increasing energy.

Electron Configurations (Sections 9.6 and 9.7)

9.43 Describe the meaning of the notation $1s^22s^22p^63s^23p^4$. How many electrons are in the atom?

9.44 Describe the meaning of the following notation: $1s^22s^22p^63s^23p^63d^54s^2$. How many electrons are in the atom?

9.45 How many electrons can be placed in all of the orbitals in the $n = 2$ shell?

9.46 How many electrons can be placed in all of the orbitals in the $n = 3$ shell?

9.47 Silicon (Si) is a nonmetallic element under carbon in Group IVA. Use the periodic table on the inside front cover of this book to write the complete electron configuration for the silicon atom.

9.48 Magnesium (Mg) is a reactive metal in Group IIA. Use the periodic table on the inside front cover of this book to write the complete electron configuration for the magnesium atom.

9.49 Arsenic (As) is a metalloid in Group VA. Use the periodic table on the inside front cover of this book to write the complete configuration for the arsenic atom.

9.50 Vanadium (V) is a transition metal used to make steel. Use the periodic table on the inside front cover of this book to write the complete configuration for the vanadium atom.

9.51 Strontium (Sr) is a metallic element similar to calcium in chemical properties. Using a periodic table, write the valence-shell configuration of strontium.

9.52 Bismuth (Bi) is a metal with a slightly pinkish color. It is in Group VA and Period 6. What is the valence-shell configuration of bismuth?

9.53 The electron configuration of lowest energy for an atom is $1s^2 2s^2 2p^6 3s^2 3p^6 3d^{10} 4s^2 4p^4$. What is the group and period of the element? What is the name of the element?

9.54 The electron configuration of lowest energy for an atom is $1s^2 2s^2 2p^6 3s^2 3p^6 3d^{10} 4s^2 4p^6$. What is the group and period of the element? What is the name of the element?

Atomic and Ionic Radii (Section 9.8)

9.55 Use a periodic table to decide which atom in each of the following pairs has the larger atomic radius.
(a) Br, F (b) K, Br

9.56 Use a periodic table to decide which atom in each of the following pairs has the larger atomic radius.
(a) S, Te (b) I, Rb

9.57 Use a periodic table to decide which ion in each of the following pairs has the smaller radius.
(a) O^{2-}, Se^{2-} (b) S^{2-}, Cl^-

9.58 Use a periodic table to decide which ion in each of the following pairs has the smaller radius.
(a) Ga^{3+}, S^{2-} (b) Ca^{2+}, Ba^{2+}

Ionization Energies (Section 9.9)

9.59 Use a periodic table to decide which atom in each of the following pairs has the greater ionization energy.
(a) Br, F (b) K, Br

9.60 Use a periodic table to decide which atom in each of the following pairs has the greater ionization energy.
(a) S, Te (b) I, Rb

Additional Problems

9.61 How many electrons are in each of the following atoms?
(a) Al (b) B (c) Ar (d) Br

9.62 How many electrons are in each of the following atoms?
(a) Cr (b) Cu (c) F (d) K

9.63 Which of the following are allowable subshells? If they are not allowed, explain.
(a) 2*p* (b) 2*d* (c) 3*f* (d) 4*f*

9.64 Which of the following are allowable subshells? If they are not allowed, explain.
(a) 1*p* (b) 3*d* (c) 4*d* (d) 0*s*

9.65 Which of the following notations are allowable for electron configurations of atoms corresponding to the lowest energy?
(a) $1s^2 2s^2 2p^3$
(b) $1s^2 2s^2 2p^4$
(c) $1s^2 2s^2 2p^7$
(d) $1s^2 2s^2 2p^2$

9.66 Which of the following notations are allowable for electron configurations of atoms corresponding to the lowest energy?
(a) $1s^2 2s^2 2p^6 3s^2 3p^6 3d^1$
(b) $1s^2 2s^2 2p^6 3s^2 3p^6 4s^1$
(c) $1s^2 2s^2 2p^6 3s^2 3p^6 3d^1 4s^2$
(d) $1s^2 2s^2 2p^6 3s^2 3p^6 4s^3$

9.67 Write the complete electron configurations for the atoms of the first four elements of Group IVA.

9.68 Write the complete electron configurations for the atoms of the first four elements of Group VIA.

9.69 What element has the electron configuration $1s^2 2s^2 2p^6 3s^2 3p^6 3d^2 4s^2$ for the neutral atom?

9.70 What element has the electron configuration $1s^2 2s^2 2p^6 3s^2 3p^6 3d^{10} 4s^2 4p^5$ for the neutral atom?

9.71 How many valence electrons are in each of the following atoms?
(a) K (b) Ba (c) Si (d) Te

9.72 How many valence electrons are in each of the following atoms?
(a) Br (b) In (c) Cs (d) Sb

9.73 Write the valence-shell configurations of the atoms of the first four elements of Group IIIA.

9.74 Write the valence-shell configurations of the atoms of the first four elements of Group VA.

9.75 Which atom has the smallest radius among the alkaline earth elements?

9.76 Which atom has the smallest radius among the halogens?

9.77 Which atom in Period 2 has the largest radius?

9.78 Which atom in Period 3 has the smallest radius?

9.79 Arrange the following elements in order of increasing atomic radius: F, S, Cl.

9.80 Arrange the following elements in order of increasing atomic radius: Se, S, As.

9.81 Arrange the following elements in order of decreasing ionization energy: Mg, S, Sr.

9.82 Arrange the following elements in order of decreasing ionization energy: Al, Cl, Na.

9.83 Arrange the following ions in order of increasing size: Br^-, Cl^-, F^-, I^-.

9.84 Arrange the following ions in order of increasing size: Ba^{2+}, Be^{2+}, Ca^{2+}, Mg^{2+}.

9.85 Arrange the following ions in order of decreasing size: Br^-, Rb^+, Se^{2-}, Sr^{2+}.

9.86 Arrange the following ions in order of decreasing size: Cl^-, P^{3-}, S^{2-}, K^+.

9.87 Which is larger, Na or Na^+?

9.88 Which is larger, Al or Al^{3+}?

9.89 Which is smaller, Cl or Cl^-?

9.90 Which is smaller, Se or Se^{2-}?

Practice in Problem Analysis

For each problem, describe the thinking you would use (the problem analysis) before doing the actual solution, but do not solve the problem.

1. What is the complete electron configuration of the bromine atom?

2. What is the valence-shell configuration of the arsenic atom?

PRACTICE EXAM

1. How many orbitals are there in the $n = 5$ shell?
(a) 5 (b) 10 (c) 15 (d) 20
(e) None of the above

2. The notation 3*d* refers to
(a) subshell 3 of a *d* type
(b) shell with $n = 3$ and *d* subshell
(c) *d* shell and subshell $n = 3$
(d) shell with $n = 2$ and subshell 3
(e) None of the above

3. How many orbitals are there in a *d* subshell?
(a) 2 (b) 3 (c) 4 (d) 5
(e) None of the above

4. How many subshells are there in the $n = 3$ shell?
(a) 2 (b) 3 (c) 4 (d) 6
(e) None of the above

5. What is the complete electron configuration of the nitrogen atom?
(a) $1s^22s^22p^2$ (b) $1s^22s^12p^4$ (c) $1s^22s^22p^3$
(d) $1s^22s^32p^2$ (e) None of the above

6. What is the complete electron configuration of the arsenic atom?
(a) $1s^22s^22p^63s^23p^63d^{10}4s^24p^2$
(b) $1s^22s^22p^63s^23p^63d^{10}4s^24p^3$
(c) $1s^22s^22p^63s^23p^63d^84s^24p^5$
(d) $1s^22s^22p^53s^23p^63d^{10}4s^24p^4$
(e) None of the above

7. What is the valence-shell configuration of the tin atom?
(a) $5s^25p^2$ (b) $6s^26p^2$ (c) $5s^15p^3$
(d) $6s^16p^3$ (e) None of the above

8. Which of the following has the largest atomic radius?
(a) Mg (b) Al (c) Si (d) P (e) Ca

9. Which of the following has the largest ionization energy?
(a) Mg (b) Al (c) Si (d) P (e) S

10. Which of the following has the largest ionization energy?
(a) Mg (b) Al (c) Si (d) P (e) Ca

Chapter

10 Chemical Bonding

A model showing the covalent bonds in buckminsterfullerene.

Atoms combine to form the various kinds of substances. But what holds them together? In other words, what is the nature of the chemical bond? Fundamentally, the forces of attraction that lead to chemical bonding between atoms are electrical in nature; oppositely charged particles attract one another. We have already used this idea in discussing the structure of atoms. Negatively charged electrons are attracted to the positively charged nucleus to form a stable, neutral atom. These same electrical forces are also responsible for chemical bonding. However, the electron structure of the atom that results from the quantum nature of matter is also significant in determining how atoms bond to one another. We investigated this electron structure and its implication for the periodic law in Chapter 9. Now we want to explore how this electron structure helps us to understand how atoms are held together to form familiar substances.

In this chapter, we examine the chemical bonding in ionic and molecular substances. The differences in the properties of these types of substances were discussed in Chapter 4. For example, we saw that ionic substances are solids, whereas many molecular substances are gases or liquids. The nature of the chemical bonds helps us to understand these differences in physical state, a topic that is explored in more detail in Chapter 12. We also look at how the shape of a molecule can be explained and predicted by considering the electron structure of the atoms within the molecule.

IONIC BONDS

Chemical bonds are of two principal types: *ionic* and *covalent.* In this section, we explore the nature of bonds in ionic substances. Later in the chapter we discuss covalent bonding. Ionic bonding seems simple enough to explain, at least in principle. Positive ions and negative ions attract one another to form electrically neutral aggregates of ions. These aggregates are regular arrays, or ionic crystals. You may be familiar with ionic crystals of some common minerals, such as halite (NaCl), fluorite (CaF_2), and calcite

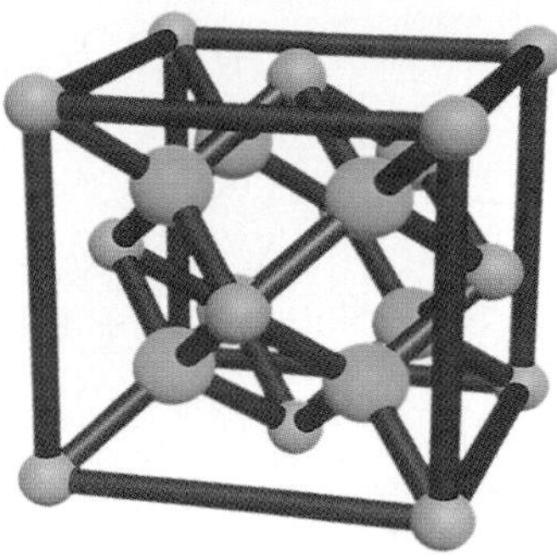

FIGURE 10.1
Crystal structure of fluorite. Fluorite (CaF_2) consists of an array of calcium and fluoride ions in which each calcium ion is surrounded by eight fluoride ions and each fluoride ion is surrounded by four calcium ions.

($CaCO_3$), and of solid kitchen chemicals, such as baking soda ($NaHCO_3$). The structure of sodium chloride crystals was discussed in Chapter 4, and the structure of fluorite is shown in Figure 10.1. It may not be obvious why the ions form in the first place, but we discuss this and the details of ionic bonding in the next two sections.

10.1 Forming an Ionic Bond from Atoms

OBJECTIVES

- Define *ionic bond.*
- Describe three steps in the formation of an ionic compound from atoms.
- Determine from the identity of the component elements whether a compound is likely to contain ionic bonds.

When sodium metal burns in chlorine gas, it forms sodium chloride, or common table salt. Sodium chloride is a crystalline compound composed of sodium ions and chloride ions. The ions in such a compound are held together by ionic bonds. An **ionic bond** is *the strong attractive force that exists between a positive ion and a negative ion in an ionic compound.* The elements sodium and chlorine are made of atoms that exchange electrons to produce ions, which in turn attract one another to yield an ionic solid (sodium chloride).

Let's look more closely at the bonding that occurs between sodium atoms and chlorine atoms when sodium chloride forms. The bonding process includes the following steps: ●

● This discussion uses the concepts of ionization energy and electron affinity, which were discussed in Section 9.9. The formation of atoms from elemental sodium and molecular chlorine is ignored in this analysis; the energy involved in these processes is small relative to that of the steps discussed here and does not modify the conclusions.

1. *Ionization of sodium atoms.* The sodium atom has one electron in its valence, or outer, shell. This electron is easily lost when a small quantity of energy (the *ionization energy*) is supplied. This ionization energy is low for the Group IA elements, as we saw in the previous chapter. We can denote this ionization process as follows (see also Figure 10.2A):

$$\text{Na atom } (1s^22s^22p^63s^1) + \text{ionization energy} \longrightarrow \text{Na}^+ \text{ ion } (1s^22s^22p^6) + e^-$$

The sodium atom loses its 3*s* electron to give the sodium ion, Na^+. Note that the electron configuration of Na^+ is the same as that of the noble gas neon. The electron released in this step is used in the next step.

2. *Formation of chloride ions.* A chlorine atom tends to gain an electron to form the chloride ion, Cl^-, releasing energy in the process. Chlorine is an element with a large *electron affinity*. The electron goes into the valence shell, filling it. We can denote this process as follows (see also Figure 10.2B):

$$\text{Cl atom } (1s^22s^22p^63s^23p^5) + e^- \longrightarrow \text{Cl}^- \text{ ion } (1s^22s^22p^63s^23p^6) + \text{energy}$$

The added electron goes into the 3*p* subshell. Note that the electron configuration of Cl^- is the same as that of the noble gas argon.

The overall process of forming ions from atoms can be written as

$$\text{Na atom } (1s^22s^22p^63s^1) + \text{Cl atom } (1s^22s^22p^63s^23p^5) \longrightarrow \text{Na}^+ \text{ ion } (1s^22s^22p^6) + \text{Cl}^- \text{ ion } (1s^22s^22p^63s^23p^6)$$

3. *Formation of NaCl(s) from ions.* Positive ions and negative ions attract one another to form an aggregate of ions, with energy being released. The optimum arrangement of Na^+ and Cl^- ions is a cubic array in which each sodium ion is surrounded by six chloride ions and each chloride ion is surrounded by six sodium ions (see Figure 4.15, Chapter 4). This cubic array forms the solid NaCl crystal.

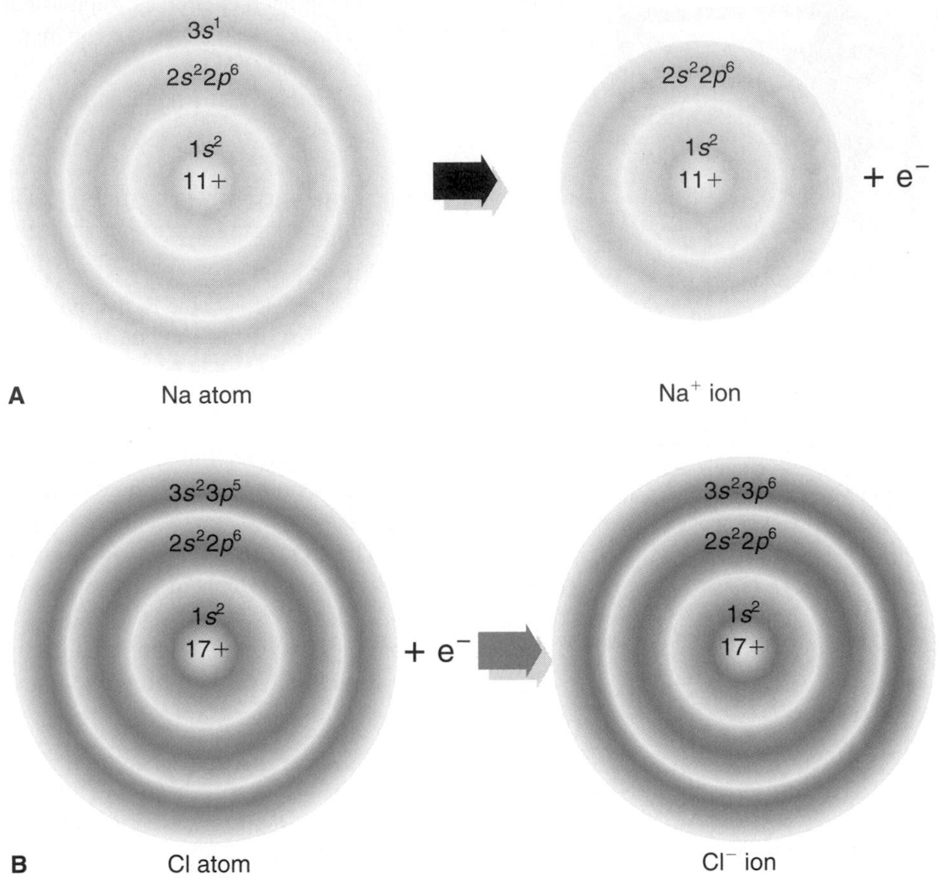

FIGURE 10.2
Formation of ions from atoms. *(A)* The sodium atom loses its valence electron to form the sodium ion. *(B)* The chlorine atom gains this electron to form the chloride ion.

Overall, the reaction of sodium and chlorine releases energy. Although it requires energy to ionize sodium atoms (step 1), the formation of Cl^- ions (step 2) and the attraction of positive and negative ions to form the solid crystal product (step 3) release more than enough energy for the ionization of sodium atoms. The overall release or loss of energy is associated with the stability of the solid product, which is at a lower energy than the two reactants. The overall reaction is like a boulder rolling down a hill. As the boulder rolls down the hill, it loses energy and becomes stabler. The boulder comes to rest at the bottom of the hill, its stablest position, where it has its lowest energy.

In general, a metal usually reacts with a nonmetal to form an ionic compound. You can see why from the previous analysis. Elements with low ionization energy (metals) tend to lose their valence electrons to form cations. Those elements having large electron affinities (the reactive nonmetals) tend to pick up electrons in their valence shells to form anions. The resulting cations and anions attract one another to give an ionic compound.

Note that main-group metal atoms tend to form cations by losing their valence electrons, giving ions with noble-gas configurations. The number of electrons usually lost in forming the ion equals the group number (the Roman numeral in the group designation) of the element. Nonmetal atoms tend to gain enough electrons to fill their valence shells to form anions with noble-gas configurations. The number of electrons gained usually equals eight minus the group number.

10.2 Describing Ionic Bond Formation with Electron-Dot Symbols

OBJECTIVES

- State the octet rule.
- Write the electron-dot symbols for some common atoms and ions.
- Describe ionic bond formation between atoms using electron-dot symbols.

In the previous section, we saw that an ionic bond forms when metal atoms lose electrons to give cations, and nonmetal atoms gain these electrons to form anions. During the process, the atoms generally form ions with noble-gas configurations. These noble-gas configurations have eight electrons in their valence shells (except for helium, which has two electrons). This tendency toward noble-gas configurations is sometimes stated as the **octet rule,** which says that *atoms tend to lose or gain electrons when bonding to give eight electrons in their valence shells.* (Lithium and beryllium atoms can lose electrons and hydrogen atoms can gain electrons to give, in each case, a single shell of two electrons, as in the noble gas helium.)

We can simplify the notation for the formation of ions from atoms if we use electron-dot symbols. An **electron-dot symbol** (or **Lewis symbol**) is *a symbol of an atom or ion in which the valence-shell electrons are represented by dots placed around the letter symbol of the element.* ● For example, the electron-dot symbols of the sodium and chlorine atoms are

● The U.S. chemist Gilbert Newton Lewis (1875–1946) devised electron-dot symbols and formulas to discuss his concept of electron pairs in bonding. He also contributed to chemical thermodynamics, a study of the relationship between heat energy and chemical reactions. In Lewis's symbols, he placed dots one at a time around the atom, as done here; but the manner in which the symbol is written is usually not crucial.

$$\text{Na}\cdot \quad \text{and} \quad :\overset{..}{\underset{..}{\text{Cl}}}\cdot$$

Sodium has one valence electron, so the electron-dot symbol has one dot placed on one side of the symbol. (The dot can be placed on any side.) Chlorine has seven valence electrons, so the electron-dot symbol has seven dots representing them. In general, you place one dot on each side of the element symbol until all four sides are occupied; then you place the dots two to a side. The exact placement of single and double dots is immaterial. For example, you can write the electron-dot symbol for the chlorine atom in any of the following ways.

$$:\overset{..}{\underset{..}{\text{Cl}}}\cdot \qquad :\overset{..}{\underset{\cdot}{\text{Cl}}}: \qquad \cdot\overset{..}{\underset{..}{\text{Cl}}}: \qquad :\overset{\cdot}{\underset{..}{\text{Cl}}}:$$

Table 10.1 shows the electron-dot symbols of some common main-group elements and the usual placement of the dots.

TABLE 10.1
Electron-Dot Symbols of Some Main-Group Atoms

IA	IIA	IIIA	IVA	VA	VIA	VIIA	VIIIA
$\text{H}\cdot$							
$\text{Li}\cdot$	$\cdot\text{Be}\cdot$	$\cdot\overset{\cdot}{\text{B}}\cdot$	$\cdot\overset{\cdot}{\underset{\cdot}{\text{C}}}\cdot$	$:\overset{\cdot}{\underset{\cdot}{\text{N}}}\cdot$	$:\overset{\cdot}{\underset{..}{\text{O}}}\cdot$	$:\overset{..}{\underset{..}{\text{F}}}\cdot$	$:\overset{..}{\underset{..}{\text{Ne}}}:$
$\text{Na}\cdot$	$\cdot\text{Mg}\cdot$	$\cdot\overset{\cdot}{\text{Al}}\cdot$	$\cdot\overset{\cdot}{\underset{\cdot}{\text{Si}}}\cdot$	$:\overset{\cdot}{\underset{\cdot}{\text{P}}}\cdot$	$:\overset{\cdot}{\underset{..}{\text{S}}}\cdot$	$:\overset{..}{\underset{..}{\text{Cl}}}\cdot$	$:\overset{..}{\underset{..}{\text{Ar}}}:$
$\text{K}\cdot$	$\cdot\text{Ca}\cdot$						

When you write the electron-dot symbol of an ion, you start with the corresponding atom, either taking away or adding dots, depending on whether you are representing a cation or an anion. For main-group elements, you take away or add dots from the atoms to give ions in accord with the octet rule. (You learned how to predict the charges of monatomic ions in Section 5.1.) If there are dots about the symbol, you add square brackets and then write the charge as a superscript, just as for the normal formula of the ion. If there are no dots left after taking the appropriate number of dots away, square brackets are not written; the symbol with the charge is written exactly as the normal formula of the ion. You write the electron-dot symbols of the sodium ion and the chloride ion as follows:

$$Na^{+} \qquad \left[:\overset{..}{\underset{..}{Cl}}:\right]^{-}$$

Now you can represent the equation for the transfer of an electron from the sodium atom to the chlorine atom to produce the sodium and chloride ions using electron-dot symbols.

$$Na\cdot + \cdot\overset{..}{\underset{..}{Cl}}: \longrightarrow Na^{+} + \left[:\overset{..}{\underset{..}{Cl}}:\right]^{-}$$

To summarize, the sodium atom loses its outer electron, which is picked up by the chlorine atom. When the sodium atom loses its outer electron, it becomes a sodium ion, which has a noble-gas configuration (neon). The chlorine atom picks up the electron, becoming a chloride ion, which also has a noble-gas configuration (argon). Once ions have formed, they attract one another to yield an ionic compound in which ions are held together by ionic bonds.

You can easily extend this electron-dot representation to other cases of ionic bonding between main-group elements. First note that ionic bonding between two elements normally occurs between a metal and a nonmetal. Also note that, to produce an electrically neutral compound, the number of electrons lost by the metal atoms must equal the number of electrons picked up by the nonmetal atoms. The following example illustrates the analysis you go through to solve this type of problem.

■ EXAMPLE 10.1 Representing Ionic Bond Formation with Electron-Dot Symbols

Use electron-dot symbols to write the equation for the formation of ions from atoms, prior to the formation of the ionic compound, for each of the following pairs of atoms: (a) Mg and F; (b) Mg and O.

Problem Analysis

Decide first what ions are formed. The metal atom becomes the cation and the nonmetal atom becomes the anion. The number of electrons lost by the metal atom equals the group number; the number of electrons gained by the nonmetal atom equals eight minus the group number. You derive the charge on an ion from the number of electrons lost or gained by the atom. To write the equation, you need to know the smallest number of cations and anions required to form the neutral compound (see Example 5.2 in Chapter 5). Write the equation with this number of ions on the right and the corresponding atoms on the left.

Solution

(a) Mg is the metal and F is the nonmetal, so Mg will lose electrons to form cations, and F will gain electrons to form anions. Because Mg is in Group IIA, it loses two electrons, forming Mg^{2+}. Because F is in Group VIIA, it picks up $8 - 7 = 1$ electron, forming F^-.

For every Mg^{2+}, you need two F^- to achieve an electrically neutral ionic compound. Write the equation using the electron-dot symbols for Mg + 2F on the left, and the electron-dot symbols for $Mg^{2+} + 2F^-$ on the right.

$$\cdot Mg \cdot + \begin{matrix} \cdot\ddot{F}: \\ \cdot\ddot{F}: \end{matrix} \longrightarrow Mg^{2+} + \begin{matrix} [:\ddot{F}:]^- \\ [:\ddot{F}:]^- \end{matrix}$$

(b) Mg is the metal and O is the nonmetal. Again, Mg forms Mg^{2+} cations. Because O is in Group VIA, it gains $8 - 6 = 2$ electrons to form O^{2-} anions. Mg^{2+} and O^{2-} occur in equal numbers, since the number of electrons lost by magnesium is exactly the number gained by oxygen. The equation for formation of ions is

$$\cdot Mg \cdot + \cdot\ddot{O}: \longrightarrow Mg^{2+} + [:\ddot{O}:]^{2-}$$

Exercise 10.1

For each of the following pairs of atoms, use electron-dot symbols to write the equation for formation of the ionic compound: (a) Al and F; (b) Ca and O.

(Try Problems 10.27 and 10.28.)

COVALENT BONDS

In the preceding sections, we looked at the formation of ionic bonds between two atoms. Now consider the bonding between two nonmetal atoms in a molecule like Cl_2. We cannot explain this bonding in ionic terms, because we would expect these nonmetal atoms to form two negative ions, which could not bond together. The bonding between nonmetal atoms must have a different explanation.

Instead of completely transferring electrons to form ions, the two chlorine atoms in Cl_2 share a pair of electrons to form a *covalent bond*. Covalent bonds tend to occur between nonmetals—those elements that have high ionization energies (meaning they tend not to give up electrons) and high electron affinities (meaning they tend to gain electrons). By sharing pairs of electrons, two nonmetal atoms effectively gain electrons without having to give up any, a case of having your cake and eating it too!

As you might expect, the sharing of a pair of electrons between atoms need not be quite equal. One atom might tend to hold the electron pair closer to it than does the other atom. The result is a *polar covalent bond*. The atom that more strongly holds electrons is the more *electronegative*.

In the next section, we discuss covalent bonding as a sharing of electron pairs. In the two following sections, we discuss electronegativity and polar covalent bonds, and a general method for writing electron-dot formulas of molecules having covalent bonds.

Animation: *s* orbitals and bonding.

10.3 Covalent Bonding as a Sharing of Electron Pairs

OBJECTIVES

- Define *covalent bond.*
- Describe the formation of a covalent bond between two hydrogen atoms in terms of orbitals.
- Represent covalent bond formation between two hydrogen atoms or two halogen atoms (such as two fluorine atoms) using electron-dot symbols for the atoms and the electron-dot formula for the molecule.
- Define *single bond, double bond,* and *triple bond.*
- Identify the bonding pairs and lone pairs in the electron-dot formula of a molecule.

The first explanation of why two nonmetal atoms might form a bond between them was put forth in 1916 by G. N. Lewis. Lewis proposed that the strong attractive force between two atoms in a molecule, which he called a **covalent bond,** is *a chemical bond formed by the sharing of a pair of electrons between the two atoms.* Often, this sharing of electrons results in a full set of eight electrons in the valence shells of the atoms so that the octet rule is followed. In other words, covalent bonding offers another way for atoms to attain noble-gas configurations of electrons, besides the outright electron transfer that occurs in ionic bonding.

Exploration: Molecular orbitals in H_2.

Although the octet rule and the idea of the stability of the noble-gas electron configuration are very useful notions, fundamentally the covalent bond results from the same forces of attraction and repulsion that exist in ionic bonds: Opposite charges attract and like charges repel. However, these forces operate within the laws of quantum mechanics, so a complete explanation of covalent bonding had to await the development of this theory in the late 1920s. The theory was first applied to the simplest case of covalent bonding: the hydrogen molecule, H_2.

Quantum mechanics explains how a covalent bond forms between two hydrogen atoms by starting with the orbital structure of each atom. Each hydrogen atom consists of an electron in a 1*s* orbital that surrounds a hydrogen nucleus. As two hydrogen atoms approach one another, the 1*s* orbitals on each atom start to overlap (Figure 10.3). With this overlap, the orbitals interact; that is, the two atomic orbitals form a new orbital, called a *molecular orbital.* Unlike individual atomic orbitals, which surround their indi-

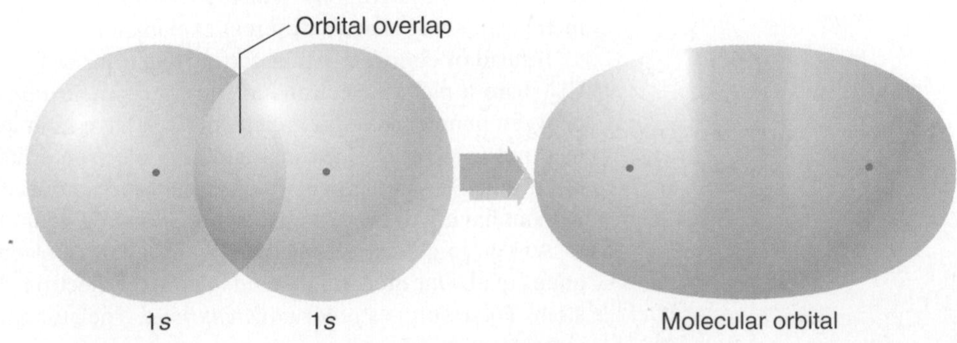

FIGURE 10.3
The formation of a covalent bond in H_2. As the hydrogen atoms approach, their orbitals overlap to form a bonding molecular orbital. In this bonding orbital, the electrons are more likely to be in the region between the two atoms.

vidual atoms, this molecular orbital encompasses both hydrogen atoms. However, the molecular orbital is similar to an atomic orbital in that it can hold a maximum of two electrons as long as they have opposite electron spins (see Section 9.4).

Once bonding occurs, the two electrons being shared by the individual atoms occupy a single molecular orbital. In this orbital, each electron occupies the region around both atoms, especially the region between the two atoms (see Figure 10.3). An electron in the region between the two atoms is attracted to both hydrogen nuclei at once, which is a more stable situation than exists in separate atoms, where an electron can be attracted to only one nucleus (Figure 10.4). As a result of this stability, when the two atoms form a covalent bond, energy is released.

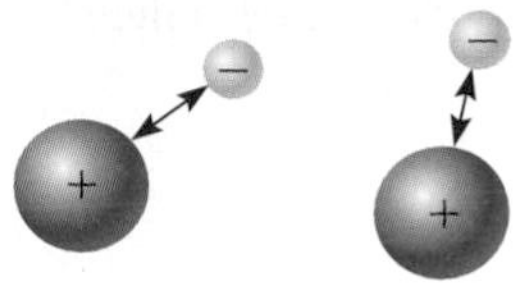

A Separate atoms

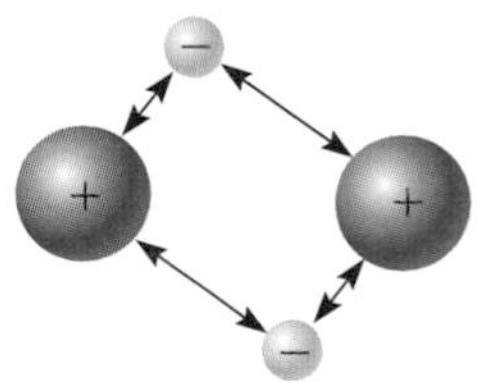

B Molecule

FIGURE 10.4

Electron-nuclear attractions. *(A)* The electrons in separate atoms are attracted to a single nucleus. *(B)* An electron in the bonding region of a molecule is attracted to both nuclei at once. This is a more stable configuration than the one in *(A)*.

Electron-Dot Formulas

We can represent the formation of a covalent bond between two hydrogen atoms by using electron-dot symbols for the atoms and an electron-dot formula for the resulting molecule. An **electron-dot formula** (also called a **Lewis formula**) is *a formula that uses pairs of dots to represent covalent bonds, as well as dots to represent the electrons on individual atoms.* Using this electron-dot notation, you would represent the formation of a covalent bond between hydrogen atoms as follows:

$$\mathrm{H}\cdot \ + \ \cdot\mathrm{H} \longrightarrow \mathrm{H} : \mathrm{H}$$

The pair of dots between the H atoms represents the covalent bond, consisting of two electrons in a molecular orbital that encompasses two atoms.

You can think of the molecular orbital in H_2 as a distorted 1*s* orbital. The molecular orbital in H_2 has its maximum number of electrons (2), just as the 1*s* orbital in the helium atom has its maximum number of electrons. In effect, each hydrogen atom in the molecule has two electrons surrounding it. You can emphasize this in the electron-dot formula of H_2 by drawing circles around each atom.

Each circle encloses two electrons. Thus, in effect, both H atoms simultaneously have a noble-gas (helium atom) type of electron configuration.

Consider the formation of a bond between two fluorine atoms to form the F_2 molecule. Using electron-dot symbols for the atom, we can represent the formation of the molecule as

$$:\ddot{\underset{\cdot\cdot}{\mathrm{F}}}\cdot \ + \ \cdot\ddot{\underset{\cdot\cdot}{\mathrm{F}}}: \longrightarrow :\ddot{\underset{\cdot\cdot}{\mathrm{F}}}:\ddot{\underset{\cdot\cdot}{\mathrm{F}}}:$$

Orbitals that are singly occupied on the F atoms overlap to form a molecular orbital (covalent bond) between the two atoms. Note that each F atom in F_2 has an octet of electrons about it. You can see this more easily perhaps if you draw circles enclosing each of the octets.

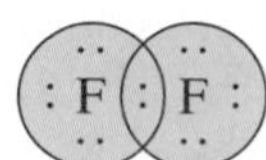

The pair of electrons shared between the two F atoms lies within both octets, and this pair of electrons constitutes the covalent bond between the atoms. Each atom now has a noble-gas (neon atom) type of electron configuration, with eight valence electrons.

As a final example, consider the formation of a more complicated molecule, NH_3.

$$3H\cdot \quad + \quad \cdot\dot{\underset{\cdot}{N}}: \longrightarrow \begin{array}{c} H \\ \ddot{} \\ H:N: \\ \ddot{} \\ H \end{array}$$

Each hydrogen atom with its singly occupied orbital overlaps a singly occupied orbital on the nitrogen atom. Note that the nitrogen atom in NH_3 has an octet of electrons about it.

Multiple Bonds

Both H_2 and F_2 have a single bond between the atoms, consisting of one pair of shared electrons. Each atom contributes one atomic orbital to the molecular orbital containing the shared pair. Double bonds and triple bonds can form when each atom contributes two or three of its orbitals to make bonding molecular orbitals so that two or three pairs of electrons are shared. We can define each of these bonds as follows:

Single bond: *a covalent bond formed by the sharing of a single pair of electrons between two atoms*

Double bond: *a covalent bond formed by the sharing of two pairs of electrons between two atoms*

Triple bond: *a covalent bond formed by the sharing of three pairs of electrons between two atoms*

But how can we use experiment to tell a double bond from, say, a single bond? We can tell from the *bond strength,* as measured by the amount of energy required to break a bond between two atoms. The greater the number of electron pairs shared between the two atoms, the greater the bond strength. This phenomenon is analogous to the difference in effort required to break a single strand of string and the effort required if the number of strands is doubled (or tripled).

Carbon dioxide (CO_2) and nitrogen (N_2), both present as components of air, are simple examples of molecules exhibiting double bonds (in CO_2) and a triple bond (in N_2). The electron-dot formulas are

$$\underset{\cdot\cdot}{\ddot{O}}::C::\underset{\cdot\cdot}{\ddot{O}} \qquad\qquad :N:::N:$$

Note that each atom has an octet of electrons about it. You can see this more clearly if you enclose each atom and its octet of electrons in a circle.

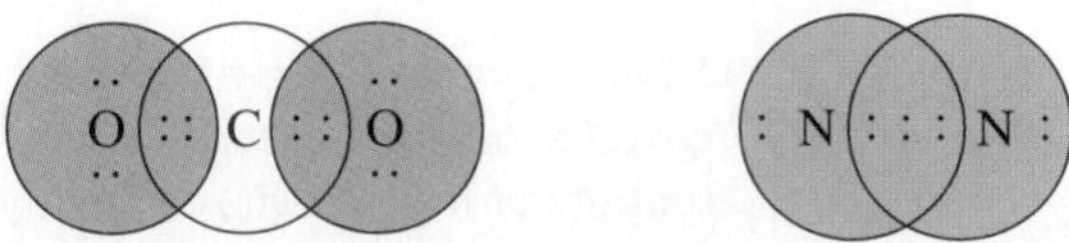

The electron pairs that are shared by atoms in each molecule are those enclosed by two circles. Double or triple bonds are often found in the compounds of the elements C, N, O, and S.

Bonding Pairs and Lone Pairs

The electron pairs in an electron-dot formula are of two kinds, bonding pairs and lone pairs. A **bonding pair** is *an electron pair that is shared between two atoms.* A **lone pair** (or **nonbonding pair**) is *an electron pair that is on one atom; it is not involved in bonding*. For example, ammonia (NH_3) has three bonding pairs and one lone pair.

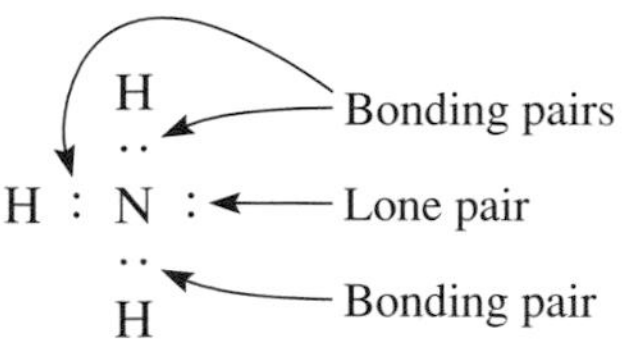

In writing electron-dot formulas, one frequently replaces a pair of dots representing a bond by a dash (—). Using dashes for bonds, the preceding electron-dot formula (or Lewis formula) for the ammonia molecule becomes

```
    H
    |
 H—N :
    |
    H
```

Carbon dioxide and nitrogen molecules, whose electron-dot formulas we used earlier to illustrate double and triple bonds, are written

```
 ..     ..
 O=C=O          :N≡N:
 ..     ..
```

In carbon dioxide, each oxygen atom has two bonding pairs and two lone pairs of electrons, whereas the carbon atom has four bonding pairs. In nitrogen, each atom has three bonding pairs and one lone pair of electrons.

10.4 Electronegativity and Polar Covalent Bonds

OBJECTIVES

- Define *electronegativity.*
- Describe the trends in electronegativity in the periodic table.
- Define *polar covalent bond.*
- Use delta notation to denote bond polarity, given any two atoms and their electronegativities.

In describing the covalent bond, we say that a pair of electrons is shared between two atoms. It is useful to imagine that the two atoms are in competition for the pair of electrons. As long as the two atoms are of the same element, as in H_2, the pair of electrons is equally shared. If atoms of two different elements are bonded together, however, this electron-pair sharing need not be equal. One atom may draw a bonding pair of electrons more strongly to itself than does the other atom of the bond. An atom that draws electrons to itself strongly is very *electronegative*. **Electronegativity** is *a measure of the ability of an atom in a covalent bond to draw bonding electrons to itself.*

In any bond, if electrons are drawn toward (or added to) one atom, they are withdrawn (or removed) from another. Therefore, the electronegativity of an atom (in a bond) should depend on two attributes of separate atoms that we discussed briefly in the previous chapter: electron affinity (a measure of the ease of adding an electron to an atom) and ionization (a measure of difficulty of removing an electron from an atom) energy. In practice, though, electronegativity values are often obtained from other data. Figure 10.5 shows a scale of electronegativities devised by Linus Pauling. ● On this scale, the most electronegative atom is fluorine, F, which has an electronegativity of 4.0.

● Linus Pauling (1901–1994) received the Nobel Prize in chemistry in 1954 for his work on the nature of the chemical bond. He received the Nobel Peace Prize in 1962.

IA	IIA	IIIB	IVB	VB	VIB	VIIB	VIIIB	VIIIB	VIIIB	IB	IIB	IIIA	IVA	VA	VIA	VIIA
								H 2.1								
Li 1.0	Be 1.5											B 2.0	C 2.5	N 3.0	O 3.5	F 4.0
Na 0.9	Mg 1.2											Al 1.5	Si 1.8	P 2.1	S 2.5	Cl 3.0
K 0.8	Ca 1.0	Sc 1.3	Ti 1.5	V 1.6	Cr 1.6	Mn 1.5	Fe 1.8	Co 1.8	Ni 1.8	Cu 1.9	Zn 1.6	Ga 1.6	Ge 1.8	As 2.0	Se 2.4	Br 2.8
Rb 0.8	Sr 1.0	Y 1.2	Zr 1.4	Nb 1.6	Mo 1.8	Tc 1.9	Ru 2.2	Rh 2.2	Pd 2.2	Ag 1.9	Cd 1.7	In 1.7	Sn 1.8	Sb 1.9	Te 2.1	I 2.5
Cs 0.7	Ba 0.9	La-Lu 1.1-1.2	Hf 1.3	Ta 1.5	W 1.7	Re 1.9	Os 2.2	Ir 2.2	Pt 2.2	Au 2.4	Hg 1.9	Tl 1.8	Pb 1.8	Bi 1.9	Po 2.0	At 2.2
Fr 0.7	Ra 0.9	Ac-No 1.1-1.7														

FIGURE 10.5
Electronegativities of the elements. The values given are those of Linus Pauling.

The least electronegative elements are the alkali metals cesium (Cs) and francium (Fr), both having an electronegativity of 0.7.

Figure 10.5 is in the form of a periodic table. (The noble gases are not listed, since few compounds are known.) But you need not have such a table in front of you to use electronegativities. All you need is to understand the general trends in electronegativities and remember some of the most electronegative and least electronegative elements. You will see that *the most electronegative elements are at the upper right in the periodic table (omitting the noble gases).* Fluorine is the most electronegative element, as we noted earlier. After it, the most electronegative elements are oxygen, then chlorine and nitrogen. *The least electronegative elements are at the lower left in the periodic table.* These are the reactive metals—the alkali metals and the alkaline earth metals. *In general, electronegativity increases in any row of the periodic table from left to right, and it decreases in going from the top of a column to the bottom.* Hydrogen, with its intermediate value of electronegativity, has been placed in the middle of the table.

The absolute difference (that is, the difference regardless of sign) between the electronegativities of the two atoms in a covalent bond is a measure of how equally or unequally the electron pair is shared. If the two atoms are alike, as in H_2, the electronegativity difference is zero, and the bonding electrons are equally shared. Consider the hydrogen chloride molecule (HCl), however. Chlorine (Cl) has an electronegativity of 3.0; hydrogen (H) has an electronegativity of 2.1. The difference of electronegativities is 0.9. Because this difference is not zero, the bonding electrons are not shared equally. Chlorine is more electronegative than hydrogen and pulls the bonding electrons toward itself, but not so far that a chloride ion forms.

The bond in the HCl molecule is a polar covalent bond. A **polar covalent bond** (or simply a **polar bond**) is *a covalent bond in which the bonding electrons spend more time near one atom than the other.* Generally, we do not try to represent a polar bond with the electron dots written an unequal distance between the two atoms. Instead, we denote the atom toward which the bonding electrons are drawn with a δ^- to indicate a slightly negative area; and we denote the other end of the bond by δ^+ to indicate a slightly positive area (δ is a lowercase Greek delta). For example, if you want to show that the HCl bond is polar, you write

$$\overset{\delta^+}{H} : \overset{\delta^-}{\underset{\cdot\cdot}{\overset{\cdot\cdot}{Cl}}} :$$

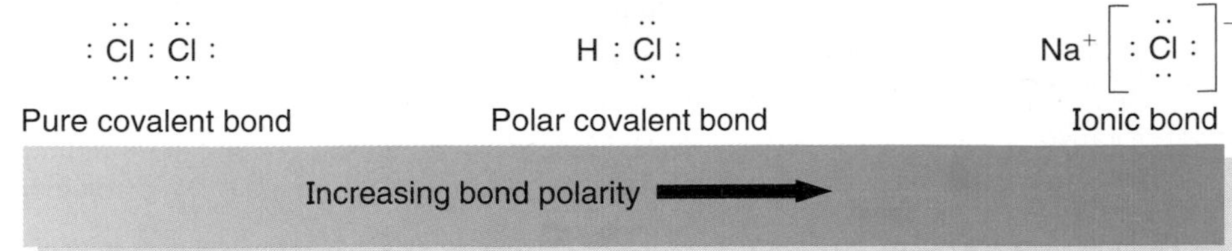

FIGURE 10.6

The continuous variation of bond polarity. Bonding between atoms varies from pure covalent bonds (where the electron pair is equally shared) to ionic bonds (where the electron pair is controlled completely by one element). Polar covalent bonds are intermediate between the two extremes.

Instead of considering chemical bonding between atoms as either strictly covalent or strictly ionic, it helps to think of bonding as varying continuously between two extremes of polarity. One extreme is represented by the purely covalent bond, as in H_2 or Cl_2, where the electronegativity difference between bonding atoms is zero. This bond has no polarity. The other extreme is found in ionic compounds such as NaCl or CsF, where the electronegativity difference is so large that the bonding pair of electrons is not shared but is instead completely transferred to the halogen atom. The resulting ions aggregate into a solid held together by electrical attraction, which you can consider to be an extreme type of polar bond. Somewhere between these two extremes of polarity are molecules such as HCl and HBr, with a moderate difference in electronegativities between elements. The atoms in these compounds still share bonding electrons, though unequally, resulting in polar molecules. Thus, a **polar molecule** is *a covalent molecule in which the centers of partial positive charge and partial negative charge are separated.* A molecule that is polar is also described as a *dipole.* Figure 10.6 illustrates the continuous variation of bond types.

10.5 General Method of Writing Electron-Dot Formulas

OBJECTIVES

- Describe three types of small molecules according to their basic arrangement of atoms.
- State the four steps used in the general method to draw electron-dot formulas.
- Draw a molecule's electron-dot formula, given its molecular formula.

It is very useful to have a general scheme for writing an electron-dot formula given the formula of a small molecule (or polyatomic ion). Rather than try to construct the electron-dot formula from atoms, as in the previous sections, you start by writing the molecule's atoms in their basic arrangement, and then you fill in the available electron dots.

Before you can begin to write an electron-dot formula for a molecule or polyatomic ion, you need to know which atoms are bonded to any given atom (without regard to whether the bonds are single, double, or triple); that is, you need to know the basic arrangement of atoms in the molecule. The molecules and ions we consider in this section belong to one of three types:

1. Many small molecules and ions consist of electronegative atoms surrounding a central atom that is less electronegative than the others. An example is nitrogen trifluoride (NF_3), in which N is the central atom and the F atoms are bonded to it.

● You can extend this to include molecules with two similar central atoms, adding that a symmetrical formula gives a symmetrical molecule. For example, S_2Cl_2 would have the arrangement Cl—S—S—Cl.

Nitrogen is the less electronegative element, and it is surrounded by the more electronegative F atoms. ●

2. Other similar molecules and ions consist of H atoms surrounding a central atom. An H atom cannot normally bond to more than one atom, so it cannot be a central atom. An example is ammonia (NH_3), in which N is the central atom and H atoms are bonded to it.
3. An *oxyacid* is a molecule that consists of a central atom (from which the acid is named), plus hydrogen and oxygen atoms. The oxygen atoms are bonded to the central atom, and one or more hydrogen atoms are bonded to the oxygen atoms. Perchloric acid ($HClO_4$) is an example. Chlorine (Cl) is the central atom, and the O atoms are bonded to it; the H atom is bonded to one of the O atoms.

To draw electron-dot formulas, you need only know the order in which atoms are bonded to one another. Once you have the arrangement of atoms in a molecule (or polyatomic ion), you can write the electron-dot formula by applying the following rules.

Steps for Writing Electron-Dot Formulas for Molecules and Polyatomic Ions

1. *Calculate the total number of valence electrons for the molecule or ion.* You do this by adding up the valence electrons for each of the atoms in the species (molecule or ion). The number of valence electrons contributed by an atom equals the group number of the element. If the species has a charge, you need to add or subtract electrons from the neutral atoms to give that charge. Add one electron for each negative charge and subtract one electron for each positive charge.
2. *Connect atoms by single bonds, using either a pair of dots or a dash.* You may decide later that one or more of these bonds is a double or triple bond.
3. *Distribute electron dots to the atoms surrounding the central atom to satisfy the octet rule for them.* Calculate the number of remaining valence electrons.
4. *Distribute the remaining valence electrons to the central atom.* If there are fewer than eight electrons around the central atom, this suggests that you need to convert one or more bonds to the central atom into multiple bonds. For instance, two electrons fewer than an octet suggests a double bond. In such a case, you should move a lone pair on a surrounding atom to the bonding region between the atoms. In this way, you obtain a double bond between the atoms, giving the central atom an octet while retaining the octet on the other atom. When you have a choice of atoms from which to obtain such a lone pair, note that C, N, O, and S atoms tend to form multiple bonds, so you take the lone pair from such an atom.

In some cases, you may be able to write more than one electron-dot formula for a molecule, in which the formulas differ by placement or type of multiple bond. For example, using the previous rules, you could obtain the following electron-dot formulas for carbonyl chloride, COCl.

```
 ..      ..      ..       ..
:Cl—C=O        Cl=C—O:
 ..  |   ..      ..  |    ..
    :Cl:            :Cl:
     ..              ..
```

As a rule, the electron-dot formula that best describes the bonding is the one in which each atom has its "usual" number of bonds (equal to eight minus the group number of the atom). Thus, the usual number of bonds to oxygen, whose group number is six, is $8 - 6 = 2$. Of the previous electron-dot formulas, only the one on the left has two bonds to oxygen, so we choose it as the best description of carbonyl chloride.

The next three examples illustrate how to apply these steps in writing electron-dot formulas for a variety of molecules and polyatomic ions. First, let's consider an example that involves only single bonds in molecules.

EXAMPLE 10.2 Writing Electron-Dot Formulas (Single Bonds)

Write the electron-dot formulas of (a) $SiCl_4$ (silicon tetrachloride, used to make silicon for transistors); (b) $HClO_2$ (chlorous acid; the sodium salt is used to bleach textiles).

Problem Analysis

Decide from the formula and the type of molecule which atom is the central atom; then arrange the other atoms around it. Add up valence electrons for the atoms (step 1), connect the central atom to the atoms around it by single bonds (step 2), and then distribute electrons to the surrounding atoms to satisfy the octet rule for them (step 3). Distribute any remaining electrons to the central atom (step 4).

Solution

(a) Silicon is much less electronegative than chlorine (see Figure 10.5). Thus, you expect the silicon tetrachloride molecule to have Si as the central atom, with the Cl atoms arranged about it.

```
    Cl
Cl  Si  Cl
    Cl
```

Step 1. Si has four valence electrons since it is in Group IVA; Cl has seven valence electrons since it is in Group VIIA. The total number of valence electrons is $(1 \times 4) + (4 \times 7) = 32$.

Step 2. Connect the atoms with pairs of dots representing single bonds.

```
      Cl
      ..
Cl : Si : Cl
      ..
      Cl
```

Step 3. Now distribute electron pairs to the Cl atoms to satisfy the octet rule for each of them. You obtain

```
         ..
        : Cl :
  ..     ..     ..
: Cl :  Si  : Cl :
  ..     ..     ..
        : Cl :
         ..
```

Counting up the dots used so far, you obtain 32, which is the total number of valence electrons available.

Step 4. There are no more electrons to distribute. Note that each atom has an octet of electrons around it. Thus the electron-dot formula is

```
        ..                              ..
      : Cl :                          : Cl :
   ..   ..   ..                    ..   |   ..
 : Cl : Si : Cl :      or        : Cl—Si—Cl :
   ..   ..   ..                    ..   |   ..
      : Cl :                          : Cl :
        ..                              ..
```

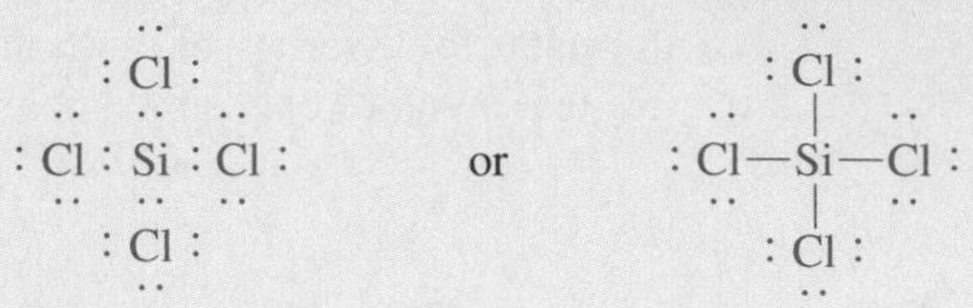

(b) $HClO_2$ is an oxyacid. Place the Cl atom in the center with the O atoms around it, and bond the H atom to one of the O atoms.

O Cl O H

Step 1. Hydrogen has one valence electron, oxygen has six valence electrons (because O is a Group VIA element), and chlorine has seven valence electrons (because it is in Group VIIA). The total number of valence electrons is $(1 \times 1) + (2 \times 6) + (1 \times 7) = 20$.

Step 2. Connect the atoms by pairs of dots.

O : Cl : O : H

Step 3. Distribute electron pairs to the O atoms to satisfy the octet rule for them.

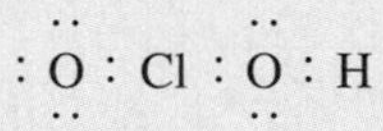

```
  ..        ..
: O : Cl : O : H
  ..        ..
```

The number of electrons remaining is $20 - 16 = 4$ electrons, or two electron pairs.

Step 4. Distribute the remaining two electron pairs to the central atom, Cl.

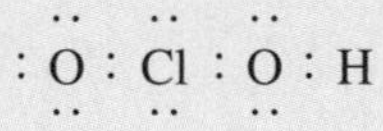

```
  ..   ..   ..
: O : Cl : O : H
  ..   ..   ..
```

Note that each atom satisfies the octet rule (with two electrons on the H atom). Therefore, the final electron-dot formula is

```
  ..   ..   ..                ..  ..  ..
: O : Cl : O : H      or    : O—Cl—O—H
  ..   ..   ..                ..  ..  ..
```

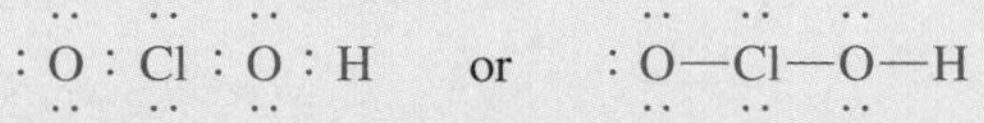

Exercise 10.2

Write electron-dot formulas of (a) H_2O; (b) $HClO_3$.

(Try Problems 10.33 and 10.34.)

Next, we consider an example with molecules that have multiple bonds. Recall that in these cases the number of valence electrons is insufficient to form single bonds between connected atoms and to satisfy the octet rule for all atoms. So you will need to redistribute some electrons to achieve the octet rule for all the atoms.

EXAMPLE 10.3 Writing Electron-Dot Formulas (Multiple Bonds)

Write electron-dot formulas of (a) CO_2 (carbon dioxide, a component of air that exists in solid form as dry ice); (b) HCN (hydrogen cyanide, used in the production of chemicals for plastics and synthetic fibers).

Problem Analysis

Decide from the formula how the atoms are arranged. Obtain the total number of valence electrons (step 1), connect atoms to the central atom by single bonds (step 2), distribute electrons to surrounding atoms to give octets (step 3), and then distribute any remaining electrons to the central atom (step 4). When there are not enough electrons to give the central atom an octet, move lone pairs on the other atoms into multiple bonds on the central atom to satisfy the octet rule. Remember that C, N, O, and S atoms often form multiple bonds.

Solution

(a) The C atom is less electronegative than the O atom, so it is the central atom. Thus, the arrangement of atoms is

$$\mathrm{O\quad C\quad O}$$

Step 1. C has four valence electrons and O has six. The total number of valence electrons is $(1 \times 4) + (2 \times 6) = 16$.

Step 2. Connect atoms by pairs of dots representing bonding electrons.

$$\mathrm{O : C : O}$$

Step 3. Add electron pairs to the O atoms (which surround the central atom) to satisfy the octet rule.

$$:\underset{\cdot\cdot}{\overset{\cdot\cdot}{\mathrm{O}}} : \mathrm{C} : \underset{\cdot\cdot}{\overset{\cdot\cdot}{\mathrm{O}}}:$$

This uses eight electron pairs, or 16 electrons, which is all the available electrons.

Step 4. Note that the C atom does not have an octet of electrons on it; it requires two more pairs of electrons. The only lone pairs of electrons are on the O atoms. We move one pair of electrons from each O atom into the bonding region to give two double bonds.

$$\underset{\cdot\cdot}{\overset{\cdot\cdot}{\mathrm{O}}} :: \mathrm{C} :: \underset{\cdot\cdot}{\overset{\cdot\cdot}{\mathrm{O}}}$$

Now the octet rule is satisfied for all atoms. Therefore, the electron-dot formula for the carbon dioxide molecule is

$$\underset{\cdot\cdot}{\overset{\cdot\cdot}{\mathrm{O}}} :: \mathrm{C} :: \underset{\cdot\cdot}{\overset{\cdot\cdot}{\mathrm{O}}} \qquad \text{or} \qquad \underset{\cdot\cdot}{\overset{\cdot\cdot}{\mathrm{O}}}{=}\mathrm{C}{=}\underset{\cdot\cdot}{\overset{\cdot\cdot}{\mathrm{O}}}$$

(It is possible to write an electron-dot formula for CO_2 in which you move two electron pairs from one O atom into the bonding region. The result is

$$:\underset{\cdot\cdot}{\overset{\cdot\cdot}{\mathrm{O}}}{-}\mathrm{C}{\equiv}\mathrm{O}:$$

However, the previous electron-dot formula better describes the bonding in CO_2 because the O atoms have their usual two bonds, whereas in this formula one O atom has one bond and the other O atom has three bonds.)

(b) C is the central atom since it is less electronegative than N. (H cannot be the central atom.) The arrangement of atoms is

H C N

Step 1. H has one valence electron, C has four valence electrons, and N has five valence electrons, giving a total of ten valence electrons.

Step 2. Connect the atoms by pairs of dots representing bonding electrons.

H : C : N

Step 3. Add electron pairs to N (an atom bonded to the central atom) to give it an octet.

$$\mathrm{H : C : \overset{\cdot\cdot}{\underset{\cdot\cdot}{N}} :}$$

This uses all ten electrons.

Step 4. Note that the C atom requires two more electron pairs to have an octet. Therefore, we move two of the lone pairs on the N atom into the bonding region, giving a triple bond.

$$\mathrm{H : C ::: N :} \quad \text{or} \quad \mathrm{H{-}C{\equiv}N :}$$

Now all atoms have octets.

Exercise 10.3

Write electron-dot formulas of (a) N_2; (b) CS_2.

(Try Problems 10.35 and 10.36.)

Finally, we will examine how to write an electron-dot formula of a polyatomic ion. The only difference from the examples you have already considered is that you must take the ionic charge into account when you count the number of valence electrons.

■ EXAMPLE 10.4 Writing Electron-Dot Formulas (Polyatomic Ions)

Write an electron-dot formula for NO_2^-, the nitrite ion. Nitrite salts are used in small amounts in the curing of meats such as bacon.

Problem Analysis

Decide from the formula how to arrange the atoms. Obtain the total number of valence electrons, noting how many electrons you must add or subtract to give the correct charge on the ion (step 1). Connect the atoms by single bonds (step 2), distribute electrons to atoms surrounding the central atom to give octets (step 3), and then distribute any remaining electrons to the central atom (step 4). If there are not enough electrons to give the central atom an octet, move lone pairs on the other atoms into bonds on the central atom (giving multiple bonds) to satisfy the octet rule.

Solution

N is less electronegative than O, so you expect N to be the central atom.

$$\mathrm{O \quad N \quad O}$$

Step 1. N has five valence electrons; O has six valence electrons. You need to add one electron to account for the negative charge on NO_2^-. Therefore, the total number of valence electrons is $5 + (2 \times 6) + 1 = 18$.

Step 2. Connect the atoms by electron pairs.

$$\mathrm{O : N : O}$$

Step 3. Distribute electron pairs to the O atoms to give them octets.

$$:\ddot{\underset{\cdot\cdot}{\mathrm{O}}}:\mathrm{N}:\ddot{\underset{\cdot\cdot}{\mathrm{O}}}:$$

This uses eight electron pairs, leaving one electron pair.

Step 4. Place this remaining electron pair on N.

$$:\ddot{\underset{\cdot\cdot}{\mathrm{O}}}:\ddot{\mathrm{N}}:\ddot{\underset{\cdot\cdot}{\mathrm{O}}}:$$

N has only six electrons around it, which suggests that you create one double bond by moving a lone pair from one of the O atoms into the bonding region.

$$\ddot{\underset{\cdot\cdot}{\mathrm{O}}}::\ddot{\mathrm{N}}:\ddot{\underset{\cdot\cdot}{\mathrm{O}}}:$$

Now all atoms have octets. Therefore, the electron-dot formula for the nitrite ion is

$$\left[\ddot{\underset{\cdot\cdot}{\mathrm{O}}}::\ddot{\mathrm{N}}:\ddot{\underset{\cdot\cdot}{\mathrm{O}}}:\right]^- \quad \text{or} \quad \left[\ddot{\underset{\cdot\cdot}{\mathrm{O}}}=\ddot{\mathrm{N}}-\ddot{\underset{\cdot\cdot}{\mathrm{O}}}:\right]^-$$

Exercise 10.4

Write an electron-dot formula for the tetrafluoroborate ion, BF_4^-.

(Try Problems 10.37 and 10.38.)

In the previous example, you could have written the electron-dot formula of the nitrite ion, NO_2^-, with the double bond on the right instead of on the left. Then you would have had two equally acceptable electron-dot formulas:

$$\left[\ddot{\underset{\cdot\cdot}{\mathrm{O}}}=\ddot{\mathrm{N}}-\ddot{\underset{\cdot\cdot}{\mathrm{O}}}:\right]^- \quad \text{and} \quad \left[:\ddot{\underset{\cdot\cdot}{\mathrm{O}}}-\ddot{\mathrm{N}}=\ddot{\underset{\cdot\cdot}{\mathrm{O}}}\right]^-$$

But which is the correct formula? The two structures differ only in that the electron arrangement around the two oxygen atoms is reversed. This problem reveals a limitation of electron-dot formulas. In these formulas, we draw pairs of electron dots so that they are either on one atom (giving a lone pair) or in the region between two atoms (giving a covalent bond between them). However, one of the pairs of electrons in the nitrite ion actually occupies a molecular orbital that encompasses all three atoms. Rather than being localized between two atoms, one bonding pair is delocalized, or spread out, over three atoms.

You can retain the simplicity of the electron-dot representation by using *resonance formulas*. You simply write both of the formulas and connect them by a double-headed arrow.

$$\left[\ddot{O}=\ddot{N}-\ddot{\underset{..}{O}}:\right]^- \longleftrightarrow \left[:\ddot{\underset{..}{O}}-\ddot{N}=\ddot{O}\right]^-$$

This notation means that *resonance* occurs; that is, one or more of the electron pairs are delocalized, occupying a region about more than two atoms.

Chemical Perspective

Explosives and Bond Strengths

Nitroglycerin is a pale yellow, oily liquid. The liquid has gained a nasty reputation for unpredictability: The mildest vibration can sometimes cause it to detonate. However, it freezes at 11°C (52°F), and as the solid (or even the cool liquid), nitroglycerin is relatively safe. Soon after its discovery in 1846, it was used as an explosive for blasting tunnels through rock and in blasting oil wells to improve their oil flow. Today, the pure liquid is seldom used as an explosive. It does provide a source of heart-pumping entertainment, though. In the award-winning French film *The Wages of Fear* (American remake in 1977 as *The Sorcerer*), four men eager for money agree to ferry two trucks loaded with nitroglycerin from a remote village in South America to fight a fire that is burning out of control on an oil rig. Along the way, the men find the road obstructed by a huge boulder that has rolled across it. They decide to use some of their explosive cargo to blast the boulder out of their way. You see one of the men gingerly pouring nitroglycerin from a thermos bottle down a stick and into a hole in the rock. Beads of sweat form on his contorted face, while he tries desperately to suspend his breathing and any extra movement that might set off the nitroglycerin.

Most chemical explosives contain nitro groups (—NO_2) or nitrate groups (—O—NO_2). The chemical structure of nitroglycerin is

```
      H           H           H
      |           |           |
   H—C———————————C———————————C—H
      |           |           |
     :O:         :O:         :O:
      |           |           |
      N           N           N
     / \\        / \\        / \\
   :O:   O:    :O:   O:    :O:   O:
```

The name nitroglycerin is a misnomer, as you can see from its structure; nitroglycerin does not contain nitro groups (—NO_2), but rather it contains nitrate groups (—O —NO_2). Its chemically correct name is *glyceryl trinitrate*.

Nitroglycerin, $C_3H_5(ONO_2)_3$, decomposes as follows:

$$4C_3H_5(ONO_2)_3(l) \longrightarrow 6N_2(g) + 12CO_2(g) + 10H_2O(g) + O_2(g)$$

Notice that the products are all gases. When nitroglycerin decomposes, it does so quickly and produces gases that occupy more than 1200 times the volume of the original nitroglycerin. This enormous, quick increase in volume is what is responsible for the explosiveness of the decomposition reaction.

Nitroglycerin decomposes because its products have lower energies and are therefore more stable. The chemical bonds of the product gases are among the strongest (that is, compounds with these bonds have low energy). Nitrogen, for example, has a very strong triple bond, and carbon dioxide has two strong double bonds.

$$:N\equiv N: \qquad \ddot{\underset{..}{O}}=C=\ddot{\underset{..}{O}}$$

By contrast, the nitrogen–oxygen bonds of nitroglycerin are relatively less strong than the bonds of the products. By slight rearrangement of the atoms in nitroglycerin, these less strong bonds are replaced with stronger ones. With just a little jostling, the atoms of nitroglycerin quickly rearrange to attain the lower energy of the products.

The dangers of handling nitroglycerin inspired investigators to seek alternatives. In 1867, the

Swedish chemist and industrialist Alfred Nobel discovered that nitroglycerin was less shock sensitive when it was absorbed on diatomaceous earth, a crumbly rock. He patented this explosive mixture, which he called *dynamite*. His explosives factories and oil interests brought him great wealth, part of which he left in trust to establish the Nobel Prizes.

Explosives often conjure up images of warfare and terrorist acts, but explosives do have many peacetime purposes, including road building, mining, and demolition (Figure 10.7).

FIGURE 10.7
Demolition of a building. Explosives were placed at predetermined positions so that when detonated, the building would collapse on itself.

SHAPES OF MOLECULES

In the preceding sections, we looked at the nature of the covalent bond and how we can represent the bonding between atoms in molecules by electron-dot formulas. An electron-dot formula describes, in an approximate way, the distribution of bonding electrons between the atoms and the placement of any lone pairs. The electron-dot formula does not generally attempt to represent the *shape of the molecule*. When an atom forms covalent bonds, however, it does so in definite directions in space, and this gives a molecule a characteristic shape. In the following sections, we describe molecular shapes and explain how you can predict the shapes from electron-dot formulas.

10.6 Molecular Structure

OBJECTIVES

- Define *bond length.*
- State the two pieces of information needed to describe molecular structure.

When two hydrogen atoms come together, the electrons move into the bonding region, where they are attracted by both hydrogen nuclei at once. A covalent bond begins to form. As the hydrogen atoms move closer, the attractions between electrons and nuclei increase, and a stronger bond forms. However, the repulsion between the two electrons and between the two nuclei also increases, and at very close distances this repulsion is much greater than the attraction between electrons and nuclei. The strongest bond forms when the two hydrogen nuclei are 74 pm apart. (The symbol *pm* stands for picometers, or 10^{-12} m; see Table 2.2 in Chapter 2.) At this distance, the attraction between electrons and nuclei is maximized; if the particles get any closer together, repulsive forces start to dominate. This distance is called the bond length in the H_2 molecule. The **bond length** is *the normal distance between nuclei whose atoms form a bond in a molecule.*

In the water (H_2O) molecule, the H—O bonds have a length of 96 pm. Moreover, as we already noted, atoms form covalent bonds in definite directions in space. The bonds

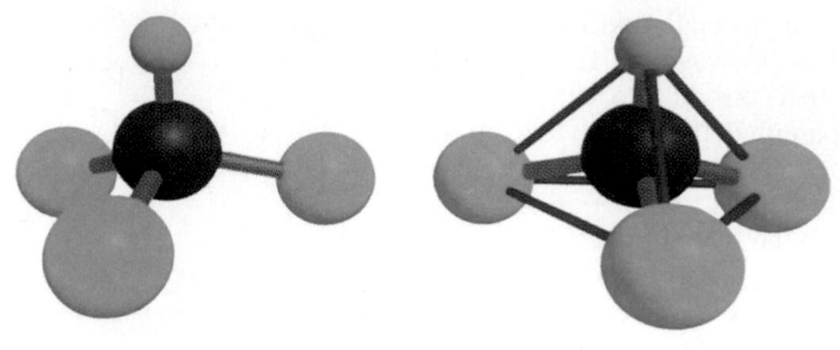

FIGURE 10.8
The chloroform molecule, $CHCl_3$. *(A)* Note the three-dimensional shape of the chloroform molecule in this ball-and-stick model. (The carbon atom is shown in black, the hydrogen atom in light blue, and the chlorine atoms in green.) *(B)* The molecule is inscribed in a tetrahedron to show how its bonds are directed toward the four tetrahedral apexes (points).

in the water molecule meet at a definite angle at the oxygen atom, and experiments show that this *bond angle* is 105°.

96 pm O 96 pm
H 105° H

You need bond lengths and bond angles to describe molecular structure. Often, though, you may just want to know the molecular shape without knowing the precise bond lengths and bond angles. In that case, you would describe H_2O in a general way as a *bent,* or *angular,* molecule. Also, the three atoms of the H_2O molecule are in a plane (or flat), so you might add that the molecule is *planar.*

Most molecules are not planar but rather extend in three dimensions. Figure 10.8A shows a ball-and-stick model of the chloroform molecule, $CHCl_3$. Note that the carbon atom (shown in black) is in the center of the molecule, with the hydrogen atom (light blue) and the chlorine atoms (green) arranged around it. If you were to draw lines connecting the atoms in $CHCl_3$ (in the direction of the bonds), they would form the outline of a *tetrahedron,* a solid figure having four points and four identical triangular faces (Figure 10.8B). You would say that the $CHCl_3$ molecule has a tetrahedral shape. In the next section, we describe how you can obtain molecular shapes from the electron-dot formulas of molecules.

10.7 The VSEPR Model of Molecular Shape

Exploration: The VSEPR theory of molecular structure.

OBJECTIVES

- Describe the valence-shell electron-pair repulsion (VSEPR) model for predicting molecular shape.
- Explain what is meant by *linear, trigonal planar,* and *tetrahedral* arrangements of electron groups about a central atom, and give one example of each arrangement.
- Describe and name a molecule's shape, given its formula.

You can predict the shape of a molecule using a very simple model. In this model, you look at the electron pairs in the valence shells, and you assume that these pairs occupy positions in space in such a way that they avoid one another as much as possible (because of mutual repulsion of electrons). This **valence-shell electron-pair repulsion (VSEPR) model** is *a model for predicting the shapes of molecules and ions, in which the valence-shell electron pairs are arranged about each atom in such a way as to keep electron pairs as far away from one another as possible.* (The acronym VSEPR is pronounced "vesper.")

We will apply this model to simple molecules consisting of a central atom surrounded by other atoms. You can predict molecular shape using a three-step approach.

1. First draw the electron-dot formula of the molecule.
2. Then, looking at the central atom, decide how the electron pairs on this atom (bonding pairs and lone pairs) would arrange themselves to minimize their repulsion.
3. Finally, look at just the bonding pairs and the positions of the atoms and decide on the shape of the molecule. (Although the lone pairs are important in deciding how the atoms are arranged in space, they are not considered part of the molecular shape.)

Arrangement of Electron Groups About the Central Atom

You already know how to write electron-dot formulas of molecules and polyatomic ions (step 1) from Section 10.5. Now we explain how you determine the arrangement of lone pairs and bonding pairs about the central atom (step 2). According to the VSEPR model, the electron pairs of lone pairs and single bonds occupy regions of space about a central atom so that the repulsion among them is minimized. Think of each lone pair or bond as a single *electron group*. Then if a central atom has just two single bonds attached to it, as in the BeF_2 molecule, each of these bonds is an electron group, and these two groups arrange themselves so that they are on opposite sides of the central atom.

$$:\ddot{\underset{..}{F}}\text{—Be—}\ddot{\underset{..}{F}}:$$

The single bonds are as far from one another as possible, and the electron repulsion between them is as low as possible.

You can extend the VSEPR model to include multiple bonds (double and triple bonds). Each lone pair, single bond, or multiple bond is a single electron group. The electrons within an electron group occupy the same general region of space. *According to the VSEPR model, the electron groups arrange themselves about the central atom to minimize the repulsion among themselves.* The electron groups spread as far apart from one another as possible, although the electrons within a single group are in the same region of space. We now look at the three possibilities in which the central atom has either two, three, or four electron groups about it.

As an example of two electron groups, let's consider the CO_2 molecule. Carbon dioxide is a component of air produced by animal respiration and consumed by plants. It has the following electron-dot formula.

$$\ddot{\underset{..}{O}}::C::\ddot{\underset{..}{O}} \quad \text{or} \quad \ddot{\underset{..}{O}}=C=\ddot{\underset{..}{O}}$$

The C atom has two electron groups about it (two double bonds). Two electron groups minimize their repulsion by moving to positions that are 180° apart.

$$\ddot{\underset{..}{O}}=\overset{180°}{C}=\ddot{\underset{..}{O}}$$

This is a *linear* arrangement of electron groups.

As an example of three electron groups, consider the nitrate ion, NO_3^-. The electron-dot formula of this ion is

$$\left[\begin{array}{c} :\ddot{O}: \\ | \\ \ddot{\underset{..}{O}}=N-\ddot{\underset{..}{O}}: \end{array}\right]^-$$

The N atom has three electron groups about it: a double bond and two single bonds. Three electron groups minimize their repulsion by moving to positions that are in a plane, with 120° angles between groups.

$$\left[\begin{array}{c} :\ddot{O}: \\ | \\ \ddot{O}{=}N{-}\ddot{O}: \end{array}\right]^{-} \quad (120^\circ \text{ between each pair of groups})$$

This is a *trigonal planar* arrangement of electron groups.

As an example of four electron groups, consider the methane molecule, CH_4. Methane is the principal component of natural gas. The electron-dot formula is

$$\begin{array}{c} H \\ H:\ddot{\underset{..}{C}}:H \\ H \end{array} \quad \text{or} \quad \begin{array}{c} H \\ | \\ H{-}C{-}H \\ | \\ H \end{array}$$

The C atom has four electron groups (four single bonds) about it. Four electron groups minimize their repulsions by moving to *tetrahedral* positions. The electron groups are directed toward the apexes (points) of a tetrahedron, and the angle between the electron groups (C—H bonds) is 109°. ●

● The angles will be somewhat different from 109° when the electron groups are different; in methane, all electron groups are identical C—H single bonds.

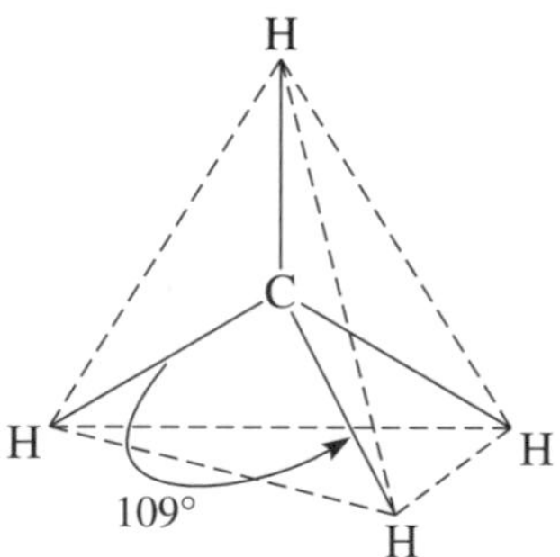

This is a *tetrahedral* arrangement of electron groups.

Predicting Molecular Shapes

Molecules have different shapes about their central atoms, which we describe as linear, trigonal planar, or tetrahedral (the same terms we used to describe the arrangements of electron groups about the central atom). However, molecular shape depends only on the arrangement of the bonds in space, which may be all or only some of the electron pairs about the central atom. You can decide this molecular shape once you know the arrangement of electron groups about the central atom in the molecule. If all of these electron groups are bonding groups and none are lone pairs, then the molecule has the same shape as the electron arrangement. But, if one or more of the electron groups are lone pairs, the molecular shape is different from the arrangement of electron groups. We leave out the lone pairs and look at just the bonds to decide the molecular shape. The different possibilities are summarized in Figure 10.9. We discuss examples of these possibilities in the remainder of this section.

In the carbon dioxide molecule (CO_2), which we considered earlier, the central atom (C) has two electron groups about it, each of which is a double bond. This is a linear

FIGURE 10.9
Molecular shapes. Shapes that are observed for central atoms with two, three, or four electron groups.

Electron Groups			Arrangement of Groups	Molecular Shape	Example
Total	Bonding	Lone			
2	2	0	Linear	Linear	CO_2 
3	3	0	Trigonal planar	Trigonal planar	NO_3^- 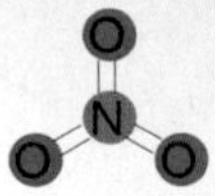
	2	1		Bent (or angular)	O_3 (Lone pair) 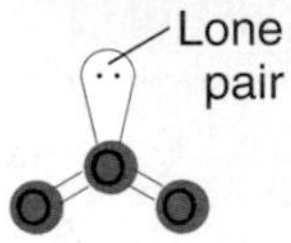
4	4	0	Tetrahedral	Tetrahedral	CH_4 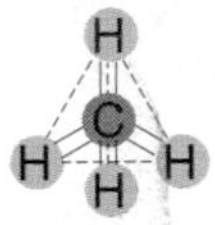
	3	1		Trigonal pyramidal	PF_3 (Lone pair) 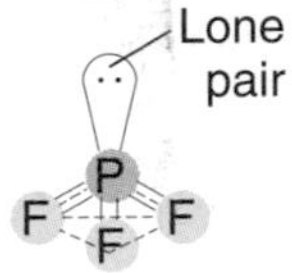
	2	2		Bent (or angular)	H_2O (Lone pairs) 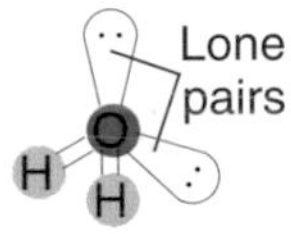

FIGURE 10.10
The carbon dioxide molecule. *(Top)* Both the electron groups about carbon are double bonds. *(Bottom)* The molecular shape is *linear;* that is, the atoms are in a straight line.

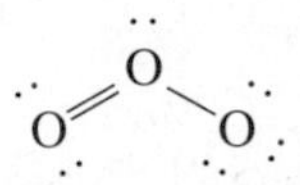

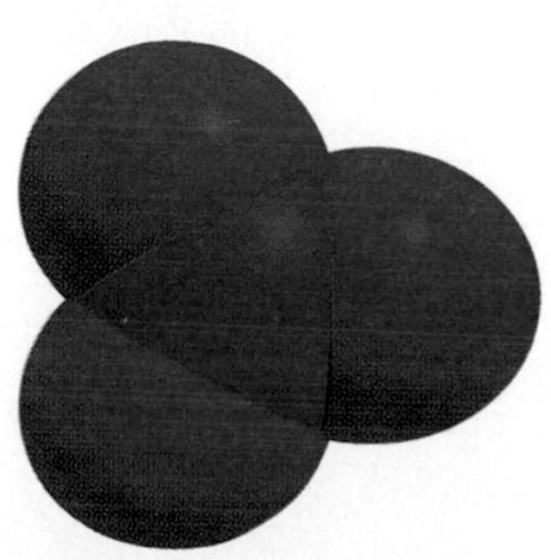

FIGURE 10.11
The nitrate ion. The nitrate ion has a *trigonal planar* shape.

arrangement of electron groups, and the molecular shape is also *linear.* (In this case, the name of the molecular shape is the same as that of the electron arrangement because there are no lone pairs.) See Figure 10.10.

In the nitrate ion (NO_3^-), which we considered earlier, the central atom (N) has three electron groups about it, two single bonds and one double bond. All three are bonding groups; there are no lone pairs on the nitrogen atom. The arrangement of electron groups is trigonal planar, and the molecular shape is also *trigonal planar,* as shown in Figure 10.11.

Now let's look at an example where the presence of a lone pair results in a molecule with a different shape from its electron arrangement. Consider ozone, which is a health hazard in the lower atmosphere, but in the stratosphere it protects us by absorbing harmful ultraviolet radiation. In the ozone molecule (O_3), the central oxygen atom has three electron groups about it; two are bonding groups (one double bond and one single bond), and the third is a lone pair. The electron-dot formula is

$$\underset{\cdot\cdot}{\ddot{O}}=\ddot{O}-\underset{\cdot\cdot}{\ddot{O}}:$$

The three electron groups form a trigonal planar arrangement about the central atom. We can redraw the previous electron-dot formula to show this arrangement of the electron groups.

$$\begin{array}{ccc} & \ddot{O} & \\ \ddot{O}\!\!\!\!\!\!\! & \diagup\!\!\!\!\!\!=\;\;\;\diagdown & \!\!\!\!\!\!\ddot{O}: \end{array}$$

FIGURE 10.12
The ozone molecule. The O_3 molecule has a *bent* (or *angular*) shape. The atoms form a plane.

Looking at just the atoms and ignoring the lone pair, we describe the molecular shape as *bent,* or *angular.* See Figure 10.12.

Now let's consider some molecules that have four electron groups about the central atom. In methane (CH_4), as we saw earlier, there are four electron groups (all single bonds) in a tetrahedral arrangement about the central carbon atom. There are no lone pairs, so the molecular shape is *tetrahedral,* the same as the arrangement of electron groups.

But what shapes do we get if there are some lone pairs? Phosphorus trifluoride (PF_3) is an example of a molecule with a central atom (P) that has four electron groups about it, three of which are bonding groups. This is shown by the electron-dot formula

```
        ..
       :F:
   ..   ..
  :F : P :
   ..   ..
       :F:
        ..
```

The arrangement of electron groups about P is tetrahedral.

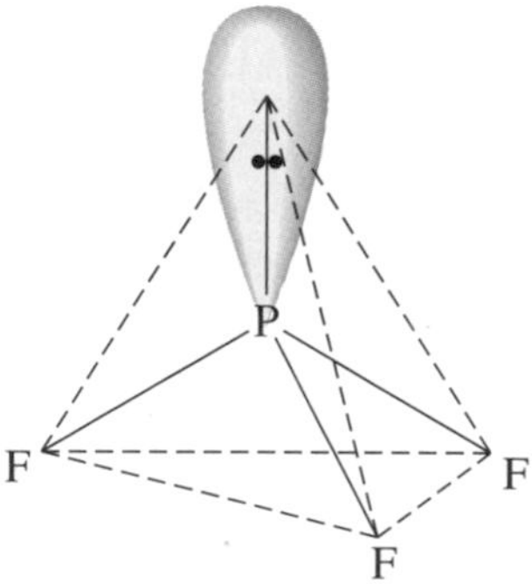

But if you look at just the atoms and imagine them connected with lines, you will see a squat pyramid with triangular faces.

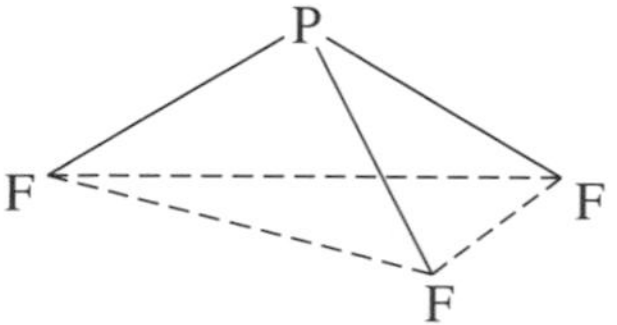

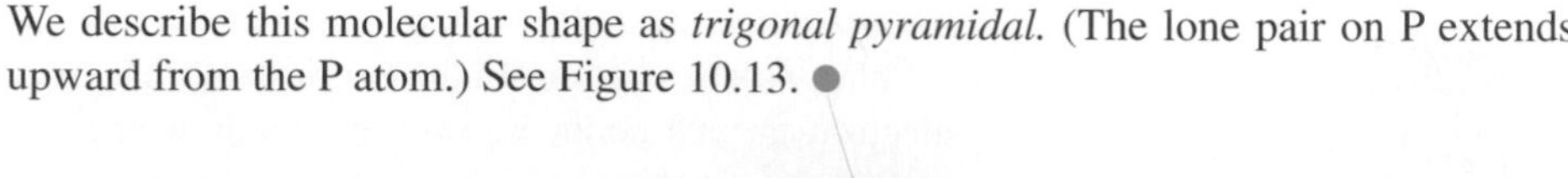

● Pyramidal is pronounced "py-ram′-i-dal."

We describe this molecular shape as *trigonal pyramidal.* (The lone pair on P extends upward from the P atom.) See Figure 10.13. ●

FIGURE 10.13
The phosphorus trifluoride molecule. *(Left)* A ball-and-stick model of the PF_3 molecule. The molecule has a *trigonal pyramidal* shape, with the P atom at the top of the pyramid and the F atoms at the base. *(Right)* A space-filling model of PF_3, showing the arrangement of the electron groups around the atoms.

FIGURE 10.14
The water molecule. The H_2O molecule has a *bent* (or *angular*) shape.

You may wonder whether there is any difference between the tetrahedral and trigonal pyramidal shapes—after all, they both have three triangular sides on a triangular base. The difference is that, in the tetrahedron, all four triangles are identical. A simple way to distinguish between these two shapes is to examine the location of the central atom. In the tetrahedral shape, the central atom is located at the center of the tetrahedron. Furthermore, in a tetrahedral molecule, there are five atoms. In the trigonal pyramidal shape, the central atom is located at one of the points of the pyramid, and there are only four atoms.

The common water molecule (H_2O) illustrates what happens to the molecular shape when there is more than one lone pair about the central atom. In H_2O, the central atom (O) has four electron groups about it, only two of which are bonding groups. The other two are lone pairs. The electron-dot formula is

$$\mathrm{H : \ddot{\underset{\cdot\cdot}{O}} : H}$$

The four electron groups are in a tetrahedral arrangement.

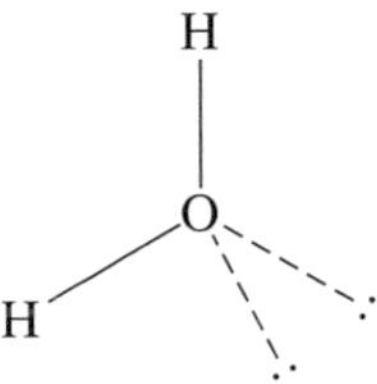

Looking at just the atoms, we describe the molecular shape as *bent*, or *angular*. See Figure 10.14. ●

● If the arrangement of electron pairs were precisely tetrahedral, the H—O—H bond angle would be 109°; the experimental value is 105°.

Now let's look at an example that further illustrates the VSEPR model for predicting molecular shape.

■ EXAMPLE 10.5 Predicting Molecular Shape with the VSEPR Model

Predict the shape of the following molecules: (a) NH_3; (b) HCN.

Problem Analysis

First draw the electron-dot formula of the molecule. Then, after noting how many electron groups there are about the central atom, decide on the arrangement of these electron groups. Finally, look at just the atoms and predict the molecular shape.

Solution

(a) The electron-dot formula of NH_3 is

$$\begin{array}{c} \mathrm{H} \\ \cdot\cdot \\ \mathrm{H : N :} \\ \cdot\cdot \\ \mathrm{H} \end{array}$$

There are four electron groups about the N atom; they will have a tetrahedral arrangement. Looking at just the atoms, you would predict that NH_3 has a trigonal pyramidal shape.

(b) The electron-dot formula of HCN is

$$\mathrm{H : C : : : N :}$$

There are two electron groups about the central atom (C), and their arrangement is linear. Because there are no lone pairs on the central atom, the molecular shape is the same as the arrangement of electron groups. The HCN molecular shape is linear.

Exercise 10.5

Predict the molecular shape of the following: (a) NH_4^+; (b) NF_3.

(Try Problems 10.39 and 10.40.)

Polarity of Molecules

Earlier we saw that for molecules that contain only two atoms their electronegativity difference could be used to decide whether the molecule is polar or nonpolar. Molecules such as HF, with the atoms having different electronegativities, are polar because the electrons spend more of their time in the vicinity of one atom (F) than the other (H). This means that centers of partial negative charge and partial positive charge are formed. In molecules such as F_2, in which the atoms have identical electronegativities, the electrons are distributed equally between the two atoms, and the molecule is nonpolar.

But what about molecules that contain more than two atoms? If the atoms are all identical, there is no separation of charges, and the molecule is nonpolar, just as for molecules with two atoms. If the atoms are not all identical, there may be a difference of electronegativities between bonding atoms so that bonds are polar. Whether or not the molecule is polar depends on its shape. Consider a molecule of the type AB_n. If the B atoms are on one side of the A atom (because the other side is occupied by lone pairs), the molecule will usually be polar. However, if the B atoms are symmetrically distributed about the A atom (because there are no lone pairs), the molecule will be nonpolar.

Consider the H_2O molecule. The electronegativity of the O atom is greater than the electronegativity of the H atom, so the bonds in H_2O are polar. Moreover, the molecule has a bent shape, with H atoms occupying two points of a tetrahedron, and lone pairs occupying the other two points. Thus, the center of positive charge is at one end of the molecule (the end occupied by the two H atoms), and the center of negative charge is at the other end (the O end). Because there is a difference between the centers of partial positive charge and partial negative charge, the molecule is polar.

Now, let's look at the CO_2 molecule. The O atom is more electronegative than the C atom, so the bonds in CO_2 are polar. Earlier, we saw that the CO_2 molecule is linear; it has two bonds arranged symmetrically and no lone pairs. In this case, since the molecule has a linear shape, the center of partial positive charge is at the C atom at the center of the molecule. The partial negative charges on the two O atoms are located opposite one another at the outside ends of the molecule. Because the two O atoms are located equal distances from the C atom, the center of partial negative charge is also located at the C atom. There is no separation of the centers of partial positive charge and partial negative charge, so the molecule is nonpolar.

We can generalize this analysis for molecules of the type AB_n. Generally the molecule will be polar if it meets the following criteria:

- The atoms A and B have different electronegativities.
- The central atom A has lone pairs of electrons.

If either of these criteria is not met, the molecule will be nonpolar. In Chapter 12, you will see that consideration of the polarity of molecules helps you understand the physical state and other properties of substances.

Chemical Perspective

Computer-Generated Molecular Modeling

Almost daily, scientists announce remarkable advances in the production of new pharmaceutical chemicals or in the manipulation of genetic material in attempts to find cures for genetic disorders. The pharmaceutical industry makes heavy use of molecular modeling based on models similar to those shown throughout this chapter. It has become commonplace for chemists to use computers to examine the structures of successful pharmaceutical drugs and to use this information to construct molecules with enhanced pharmaceutical properties. Rather than systematically preparing and examining the hundreds of molecules that might follow the desired chemistry, they use computers to decide which molecule is most likely to have the desired properties. Chemists can then concentrate on synthesizing the most promising molecules.

But chemists have not always had such elegant computer-based tools. In the past they used various models of wood, metal, plastic, and even paper to help them visualize the three-dimensional shapes of molecules. These physical models usually consisted of balls or similar objects to represent atoms and sticks or rods to represent bonds. These models worked because atoms of a given element tend to bond at definite angles and to have definite bond lengths. By constructing a set of atoms with specific bond angles (such as tetrahedral or trigonal planar) and specific bond lengths, one can "build" molecules having various shapes.

One of the more famous molecular models was that of James Watson and Francis Crick, who in 1953 deduced the structure of deoxyribonucleic acid (DNA) by playing with molecular models constructed of paper and wire, trying to find the arrangement of atoms that was consistent with the results of x-ray experiments (Figure 10.15). In these experiments, x rays strike a crystal of a molecular substance, generating a pattern of reflections. By trying to find the arrangement of atoms that will exactly reproduce this pattern, chemists can deduce the molecular structure. In the days before high-speed computers, it was tedious work. Once Watson and Crick were certain of the structure of DNA, they could explain how DNA molecules could carry the hereditary information of living organisms and could duplicate that information during cell division.

The physical construction of models of biological molecules can quickly exhaust one's patience. Most biological molecules are proteins, and these are very large molecules, each containing several thousand atoms. Even when you know the atoms that make up the molecule and you know how the *(continued)*

A

B

FIGURE 10.15
A model of a fragment of a DNA molecule. *(A)* James Watson and Francis Crick discovered how DNA contains its genetic information using paper and wire molecular models. *(B)* A computer-generated molecular model provides a more accurate view of the structure of DNA.

atoms are bonded to one another, it is still a considerable feat to try to deduce the shape of a protein molecule by constructing a molecular model of it.

Today scientists are solving this problem by using high-speed computers to generate screen images of the molecules. These computer-generated images can rotate on-screen to show the shape of the molecule in three dimensions. Photographs of some of these computer-generated models appear at various places in this book. For small molecules, space-filling models are often convenient (Figure 10.16A), but in trying to visualize very large molecules, bonds and groups of atoms are frequently reduced to lines, coils, and similar graphic symbols (Figure 10.16B). More and more, computer-generated molecular modeling is helping biochemists solve questions of how biological molecules function in living organisms.

A

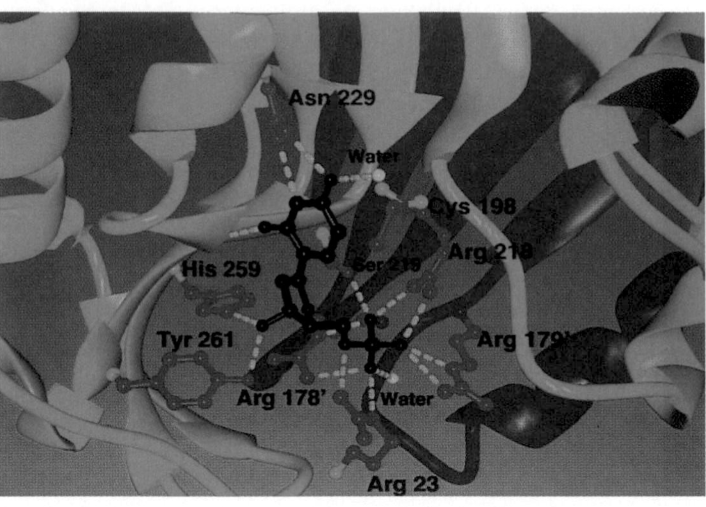

B

FIGURE 10.16

Different forms of computer-generated molecular models. *(A)* A space-filling model of methionine, an amino acid. Amino acids are the building blocks that biological organisms use to construct protein molecules. *(B)* In large molecules, groups of atoms are represented by rings, lines, coils, or dots. This image shows the active site in the enzyme thymidylate synthase as a ball-and-stick model, while the surrounding protein is shown as ribbons. This reaction produces thymidylate, used in DNA synthesis.

CHAPTER REVIEW

Key Words

ionic bond *(p. 291)*
octet rule *(p. 293)*
electron-dot symbol (Lewis symbol) *(p. 293)*
covalent bond *(p. 296)*
electron-dot formula (Lewis formula) *(p. 297)*
single bond *(p. 298)*
double bond *(p. 298)*
triple bond *(p. 298)*
bonding pair *(p. 298)*
lone pair (nonbonding pair) *(p. 298)*
electronegativity *(p. 299)*
polar covalent bond (polar bond) *(p. 300)*
polar molecule *(p. 301)*
bond length *(p. 309)*
valence-shell electron-pair repulsion (VSEPR) model *(p. 310)*

Summary

Chemical bonds are of two principal types: ionic and covalent. In the formation of an *ionic bond* such as in sodium chloride, the sodium atoms and chlorine atoms first exchange electrons to yield ions; the sodium atom loses an electron and the chlorine atom gains this electron. The resulting ions attract one another to give the ionic compound. In general, atoms tend to lose or gain electrons to give a noble-gas structure, a configuration having eight electrons in the valence shell. This tendency toward an outer shell of eight electrons is referred to as the *octet rule.* The transfer of electrons between atoms is conveniently denoted by *electron-dot symbols.*

A *covalent bond* is formed when two atoms share a pair of electrons. Molecules with covalent bonding are conveniently represented by *electron-dot formulas.* The electron pairs in a molecule can be *bonding pairs* or *lone pairs.* Bonding pairs may be unequally shared, because the atoms of the bond may have different *electronegativities.* This unequal sharing of electrons results in a *polar covalent bond.* Before writing an electron-dot formula for a molecule, you need to know the basic arrangement of atoms. Many small molecules consist of a less electronegative atom surrounded by atoms of higher electronegativities. Once you have the basic arrangement of atoms, you draw the electron-dot formula by connecting the atoms by electron pairs, distributing electrons to satisfy the octet rule for the surrounding atoms, and finally distributing remaining pairs to the central atom. If the central atom has fewer than eight electrons about it, you move one or more lone pairs on surrounding atoms into the bonding region to form a multiple bond. A covalent bond may be a *single bond,* a *double bond,* or a *triple bond,* depending on whether the two atoms share two, four, or six electrons.

The structures of molecules are characterized by *bond lengths* and bond angles, which determine the *molecular shapes.* Molecular shapes can be predicted by using the *valence-shell electron-pair repulsion (VSEPR) model.* The total number of electron groups (bonding and nonbonding) determines the arrangement of electrons about the central atom. The number of bonding groups (and bonded atoms) determines the molecular shape. *Polar molecules* result when the centers of partial positive and partial negative charges are separated. This condition occurs when there is both a difference in the electronegativities of the bonded atoms and the presence of lone pairs of electrons on the central atom.

Problem-Solving Skills

1. **Representing Ionic Bond Formation with Electron-Dot Symbols:** Given a metal atom and a nonmetal atom, use electron-dot symbols to represent the formation of ions prior to the formation of an ionic compound (Example 10.1).
2. **Writing Electron-Dot Formulas:** Given a molecular formula or the formula of a polyatomic ion, write the electron-dot formula (Examples 10.2, 10.3, and 10.4).
3. **Predicting Molecular Shape with the VSEPR Model:** Given a molecular formula, predict the molecular shape (Example 10.5).

Questions to Test Your Reading

10.1 What are the two principal types of chemical bonds?

10.2 Define the term *ionic bond.*

10.3 What types of forces are involved in ionic bonding?

10.4 Describe the process that occurs when ions form from a lithium atom and a fluorine atom.

10.5 What energy term is associated with the formation of cations from atoms? of anions from atoms?

10.6 Which types of elements form ionic compounds?

10.7 Describe a method for using the periodic table to predict the charges on cations and anions.

10.8 Consider each of the following pairs of atoms. Label each pair as ionic or covalent, depending on the type of bond you expect to form between them.
(a) K and O (b) C and O (c) Ca and Cl
(d) N and F

10.9 What is the formula of the ion formed from each of the following atoms?
(a) Cs (b) Ba (c) S (d) I

10.10 State the octet rule.

10.11 Draw electron-dot symbols for the following.
(a) Rb (b) Ba (c) Se (d) I

10.12 Define the term *covalent bond.* How does a covalent bond differ from an ionic bond?

10.13 What is the difference between an electron-dot symbol and an electron-dot formula?

10.14 Use electron-dot symbols and an electron-dot formula to represent the formation of a covalent bond between two chlorine atoms.

10.15 Illustrate each of the following by an electron-dot formula.
(a) a single bond (b) a double bond
(c) a lone pair

10.16 What is meant by a multiple bond?

10.17 Define the term *electronegativity.*

10.18 Describe the trends in electronegativity across the periodic table.

10.19 Looking only at the periodic table on the inside front cover of this book, decide which one of the following pairs of elements is the more electronegative.
(a) K, As (b) S, Po (c) Ge, Br (d) I, F

10.20 Use the delta notation (δ) to indicate the positive and negative ends of each of the following bonds.
(a) C—O (b) H—S (c) Cl—C (d) F—N

10.21 For each of the following molecules, indicate the central atom.
(a) H_2Te (b) CCl_4 (c) OF_2 (d) HOCl

10.22 Calculate the number of valence electrons available to draw electron-dot formulas for each of the following molecules.
(a) BF_4^- (b) SCl_2 (c) H_2SO_4 (d) PCl_3

10.23 Define the term *bond length.*

10.24 What are the quantities required to describe molecular structure?

10.25 Describe the valence-shell electron-pair repulsion model for predicting molecular shapes.

10.26 What two conditions are necessary for a molecule to be polar?

Practice Problems

Ionic Bonds (Section 10.2)

10.27 For each of the following pairs of atoms, use electron-dot symbols to write the equation for the formation of ions prior to the formation of an ionic compound.
(a) Ca and S (b) Na and O (c) K and Cl
(d) Al and O

10.28 For each of the following pairs of atoms, use electron-dot symbols to write the equation for the formation of ions prior to the formation of an ionic compound.
(a) Li and N (b) Ca and Cl (c) Li and Br
(d) Li and S

Identifying Types of Electron Pairs (Section 10.3)

10.29 In each of the following electron-dot formulas, identify bonding pairs and lone pairs. For each of the bonding pairs, identify the bond as a single, double, or triple bond.

(a) :N≡N: (b) H—Ö—C(=Ö:)—H (with :O: double-bonded above C; O of H—O—C bearing two lone pairs)

10.30 In each of the following electron-dot formulas, identify bonding pairs and lone pairs. For each of the bonding pairs, identify the bond as a single, double, or triple bond.

(a) H—P:(—H)(—H) (H above and below P) (b) Ö=S̈—Ö: (with lone pairs as dots)

Electronegativity and Polar Covalent Bonds (Section 10.4)

10.31 For each of the following bonds, calculate the absolute value (value without sign) of the difference in electronegativities of the atoms. Arrange all four bonds in order, left to right, from pure covalent to most ionic.
(a) F—Na (b) C—Cl (c) F—C
(d) C—C

10.32 For each of the following bonds, calculate the absolute value (value without sign) of the difference in electronegativities of the atoms. Arrange all four bonds in order, left to right, from pure covalent to most ionic.
(a) Li—F (b) N—N (c) F—N
(d) Cl—N

Writing Electron-Dot Formulas (Section 10.5)

10.33 Write electron-dot formulas for each of the following.
(a) GeF_4 (b) AsF_3 (c) AsH_3 (d) HOCl

10.34 Write electron-dot formulas for each of the following.
(a) H_2Te (b) PBr_3 (c) CHI_3 (d) $HClO_4$

10.35 For each of the following, draw electron-dot formulas.
(a) $COCl_2$ (b) CO (c) BrCN

10.36 For each of the following, draw electron-dot formulas.
(a) HNO_2 (b) CSe_2 (c) NSF

10.37 Write electron-dot formulas for each of the following ions.
(a) ClO^- (b) NO_3^-

10.38 Write electron-dot formulas for each of the following ions.
(a) $GeCl_3^-$ (b) CO_3^{2-}

Predicting Molecular Shapes (Section 10.7)

10.39 Predict the shape of the following molecules or ions.
(a) SiF_4 (b) AsF_3 (c) CO_3^{2-}

10.40 Predict the shape of the following molecules or ions.
(a) H_2Te (b) BrO_2^- (c) $GeCl_4$

Additional Problems

10.41 Compare lithium chloride (LiCl) and chlorine (Cl_2). Which is ionic and which is covalent? Describe the formation of a bond in each of these substances using electron-dot symbols and formulas.

10.42 Compare cesium bromide (CsBr) and bromine (Br_2). Which is ionic and which is covalent? Describe the formation of a bond in each of these substances using electron-dot symbols and formulas.

10.43 Lead forms ions by losing only part of its valence electrons. The normal lead ion is Pb^{2+}. What is the electron configuration of this ion?

10.44 Tin forms Sn^{2+} ions by losing only part of its valence electrons. What is the electron configuration of the tin(II) ion, Sn^{2+}?

10.45 Which bond in each of the following pairs is more polar?
(a) Na—O, O—O (b) O—O, H—O
(c) F—O, F—H

10.46 Which bond in each of the following pairs is more polar?
(a) H—F, H—Br (b) Cl—Cl, Li—Cl
(c) N—Cl, N—Br

10.47 Write electron-dot formulas for the following ionic compounds.
(a) NaBr (b) CaF_2 (c) Na_2S

10.48 Write electron-dot formulas for the following ionic compounds.
(a) $MgBr_2$ (b) CaO (c) Na_2O

10.49 Write electron-dot formulas for each of the following molecules or ions.
(a) CH_2ClF (b) SO_2 (c) HNO_3 (d) SO_4^{2-}

10.50 Write electron-dot formulas for each of the following molecules.
(a) $SiHF_3$ (b) H_2SO_4 (c) SO_3 (d) PBr_3

10.51 What is the arrangement of electron groups around the atom underlined in each of the following molecules or ions?
(a) $\underline{S}O_2$ (b) $\underline{Se}Cl_2$ (c) $\underline{Cl}O_4^-$ (d) $\underline{P}H_3$

10.52 What is the arrangement of electron groups around the atom underlined in each of the following molecules?
(a) $\underline{As}Br_3$ (b) $H\underline{O}Br$ (c) $Cl\underline{C}N$ (d) $Cl_2\underline{O}$

10.53 For each of the molecules listed in Problem 10.49, predict the molecular shape. Draw a rough sketch showing this shape.

10.54 For each of the molecules listed in Problem 10.50, predict the molecular shape. Draw a rough sketch showing this shape.

10.55 Decide which of the following bonds is least polar on the basis of the electronegativities of the atoms: H—Cl, P—Cl, N—Cl.

10.56 Decide which of the following bonds is most polar on the basis of the electronegativities of the atoms: H—Cl, P—Cl, N—Cl.

10.57 For each of the following pairs of elements, state whether the compound formed is likely to be ionic or covalent.
(a) Ba, O (b) C, F (c) In, Cl (d) P, Br

10.58 For each of the following pairs of elements, state whether the compound formed is likely to be ionic or covalent.
(a) Na, S (b) N, Cl (c) Si, Br (d) Ca, Se

10.59 Select the molecule from each pair that is polar.
(a) HF, F_2 (b) CF_4, NF_3 (c) SiH_4, H_2O

10.60 Select the molecule from each pair that is polar.
(a) H_2, HCl (b) CS_2, SO_2 (c) CH_4, PH_3

Practice in Problem Analysis

For each problem, describe the thinking you would use (the problem analysis) before doing the actual solution, but do not solve the problem.

1. Hydrogen sulfide, H_2S, is a poisonous gas with a foul odor. What is the electron-dot formula of H_2S?

2. Phosgene, $COCl_2$, is used in the manufacture of polyurethane plastics. What is the electron-dot formula for $COCl_2$?

PRACTICE EXAM

1. How many electrons are there in the valence shell of the sodium atom?
(a) 1 (b) 2 (c) 4 (d) 7
(e) None of the above

2. How many electrons are there in the valence shell of the chlorine atom?
(a) 1 (b) 2 (c) 4 (d) 7
(e) None of the above

3. How many electrons are there in the valence shell of the sodium ion?
(a) 1 (b) 2 (c) 4 (d) 7
(e) None of the above

4. How many electrons are there in the valence shell of the chloride ion?
(a) 1 (b) 2 (c) 4 (d) 7
(e) None of the above

5. An electron pair that is shared between two atoms is called
(a) a rebonding pair (b) a bonding pair
(c) a lone pair (d) a double bond
(e) None of the above

6. An electron pair that is entirely on one atom in a molecule is called
(a) a rebonding pair (b) a bonding pair
(c) a lone pair (d) a double bond
(e) None of the above

7. Which one of the following elements is more electronegative than sulfur, S?
(a) Te (b) Al (c) Si (d) P
(e) None of the above

8. Which one of the following elements is more electronegative than oxygen, O?
(a) C (b) N (c) Cl (d) F
(e) None of the above

9. Which of the following is the correct electron-dot formula for phosphine, PH_3?

```
         H               H                H
        ..              ..               ..
(a) H : P :     (b) H : P        (c) H : : P :
        ..              ..               ..
         H               H                H

          H
          ..
(d) : H : P      (e) None of the above
          ..
          H
```

10. Which of the following is the correct electron-dot formula for phosgene, $COCl_2$?

```
            :O:                         ..
            ..                         :O:
      ..    ..    ..             ..    ..    ..
(a) : Cl  : C  : Cl :      (b) : Cl  : C  : Cl :
      ..    ..    ..             ..          ..

            :O:                          ..
            ..                          :O:
      ..    ..    ..              ..    ..    ..
(c) : Cl  : C  : Cl :      (d) : Cl  : C  : Cl :
      ..          ..              ..    ..    ..
```

(e) None of the above

11. The molecular shape of the phosphine molecule, PH_3, is
(a) linear (b) angular (c) trigonal planar
(d) trigonal pyramidal (e) None of the above

12. The molecular shape of the carbon disulfide molecule, CS_2, is
(a) linear (b) angular (c) trigonal planar
(d) trigonal pyramidal (e) None of the above

13. The molecular shape of the silane molecule, SiH_4, is
(a) linear (b) angular (c) trigonal planar
(d) trigonal pyramidal (e) None of the above

14. The molecular shape of the hydrogen sulfide molecule, H_2S, is
(a) linear (b) angular (c) trigonal planar
(d) trigonal pyramidal (e) None of the above

15. Which of the following is a polar molecule?
(a) CO_2 (b) NH_3 (c) CH_4 (d) SiF_4
(e) None of the above

Chapter

11 The Gaseous State

Neon and argon gases are used in advertising signs.

Chapter Outline

● Gases, liquids, and solids were discussed briefly in Chapter 3. Solids and liquids also are the subjects of Chapter 12.

You are surrounded by the three states of matter: gases, liquids, and solids. ● It is hard to overlook the solids and liquids. If you are reading this book at a desk while sipping a drink, you are well aware of the solid chair you are sitting on and the liquid you are drinking. But you are also immersed in a sea of invisible gases. Because most gases are colorless and because they cannot be picked up and handled, their presence can go unnoticed.

Nevertheless, gases are enormously important. We must breathe oxygen to survive. Gases make up the air of earth's atmosphere, and without them this planet would be a very dead place, like the moon (Figure 11.1).

The behavior of gases is actually simpler and easier to predict than the behavior of liquids and solids because the particles (atoms or molecules) of a gas are far apart and do not interact with each other very much. In contrast, the particles of liquids and solids are so close together that complicated interactions occur. Chemists quantified the behavior of gases beginning some 300 years ago—in the seventeenth and eighteenth centuries—but an understanding of the behavior of liquids and solids had to wait for the

FIGURE 11.1
Gases make up the atmosphere. Earth's atmosphere (shown here with the moon behind it) consists of gases that provide the oxygen and pressure necessary to support life as we know it. In contrast, the moon (shown more clearly in the inset) has no atmosphere.

TABLE 11.1
The Properties of Some Gases

Name	Formula	Color	Odor
Ammonia	NH_3	Colorless	Penetrating
Carbon dioxide	CO_2	Colorless	Odorless
Carbon monoxide	CO	Colorless	Odorless
Chlorine	Cl_2	Pale yellow	Irritating
Helium	He	Colorless	Odorless
Methane	CH_4	Colorless	Odorless
Neon	Ne	Colorless	Odorless
Nitrogen	N_2	Colorless	Odorless
Nitrogen dioxide	NO_2	Red brown*	Irritating
Oxygen	O_2	Colorless	Odorless

*Brownish-yellow when dilute.

theories and technology of the twentieth century. In this chapter, we present some simple laws governing the behavior of gases.

GASES AND THEIR BEHAVIOR

Some of the elements and many compounds are gases under ordinary, everyday conditions. Table 11.1 lists some common gases along with their formulas, colors, and odors. As you can see, most gases are colorless, but chlorine and nitrogen dioxide are exceptions. The brownish-yellow color of smoggy air (Figure 11.2) is due principally to nitrogen dioxide.

Unpolluted air is a mixture of gases; at sea level it consists of 78% elemental nitrogen (N_2) and 21% elemental oxygen (O_2). The remaining 1% is mostly argon, but trace amounts of other gases such as CO_2 and H_2O (in vapor form) are also present. The composition of air changes with altitude, however.

We begin our study of gases by examining their properties and the meaning of gas pressure. We then describe the *kinetic molecular theory,* which explains how gas particles behave and how they interact with each other and with the walls of containers. This theory successfully accounts for the properties of gases and allows us to predict their behavior.

FIGURE 11.2
Smog over Los Angeles. Although this smog contains a mixture of conponents, nitrogen dioxide (NO_2) is the principal cause of its color.

11.1 The Nature of Gases

OBJECTIVE

- Describe some of the properties shared by all gases.

The following properties are common to all gases:

1. *Gases fill a container completely and uniformly.* When you inflate a tire, all parts of the tire are uniformly filled. In contrast, a liquid fills only the bottom of a container and a solid has a fixed shape.
2. *Gases are compressible.* Although the volume enclosed by a hard rubber tire is essentially fixed, we can force more and more air into the tire using a pump. Thus more and more air is compressed into the same volume. Liquids and solids cannot be compressed, at least not to any great degree.

3. *Gases have low densities.* The density of a typical gas is about 1 g/L, whereas the density of a typical solid may be about 2000 g/L (or 2 g/mL) or even greater. Thus a solid may be about 2000 times more dense than a gas. A typical liquid may be 800 to 1000 times more dense than a gas.
4. *Gases exert a uniform pressure on all inner surfaces of a container because they completely fill their containers.* Solids and liquids exert pressure only on those areas of a container that they touch.

Some of these properties occur because gas particles (atoms or molecules, depending on the substance) are far apart compared with the size of the particles themselves. For example, gases are compressible because the widely spaced particles can be squeezed together. When we compress a gas, its pressure increases. In the next section, we define gas pressure and describe how gas pressure is measured.

11.2 Gas Pressure

OBJECTIVES

- Define *pressure*.
- Describe atmospheric pressure and its measurement with a barometer.
- Give units for pressure.
- Describe the measurement of gas pressure with a manometer.

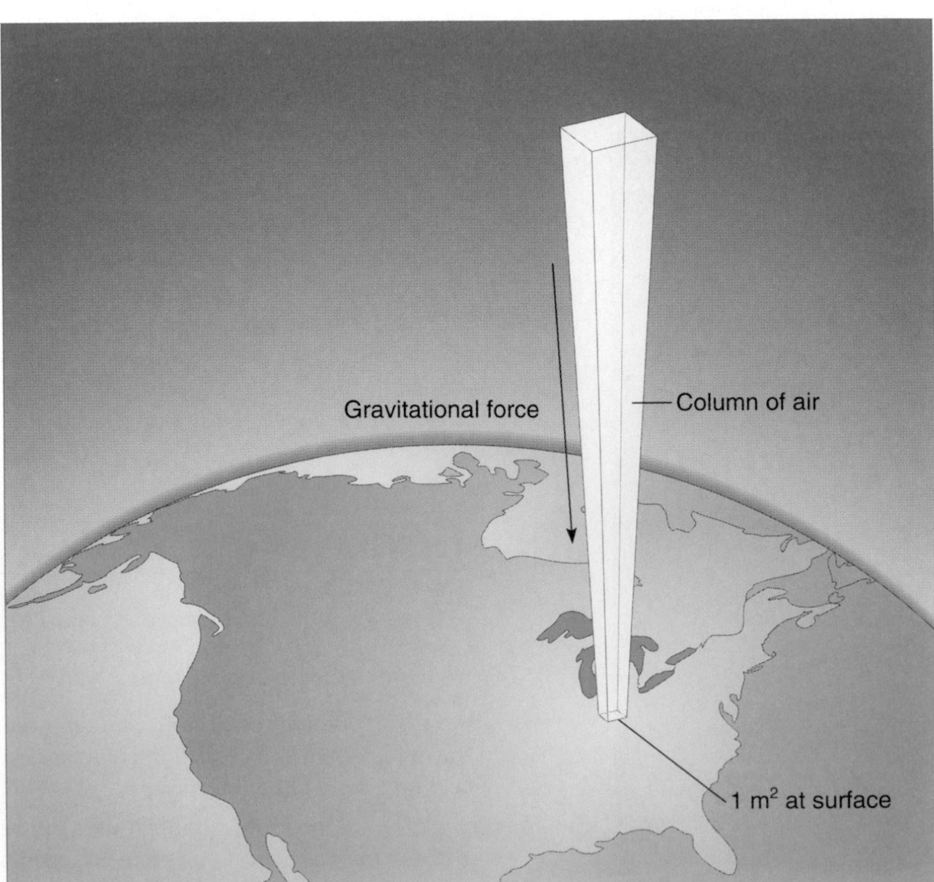

FIGURE 11.3
Atmospheric pressure. Gravitational attraction between a column of air and the earth causes the atmosphere to exert a pressure on a particular area (in this case, 1 m^2).

Pressure is a familiar physical property. For example, when you press a finger against your nose, you feel the pressure from your finger. The concept of atmospheric pressure is not very different. The atmosphere around our earth contains an estimated 5×10^{18} kg of gaseous substances. This mass exerts a force on every part of the earth's surface because of the earth's gravitational attraction. The atmosphere pushes on the earth like your finger pushes on your nose. If we look at the force from the atmosphere on some particular area of the earth—say, 1 square meter—we are considering pressure. **Pressure** is *the force exerted on a unit area.* A column of the atmosphere is pressing downward on every square meter of the earth's surface (Figure 11.3). The pressure exerted by the atmosphere on the earth is called *atmospheric pressure.* Atmospheric pressure is greatest near the earth's surface and decreases with increasing altitude. As a result, the atmospheric pressure in Denver, Colorado (in the Rocky Mountains), is considerably lower than the atmospheric pressure at the seaside in Asbury Park, New Jersey. ●

● Atmospheric pressure also depends on weather conditions. The "highs" and "lows" that a TV meteorologist talks about refer to high- and low-pressure air masses. A high usually signals fair weather, whereas a low can indicate an approaching storm.

Barometers

A **barometer** is *a device used to measure atmospheric pressure.* The mercury barometer shown in Figure 11.4 is constructed by filling a glass tube (about 1 meter long but having any diameter) with mercury, a metal that is a liquid under ordinary conditions. The tube is then inverted in a dish of mercury. ● As you can see, some of the mercury drains from the tube into the dish but a considerable portion of it remains in the tube. A vacuum (empty space) forms above the surface of the mercury in the tube because some of the mercury has left the tube and air has not been able to replace it. At sea level, the

● Any liquid can be used in a barometer, but the liquid's density affects the height of the liquid in the glass tube. If water is used, the height of water will be 13.6 times greater than the height of the mercury column because mercury's density is 13.6 times greater than that of water.

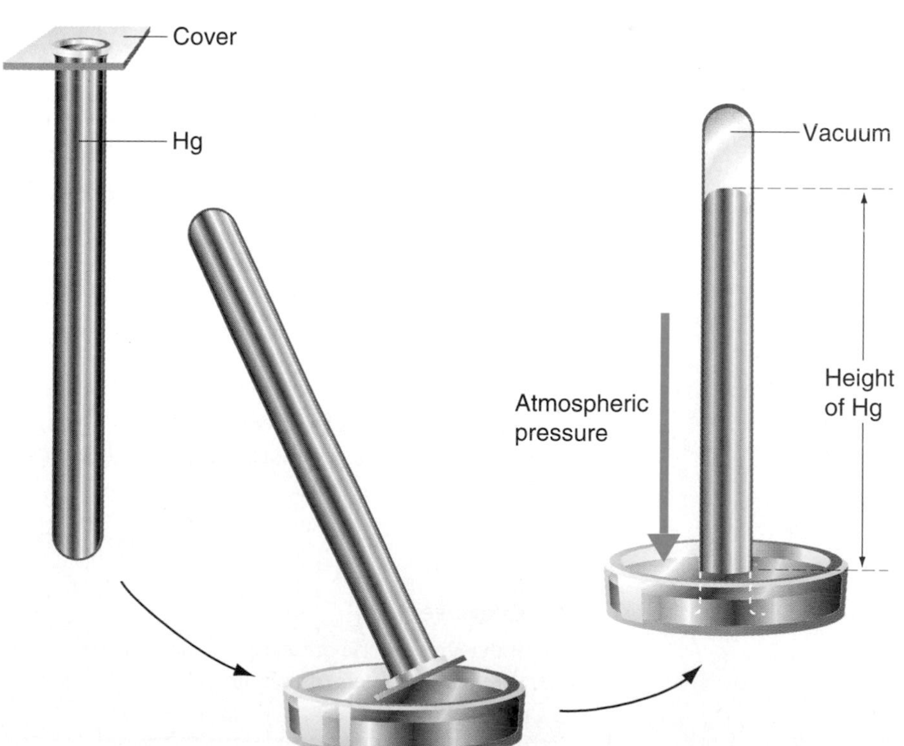

FIGURE 11.4
A mercury barometer. A column is filled with mercury (Hg), covered, inverted in a dish of mercury, and uncovered. Some of the mercury drains from the column, leaving a vacuum in its place. The height of mercury in the column reflects the atmospheric pressure. The height increases as the atmospheric pressure increases and decreases as the atmospheric pressure decreases.

average height of the mercury in the tube is 760 mm, but the height varies as the atmospheric pressure changes. This pressure is exerted downward on the surface of the liquid, as you can see in the figure. It is also transmitted through the liquid and exerts a pressure upward at the base of the column of mercury. The upward pressure supports the column. Thus, the height of the mercury within the column is a direct measure of atmospheric pressure because the mercury rises when atmospheric pressure increases or falls when it decreases.

Units of Pressure

The most frequently used units of pressure are based on the height of the mercury in such a column. The unit **millimeters of mercury (mmHg),** also called the **torr** (after Evangelista Torricelli, who invented the barometer in 1643), is *a unit of pressure equal to that exerted by a column of mercury that is exactly 1 millimeter high.* One standard **atmosphere (atm)** is *another unit of pressure equal to exactly 760 mmHg.* Exactly 1 atm is also equal to 1.01325×10^5 pascals. The **pascal (Pa)** is *the SI unit of pressure.* The relationship between these units is given by

$$1 \text{ atm} = 760 \text{ mmHg} = 760 \text{ torr} = 1.01325 \times 10^5 \text{ Pa}$$

We use millimeters of mercury or atmospheres as the units of gas pressure in this book.

■ EXAMPLE 11.1 Converting One Pressure Unit to Another

The height of mercury in a barometer in Indiana reads 746 mmHg. What is the atmospheric pressure in atmospheres?

Problem Analysis

The atmospheric pressure in millimeters of mercury is equal to the height of the column of mercury in the barometer. Use the conversion factor

$$\frac{1 \text{ atm}}{760 \text{ mmHg}}$$

to convert millimeters of mercury to atmospheres.

Solution

Since the height of the mercury column is 746 mm, the atmospheric pressure is 746 mmHg. When this pressure is expressed in atmospheres, it becomes

$$746 \cancel{\text{mmHg}} \times \frac{1 \text{ atm}}{760 \cancel{\text{mmHg}}} = 0.982 \text{ atm}$$

Exercise 11.1

If the atmospheric pressure in a mountainous region is 0.712 atm, what is the pressure in millimeters of mercury?

(Try Problems 11.21 and 11.22.)

Manometers

All gases exert a pressure on any surface that they contact. For example, the gas inside an inflated child's balloon exerts a pressure and pushes the walls of the balloon outward. Assuming the balloon is perfectly elastic, the outward movement stops when the gas pressure within the balloon equals the atmospheric pressure. Thus by measuring the atmospheric pressure with a barometer, you have also determined the gas pressure within the balloon.

A gas on the inside of a closed, nonelastic glass or metal container also exerts a pressure, but we need another way to measure it. In the laboratory, we use a **manometer,** *a device that measures a gas's pressure in a container.* Figure 11.5 shows a manometer with an open-ended U-tube containing mercury. The difference in the heights of mercury in the two arms of the U-tube allows us to compare the gas pressure in the container to the atmospheric pressure. If the pressure of the gas is identical to atmospheric pressure, the mercury levels in the two arms of the U-tube are equal because the gas and the atmosphere are pushing on the mercury with the same amount of force, as in Figure 11.5A. If the gas's pressure is greater than that of the atmosphere, the mercury is forced higher in the arm exposed to the atmosphere, as in Figure 11.5B. The difference between the heights of mercury in the two arms equals the pressure difference (in mmHg) between the gas and the atmosphere. For example, suppose the atmospheric pressure is 743 mmHg (measured with a barometer) and the height of mercury in the arm open to the atmosphere is 182 mm greater than in the end open to the gas (measured with a ruler). The gas pressure (P) must be 182 mmHg *greater* than the pressure of the atmosphere, so

$$P = 743 \text{ mmHg} + 182 \text{ mmHg} = 925 \text{ mmHg}$$

Conversely, if the gas's pressure is less than atmospheric pressure, the mercury level is higher in the arm open to the gas, as in Figure 11.5C. Again, the difference between the heights of mercury in the two arms equals the pressure difference (in mmHg) between

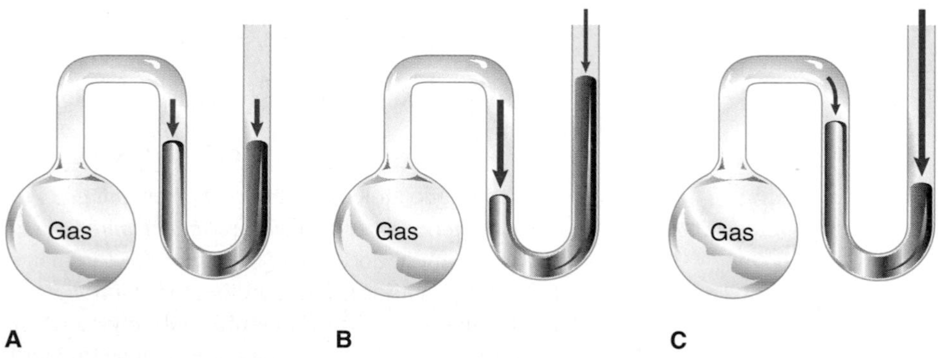

FIGURE 11.5

A mercury manometer. *(A)* The levels of mercury in the two arms of the U-tube are equal because the pressure of the gas is equal to the atmospheric pressure. *(B)* The level of the mercury in the arm open to the atmosphere is higher than that in the other arm because the pressure of the gas is greater than the atmospheric pressure. *(C)* The level of the mercury in the arm open to the gas is lower than that in the other arm because the atmospheric pressure is greater than the pressure of the gas.

Chemical Perspective

Blood Pressure

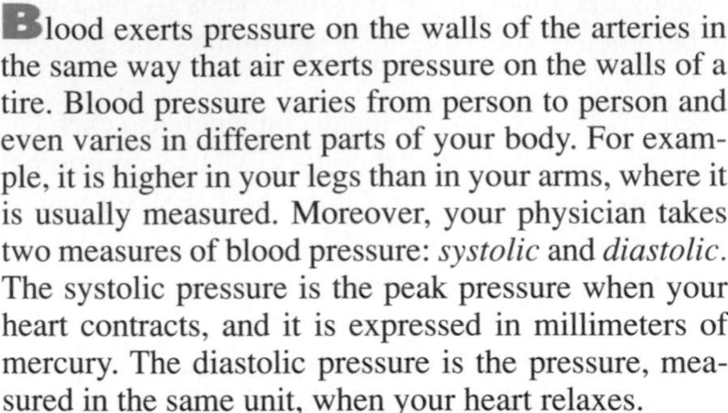

Blood exerts pressure on the walls of the arteries in the same way that air exerts pressure on the walls of a tire. Blood pressure varies from person to person and even varies in different parts of your body. For example, it is higher in your legs than in your arms, where it is usually measured. Moreover, your physician takes two measures of blood pressure: *systolic* and *diastolic*. The systolic pressure is the peak pressure when your heart contracts, and it is expressed in millimeters of mercury. The diastolic pressure is the pressure, measured in the same unit, when your heart relaxes.

If your physician tells you that your blood pressure is 120 over 80, he or she means that your systolic pres-

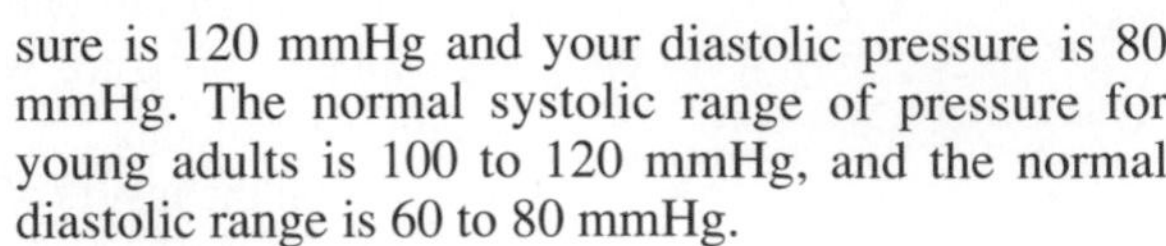

sure is 120 mmHg and your diastolic pressure is 80 mmHg. The normal systolic range of pressure for young adults is 100 to 120 mmHg, and the normal diastolic range is 60 to 80 mmHg.

Even if only one of the two pressures is high, but especially the diastolic pressure, you may have high blood pressure. If so, your risk of having a heart attack or a stroke is increased. One treatment for this malady is a group of drugs called *diuretics*. These drugs help to reduce your blood pressure by expelling fluid from your body through frequent urination. Because the volume of blood is reduced, your blood pressure will usually decrease.

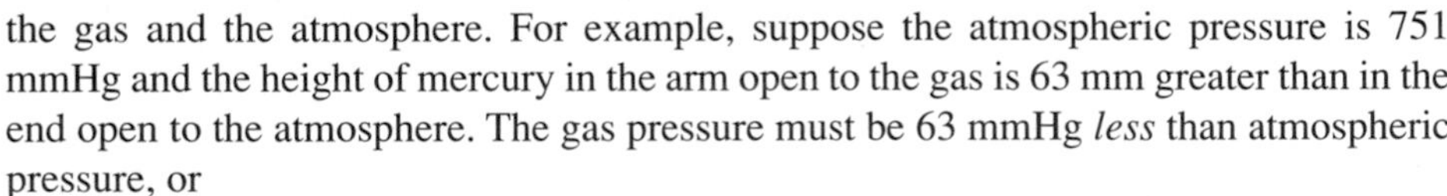

the gas and the atmosphere. For example, suppose the atmospheric pressure is 751 mmHg and the height of mercury in the arm open to the gas is 63 mm greater than in the end open to the atmosphere. The gas pressure must be 63 mmHg *less* than atmospheric pressure, or

$$P = 751\text{ mmHg} - 63\text{ mmHg} = 688\text{ mmHg}$$

Animations: Kinetic molecular theory and heat transfer; Visualizing molecular motion (single molecule); Visualizing molecular motion (many molecules).

11.3 The Kinetic Molecular Theory of Gases

OBJECTIVE

- Name the theory that explains the behavior of real gases under moderate conditions, and list its four postulates.

The **kinetic molecular theory of gases,** developed during the eighteenth and nineteenth centuries by several scientists, *explains the behavior of gases*. As we will see in Chapter 12, part of this theory can also be used to explain the behavior of liquids and solids.

The kinetic molecular theory applies to *a gas that obeys a simple mathematical relationship between its pressure, volume, temperature, and amount.* ● This gas is called an **ideal gas** and it does not really exist. We can make mathematical predictions about the behavior of an ideal gas, and these predictions will also apply to a real gas as long as moderate conditions of pressure and temperature are maintained. Moderate conditions are pressures that are less than or not much greater than 1 atm and temperatures that are considerably greater than those at which the gases liquefy. We cannot be more specific about these conditions; they vary from gas to gas because the properties of one substance will always differ in many respects from those of another substance. If a real gas is subjected to more extreme conditions, namely, high pressure or low temperature, then deviations from ideal behavior are noticeable. Throughout this chapter, we assume ideal behavior due to moderate conditions. Let's consider the postulates in this theory.

● This relationship, called the *ideal gas law*, is introduced in Section 11.8.

Postulates from the Kinetic Molecular Theory

The kinetic molecular theory can be summarized by four postulates.

1. A gas is composed of particles (atoms or molecules) whose dimensions are very small compared with the very large average distances between the particles. ●
2. These particles are in constant, random motion, moving in straight lines unless they collide with another particle or the walls of the container.
3. Attractions (or repulsions) between gas particles are negligible.
4. The average kinetic energy (the energy of motion) of these particles is directly proportional to the Kelvin temperature. ●

● The noble gases (He, Ar, etc.) are composed of atoms. Gases consisting of molecules include oxygen (O_2), nitrogen (N_2), carbon dioxide (CO_2), and water vapor (steam).

● The Kelvin (absolute) temperature scale was introduced in Section 2.6. The relationship between Kelvin and Celsius temperatures is $K = °C + 273$.

Thus, a gas is a collection of particles moving around randomly, bouncing off the walls of the container and occasionally hitting each other. Although the third postulate states that attractions between gas particles are negligible, attractions must exist or gases would never condense into the liquid state. Attractive forces are responsible for holding the liquid state together. Clearly, the assumption that these forces are absent in a gas is a simplification. However, attractive forces between gas particles are negligible under moderate conditions of pressure and temperature.

The fourth postulate says that the average kinetic energy of gas particles is directly proportional to the Kelvin temperature. The term *directly proportional* means that the average kinetic energy will increase in a specific way when the Kelvin temperature is increased, and decrease in the same way when the Kelvin temperature is decreased. For example, the average kinetic energy will double if the Kelvin temperature is doubled, or it will be halved if the Kelvin temperature is halved. This postulate means that the motions of gas particles become more violent and chaotic as the temperature increases and more restrained as the temperature decreases. ●

● The kinetic energy of any object (including a gas particle) is related to the object's mass and speed by the equation

$$\text{Kinetic energy} = \frac{1}{2} \times \text{mass} \times (\text{speed})^2$$

Because of this relationship and because the average kinetic energy of gas particles is directly proportional to the Kelvin temperature, changes in temperature affect the average speed of the particles.

Let's see how the kinetic molecular theory explains the behavior noted in Section 11.1. Gases fill a container completely and uniformly because the particles move constantly and rapidly and have virtually no attraction to one another. Gases are easily compressed and have low densities because of the large average distances between the particles. In fact, most of the space in any gas sample under moderate conditions is empty space (without any gas particles). Finally, gases exert a pressure on any surface that they touch because the gas particles collide with the surface and push against it.

THE GAS LAWS AND STOICHIOMETRY

Beginning in the mid-seventeenth century, several scientists studied the quantitative relationships between a gas's pressure, volume, temperature, and amount. For instance, they were able to predict the volume that a certain amount of gas (in moles) would occupy at a specified temperature and pressure. All gases, *regardless of chemical composition,* obeyed these mathematical relationships as long as the gases were under moderate conditions. In the next six sections, we describe six laws that arose from these experimental studies. Because these gas laws led to the formulation of the kinetic molecular theory, we examine them through contemporary eyes and see how the theory explains the behavior that gas laws merely describe.

Animations: Boyle's law; Microscopic illustration of Boyle's law.

11.4 Boyle's Law (Pressure and Volume)

OBJECTIVES

- Show how the pressure and volume of a certain amount of a gas are related at constant temperature.
- Given the pressure and volume of a certain amount of a gas at a constant temperature, calculate the volume of the gas at a new pressure or the pressure at a new volume.
- Explain Boyle's law using the kinetic molecular theory of gases.

The English chemist Robert Boyle (1627–1691) experimented with the compressibility of air and discovered the law that is now named after him. **Boyle's law** states that *the volume of a fixed amount of a gas at a given temperature is inversely proportional to the applied pressure*. The term *inversely proportional* means that one of these quantities will increase in a particular way when the other decreases and vice versa. For example, the volume of a gas will double when the applied pressure is halved. The mathematical equation for Boyle's law is

$$\text{Pressure} \times \text{volume} = \text{constant}$$

$$PV = a$$

where a is a constant for a given temperature and a given amount of the gas.

To visualize the relationship between pressure and volume described by Boyle's law, consider Figure 11.6. We assume that the temperature and the amount of the gas do not change during the experiment shown in the figure. As we begin, the pressure is 1.0 atm and the volume is 4.0 L. When the pressure is doubled by increasing it to 2.0 atm, the volume is halved so that it is 2.0 L. Next, the pressure is doubled again, and the volume is halved once again. This demonstrates the inverse proportionality between the pressure and volume of a gas, which you can also see graphically in Figure 11.7. If you know the pressure of a gas as shown on the horizontal axis, you can read the corresponding volume on the vertical axis.

Figure 11.6 also shows that a constant (4.0 atm · L) is obtained when we multiply each pressure by its corresponding volume. Let us identify the first pressure and volume in the figure as P_1 and V_1 and the second pressure and volume as P_2 and V_2. We can write

$$P_1V_1 = 4.0 \text{ atm} \cdot \text{L} = P_2V_2$$

since

$$1.0 \text{ atm} \times 4.0 \text{ L} = 4.0 \text{ atm} \cdot \text{L} = 2.0 \text{ atm} \times 2.0 \text{ L}$$

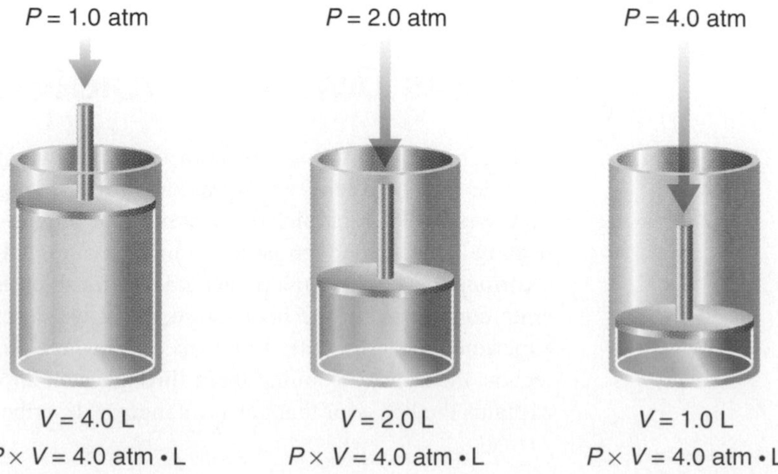

FIGURE 11.6
The effect of pressure on the volume of a gas. Consider a cylinder with a movable piston. Each time the pressure on a gas enclosed in the cylinder is doubled, the volume of the gas is cut in half.

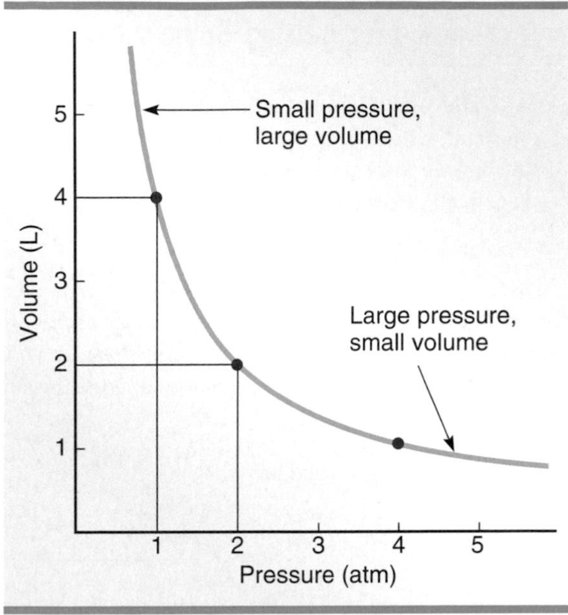

FIGURE 11.7
Volume of an ideal gas at various pressures. The volume decreases with increasing pressure. The data on this curve are from Figure 11.6. When the pressure is doubled (for example, from 1.0 to 2.0 atm), the volume is halved (from 4.0 to 2.0 L).

(a relationship that occurs throughout Figure 11.6). Thus, we can write

$$P_1V_1 = P_2V_2$$

and *any units for pressure and volume can be used.* (However, the units of both pressures must be the same, just as the units for both volumes must be identical.) This equation is an equivalent way of stating Boyle's law, and it puts Boyle's law in a form that is practical for solving problems.

Using Boyle's law in this form allows us to predict the new volume of a gas if the pressure changes. If P_1 and V_1 are the old pressure and volume (both known) and P_2 is the new pressure (again known), you need to arrange the equation $P_1V_1 = P_2V_2$ so that it begins with $V_2 =$. You can put the equation in that form by dividing both sides by P_2.

$$\frac{P_1V_1}{P_2} = \frac{\cancel{P_2}V_2}{\cancel{P_2}}$$

The cancellation on the right side leads to

$$\frac{P_1V_1}{P_2} = V_2 \qquad \text{or} \qquad V_2 = V_1 \times \frac{P_1}{P_2}$$

This equation indicates that you can calculate the new volume by multiplying the old volume by a ratio of the pressures.

Similarly, you can predict the new pressure of a gas if the volume changes. When you rearrange the equation $P_1V_1 = P_2V_2$, you obtain

$$P_2 = P_1 \times \frac{V_1}{V_2}$$

We illustrate the use of these equations in the next two examples.

■ EXAMPLE 11.2 Using Boyle's Law to Calculate a New Volume

At sea level in Maine, a balloon has a volume of 635 mL. If the temperature remains constant, what will be the volume of the balloon if it is moved to Colorado, where the new atmospheric pressure on a particular day is 0.803 atm? Assume the balloon is perfectly elastic.

Problem Analysis

Identify each quantity before you begin. Assume that the atmospheric pressure at sea level on a particular day has its customary average value of 1.00 atm. *Remember that your answer to this problem must show that the volume increases because the pressure decreases.* Use this expectation to judge the soundness of your answer.

Solution

The quantities you need are

$P_1 = 1.00$ atm
$V_1 = 635$ mL
→
$P_2 = 0.803$ atm
$V_2 = ?$

When these quantities are inserted into the equation

$$V_2 = V_1 \times \frac{P_1}{P_2}$$

you obtain

$$V_2 = 635 \text{ mL} \times \frac{1.00 \cancel{\text{atm}}}{0.803 \cancel{\text{atm}}} = 791 \text{ mL}$$

The decrease in the pressure leads to an increased volume, so the balloon expands, just as we predicted.

Exercise 11.2

Suppose a quantity of gaseous carbon dioxide (CO_2) at 1.00 atm and 23°C occupies 20.0 L. What will be the volume at 2.10 atm and 23°C?

(Try Problems 11.23, 11.24, 11.25, and 11.26.)

■ EXAMPLE 11.3 Using Boyle's Law to Calculate a New Pressure

A cylinder contains a gas with an initial volume of 0.669 L. The gas exerts an initial pressure of 1.00 atm. Calculate the pressure of the gas when it is compressed to 0.223 L at constant temperature.

Problem Analysis

Identify the quantities you need. Notice that you need to solve for a new pressure rather than a new volume as in Example 11.2. *Remember that the pressure must increase because the volume decreases.* Make sure that your answer makes sense.

Solution

The quantities you are given are

$$P_1 = 1.00 \text{ atm} \quad V_1 = 0.669 \text{ L} \longrightarrow P_2 = ? \quad V_2 = 0.223 \text{ L}$$

After you put these quantities into the equation

$$P_2 = P_1 \times \frac{V_1}{V_2}$$

you can solve the problem by writing

$$P_2 = 1.00 \text{ atm} \times \frac{0.669 \not{\text{L}}}{0.223 \not{\text{L}}} = 3.00 \text{ atm}$$

This result is reasonable because the problem tells you that the gas is compressed and, indeed, you have found that the pressure has increased.

Exercise 11.3

A perfectly elastic balloon has a volume of 1.32 L at an atmospheric pressure of 745 mmHg and a temperature of 18°C. When the balloon ascends to a particular altitude, its volume is 1.89 L. What is the atmospheric pressure at that altitude assuming the temperature remains constant?

(Try Problems 11.27, 11.28, 11.29, and 11.30.)

The kinetic molecular theory of gases explains Boyle's law. According to the second postulate of the theory, gas particles are in constant, random motion, moving in straight lines until they collide with another particle or the walls of the container. As we have seen, the pressure a gas exerts in a container is a result of the forces exerted by the collision of gas particles with the walls of the container. In Figure 11.8, we consider what happens to the movement of gas particles and the pressure when the volume of the container decreases at constant temperature. The average kinetic energy of these particles is directly proportional to the Kelvin temperature, according to the fourth postulate given earlier. As a result, the average energy of a collision when one of the gas particles hits a

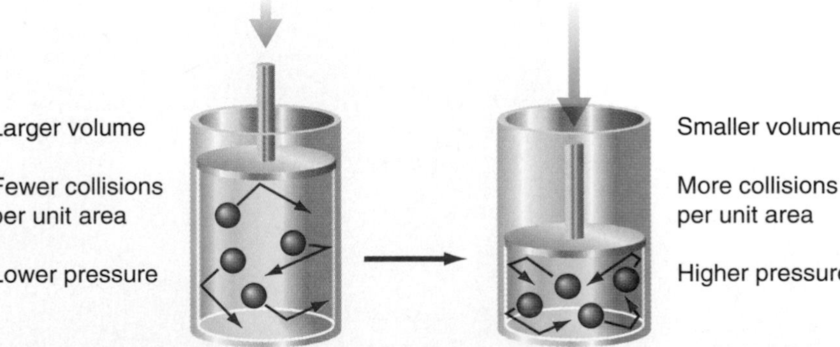

FIGURE 11.8
The pressure and volume of a gas are inversely proportional. Decreasing the volume increases the pressure by increasing the number of collisions per unit area on the walls of the container.

wall of the container will not change when the volume decreases at constant temperature. However, the pressure will *increase* when the volume *decreases* because there will be *more* collisions with the container per unit area. Thus pressure and volume should be *inversely* related, as Boyle's law states.

Animations: Charles's law; Microscopic illustration of Charles's law.

11.5 Charles's Law (Volume and Temperature)

OBJECTIVES

- Show how the volume and temperature of a certain amount of a gas are related at constant pressure.
- Explain the origin of the Kelvin temperature scale.
- Given the volume and temperature of a certain amount of a gas at constant pressure, calculate the volume at a new temperature or the temperature at a new volume.
- Explain Charles's law using the kinetic molecular theory of gases.

After Boyle completed his studies, chemists and other scientists continued to examine gases and their properties. Jacques Charles (1746–1823), a gifted lecturer and inventor, was the first to inflate a balloon with hydrogen and make an ascent with it. Hot-air balloons are a popular hobby today (Figure 11.9), but hydrogen is no longer used because it will explode if a spark occurs.

Charles found that the volume of a fixed amount of gas increased when he increased its temperature, as long as the pressure remained constant. If he cooled the gas under the same conditions, the volume decreased. We can demonstrate this phenomenon with the balloons shown in Figure 11.10. The balloons are filled with air and cooled in liquid nitrogen (−196°C). The chilled air contracts, making the balloons shrink. When the balloons and their contents return to room temperature again, the air in each one expands

FIGURE 11.9
Hot-air balloons. A propane gas burner heats the air within each balloon. The heated air expands so that it occupies a larger volume. Since the hot air has a lower density than the surrounding air, the balloon rises.

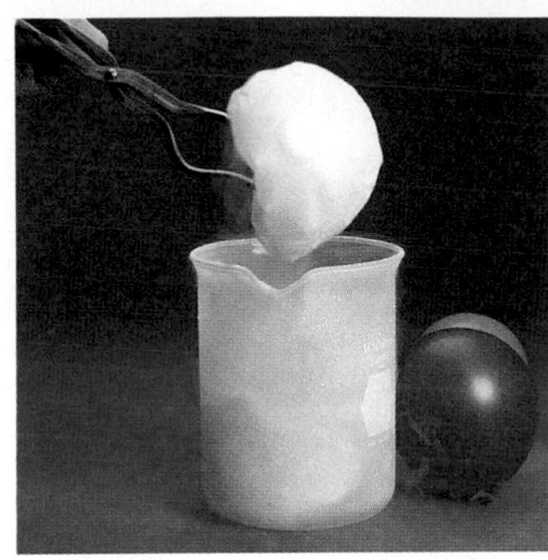

FIGURE 11.10

The effect of temperature on the volume of a gas. When balloons are immersed in liquid nitrogen (−196°C), they shrink because the volume of the air within them decreases. After the balloons are removed from the liquid nitrogen, the air inside warms up and expands, returning each one to its original size.

Animation: Balloon upon submersion in liquid N_2.

Video: Liquid nitrogen and balloons.

until it has the original volume. The pressure is constant at atmospheric pressure throughout these events.

The volume–temperature relationship for a gas is shown in the graph in Figure 11.11A. You can see the straight lines that result when we plot the volumes of two different gases at constant pressure against their Celsius temperatures. When we extend the straight lines from the last experimental point toward lower temperatures—that is, when we *extrapolate* the straight lines—we find something rather surprising. Both lines intersect the horizontal axis (where the graph indicates that the volume of the gas is zero) at the same temperature, −273°C. Because a negative volume is not physically possible, our results suggest that this temperature is the lowest possible. To accommodate this conclusion, we can create a new temperature scale in which the lowest possible temperature becomes the zero point. We do so by adding 273 to any Celsius temperature. The resulting temperature scale is called the *Kelvin scale* in honor of William Thompson, Lord Kelvin (1824–1907), the Scottish physicist and nobleman who first proposed it. The relationship between the Kelvin scale and the Celsius scale is

$$\mathrm{K} = {}^{\circ}\mathrm{C} + 273$$

as introduced in Section 2.6. Recall that "degrees" on the Kelvin scale are called kelvins and they are written with a capital K without a degree symbol.

When we replot the volumes of the gases in Figure 11.11A against Kelvin temperatures rather than Celsius temperatures, we obtain the results shown in Figure 11.11B. Now the straight lines extrapolate to 0 K, where the graph indicates the volume is also equal to zero. Note that the graph seems to say that matter disappears at absolute zero (a term for 0 K). However, when real gases are cooled, they liquefy before their temperatures reach absolute zero. Liquids and solids do not obey the linear (straight-line) relationship between volume and temperature shown in Figure 11.11 because their particles (atoms or molecules) are already packed quite close together.

The straight-line relationship between volume and temperature is summarized in a statement now known as **Charles's law:** *The volume of a fixed amount of gas is directly proportional to its Kelvin temperature as long as the pressure is kept constant.* Charles's law can be represented by the equation

$$\text{Volume} = \text{constant} \times \text{Kelvin temperature}$$

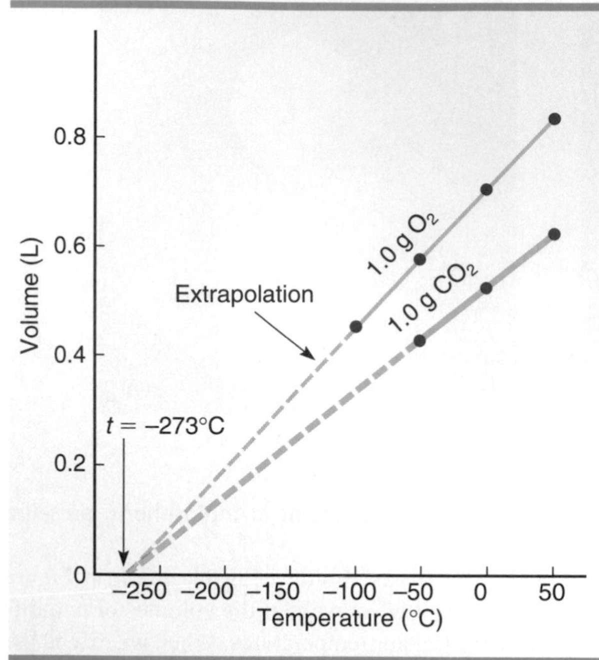

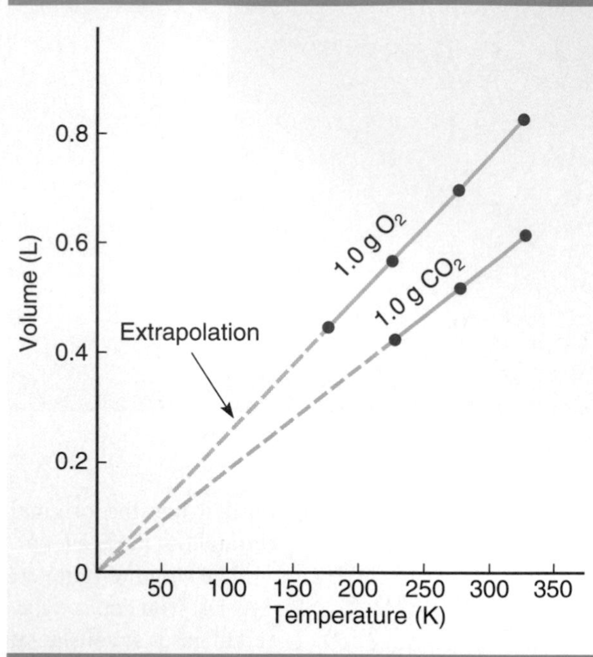

FIGURE 11.11

Volume–temperature relationship for gases at constant pressure. *(A)* The graph shows volumes of 1.0 g of two gases at various temperatures and 1.00 atm pressure. Both lines extrapolate to −273°C at zero volume. *(B)* The volumes of the same gases plotted against Kelvin temperatures. Both lines now extrapolate to 0 K.

If we write T for the temperature on the Kelvin scale, our equation becomes

$$V = b \times T$$

where b is the constant. When we divide both sides of this equation by T, we obtain

$$\frac{V}{T} = \frac{b \times \cancel{T}}{\cancel{T}}$$

After cancellation, the equation becomes

$$\frac{V}{T} = b$$

This equation says that the volume divided by the Kelvin temperature is a constant as long as the pressure and the amount of gas are fixed.

If you are considering a sample of gas whose volume and temperature are changing, you can also write this equation in terms of two different conditions,

$$\frac{V_1}{T_1} = b = \frac{V_2}{T_2}$$

or

$$\frac{V_1}{T_1} = \frac{V_2}{T_2}$$

This is an equivalent way of stating Charles's law that can be very practical for problem solving. The equation means that, if you know the volume and temperature of a certain amount of a gas at a particular pressure, you can predict the volume of the gas at a new temperature. Because you need the equation in a form that begins with $V_2 =$, you multiply both sides by T_2 and cancel where possible.

$$T_2 \times \frac{V_1}{T_1} = \frac{V_2}{\cancel{T_2}} \times \cancel{T_2}$$

The result is

$$V_2 = V_1 \times \frac{T_2}{T_1}$$

You can also predict the new temperature of a gas if the volume changes without changing the pressure or the quantity of the gas. After rearranging the equation

$$\frac{V_1}{T_1} = \frac{V_2}{T_2}$$

you obtain

$$T_2 = T_1 \times \frac{V_2}{V_1}$$

We use these equations in the next two examples. Note that *you can use volume in any unit (provided both V_1 and V_2 are in the same unit)*, but you must express temperatures in kelvins.

■ EXAMPLE 11.4 Using Charles's Law to Calculate a New Volume

The air in a perfectly elastic balloon occupies 785 mL during the fall when the temperature is 21°C. During the winter, the temperature on a particular day is 0°C. Calculate the new volume if the pressure is unchanged.

Problem Analysis

Convert each temperature from degrees Celsius to kelvins. Identify each quantity in the problem by labeling the initial Kelvin temperature and volume as T_1 and V_1, and the final Kelvin temperature and volume as T_2 and V_2. Decide how to write the ratio of temperatures, *keeping in mind that the volume must decrease because the temperature decreases.*

Solution

Convert the temperatures to the Kelvin scale using K = °C + 273.

$$\text{Initial temperature in kelvins} = 21 + 273 = 294\ \text{K} = T_1$$

$$\text{Final temperature in kelvins} = 0 + 273 = 273\ \text{K} = T_2$$

The given quantities and the unknown quantity for this problem are

$V_1 = 785$ mL $T_1 = 294$ K	→	$V_2 = ?$ $T_2 = 273$ K

Substituting the given quantities into the equation

$$V_2 = V_1 \times \frac{T_2}{T_1}$$

you get

$$V_2 = 785\text{ mL} \times \frac{273\text{ K}}{294\text{ K}} = 729\text{ mL}$$

You expected the volume to decrease, so this result is reasonable.

Exercise 11.4

If a reaction produces 438 mL of hydrogen (H_2), at a pressure of 742 mmHg and 19°C, what volume will you expect at the same pressure if the temperature is 25°C?

(Try Problems 11.31, 11.32, 11.33, and 11.34.)

EXAMPLE 11.5 Using Charles's Law to Calculate a New Temperature

Old-fashioned gas thermometers were calibrated to measure temperature based on the expansion and contraction of gases. Let's see how one of these thermometers worked. Suppose a gas occupies 4.23 L inside a chilly room at 16°C and 0.985 atm. The gas sample is carried outside, and the volume of the gas is now 3.84 L at the same pressure. What is the temperature outside on the Celsius scale?

Problem Analysis

First, convert the initial temperature to kelvins. Then, identify the given quantities and the one you will want. Notice that this problem (unlike Example 11.4) requires you to calculate a new temperature rather than a new volume. The equation you want is

$$T_2 = T_1 \times \frac{V_2}{V_1}$$

Remember that Charles's law tells you that the temperature must decrease because the volume decreases.

Solution

Your first task is to convert the initial temperature to the Kelvin scale.

$$\text{Initial Kelvin temperature} = 16 + 273 = 289\text{ K} = T_1$$

Next, identify and label all the quantities.

$V_1 = 4.23$ L	→	$V_2 = 3.84$ L
$T_1 = 289$ K		$T_2 = ?$

You can solve the problem by writing

$$T_2 = 289 \text{ K} \times \frac{3.84 \not{L}}{4.23 \not{L}} = 262 \text{ K}$$

So the temperature outside is 262 K, but you were asked for the temperature on the Celsius scale. To get this, you convert using

$$\text{K} = {}^\circ\text{C} + 273$$

by subtracting 273 from each side.

$$\text{K} - 273 = {}^\circ\text{C} + 273 - 273 = {}^\circ\text{C}$$

or

$${}^\circ\text{C} = \text{K} - 273 = 262 - 273 = -11^\circ\text{C}$$

The decrease in the volume of the gas tells you that the temperature is -11°C.

Exercise 11.5

Suppose a sample of gas occupies 10.0 mL at 50.0°C and 736 mmHg. At what temperature (on the Celsius scale) will this gas sample occupy 20.0 mL if the pressure remains unchanged?

(Try Problems 11.35, 11.36, 11.37, and 11.38.)

The kinetic molecular theory explains Charles's law. Suppose that a gas is enclosed in an elastic container, such as a balloon (see Figure 11.10 again). What happens if we cool the balloon (decrease the temperature)? Because the pressure of the gas inside an elastic balloon must always be identical to that of the atmosphere, the gas pressure will not change when the temperature is decreased. However, the average kinetic energy of the gas particles will decrease in accordance with the fourth postulate of the theory. Less powerful collisions with the balloon's walls will occur, so that the gas particles exert less force on those walls. Remember that pressure is force per unit area. To keep the pressure constant at atmospheric pressure, the area of the balloon's walls must decrease if the collision force decreases. As a result, the balloon contracts. Clearly, when the temperature *decreases* at constant pressure, the volume must *decrease* (and vice versa). In agreement with Charles's law, the kinetic molecular theory predicts that the volume of a gas and its temperature must be directly proportional.

11.6 The Combined Gas Law (Pressure, Volume, and Temperature)

OBJECTIVE

- Show that the equations from Boyle's law and Charles's law can be combined into a single equation.

To use Boyle's law, you must have constant temperature. To use Charles's law, you must have constant pressure. But sometimes temperature and pressure change at the same time, and both of these affect gas volume. In such a situation, you can use the **combined gas law,** *the combination of Boyle's law and Charles's law* for a constant amount of gas.

The equation for this law is

$$\frac{\text{Pressure} \times \text{volume}}{\text{Kelvin temperature}} = \text{constant}$$

or

$$\frac{PV}{T} = c$$

where c is the constant. If you write this equation for two sets of pressure–volume–temperature conditions, you obtain

$$\frac{P_1V_1}{T_1} = \frac{P_2V_2}{T_2}$$

This equation can be solved algebraically for any one of the six quantities. For instance, if you wanted a new gas volume when both pressure and temperature are changing, you would find that

$$V_2 = V_1 \times \frac{P_1}{P_2} \times \frac{T_2}{T_1}$$

We use the combined gas law in the next section.

11.7 Avogadro's Law (Volume and Moles)

OBJECTIVES

- Show how the volume and amount of a gas in moles are related if the pressure and temperature are held constant.
- Define standard temperature and pressure (STP).
- Give the molar volume of a gas at STP, and calculate the molar mass of a gas under any pressure or temperature.
- Explain Avogadro's law using the kinetic molecular theory of gases.

Boyle's law states the relationship between the pressure of a gas and its volume (at constant temperature and amount), and Charles's law states the relationship between the volume of a gas and its temperature (at constant pressure and amount). There is also a simple relationship between the volume of a gas and its amount. According to **Avogadro's law,** *equal volumes of any two gases at the same temperature and pressure contain equal moles of molecules.* This relationship is named for Italian chemist Amedeo Avogadro (1776–1856), who postulated it in 1811 during his studies of gas volumes. ●

● Recall from Chapter 7 that 1 mol of a gas contains 6.022×10^{23} atoms or molecules (Avogadro's number).

We can also state Avogadro's law in another way: *The volume of a gas is directly proportional to its amount in moles as long as the temperature and pressure are constant.* Thus, you can write

$$V = d \times n$$

where d is a constant and n is the number of moles. If you divide both sides of this equation by n and cancel, you get

$$\frac{V}{n} = \frac{d \times \cancel{n}}{\cancel{n}} = d$$

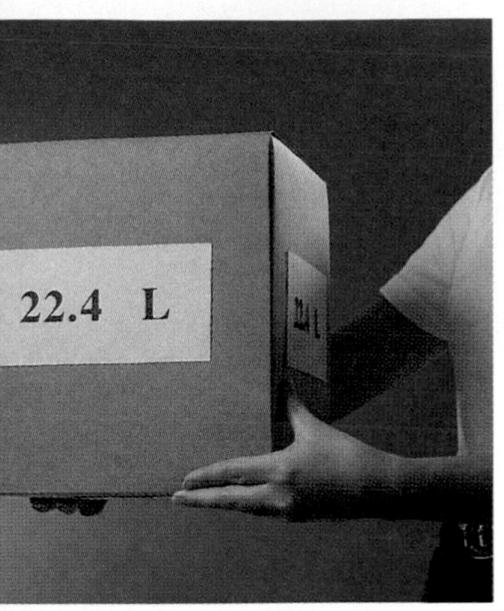

FIGURE 11.12
The molar volume of a gas. The molar volume of any gas at STP is 22.4 L. The molar volume is the volume of exactly 1 mol. Thus the cube will hold 2.0 g of H_2, 16.0 g of CH_4, 32.0 g of O_2, 44.0 g of CO_2, or 71.0 g of Cl_2 at 0°C and a pressure of 1.00 atm.

or

$$\frac{V}{n} = d$$

When n is exactly 1 mol, the equation becomes

$$\frac{V_M}{1} = d \qquad \text{or} \qquad V_M = d$$

where V_M is the molar volume (the volume of 1 mol of a gas). One mole of anything (solid, liquid, or gas) always contains 6.022×10^{23} particles (Avogadro's number), and *the equation* $V_M = d$ *tells you that the molar volume of any gas is a constant for a given temperature and pressure, independent of chemical composition.* In other words, 1 mol of hydrogen and 1 mol of radon occupy equal volumes, provided both are at the same temperature and pressure. You can use this fact to find the molar mass of any gas, as discussed in the next paragraph.

Volumes of gases are often compared at **standard temperature and pressure (STP),** *arbitrarily chosen to be exactly 0°C and 1 atm (760 mmHg) pressure. At STP, the volume of 1 mol of any gas has been shown by experiment to be 22.4 L* (Figure 11.12).

$$V_M = 22.4\ \text{L} \qquad \text{(at STP)}$$

If you weigh 22.4 L of any gas at STP, you will know its molar mass because 22.4 L contains 1 mol of gas particles under those conditions. As a result, you can use this relationship to obtain the molar mass of an unknown gas. Notice, however, that you do not need 22.4 L of a gas to make this determination; any lesser volume will suffice. For example, (22.4/2) L of any gas at STP will weigh one-half of its molar mass, (22.4/4)L will weigh one-quarter of its molar mass, and so forth. All you need is this conversion factor: 22.4 L/1 mol. Let's look at an example.

■ EXAMPLE 11.6 Calculating the Molar Mass of a Gas at STP

An amateur chemist working in Newport, a city on the seacoast of Rhode Island, obtains an unknown gas on a winter day during her studies of acid rain. She thinks the gas is nitrogen monoxide (NO). Suppose she finds that the volume of the gas at 0°C is 1.15 L and its mass is 1.54 g. Does the molar mass that she calculates from these data agree with her guess about the identity of the gas?

Problem Analysis

Since the temperature is 0°C and she is working on the sea coast, you may assume she is working at STP. You know that 22.4 L of this gas at STP will contain the molar mass. Therefore, you calculate the mass of 22.4 L of this gas, using the conversion factor 22.4 L/1 mol and knowing that the mass of 1.15 L is 1.54 g.

Solution

Calculate the mass of 22.4 L from

$$\frac{1.54\ \text{g}}{1.15\ \cancel{\text{L}}} \times \frac{22.4\ \cancel{\text{L}}}{1\ \text{mol}} = 30.0\ \text{g/mol}$$

Thus the molar mass of the gas is 30.0 g. By expressing the formula weight of NO in grams, you can show that the molar mass of nitrogen monoxide is 30.01 g, so the amateur chemist may be correct about the identity of the gas. The answer is yes.

Exercise 11.6

If 5.9 g of CO_2 occupies 3.0 L at STP, what is the molar mass of this gas?

(Try Problems 11.39 and 11.40.)

If a gas sample of known volume and mass is not at STP, you can use the combined gas law to calculate the volume the gas would have at STP. Once you calculate this volume, the calculation of molar mass proceeds as before.

■ EXAMPLE 11.7 Calculating the Molar Mass of a Gas Under Any Conditions

Radon, a radioactive gas that comes from rocks and soils containing uranium, seeps into many houses in the United States. Public health officials are concerned about its presence because excessive exposure to radioactivity from any source can cause cancer. Calculate the molar mass of radon if 12.1 g occupies 1.27 L at 16°C and 773 mmHg.

Problem Analysis

Convert the volume of radon at the given conditions (16°C, 773 mmHg) to STP (0°C, 760 mmHg). Begin by converting the temperatures to kelvins. Then, identify the quantities you will need to calculate the volume this gas would occupy at STP using the combined gas law,

$$\frac{P_1V_1}{T_1} = \frac{P_2V_2}{T_2}$$

You can rearrange this equation to give an equation that begins with $V_2 =$, as you saw earlier.

$$V_2 = V_1 \times \frac{P_1}{P_2} \times \frac{T_2}{T_1}$$

Finally, you know that 22.4 L of this gas at STP will contain the molar mass. So, you calculate the mass of 22.4 L of this gas, knowing that the mass of this sample is 12.1 g.

Solution

Convert the temperatures to the Kelvin scale using K = °C + 273.

$$\text{Initial temperature in kelvins} = 16 + 273 = 289\text{ K} = T_1$$

$$\text{Final temperature in kelvins} = 0 + 273 = 273\text{ K} = T_2$$

The quantities you need for this problem are

$P_1 = 773$ mmHg		$P_2 = 760$ mmHg
$V_1 = 1.27$ L	⟶	$V_2 = ?$
$T_1 = 289$ K		$T_2 = 273$ K

Using the combined gas law, you can calculate the volume under the final conditions of pressure and temperature.

$$V_2 = 1.27\text{ L} \times \frac{773\text{ }\cancel{\text{mmHg}}}{760\text{ }\cancel{\text{mmHg}}} \times \frac{273\text{ }\cancel{\text{K}}}{289\text{ }\cancel{\text{K}}} = 1.22\text{ L}$$

Then you can calculate the mass of 22.4 L from

$$\frac{12.1\text{ g}}{1.22\text{ }\cancel{\text{L}}} \times \frac{22.4\text{ }\cancel{\text{L}}}{1\text{ mol}} = 222\text{ g/mol}$$

Thus the molar mass of radon is 222 g.

Exercise 11.7

Calculate the molar mass of chlorine if 1.00 g occupies 119 mL at 255°C and 5.14 atm.

(Try Problems 11.41 and 11.42.)

You may have been surprised to learn that 1 mol of hydrogen and 1 mol of radon occupy equal volumes, provided both are the same temperature and pressure. This is certainly not true for liquids and solids, as you can see from Figure 7.6 in Chapter 7. One mole each of copper, mercury, carbon, water, and sodium chloride certainly do not occupy equal volumes when all are at the same temperature and pressure. Avogadro's law for gases is explained by the first postulate of the kinetic molecular theory, which says that the dimensions of gas particles are very small compared with the very large average distances between the particles. Although radon atoms are larger (and heavier) than hydrogen molecules, the difference in size does not affect the molar volumes because each gas is mostly empty space.

Animation: The ideal gas law, $PV = nRT$

11.8 The Ideal Gas Law

OBJECTIVES

- Combine Boyle's law, Charles's law, and Avogadro's law into a single law describing the behavior of an ideal gas.
- Given any three of the quantities pressure, volume, temperature, and amount of gas in moles, calculate the fourth.

The **ideal gas law** *combines all the information from the laws of Boyle, Charles, and Avogadro.* The equation for this law is

$$PV = nRT$$

Exploration: Properties of gases.

where R is the molar gas constant. The value of this constant is

$$R = 0.08206 \text{ L} \cdot \text{atm/(K} \cdot \text{mol)}$$

All gases behave ideally and conform to the equation for the ideal gas law provided that the pressures are relatively low and the temperatures are relatively high.

We now show how the ideal gas equation contains all of the information from the other gas laws.

1. Suppose the temperature T of a gas and its molar amount n are fixed. Then, since R is also a constant, each quantity on the right side of the ideal gas equation is a constant. When you multiply these three constants, you get another constant.

   $$PV = nRT = \text{constant}$$

   or

   $$PV = \text{constant}$$

   This is the equation for Boyle's law.

2. Suppose the pressure P and molar amount n of a gas are constants.

   $$\underbrace{P}_{}V = \underbrace{nR}_{}T$$

   Constants

   When you rearrange the equation so that all of the constants are on the right side, you get

   $$\frac{V}{T} = \frac{nR}{P} = \text{constant}$$

   or

   $$\frac{V}{T} = \text{constant}$$

   This is Charles's law.

3. Similarly, suppose that the pressure P and temperature T of a gas are fixed. Then the constants in the ideal gas equation become

   $$PV = nRT$$

   Constants

   When you collect all of the constants on the right side of the equation, you obtain

   $$\frac{V}{n} = \frac{RT}{P} = \text{constant}$$

   or

   $$\frac{V}{n} = \text{constant}$$

   This is Avogadro's law.

When you worked with Boyle's law and Charles's law, the problems always described a change in some condition (P, V, or T) while you calculated the change in another condition. With the combined gas law, you calculate the change in one of these conditions as two others change. The ideal gas law allows you to solve another type of problem. If you are given any three of the quantities P, V, n, and T, you can calculate the fourth, as seen in the following two examples. Moreover, you can use the equation from the ideal gas law to replace the equations from the laws of Boyle, Charles, and Avogadro.

■ EXAMPLE 11.8 Using the Ideal Gas Law to Calculate Pressure

Dinitrogen monoxide (N_2O), a colorless substance sometimes called nitrous oxide and sometimes called laughing gas, is used by dentists as an anesthetic. If 2.86 mol of this gas occupies a 20.0-L tank at 23°C, what is the pressure in the tank?

Problem Analysis

First convert the Celsius temperature to kelvins. Rearrange the equation for the ideal gas law so that pressure P is by itself on the left side of the equation: $P =$. Substitute the given values for n, T (in kelvins), and V into the equation. (The molar gas constant R is always 0.08206 L · atm/(K · mol); you will find it listed on the inside back cover of this book.)

Solution

Changing the temperature to the Kelvin scale, you obtain

$$\text{Kelvin temperature} = 23 + 273 = 296 \text{ K} = T$$

Next, rearrange the ideal gas equation so that you can solve for the pressure by dividing both sides by V.

$$\frac{P\cancel{V}}{\cancel{V}} = \frac{nRT}{V}$$

or

$$P = \frac{nRT}{V}$$

The quantities to use in the equation are

$P = ?$	$n = 2.86$ mol
$V = 20.0$ L	$R = 0.08206$ L · atm/(K · mol)
	$T = 296$ K

Substitution leads to

$$P = \frac{2.86 \cancel{\text{mol}} \times 0.08206 \cancel{\text{L}} \cdot \text{atm}/(\cancel{\text{K}} \cdot \cancel{\text{mol}}) \times 296 \cancel{\text{K}}}{20.0 \cancel{\text{L}}} = 3.47 \text{ atm}$$

Exercise 11.8

Calculate the pressure of helium at 17°C if a 2.00-g sample occupies 3.06 L.

(Try Problems 11.43, 11.44, 11.45, and 11.46.)

In addition to calculating an unknown pressure as you did in the last example, you can also use the equation to calculate either the volume of the gas, the amount of the gas in moles, or the gas's temperature. In the next example, you calculate n, the amount in moles.

■ EXAMPLE 11.9 Using the Ideal Gas Law to Calculate Molar Amount

Nitrogen monoxide (NO) is formed in our atmosphere during lightning storms. If a 1.42-L sample of this gas is collected at 734 mmHg and 28°C, how many moles of the gas are present? How many grams?

Problem Analysis

Begin by converting the given temperature and pressure conditions into units compatible with those for the molar gas constant R, which is in L · atm/(K · mol). The temperature must be in kelvins and the pressure must be in atmospheres (1 atm = 760 mmHg). Rearrange the ideal gas equation so that molar amount n is by itself on the left side of the equation: $n =$. Substitute the given values for P (in atm), V, and T (in kelvins) into this equation and calculate n. Compute the mass of this sample using the molar mass of NO (30.01 g/mol).

Solution

When you convert the pressure to atmospheres and the temperature to kelvins, you get

$$P = 734\ \cancel{\text{mmHg}} \times \frac{1\ \text{atm}}{760\ \cancel{\text{mmHg}}} = 0.966\ \text{atm}$$

$$T = \text{Kelvin temperature} = 28 + 273 = 301\ \text{K}$$

Since you need to solve for moles, divide both sides of the ideal gas equation by RT.

$$\frac{PV}{RT} = \frac{n\cancel{RT}}{\cancel{RT}} = n$$

or

$$n = \frac{PV}{RT}$$

The quantities you need to use are

$P = 0.966$ atm	$n = ?$
$V = 1.42$ L	$R = 0.08206$ L · atm/(K · mol)
	$T = 301$ K

Substituting these quantities into the equation, you obtain

$$n = \frac{0.966\ \cancel{\text{atm}} \times 1.42\ \cancel{\text{L}}}{0.08206\ \cancel{\text{L}} \cdot \cancel{\text{atm}}/(\cancel{\text{K}} \cdot \text{mol}) \times 301\ \cancel{\text{K}}} = 0.0555\ \text{mol}$$

Calculate the mass of the sample from

$$0.0555 \cancel{\text{mol}} \times \frac{30.01 \text{ g}}{1 \cancel{\text{mol}}} = 1.67 \text{ g}$$

Exercise 11.9

The pressure, volume, and temperature of a sample of carbon dioxide are 0.886 atm, 0.452 L, and 298 K, respectively. Calculate how many moles are present. What is the mass of this gas?

(Try Problems 11.51, 11.52, 11.53, and 11.54.)

Real gases behave ideally (that is, they obey the ideal gas law) provided that pressures are relatively low and temperatures relatively high. These conditions are not too far from STP. At such moderate conditions, the volumes of gas particles and their intermolecular attractions can be ignored according to the kinetic molecular theory.

Real gases are often mixtures of pure gaseous substances. The most familiar example is air, which is mostly nitrogen, oxygen, and argon with small amounts of other gases. If ideal gas behavior holds so that particle volumes and intermolecular attractions are negligible, then it should not matter whether the gas particles in a sample are all the same or whether the sample contains a mixture—the same gas laws should hold.

John Dalton, who is credited with atomic theory, studied the pressure relationships among components of a mixture of gases. In the next section, we present his results.

11.9 Dalton's Law of Partial Pressures

OBJECTIVES

- Define *partial pressure.*
- Given the partial pressure of each component in a gas mixture, find the total pressure.

Gases exert a pressure on any surface they touch because the gas particles collide with the surface and push against it. For example, air exerts pressure on all sides of an exposed object because gas particles in the air collide with it. Since air is a mixture of nitrogen, oxygen, and argon (plus other gases in trace amounts), each substance exerts its own pressure. The collision of nitrogen molecules (N_2) with objects on the earth's surface therefore creates a pressure due to nitrogen, and similar pressures are created by the impact of oxygen molecules (O_2) and argon atoms (Ar). How do these pressures relate to the total pressure exerted by air?

According to John Dalton, the components of a mixture of gases act independently insofar as their pressures are concerned, and the total pressure is the sum of the pressures due to the components. *The pressure that each component of a gas mixture exerts* is called the **partial pressure** of that component. According to **Dalton's law of partial pressures,** *the total pressure exerted by a gaseous mixture is the sum of the partial pressures of the components of the mixture.* Thus the total pressure (P_{Total}) of air is given principally by

$$P_{\text{Total}} = P_{\text{Nitrogen}} + P_{\text{Oxygen}} + P_{\text{Argon}}$$

where P_{Nitrogen}, P_{Oxygen}, and P_{Argon} are the partial pressures of nitrogen, oxygen, and argon, respectively.

Let's summarize Dalton's law of partial pressures. In general, if P_A, P_B, P_C, . . . are the partial pressures for a gas mixture containing gases denoted by A, B, C, . . . , then the total pressure will be given by

$$P_{\text{Total}} = P_A + P_B + P_C + \cdots$$

This law exists because the partial pressure exerted by any one of the gases in a mixture is the same as it would be if the gas were alone. We use this law in the following example.

■ EXAMPLE 11.10 Finding the Partial Pressures and the Total Pressure of a Mixture of Gases

Nitrogen in air dissolves in the blood at the high pressures experienced by scuba divers who are under several hundred feet of water. This gas will bubble out of the blood if a diver returns to the surface too quickly, causing the dangerous and excruciatingly painful condition known as "the bends." Thus, these divers cannot breathe ordinary air from scuba tanks. Instead, they breathe a mixture of oxygen and helium. Helium is lightweight, inert, and does not dissolve in blood. Suppose a 5.00-L scuba tank contains 1.05 mol of oxygen and 0.418 mol of helium at 25°C. What is the partial pressure of each substance, and what is the tank's total pressure?

Problem Analysis

Begin by recalling the basis for Dalton's law of partial pressures: The pressure exerted by any one of the gases in the mixture is the same as it would be if the gas occupied the tank by itself. Calculate each partial pressure from the ideal gas law as if each substance were by itself. Then add the partial pressures to get the total pressure.

Solution

The Kelvin temperature is

$$T = 25 + 273 = 298 \text{ K}$$

Next, you need to calculate the pressure of each component of the mixture in the tank by the same method you used in Example 11.8. The correct form of the ideal gas equation is

$$P = \frac{nRT}{V}$$

The quantities that you need for oxygen are

$P_{\text{Oxygen}} = ?$	$n = 1.05$ mol
$V = 5.00$ L	$R = 0.08206$ L · atm/(K · mol)
	$T = 298$ K

Substitution leads to

$$P_{\text{Oxygen}} = \frac{1.05 \cancel{\text{mol}} \times 0.08206 \cancel{\text{L}} \cdot \text{atm}/(\cancel{\text{K}} \cdot \cancel{\text{mol}}) \times 298 \cancel{\text{K}}}{5.00 \cancel{\text{L}}} = 5.14 \text{ atm}$$

The quantities that you need for helium are

$P_{\text{Helium}} = ?$	$n = 0.418$ mol
$V = 5.00$ L	$R = 0.08206$ L · atm/(K · mol)
	$T = 298$ K

When these quantities are inserted in the ideal gas equation, you obtain

$$P_{\text{Helium}} = \frac{0.418 \cancel{\text{mol}} \times 0.08206 \cancel{\text{L}} \cdot \text{atm}/(\cancel{\text{K}} \cdot \cancel{\text{mol}}) \times 298 \cancel{\text{K}}}{5.00 \cancel{\text{L}}} = 2.04 \text{ atm}$$

The total pressure in the scuba tank is given by

$$P_{\text{Total}} = P_{\text{Oxygen}} + P_{\text{Helium}} = 5.14 \text{ atm} + 2.04 \text{ atm} = 7.18 \text{ atm}$$

Exercise 11.10

A 2.00-L flask contains 0.0200 mol of carbon dioxide and 0.0400 mol of oxygen at 25°C. What are the partial pressures of carbon dioxide and oxygen? What is the total pressure within the flask?

(Try Problems 11.59 and 11.60.)

● We return to Dalton's law of partial pressures in Chapter 12.

The kinetic molecular theory of gases explains Dalton's law of partial pressures. Because the attractions (or repulsions) between gas particles are negligible, the components of a gaseous mixture act independently, causing partial pressures that are proportional to their molar quantities. ●

11.10 Stoichiometry of Reactions Involving Gases

OBJECTIVE

- Given the masses of reactants, calculate the volume of a gas released during a reaction.

In discussing the stoichiometry of reactions in Chapter 8, we showed how to calculate the amount of a product from the amount of a reactant. If the reactant amount was given in grams, we used the molar mass of that reactant to convert the quantity to moles. Then, using the balanced chemical equation for the reaction, we constructed a mole ratio that converted moles of reactant to moles of product. If desired, we could then convert the product amount in moles to an amount in grams, using the product's molar mass.

If the product is a gas, however, we are often more interested in its volume than its mass. (Gas volumes are easy to observe.) If we know the pressure and temperature of a certain molar amount of gas, we can calculate its volume using one of the gas laws. Consider the following historical example involving a rather large gas volume.

■ EXAMPLE 11.11 Finding the Volume of a Gas Generated from Known Quantities of Reactants

Jacques Charles, who discovered Charles's law, made the first ascent in a balloon filled with hydrogen (Figure 11.13). He obtained the hydrogen from the reaction of iron with sulfuric acid:

$$Fe(s) + H_2SO_4(aq) \longrightarrow FeSO_4(aq) + H_2(g)$$

According to Will Hayes, the author of *The Complete Ballooning Book* (World Publications, Mountain View, CA), Charles used about 1000 lb of iron filings for this reaction. How many liters of hydrogen were generated? Let's assume that Charles used exactly 1000 lb (four significant figures) of iron. We also assume that iron was the limiting reactant (the reactant that controls the amount of product formed; see Section 8.4 in Chapter 8). Moreover, we assume that the pressure was 1.00 atm and the temperature was 0°C (since the ascent was made on December 1, 1783). Thus the gas was at STP.

FIGURE 11.13
Charles making his ascent. On December 1, 1783, Jacques Charles rose from the Tuileries gardens in Paris, France, in the first hydrogen-filled balloon. He obtained hydrogen from the reaction of iron filings with sulfuric acid. The balloon was 27 feet in diameter and consisted of silk coated with a thin film of rubber.

Problem Analysis

Convert pounds of iron to grams. The next step of your strategy will resemble that for Example 8.9 in Chapter 8 except that you do not need to find the mass of hydrogen. Write down the desired conversion: mass iron $\longrightarrow$ moles iron $\longrightarrow$ moles hydrogen. Use the molar mass of Fe and the proper mole ratio from the balanced chemical equation to find the amount of hydrogen in moles.

Convert the temperature to kelvins. Use the equation for the ideal gas law to calculate the volume from the amount of hydrogen in moles, the temperature, and the pressure. The molar gas constant R is 0.08206 L · atm/(K · mol).

Solution

Convert pounds to grams using the conversion factor listed on the inside back cover of this book (1 lb = 453.6 g).

$$\text{Mass of iron} = 1000.\ \cancel{\text{lb}} \times \frac{453.6\ \text{g}}{1\ \cancel{\text{lb}}} = 4.536 \times 10^5\ \text{g Fe}$$

The molar mass of Fe is its atomic weight expressed in grams, or 55.85 g. According to the balanced chemical equation, each mole of iron used in the reaction generates 1 mol of hydrogen. So the amount of hydrogen in moles is

$$4.536 \times 10^5\ \cancel{\text{g Fe}} \times \frac{1\ \cancel{\text{mol Fe}}}{55.85\ \cancel{\text{g Fe}}} \times \frac{1\ \text{mol H}_2}{1\ \cancel{\text{mol Fe}}} = 8.122 \times 10^3\ \text{mol H}_2$$

The temperature of this gas in kelvins is 0 + 273 or 273 K.

Next, rearrange the ideal gas equation so that you can solve for the volume by dividing both sides by P.

$$\frac{\cancel{P}V}{\cancel{P}} = \frac{nRT}{P}$$

or

$$V = \frac{nRT}{P}$$

The quantities to use in the equation are

$P = 1.00$ atm	$n = 8.122 \times 10^3$ mol
$V = ?$	$R = 0.08206$ L · atm/(K · mol)
	$T = 273$ K

Substitution leads to

$$V = \frac{8.122 \times 10^3\ \cancel{\text{mol}} \times 0.08206\ \text{L} \cdot \cancel{\text{atm}}/(\cancel{\text{K}} \cdot \cancel{\text{mol}}) \times 273\ \cancel{\text{K}}}{1.00\ \cancel{\text{atm}}} = 1.82 \times 10^5\ \text{L}$$

Exercise 11.11

How many liters of carbon dioxide will be obtained at 745 mmHg and 27°C from the decomposition of 15.0 g of calcium carbonate? The chemical equation for the reaction is

$$CaCO_3(s) \longrightarrow CaO(s) + CO_2(g)$$

(Try Problems 11.61, 11.62, 11.63, and 11.64.)

Chemical Perspective

Holes in the Ozone Layer

In 1985, scientists from the British Antarctic Survey reported that the ozone layer over the South Pole had thinned each September and October since the late 1970s. An expedition to the North Pole in 1989 showed that less severe ozone depletion was also occurring over the North Pole. Another unexpected shock came in 1991 when analysis of satellite data indicated that ozone was being depleted at an accelerating rate over heavily populated areas. Since that time, there have been numerous warnings about the dangers presented by this depletion. Is this a real threat, or is it like Chicken Little's warning that the sky is falling? Some understanding of how ozone is formed will help answer this question.

A layer of gaseous ozone surrounds the earth at an average altitude of close to 30 km (about 20 miles). Ultraviolet radiation from the sun provides the energy for its formation from O_2. The process begins with the formation of oxygen atoms,

$$O_2 \xrightarrow[\text{radiation}]{\text{Ultraviolet}} 2O$$

Since the atmospheric pressure at the altitude of the ozone layer is very low, the oxygen atoms from this reaction tend to separate and remain very far apart. The great distances between these atoms mean that they rarely encounter one another to recombine to form O_2 molecules. However, when one of these atoms encounters a more plentiful O_2 molecule, ozone is formed. The reaction is described by

$$O + O_2 \longrightarrow O_3$$

This reaction is reversed when even more ultraviolet radiation (at another wavelength) causes the rupture of the ozone molecule and the regeneration of an oxygen molecule and an oxygen atom.

$$O_3 \xrightarrow[\text{radiation}]{\text{Ultraviolet}} O_2 + O$$

Notice that a cyclic process is occurring: Ozone formed in one reaction disappears in another reaction. As long as the process is undisturbed, a near-steady amount of ozone will be maintained in our atmosphere.

Atmospheric ozone has a valuable and essential function. Certain wavelengths of ultraviolet radiation can cause three different kinds of skin cancers. Since these wavelengths are the ones that are absorbed by the ozone molecule in its natural destruction, the ozone layer serves as an invisible protective screen around the earth. This ultraviolet screen is so valuable that scientists are very concerned about any atmospheric chemical that will react with ozone, throwing the natural cyclic process out of balance and destroying ozone faster than it can be regenerated. The chief culprits in the recent depletion of ozone are CFCs (*c*hloro*f*luoro*c*arbons), a family of substances used as aerosol propellants and refrigerants. These compounds decompose in sunlight just like ozone. Chlorine atoms from the decomposition as well as other compounds have reacted with the atmospheric ozone layer so completely that the well-known and well-documented "ozone holes" over the North and South Poles have appeared (Figure 11.14). The threat to human health from an increased incidence of skin cancer is real.

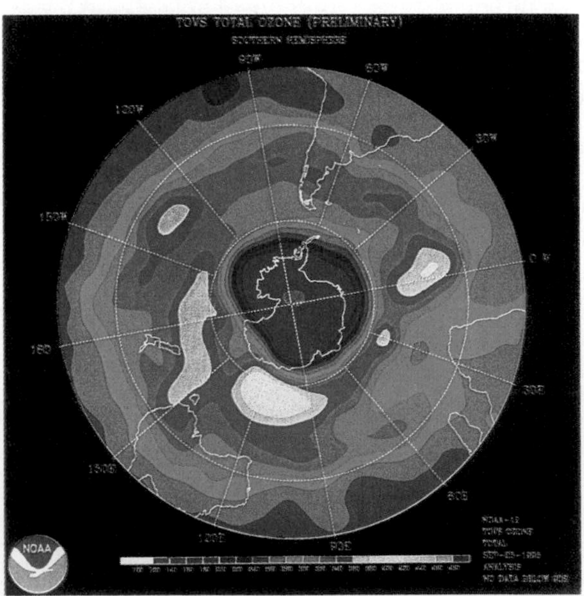

FIGURE 11.14

The ozone hole. This photo shows an obvious depletion of the ozone layer over Antarctica (represented by the red spot) where CFCs (chlorofluorocarbons) accumulate. All colors have been assigned by computer.

International efforts to protect the ozone layer from these chemicals resulted in a meeting of diplomats from 31 nations in Montreal, Canada, in 1987. The Montreal Protocol (as the significant result of this meeting is called) requires a sharp restriction in the production of CFCs by the end of this century. Moreover, efforts to curb the production of CFCs completely by the year 2000 are gaining considerable momentum.

CHAPTER REVIEW

Key Words

pressure *(p. 327)*
barometer *(p. 327)*
millimeters of mercury (mmHg) *(p. 328)*
torr *(p. 328)*
atmosphere (atm) *(p. 328)*
pascal (Pa) *(p. 328)*
manometer *(p. 329)*
kinetic molecular theory of gases *(p. 330)*
ideal gas *(p. 330)*
Boyle's law *(p. 332)*
Charles's law *(p. 337)*
combined gas law *(p. 341)*
Avogadro's law *(p. 342)*
standard temperature and pressure (STP) *(p. 343)*
ideal gas law *(p. 345)*
partial pressure *(p. 349)*
Dalton's law of partial pressures *(p. 349)*

Key Equations

$$1 \text{ atm} = 760 \text{ mmHg} = 760 \text{ torr} = 1.01325 \times 10^5 \text{ Pa}$$

$$P_1V_1 = P_2V_2 \quad \text{(constant } n, T)$$

$$\text{K} = {}^\circ\text{C} + 273$$

$$\frac{V_1}{T_1} = \frac{V_2}{T_2} \quad \text{(constant } n, P)$$

$$\frac{P_1V_1}{T_1} = \frac{P_2V_2}{T_2} \quad \text{(constant } n)$$

$$V_M = 22.4 \text{ L} \quad \text{(at STP)}$$

$$PV = nRT$$

$$P_{\text{Total}} = P_A + P_B + P_C + \cdots$$

Summary

Pressure is force exerted on a unit area of a surface. Atmospheric pressure is the result of about 5×10^{18} kg of gaseous substances exerting a force on the earth's surface. Atmospheric pressure can be measured with a *barometer.* The most commonly used units of pressure are *millimeters of mercury* (*mmHg*) and *atmospheres* (*atm*). Any enclosed gas exerts a pressure on the walls of its container, and this pressure can be measured with a *manometer.*

The *kinetic molecular theory of gases* is an explanation of *ideal gas* behavior. The theory is based on four postulates concerning an ideal gas. A gas consists of tiny, noninteracting particles that are very far apart. They are in constant random motion. When they collide with the walls of a container, pressure is produced. The average kinetic energy of these particles is proportional to the Kelvin temperature.

Under moderate conditions, all gases obey the same simple relationships, or gas laws. Thus, the volume of a gas varies inversely with pressure for a given molar quantity and a constant temperature (*Boyle's law*). It is also true that the volume of a gas varies directly with the Kelvin temperature when the molar quantity and the pressure are constant (*Charles's law*). Finally, equal volumes of gases contain the same number of molecules (*Avogadro's law*). At *standard temperature and pressure* (*STP*), conditions arbitrarily chosen to be exactly 0°C and 1 atm pressure, the volume of one mole of any gas has been shown by experiment to be 22.4 L. The three gas laws can be formulated as one equation, $PV = nRT$ (*ideal gas law*). This equation gives the relationship between the pressure, the volume, the quantity, and the temperature of a gas. It also relates these quantities for each component of a gas mixture. The total pressure is the sum of the partial pressures of each component (*Dalton's law of partial pressures*).

The volume of a gas produced in a chemical reaction can be calculated from the quantities of the reactants, keeping the stoichiometry of the reaction in mind, and the equation from the ideal gas law.

Problem-Solving Skills

1. **Converting One Pressure Unit to Another:** Given a pressure in either millimeters of mercury or atmospheres, convert the pressure to the other unit (Example 11.1).
2. **Using Boyle's Law:** Given an initial volume, use Boyle's law to calculate the final volume when the pressure is changed (Example 11.2); or, given an initial pressure, calculate the final pressure when the volume is changed (Example 11.3).
3. **Using Charles's Law:** Given an initial volume, use Charles's law to calculate the final volume when the temperature is changed (Example 11.4); or, given an initial temperature, calculate the final temperature when the volume is changed (Example 11.5).
4. **Calculating the Molar Mass of a Gas:** Given the volume and mass of a gas, calculate the molar mass at STP (Example 11.6) or at any other conditions (Example 11.7).
5. **Using the Ideal Gas Law:** Given any three of the variables P, V, T, and n, use the ideal gas law to calculate the fourth one (Examples 11.8 and 11.9).
6. **Finding the Partial Pressures and the Total Pressure of a Mixture of Gases:** Given the molar amounts of each gas in a mixture of gases, and the temperature and volume, calculate their partial pressures and the total pressure (Example 11.10).
7. **Solving Stoichiometry Problems Involving Gas Volumes:** Given the mass of one substance in a reaction, calculate the volume of a gas produced in the reaction (Example 11.11).

Questions to Test Your Reading

11.1 What are the three states of matter?

11.2 Name four properties of all gases.

11.3 Define pressure. What is atmospheric pressure, and how is it measured?

11.4 Why does the atmosphere exert pressure on the earth?

11.5 What is the average atmospheric pressure at sea level?

11.6 What is a manometer, and how does it work?

11.7 If the atmospheric pressure is 1 atm, what is the pressure of a gas inside a perfectly elastic balloon?

11.8 According to the kinetic molecular theory of gases, what is the origin of gas pressure?

11.9 What is Boyle's law?

11.10 Would an inflated balloon be larger in Denver, Colorado, or in Asbury Park, New Jersey, if the temperature is unchanged? Explain.

11.11 Explain Boyle's law in terms of the kinetic molecular theory.

11.12 What is Charles's law?

11.13 Will a balloon be larger on a cold winter day or on a hot summer day if the atmospheric pressure is constant? Explain.

11.14 Explain Charles's law in terms of the kinetic molecular theory.

11.15 What is STP?

11.16 What is Avogadro's law?

11.17 Explain Avogadro's law in terms of the kinetic molecular theory.

11.18 Give the equation for the ideal gas law.

11.19 What is Dalton's law of partial pressures?

11.20 Explain Dalton's law of partial pressures in terms of the kinetic molecular theory.

Practice Problems

Pressure (Section 11.2)

11.21 Convert 1128 mmHg to atmospheres.

11.22 Convert 0.912 atm to millimeters of mercury.

Boyle's Law (Section 11.4)

11.23 A helium-filled balloon has a volume of 2.60 L at 21°C and 768 mmHg. If the balloon ascends to an altitude where the helium pressure is 614 mmHg at the same temperature, what is the volume of the gas?

11.24 A 6.50-L sample of oxygen at 24°C is confined in a perfectly elastic balloon under water at 1.50 atm. If the balloon is moved to a lower depth at the same temperature where the pressure is 2.50 atm, what is the volume of the gas?

11.25 Nitrogen gas is stored in a tank at 21.6 atm pressure and 21°C. If the gas is allowed to expand to a pressure of 1.00 atm, it occupies 849 L at the same temperature. Calculate the volume of the tank.

11.26 Carbon dioxide gas is stored in a cylinder at 6.50 atm and 25°C. If the gas is allowed to expand to a pressure of 1.05 atm, it occupies 24.8 L at the same temperature. What is the volume of the cylinder?

11.27 A sample of argon gas in a cylinder occupies 845 mL at 33°C and a pressure of 763 mmHg. After it is compressed with a piston at the same temperature, this gas occupies 643 mL. What is the new pressure?

11.28 A sample of neon gas in a cylinder occupies 1.05 L at 20°C and a pressure of 1.90 atm. After it is allowed to expand at the same temperature, this gas occupies 3.45 L. What is the new pressure?

11.29 A meteorological balloon containing helium has a volume of 145 L at 18°C and 1.00 atm pressure. When the balloon ascends to a certain altitude, the volume is 165 L at the same temperature. What is the new pressure?

11.30 If 345 mL of oxygen gas is compressed to 208 mL at 3.08 atm without a change in temperature, what is the original pressure of the gas?

Charles's Law (Section 11.5)

11.31 A perfectly elastic balloon filled with helium has a volume of 1.33×10^3 L at 1.00 atm and 25°C. What volume would the gas occupy at 0°C and the same pressure?

11.32 A bacterial culture isolated from sewage produces 41.3 mL of methane (CH_4) gas at 31°C and 763 mmHg. What volume will the gas occupy at 0°C and the same pressure?

11.33 Oxygen from the decomposition of potassium chlorate ($KClO_3$) occupies 125 mL at 26°C and 0.944 atm. Calculate the volume of the gas at 14°C and the same pressure.

11.34 A perfectly elastic balloon filled with hydrogen has a volume of 222 mL at 51°C and 1.00 atm. Calculate the new volume if the balloon is cooled to −111°C at the same pressure.

11.35 A gas occupies 2.22 L at 0°C. To what temperature (on the Celsius scale) must this sample be heated for it to occupy 3.22 L at the same pressure?

11.36 A gas occupies 2.22 L at 0°C. To what temperature (on the Celsius scale) must this sample be cooled for it to occupy 1.22 L at the same pressure?

11.37 Suppose a sample of gas occupies 2.00 L at 30°C and 760 mmHg. At what temperature (on the Celsius scale) will this gas sample occupy 4.00 L if the pressure remains unchanged?

11.38 Suppose a sample of gas occupies 2.00 L at 30.0°C and 760 mmHg. At what temperature (on the Celsius scale) will this gas sample occupy 1.00 L if the pressure remains unchanged?

Avogadro's Law (Section 11.7)

11.39 If 15.3 g of carbon monoxide occupies 12.2 L at STP, what is the molar mass of this gas?

11.40 If 1.00 g of oxygen occupies 0.700 L at STP, what is the molar mass of this gas?

11.41 An experiment shows that 1.83 g of argon at 25°C and 2.00 atm occupies 0.561 L. What is the molar mass of this gas?

11.42 A chemist finds that 3.54 g of an unknown gas at 25°C and 1.56 atm occupies 1.74 L. What is the molar mass of this gas?

Ideal Gas Law (Section 11.8)

11.43 A cylinder has a volume of 9.65 L. If the cylinder contains 3.03 mol of neon at 23°C, what is the pressure in the cylinder?

11.44 If 7.99 g of argon is compressed into a heavy-walled 6.00-L flask at 19°C, what is the final pressure of the gas?

11.45 What is the pressure of a 0.200-mol sample of radon if it occupies 5.00 L at 100°C?

11.46 What is the pressure of 5.12 g of methane (CH_4) at 24°C in an 8.45-L cylinder?

11.47 An experiment requires 2.50 moles of carbon monoxide (CO) at 55°C and 1.50 atm. What is the volume of this sample?

11.48 If calculations show that a reaction should yield 94.5 mg of carbon dioxide (CO_2), what volume of this gas should be expected at 26°C and 738 mmHg?

11.49 What is the volume of 0.125 mol of hydrogen at 75°C and a pressure of 775 mmHg?

11.50 What is the volume of 10.6 g of oxygen at 22°C and 2.00 atm?

11.51 A chemical reaction produces 125 mL of nitrogen at 26°C and 745 mmHg. How many moles are produced?

11.52 When a sample of calcium carbonate is heated to 600°C, 2.68 L of carbon dioxide (CO_2) is produced at 1.12 atm. How many moles of carbon dioxide are formed? How many grams?

11.53 How many moles of helium will occupy 545 mL at 760 mmHg and 20°C?

11.54 How many moles of xenon are in a 2.00-L sample at STP? How many grams?

11.55 The maximum safe pressure that a certain 4.00-L flask can hold is 3.50 atm. If the flask contains 0.410 mol of a gas, what is the maximum temperature (in degrees Celsius) to which this flask can be subjected?

11.56 A 2.50-L flask was used to collect 5.63 g of gaseous propane (C_3H_8). After the gas was collected, its pressure was found to be 956 mmHg. What was the temperature (in degrees Celsius) of the gas?

11.57 What is the temperature (on the Celsius scale) of 28.0 g of nitrogen if it occupies 10.0 L at 755 mmHg?

11.58 What is the temperature (on the Celsius scale) of 98.9 mg of argon if it occupies 253 mL at 1.11 atm?

Dalton's Law of Partial Pressures (Section 11.9)

11.59 A 2.00-L flask contains 0.00125 mol O_2 and 0.00325 mol He at 15°C. What are the partial pressures of these gases, and what is the total pressure?

11.60 A 5.00-L flask contains 1.00 mol of nitrogen and 2.00 mol of hydrogen at 15°C. What are the partial pressures of these gases, and what is the total pressure?

Gas Stoichiometry (Section 11.10)

11.61 How many liters of oxygen will be obtained from the decomposition of 1.00 g of hydrogen peroxide (in an aqueous solution) if the gas is collected at 28°C and 0.867 atm? The chemical equation for this reaction is

$$2H_2O_2(aq) \longrightarrow 2H_2O(l) + O_2(g)$$

11.62 How many liters of oxygen will be obtained from the decomposition of 2.00 g of solid mercury(II) oxide if the gas is collected at −2°C and 0.684 atm? The chemical equation for this reaction is

$$2HgO(s) \longrightarrow 2Hg(l) + O_2(g)$$

11.63 How many liters of oxygen at 23°C and 1.12 atm will be required for the combustion of 10.0 g of methane? The chemical equation for this reaction is

$$CH_4(g) + 2O_2(g) \longrightarrow CO_2(g) + 2H_2O(g)$$

11.64 How many liters of oxygen at 16°C and 0.756 atm will be required for the combustion of 10.0 g of ethane? The chemical equation for this reaction is

$$2C_2H_6(g) + 7O_2(g) \longrightarrow 4CO_2(g) + 6H_2O(g)$$

11.65 When magnesium burns in air, the reaction is described by

$$2Mg(s) + O_2(g) \longrightarrow 2MgO(s)$$

What mass of magnesium will react with 10.8 L of oxygen if the temperature and pressure of the gas before the reaction are −12°C and 664 mmHg?

11.66 Iron reacts with fluorine according to the chemical equation

$$2Fe(s) + 3F_2(g) \longrightarrow 2FeF_3(s)$$

What mass of iron will react with 5.88 L of fluorine if the temperature and pressure of the gas before the reaction are 22°C and 787 mmHg?

11.67 Sulfur dioxide can be removed from other gases by its reaction with calcium oxide. The chemical equation for the reaction is

$$CaO(s) + SO_2(g) \longrightarrow CaSO_3(s)$$

What mass of calcium oxide is required to react with 102 mL of sulfur dioxide at STP?

11.68 Water vapor will react with hot copper according to the chemical equation

$$Cu(s) + H_2O(g) \longrightarrow CuO(s) + H_2(g)$$

What mass of copper is necessary to react with 80.8 mL of water vapor at 0.987 atm and 503°C?

11.69 When solid barium carbonate is heated strongly, solid barium oxide and gaseous carbon dioxide are formed. Write the balanced chemical equation. Calculate the mass of barium carbonate that will produce 19.4 L of carbon dioxide if the pressure is 0.955 atm and the temperature is 802°C.

11.70 When solid potassium chlorate is heated strongly, solid potassium chloride and oxygen gas are produced. Write the balanced equation. Calculate the mass of potassium chlorate that will produce 13.7 L of oxygen gas when it is collected at a pressure of 0.973 atm and a temperature of 24°C.

Additional Problems

11.71 A radioactive metal atom decays (changes to another kind of atom) by emitting a single alpha particle (He^{2+} ion). After these particles are converted to gaseous atomic helium, they can be collected like any other gas. If 12.5 mL is obtained at 765 mmHg and 23°C, how many metal atoms decayed to obtain this sample?

11.72 A person exhales about 5.8×10^2 L of carbon dioxide per day (at STP). This quantity of this gaseous substance can be a problem in a space shuttle. However, it can be absorbed by lithium hydroxide according to the chemical reaction

$$2LiOH(s) + CO_2(g) \longrightarrow Li_2CO_3(s) + H_2O(l)$$

How many grams of lithium hydroxide would be required for each astronaut on each day of the flight?

11.73 A sealed diving bell contains oxygen and helium. If the partial pressure of oxygen is 0.200 atm and the total pressure is 3.00 atm, calculate the number of moles of helium in 1.00 L of the gas mixture at 23°C.

11.74 A gas mixture contains 1.50 mol of oxygen and 2.00 mol of helium at 755 mmHg and 25°C. What is the volume of this mixture?

Practice in Problem Analysis

For each problem, describe the thinking you would use (the problem analysis) before doing the actual solution, but do not solve the problem.

1. A sample of calcium carbonate ($CaCO_3$) is heated until no more carbon dioxide is released. This gas is collected at 23°C and a pressure of 743 mmHg. The gas is then compressed at the same temperature by increasing the pressure to 1158 mmHg. How much smaller is the volume of the gas after the compression?

2. A sample of hydrogen gas (41.4 L at 756 mmHg and 23°C) is burned in oxygen. The water from this reaction is allowed to react with P_4O_{10} to form phosphoric acid. Assuming water is the limiting reactant, how many grams of phosphoric acid are formed?

PRACTICE EXAM

1. Which statement is true?
   (a) Heavy gases will fill only the bottom of a container.
   (b) Some gases are not compressible.
   (c) Gases have densities that are similar to those of liquids and solids.
   (d) Gases exert a uniform pressure on all inner surfaces of a container.
   (e) None of the above

2. Unpolluted air is
   (a) mostly oxygen
   (b) 50% nitrogen and 50% oxygen
   (c) 78% nitrogen and 21% oxygen
   (d) mostly nitrogen
   (e) None of the above

3. Atmospheric pressure can be measured with
(a) a kelvinometer (b) an inclinometer
(c) a manometer (d) a barometer
(e) None of the above

4. The pressure of a gas within a container can be compared to atmospheric pressure with
(a) a kelvinometer (b) an inclinometer
(c) a manometer (d) a barometer
(e) None of the above

5. The height of mercury in a barometer is 752 mmHg. The atmospheric pressure in atmospheres is
(a) 0.989 atm (b) 0.898 atm (c) 1.01 atm
(d) 1.11 atm (e) None of the above

6. Which of the following statements about the kinetic molecular theory of matter is true?
(a) A gas is composed of particles whose dimensions may be larger than the distances between the particles.
(b) Gas particles rarely move.
(c) Gas particles experience considerable attraction for one another.
(d) The average kinetic energy of a collection of gas particles is directly proportional to the Kelvin temperature.
(e) None of the above

7. A sample of nitrogen has a certain volume at a particular pressure and temperature. How must the pressure be changed to double the volume at constant temperature?
(a) The pressure must be quadrupled.
(b) The pressure must be doubled.
(c) The pressure must be halved.
(d) The pressure must be quartered.
(e) None of the above

8. A fixed quantity of a gas occupies a closed, perfectly elastic 3-L balloon at 1 atm and 294 K. If the pressure is increased to 3 atm without changing the temperature, the new volume is
(a) unchanged (b) 1 L (c) 6 L
(d) 9 L (e) None of the above

9. A fixed quantity of a gas occupies a closed, perfectly elastic 3-L balloon at 1 atm and 294 K. If the temperature is doubled without changing the pressure, the new volume is
(a) unchanged (b) 1 L (c) 6 L
(d) 9 L (e) None of the above

10. The lowest possible temperature is
(a) −273°C (b) −273 K (c) 273°C
(d) 273 K (e) None of the above

11. Which gas will occupy the largest volume at STP?
(a) 1 mol of H_2 (b) 0.5 mol of He
(c) 0.3 mol of Cl_2 (d) 0.2 mol of Ar
(e) 0.1 mol of Xe

12. Which statement about a closed, inflated, perfectly elastic balloon containing a given quantity of a gas is true? The balloon will be larger if
(a) the pressure and Kelvin temperature are doubled.
(b) the pressure and Kelvin temperature are halved.
(c) the pressure is doubled and the Kelvin temperature is halved.
(d) the pressure is halved and the Kelvin temperature is doubled.
(e) None of these changes will affect the volume.

13. A 12-L cylinder contains oxygen and helium in a mole ratio of 1:4. Which of the following statements about the contents of the cylinder is true?
(a) The partial pressure and volume of helium must be 40% of the total pressure and volume.
(b) The partial pressure and volume of helium must be 80% of the total pressure and volume.
(c) The partial pressure of helium must be 40% of the total pressure, but its volume is 12 L.
(d) The partial pressure of helium must be 80% of the total pressure, but its volume is 12 L.
(e) None of the above

14. The volume occupied by 19.6 g of gaseous methane at 27°C and 1.59 atm is
(a) 18.9 L (b) 18.9 mL (c) 27.7 L
(d) 27.7 mL (e) None of the above

15. When potassium nitrate is heated, it decomposes according to the chemical equation

$$2KNO_3(s) \longrightarrow 2KNO_2(s) + O_2(g)$$

If the oxygen from the decomposition of 5.06 g of potassium nitrate is collected at STP, its volume will be
(a) 22.4 L (b) 5.60 L (c) 1.12 L
(d) 0.560 L (e) None of the above

Chapter

12 Liquids, Solids, and Attractions Between Molecules

Even rock can be a liquid.

Chapter Outline

The steam erupting from Old Faithful, the water pouring over Niagara Falls, and the ice and snow blanketing the vast expanse of the Arctic are the gas, liquid, and solid forms, respectively, of the same substance—water (Figure 12.1). Although the three states of water have identical chemical properties, their physical properties differ. Certainly, they do not look the same or feel the same. Perhaps the difference we notice most is their temperatures. Steam feels much hotter than unheated water, and ice feels much colder than either of them. Indeed, temperature determines which state is present. But how does temperature influence which state is present at any particular time, and how does it affect the physical properties of matter?

In this chapter, we study liquids and solids by extending the kinetic molecular theory that we introduced in Chapter 11 to explain the behavior of gases. As we describe the

FIGURE 12.1

The three states of water. *(A)* The steam coming from Old Faithful (a geyser in Yellowstone National Park, Wyoming), *(B)* the liquid water pouring over Niagara Falls, and *(C)* ice and snow covering the Arctic are examples of the three states of matter (gas, liquid, and solid).

FIGURE 12.2
The liquid state of iron. Iron is a liquid between 1535°C and 2750°C. It is a solid below 1535°C and a gas above 2750°C at a pressure of 1 atm.

predictions of the theory, you will see why temperature is so influential in determining the state of a substance. Specifically, you will see how the theory explains each state of matter by examining the balance of the attractive forces between the particles of matter (atoms or molecules depending on the substance) and the temperature-dependent motions of these particles. As a background for these discussions, we begin with a thorough description of the states of matter.

MACROSCOPIC DESCRIPTION OF THE STATES OF MATTER

In the next few sections, we describe gases, liquids, and solids macroscopically. The term *macroscopic* means large enough to be observed and examined by the naked eye. You will find that you are already familiar with many differences in the states of matter.

12.1 The States of Matter and Changes in These States

OBJECTIVES

- Describe the states of matter.
- Name the changes in the states of matter.
- Explain the role of heat and temperature in each change of state.

Although we have identified the three states of water, other substances have three states also (Figure 12.2).

A **gas** (sometimes called a *vapor*) has *neither a fixed volume nor a fixed shape.* A gas assumes the shape and volume of any container in which it is introduced. The same quantity of a gas can be compressed so that it occupies a very small container or it can be allowed to expand to fit a much larger one. As you saw in Chapter 11, the gas laws summarize the relationships between the pressure, volume, quantity, and temperature of a gas.

A **liquid** has *a fixed volume but does not have a specific shape.* It assumes the shape of as much of a container as it is able to fill. A liquid can be compressed only slightly by applying pressure, and it will not expand to fill a large container. We cannot describe the properties of a liquid by simple mathematical laws as we can for gases.

A **solid** has *a fixed volume and a fixed shape.* Unlike gases and liquids, a solid is firm when we touch it. Solids do not allow much compression under applied pressure, nor do they expand when pressure is removed.

Temperature changes can trigger changes in these states. For example, water exists as a solid below 0°C, but it is a liquid above that temperature (at normal pressures). Each change of state has a specific name. **Melting** (also called **fusion**) is *the change of a solid to a liquid.* The melting of ice is a common example.

$$H_2O(s) \longrightarrow H_2O(l)$$

Ice — Liquid water

Freezing is *the change of a liquid to a solid.* For example,

$$H_2O(l) \longrightarrow H_2O(s)$$

Liquid water — Ice

Exploration: Changes of state.

TABLE 12.1
Types of Changes of State

Change	Name	Example
Solid ⟶ liquid	Melting (fusion)	Melting of ice
Liquid ⟶ solid	Freezing	Freezing of water
Liquid ⟶ gas	Vaporization	Evaporation of water
Gas ⟶ liquid	Condensation	Condensation of steam
Gas ⟶ solid	Deposition	Formation of frost
Solid ⟶ gas	Sublimation	Evaporation of snow

Vaporization (evaporation) is *the change of a liquid to a gas,* as in

$$H_2O(l) \longrightarrow H_2O(g)$$

Liquid water   Steam

Condensation is *the change of a gas to a liquid.* For example,

$$H_2O(g) \longrightarrow H_2O(l)$$

Steam   Liquid water

Deposition is *the change of a gas to a solid.* An example occurs when frost forms from water vapor.

$$H_2O(g) \longrightarrow H_2O(l)$$

Steam   Ice

Sublimation is *the change of a solid to a gas,* as in

$$H_2O(s) \longrightarrow H_2O(g)$$

Ice   Steam

This is the process that makes ice and snow slowly disappear without melting. We smell moth balls because they sublime easily. Table 12.1 summarizes the changes of state we have encountered.

■ EXAMPLE 12.1 Identifying a Change of State

The water level in an aquarium decreases slowly even though the tank does not leak. What change of state is occurring?

Problem Analysis

Identify the state of the substance before and after the change.

Solution

Liquid water must be changing into gaseous water. Thus, vaporization is occurring.

Exercise 12.1

Frost forms on a cold, clear morning. What change of state has occurred?

(Try Problems 12.27 and 12.28.)

Ice cannot melt without the addition of heat (at least under ordinary circumstances), nor can liquid water freeze unless heat is removed from it. In general, melting and vaporization require the addition of heat, whereas freezing and condensation require cooling to remove heat. Let's consider what happens to the three states of water when we add heat at a constant rate to a very cold sample of ice. Figure 12.3 shows the *heating curve* (a graph of temperature versus heat added) for this sample. Starting at the left of the figure, you can see that the input of heat causes the temperature of the sample of ice to rise. When the sample reaches 0°C, it melts. Notice that the temperature does not change while the ice is melting even though heat is still being added. This heat makes the ice melt; it does not raise the temperature of the liquid water being formed. If we stopped adding heat and if we prevented heat from escaping, the ice and the liquid water would coexist indefinitely. Melting would continue but freezing would also occur, and at a rate equal to that of melting.

The temperature at which any solid melts depends slightly on the pressure exerted on the solid (usually the pressure of the atmosphere). *At 1 atm pressure, the constant temperature at which the solid changes into a liquid* is called the **normal melting point** of the solid and the **normal freezing point** of the liquid. For a given substance, the normal

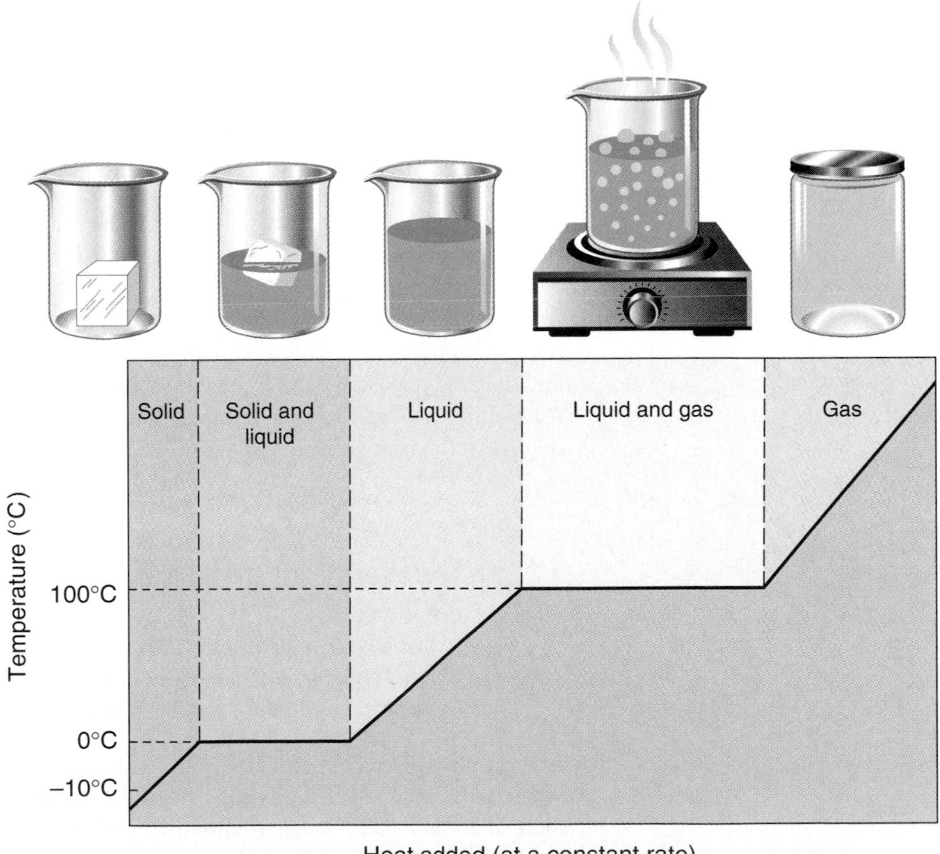

FIGURE 12.3
Heating (or cooling) curve for water. The addition (and removal) of heat affects the states of water.

TABLE 12.2
Normal Melting Points and Normal Boiling Points of Several Substances

Substance	Normal Melting Point (°C)	Normal Boiling Point (°C)
Neon (Ne)	−249	−246
Hydrogen sulfide (H_2S)	−86	−61
Mercury (Hg)	−39	357
Water (H_2O)	0	100
Sodium (Na)	98	883
Sodium chloride (NaCl)	801	1413
Quartz (SiO_2)	1610	2230

melting point of the solid and the normal freezing point of the liquid are always identical. For example, the normal melting point and the normal freezing point of water are both 0°C.

After all of the ice has melted, the temperature of the sample, which is now entirely liquid, begins to rise once again. At 100°C, the liquid begins to boil, as shown in Figure 12.3. The temperature does not change while the water is boiling even though heat is still being added. This heat makes the water vaporize; it does not raise the temperature of the gaseous water that is being formed. If we stopped adding heat and prevented heat from escaping, the liquid water and the steam would coexist indefinitely. The liquid would continue to vaporize, but the steam would also condense, and at a rate equal to that of the vaporization. The **normal boiling point** of the liquid is *the constant temperature at 1 atm pressure at which the liquid changes into a gas.* The normal boiling point of water is 100°C. When all of the liquid water has changed to steam, the temperature of the sample begins to rise once again. The normal melting and normal boiling points of several substances, including water, are listed in Table 12.2. ●

● The boiling point of a substance is usually more dependent on the pressure than is the melting point.

When the entire process is reversed, the same curve, now called a *cooling curve,* is followed in the opposite direction and heat is released from the sample. The heat released at the condensation point (the temperature at which a gas changes into a liquid) and the heat released at the freezing point are identical to the quantities of heat required for boiling and melting the sample. We examine these quantities of heat more thoroughly in the next section.

12.2 The Energy for a Change of State

OBJECTIVES

- Define the molar heat of fusion and the molar heat of vaporization.
- Given the molar heat of fusion or the molar heat of vaporization, calculate the heat required to melt or vaporize a known mass of a substance.

As you have already seen, energy in the form of heat is required to melt a solid or to vaporize a liquid. *The energy required to melt (fuse) 1 mol of a solid* is called the **molar heat of fusion.** *The energy required to vaporize 1 mol of a liquid* is called the **molar heat of vaporization.** The molar heats of fusion and vaporization of several substances are compared in Table 12.3. The next two examples demonstrate the use of these quantities.

TABLE 12.3
Some Molar Heats of Fusion and Vaporization

Substance	Molar Heat of Fusion (kJ/mol)	Molar Heat of Vaporization (kJ/mol)
Hydrogen (H_2)	0.12	0.9
Methane (CH_4)	0.94	10.4
Water (H_2O)	6.01	40.7
Mercury (Hg)	2.3	59.3

■ EXAMPLE 12.2 Calculating the Energy for a Change of State: Solid to Liquid

How much energy in the form of heat is required to melt 10.0 g of ice cubes at 0°C? The molar heat of fusion for water is 6.01 kJ/mol.

Problem Analysis

Calculate the quantity of water in moles and multiply it by water's molar heat of fusion.

Solution

Begin by calculating moles of water. Because the molecular weight of water is (2 × 1.008 + 16.00) amu or 18.02 amu, the molar mass is 18.02 g/mol, and

$$\text{Moles of water} = 10.0\ \cancel{g} \times \frac{1\ \text{mol}}{18.02\ \cancel{g}} = 0.555\ \text{mol}$$

Next, calculate the heat required to melt this quantity of ice.

$$0.555\ \cancel{\text{mol}} \times \frac{6.01\ \text{kJ}}{1\ \cancel{\text{mol}}} = 3.34\ \text{kJ}$$

Thus 3.34 kJ of heat must be added to melt 10.0 g of ice at 0°C.

Exercise 12.2

Mercury freezes at −39°C. Calculate the heat that must be removed from 1.0 g of mercury at the freezing point so that the sample will freeze. The molar heat of fusion is 2.3 kJ/mol.

(Try Problems 12.29, 12.30, 12.31, and 12.32.)

■ EXAMPLE 12.3 Calculating the Energy for a Change of State: Liquid to Gas

How much energy is required to convert 10.0 g of liquid water at 100°C to steam at the same temperature? The molar heat of vaporization at 100°C is 40.7 kJ/mol.

Problem Analysis

Obtain the quantity of water in moles and multiply it by water's heat of vaporization.

Solution

From Example 12.2 you know that there is 0.555 mol in 10.0 g of water. The heat required to vaporize this quantity of water is

$$0.555\ \cancel{\text{mol}} \times \frac{40.7\ \text{kJ}}{1\ \cancel{\text{mol}}} = 22.6\ \text{kJ}$$

Thus 22.6 kJ of heat must be added to vaporize 10.0 g of water at 100°C.

Exercise 12.3

Calculate the heat required to vaporize 1.0 g of mercury at its normal boiling point (357°C). The molar heat of vaporization of mercury at the boiling point is 59.3 kJ/mol.

(Try Problems 12.33, 12.34, 12.35, and 12.36.)

Example 12.2 showed that 3.34 kJ of heat is required to melt 10.0 g of ice while maintaining the temperature at 0°C. What happens if we now freeze this liquid again? Clearly, 3.34 kJ of heat will be released. Similarly, you saw in Example 12.3 that vaporizing 10.0 g of water at 100°C requires 22.6 kJ of heat. These calculations also imply that 22.6 kJ of heat will be released if we condense 10.0 g of steam at 100°C.

Suppose you want to know how much energy in the form of heat is required to raise the temperature of a certain quantity of ice at its normal freezing point to steam at the normal boiling point. You are faced with three problems, as the next example shows.

■ EXAMPLE 12.4 Calculating the Energy for a Change of State: Solid to Liquid to Gas

How much heat is required to melt 10.0 g of ice at 0°C, raise the temperature of the liquid water from 0°C to 100°C, and then vaporize the liquid at 100°C? The specific heat of water is 4.18 J/(g · °C).

Problem Analysis

Divide this problem into three separate parts.

Total heat = Heat needed for melting at 0°C + Heat needed to raise temperature from 0°C to 100°C + Heat needed for vaporizing at 100°C

You calculated answers for the first and third parts in Examples 12.2 and 12.3. The second part requires the specific heat of water. Review the topic of specific heat in Section 3.7 in Chapter 3. Then calculate the heat required to raise the temperature from 0°C to 100°C using the specific heat and the mass expressed in grams from

$$\text{Heat} = \text{specific heat} \times \text{mass} \times \text{temperature change}$$

Solution

The specific heat is the heat needed to raise the temperature of 1 g of water by 1°C. Thus you need to multiply the specific heat by the mass in grams (10.0 g) and the temperature change (100°C − 0°C = 100°C). Thus

$$\text{Heat} = \frac{4.18\ \text{J}}{1\cancel{\text{g}} \cdot 1\cancel{^\circ\text{C}}} \times 10.0\ \cancel{\text{g}} \times 100\cancel{^\circ\text{C}} = 4.18 \times 10^3\ \text{J}$$

and

$$4.18 \times 10^3\ \cancel{\text{J}} \times \frac{1\ \text{kJ}}{1000\ \cancel{\text{J}}} = 4.18\ \text{kJ}$$

The total heat required is the sum of the heat requirements from the three steps, or

$$3.34\ \text{kJ} + 4.18\ \text{kJ} + 22.6\ \text{kJ} = 30.1\ \text{kJ}$$

Exercise 12.4

How much heat must be removed from 5.0 g of gaseous mercury at 357°C (the normal boiling point) to cause condensation, a decrease in the temperature of the liquid to −39°C (the normal freezing point), and then freezing at that temperature? The specific heat of mercury is 0.14 J/(g · °C). The molar heats of fusion and vaporization are 2.3 and 59.3 kJ/mol, respectively.

(Try Problems 12.37, 12.38, 12.39, and 12.40.)

12.3 Vapor Pressure and Evaporation

OBJECTIVES

- Discuss evaporation in terms of vapor pressure.
- Given the vapor pressure of water and Dalton's law of partial pressures, calculate the partial pressure of a gas that has displaced water and its molar amount.

Liquids (and some solids) are constantly evaporating (vaporizing) *even when the temperature is less than the boiling point.* We know, for example, that we need to water indoor plants often. Although some of the water is taken up by the plants, a significant portion evaporates from the soil. Similarly, water in an open container evaporates until no more of the water remains (Figure 12.4). If the container is covered, however, a small, almost undetectable quantity evaporates, and then evaporation appears to stop. The reason is simple: Because gaseous water in the closed container is always in contact with the surface of the liquid, it can condense (and often does). When the rates of condensation and evaporation become equal, the quantity of the liquid will no longer change. The liquid and gas are in *equilibrium.* Chapter 14 presents a more complete discussion of equilibrium.

What is the difference between evaporation below the boiling point of a liquid and evaporation at the boiling point? Below the boiling point, the liquid evaporates principally from its surface. Although evaporation from the surface occurs at the boiling point, evaporation also takes place within the liquid so that bubbles of gas appear and escape by rising to the surface.

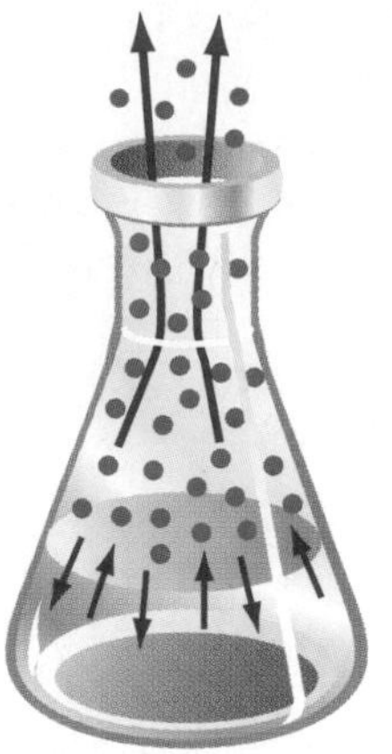

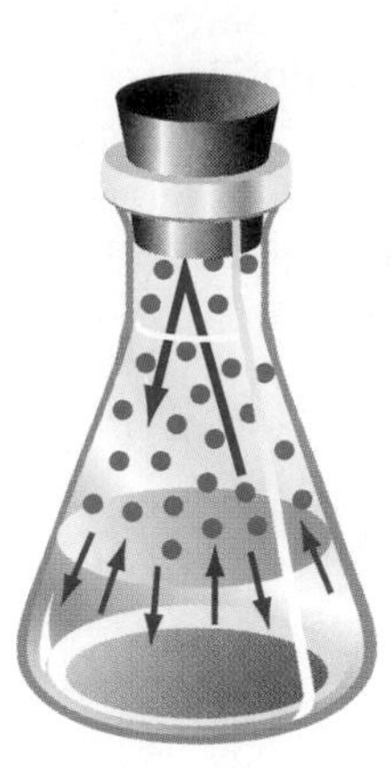

FIGURE 12.4
Behavior of a liquid in an open and closed container. The liquid on the left evaporates readily because the gaseous molecules can leave the container. The liquid on the right cannot evaporate because the stopper prevents molecules from leaving the container. With enough time, the rate of molecules leaving the liquid will be equaled by the rate of molecules returning to it, so the quantity of the liquid will not change.

The vapor above the liquid exerts a pressure like any other gas. The **vapor pressure** of a liquid is *the pressure exerted by the liquid's vapor at equilibrium.* Table 12.4 gives the vapor pressure of water for several temperatures. Note that the vapor pressure increases smoothly as the temperature increases.

The temperature at which the vapor pressure of a liquid equals the atmospheric pressure is another definition for the **boiling point** of the liquid. The vapor pressure of water

TABLE 12.4
Vapor Pressure of Water at Various Temperatures

Temperature (°C)	Pressure (mmHg)
0	4.6
5	6.5
10	9.2
15	12.8
20	17.5
21	18.7
22	19.8
23	21.1
24	22.4
25	23.8
26	25.2
27	26.7
28	28.3
29	30.0
30	31.8
40	55.3
50	92.5
60	149.4
70	233.7
80	355.1
90	525.8
100	760.0

is 760 mmHg, or exactly 1 atm, when the temperature is 100°C. This is the *normal boiling point,* defined earlier—the temperature at which the vapor pressure of a liquid equals 1 atm pressure. However, the atmospheric pressure affects the boiling point of water (and other liquids). For example, the vapor pressure of water at 70°C is about 234 mmHg. If the atmospheric pressure were also 234 mmHg, then this sample of water would boil. A liquid will boil when its vapor pressure is equal to the atmospheric pressure. Thus the boiling point of water on the top of Mt. Everest is usually about 70°C because the atmospheric pressure on top of the mountain is usually close to 234 mmHg. ●

● Because of this phenomenon, hard-boiling an egg in the mountains takes much more time than cooking it at sea level.

Collecting Gases over Water

The information in Table 12.4 has at least two other uses. It is used by meteorologists to calculate the relative humidity, as described in the Chemical Perspective that follows this section. The data are also used when scientists want to measure the quantity of a gas produced in a chemical reaction by allowing the gas to displace water (provided the gas will not react with water or dissolve appreciably in it). An example of this procedure is shown in Figure 12.5. Hydrogen produced by a chemical reaction in the flask on the left is collected by leading it to an inverted tube where it displaces the water that initially filled the tube. The tube is graduated so that the volume of the gas can be measured. The hydrogen is not pure, however, because the gas bubbles moving through the water in the tube pick up water vapor (gaseous water). Thus the gas collected is really a mixture of hydrogen and water vapor. Because both hydrogen and water molecules collide with the

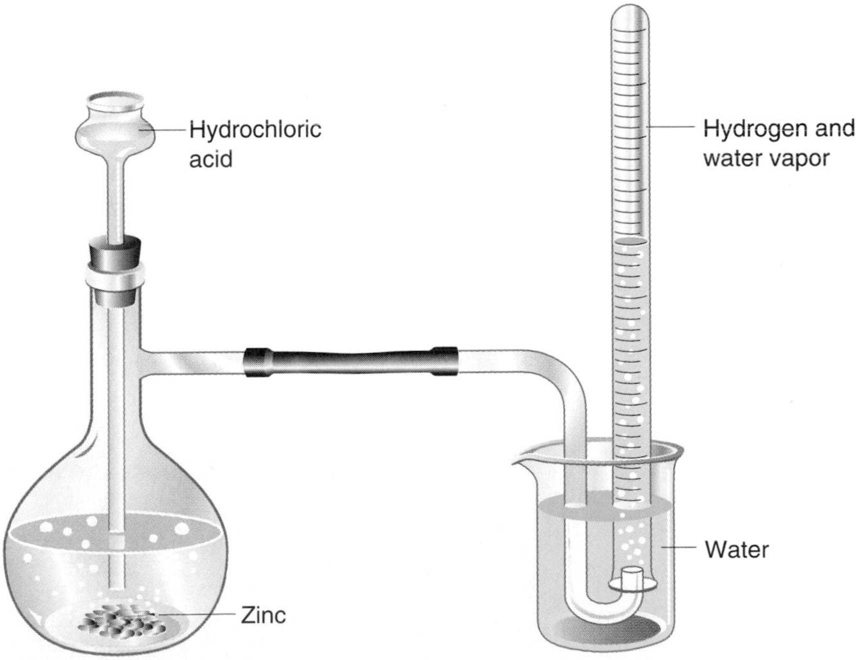

FIGURE 12.5
Collection of a gas over water. Hydrogen, prepared by the reaction of zinc with hydrochloric acid, is collected by displacing water that was initially in the tube.

walls of the tube, both exert a pressure. According to Dalton's law of partial pressures (see Section 11.9 in Chapter 11), the total pressure is the sum of the pressure exerted by hydrogen and the vapor pressure of water, or

$$P_{\text{Total}} = P_{\text{Hydrogen}} + P_{\text{Water vapor}}$$

(The total pressure is also equal to the atmospheric pressure if the water level inside the tube is adjusted so that it is equal to the water level outside the tube.) The following example uses this information to find the amount of oxygen generated during the decomposition of hydrogen peroxide.

EXAMPLE 12.5 Using Dalton's Law of Partial Pressures to Calculate the Amount of Gas Produced

Hydrogen peroxide (H_2O_2), the substance used for bleaching hair, decomposes slowly to give water and molecular oxygen. Suppose the oxygen that evolves from this reaction is collected over water. The volume of the gas is 128 mL at 24°C and 762 mmHg. Calculate the amount of oxygen obtained in moles.

Problem Analysis

Find the vapor pressure of water at 24°C from Table 12.4. Next, use Dalton's law of partial pressures to calculate the partial pressure of oxygen by subtracting the vapor pressure of water from the total pressure of the gas (762 mmHg). At this point, you know the pressure, volume, and temperature of the oxygen gas, so you can use the ideal gas law to calculate the amount of oxygen in moles (see Section 11.8).

Solution

According to Table 12.4, the vapor pressure of water at 24°C is 22.4 mmHg. You know that the total pressure is given by

$$P_{\text{Total}} = P_{\text{Oxygen}} + P_{\text{Water vapor}}$$

Solving for P_{Oxygen}, you obtain

$$P_{\text{Oxygen}} = P_{\text{Total}} - P_{\text{Water vapor}}$$

Thus

$$P_{\text{Oxygen}} = 762 \text{ mmHg} - 22.4 \text{ mmHg} = 740 \text{ mmHg}$$

Convert this result to atmospheres, the volume to liters, and the temperature to kelvins because R from the ideal gas equation contains these units.

$$P_{\text{Oxygen}} = 740 \cancel{\text{mmHg}} \times \frac{1 \text{ atm}}{760 \cancel{\text{mmHg}}} = 0.974 \text{ atm}$$

$$V = 128 \cancel{\text{mL}} \times \frac{1 \text{ L}}{1000 \cancel{\text{mL}}} = 0.128 \text{ L}$$

$$T = 24 + 273 = 297 \text{ K}$$

Collecting your data, you obtain

$P_{Oxygen} = 0.974$ atm	$n = ?$
$V = 0.128$ L	$R = 0.08206$ L · atm/(K · mol)
	$T = 297$ K

Solve for moles of oxygen by using the form of the ideal gas equation in Example 11.9.

$$n = \frac{PV}{RT} = \frac{0.974\ \cancel{\text{atm}} \times 0.128\ \cancel{\text{L}}}{0.08206\ \cancel{\text{L}} \cdot \cancel{\text{atm}}/(\cancel{\text{K}} \cdot \text{mol}) \times 297\ \cancel{\text{K}}} = 0.00512 \text{ mol}$$

Exercise 12.5

If 203 mL of a gas is collected over water at 15°C and 735 mmHg, how many moles of dry gas were collected?

(Try Problems 12.41 and 12.42.)

Chemical Perspective

Relative Humidity

Water vapor in the air is not usually in equilibrium with liquid water below it (except under special conditions encountered in a laboratory). Instead, the amount of water vapor in the air varies from trace quantities to about 4% by volume. The *relative humidity* expresses the water vapor content as a percentage of the maximum possible amount. The maximum possible amount at a given temperature is the equilibrium vapor pressure.

$$\text{Relative humidity} = \frac{\text{actual partial pressure}}{\text{equilibrium vapor pressure}} \times 100\%$$

For example, suppose the partial pressure of water vapor on a particular evening in New York is 12.8 mmHg, and the temperature is 20°C. What is the relative humidity? The equilibrium vapor pressure at 20°C is 17.5 mmHg, so

$$\text{Relative humidity} = \frac{12.8\ \cancel{\text{mmHg}}}{17.5\ \cancel{\text{mmHg}}} \times 100\% = 73.1\%$$

However, if the air temperature drops to 15°C, as the evening turns into night, without any change in the partial pressure of water vapor, the relative humidity becomes 100%, because the equilibrium vapor pressure at 15°C is 12.8 mmHg.

$$\text{Relative humidity} = \frac{12.8\ \cancel{\text{mmHg}}}{12.8\ \cancel{\text{mmHg}}} \times 100\% = 100\%$$

The air is now saturated with water vapor. Any further temperature drop will result in the formation of dew because the air cannot hold any more water. The drops of water that sometimes form on a glass of cold water are caused by this phenomenon. The temperature at which water just begins to condense from the air is called the *dew point*, a term that you may have heard mentioned on a TV weather report.

MOLECULAR EXPLANATIONS

The *kinetic molecular theory* introduced in Chapter 11 uses the behavior of gas molecules to explain the properties of gases and the temperature, pressure, and volume relationships summarized in the gas laws. The theory also helps us to understand why liquids and solids form, and why they possess their characteristic properties.

12.4 A Kinetic Molecular Description of Gases, Liquids, and Solids

OBJECTIVE

- Explain the differences between the three states of matter and the factors that lead to changes of state in terms of attractions between molecules and their temperature-dependent motions.

The kinetic molecular theory of gases indicates that the average distances between the molecules (or atoms in the case of the Group VIIIA elements) that compose a gas are so large that most of the space within a gas is empty. It is easy to compress a gas by increasing the pressure because the space occupied by the gas is so thinly populated with gas molecules. The theory also indicates that the molecules in a gas are in constant, random, and chaotic motion, and that their average kinetic energy—that is, their energy of motion—increases as the temperature increases and decreases as the temperature decreases. Attraction between the gas molecules is negligible because the kinetic energy of a gas under moderate conditions overcomes the forces of attraction between the molecules. The result is that the molecules are free to wander in any direction. They expand to fill the entire container, collectively adopting its shape (Figure 12.6A). Clearly, a gas has neither a fixed shape nor a fixed volume.

When the gas in Figure 12.6A is cooled, the average kinetic energy of the molecules decreases, and they move more slowly. When the temperature of the gas has dropped sufficiently, the attractions between the molecules begin to be important because of the molecules' reduced kinetic energy. (We discuss the nature of these attractions in the next section.) The attractions cause the molecules to come so close together that many of them touch. As a result, the gas condenses and a liquid appears. However, because random (albeit restricted) movement of the molecules continues, we can say only that the molecules of a liquid are *relatively* closely packed in a random arrangement (Figure 12.6B).

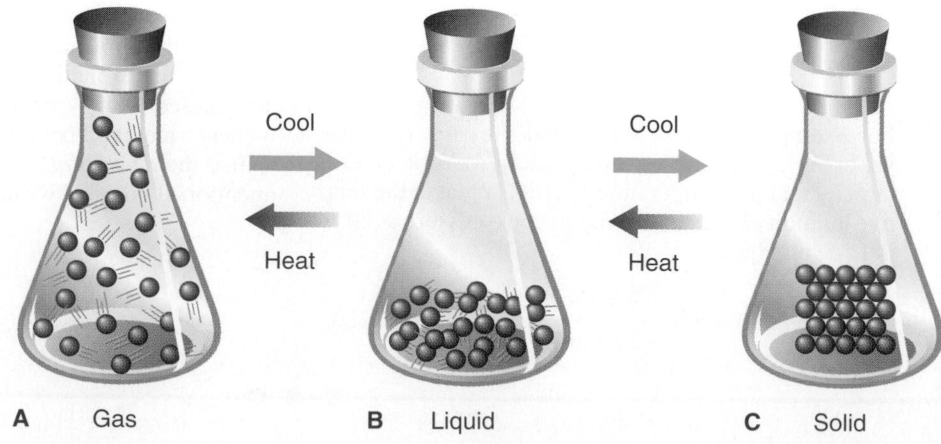

FIGURE 12.6

Heating and cooling affect the states of matter. Cooling a gas *(A)* decreases chaotic molecular motion and eventually converts the gas to a partially ordered liquid *(B)* and finally to an ordered solid *(C)*. Heating increases molecular motion and reverses the process.

FIGURE 12.7
Surface tension. *(A)* An insect walks on water and *(B)* water droplets are spherical because of surface tension. *(C)* Surface tension occurs because a molecule on the surface of a liquid experiences a net attraction toward the interior, whereas a molecule in the interior experiences balanced attraction from all sides.

Because many of the molecules are touching, liquids cannot be compressed significantly. Although the molecules in a liquid still move about, the attractions between them are strong enough to keep them from wandering in all directions. Thus a liquid has a fixed volume. However, these attractions are not strong enough to prevent the molecules from flowing past one another. As a result, a liquid assumes the shape of as much of a container as it is able to fill.

If we continue to cool the liquid, motions of the molecules will diminish until they no longer have enough kinetic energy to move past each other. At that point, the liquid freezes and a solid forms (Figure 12.6C). The attractive forces keep the molecules fixed in position (although they vibrate). As a result, the molecules in solids possess considerable order and little chaotic movement. They are unable to flow past one another. Thus solids are rigid and have a fixed volume and shape. ●

● The average distance between molecules and the tendency for chaotic motion follow the trend

gas > liquid > solid

The trend in attractions between molecules is

gas < liquid < solid

The kinetic molecular theory also allows us to explain other macroscopic observations. For example, we have seen that the vapor pressure of a liquid changes as the temperature changes. The theory tells us that the vapor pressure at a particular temperature depends on the ease with which the molecules of the liquid are able to escape into the gas phase. Thus the vapor pressure is dependent on the strength of the attractions within the liquid. As the temperature increases, molecular motions become greater and more chaotic, more molecules are able to escape from a liquid's surface, and the vapor pressure increases.

Liquids also have another important property, one that explains why insects can walk on water and why water droplets are spherical (Figure 12.7). As Figure 12.7C shows, a molecule within the liquid tends to be attracted to other molecules in all directions. A molecule on the surface, however, has no molecules above it to counteract the pulls and tugs from those below. The unbalanced attractions cause the layer of molecules on the surface to behave like an elastic skin. Like any skin, it resists stretching. The energy required to stretch a unit area of this skin is called *surface tension*; strong attractions lead to a high surface tension. The insect in Figure 12.7A can walk on water because of this skin. This same surface tension also causes the skin to make itself as small as possible. Water droplets like those in Figure 12.7B are spherical because a sphere has the smallest area per unit volume.

Although we have mentioned the importance of attractions in liquids and solids, we have not yet shown what causes them. We look at the causes of these attractions in Sections 12.5 and 12.6.

12.5 The Liquid State

Exploration: Intermolecular forces.

OBJECTIVE

- Describe the nature of the forces between the molecules of a liquid.

The boiling point of a liquid is a rough measure of the strength of the attractive forces between the molecules in the liquid. Liquids with high boiling points require considerable energy in the form of heat to overcome the attractive forces that keep the molecules from escaping into the gaseous state. The three types of attractive forces between molecules are *dispersion forces, dipole–dipole forces,* and *hydrogen bonding*.

Dispersion Forces

Because helium, neon, argon, krypton, and xenon (the noble gases) have very low boiling points (all of them are lower than −100°C), we know that the attractive forces in their liquid states are exceedingly weak. Let's examine the attractive forces in neon as an example. The ten electrons of a neon atom move about the nucleus so that over a period of time they behave as if they had been smeared on the surface of a spherical ball. The electrons spend as much time on one side of the nucleus as the other. Nevertheless, at any one instant, there may be more electrons on one side than on the other. When this occurs (and it probably occurs very often), one side of the atom will have more negative charges than the other. For a brief instant, one side of the atom has a small negative charge, $\delta-$, while the other side has a partial positive charge, $\delta+$, as shown in Figure 12.8A. The neon atom has become polarized.

Next, imagine that another neon atom is near the negative end of the polarized atom. (If it were near the positive end, it would not change our conclusion.) The partial negative charge on the first atom repels the negatively charged electrons of the second atom, so that it too becomes polarized for a brief instant (Figure 12.8B). The two polarized atoms are now oriented so that the partial negative charge of one is next to the partial positive charge of the other. An attractive force exists between these atoms for a brief

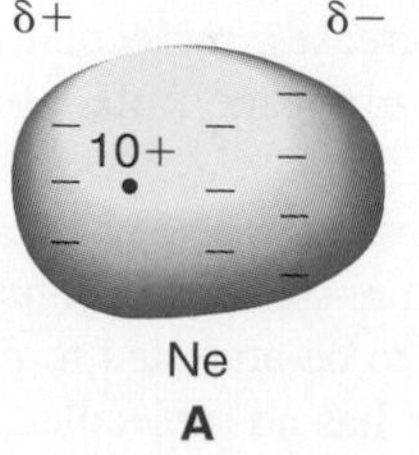

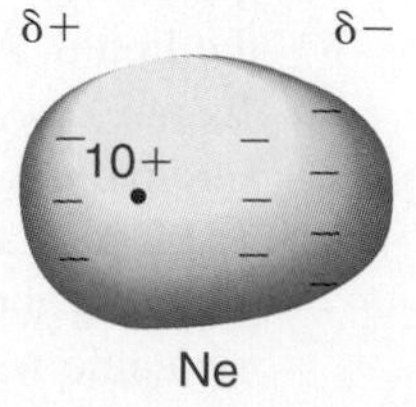

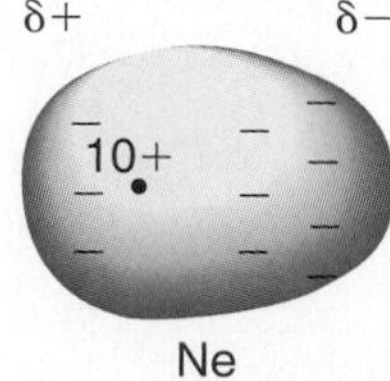

FIGURE 12.8
Origin of dispersion forces in neon. *(A)* At a particular instant, there may be more electrons (shown as negative charges) on one side of a neon atom's nucleus (shown as 10+ over the black dot) than the other. The atom becomes polar. *(B)* If this polarized atom is near another neon atom, it also becomes polarized, and an instantaneous attractive force exists.

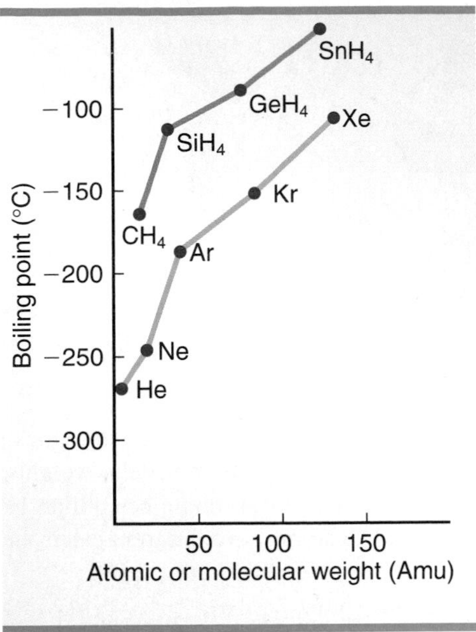

FIGURE 12.9
Boiling point versus atomic weight or molecular weight. Graphs for noble gases and for a series of nonpolar molecules; both show a fairly smooth increase of boiling point with atomic weight due to increasing dispersion forces.

moment, and it is only one of the many temporary attractive forces of this type in any sample of neon. Although these forces come and go, a net attractive force is always present in the liquid.

Thus we see that **dispersion forces** are *the weak attractive forces between temporarily polarized atoms (or molecules) caused by the varying positions of the electrons during their motion about nuclei.* Dispersion forces occur in the liquid states of all molecules, both polar and nonpolar, but they are solely responsible for holding otherwise nonpolar molecules in the liquid state. A nonpolar molecule cannot pass into the gaseous state during boiling until the dispersion forces have been overcome.

Figure 12.9 shows the boiling points of the noble gases as a function of their atomic weights. Notice that the boiling points increase as the atomic weights increase. Thus dispersion forces must increase as atomic weights increase. The reason for this dependence is simple. Atoms with larger atomic weights have more electrons, and dispersion forces increase in strength with the number of electrons. For the same reason, dispersion forces increase with the molecular weights of nonpolar molecules, as also shown in Figure 12.9. Methane (CH_4), with a molecular weight of 16.04 amu, boils at −164°C; silane (SiH_4), with a molecular weight of 32.12 amu, boils at −112°C; germane (GeH_4), with a molecular weight of 76.64 amu, boils at −89°C; and stannane (SnH_4), with a molecular weight of 122.74 amu, boils at −52°C.

Dipole–Dipole Forces

Although the molecules in *any* liquid experience dispersion forces, polar molecules also experience another type of intermolecular force. A polar molecule is attracted to another polar molecule through dipole–dipole forces (Figure 12.10). A **dipole–dipole force** is *a permanent attractive intermolecular force resulting from the interaction of the positive end of one molecule with the negative end of another.* Dipole–dipole forces are generally stronger than dispersion forces, at least in molecules with approximately equal molecular weights.

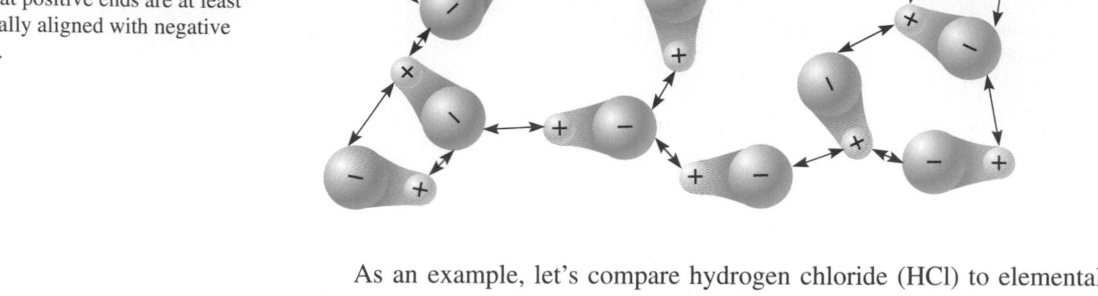

FIGURE 12.10
Alignment of the polar molecules in liquid HCl. Even with the random motions that occur, molecules of liquid HCl tend to be oriented so that positive ends are at least partially aligned with negative ends.

As an example, let's compare hydrogen chloride (HCl) to elemental fluorine (F_2). Because HCl and F_2 have similar molecular weights (36.46 amu for HCl and 38.00 amu for F_2), their dispersion forces should be similar. However, a molecule of HCl is polar because the H and Cl atoms have different electronegativities. ●

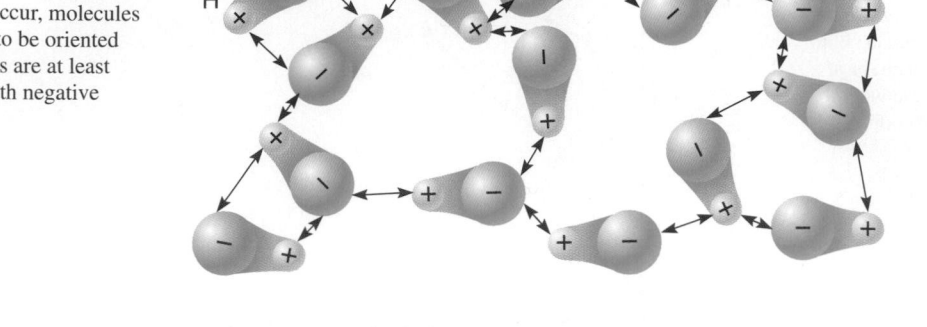

● The electronegativity of chlorine is greater than that of hydrogen.

$$\overset{\delta+}{\text{H}}-\overset{\delta-}{\text{Cl}}$$

In contrast, a molecule of F_2 is nonpolar because both of the atoms in the molecule are identical. HCl molecules experience both dispersion forces and dipole–dipole forces, whereas F_2 molecules experience only dispersion forces. As a result, polar HCl boils at −85°C, but nonpolar F_2 boils at −188°C. ●

● Here's another example: Polar AsH_3 (a triangular pyramidal molecule) with a molecular weight of 77.94 amu boils at −55°C, but nonpolar GeH_4 (a tetrahedral molecule) with a molecular weight of 76.62 amu boils at −90°C.

Hydrogen Bonding

Molecules that contain a hydrogen atom covalently bonded to a small, electronegative atom are subject to yet another attractive force, called hydrogen bonding. A **hydrogen bond** is *a dipole–dipole attractive force that exists between two polar molecules containing a hydrogen atom covalently bonded to an atom of nitrogen, oxygen, or fluorine.* Water is an example of a substance with extensive hydrogen bonding. Examine the pattern of hydrogen bonding in liquid water in Figure 12.11. A hydrogen bond is repre-

FIGURE 12.11
Hydrogen bonding in water. The hydrogen bonds are represented by a series of dots between a hydrogen atom of one molecule and the oxygen atom of another one.

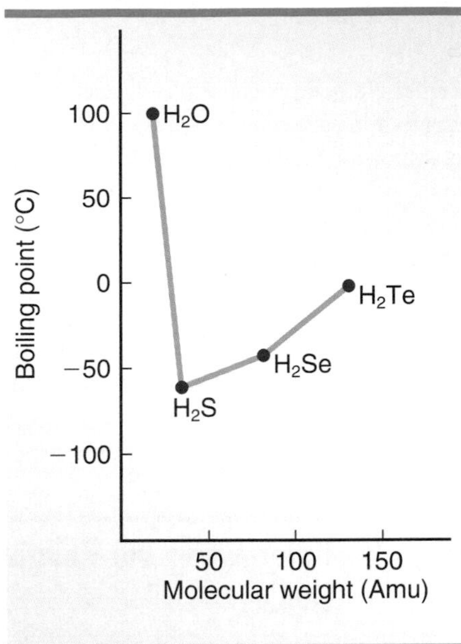

FIGURE 12.12
Boiling point versus molecular weight for the polar hydrogen compounds of Group VIA elements. The boiling points of H_2S, H_2Se, and H_2Te increase regularly, suggesting the importance of dispersion forces. However, H_2O has a much higher boiling point than any other member of this series, indicating additional attractions between molecules due to hydrogen bonding.

sented by the series of dots, and the normal covalent bonds are shown as dashes. As you can see, water contains two kinds of bonds: the normal covalent bonds within each molecule and the hydrogen bonds between a hydrogen atom in one molecule and the oxygen atom in another molecule. Hydrogen bonding is usually stronger than other dipole–dipole attractive forces, but it is considerably weaker than the covalent bonding within molecules.

Figure 12.12 compares the boiling point of H_2O to those of H_2S, H_2Se, and H_2Te. This comparison is of interest because oxygen, sulfur, selenium, and tellurium belong to the same group in the periodic table. If only dispersion forces prevailed, we would expect the boiling points to follow the same trend as the molecular weights, or $H_2Te > H_2Se > H_2S > H_2O$. Although this trend occurs for H_2Te, H_2Se, and H_2S, it is easy to see that water has a boiling point that does not follow this trend. In fact, we would be puzzled by water's high boiling point if we did not know about hydrogen bonding.

The next three examples show you how to identify forces between molecules and how these forces affect the relative vapor pressures and boiling points of liquids.

EXAMPLE 12.6 Identifying Forces Between Molecules

What types of attractive forces do you expect to occur between the molecules in the liquid forms of HF and HCl?

Problem Analysis

First, remember that dispersion forces occur in all liquids. Next, look for polar molecules, and then look for molecules that can participate in hydrogen bonding. Recall that hydrogen bonds occur between polar molecules containing a hydrogen atom bonded to an atom of nitrogen, oxygen, or fluorine.

Solution

Both molecules are polar because the electronegativity of hydrogen differs from that of the other atom in each molecule. Liquid HF also contains hydrogen bonds because HF is a polar molecule containing a hydrogen atom bonded to a fluorine atom. Thus you should expect hydrogen bonding in liquid HF in addition to dispersion forces. Liquid HCl is held together by both dispersion forces and dipole–dipole forces.

Exercise 12.6

List the attractive forces that you would expect to occur between molecules in the liquid forms of each of the following compounds: hydrogen (H_2), bromine (Br_2), and hydrogen bromide (HBr).

(Try Problems 12.43, 12.44, 12.45, and 12.46.)

EXAMPLE 12.7 Determining the Relative Vapor Pressures of Liquids

Does liquid HF or liquid HCl have the higher vapor pressure at a particular temperature? Explain your reasoning.

Problem Analysis

Identify the forces between molecules in the liquid forms of the compounds (see Example 12.6). Determine which one has the weaker forces. The compound with weaker attractions between the molecules has the higher vapor pressure.

Solution

Remember that dispersion forces increase with molecular weight. Since the molecular weights of HF and HCl are 19.01 and 36.46 amu, respectively, the dispersion forces between HCl molecules should be greater than those between HF molecules. However, liquid HF also contains hydrogen bonds, and the attractions coming from hydrogen bonding are much stronger than those resulting from dispersion forces. Thus liquid HCl should have weaker attractions between the molecules and a higher vapor pressure (at any temperature).

Exercise 12.7

Arrange the following nonpolar compounds in order of increasing vapor pressure at a particular temperature: CH_4, C_2H_6, and C_3H_8. Explain your reasoning.

(Try Problems 12.47 and 12.48.)

EXAMPLE 12.8 Determining the Relative Boiling Points of Liquids

Does liquid HF or liquid HCl have the higher normal boiling point? Explain your reasoning.

Problem Analysis

Identify the forces between molecules in the liquid forms of the compounds (see Example 12.6). Determine which one has the lower vapor pressure (see Example 12.7). The compound with the lower vapor pressure has a higher boiling point.

Solution

Recall that the normal boiling point is the temperature at which the vapor pressure of a liquid is identical to an atmospheric pressure of 760 mmHg. You saw in Example 12.7 that the vapor pressure of HCl should be higher than that of HF at any particular temperature. Suppose, then, that you observe the vapor pressures of samples of these substances as you raise the temperature. Clearly, the vapor pressure of HCl reaches 760 mmHg at a lower temperature, whereas a higher temperature is required for HF. Thus the normal boiling point of HF is higher than that of HCl. (In fact, the normal boiling points of HF and HCl are 19.5°C and −84.9°C, respectively.)

Exercise 12.8

Does nonpolar methane (CH_4) or nonpolar silane (SiH_4) boil at a higher temperature? Explain your reasoning.

(Try Problems 12.49, 12.50, 12.51, and 12.52.)

12.6 The Solid State

OBJECTIVE

- Describe the nature of the forces between atoms, molecules, or ions in solids.

Solids such as ice, copper, table salt, and table sugar are pure solids that we encounter regularly. Other solids, such as wood, cement, and paper are mixtures of various substances. A pure solid consists of an array of structural units—atoms, molecules, or ions—that are packed more or less tightly together because of strong attractions between the units. Solids can be crystalline or amorphous. A **crystalline solid** is *a solid that is composed of one or more crystals with each crystal having a well-defined, ordered arrangement of structural units.* Examples are sodium chloride (table salt) and sucrose (table sugar). In contrast, an **amorphous solid** is *a solid that lacks order and, as a result, does not have a well-defined arrangement of structural units.* Window glass is an example. Our discussion here focuses entirely on crystalline solids. ●

● Many of the common solids, such as wood and cement, are not pure; they are mixtures.

We classify crystalline solids according to the five types of structural units found in the solids:

1. An **atomic solid** is *a solid containing atoms from a nonmetallic element.* Examples are neon and the remainder of the noble gases (Group VIIIA).
2. A **metallic solid** is *a solid containing atoms from a metallic element.* Examples are copper, silver, and gold.
3. A **covalent network solid** is *a solid containing atoms held in large networks or chains by covalent bonds.* Examples are carbon and silicon dioxide.
4. A **molecular solid** is *a solid consisting of molecules.* Chlorine (Cl_2), water (ice), carbon dioxide (dry ice), and sucrose (table sugar) are examples.
5. An **ionic solid** is *a solid consisting of ions.* Sodium chloride (table salt) is an example.

Let's examine each of these types in more detail.

Atomic Solids

Solids of frozen noble gases consist of single atoms that are fixed in an ordered, well-defined array. The atoms are held in place by dispersion forces. These forces are exerted equally in all directions. Thus the greatest attraction occurs when each atom is surrounded by the largest possible number of other atoms. This type of arrangement is called a *close-packed structure.* These structures are analogous to those you would obtain if you carefully filled a container with marbles so that it would contain the largest possible number of marbles. Atomic solids have low melting points because the attractions between the atoms are weak dispersion forces.

Metallic Solids

Metallic solids also consist of single atoms in an ordered, well-defined array, but the atoms are held in place by metallic bonding. In this type of bonding, the atoms have lost one or more of their outer electrons so that only positively charged cores of the atoms remain. These cores occupy fixed positions and are held together by a surrounding "sea" of the outer electrons. These electrons are responsible for a metal's ability to conduct electricity.

Many metals crystallize in a close-packed structure similar to those found with the noble gases. Examples are copper, silver, and gold (Figure 12.13). Although the melting points of metals are often very high, they exhibit considerable variability. For example, mercury melts at −39°C (mercury is the only metal that is a liquid at room temperature), but tungsten melts at 3410°C (the highest melting point of any of the metals).

Covalent Network Solids

Solids of some nonmetals contain large networks of atoms that are held together by strong covalent bonds. Individual molecules do not exist. Diamond, one form of carbon, is an excellent example. Every carbon atom is covalently bonded to four others (Figure 12.14). As you can see, these four atoms are arranged in the shape of a regular tetrahedron. Clearly, a single diamond crystal (such as one in a ring) might be considered a giant molecule. When covalent network solids melt, we must break the strong covalent

FIGURE 12.13
The structure of copper, a metallic solid. Although the copper atoms actually touch, this view allows the ordered arrangement of the close-packed atoms to be seen easily.

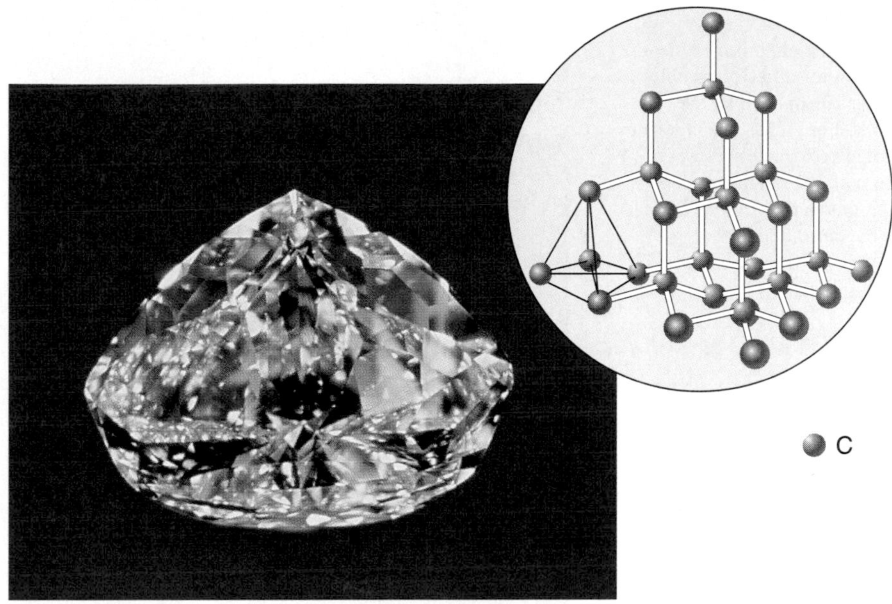

FIGURE 12.14
The structure of diamond, a covalent network solid. The *Centenary* shown here is the largest modern-cut diamond in the world. Like any other diamond, it is a giant molecule. Every carbon atom in it is covalently bonded to four other carbon atoms.

bonds. For that reason, their melting points are relatively high. Diamond, for example, melts at 3550°C. Solid silicon and germanium have structures that mimic that of diamond. Graphite, another form of carbon, is a different type of covalent network solid. Silicon dioxide (SiO_2) is another substance that forms a solid with this type of structure.

Molecular Solids

Nonmetals to the right of carbon in the periodic table form molecular solids (excluding the noble gases). For example, we find N_2, P_4, O_2, and F_2 in molecular solids. A third form of carbon, the recently discovered buckminsterfullerene, belongs in this category because it consists of C_{60} molecules. In addition, molecular solids include compounds such as water (Figure 12.15), carbon dioxide, and sucrose. The forces that hold the

FIGURE 12.15
The structure of snow and ice, a molecular solid. Every water molecule is connected to four other water molecules by hydrogen bonds to give a very ordered array of molecules.

FIGURE 12.16
The structure of sodium chloride, an ionic solid. Every chloride ion is surrounded by six smaller sodium ions. The chloride ions are approximately close-packed with the much smaller sodium ions in the cavities between the chloride ions.

ordered array of molecules together in a molecular solid always include dispersion forces, but dipole–dipole forces or hydrogen bonding may also contribute. As a result, the melting points of molecular solids reflect the same trends that we found in the boiling points of liquids. Molecular solids with molecular weights of less than 200 amu are often liquids (such as water) or even gases (such as carbon dioxide) at room temperature.

Ionic Solids

Ionic solids such as sodium chloride are held together by the strong attractive forces that exist between oppositely charged ions. These forces are much stronger than those between the partial positive and negative charges found in molecular solids with dipole–dipole forces. The stronger attractive forces cause higher melting points, much higher than the melting points of molecular solids (sodium chloride melts at 801°C). Unlike molecular solids, ionic substances are always solids at room temperature.

Ionic forces are exerted equally in all directions. As a result, each ion in an ionic solid can be surrounded by the largest possible number of oppositely charged ions. Sodium chloride is a good example, as you will see when you study Figure 12.16 carefully. Several other solids, such as magnesium oxide (MgO), share this arrangement of ions. However, other arrangements of ions occur in some other ionic compounds.

The next two examples show you how to identify types of solids and how the forces in them affect their relative melting points.

■ EXAMPLE 12.9 Identifying Types of Solids

Sulfur dioxide (SO_2) is an air pollutant; potassium iodide (KI) is added to common table salt by the distributor to provide iodine in our diets. Identify the type of solid you expect for each substance.

Problem Analysis

Notice that both substances are compounds. Distinguish between ionic and covalent compounds.

Solution

Sulfur dioxide is a molecular substance (both sulfur and oxygen are nonmetals); therefore, it freezes as a molecular solid. Potassium iodide is an ionic substance (potassium is a metal, and iodine is a nonmetal); it exists as an ionic solid.

Exercise 12.9

Classify each of the following solids according to its type: zinc (Zn), sodium bromide (NaBr), and methane (CH_4).

(Try Problems 12.53, 12.54, 12.55, and 12.56.)

■ EXAMPLE 12.10 Determining Relative Melting Points

Does sulfur dioxide or potassium iodide have a higher melting point? Explain your reasoning.

Problem Analysis

Identify the type of solid expected for each substance (see Example 12.9). Determine the types of attractions in each solid and which one has the stronger forces. The compound with stronger attractions between the molecules has a higher melting point.

Solution

As you saw in Example 12.9, sulfur dioxide is a molecular solid, so it is held together by dispersion forces. Potassium iodide is an ionic compound, so it is held together by electrical attractions between oppositely charged ions. Because ionic attractions are stronger than dispersion forces, potassium iodide will have a higher melting point than sulfur dioxide.

Exercise 12.10

Name the type of solid that is formed for each of the following substances: $MgCl_2$, CH_3OH, and Ar. On the basis of the type of solid that you expect, arrange these substances in order of increasing melting point.

(Try Problems 12.57, 12.58, 12.59, and 12.60.)

Chemical Perspective

Why Does Ice Float?

As our population grows and more people reach for the water faucet, we ask more and more questions about water. The answers to some of these questions depend on the structure of water molecules and the influence that they exert on each other. Let's illustrate by asking a simple question: "Why does ice float?"

As most liquids freeze, the solid part falls to the bottom because it is more dense than the liquid; this is not the case with water. Extensive hydrogen bonding occurs in ice because water contains two O—H bonds. As you saw in Figure 12.15, each oxygen atom interacts with four hydrogen atoms. Two of these hydrogen atoms are covalently bonded to oxygen, and the other two interactions are the result of hydrogen bonding. As a result, the structure of ice consists of an interlocking array of six-membered rings. When ice melts, some (but not all) of the hydrogen bonds that maintain the open structure in Figure 12.15 break. This partial collapse causes liquid water to have a greater density than ice because a given mass of water molecules will occupy less volume after the structure has collapsed. Because of the difference in densities, ice floats on liquid water.

If ice did not float, many northern lakes would freeze completely during the winter and much of the aquatic life in those waters would not survive. Instead, a layer of ice floats on the surface, insulating the water below from further cooling and freezing and protecting the lake from freezing solid (Figure 12.17).

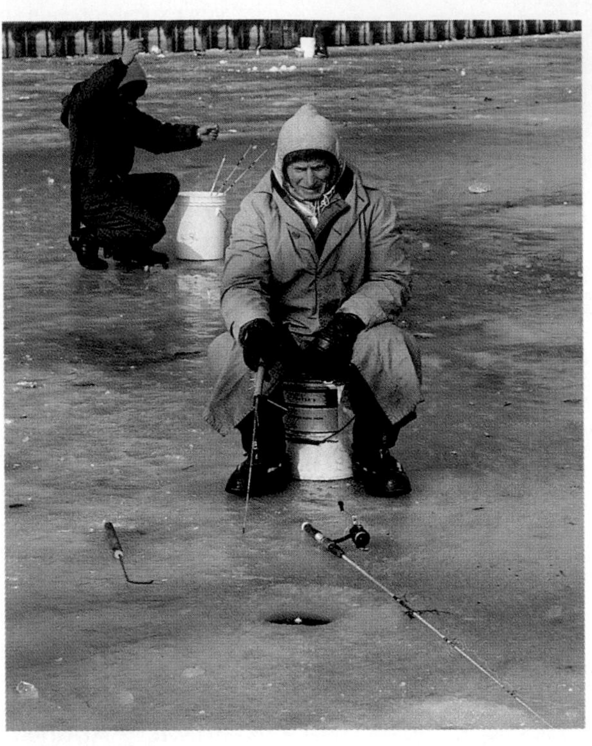

FIGURE 12.17
Ice is less dense than liquid water. The aquatic life in this lake is able to withstand the winter because the ice does not sink; instead, it insulates the water from the cold.

CHAPTER REVIEW

Key Words

gas *(p. 363)*
liquid *(p. 363)*
solid *(p. 363)*
melting (fusion) *(p. 363)*
freezing *(p. 363)*
vaporization (evaporation) *(p. 364)*
condensation *(p. 364)*
deposition *(p. 364)*
sublimation *(p. 364)*
normal melting point *(p. 365)*
normal freezing point *(p. 365)*
normal boiling point *(p. 366)*
molar heat of fusion *(p. 366)*
molar heat of vaporization *(p. 366)*
vapor pressure *(p. 370)*
boiling point *(p. 370)*
dispersion forces *(p. 377)*
dipole–dipole force *(p. 377)*
hydrogen bond *(p. 378)*
crystalline solid *(p. 381)*
amorphous solid *(p. 381)*
atomic solid *(p. 381)*
metallic solid *(p. 381)*
covalent network solid *(p. 381)*
molecular solid *(p. 381)*
ionic solid *(p. 381)*

Summary

The three states of matter can be distinguished by the following macroscopic properties: a *gas* has neither a fixed volume nor a fixed shape; a *liquid* has a fixed volume but it does not have a specific shape; and a *solid* has a fixed volume and a fixed shape.

Any state of matter can change into another state. These changes of state occur during melting, freezing, vaporization, condensation, deposition, and sublimation. Every change of state involves energy in the form of heat. For example, if a sample of water is to melt, heat must be supplied. The *normal melting point* and *normal boiling point* are characteristic macroscopic properties of a substance that involve changes of state and energy in the form of heat. The *molar heat of fusion* of a solid is the energy required to melt (fuse) 1 mol of the solid. The *molar heat of vaporization* of a liquid is the energy required to vaporize 1 mol of the liquid. The *vapor pressure* of a liquid at a particular temperature is another characteristic macroscopic property of a liquid. If a gas is collected over water, the total pressure includes a contribution from the vapor pressure of water.

The *kinetic molecular theory* provides insight into the reasons for the macroscopic properties of the three states of matter. For example, the theory indicates that gases are composed of molecules in constant random motion throughout mostly empty space. Attractions between the molecules are overwhelmed by the chaotic motions of the molecules. Thus gases can be compressed easily and occupy any container entirely. The molecules of a liquid are also in constant but lessened random motion, and so are more tightly packed than in a gas. Attractions between the molecules have become important because the random motions are weaker. As a result, liquids have a fixed volume, but they cannot be compressed significantly. Solids are composed of atoms, molecules, or ions that are in close contact and that occupy fixed positions because the kinetic energy of these particles is not enough to overwhelm the attractions between the particles. Thus solids are rigid and cannot be compressed.

The three attractive forces are *dispersion forces, dipole–dipole forces,* and *hydrogen bonding.* Dispersion forces are weak attractive forces found in all molecules. Dipole-dipole forces are the attractive forces coming from the attraction of the positive end of one polar molecule to the negative end of another. Hydrogen bonding occurs in substances containing hydrogen atoms covalently bonded to nitrogen, oxygen, or fluorine atoms.

Solids are either *crystalline* or *amorphous.* Crystalline solids can also be classified according to the type of structural unit in the solid. Thus we have *atomic, metallic, molecular, covalent network,* and *ionic solids* in which the structural units are atoms, molecules, or ions.

Problem-Solving Skills

1. **Identifying a Change of State:** Given a description of a sample of matter before and after a change of state, identify the change of state (Example 12.1).
2. **Calculating the Energy for a Change of State (Solid to Liquid and Liquid to Gas):** Given the mass of a substance and either its molar heat of fusion or its molar heat of vaporization, calculate the energy required for a change of state (Examples 12.2 and 12.3).
3. **Calculating the Energy for a Change of State (Solid to Liquid to Gas):** Given the mass of a substance, its molar heat of fusion, its molar heat of vaporization, and the specific heat of the liquid, calculate the energy required to change a solid to a liquid and then to a gas (Example 12.4).
4. **Using Dalton's Law of Partial Pressures to Calculate the Amount of Gas Produced**: Given the volume, total pressure, and temperature of a gas collected over water, calculate the molar quantity of the dry gas (Example 12.5).
5. **Identifying Forces Between Molecules:** Given a chemical formula, identify the expected attractive force or forces between molecules (Example 12.6).
6. **Determining the Relative Vapor Pressures of Liquids:** Given two liquids of known composition, decide on the basis of attractive forces between molecules which has the larger (or smaller) vapor pressure at the same temperature (Example 12.7).
7. **Determining the Relative Boiling Points of Liquids:** Given two liquids of known composition, decide on the basis of attractive forces between molecules which has the higher (or lower) normal boiling point (Example 12.8).
8. **Identifying Types of Solids:** Given a solid of known composition, classify it as an atomic, metallic, covalent network, molecular, or ionic solid (Example 12.9).
9. **Determining Relative Melting Points:** Given two (or more) solids of known composition, decide on the basis of attractive forces between molecules which has the higher (or lower) melting point (Example 12.10).

Questions to Test Your Reading

12.1 Describe two macroscopic differences between gases, liquids, and solids.

12.2 What is a change of state? Give an example.

12.3 How does temperature affect the states of matter? Give an example.

12.4 List the different changes of state and give an example of each.

12.5 Name two substances that are liquids at 25°C. Name two others that are solids at this temperature.

12.6 Can ice and liquid water coexist indefinitely at 0°C and an atmospheric pressure of 1 atm? Explain. Can liquid water and steam coexist indefinitely at 120°C and an atmospheric pressure of 1 atm? Explain.

12.7 Does the melting point of a solid differ from the freezing point of the liquid form of the same substance? Explain.

12.8 What is the *normal melting point* of a solid?

12.9 What is the *normal boiling point* of a liquid?

12.10 What are the normal freezing point and the normal boiling point of water?

12.11 What are the *molar heat of fusion* and the *molar heat of vaporization* of a substance?

12.12 Why will 18 g of steam at 100°C melt more ice than 18 g of liquid water at 100°C?

12.13 Why would evaporation of some part of a liquid lead to cooling of the remaining liquid?

12.14 What is *vapor pressure*? What variable will affect the vapor pressure of a liquid?

12.15 Under what circumstances will water boil at 50°C?

12.16 Explain evaporation of a liquid below its boiling point at the molecular level.

12.17 Why are raindrops almost spherical?

12.18 Explain the origin of dispersion forces between two molecules. Will these forces exist in liquid nitrogen (N_2) or in water? Explain.

12.19 Explain the origin of dipole–dipole forces between two molecules. Give an example of a liquid in which this type of force is important.

12.20 When will hydrogen bonding occur? Give an example of a liquid (other than water) in which this type of force is important.

12.21 If a molecule of water had a linear shape, would the molecule be polar? Explain.

12.22 Describe the distinguishing characteristics of a crystalline solid and an amorphous solid.

12.23 Describe the distinguishing characteristics of an atomic solid, a metallic solid, a covalent network solid, a molecular solid, and an ionic solid. Give an example of each.

12.24 Explain why ionic solids always have relatively high melting points but the melting points of atomic solids are relatively low.

12.25 Explain why covalent network solids have relatively high melting points.

12.26 Explain why molecular solids can have either relatively low or relatively high melting points.

Practice Problems

Changes of State (Sections 12.1 and 12.2)

12.27 Identify the change of state occurring in each of the following events.
(a) When chlorine gas is admitted into an extremely cold tube, a yellow liquid forms.
(b) Molten lava from a volcano cools and turns into rock.

12.28 Identify the change of state occurring in each of the following events.
(a) Mothballs slowly disappear.
(b) A cold windshield becomes covered with ice when struck by raindrops.

12.29 How much heat is needed to melt 75.0 g of ice at 0°C? The molar heat of fusion for water is 6.01 kJ/mol.

12.30 How much heat is produced when 75.0 g of liquid water freezes at 0°C? The molar heat of fusion for water is 6.01 kJ/mol.

12.31 How much heat is required to melt 26.4 g of solid ammonia at its melting point (−78°C)? The molar heat of fusion of this substance is 5.65 kJ/mol.

12.32 How much heat is released when 26.4 g of liquid ammonia freezes at −78°C? The molar heat of fusion of this substance is 5.65 kJ/mol.

12.33 How much heat is required to convert 75.0 g of liquid water at 100°C to steam at the same temperature? The molar heat of vaporization at 100°C is 40.7 kJ/mol.

12.34 How much heat is produced when 33.0 g of steam at 100°C is converted to liquid water at the same temperature? The molar heat of vaporization at 100°C is 40.7 kJ/mol.

12.35 How much heat is released when 26.4 g of gaseous gold is converted to liquid at its normal boiling point (2660°C)? The molar heat of vaporization at this temperature is 310. kJ/mol.

12.36 How much heat is required to vaporize 1.0 g of liquid gold at the normal boiling point (2660°C)? The molar heat of vaporization at this temperature is 310. kJ/mol.

12.37 How much heat is required to change 12.2 g of ice at its normal freezing point to steam at 100°C? The specific heat of water is 4.184 J/(g · °C), the molar heat of fusion at 0°C is 6.01 kJ/mol, and the molar heat of vaporization at 100°C is 40.7 kJ/mol.

12.38 How much heat is liberated if 53 g of steam at the normal boiling point of water is converted to ice at the normal freezing point? The specific heat of water is 4.184 J/(g · °C), the molar heat of fusion at 0°C is 6.01 kJ/mol, and the molar heat of vaporization at 100°C is 40.7 kJ/mol.

12.39 How much heat is needed to change 33.3 g of water at 37°C to steam at 100°C, the normal boiling point? The specific heat of water is 4.184 J/(g · °C), and the molar heat of vaporization at 100°C is 40.7 kJ/mol.

12.40 How much heat is liberated when 45 g of steam at the normal boiling point of water is converted to liquid water at 43°C? The specific heat of water is 4.184 J/(g · °C), and the molar heat of vaporization at 100°C

Vapor Pressure and Dalton's Law of Partial Pressures (Section 12.3)

12.41 A 232-mL sample of hydrogen is collected over water at 30°C. The total pressure of the gas mixture is 752 mmHg. What is the partial pressure of the hydrogen? How many moles of this substance have been collected?

12.42 Some carbon monoxide (4.52 L) is collected over water at 45°C. If the total pressure is 1.05 atm, what is the partial pressure of carbon monoxide? How many moles of this substance have been collected? How many grams?

Attractive Forces Between Molecules in Liquids (Section 12.5)

12.43 For each of the following substances, list the kinds of attractive forces between atoms or molecules that are expected.
(a) radon (Rn)
(b) hydrogen fluoride (HF)
(c) hydrogen iodide (HI)

12.44 For each of the following substances, list the kinds of attractive forces between molecules that are expected.
(a) hydrogen sulfide (H_2S)
(b) ammonia (NH_3)
(c) iodine (I_2)

12.45 Arrange the following substances in order of increasing dispersion forces: $SiCl_4$, CCl_4, and $GeCl_4$. Explain.

12.46 Arrange the following substances in order of increasing dispersion forces: Rn, He, and Ar. Explain.

12.47 Tetrahedral methane (CH_4) reacts with elemental chlorine to form a series of chlorinated substances. One of these is tetrahedral carbon tetrachloride (CCl_4). Both compounds (CH_4 and CCl_4) are nonpolar. Which of these two substances is expected to have the lower vapor pressure at a given temperature? Explain.

12.48 The halogens form a series of interesting compounds with each other. Two of these are bromine monochloride (BrCl) and bromine monofluoride (BrF). Which of these two substances is expected to have the lower vapor pressure at a given temperature? Explain.

12.49 Arrange the following nonpolar substances in order of increasing boiling point: $SiCl_4$, CCl_4, and $GeCl_4$. Explain your reasoning.

12.50 Arrange the following substances in order of increasing boiling point: Rn, He, and Ar. Explain your reasoning.

12.51 Will water or hydrogen selenide (H_2Se) have the higher boiling point? Both molecules have a bent shape. Explain your answer.

12.52 Will hydrogen selenide (H_2Se) or hydrogen telluride (H_2Te) have the higher boiling point? Both molecules have a bent shape. Explain your answer.

Solids (Section 12.6)

12.53 One form of elemental phosphorus occurs as tetrahedral P_4 molecules. What kind of solid is expected for this substance?

12.54 The most abundant form of elemental oxygen occurs as O_2 molecules. What kind of solid is expected for this substance?

12.55 What kind of solid is expected for barium chloride ($BaCl_2$)?

12.56 What kind of solid is expected for titanium (Ti)?

12.57 Will water or hydrogen selenide (H_2Se) have the higher melting point? Both molecules have a bent shape. Explain your answer.

12.58 Will hydrogen selenide (H_2Se) or hydrogen telluride (H_2Te) have the higher melting point? Both molecules have a bent shape. Explain your answer.

12.59 Will elemental nitrogen (N_2) or elemental phosphorus (P_4) have the higher melting point? Explain your reasoning.

12.60 There are two forms of elemental oxygen: O_2 and O_3. Which one will have the higher melting point? Explain.

Additional Problems

12.61 A particular refrigerator cools by evaporating liquefied dichlorodifluoromethane (CCl_2F_2), a substance whose molar heat of vaporization is 17.4 kJ/mol. How many grams of this compound must evaporate to freeze 525 g of water at 0°C to ice at the same temperature?

12.62 The molar heat of vaporization of ammonia is 23.4 kJ/mol. How many grams of water at 0°C can be frozen to ice at 0°C by the vaporization of 1.00 kg of ammonia?

12.63 The heats of vaporization of liquid Cl_2 and liquid H_2 are 20.4 and 0.9 kJ/mol, respectively. Are the relative values in line with your expectations? Explain.

12.64 The heats of vaporization of liquid Ne and liquid CH_3OH are 1.8 kJ/mol and 34.5 kJ/mol, respectively. Are the relative values in line with your expectations? Explain.

Practice in Problem Analysis

For each problem, describe the thinking you would use (the problem analysis) before doing the actual solution, but do not solve the problem.

1. Could we use water as a replacement for chlorofluorocarbons (CFCs) as a refrigerator coolant?

2. Sometimes a physician will prescribe an alcohol rub to reduce fever. How does this procedure work?

PRACTICE EXAM

1. Which statement is true?
   (a) Ice will not melt without the addition of heat.
   (b) Steam will not condense without the addition of heat.
   (c) Water will not freeze without the addition of heat.
   (d) Frost will not form without the addition of heat.
   (e) None of the above

2. A gas that does not dissolve in water is collected over water at 26°C. After the pressure is adjusted to atmospheric pressure (747 mmHg), the pressure due to the gas is
   (a) 747 mmHg (b) 772 mmHg
   (c) 722 mmHg (d) unpredictable
   (e) None of the above

3. Which statement is true?
   (a) The vapor pressure of a liquid is the partial pressure of the gas from the liquid at equilibrium.
   (b) The boiling point of a liquid is the temperature at which the vapor pressure of the liquid equals the atmospheric pressure.
   (c) The vapor pressure of a liquid increases as the temperature increases.
   (d) The boiling point of a liquid decreases as the atmospheric pressure decreases.
   (e) All of the above

4. Which of the following liquid compounds has the highest boiling point?
   (a) CF_4 (b) SiF_4 (c) GeF_4 (d) SnF_4
   (e) PbF_4

5. Which of the following liquid compounds has a boiling point greater than 50°C?
   (a) H_2O (b) H_2S (c) H_2Se (d) H_2Te
   (e) None of the above

6. Which of the following liquid compounds has the highest vapor pressure at a particular temperature?
   (a) CF_4 (b) SiF_4 (c) GeF_4 (d) SnF_4
   (e) PbF_4

7. Which of the following liquid compounds has a vapor pressure of 760 mmHg at 100°C?
   (a) H_2O (b) H_2S (c) H_2Se (d) H_2Te
   (e) None of the above

8. Which of the following solid compounds has the highest melting point?
   (a) CF_4 (b) SiF_4 (c) GeF_4 (d) SnF_4
   (e) PbF_4

9. Which of the following liquid compounds has a melting point greater than 0°C?
   (a) H_2O (b) H_2S (c) H_2Se (d) H_2Te
   (e) None of the above

10. Which of the following substances will have dispersion forces?
    (a) NH_3 (b) PH_3 (c) AsH_3 (d) SbH_3
    (e) All of the above

11. Which of the following substances will have hydrogen bonding?
    (a) NH_3 (b) PH_3 (c) AsH_3 (d) SbH_3
    (e) All of the above

12. Which of the following solid compounds has the highest melting point?
    (a) HCl (b) NaCl (c) Cl_2 (d) CCl_4
    (e) Ar

13. Which of the following solid substances is a metallic solid?
    (a) $BaCl_2$ (b) Ti (c) Cl_4 (d) Ar
    (e) None of the above

14. Which of the following solid substances is an ionic solid?
    (a) $BaCl_2$ (b) Ti (c) CCl_4 (d) Ar
    (e) None of the above

15. Which of the following solid substances is a molecular solid?
    (a) $BaCl_2$ (b) Ti (c) CCl_4 (d) Ar
    (e) None of the above

Chapter

13 Solutions

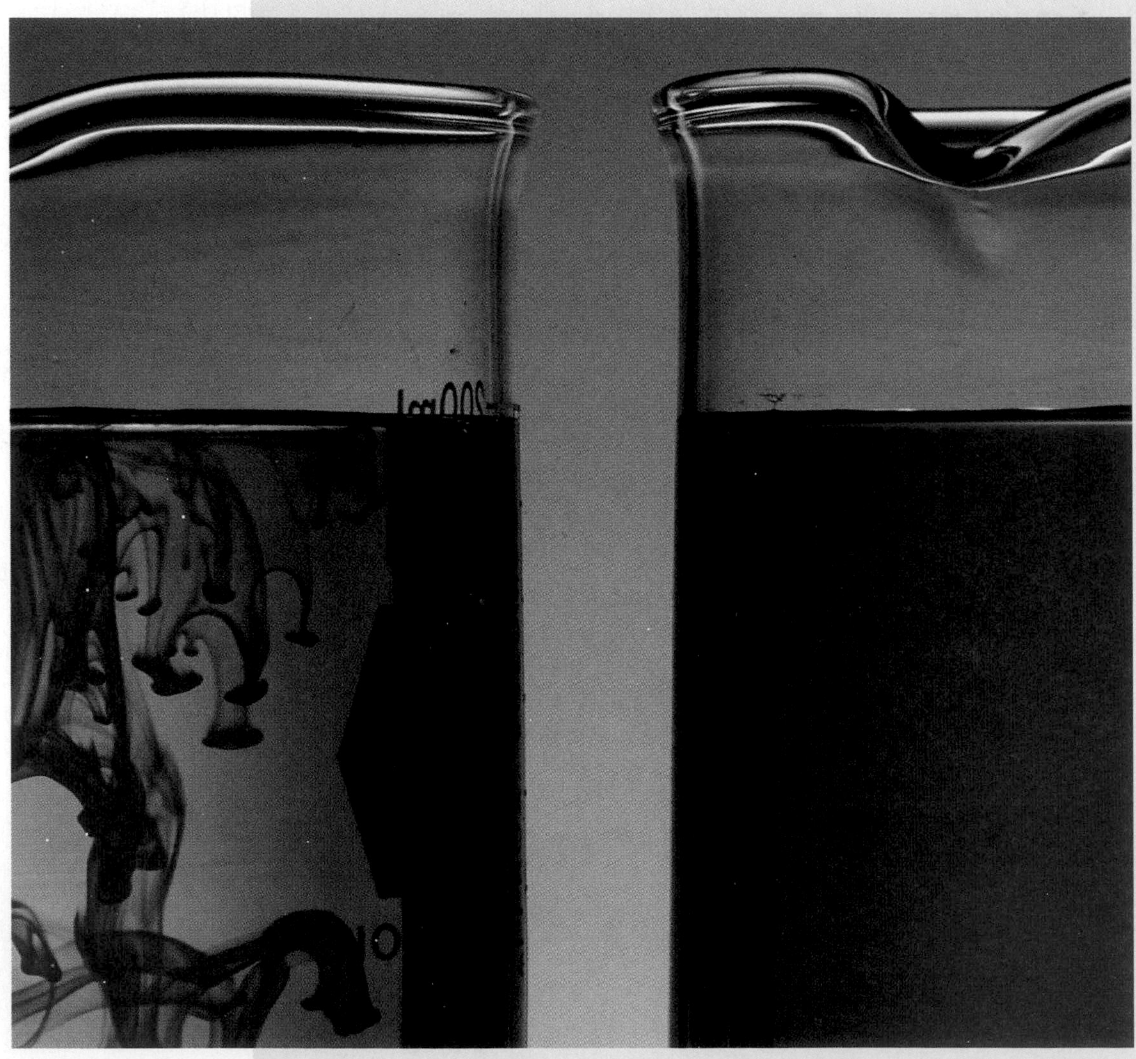

Red dye in solution.

Chapter Outline

Water as it flows from the kitchen tap is not a pure substance; it is a solution. You may recall from Chapter 3 that a solution is a homogeneous mixture of substances, that is, a mixture that has the same properties throughout (unlike beach sand, in which you might see specks of various shells, as well as different kinds of rock particles). As you look at most tap water, it appears clear, without noticeable bits of other substances in it. If you evaporate the water, however, you will see a residue of mineral substances that were previously dissolved in the water. These minerals were dispersed uniformly through the solution as ions. Tap water also contains dissolved gases from the atmosphere, which are dispersed uniformly through the solution as molecules. When cool tap water warms up, some of these dissolved gases come out of solution to form small bubbles. You can see these bubbles form when you set aside a glass of cold tap water.

Solutions are very common. Many, like tap water and commercial products such as window cleaners, are liquids. Others are gases and solids. Air is a solution of nitrogen, oxygen, and other gases. Like any solution, air is a homogeneous mixture; all parts of the gaseous solution appear the same. Jewelry gold is an example of a solid solution (Figure 13.1). It is a homogeneous mixture of silver dissolved in gold.

We often prepare a solution because of some special property it has. One of the reasons for adding silver to gold is to prepare a harder metal that is more easily worked. Pure gold is very soft, and as an item of jewelry would not withstand everyday wear. Also, a solid solution of gold containing some silver melts at a lower temperature than

FIGURE 13.1
Jewelry gold. Jewelry gold is a solid solution of gold containing some silver.

pure gold, so it is easier to cast. (Generally, a solution has a lower melting point than a pure substance; that is, the melting point of a substance is depressed by dissolving another substance in it.)

In this chapter, we explore a number of important questions relating to solutions. For example, how much salt will dissolve in water at 20°C? In other words, what is the solubility of salt at this temperature? How do we express the amount of salt that is dissolved in a solution? What is the concentration of salt in the solution? What is the freezing point of a solution of salt in water? (By how much is the freezing point of water depressed by adding a certain amount of salt to it?) Before answering these questions, we need an introduction to solutions.

AN INTRODUCTION TO SOLUTIONS

You can see from the examples mentioned that the characteristics of solutions vary. To begin our discussion of solutions, we need a basic vocabulary to describe them, which we discuss in the next section. Then, in Section 13.2, we look at examples of the different types of solutions. In the final section of this introduction, we examine some general properties of solutions.

13.1 Some Terms Used to Describe Solutions

OBJECTIVE

- Define *solvent, solute, miscible,* and *immiscible.*

To discuss solutions, we need to define a few basic terms. A **solution** is *a homogeneous mixture of two or more substances*. When one of these substances is a liquid and the others are solids or gases, we tend to think of the liquid as the substance that dissolves the solid or gas. The liquid is referred to as the *solvent* and the solids and gases as the *solutes* (Figure 13.2). In the case of a solution of sucrose (table sugar) in water, sucrose is the solute and water is the solvent. You make up the solution by dissolving sucrose (the solute) in water (the solvent). Similarly, you dissolve air (the solute) in water (the solvent) by bubbling air into water.

When two gases or two liquids are mixed, the distinction between the substance that dissolves and the substance that is dissolved is not so clear. In those cases, we refer to

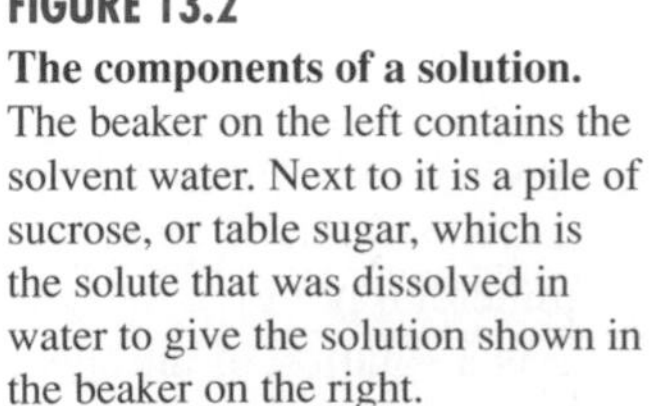

FIGURE 13.2
The components of a solution. The beaker on the left contains the solvent water. Next to it is a pile of sucrose, or table sugar, which is the solute that was dissolved in water to give the solution shown in the beaker on the right.

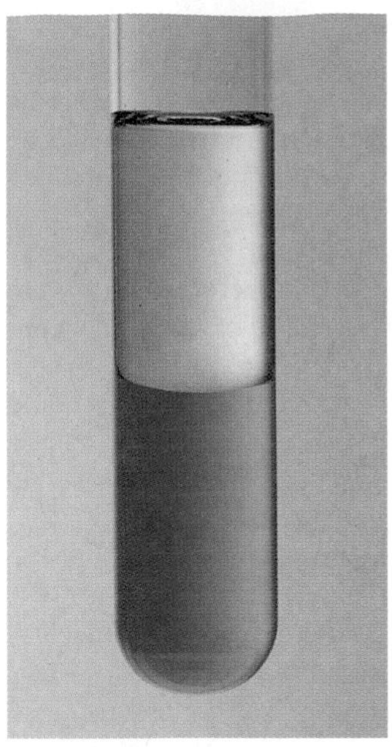

FIGURE 13.3
Immiscible liquids. The test tube contains two immiscible liquids, water and vegetable oil, which form two layers. Water is the bottom layer. (Food dye has been added to the water to make the layers show up more distinctly.)

the substance in greater amount as the solvent; the other substance is the solute. In summary, we define the terms *solvent* and *solute* as follows: The **solvent** is *the substance in a solution that dissolves another substance; if it is not clear which substance does the dissolving, it is the substance in the solution that is in greater amount.* A **solute** is *either a substance in a solution considered to have been dissolved by a solvent, or else a substance in a solution that is in smaller amount.*

Generally, when you mix substances to make a solution, you find that you can dissolve only a limited amount of one substance in another. As you dissolve more and more sucrose in water, you reach a point when no more sucrose dissolves no matter how much you stir the solution. When you mix two liquids together, however, you may find that they dissolve completely in one another, whatever proportions you take. Suppose, for example, that you have 100 mL of water in a beaker, and you begin adding ethyl alcohol to it. You find that no matter how much alcohol you add to the water, the liquids mix to form a homogeneous mixture, or solution. Eventually, you will have added so much alcohol that the solution is essentially alcohol with some water in it. *Two liquids that mix completely in one another to form a solution, no matter what the proportions of liquids,* are said to be **miscible.**

On the other hand, the two liquids might not mix appreciably. Consider what happens when you try to mix oil in water. You find that, rather than mixing, the oil forms a layer that floats on the water (Figure 13.3). Similarly, if you try to mix gasoline and water, you obtain two liquid layers. The bottom layer is mostly water with a small amount of gasoline dissolved in it, whereas the top layer is mostly gasoline with a small amount of water dissolved in it. (If the gasoline in your car has water dissolved in it, the water may separate out and freeze in the winter if it is cold enough, blocking the gasoline intake line.) *Two liquids that do not mix together or dissolve in one another in significant amounts, but rather tend to separate into two distinct layers,* are said to be **immiscible.** Now with this basic vocabulary, we can look at examples of various types of solutions.

13.2 Types of Solutions

OBJECTIVES

- Define aqueous and nonaqueous solutions.
- List examples of solutions in all three states of matter.

Solutions of solutes in water are generally referred to as *aqueous* solutions. (The word *aqueous* derives from the Latin *aqua,* meaning "water.") Aqueous solutions are frequently used when we want to react solid solutes in preparing other chemical substances. Solids generally react slowly when mixed because the reactant molecules (or ions, if they are ionic solids) have difficulty coming together to react. When the solid reactants are dissolved in water, however, the reactant molecules (or ions) readily move about the solution, where they come in contact with one another and react. Aqueous solutions are also frequently used to dispense substances, such as medicines.

Gasoline is a common example of a *nonaqueous* liquid solution. It is a mixture of hydrocarbons, which are compounds of hydrogen and carbon. Many other examples of nonaqueous solutions can be found. For example, many paint removers are solutions of various substances in dichloromethane (CH_2Cl_2).

Although aqueous and nonaqueous are useful labels for classifying some solutions, we need broader criteria to include all solutions. As we saw in the introduction, solutions can exist in any one of the three states of matter: gas, liquid, or solid. Thus, we classify solutions as to their physical state. Solutions in these three states may be further subdivided, depending on the physical states of the solutes and solvents from which they are formed.

Let's begin by looking at liquid solutions, particularly solutions in which water is the solvent. Water dissolves a broad range of substances. It dissolves many minerals in small amounts, including limestone, which is a common component of well water and water obtained from other natural sources. ● It dissolves small amounts of air, giving a solution whose oxygen content is important for the survival of fish and other aquatic animals. Carbon dioxide gas dissolves quite readily in water, a property that is used to make carbonated beverages. Water also dissolves many liquids. We have already mentioned alcohol–water solutions as an example.

● Limestone is not very soluble in water, but it does dissolve more easily if the water contains some dissolved carbon dioxide, from air or from decaying vegetation.

Liquid solutions are usually obtained by dissolving one or more solutes in a liquid solvent. However, sometimes when you mix the appropriate quantities of solids, such as sodium metal and potassium metal, you obtain a liquid solution. This is because the freezing point of a substance can be lowered by dissolving other substances in it, as discussed later in the chapter. By dissolving potassium in sodium, you lower the freezing point sufficiently that the solution melts to give a liquid solution.

Air is an example of a gaseous mixture. It contains about 78% nitrogen molecules (N_2) and about 21% oxygen molecules (O_2), with small amounts of other gases, such as Ar (argon), CO_2, and H_2O. Gaseous mixtures are commonly used by the chemical industry to prepare chemical substances. As in liquid solutions, the molecules in a gas are free to move about, and so can react to form products. A gaseous mixture of nitrogen (N_2), from air, and hydrogen (H_2), which is prepared from natural gas, is used to prepare ammonia (NH_3) for fertilizer and other uses. ●

● Natural gas contains methane, CH_4. It reacts with steam to produce hydrogen and carbon monoxide.

Although when we talk about solutions we often have liquid solutions in mind, solid solutions are quite common. Most commercial metals, for example, are mixtures of various metals, called *alloys,* and frequently these are homogeneous mixtures, or solutions. We have already mentioned jewelry gold. Another example is the alloy that is used to fill cavities in teeth, a solid solution of silver and other metals in mercury (a liquid metal). Many naturally occurring minerals and gemstones are solid solutions. Ruby, for example, is a solid solution of a small amount of chromium(III) oxide dissolved in aluminum oxide.

Table 13.1 lists examples of eight common types of solutions. They have been classified by the state (gas, liquid, solid) of the solution and the various states of the components, or constituents, of the solution.

TABLE 13.1
Examples of Solutions

Physical State of:			
Solution	**Solute**	**Solvent**	**Examples**
Gas	Gas	Gas	Air, auto exhaust fumes
Liquid	Gas	Liquid	Oxygen in water, carbonated beverages
Liquid	Liquid	Liquid	Ethanol in water, bromine in water
Liquid	Solid	Liquid	Sodium chloride in water (brine)
Liquid	Solid	Solid	Potassium-sodium alloy
Solid	Gas	Solid	Hydrogen in palladium
Solid	Liquid	Solid	Mercury in silver (dental alloys)
Solid	Solid	Solid	Alloys, such as copper in gold

13.3 General Properties of Solutions

OBJECTIVE

- List some general properties of solutions.

Although solutions vary widely in their properties, they do have a number of general characteristics. As we noted earlier, solutions are homogeneous, a property they share with pure substances (elements and compounds). Unlike a substance, however, a solution can vary in its composition. Thus, the percentage of sodium chloride in seawater (a solution), although approximately constant, is variable. The percentage of sodium chloride is lower in a sample of seawater obtained near the mouth of a river than it is further out in the ocean. On the other hand, the elemental analysis of sodium chloride (a compound) always gives 39.3% sodium by mass and 60.7% chlorine by mass, whether the sample is obtained from seawater, from an underground mine, or from any other source. ●

● We introduced compounds and mixtures in Chapter 3.

Also, unlike compounds, solutions can generally be separated by physical processes. In Chapter 3, we described how you can separate an aqueous solution of sodium chloride into its components, sodium chloride and water, by the physical process of distillation. Sodium chloride itself, however, cannot be separated by physical processes into elements; this can only be done by chemical reactions.

Another characteristic of solutions is that they remain evenly distributed over time. If you vigorously stir sand into water in a glass, you momentarily obtain an apparently uniform mixture of sand and water. But the uniformity does not last—soon the sand settles to the bottom of the glass. Sand and water form a heterogeneous mixture, not a solution. But suppose you add a drop of food coloring to a glass of pure water. At first the dye forms a stream of color as it descends through the water. As time passes, the dye diffuses through the water and becomes uniformly mixed through the solution (Figure 13.4). You can speed up this process by vigorous stirring, as you did with the sand and water, but however you add the dye, the result is a true solution. Unlike the uniform mixture of sand and water, the dye molecules do not settle out in time. The solute remains uniformly distributed.

FIGURE 13.4

Continued uniform distribution of a dye in a solution. In the beaker at the left, a concentrated dye solution has been added. On the right, a similar dye solution has already diffused uniformly through the solution. The dye molecules do not settle out, but rather remain uniformly distributed.

The reason for this difference has to do with the size of the particles. An aqueous solution and a uniform mixture of sand in water both consist of particles mixed through water. In an aqueous solution, the solute particles are molecules or ions, which are normally a few hundred picometers in diameter. Sand particles, on the other hand, are tenths of a millimeter or larger in size. Therefore, the force of gravity acting on the sand particles is relatively large, causing the particles to settle to the bottom of the glass. In a solution, the force of gravity acting on solute molecules or ions is inconsequential compared with the forces derived from constant buffeting of the solute particles by the solvent molecules. This buffeting keeps the solvent molecules and other solute particles in constant, random motion and they do not settle out. You can view the solution as a mixture of solute particles moving randomly and constantly through a swarm of solvent molecules.

This brings up the question of what we mean by the term *homogeneous* in describing a solution as a homogeneous mixture. Our theoretical view of a solution such as an aqueous solution of sodium chloride is that it consists of a mixture of sodium ions and chloride ions moving randomly through a collection of water molecules. If you look at a series of normal-size or even microscopic-size samples of the solution, the composition of each sample will be the same. Other properties of each sample will also be the same. However, if you could look at the solution at the molecular level, the solution will appear "grainy," or inhomogeneous (you will see separate molecules and ions). Thus, when we say a solution is homogeneous, we mean for samples of normal or even microscopic size, but not at the molecular level.

We can now summarize what we have discussed in this section.

1. Solutions are like compounds in terms of their homogeneity (except at the molecular level, where they appear grainy), but they are unlike compounds in that they have variable compositions.
2. A solution can generally be separated into its component parts by physical processes, unlike a compound, which can only be separated into its elements by chemical processes.
3. The solute in a solution is uniformly distributed and does not settle out over a period of time.

CONCEPT OF SOLUBILITY

When you stir a small quantity of sodium chloride (ordinary salt) into water, it dissolves to form a solution. If you continue adding similar samples of sodium chloride to the solution, the first samples dissolve as before, but eventually you reach a point where any additional salt that you add remains undissolved at the bottom of the container. Continued investigation would show you that there is a definite maximum amount of salt that you can dissolve in a given quantity of water. A 100-mL volume of water dissolves a maximum of 36 g of sodium chloride at 20°C. Therefore, we say that the *solubility* of sodium chloride at 20°C is 36 g per 100 mL of water. In the next section, we look at this concept of solubility more closely.

13.4 Saturated, Unsaturated, and Supersaturated Solutions

OBJECTIVES

- Define *solubility.*
- Distinguish between a saturated solution and an unsaturated solution.
- Distinguish between a saturated solution and a supersaturated solution.

To explore the concept of solubility in more detail, we will use sodium thiosulfate ($Na_2S_2O_3$) as our example. This is a white, crystalline compound used in photography to fix the image on a negative. (In this application, it is often referred to as "hypo," from an older name, sodium hyposulfite.) ●

● The thiosulfate ion combines with the silver ion to give a soluble silver compound, which is washed from the negative.

Definition of Solubility

Suppose you add 60 g of sodium thiosulfate to 100 mL of water contained in a flask at 20°C. After stirring vigorously, you find that 50 g of the crystalline compound dissolves, leaving 10 g at the bottom of the flask. Thus, the maximum amount of the compound that you can dissolve in this quantity of water at this temperature is 50 g. **Solubility** is *the maximum amount of substance that dissolves in a given volume of solvent at a specified temperature*. Thus, the solubility of sodium thiosulfate at 20°C is 50 g in 100 mL of water.

It is important to understand that it takes time for a substance to dissolve. It is also important to understand the factors that determine the rate, or speed, of dissolution when trying to determine the solubility of a substance. Fine crystals dissolve faster than large crystals. Stirring also accelerates the solution process. When you use fine crystals of sucrose, they dissolve readily in water with only a little stirring. When you try to dissolve a large chunk of sucrose, however, it dissolves very slowly. If you are not careful to stir thoroughly and to wait long enough, you might think that the maximum amount of sucrose has dissolved when it has not. You would then obtain the wrong value for the solubility. The solubility of sucrose is 200 g in 100 mL of water at 20°C, independent of the initial size of the sucrose crystals and independent of how vigorously you stir the solution. It simply takes longer for the crystals to dissolve if they are large or if you do not stir the solution vigorously.

Saturated and Unsaturated Solutions

The solution you obtain after you have dissolved the maximum amount of solute in a solvent at a given temperature is a **saturated solution.** Thus, the solution of 100 mL water containing 50 g of sodium thiosulfate at 20°C is a saturated solution of sodium thiosulfate. *Any solution that contains less solute in a given volume of solvent than is contained in the saturated solution* is an **unsaturated solution.** If you take a small quantity of sodium thiosulfate and add it to an unsaturated solution, it will dissolve. On the other hand, if you add the same quantity of sodium thiosulfate to a saturated solution, the crystals will simply fall to the bottom of the vessel, without dissolving. Figure 13.5 compares an unsaturated solution of sodium thiosulfate with a saturated solution.

Supersaturated Solutions

Under certain circumstances, it is possible to prepare a **supersaturated solution,** which is *a solution that contains more of a solute than is contained in the saturated solution.* Suppose you dissolve sodium thiosulfate in water at its boiling point (100°C). The solubility of sodium thiosulfate at this temperature is 231 g of the compound in 100 mL of water, which is much more than you can dissolve at room temperature. Now you allow the solution to cool slowly to 20°C. If you do this carefully, you will obtain a solution

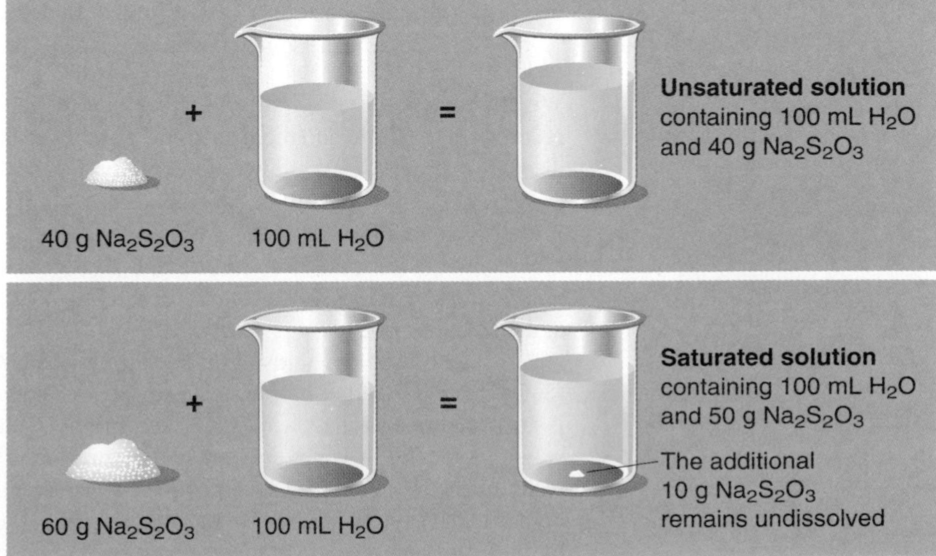

FIGURE 13.5

Saturated and unsaturated solutions. *Top:* A pile of 40 g $Na_2S_2O_3$ dissolves completely in 100 mL of water at 20°C, giving an unsaturated solution. *Bottom:* When a pile of 60 g $Na_2S_2O_3$ is added to 100 mL of water at 20°C, 50 g dissolves and 10 g remains undissolved. The solution is saturated.

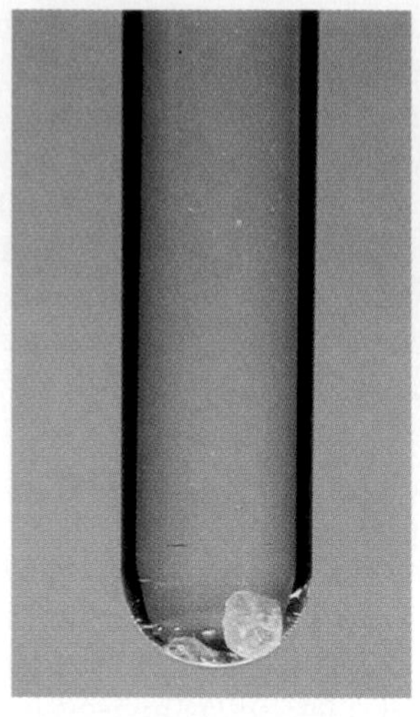
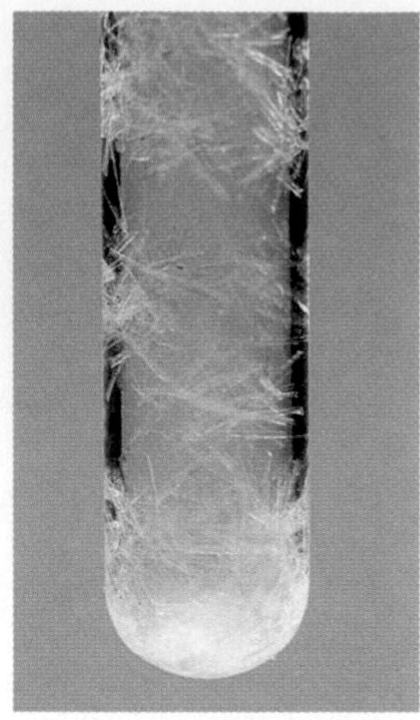

FIGURE 13.6
Crystallization from a supersaturated solution. A supersaturated solution of sodium thiosulfate ($Na_2S_2O_3$) is prepared. When a small crystal of $Na_2S_2O_3$ is added, rapid crystallization occurs.

containing 231 g of sodium thiosulfate in 100 mL of water at 20°C. This is a supersaturated solution, because it contains much more solute (231 g) than would be in the saturated solution (50 g), the amount that you would be able to dissolve in water at 20°C.

How do you know whether a solution is saturated or supersaturated? A supersaturated solution is not stable in the presence of crystals of solute. If you add a crystal of sodium thiosulfate to a saturated solution, it merely drops to the bottom of the vessel, without dissolving, as we noted earlier. However, if you add a crystal of sodium thiosulfate to a supersaturated solution, it initiates a dramatic process in which crystals of sodium thiosulfate come out of the solution (Figure 13.6). When this *crystallization* has finished, you have a saturated solution in the presence of sodium thiosulfate crystals. The quantity of sodium thiosulfate that crystallizes out is 231 g − 50 g, or 181 g.

13.5 The Solution Process

Animation: The dissolution of a solid in a liquid.

OBJECTIVES

- Describe the solution process in terms of the dissolution of solute in the solution and the crystallization of solute from the solution.
- Describe the solution process as an *equilibrium*.

When you first add a few crystals of sodium thiosulfate to water, the crystals dissolve readily. Later, as you add more and more crystals to the flask containing the solution, the dissolving process slows down. Eventually, when the solution becomes saturated and crystals of sodium thiosulfate remain at the bottom of the flask, it appears that nothing further is happening. However, if you imagine seeing the solution process at the level of molecules and ions, the story is different. You see the solution process as one of continuous activity, even when the solution is saturated. Let's take such a submicroscopic look at the solution process.

Crystals of sodium thiosulfate consist of sodium ions (Na^+) and thiosulfate ions ($S_2O_3^{2-}$) held together by the strong attractions between charges of opposite sign (that is, by the attractions between cations and anions). When you add one of these crystals to water, molecules of water are attracted to ions in the crystal. These attractions between water molecules and ions can sometimes overpower the attractions between cations and anions in the crystal. The ions at the corners of a crystal are particularly easy to dislodge in this way, because each of these ions is held to the crystal by only a few ions of opposite charge. Once an ion has been loosened from its crystal, it moves off into the body of the solvent, where it is surrounded by water molecules. As more and more ions leave the crystal, the crystal dissolves. We represent this process of dissolving of sodium thiosulfate as

$$Na_2S_2O_3(s) \longrightarrow 2Na^+(aq) + S_2O_3^{2-}(aq)$$

The solution process, however, does not consist simply of ions leaving crystals to go into the solution. As crystals continue to dissolve, the solution becomes more and more crowded with ions. Some of these ions in the solution collide, through random motions, with crystals at the bottom of the flask. Once an ion collides with a crystal, it may be so strongly attracted to ions of opposite charge in that crystal that it sticks. In other words, the solution process actually consists of both dissolution (in which ions leave the crystal) and crystallization (in which ions in the solution attach themselves to a crystal). We represent these two parts of the solution process by writing an equation with a double arrow.

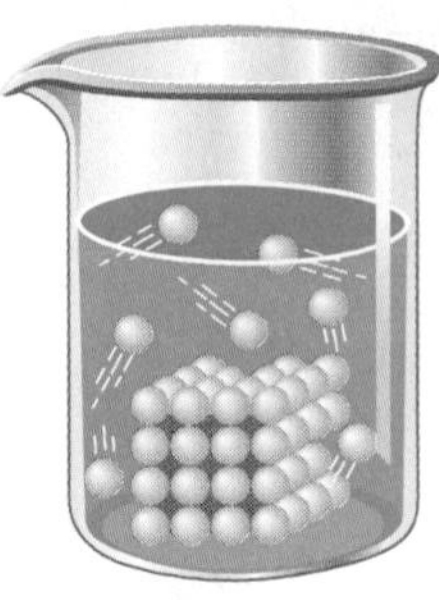

FIGURE 13.7
Solubility equilibrium. At equilibrium, as many particles (molecules or ions) leave the solid as return to the solid from the solution. No more solid appears to dissolve.

$$Na_2S_2O_3(s) \rightleftharpoons 2Na^+(aq) + S_2O_3^{2-}(aq)$$

These two parts of the solution process, dissolution and crystallization, continue even in the saturated solution in contact with the solid solute. When you dissolve the first quantity of sodium thiosulfate in water, the solution contains so few ions that the rate (or speed) at which ions in the solution return to the crystals is very low. At the same time, the rate at which ions leave the crystals is relatively high, so you see crystals of sodium thiosulfate dissolving readily. But as you dissolve more and more solute, the rate at which ions come back to the crystals from the solution increases. You see the solution process slowing down. Eventually, the rate at which ions in the solution come back and reattach themselves to crystals becomes equal to the rate at which ions are leaving the crystals (Figure 13.7). You find that no more solute dissolves; the solution is now saturated. When dissolution and crystallization are proceeding *at the same rate* so that the solution process appears to have stopped, a solution *equilibrium* has been reached, and the solid solute is in equilibrium with the saturated solution. ●

● We introduced equilibrium in Chapter 12, where we discussed vapor pressure. Equilibrium will be discussed thoroughly in Chapter 14.

13.6 Factors That Affect Solubility

OBJECTIVES

- List the three factors that affect solubility.
- Describe how the different ends of the polar molecule water are attracted to cations and anions.
- Explain why a substance like sodium chloride is soluble in water and a substance like calcium phosphate is insoluble.
- Apply the general rule that "like dissolves like" to predict solubility.
- State the effect that increasing temperature has on the solubility of most (but not all) ionic substances.
- Predict the effect of pressure on the solubility of a gas.

Substances vary widely in their solubilities in water and other solvents. Sucrose (table sugar) is quite soluble in water (approximately 200 g in 100 mL of water at 20°C), whereas octane, C_8H_{18} (a component of gasoline), is hardly soluble in water at all. Why is sucrose very soluble in water, but octane is not? Can we say anything about the factors that affect the solubility of a solute in a solvent?

As you might expect, the natures of both the solute and the solvent influence the solubility of the solute in the solvent. Temperature also affects the solubility in most cases,

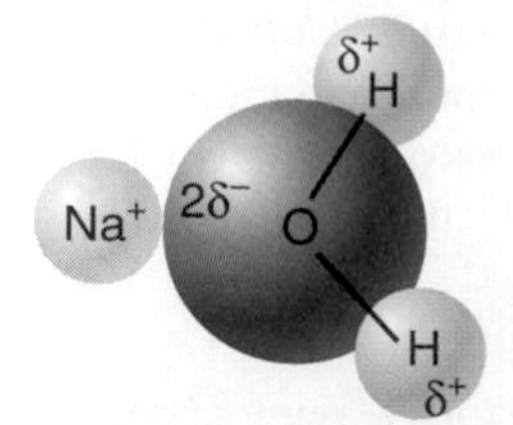

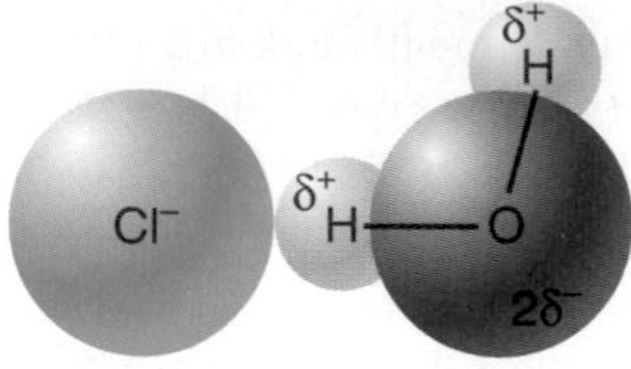

FIGURE 13.8

Attraction of ions for water molecules. The O end of the H_2O molecule is slightly negative and is attracted to cations, such as Na^+. The other end of the H_2O molecule is slightly positive and is attracted to anions, such as Cl^-.

● The notation δ^+ and δ^- for partial charges in polar molecules was introduced in Section 10.4.

and pressure has an effect on the solubility of gaseous solutes. These then are the three main factors affecting the solubility of a solute:

1. Natures of the solute and solvent
2. Temperature
3. Pressure

Natures of the Solute and Solvent

The natures of the solute and solvent determine the attractive forces that exist between particles in solution, and it is these attractive forces that dictate the extent to which a solute dissolves in a solvent. You can see these attractive forces at play when sodium chloride dissolves in water. The water molecule is a *polar molecule,* which means that the molecule has a slightly positively charged end (the hydrogen atoms) and a slightly negatively charged end (the oxygen atom). Figure 13.8 shows these partial charges (indicated in the diagram by the symbols δ^+ and δ^-) on the water molecule. ● The negative end of a water molecule is attracted to sodium ions (the positive ions) in the solute, and the positive end of the water molecule is attracted to chloride ions (the negative ions). These attractive forces between water molecules and the ions in sodium chloride are sufficient to overcome the strong attractions that exist between ions in the solid solute. Thus, the ions tend to leave the sodium chloride crystals and go into the aqueous solution. We find that sodium chloride dissolves readily. Figure 13.9 shows the dissolving of sodium chloride in water.

Compare this situation with the dissolution of calcium phosphate, $Ca_3(PO_4)_2$, in water. In this case, the attractions between water molecules and ions in the solute are not strong enough to overcome the attractions between ions in the solid. The ions now have multiple charges (2+ in the case of the calcium ion, and 3− in the case of the phosphate ion), and these multiple charges produce very strong attractions between ions. The ions tend to stay in the solid rather than go into the water. As a result, calcium phosphate is

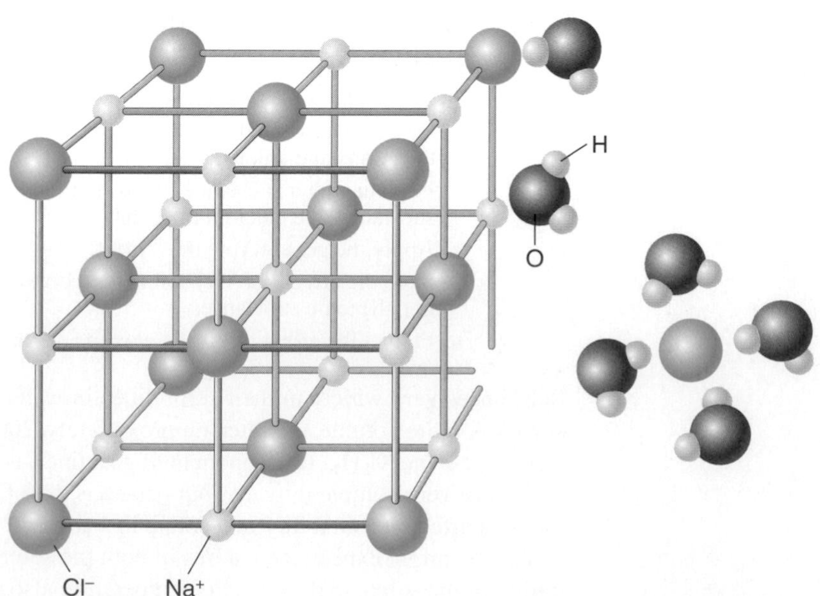

FIGURE 13.9

The dissolving of sodium chloride in water. Ions on the surface of the NaCl crystal can associate with water molecules. This association with water loosens the ions, which then drift off into the body of the solvent.

not very soluble in water. From these two examples, you might expect ionic substances to vary widely in their solubilities, as they do. ●

● Note that in Table 6.3 many of the insoluble compounds contain anions with two or more charges.

There is a general rule of solubility that says *"like dissolves like."* Polar substances (substances composed of polar molecules, such as water) tend to dissolve in one another. Similarly, nonpolar substances dissolve in one another. But polar substances tend not to dissolve in nonpolar substances.

Water and ethyl alcohol, which are both polar molecules (both have polar —O—H bonds in them), are miscible in one another. Similarly, sucrose, whose molecule has many —O—H bonds, is very soluble in water. Gasoline and oil, which have similar nonpolar molecules, are soluble in one another. On the other hand, water and oil tend not to mix; and gasoline is hardly soluble in water. The attraction between a water molecule and an oil molecule (or octane molecule in gasoline) is slight and not sufficient to overcome the mutual attraction between water molecules. There is little incentive for a water molecule to leave the water layer, where it is attracted to other molecules, and go into the oil or gasoline layer, where the attraction is slight.

Temperature

Temperature generally has a profound effect on the solubility of a substance. Consider potassium nitrate (KNO_3), which is used in gunpowder and matches. Its solubility varies from 13 g per 100 mL of water at 0°C to 246 g per 100 mL of water at 100°C. Usually, the solubility of an ionic substance increases with temperature, although sometimes the opposite is true. For example, strontium acetate, $Sr(C_2H_3O_2)_2$, has a solubility of 43.0 g per 100 mL of water at 10°C, but a solubility of only 36.4 g per 100 mL of water at 100°C. Figure 13.10 shows the variation of solubility of some ionic substances with temperature.

In most cases, gases are more soluble in a liquid solvent at lower temperatures than they are at higher temperatures. When you first begin to heat a pot of water, you may see small bubbles form on the sides of the vessel well before the solution boils. These bubbles are composed of air. Because air is less soluble in hot water than it is in cool water, air comes out of the heated solution, forming small bubbles. Similarly, carbonated beverages contain carbon dioxide gas, which comes out of solution as the cold beverage warms to room temperature. The fish in a home aquarium on a hot day may show signs of stress because of the lack of oxygen in the water; less oxygen from air dissolves in the warm water than dissolves in cooler water.

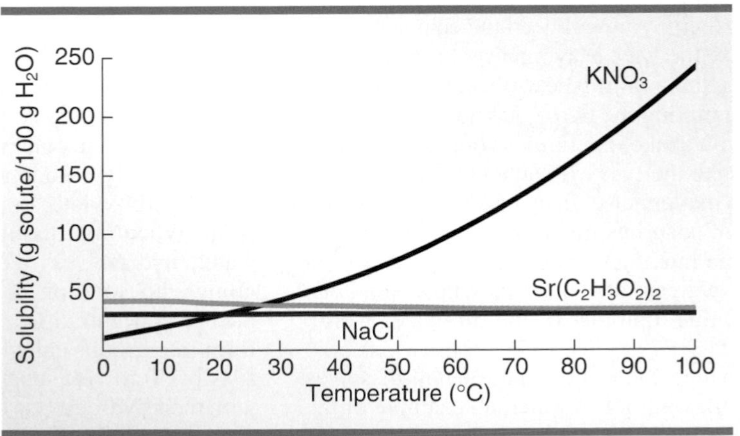

FIGURE 13.10

Solubilities of some ionic compounds at different temperatures. The solubilities of KNO_3 and NaCl rise with temperature. (The solubility of NaCl varies only slightly.) The solubility of $Sr(C_2H_3O_2)_2$ decreases with temperature.

FIGURE 13.11
Effect of decreasing gas pressure on gas solubility. The bottle of carbonated beverage was recently opened, decreasing the gas pressure over the solution. Because the solubility of a gas decreases with decreasing gas pressure, excess gas bubbles from the solution.

Pressure

Generally, pressure has a significant effect only on the solubility of a gas in a liquid or solid solvent. In these cases, the solubility is directly proportional to the pressure of the gas. To prepare a carbonated beverage, you dissolve carbon dioxide gas under pressure (at about 6 atm) in the beverage solution. This pressure is maintained over the solution while it is in the bottle. But when you open the bottle, the pressure of the carbon dioxide gas is reduced and it fizzes out of the solution (Figure 13.11). As the container sits on a table, the beverage eventually goes "flat." ●

● Divers must be careful about descending to great depths, because the increased pressure increases the solubility of gases such as nitrogen in the blood. If a diver ascends too quickly, these gases come out of the blood as small bubbles that cause great pain and injury, a condition known as "the bends."

Chemical Perspective

Soap, the Molecular Diplomat

Oil and water do not mix. Figuratively, this is used as a cliché to express the difficulty of bringing people with grudges together. When it seems necessary to do that (say, for a wedding or funeral), it helps to have someone with diplomatic skills to keep the parties from brawling. What about the real substances oil and water? To mix them, you will need the equivalent of a diplomat to ease the way. Mayonnaise, for instance, is principally vegetable oil mixed with water; but if you have ever tried mixing the two, you know that no matter how hard you shake, the liquids soon separate into two layers. Shake the two with an egg yolk, however, and—voila!—mayonnaise. In using soap to clean fabrics or the skin, soap has much the same function as egg yolk does in making mayonnaise; it allows oil (in the form of dirty grease spots) to mix with water. Let's see how soap, this diplomat of the oil–water world, works.

Oil, whether vegetable oil or petroleum, dissolves easily in gasoline, but not in water, an example of the rule that like dissolves like. Oil and gasoline are principally hydrocarbons (compounds of hydrogen and carbon) and are nonpolar materials. They are similar types of compounds and are mutually soluble. On the other hand, dissimilar compounds, say, a nonpolar substance and a polar one, might be expected to be insoluble, as are oil and water.

Soaps are salts of fatty acids. You make soap by boiling fat with sodium or potassium hydroxide. Soap anions, those diplomats of the oil–water world, are *amphipathic*. This means that they have both polar and nonpolar character. A typical soap anion is shown in Figure 13.12. It consists of a hydrocarbon portion, about 16 to 18 carbon atoms long, which is nonpolar, and a carboxylate ion group ($—CO_2^-$), which is polar. When you dissolve a small amount of soap in water, it dissolves like any other ionic compound, going into solution as separate ions. At somewhat greater concentrations, though, soap anions begin to form into groups called *micelles*. A soap micelle consists of 50 to 100 soap anions arranged into a sphere with their hydrocarbon tails tucked inside the sphere to form an "oily" center, with the ion charges of the carboxylate groups on the surface of the sphere, where

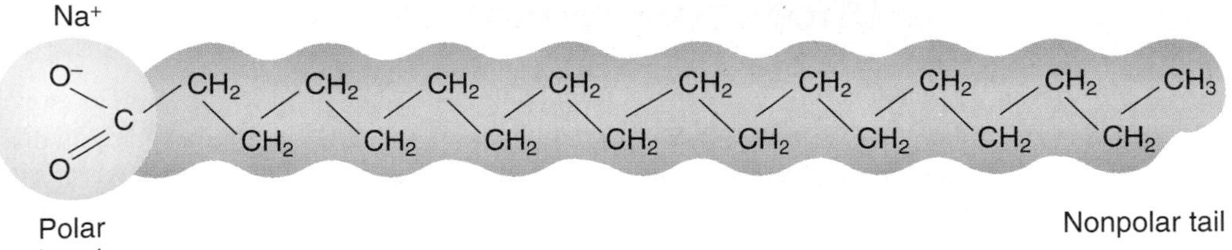

FIGURE 13.12

A soap anion. This is the anion of stearic acid ($C_{17}H_{35}CO_2^-$), a typical soap anion. The ionic end is polar and therefore attracted to water. (The charge is balanced by a metal cation, usually Na^+.) The long hydrocarbon tail is nonpolar.

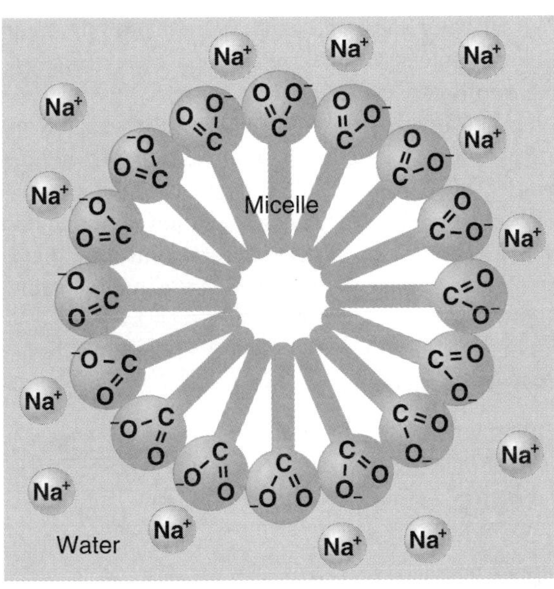

FIGURE 13.13

A soap micelle. Soap anions associate into spherical groups, called *micelles*, in which the hydrocarbon tails of the soap anions are tucked inside the sphere to form an "oily" center, with the ion charges on the surface of the sphere. Sodium ions in the water solution counter the charges of the anions.

they are attracted to the water, a polar solvent (see Figure 13.13). The negative ion charges on the micelle are countered by positive sodium ions in the water solution surrounding the micelles. By forming micelles, soap anions solve the problem of what to do with their nonpolar groups in a polar solvent like water.

Figure 13.14 depicts the cleansing action of soap. The hydrocarbon ends of soap anions are attracted to grease stains on a fabric. Once surrounded by soap anions, a fragment of dirty grease becomes entrapped in a soap micelle, where it can be washed away from the fabric.

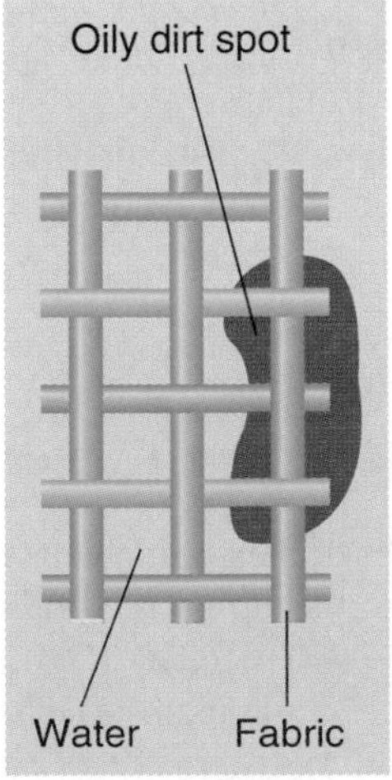

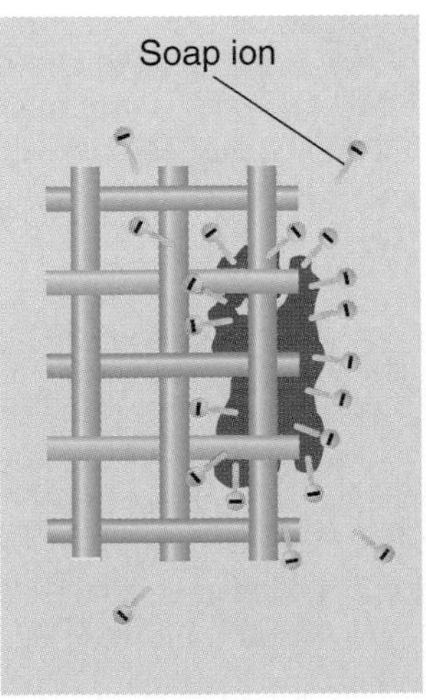

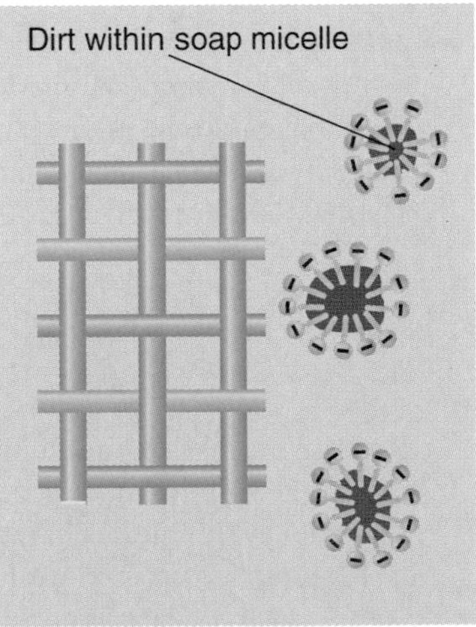

FIGURE 13.14

The cleansing action of soap. Dirt spots on a fabric are usually oily spots containing dirt. The nonpolar, hydrocarbon (oily) tails of soap anions tend to associate with the oily dirt spots on the fabric. Once soap anions surround a dirt spot, pieces of the oily dirt spot are effectively trapped inside of soap micelles, which can then float away in the wash water.

SOLUTION CONCENTRATION

Concentration is a general term that refers to the quantity of solute dissolved in a given quantity of solution or a given quantity of solvent. The more sugar you dissolve in your cup of tea, the higher the concentration of sugar in the solution. We use the qualitative terms *dilute* and *concentrated* to refer to the relative concentrations of a solution. A **dilute solution** is *a solution whose concentration is relatively low.* A **concentrated solution** is *a solution whose concentration is relatively high.* Often these terms are used comparatively. You would say that a solution containing 15 g of sucrose in 100 mL of water is more concentrated than one containing 5 g of sucrose in 100 mL of water.

These terms are also used in referring to solutions of common acids and bases. Commercially available solutions that contain the maximum available concentration of solute are referred to as concentrated solutions. For example, concentrated hydrochloric acid contains about 37 g HCl in 100 g of solution, which is the maximum quantity of HCl found in the commercially available solution. If a solution contains substantially less than this, it is a dilute solution of hydrochloric acid.

You obtain a unit of concentration by specifying the unit for the quantity of solute and the unit for the quantity of solution or the quantity of solvent. In the previous paragraph, we defined the concentration of HCl in a solution in terms of the mass of solute in a given mass of solution. We could also specify the concentration of this acid in terms of the mass of HCl dissolved in a given mass of solvent (water). Thus, the solution contains 37 g HCl in 63 g of water. (You obtain this by subtracting the mass of solute, 37 g, from the mass of solution, 100 g.)

You can specify the quantity of substance in terms of mass, moles, or volume. In the next three sections, we discuss several concentration units.

13.7 Mass Percent of Solute

OBJECTIVES

- Define the concentration unit *mass percent of solute.*
- Use the defining equation for mass percent of solute to relate this concentration unit to mass of solute and mass of solution (or mass of solvent).

Mass percent of solute is *a unit that expresses the concentration of solute as a percentage of the mass of solution.* (Mass percent is also referred to as *weight percent.*) It is equivalent to the grams of solute in 100 g of solution. Earlier, we noted that concentrated hydrochloric acid contains 37 g HCl in 100 g of solution. Alternatively, we could say that this acid is 37% HCl by mass. ●

● The concept of mass percent was discussed in Chapter 7.

$$\text{Mass percent of solute} = \frac{\text{mass of solute}}{\text{mass of solution}} \times 100\%$$

Because the mass of any solution equals the sum of the masses of the components making up the solution, the percentages of these components always add up to 100%. For example, hydrochloric acid that is 37% HCl must be 63% H_2O (37 + 63 = 100).

As long as you mix a solution to give a total mass of 100 g, the mass of solute in grams (such as 37 g) equals the numerical value (37) of the mass percent of solute. Because you usually do not have 100 g, however, you must use the previous equation to calculate the mass percent of solute. We refer to this as the *defining equation* for the mass percent of solute. For example, a solution weighing 55 g that contains 15 g of HCl has the following mass percent of solute.

$$\text{Mass percent of solute} = \frac{15\ \cancel{g}}{55\ \cancel{g}} \times 100\% = 27\%$$

You can choose any units for the masses of solute and solution, as long as they are the same, because then the units of mass cancel. The following three examples show several types of problems that use the defining equation for the mass percent of solute. The first one is a direct application of the previous equation.

■ EXAMPLE 13.1 Calculating Mass Percent of Solute (from Solute and Solution Masses)

Potassium permanganate is used to bleach cotton, silk, and other fabrics and is used as a laboratory reagent. A laboratory worker prepares a solution by dissolving 4.5 g of potassium permanganate ($KMnO_4$) in enough water to give 68.0 g of solution (Figure 13.15). What is the mass percent of potassium permanganate in the solution?

FIGURE 13.15
Preparation of potassium permanganate solution. Potassium permanganate is a dark blue-black compound; when it dissolves in water, it forms a bright purple solution.

Problem Analysis

The mass of solute and the mass of solution are given, so you simply substitute into the defining equation to find the mass percent of solute.

Solution

Insert the quantities into the defining equation for mass percent of solute.

$$\text{Mass percent of solute} = \frac{\text{mass of solute}}{\text{mass of solution}} \times 100\% = \frac{4.5\ \cancel{g}}{68.0\ \cancel{g}} \times 100\% = 6.6\%$$

Exercise 13.1

You prepare a solution by dissolving 5.4 g of potassium sulfate (K_2SO_4) in water to give 83.5 g of solution. What is the mass percent of potassium sulfate in the solution?

(Try Problems 13.33 and 13.34.)

Usually you know the mass of solute and the mass of solvent, rather than the mass of solution. In these cases the mass of solution can be obtained by adding the masses of solute and solvent.

■ EXAMPLE 13.2 Calculating Mass Percent of Solute (from Solute and Solvent Masses)

You make up a solution by dissolving 5.5 g of glucose ($C_6H_{12}O_6$) in 20.0 g of water. What is the mass percent of glucose in the aqueous solution?

Problem Analysis

To use the defining equation for the mass percent of solute, you need the mass of solute (5.5 g) and the mass of solution (which is not given). Note, however, that the mass of the solution equals the sum of the masses of solute (5.5 g) and solvent (20.0 g), so you can calculate it.

Solution

You first calculate the mass of the solution.

$$\text{Mass of solution} = \text{mass of solute} + \text{mass of solvent} = 5.5\text{ g} + 20.0\text{ g} = 25.5\text{ g}$$

Then you calculate the mass percent of the solute.

$$\text{Mass percent of solute} = \frac{\text{mass of solute}}{\text{mass of solution}} \times 100\% = \frac{5.5\ \cancel{\text{g}}}{25.5\ \cancel{\text{g}}} \times 100\% = 22\%$$

Exercise 13.2

You prepare a solution by dissolving 5.4 g of potassium sulfate (K_2SO_4) in 83.5 g of water. What is the mass percent of potassium sulfate in the solution?

(Try Problems 13.35 and 13.36.)

If you know the total mass of a solution and the mass percent of solute that is needed, you can use the defining equation for mass percent to calculate the amounts of solute and solvent that must be used to make up the solution.

■ EXAMPLE 13.3 Preparing a Solution of Given Mass Percent of Solute

A physiological saline solution consists of 0.85% by mass of sodium chloride. It is sometimes administered intravenously to people suffering from severe sodium depletion. You wish to prepare 500 g of such a saline solution. How many grams of sodium chloride and how many grams of water do you need to make up the solution?

Problem Analysis

The defining equation for the mass percent of solute relates this mass percent (given in the problem statement) to the mass of solute (wanted) and the mass of solution (given). You can rearrange the defining equation for mass percent of solute to give the mass of solute. Then you can subtract the mass of solute from the mass of solution (500 g) to obtain the mass of solvent (water).

Solution

You start with the equation

$$\text{Mass percent of solute} = \frac{\text{mass of solute}}{\text{mass of solution}} \times 100\%$$

and rearrange it to give

$$\text{Mass of solute} = \frac{\text{mass \% of solute} \times \text{mass of solution}}{100\%}$$

Now, you substitute into this equation to obtain the mass of solute.

$$\text{Mass of solute} = \frac{0.85 \times 500 \text{ g}}{100} = 4.3 \text{ g of sodium chloride}$$

Recall that the mass of solution equals the sum of the masses of solute and solvent, so

$$\begin{aligned}\text{Mass of solvent} &= \text{mass of solution} - \text{mass of solute} \\ &= 500 \text{ g} - 4.3 \text{ g} = 496 \text{ g of water}\end{aligned}$$

Exercise 13.3

Potassium dichromate ($K_2Cr_2O_7$) is a red-orange compound. A chemist wants to prepare 42.5 g of a 5.0% aqueous solution of potassium dichromate. How many grams of potassium dichromate and how many grams of water are needed to make up this solution?

(Try Problems 13.37 and 13.38.)

Very dilute solutions are sometimes expressed in parts per million (that is, parts solute per million parts solution, by mass). This is a unit similar to percent by mass (which is equivalent to parts per hundred). The unit *parts per million* (abbreviated *ppm*) is defined as

$$\text{Parts per million} = \frac{\text{mass of solute}}{\text{mass of solution}} \times 1{,}000{,}000$$

For example, the maximum allowable concentration of lead in drinking water, established by the U.S. Public Health Service, is 0.05 ppm. This is 0.05 mg of Pb in 1000 g of solution, or 0.05 mg per kg of solution.

The concentration of a liquid solute in a liquid solution is sometimes expressed in a unit called *volume percent*. It is defined by an equation similar to mass percent.

$$\text{Volume percent of solute} = \frac{\text{volume of solute}}{\text{volume of solution}} \times 100\%$$

Thus, a solution that is made from 30.0 mL of ethyl alcohol with enough water to give 97.3 mL of solution contains (30.0 mL/97.3 mL) × 100% = 30.8% by volume of ethyl alcohol.

13.8 Molarity and Normality

OBJECTIVES

- Define the concentration unit *molarity.*
- Use the defining equation for molarity to calculate the molar concentration from moles (or mass) of solute and volume of solution.
- Use molarity as a conversion factor to convert between moles of solute and volume of solution.
- Use molarity as a conversion factor in stoichiometry problems involving volumes of solutions of given molarity.
- Define *normality* and *equivalent.*
- Calculate the mass of one reactant from the normality and volume of another reactant.
- Relate the final volume and molarity to the initial volume and molarity in a dilution.

Instead of expressing the concentration of a solution in terms of the mass of solute in a given quantity of solution, as we did in the previous section, we could express the concentration in terms of moles of solute in a given quantity of solution. **Molarity** is *a concentration unit equal to the moles of solute dissolved in a liter of solution.* It is defined by the equation

$$\text{Molarity} = \frac{\text{moles of solute}}{\text{liters of solution}}$$

Note that the unit of molarity is moles per liter (of solution). It is abbreviated by the capital letter M. Concentrated hydrochloric acid is 12.1 M HCl (read as "12.1 molar HCl"). This means that there are 12.1 mol HCl in each liter of solution.

Molarity is an important unit because it allows you to dispense molar amounts of substances by measuring volumes of solutions, rather than masses of materials, and measuring volumes is relatively simple. If you want 12.1 mol HCl, you pour out 1.00 L of solution. If you want 1.21 mol HCl, you need to pour out 0.100 L of solution.

You can understand the concept of molarity by looking at the procedure used to make such a solution. You begin with a volumetric flask (Figure 13.16). This type of flask has

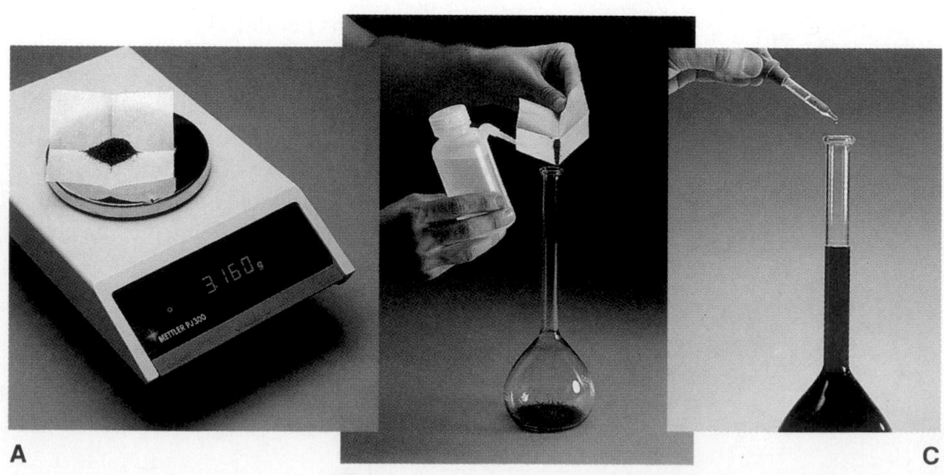

FIGURE 13.16
Preparing a solution of given molarity. *(A)* To prepare a 0.400 M solution of $KMnO_4$, you first weigh out 0.200 mol (or 3.16 g) of $KMnO_4$. *(B)* Then you add the solid to the volumetric flask and add some water to dissolve it. *(C)* You keep adding water until the volume of solution rises to the line on the neck of the flask; then you mix the solution.

a long neck with an etched line to indicate the precise volume when the flask is filled to this line. Suppose you use a flask having a volume of 500 mL (or 0.500 L) when filled to this line. You add 0.200 mol $KMnO_4$ (potassium permanganate) to the flask, and then add a quantity of water to dissolve the solid. You add more water until the volume of solution rises to the line on the neck of the flask, and you shake the flask to mix the solution. The volume of solution is 500 mL, which is 0.500 L, and the amount of solute is 0.200 mol. Therefore,

$$\text{Molarity} = \frac{0.200 \text{ mol}}{0.500 \text{ L}} = 0.400\ M \qquad \text{(or 0.400 mol/L)}$$

Note that you must express the volume of solution in liters. The next two examples illustrate the calculation of the molarity of a solution from given quantities of solute and solution. First we consider a case in which you are given the moles of solute and volume of solution.

■ EXAMPLE 13.4 Calculating the Molarity of a Solution

Potassium hydroxide (KOH) is used in the manufacture of liquid soaps and in the preparation of other potassium compounds. (a) An experimenter makes up a solution by dissolving 0.0150 mol KOH in enough water to give 25.0 mL of solution. What is the molarity of the solution? (b) You dissolve 0.350 g of KOH in 50.0 mL of solution. What is the molarity of the solution?

Problem Analysis

You use the defining equation for molarity and substitute the moles of solute and volume of solution in liters into the solution. If the mass rather than the moles of solute is given, you must first convert this mass to moles, then use the defining equation for molarity.

Solution

In these problems, be careful to express the volume of solution in liters.

(a) The volume is 25.0 mL, which is 0.0250 L. Therefore,

$$\text{Molarity} = \frac{0.0150 \text{ mol}}{0.0250 \text{ L}} = 0.600\ M$$

(b) The molar mass of KOH is 56.1 g. Therefore,

$$0.350 \text{ g } \cancel{\text{KOH}} \times \frac{1 \text{ mol KOH}}{56.1 \text{ g } \cancel{\text{KOH}}} = 0.00624 \text{ mol KOH}$$

Now you substitute into the defining equation for molarity.

$$\text{Molarity} = \frac{0.00624 \text{ mol}}{0.0500 \text{ L}} = 0.125\ M$$

Exercise 13.4

Solutions of nickel(II) sulfate ($NiSO_4$) are bright green in color. A solution whose volume is 61.2 mL contains 1.58 g $NiSO_4$. What is the molarity of this solution?

(Try Problems 13.45 and 13.46.)

Molarity as a Conversion Factor

You can think of molarity as a conversion factor for converting moles of solute to liters of solution, and vice versa. For example, a 0.200 *M* hydrochloric acid solution contains 0.200 mol HCl in 1 L of solution. To find out how many moles of HCl there are in 1.50 L of this hydrochloric acid solution, you perform the following conversion factor calculation.

$$1.50\ \cancel{\text{L HCl solution}} \times \frac{0.200\ \text{mol HCl}}{1\ \cancel{\text{L HCl solution}}} = 0.300\ \text{mol HCl}$$

You write the conversion factor (0.200 mol HCl/1 L HCl solution) directly from the molarity (0.200 *M*). It converts liters of solution to moles HCl.

If you invert this conversion factor, you obtain

$$\frac{1\ \text{L HCl solution}}{0.200\ \text{mol HCl}}$$

which converts moles of HCl to liters of solution. Suppose you want to know the volume of 0.200 *M* HCl needed to give 0.150 mol HCl. You simply convert 0.150 mol HCl to volume.

$$0.150\ \cancel{\text{mol HCl}} \times \frac{1\ \text{L HCl solution}}{0.200\ \cancel{\text{mol HCl}}} = 0.750\ \text{L HCl solution}$$

Thus, you need to measure out 750 mL of 0.200 molar acid to give 0.150 mol HCl. The following example illustrates the use of a similar conversion factor.

■ EXAMPLE 13.5 Preparing a Solution of Given Volume and Molarity

You are asked to prepare 100. mL of 0.150 *M* KOH (potassium hydroxide) solution. How many grams of KOH will you need?

Problem Analysis

You use the molarity (0.150 *M*) as a conversion factor to convert the volume of solution (100 mL or 0.100 L) to moles of solute; then you use the formula weight to convert moles to mass of solute.

Volume of solution —Convert to→ Moles of solute —Convert to→ Mass of solute

Solution

The conversion of volume to moles of solute is

$$0.100\ \cancel{\text{L KOH solution}} \times \frac{0.150\ \text{mol KOH}}{1\ \cancel{\text{L KOH solution}}} = 0.0150\ \text{mol KOH}$$

Now you use the formula weight of KOH (56.10 amu) to write the factor to convert moles KOH to grams KOH.

$$0.0150\ \cancel{\text{mol KOH}} \times \frac{56.10\ \text{g KOH}}{1\ \cancel{\text{mol KOH}}} = 0.842\ \text{g KOH}$$

You can, of course, put both of these conversions together on the same line. The previous calculation would then be written as

$$0.100\ \cancel{\text{L KOH solution}} \times \frac{0.150\ \cancel{\text{mol KOH}}}{1\ \cancel{\text{L KOH solution}}} \times \frac{56.10\ \text{g KOH}}{1\ \cancel{\text{mol KOH}}} = 0.842\ \text{g KOH}$$

Exercise 13.5

What mass (in grams) of silver nitrate ($AgNO_3$) would you need to prepare 250 mL of 0.150 *M* $AgNO_3$?

(Try Problems 13.47 and 13.48.)

Stoichiometry Involving Solutions

Molarity, as we discussed, allows you to convert between volume of solution and moles of solute. Because of this relationship between volume and moles, it is possible to use volumes in place of moles or mass in stoichiometry problems. The next two examples illustrate this type of stoichiometry problem. ●

● You may wish to review the principles of stoichiometry from Chapter 8.

■ EXAMPLE 13.6 Calculating a Stoichiometric Quantity Using Molarity

A flask contains a solution with an unknown amount of sodium hydroxide, NaOH. A chemist adds 0.154 *M* H_2SO_4 (sulfuric acid) to the sodium hydroxide until the following reaction is complete.

$$H_2SO_4(aq) + 2NaOH(aq) \longrightarrow Na_2SO_4(aq) + 2H_2O(l)$$

If a total of 48.6 mL of sulfuric acid is required to complete the reaction, how many grams of sodium hydroxide were in the flask?

Problem Analysis

This is a stoichiometry problem, except that you are asked to obtain the mass of NaOH given the volume and molarity of H_2SO_4, instead of the moles of H_2SO_4. To do the stoichiometry problem, you need to know the moles of H_2SO_4, which you obtain by converting volume H_2SO_4 to moles H_2SO_4. Then, as in the usual stoichiometry problem, you convert moles H_2SO_4 to moles NaOH and finally to mass NaOH.

Solution

To obtain the moles of H_2SO_4, use the molarity of H_2SO_4 to convert the volume to moles.

$$0.0486\ \cancel{\text{L H}_2\text{SO}_4\ \text{solution}} \times \frac{0.154\ \text{mol H}_2\text{SO}_4}{1\ \cancel{\text{L H}_2\text{SO}_4\ \text{solution}}} = 0.00748\ \text{mol H}_2\text{SO}_4$$

The question now is, how many grams of NaOH react with 0.00748 moles H_2SO_4?

To answer this, you convert moles H_2SO_4 to moles NaOH, then to mass NaOH.

Moles of H_2SO_4 → Convert to → Moles of NaOH → Convert to → Mass of NaOH

You obtain the mole ratio that converts moles of H_2SO_4 to moles of NaOH from the balanced chemical equation, just as you did in Chapter 8. (Review Example 8.1 if necessary.) The molar mass of NaOH converts moles of NaOH to grams of NaOH.

$$0.00748 \cancel{\text{mol } H_2SO_4} \times \frac{2 \cancel{\text{mol NaOH}}}{1 \cancel{\text{mol } H_2SO_4}} \times \frac{40.00 \text{ g NaOH}}{1 \cancel{\text{mol NaOH}}} = 0.598 \text{ g NaOH}$$

Therefore, the original solution in the flask contained 0.598 g NaOH.

Exercise 13.6

Vinegar contains acetic acid ($HC_2H_3O_2$), which is responsible for its sour taste and characteristic odor. A 10.0-mL sample of vinegar is placed in a flask and enough 0.148 *M* NaOH solution added to react completely with all of the acetic acid in the vinegar. A total of 48.1 mL of the NaOH solution is required to react completely with the acetic acid by the reaction

$$HC_2H_3O_2(aq) + NaOH(aq) \longrightarrow NaC_2H_3O_2(aq) + H_2O(l)$$

How many grams of acetic acid are in the vinegar sample?

(Try Problems 13.51 and 13.52.)

The preceding example illustrates a typical problem in *titration*. Suppose substance A reacts with substance B, and you want to know the mass of substance A contained in a flask. You can find the mass of A by determining the moles of B that react completely with all of A. You can determine the moles of B if you know the volume and molarity of B. Thus, in Example 13.6, you were given the volume and molarity of H_2SO_4 that reacts completely with the NaOH in the flask. From this information, you calculated the mass of NaOH. The calculations follow this sequence of conversions:

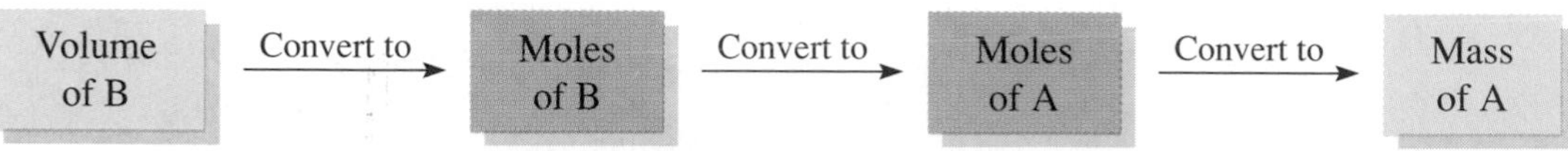

Titration is *a procedure for determining the amount of one substance* (such as NaOH) *by determining the volume of a solution of known molarity of another substance* (H_2SO_4) *that reacts completely with the first substance* (NaOH). Figure 13.17 shows the titration of a hydrochloric acid solution by a sodium hydroxide solution. In this titration, a special dye called an *indicator* was added to indicate the completion of the reaction. The indicator dye (phenolphthalein) is colorless in hydrochloric acid but turns a light pink at the end of the reaction. ●

● Various vegetable and fruit dyes change colors in acidic and basic solutions. Red cabbage juice, for example, changes from red in acidic solution to green in basic solution.

Normality and Equivalents

Normality is a concentration unit related to molarity. Chemists often use it when dealing with acid–base titrations. The unit is based on the concept of an equivalent (abbreviated

A

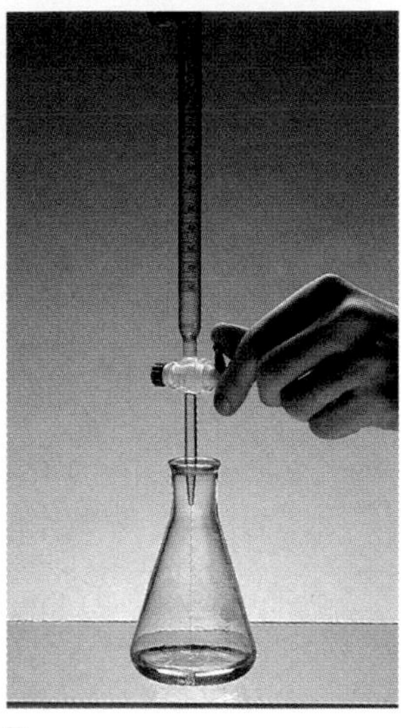
B

C

FIGURE 13.17
Titration of hydrochloric acid with sodium hydroxide. *(A)* Hydrochloric acid is pipetted into the flask to which an indicator (phenolphthalein) has been added. *(B)* This solution is then titrated with sodium hydroxide solution from a buret (a graduated glass tube with a stopcock). *(C)* At the end of the titration, the indicator shows a slight pink color.

equiv) of an acid or base. An **equivalent of an acid** is *the amount of acid that yields one mole of hydrogen ion,* H^+. An **equivalent of a base** is *the amount of base that yields one mole of hydroxide ions,* OH^-. Because one mole of hydrogen ion reacts with one mole of hydroxide ion (to produce one mole of water), one equivalent of any acid reacts with one equivalent of any base.

Let's look at some common acids. One mole of hydrochloric acid, HCl, provides one mole of hydrogen ion, H^+. Thus, one mole of HCl equals one equivalent of acid. On the other hand, one mole of sulfuric acid, H_2SO_4, provides two moles of hydrogen ion and, therefore, two equivalents of acid. Similarly, one mole of phosphoric acid, H_3PO_4, provides three moles of hydrogen ion and, therefore, three equivalents of acid. In summary:

1 mol HCl equals 1 equiv HCl

1 mol H_2SO_4 equals 2 equiv H_2SO_4

1 mol H_3PO_4 equals 3 equiv H_3PO_4

The reasoning for bases is similar. Thus, one mole of sodium hydroxide, NaOH, provides one mole of hydroxide ion, OH^-, and therefore one equivalent of base. One mole of magnesium hydroxide, $Mg(OH)_2$, provides two moles of hydroxide ion, and therefore two equivalents of base.

1 mol NaOH equals 1 equiv NaOH

1 mol $Mg(OH)_2$ equals 2 equiv $Mg(OH)_2$

The **normality** (N) of an acid or base solution is *the number of equivalents of acid or base solute in one liter of solution.*

$$\text{Normality} = \frac{\text{equivalents of solute}}{\text{liters of solution}}$$

A solution that is 1 N HCl contains 1 equiv HCl in one liter of solution.

Suppose you have a 0.10 M H_2SO_4 solution. What is its normality? One liter of the solution contains 0.10 mol H_2SO_4. Because 1 mol H_2SO_4 equals 2 equiv H_2SO_4, one liter of the solution contains 0.20 (= 2×0.10) equiv H_2SO_4, and so is 0.20 $N H_2SO_4$. In general, the normality of an acid or base solution is some multiple of its molarity.

You can rearrange the preceding equation to give a formula for the equivalents in a solution.

$$\text{Equivalents} = \text{normality} \times \text{liters of solution}$$

Or, using N for normality and V for the volume of the solution in liters,

$$\text{Equiv} = N \times V$$

You can use this equation to calculate the equivalents of a solute in a given volume of solution. For instance, 40.0 mL of 0.10 N H_2SO_4 contains

$$N \times V = 0.10\ (\text{equiv } H_2SO_4\,/\text{L}) \times 0.0400\ \text{L} = 0.0040\ \text{equiv } H_2SO_4$$

As we noted earlier, one equivalent of acid reacts with one equivalent of base. This fact provides the main advantage of using normality in acid–base calculations. The next example is a repeat of Example 13.6, but it uses the concepts of equivalents and normality in its solution.

■ EXAMPLE 13.7 Calculating the Mass of a Reactant Using Normality

A flask contains a solution with an unknown amount of sodium hydroxide, NaOH. A chemist adds 0.308 N (0.154 M) H_2SO_4 to the sodium hydroxide until the following reaction is complete.

$$H_2SO_4(aq) + 2NaOH(aq) \longrightarrow Na_2SO_4(aq) + 2H_2O(l)$$

If a total of 48.6 mL of sulfuric acid is required to complete the reaction, how many grams of sodium hydroxide were in the flask?

Problem Analysis

You are asked to obtain the mass of NaOH given the volume and normality of H_2SO_4. The equivalents of sulfuric acid that react equal the normality of the acid times its volume. This also equals the equivalents of NaOH. From this, you obtain the moles NaOH, then mass NaOH.

Solution

The equivalents of sulfuric acid that react are

$$N \times V = 0.308\ (\text{equiv/L}) \times 0.0486\ \text{L} = 0.0150\ \text{equiv}$$

The equivalents of NaOH initially in the flask are the same: 0.0150 equiv NaOH. Since 1 equiv NaOH equals 1 mol NaOH, the flask contains 0.0150 mol NaOH. You now convert this to grams NaOH.

$$0.0150\ \text{mol NaOH} \times \frac{40.00\ \text{g NaOH}}{1\ \text{mol NaOH}} = 0.600\ \text{g NaOH}$$

The original solution in the flask contained 0.600 g NaOH. (*Note:* This answer is the same as the one in Example 13.6, within the error of the calculation.)

Exercise 13.7

Vinegar contains acetic acid ($HC_2H_3O_2$). A 10.0-mL sample of vinegar is placed in a flask and enough 0.148 N NaOH solution added to react completely with all of the acetic acid in the vinegar. A total of 48.1 mL of the NaOH solution is required to react completely with the acetic acid by the reaction

$$HC_2H_3O_2(aq) + NaOH(aq) \longrightarrow NaC_2H_3O_2(aq) + H_2O(l)$$

How many grams of acetic acid are in the vinegar sample?

(Try Problems 13.57 and 13.58.)

Diluting Solutions

Often solutions are available commercially or are first prepared as concentrated solutions that are later diluted to the needed concentrations. Hydrochloric acid, as we noted earlier, is available commercially as a concentrated acid, which has a concentration of 12.1 *M*. If you want a 0.10 *M* solution, you take a quantity of the concentrated acid and add it to sufficient water so that it has the desired concentration (that is, you dilute the solution). The amount of HCl in the solution does not change during this dilution process, only the concentration of HCl. You can use this fact to prepare a solution of given concentration from a more concentrated solution. This is illustrated in the next example (13.8) using the conversion-factor method. Example 13.9 solves the same problem as Example 13.8 using an algebraic approach. (Ask your instructor if he or she has a preferred method.)

EXAMPLE 13.8 Preparing a Solution by Dilution (Conversion-Factor Method)

You want to prepare 1.50 L of 0.150 *M* HCl from a stock solution of the commercial acid, which is 12.1 *M*. What volume of the concentrated acid do you require?

Problem Analysis

During the dilution step, the amount of HCl remains constant. Therefore, you know that the moles of HCl in the dilute solution (1.50 L of 0.150 *M* HCl) equal the moles of HCl in the concentrated acid. You can calculate the moles of HCl from the dilute solution, then convert this amount of HCl to the volume of concentrated hydrochloric acid by using the molarity of this acid as a conversion factor.

Solution

The amount of HCl in the dilute solution is

$$1.50\ \text{L}\ \cancel{\text{HCl solution}} \times \frac{0.150\ \text{mol HCl}}{1\ \text{L}\ \cancel{\text{HCl solution}}} = 0.225\ \text{mol HCl}$$

This then is the amount of HCl in the concentrated solution. You can convert this amount of HCl to the volume of the concentrated acid by using the molarity of this acid (12.1 *M*) as a conversion factor.

$$0.225\ \cancel{\text{mol HCl}} \times \frac{1\ \text{L HCl solution}}{12.1\ \cancel{\text{mol HCl}}} = 0.0186\ \text{L HCl solution}$$

Thus, you take 18.6 mL of concentrated HCl and dilute it to 1.50 L; this gives a solution that is 0.150 *M* HCl.

Exercise 13.8

A stock solution of silver nitrate is 2.5 *M* $AgNO_3$. How many milliliters of this solution should you use to prepare 1.00 L of 0.10 *M* $AgNO_3$?

(*Try Problems 13.59 and 13.60.*)

Instead of using the conversion-factor method, as in the previous example, you could derive a general formula by approaching the problem algebraically. Let V_f be the volume of the diluted solution (the final solution), and let M_f be the molarity of this final solution. From the defining equation for molarity, you have

$$M_f = \frac{\text{moles of solute}}{V_f}$$

Thus, the amount of solute in the solution is

$$\text{Moles of solute} = M_f V_f$$

A similar result holds for the concentrated solution (the initial solution). In this case, you write M_i for the molarity of the initial solution and V_i for its volume (initial volume).

$$\text{Moles of solute} = M_i V_i$$

By combining the two equations, you get a formula that relates molarities and volumes for the concentrated and dilute solutions.

$$M_i V_i = M_f V_f$$

This is a general result, and you can use it in any dilution problem. We illustrate its use in the next example, using the same problem as in Example 13.8.

■ EXAMPLE 13.9 Preparing a Solution by Dilution (Algebraic Method)

You want to prepare 1.50 L of 0.150 *M* HCl from a stock solution of the commercial acid, which is 12.1 *M*. What volume of the concentrated acid do you require?

Problem Analysis

Use the equation $M_i V_i = M_f V_f$, substituting values for M_i (molarity of the concentrated acid = 12.1 *M*), M_f (molarity of the dilute acid = 0.150 *M*), and V_f (volume of the dilute acid = 1.50 L). Then solve for V_i (volume of concentrated acid).

Solution

After substituting into the dilution equation, you obtain

$$12.1 \text{ mol/L} \times V_i = 0.150 \text{ mol/L} \times 1.50 \text{ L}$$

Solving for V_i gives

$$V_i = \frac{0.150 \cancel{\text{mol/L}} \times 1.50 \text{ L}}{12.1 \cancel{\text{mol/L}}} = 0.0186 \text{ L} \qquad \text{(or 18.6 mL)}$$

You start with 18.6 mL of the concentrated acid and dilute it to 1.50 L.

Exercise 13.9

A stock solution of barium nitrate is 1.2 *M* $Ba(NO_3)_2$. How many milliliters of this solution should you use to prepare 0.50 L of 0.25 *M* $Ba(NO_3)_2$?

(Try Problems 13.61 and 13.62.)

13.9 Molality

OBJECTIVES

- Define the concentration unit *molality.*
- Calculate the molality of a solution from masses of solute and solvent.

Although molarity is the preferred unit in many situations (for example, when you are interested in dispensing a given amount of solute by volume), in certain circumstances it is an inappropriate unit. Because molarity is defined as the moles of solute in a given volume of solution, molarity changes with temperature, even though the composition remains fixed. *As you raise the temperature, the volume of solution normally increases, and therefore the molarity decreases.* Section 13.10 demonstrates how the freezing point and boiling point of a liquid change as you add a solute to it. However, to consider these changes, you need a concentration unit that does not change with temperature for a given composition of solution.

Molality is *a concentration unit equal to the moles of solute dissolved in a kilogram of solvent.* We define this unit by the following equation.

$$\text{Molality} = \frac{\text{moles of solute}}{\text{kilograms of solvent}}$$

Note that the molality of a solution does not change with temperature because neither moles of solute nor kilograms of solvent change.

Molality is abbreviated by a lowercase *m*. (A capital *M* is used for molarity; note also the similarity in the names of the units—they differ only by a change of the letter *r* to an *l*.) A solution that contains 0.100 mol NaOH dissolved in 1.00 kg of water has a concentration of 0.100 *m* NaOH (read "0.100 molal NaOH"). Note that this unit differs from molarity by specifying the mass of solvent rather than volume of solution. You would make up a solution of given molality by weighing out solute and solvent. For example, you would make up a 0.100 *m* NaOH solution by weighing out 4.00 g NaOH (0.100 mol) into a flask and adding 1.00 kg of water. The molality of the solution remains constant when the temperature changes, because this change cannot affect the masses of solute and solvent. The next example illustrates how to calculate the molality of a solution from the masses of solute and solvent.

EXAMPLE 13.10 Calculating the Molality of a Solution

Urea (NH_2CONH_2) is used in fertilizers and animal feeds. What is the molality of a solution that you make up by dissolving 2.35 g of urea in 53.0 g of water?

Problem Analysis

To obtain the molality, you need the moles of solute (not given directly) and kilograms of solvent (equal to 0.0530 kg). Although the number of moles of solute is not given directly, you can calculate it from the mass (2.35 g) and molecular weight of the solute.

Solution

The molecular weight of urea (NH_2CONH_2) is 60.06 amu. Therefore, to convert 2.35 g urea to moles, you multiply the mass by the conversion factor (1 mol urea/60.06 g urea).

$$2.35 \cancel{\text{g urea}} \times \frac{1 \text{ mol urea}}{60.06 \cancel{\text{g urea}}} = 0.0391 \text{ mol urea}$$

The molality of the solution is

$$\text{Molality} = \frac{\text{moles of solute}}{\text{kilograms of solvent}} = \frac{0.0391 \text{ mol}}{0.0530 \text{ kg}} = 0.738\ m$$

Exercise 13.10

Ethylene glycol, CH_2OHCH_2OH (the molecular formula is $C_2H_6O_2$), is used in antifreeze solutions. If a person makes up a solution containing 34.2 g of ethylene glycol and 250.0 g of water, what is the molality of this solution with respect to ethylene glycol?

(Try Problems 13.65 and 13.66.)

COLLIGATIVE PROPERTIES

In the remainder of the chapter, we look at certain colligative properties. A **colligative property** is *a property of a solution that depends only on the number of solute particles (molecules and ions) in a given quantity of solution and not on the particular characteristics of the solute particles.* Some examples of colligative properties are freezing-point depression, boiling-point elevation, and osmotic pressure. ●

● The term *colligative* stems from the Latin *colligare,* meaning "to bind." Colligative properties are bound together by a similar characteristic (that they depend on the number of particles in the solution).

13.10 Freezing-Point Depression and Boiling-Point Elevation

OBJECTIVES

- Explain what is meant by *freezing-point depression* and *boiling-point elevation.*
- Calculate the freezing point and boiling point of a solution of given molality.

When you add a nonvolatile solute (a solute with very low vapor pressure) to a liquid, you find that the liquid's freezing point is lowered, or depressed, and its boiling point is elevated, as the following discussions explain. ●

● Nonvolatile substances are those that do not evaporate easily; volatile ones do.

Freezing-Point Depression

Normally, water freezes at exactly 0°C. But if you dissolve salt in the water, the solution freezes at lower temperatures. (The exact freezing point depends on the concentration of salt added.) If you have ever spread salt on a snowy walkway to melt the snow, you have taken advantage of this freezing-point-depression phenomenon. The salt dissolves in any moisture present to give a solution. As long as the freezing point of the salt solution is below the temperature of the snow it touches, the snow will melt. This phenomenon does not occur with just salt; the addition of a solute to any liquid will depress, or lower, the liquid's freezing point.

Let's look at this phenomenon quantitatively. We restrict ourselves to nonvolatile solutes. We define the **freezing-point depression,** ΔT_f, to be *the colligative property of a solution equal to the freezing point of the pure solvent minus the freezing point of the solution.* (This gives a positive number.) The Greek letter delta (Δ) is used to indicate a change in the quantity that follows it.

$$\Delta T_f = \text{freezing point of pure solvent} - \text{freezing point of solution}$$

The freezing-point depression of a liquid is proportional to the molal concentration of solute particles in the solution. This means that if you double the number of solute particles in the solution, you double the freezing-point depression. For a molecular solute that dissolves as molecules and does not give ions, the molal concentration of solute particles equals the molal concentration of solute, m. Mathematically, then,

$$\Delta T_f \propto m$$

where $\propto$ means "is proportional to."

We can express this proportionality as an equation by introducing a proportionality constant, K_f.

$$\Delta T_f = K_f m \text{ (for a nonvolatile, nonionic solute)}$$

The quantity K_f is called the *freezing-point depression constant.* It depends only on the solvent. For water, K_f equals 1.86°C/m, which means that a 1 m aqueous solution has a freezing-point depression of 1.86°C (ΔT_f = 1.86°C/m × 1 m).

The preceding equation applies only if the solute does not form ions in solution. When the solute forms ions, it gives a solution that contains more particles in solution than would be indicated by its molality. For example, a 0.010 m NaCl solution would contain 0.010 mol Na^+ and 0.010 mol Cl^- per kilogram of water, or a total of 0.020 mol of ions per kilogram of water. The freezing-point depression is about twice that calculated by the preceding equation. ●

● The freezing point is not quite twice that calculated from the equation; it is actually about 1.93 times this value, because strong interactions between ions affect the freezing-point depression.

The next example illustrates the use of this equation to calculate the freezing point of a solution.

■ EXAMPLE 13.11 Calculating the Freezing Point of a Solution

What is the freezing point of an aqueous solution that is 0.15 m glucose ($C_6H_{12}O_6$)? K_f equals 1.86°C/m; the freezing point of pure water is 0.00°C.

Problem Analysis

Relate the freezing point of the solution to the freezing point depression (ΔT_f) and the freezing point of pure water by rearranging the equation

$$\Delta T_f = \text{freezing-point of pure water} - \text{freezing point of solution}$$

to

$$\text{Freezing point of solution} = \text{freezing point of pure water} - \Delta T_f$$

Calculate ΔT_f from the freezing-point depression constant for water (1.86°C/m) and the molality of the solution (0.15 m).

Solution

The freezing-point depression is

$$\Delta T_f = K_f m = 1.86°\text{C}/m \times 0.15\ m = 0.28°\text{C}$$

The freezing point of the solution is 0.00°C − 0.28°C = −0.28°C.

Exercise 13.11

A solution was made up by dissolving 23.4 g of sucrose, $C_{12}H_{22}O_{11}$ (table sugar), in 125 g of water. What is the freezing point of this solution?

(Try Problems 13.69 and 13.70.)

Boiling-Point Elevation

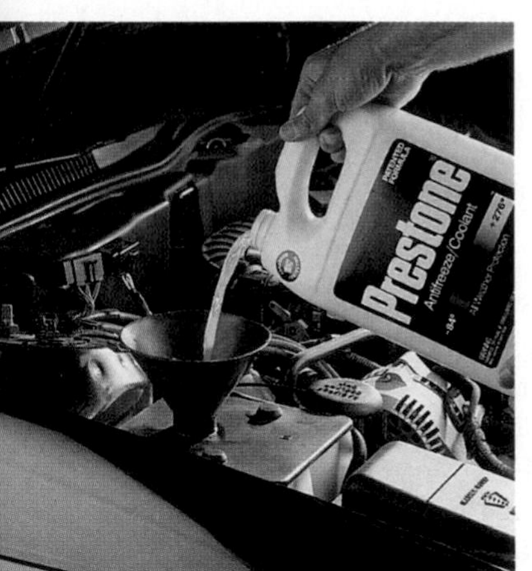

FIGURE 13.18
Automobile antifreeze. Antifreeze consists mainly of ethylene glycol (CH_2OHCH_2OH) with perhaps a dye added.

If you add a nonvolatile solute to a liquid, you not only depress its freezing point, but you also elevate its boiling point. Most antifreeze liquids contain a compound called *ethylene glycol* (Figure 13.18). The addition of this liquid to the water in your auto's radiator coolant not only lowers the freezing point of the water solution so it does not freeze in the winter, but it also elevates the boiling point. This helps prevent the radiator coolant from boiling over in the summer. If you add enough nonvolatile solute to water to depress its freezing point by 1.86°C, you find that its boiling point has been elevated by 0.52°C. Like freezing-point depression, the boiling-point elevation of a solution is a general phenomenon.

We define the **boiling-point elevation,** ΔT_b, to be *the colligative property of a solution equal to the boiling point of the solution minus the boiling point of the pure solvent.* (This gives a positive quantity.)

$$\Delta T_b = \text{boiling point of solution} - \text{boiling point of pure solvent}$$

The boiling-point elevation is proportional to the molal concentration of the nonvolatile solute, just like the freezing-point depression. Mathematically, you can write

$$\Delta T_b = K_b m \qquad \text{(for a nonvolatile, nonionic solute)}$$

The quantity K_b is called the *boiling-point elevation constant.* Like K_f, it depends only on the solvent. For water, K_b equals 0.52°C/m. The next example illustrates the use of this equation.

■ EXAMPLE 13.12 Calculating the Boiling Point of a Solution

What is the boiling point of an aqueous solution that is 0.15 m glucose ($C_6H_{12}O_6$)? K_b for water equals 0.52°C/m; the boiling point of water is 100.00°C.

Problem Analysis

Relate the boiling point of the solution to the boiling-point elevation (ΔT_b) and the boiling point of pure water by rearranging the equation

$$\Delta T_b = \text{boiling point of solution} - \text{boiling point of pure water}$$

to

$$\text{Boiling point of solution} = \Delta T_b + \text{boiling point of pure water}$$

Calculate ΔT_b from the boiling-point elevation constant for water (0.52°C/*m*) and the molality of the solution (0.15 *m*).

Solution

The boiling-point elevation is

$$\Delta T_b = K_b m = 0.52°\text{C}/m \times 0.15\ m = 0.078°\text{C}$$

The boiling point of the solution is 100.00°C + 0.078°C = 100.08°C.

Exercise 13.12

The solution described in Exercise 13.11 contains 23.4 g of sucrose, $C_{12}H_{22}O_{11}$, in 125 g of water. What is the boiling point of this solution?

(Try Problems 13.71 and 13.72.)

Animation: Osmosis.

13.11 Osmotic Pressure

OBJECTIVES

- Describe the phenomenon of osmosis.
- Define the colligative property *osmotic pressure*.
- Explain the importance of osmotic pressure in biological processes.

When you place a carrot in a concentrated sugar solution, you find that after a few hours the carrot shrinks significantly in size (Figure 13.19). What is happening? Why does the carrot shrink?

To understand why this happens, imagine the simpler situation sketched in Figure 13.20. The diagram shows two aqueous solutions of glucose (blood sugar) separated by

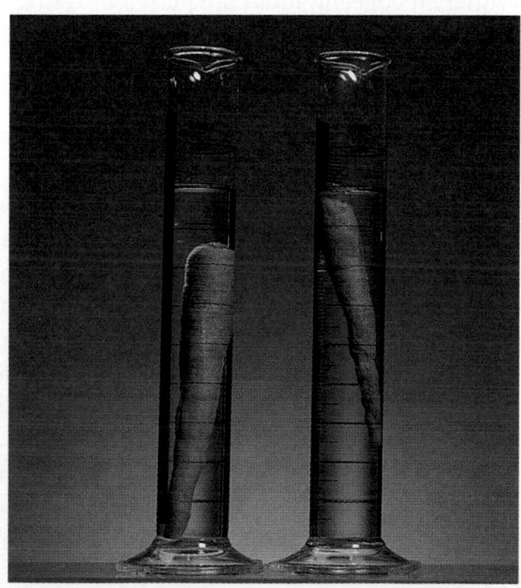

FIGURE 13.19
Effect of concentrated sugar solution on a carrot. The graduated cylinders originally contained carrots of about the same size. The cylinder on the right, however, contains a concentrated sugar solution. Note how the carrot in this cylinder has shrunk.

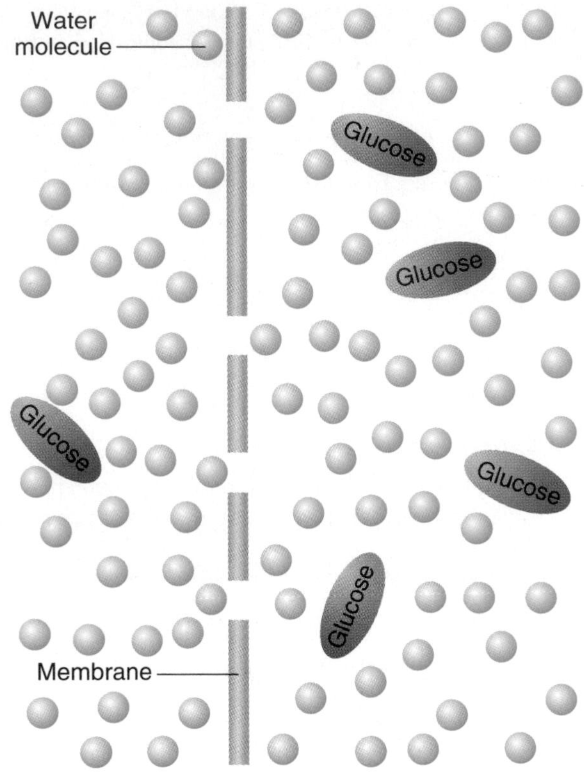

FIGURE 13.20
A semipermeable membrane separating glucose solutions. The solution on the left is less concentrated than the one on the right. As a consequence, water molecules pass through the membrane from left to right.

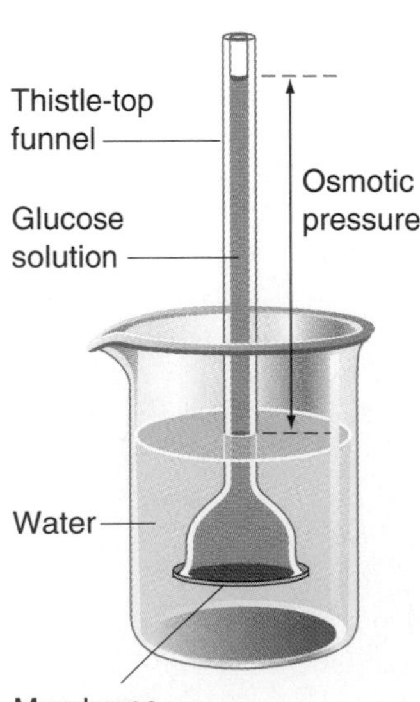

FIGURE 13.21
An experiment in osmosis. Water passes through the membrane into the glucose solution inside the funnel. The flow of water ceases when the liquid in the funnel stem exerts sufficient downward pressure.

a membrane of cellophane. The cellophane is a *semipermeable membrane* because it allows certain molecules (water molecules in this case) to pass through it, but not others (the glucose molecules cannot pass through the membrane). Water molecules pass through the cellophane from the dilute solution to the concentrated solution until the solutions on the two sides of the membrane become equal in glucose concentration. The net effect is as if the two solutions had been placed in contact. Normally, when two solutions of different concentrations are placed in contact, their molecules tend to diffuse in such a way that the solutions become equal in their concentrations. In osmosis, however, only the solvent molecules can move between the two solutions (the solute molecules cannot pass through the membrane).

A carrot is made up of cells, each of which consists of a cell membrane surrounding cell constituents and cell solution. When the carrot is placed in a concentrated sugar solution, water passes from the cells through their membranes into the sugar solution, in order to dilute the concentrated sugar solution and equalize the concentrations on the two sides of the membranes. As a result, the cells of the carrot shrink. You can see a similar effect when you have had your hands in soapy water for a while; your hands become wrinkled as water passes from your hands into the more concentrated soapy water.

Osmosis is *the process of solvent flow through a semipermeable membrane in order to equalize the concentrations of solutes on the two sides of the membrane.* Figure 13.21 shows an experiment in osmosis. In this experiment, a glucose solution is placed in a thistle-top funnel whose mouth is covered with a semipermeable membrane (cellophane, for example). The funnel and its glucose solution are now inverted and placed in a beaker of distilled water. Water flows from the beaker into the funnel containing the

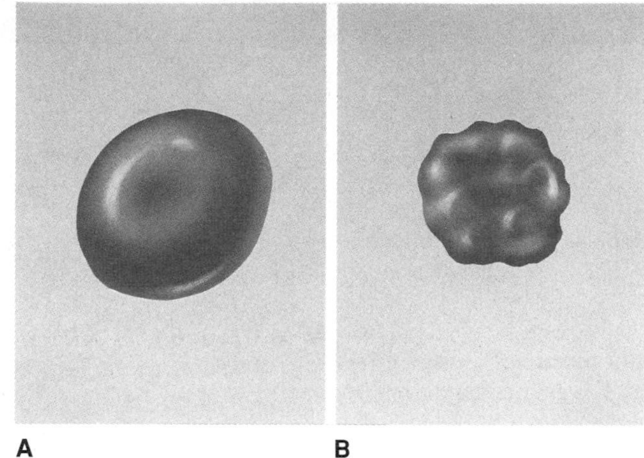

FIGURE 13.22
The importance of osmotic pressure of a solution on red blood cells. *(A)* When red blood cells are placed in pure water, the cells enlarge. *(B)* When the cells are placed in a 2.0% by mass NaCl solution, the cells shrink.

concentrated solution of glucose, in an effort to equalize the concentrations of the two solutions. As the water flows into the glucose solution diluting it, the solution rises up the funnel tube.

Any solution in the funnel tube that is above the level of water in the beaker exerts a pressure downward due to gravity. As osmosis continues, the pressure downward increases. Eventually, the level of solution in the funnel tube exerts sufficient pressure downward to counteract the upward flow of solvent molecules across the membrane into the solution in the funnel. Osmosis then stops.

Osmotic pressure is *the pressure that must be exerted on a solution to stop osmosis.* It is a colligative property because it is proportional to the concentration of solute particles in the solution. The osmotic pressure of a solution can be quite large. A 0.010 *M* aqueous solution of a nonionizable solute has an osmotic pressure of 186 mmHg. In the experiment that we sketched in Figure 13.21, the solution in the funnel stem would have to be 2.53 m (8.30 ft) high to stop osmosis! ●

● Because osmotic pressure is so high, it can be measured for quite dilute solutions. As a result, osmotic pressure is often used to measure the very high molecular weights (often greater than 10^6) of polymers, which are generally not very soluble. Values of freezing-point depression and boiling-point elevation are not large enough for such dilute solutions to provide a means of measuring the molecular weights of polymers.

Osmosis is very important in many biological processes. A biological cell, as we noted earlier, consists of cell constituents and cell solution enclosed in a membrane. If the cell solution has a concentration that is different from that of the solution surrounding the cell, osmosis will occur. When you place red blood cells in pure water, water flows through the cell membrane into the cell solution, and the cells enlarge (Figure 13.22A). If allowed to remain in this water, the red blood cells would eventually burst. On the other hand, when red blood cells are placed in an aqueous solution that is 2.0% by mass sodium chloride, water from the cell passes through the membrane into this sodium chloride solution. The red blood cells shrink (Figure 13.22B).

For this reason, a solution that is given intravenously to a patient must have an osmotic pressure that is nearly equal to that of blood plasma, to ensure that the blood cells are not adversely affected by osmosis. When a solution has an osmotic pressure equal to that of blood plasma, it is said to be *isotonic* with blood plasma. A solution that is 0.85% by mass of sodium chloride is isotonic with blood plasma. Such a solution is known as *isotonic saline* or *physiological saline* solution. It is given to patients suffering from severe dehydration. A solution more concentrated than this is said to be hypertonic, whereas a less concentrated solution is called hypotonic.

Chemical Perspective

Water, the (Somewhat) Universal Solvent

The "universal solvent" sought by the alchemists, those speculative philosophers of the Middle Ages, was supposedly a liquid that would dissolve all substances. Early chemists took up the search, but abandoned it when they realized that no vessel could contain a substance that dissolved all things. Water, though not the universal solvent, nevertheless does dissolve many materials. Most of the water that we encounter contains some dissolved substances. Very pure water is not even desirable under all circumstances. For example, if an in-ground swimming pool were filled with pure water, the plaster lining of the pool would dissolve because pure water is such a good solvent.

The ability of water to dissolve a wide variety of substances stems from two properties: the polarity of the water molecule and the ability of the molecule to form hydrogen bonds. Water molecules are strongly attracted to ions, to other polar molecules, and to molecules with which water can form hydrogen bonds. If these attractions are strong enough, they overcome the attractions between the molecules (or ions) of the other substance and that substance dissolves. Thus, water-soluble substances include a wide range of substances: ionic solids, polar substances, and hydrogen-bonded compounds.

Water, with its solvent properties, is central to the daily changes we see on earth. As water from rain and snow flows over rocks and through soil, mineral substances dissolve, and the water becomes a solution of inorganic materials from minerals. This dilute aqueous solution, or fresh water, eventually finds its way to lakes and streams and finally to the oceans, which have become concentrated solutions of the most soluble mineral substances (especially sodium chloride). The cycle continues as water from the oceans evaporates, forming clouds, then rain or snow. Freshwater sources, like the lakes and rivers, are renewed. This natural cycle of water from the oceans to freshwater sources and back to the oceans is called the *hydrologic cycle* (Figure 13.23).

The fresh water we drink and use for our daily chores or for manufacturing processes is a dilute solution containing a number of metal ions, such as Ca^{2+} and Mg^{2+}. When these ions are in sufficient concentration, the water is said to be *hard.* These ions in hard water react with soaps, which are sodium or potassium salts of so-called fatty acids found in fats and vegetable oils. The product is the curdy material you see in soapy hard water that adheres to clothing and bathtubs, forming "bathtub ring."

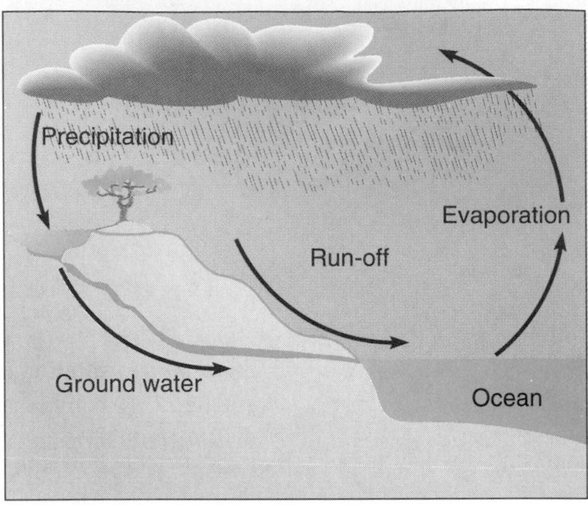

FIGURE 13.23
The hydrologic cycle. Water evaporates from the ocean to form clouds. Then it rains or snows, replenishing freshwater sources. Water from these sources eventually returns to the oceans via rivers, runoff, or groundwater.

The process of removing such ions as Ca^{2+} is referred to as *water softening.* Water is often softened in the home and in some commercial applications by using *ion-exchange resins.* These are insoluble, macromolecular materials (substances consisting of giant molecules) to which negatively charged groups are chemically bonded. The negative charges are counterbalanced by cations, such as Na^+. When hard water containing calcium ions passes through a column of this ion-exchange resin, the sodium ions in the resin are replaced by calcium ions (Figure 13.24). The water that passes through the column now contains sodium ions in place of calcium ions, so it is no longer hard. The result is *soft water.* The column of ion-exchange resin can be regenerated by passing salt (sodium chloride) solution through the column to replace the calcium ions on the resin with sodium ions. Now the column can be used again to soften hard

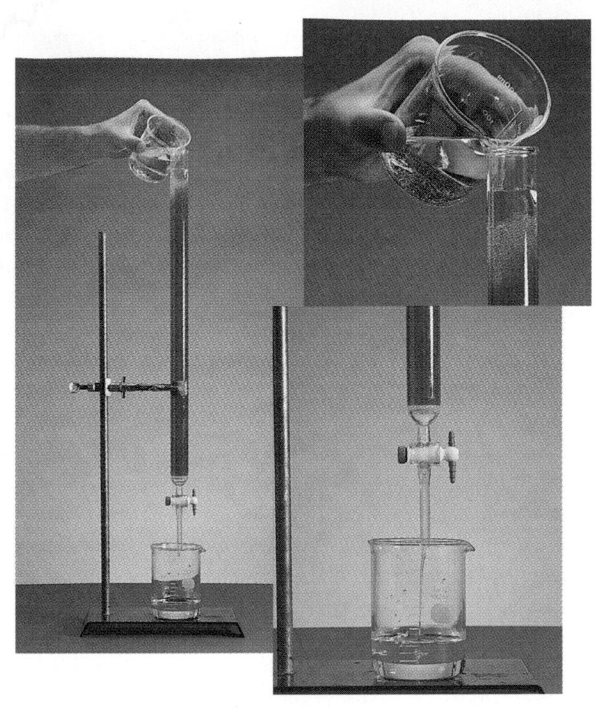

water. Because soft water produces fewer curds with soap, it is preferred to hard water for taking showers and washing clothes. If you are on a low-salt diet, you should not drink softened water because of its sodium ion content. And if you do not want to damage the plaster lining in your swimming pool, don't fill it with softened water!

FIGURE 13.24

Softening of water by ion exchange. The column contains an ion-exchange resin. Hard water, containing Ca^{2+}, is poured in at the top of the column and water containing Na^{+} is removed at the bottom.

CHAPTER REVIEW

Key Words

solution *(p. 394)*
solvent *(p. 395)*
solute *(p. 395)*
miscible *(p. 395)*
immiscible *(p. 395)*
solubility *(p. 398)*
saturated solution *(p. 399)*
unsaturated solution *(p. 399)*
supersaturated solution *(p. 399)*
dilute solution *(p. 406)*
concentrated solution *(p. 406)*
mass percent of solute *(p. 406)*
molarity *(p. 410)*
titration *(p. 414)*
equivalent of an acid *(p. 415)*
equivalent of a base *(p. 415)*
normality *(p. 415)*
molality *(p. 419)*
colligative property *(p. 420)*
freezing-point depression *(p. 421)*
boiling-point elevation *(p. 422)*
osmosis *(p. 424)*
osmotic pressure *(p. 425)*

Key Equations

$$\text{Mass percent of solute} = \frac{\text{mass of solute}}{\text{mass of solution}} \times 100\%$$

$$\text{Molarity} = \frac{\text{moles of solute}}{\text{liters of solution}}$$

$$\text{Normality} = \frac{\text{equivalents of solute}}{\text{liters of solution}}$$

$$M_iV_i = M_fV_f$$

$$\text{Molality} = \frac{\text{moles of solute}}{\text{kilograms of solvent}}$$

$\Delta T_f = K_f m$ (for a nonvolatile, nonionic solute)

$\Delta T_b = K_b m$ (for a nonvolatile, nonionic solute)

Summary

A *solution* is a homogeneous mixture of one or more substances (*solutes*) dissolved in another substance (the *solvent*). A solution may be a gas, liquid, or solid. Aqueous solutions are solutions of solutes dissolved in water. The solute in a solution does not settle out because the solute particles (molecules or ions) are being constantly hit by other particles and so are in constant random motion throughout the solution.

Solubility is the maximum amount of substance that dissolves in a given volume of solvent at a specified temperature. The solution you obtain when you dissolve the maximum amount of solute is *saturated*; otherwise, it is *unsaturated*. If more than this maximum amount is dissolved, the solution is *supersaturated*. In a saturated solution in contact with a solid solute, the processes of dissolution and crystallization occur simultaneously and at the same rate, so that the solution process appears to have stopped. The solution has reached equilibrium. Three main factors affect solubility: the natures of the solute and solvent, the temperature, and the pressure (in the case of gases dissolving in a liquid or solid).

Concentration\ is a general term used to describe the quantity of solute dissolved in a given quantity of solution or solvent. Although solutions may be called *concentrated* or *dilute*, a more quantitative expression of concentration is generally used. Four concentration units were discussed in this chapter: *mass percent of solute, molarity, normality,* and *molality.*

Molarity, which is the moles of solute per liter of solution, is a convenient unit because you can use it to convert between moles of solute and volume of solution. This allows you to do stoichiometry using solutions of reactants. It is especially useful in *titration,* which is used to determine the amount of a substance in a sample. Normality is a concentration unit useful for acid–base work; it is equal to equivalents of acid or base per liter of solution. (An equivalent of acid is the amount that provides one mole of hydrogen ion; an equivalent of base is the amount that provides one mole of hydroxide ion.) Molality equals moles of solute per kilogram of solvent. The molality does not change with temperature because the solution is made up from given masses of substances. (By contrast, molarity varies with temperature because molarity depends on the volume of solution.) This makes molality a convenient unit when discussing properties that involve a change of temperature (such as freezing-point depression).

The *colligative properties* of a solution depend only on the number of solute particles in the solution and not on the particular characteristics of the solute particles. Some examples are *freezing-point depression, boiling-point elevation,* and *osmosis*. Thus, when you add a nonvolatile solute to a solvent, the solute depresses the freezing point and elevates the boiling point of the solution. The addition of a solute also increases the *osmotic pressure* of the solution.

Problem-Solving Skills

1. **Calculating Mass Percent of Solute:** Calculate the mass percent of solute given the masses of solute and solution (Example 13.1) or given the masses of solute and solvent (Example 13.2).
2. **Preparing a Solution of Given Mass Percent of Solute:** Given the mass percent of solute and mass of solution, calculate the mass of solute and solvent required to prepare the solution (Example 13.3).
3. **Calculating the Molarity of a Solution:** Calculate the molarity of a solution given the moles of solute and volume of solution or given the mass of solute and volume of solution (Example 13.4).
4. **Preparing a Solution of Given Molarity:** Given the volume and molarity of a solution, calculate the mass of solute needed to prepare the solution (Example 13.5).
5. **Calculating a Stoichiometric Quantity Using Molarity:** Given the volume and molarity of a reactant (or product), calculate the mass of another reactant or product (Example 13.6).
6. **Calculating the Mass of a Reactant Using Normality:** Given the volume and molarity of an acid (or a base), calculate the mass of base (or acid) that reacts completely with that acid (or base) (Example 13.7).
7. **Preparing a Solution by Dilution:** Given the volume and molarity of a solution that you wish to make up and the molarity of a more concentrated stock solution, calculate the volume required of the stock solution (Example 13.8 and Example 13.9).
8. **Calculating the Molality of a Solution:** Given the masses of solute and solvent, calculate the molality (Example 13.10).
9. **Calculating the Freezing Point of a Solution:** Given the molality of a solution and the freezing point and K_f for the solvent, calculate the freezing point of the solution (Example 13.11).
10. **Calculating the Boiling Point of a Solution:** Given the molality of a solution and the boiling point and K_b for the solvent, calculate the boiling point of the solution (Example 13.12).

Questions to Test Your Reading

13.1 Define the following terms: *solution, solute, solvent.* Apply these terms to the result of dissolving sodium chloride in water.

13.2 Some rubbing alcohol solutions are mixtures of isopropyl alcohol in water. Isopropyl alcohol and water are miscible in one another. What does the term *miscible* mean?

13.3 A chemist tries to mix two liquids together by vigorously stirring them. However, they seem to quickly separate into two different layers. Does this mean that the chemist has not stirred the liquids together enough and that they would eventually dissolve in one another? What has happened?

13.4 Give examples of the following solutions.
(a) a gaseous solution
(b) a liquid solution of a liquid solvent and gaseous solute
(c) a liquid solution obtained by mixing solids
(d) a solid solution
(e) a nonaqueous solution

13.5 What are the physical states of the solute and solvent in each of the following solutions?
(a) a breathing mixture of helium and oxygen (used by deep-sea divers)
(b) brass (contains copper and zinc)
(c) antifreeze solution in your car's radiator
(d) brine (salt in water)
(e) drinking water containing chlorine (as a disinfectant)

13.6 Compare the general properties of compounds and solutions.

13.7 Explain why the solute molecules in a solution do not settle out like the particles of sand that have been vigorously stirred into a pail of water.

13.8 The solubility of lithium hydroxide in water at 20°C is 12.4 g per 100 mL of water. What does this mean?

13.9 What is a saturated solution? How would you differentiate a saturated solution from an unsaturated solution?

13.10 A solution contains 11.0 g of LiOH per 100 mL of water at 20°C. Is the solution saturated or unsaturated? Explain. (See Question 13.8.)

13.11 What is a supersaturated solution? How would you differentiate such a solution from a saturated solution?

13.12 Potassium dichromate ($K_2Cr_2O_7$) is a red-orange compound. Its solubility is 11.7 g in 100 mL of water. Suppose you add 15.0 g of potassium dichromate to 100 mL of water and stir. How many grams of $K_2Cr_2O_7$ remain undissolved when the solution is saturated with this compound?

13.13 Describe the process of equilibrium that exists between potassium dichromate in a saturated solution and the solid solute. The amount of potassium dichromate in the solution remains constant (at a given temperature). Does this mean that the process of dissolution of solid solute has stopped? Explain.

13.14 Ionic compounds vary widely in their solubilities in water, some being very soluble, others insoluble. Give one reason that this is so.

13.15 Sketch a potassium ion with a couple of water molecules about it. How does this differ from a similar sketch of a fluoride ion with water molecules about it?

13.16 Carbon tetrachloride is a nonpolar liquid. What does that mean? In which liquid do you expect carbon tetrachloride to be more soluble, water or gasoline? Explain.

13.17 What do you expect to happen if you warm up a carbonated beverage? Explain.

13.18 A saturated solution of potassium nitrate at 20°C is heated to 50°C. Do you expect the solution to continue to be saturated? Explain.

13.19 Divers who ascend to the surface too quickly sometimes get "the bends," which is a painful affliction caused by small nitrogen gas bubbles that come out of the blood. Explain what causes this gas-bubble formation.

13.20 State the meaning of each of the following.
(a) An aqueous solution is 4.0% by mass of sodium carbonate.
(b) A concentrated ammonia solution is 14.5 *M*.
(c) A solution is 0.15 *m* sucrose in water.

13.21 A solution is 0.18 *M* H_2SO_4. Write a conversion factor that converts from liters of solution to moles of H_2SO_4.

13.22 Molality can be used to obtain a conversion factor that converts from kilograms of solvent to moles of solute. Write such a conversion factor for a solution that is 0.18 *m* H_2SO_4.

13.23 In what circumstance is molarity a convenient unit of concentration?

13.24 Why can the molarity of a solution change with temperature, but its molality cannot?

13.25 Titration is a method commonly used to find the amount of a compound in a sample. For example, you could determine the quantity of citric acid in a sample of lemon juice by titrating the sample with a solution of sodium hydroxide. Describe in detail how this is done, and explain the calculations you would have to do.

13.26 Describe the procedure you would use to prepare a solution of specific molarity from a more concentrated solution of known molarity.

13.27 A solution of 1.0 L of 2.0 *M* HCl contains how many moles of HCl? This solution is diluted to 1.0 *M* HCl. What is the final volume of the solution?

13.28 What is meant by the freezing-point depression of a solution? What is the freezing-point depression of a 1.0 *m* aqueous solution of a nonionizable solute? What is the freezing-point depression of a 2.0 *m* aqueous solution of this solute?

13.29 What is the freezing point of 0.10 *m* glucose in water? What is the freezing point of 0.10 *m* sucrose in water? What are the boiling points of these two solutions?

13.30 Why does the freezing-point depression of an aqueous solution that is 1.0 *m* NaCl differ from the freezing-point depression of a similar solution that is 1.0 *m* urea (a molecular solute)?

13.31 You cut up some salad greens in a large bowl and store the bowl in the refrigerator until dinner several hours later. What would happen to the greens if you added a dressing of vinegar and other ingredients at this time instead of waiting until just before serving them?

13.32 What is meant by a saline solution that is isotonic with blood plasma? What happens to red blood cells that are placed in a solution that is more concentrated than an isotonic saline solution?

Practice Problems

Mass Percent of Solute (Section 13.7)

13.33 What is the mass percent of solute in each of the following?
(a) a solution that contains 15 g $Pb(NO_3)_2$ in a total solution weighing 45 g
(b) a solution that contains 13.6 g Na_2SO_4 in enough water to give 125 g of solution

13.34 Calculate the mass percent of solute in each of the following solutions.
(a) a chemist dissolves 0.236 g NH_4NO_3 in water to give 24.1 g of solution
(b) a chemist dissolves 7.1 g of acetic acid ($HC_2H_3O_2$) in water to give a total of 134 g

13.35 An experiment calls for a solution that you make up by dissolving 2.34 g of lead(II) nitrate in 255 g of water. What is the mass percent of lead(II) nitrate in the solution?

13.36 A solution of sodium chloride contains 5.0 g NaCl dissolved in 85 g of water. What is the mass percent of NaCl in the solution?

13.37 You want to make an aqueous solution that is 5.0% by mass of copper(II) sulfate, $CuSO_4$. If you prepare 425 g of solution, how many grams of copper(II) sulfate and how many grams of water do you need?

13.38 The label on a bottle says that it contains a 9.5 mass percent solution of ammonium chloride, NH_4Cl. You find that the bottle contains 124 g of this solution. How many grams of ammonium chloride and how many grams of water are in the solution?

13.39 A solution is 7.35% KCl by mass. An experiment requires 5.0 g KCl. How many grams of the solution do you need for this experiment?

13.40 By analysis, a chemist determines that a sample of solution contains 0.257 g $AgNO_3$ (silver nitrate). If the sample was a 5.00% solution (by mass), how many grams of the solution were in the sample?

13.41 You make up a solution by dissolving 0.145 mol NaOH in 200 g of water. What is the mass percent of sodium hydroxide in the solution?

13.42 A solution contains 0.100 mol copper(II) sulfate ($CuSO_4$) in 1.00 kg of water. What is the mass percent of $CuSO_4$ in the solution?

Molarity (Section 13.8)

13.43 Calculate the molar concentration of solute in each of the following.
(a) a solution containing 0.0450 mol $Cu(NO_3)_2$ in 250 mL of solution
(b) a solution of 0.250 mol NH_4NO_3 dissolved in enough water to give 455 mL of solution

13.44 What is the molarity when each of the following amounts of solute is made up to give 50.0 mL of solution?
(a) 0.0235 mol $NaNO_3$
(b) 0.178 mol $CuCl_2$

13.45 Calculate the molarity of each of the following solutions.
(a) 0.418 g $Ba(NO_3)_2$ in 25.0 mL of solution
(b) 23.4 mg $NaHSO_4$ in 50.0 mL of solution

13.46 Calculate the molarity of each of the following solutions.
(a) 28.5 g $CaCl_2$ in 755 mL of solution
(b) 85.6 mg $AgNO_3$ in 25.0 mL of solution

13.47 How many grams of potassium chromate (K_2CrO_4) would you need to prepare 385 mL of 0.0750 *M* solution?

13.48 An experiment calls for 25.0 mL of 0.300 *M* potassium permanganate, $KMnO_4$. How many grams of $KMnO_4$ are required for this solution?

13.49 How many milliliters of 0.265 *M* sodium carbonate (Na_2CO_3) solution do you need to give 0.150 mol Na_2CO_3?

13.50 What volume (in milliliters) of 0.0850 *M* sulfuric acid (H_2SO_4) must one have to equal 0.0450 mol H_2SO_4?

Stoichiometry Involving Solutions (Section 13.8)

13.51 Limestone is a rock composed primarily of calcium carbonate, $CaCO_3$. $CaCO_3$ reacts with hydrochloric acid (HCl) to give carbon dioxide gas.

$$CaCO_3(s) + 2HCl(aq) \longrightarrow CaCl_2(aq) + H_2O(l) + CO_2(g)$$

If a total of 38.9 mL of 0.115 *M* HCl is required to react with the calcium carbonate in a sample of limestone, how many grams of calcium carbonate are in the sample?

13.52 Aqueous ammonia, $NH_3(aq)$, is sold in grocery stores as a household cleaner. Ammonia reacts with sulfuric acid, H_2SO_4, to produce the salt ammonium sulfate.

$$2NH_3(aq) + H_2SO_4(aq) \longrightarrow (NH_4)_2SO_4(aq)$$

A sample of ammonia solution requires 25.8 mL of 0.283 *M* H_2SO_4 to react completely with the ammonia in the sample. How many grams of NH_3 are in the sample?

13.53 How many grams of sodium carbonate (Na_2CO_3) are needed to react with 35.9 mL of 0.0876 *M* nitric acid (HNO_3)? How many grams of sodium nitrate ($NaNO_3$) are produced? The reaction is

$$Na_2CO_3(s) + 2HNO_3(aq) \longrightarrow 2NaNO_3(aq) + H_2O(l) + CO_2(g)$$

13.54 How many grams of lead(II) chromate, $PbCrO_4$, are produced from 154 mL of 0.180 *M* $Pb(NO_3)_2$ according to the following reaction? How many grams of potassium nitrate, KNO_3, can you produce?

$$Pb(NO_3)_2(aq) + K_2CrO_4(aq) \longrightarrow 2KNO_3(aq) + PbCrO_4(s)$$

13.55 The sour taste of vinegar is due to acetic acid, $HC_2H_3O_2$. A 10.0-mL sample of vinegar is titrated with 0.192 *M* NaOH solution. The acetic acid reacts completely, as shown by the following equation, when 43.2 mL of NaOH solution is added.

$$HC_2H_3O_2(aq) + NaOH(aq) \longrightarrow NaC_2H_3O_2(aq) + H_2O(l)$$

What is the molarity of acetic acid in the vinegar?

13.56 Some toilet bowl cleaners contain hydrochloric acid, HCl. A 1.00-mL sample of cleaner is titrated with a 0.248 *M* NaOH solution. The hydrochloric acid reacts completely when 24.2 mL of the NaOH solution is added. The reaction is

$$HCl(aq) + NaOH(aq) \longrightarrow NaCl(aq) + H_2O(l)$$

What is the molarity of hydrochloric acid in the cleaner?

13.57 Limewater is an aqueous solution of calcium hydroxide, $Ca(OH)_2$. Calcium hydroxide reacts with sulfuric acid, H_2SO_4, to produce the salt calcium sulfate.

$$Ca(OH)_2(aq) + H_2SO_4(aq) \longrightarrow CaSO_4(s) + 2H_2O(l)$$

A sample of limewater requires 48.3 mL of 0.184 *N* H_2SO_4 to react completely with the calcium hydroxide in the sample. How many grams of $Ca(OH)_2$ are in the sample?

13.58 Aqueous ammonia, $NH_3(aq)$, is sold in grocery stores as a household cleaner. Ammonia reacts with sulfuric acid, H_2SO_4, to produce the salt ammonium sulfate.

$$2NH_3(aq) + H_2SO_4(aq) \longrightarrow (NH_4)_2SO_4(aq)$$

A sample of ammonia requires 25.8 mL of 0.566 *N* H_2SO_4 to react completely with the ammonia in the sample. How many grams of NH_3 are in the sample?

Diluting Solutions (Section 13.8)

13.59 Concentrated nitric acid is 15.9 *M* HNO_3. How many milliliters of this acid do you require to produce 150 mL of 0.15 *M* HNO_3?

13.60 Concentrated ammonia is 14.5 *M* NH_3. How many milliliters of this solution do you require to produce 185 mL of 1.5 *M* NH_3?

13.61 A solution of copper(II) chloride is 2.75 *M* $CuCl_2$. What volume of this solution, in milliliters, is needed to prepare 405 mL of 0.80 *M* $CuCl_2$?

13.62 A solution of sodium sulfide is 0.25 *M* Na_2S. What volume of this solution, in milliliters, is needed to prepare 125 mL of 0.20 *M* Na_2S?

13.63 A solution is 2.85 *M* $BaCl_2$. What is the molarity of the solution obtained by diluting 50.0 mL of the original solution to 250 mL?

13.64 Concentrated hydrochloric acid, which is 12.1 *M* HCl, is diluted from 25.0 mL to 500 mL. What is the molarity of the resulting solution?

Molality (Section 13.9)

13.65 What is the molality of ethanol (C_2H_5OH) in a solution containing 2.56 g of ethanol and 38.6 g of water?

13.66 A solution contains 4.81 g of sucrose ($C_{12}H_{22}O_{11}$) and 120.0 g of water. What is the molality of sucrose in this solution?

13.67 A solution is 0.34 *m* urea, NH_2CONH_2. How many grams of solution do you need to provide 0.10 mol of urea?

13.68 A solution is 0.47 *m* glucose, $C_6H_{12}O_6$. How many grams of solution do you need to provide 0.25 mol of glucose?

Freezing-Point Depression and Boiling-Point Elevation (Section 13.10)

13.69 Glycerol ($C_3H_8O_3$) is a syrupy, sweet-tasting liquid used in cosmetics and candy. It is a nonvolatile, nonionizable compound. What is the freezing point of an aqueous solution that is 0.25 *m* glycerol?

13.70 Urea (NH_2CONH_2) is a white solid produced commercially as a fertilizer and starting material for plastics. It is a nonvolatile, nonionizable compound. What is the freezing point of an aqueous solution that is 0.34 *m* urea?

13.71 What is the boiling point of the aqueous solution described in Problem 13.67?

13.72 What is the boiling point of the aqueous solution described in Problem 13.68?

13.73 What are the freezing point and boiling point of an aqueous solution that is 0.015 *m* KBr?

13.74 What are the freezing point and boiling point of an aqueous solution that is 0.015 *m* Na_2SO_4?

13.75 Cholesterol is an important compound in the body, but in excess has been implicated as a cause of heart disease. A solution of 4.82 g of cholesterol in 112 g of benzene (C_6H_6) has a freezing point of 4.886°C. The freezing point of pure benzene is 5.455°C; the K_f for benzene is 5.12°C/*m*. What is the molecular weight of cholesterol?

13.76 Vitamin K_1 is a substance found in green leafy vegetables; it is needed by the body to produce a blood-clotting factor. A solution of 55.8 mg of vitamin K_1 in 1.048 g of benzene (C_6H_6) has a freezing point of 4.850°C. The freezing point of pure benzene is 5.455°C; the K_f for benzene is 5.12°C/*m*. What is the molecular weight of vitamin K_1?

Additional Problems

13.77 An aqueous solution is 9.30% ammonium chloride (NH_4Cl) by mass. The density of the solution is 1.04 g/mL. What are the molarity and molality of this solution?

13.78 An aqueous solution is 6.25% calcium chloride ($CaCl_2$) by mass. The density of the solution is 1.03 g/mL. What are the molarity and molality of this solution?

13.79 Which aqueous solution has the lower freezing point, 0.10 *m* $BaCl_2$ or 0.10 *m* KCl?

13.80 Which aqueous solution has the lower freezing point, 0.22 *m* NaCl or 0.22 *m* $AlCl_3$?

13.81 A liquid bleach contains sodium hypochlorite (NaClO) dissolved in water. The freezing point of the solution is −2.78°C. What is the molality of sodium hypochlorite in the solution? What is the mass percent of solute?

13.82 An antiseptic solution contains hydrogen peroxide (H_2O_2) dissolved in water. The freezing point of the solution is −1.63°C. What is the molality of hydrogen peroxide in the solution? What is the mass percent of solute?

13.83 An aqueous solution of a compound freezes at −4.36°C. At what temperature will this solution boil?

13.84 An aqueous solution of a compound boils at 102.4°C. At what temperature will this solution freeze?

13.85 Phosphoric acid (H_3PO_4) is used to give an acidic taste to some cola beverages. A 2.50-mL sample of a phosphoric acid solution is titrated with a 0.0152 *M* $Ba(OH)_2$ solution. The reaction is

$$2H_3PO_4(aq) + 3Ba(OH)_2(aq) \longrightarrow Ba_3(PO_4)_2(s) + 6H_2O(l)$$

The reaction is complete when 34.7 mL of barium hydroxide is added. What is the molarity of the phosphoric acid solution? What mass, in grams, of solid $Ba_3(PO_4)_2$ is formed by this reaction?

13.86 Seawater contains considerable quantities of sodium chloride, NaCl. A 25.0-mL sample of seawater is titrated with 0.102 *M* $AgNO_3$ (silver nitrate) solution. The following reaction is complete when 37.1 mL of $AgNO_3$ solution has been added.

$$NaCl(aq) + AgNO_3(aq) \longrightarrow AgCl(s) + NaNO_3(aq)$$

What is the molarity of NaCl in the seawater sample? What mass, in grams, of solid AgCl is formed by this reaction?

13.87 Water from an alkali pond contains 0.0145 *M* $Ca(OH)_2$ (calcium hydroxide). How much 2.50 *M* HCl should be added to 1.00 L of water from this pond to remove completely the calcium hydroxide according to the following equation?

$$Ca(OH)_2(aq) + 2HCl(aq) \longrightarrow CaCl_2(aq) + 2H_2O(l)$$

13.88 Water from an acidic lake contains 0.00102 *M* H_2SO_4 (sulfuric acid). How much 0.0208 *M* $Ca(OH)_2$ should be added to 5.00 L of water from this lake to remove completely the sulfuric acid according to the following equation?

$$H_2SO_4(aq) + Ca(OH)_2(aq) \longrightarrow CaSO_4(s) + 2H_2O(l)$$

13.89 A solution is prepared by adding 148 mL of 3.00 *M* HCl to 252 mL of 6.25 *M* HCl. What is the molarity of HCl in this solution? (Assume that the combined volume is the sum of the separate volumes.)

13.90 A solution is prepared by adding 2.05 L of 0.250 *M* NaOH to 1.25 L of 0.120 *M* NaOH. What is the molarity of NaOH in this solution? (Assume that the combined volume is the sum of the separate volumes.)

13.91 U.S. silver coins, which contain 10% Cu and 90% Ag by mass, melt (and freeze) at 890°C. Pure silver melts (and freezes) at 961°C. What is the freezing-point depression constant of silver, in units of °C/*m*?

13.92 Yellow brass contains 33% Zn and 67% Cu by mass and melts (and freezes) at 940°C. Pure copper melts (and freezes) at 1083°C. What is the freezing-point depression constant of copper, in units of °C/*m*?

Practice in Problem Analysis

For each problem, describe the thinking you would use (the problem analysis) before doing the actual solution, but do not solve the problem.

1. What is the mass percent of solute in a solution that you prepare by dissolving 7.8 g of magnesium hydroxide, $Mg(OH)_2$, in 87.6 g of water?

2. You are given a sample solution containing an unknown mass of magnesium hydroxide. You titrate magnesium hydroxide solution with 0.23 *M* H_2SO_4. If the completed reaction requires 35 mL of sulfuric acid, what is the mass of magnesium hydroxide in the sample?

PRACTICE EXAM

1. A solution contains 5.8 g $CaCl_2$ in 95.0 g of solution. What is the mass percent of $CaCl_2$ in the solution?
(a) 0.30 (b) 0.61 (c) 1.2 (d) 6.1
(e) 12

2. You make up a solution by dissolving 9.8 g of $CaCl_2$ in 78.5 g of water. What is the mass percent of $CaCl_2$ in the solution?
(a) 2.8 (b) 5.6 (c) 11 (d) 22 (e) 33

3. An experimenter wants to prepare 200. g of an aqueous solution that contains 5.6% by mass of calcium chloride. How many grams of calcium chloride does the experimenter need?
(a) 5.6 (b) 11 (c) 15 (d) 21 (e) 25

4. How many grams of water does the experimenter require for the solution described in Question 3?
(a) 148 (b) 168 (c) 174 (d) 189
(e) 194

5. A student prepares a solution by dissolving 0.672 mol Na_2SO_4 in 565 mL of solution. What is the molarity of this solution?
(a) 1.19 (b) 1.29 (c) 1.39 (d) 1.49
(e) 1.59

6. You prepare a solution by dissolving 56.0 g of Na_2SO_4 in 498 mL of solution. What is the molarity of Na_2SO_4 in this solution?
(a) 0.505 (b) 0.792 (c) 1.02 (d) 1.51
(e) 2.15

7. A chemist wants to prepare 14.5 mL of 0.120 *M* NaOH solution. How many moles of NaOH will she need?
(a) 0.00174 (b) 0.0153 (c) 0.153
(d) 0.187 (e) 0.293

8. You want 125 mL of 0.250 *M* KCl. How many grams of KCl will you need?
(a) 0.0155 (b) 0.281 (c) 0.571 (d) 1.68
(e) 2.33

9. A flask contains an aqueous mixture with an unknown amount of $Mg(OH)_2$. A chemist titrates this mixture with 0.156 *M* HCl. The reaction is

$$2HCl(aq) + Mg(OH)_2(s) \longrightarrow MgCl_2(aq) + 2H_2O(l)$$

If a total of 36.5 mL of this hydrochloric acid is required to complete this reaction, how many grams of magnesium hydroxide were in the flask?
(a) 0.102 (b) 0.166 (c) 0.325 (d) 0.581
(e) 0.781

10. You can prepare lead(II) iodide by reacting lead(II) nitrate with potassium iodide:

$$Pb(NO_3)_2(aq) + 2KI(aq) \longrightarrow PbI_2(s) + 2KNO_3(aq)$$

How many grams of PbI_2 can you prepare from 54.6 mL of 0.350 *M* KI?
(a) 0.752 (b) 1.09 (c) 2.20 (d) 3.51
(e) 4.40

11. A solution containing an unknown quantity of calcium hydroxide, $Ca(OH)_2$, was titrated with 0.150 *N* HCl. If the titration required 48.6 mL of HCl, how many grams of calcium hydroxide were in the solution?
(a) 0.135 (b) 0.270 (c) 0.520 (d) 0.871
(e) 1.08

12. How many milliliters of 2.0 *M* HCl do you need to prepare 50.0 mL of 0.45 *M* HCl?
(a) 5.65 (b) 11.3 (c) 22.6 (d) 33.9
(e) 45.2

13. What is the molality of a solution of 1.28 g glucose, $C_6H_{12}O_6$, dissolved in 67.0 g of water?
(a) 0.106 (b) 0.124 (c) 0.232 (d) 0.341
(e) 0.415

14. What is the freezing point of the solution described in Question 13? K_f for water is 1.86°C/*m*.
(a) 0.197 (b) 0.231 (c) −0.197
(d) −0.231 (e) −0.432

15. What is the boiling point of the solution described in Question 13? K_b for water is 0.52°C/*m*.
(a) 0.055 (b) 0.64 (c) 99.9 (d) 100.06
(e) 105.5

Chapter

14 Reaction Rates and Chemical Equilibrium

These crystals of alum will stop growing when equilibrium is reached.

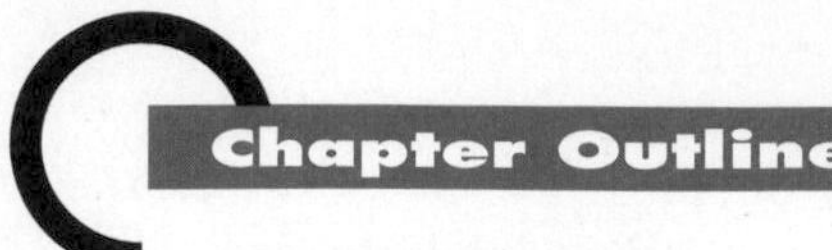

A mountain climber who attempts to scale a very tall mountain must cope not only with cold, wind, and storms but also with the restlessness, confused thinking, and even unconsciousness caused by the rarefied atmosphere around higher peaks (Figure 14.1). Even the most conditioned climber must use an oxygen mask and tank at high altitudes to ward off the deadly effects of oxygen deprivation. Passengers in modern, high-flying commercial airplanes would face the same problems if airline cabins were not pressurized. Because hemoglobin in the blood carries oxygen to all tissues, symptoms of oxygen deficiency are felt everywhere in the body, but particularly in the brain.

Oxygen from the air reacts with hemoglobin (Hb) in red blood cells within the lungs (recall the Chemical Perspective in Chapter 6). The equation for this reaction is

$$Hb + 4O_2 \longrightarrow Hb(O_2)_4$$

This equation suggests that the amount of oxygenated hemoglobin that can form from a fixed amount of hemoglobin depends solely on the number of oxygen molecules present. The experience of a mountain climber who does not have supplemental oxygen, however, indicates that this interpretation is not accurate. At sea level, essentially all of the hemoglobin is converted to $Hb(O_2)_4$ as the blood passes through the lung capillaries in its circuit through the body. But less and less of the hemoglobin is converted as a climber ascends the mountain, even though there are more than enough oxygen molecules in the climber's vicinity to oxygenate all of the hemoglobin. Why? As the elevation increases, the air "thins," and the concentration of oxygen decreases. For example, the concentration of O_2 at the top of Mt. Everest, the world's highest mountain (29,028 feet), is about one-third of its concentration at sea level. The concentration of oxygen, rather than the total number of oxygen molecules, governs how much $Hb(O_2)_4$ is formed in the lung capillaries at any elevation.

In this chapter, we show that many reactions seem to stop before they reach their theoretical yields, and concentrations rather than numbers of molecules determine when

FIGURE 14.1
Mountain climbers experience many problems. Oxygen deprivation at high altitudes can be the most serious problem of mountain climbing if a climber lacks an oxygen mask and tank.

each stopping point occurs. These reactions appear to stop because they have reached *chemical equilibrium. All* reactions proceed until chemical equilibrium has been achieved—as long as none of the products are removed—and then they continue indefinitely even though they seem to have stopped. In this chapter we also reveal some of the factors that affect the rate (speed) of a reaction on its way to equilibrium (and at equilibrium). Finally, we look at how other factors influence the concentrations of reactants and products in a reaction at equilibrium.

REACTION RATES

Many chemical reactions occur when we prepare food, and many of our recipes require heating to make those reactions occur. For example, we cannot prepare an egg without heating it because the necessary chemical reactions will not happen. Moreover, the length of time that we cook the egg determines the extent of the chemical reactions in it. We can cook a soft-boiled egg in 2 or 3 minutes, but we need more cooking time for a hard-boiled egg. We can extend these observations to make a broader statement about all chemical reactions: Chemical reactions require many different types of conditions—added heat is often one of them—and some reactions occur very rapidly, whereas others take hours, days, or even years. The length of time required for the completion of a reaction depends on the following factors:

1. The identity of the reactants
2. The concentration of the reactants
3. The temperature of the reaction mixture
4. The presence of a catalyst

● The term *reaction rate* refers to a reaction's speed.

In the next two sections, we describe a single theory (collision theory) that accounts for the way a reaction's rate depends on each of these factors. ●

14.1 Collision Theory and Activation Energy

OBJECTIVES

- Explain how molecules must collide before they can react.
- Define *activation energy*.
- Relate the magnitude of the activation energy to a reaction's rate.

We know intuitively that a molecule in Indiana is not likely to react with a molecule in Arizona. **Collision theory** goes a step further when it says that *reacting molecules must come so close that they collide.* This theory also says that the *energy of the collision must be greater than a certain minimum value.* ●

● Collision theory also says that reactant molecules must be oriented correctly if a collision is to lead to a reaction. We do not deal with this aspect of the theory.

Molecules in gases and liquids will collide if they are in the same container because they are in constant, chaotic motion. Because of these motions, they cannot avoid encountering one another. A collision is necessary for a reaction to occur, but every collision will not result in a reaction. Although temperature dictates the molecules' average kinetic energy and average speed, some molecules have energies and speeds that are less than the average and others have greater energies and speeds. Therefore, some collisions occur with greater energies than others. The *minimum energy of a collision that leads to a reaction* is called the **activation energy.** A chemical reaction has an activation energy because some of the bonds in the reactant molecules are broken and others are formed. Bond breaking and rearrangement require considerable energy, and this energy comes from the collision.

Consider the reaction of ozone (O_3) with chlorine atoms, a reaction that occurs in the lower stratosphere. (The chlorine atoms come from the decomposition of chlorofluorocarbons, better known as CFCs.) The equation is

$$O_3(g) + Cl(g) \longrightarrow O_2(g) + OCl(g)$$

The collision between the reactants is diagrammed in Figure 14.2. If a reaction is to occur after a collision, a bond between two oxygen atoms in an ozone molecule must be broken and a bond between oxygen and chlorine must be formed. When the reactants collide with insufficient energy, a reaction cannot occur because there is not enough energy in the collision to break the bond and to rearrange the atoms. Only collisions with sufficient energy can do this. The activation energy is the minimum energy that is

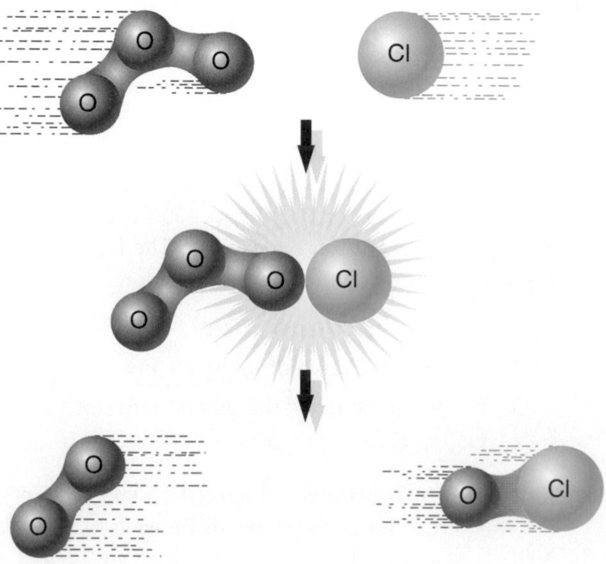

FIGURE 14.2
Ozone reacting with a chlorine atom. An ozone molecule (O_3) collides with a chlorine atom, causing the bond between two oxygen atoms to break. The result is an oxygen molecule and OCl.

FIGURE 14.3
How fast can the boulders be pushed over the hills? More boulders can be pushed over a low hill in a given amount of time than over a higher hill.

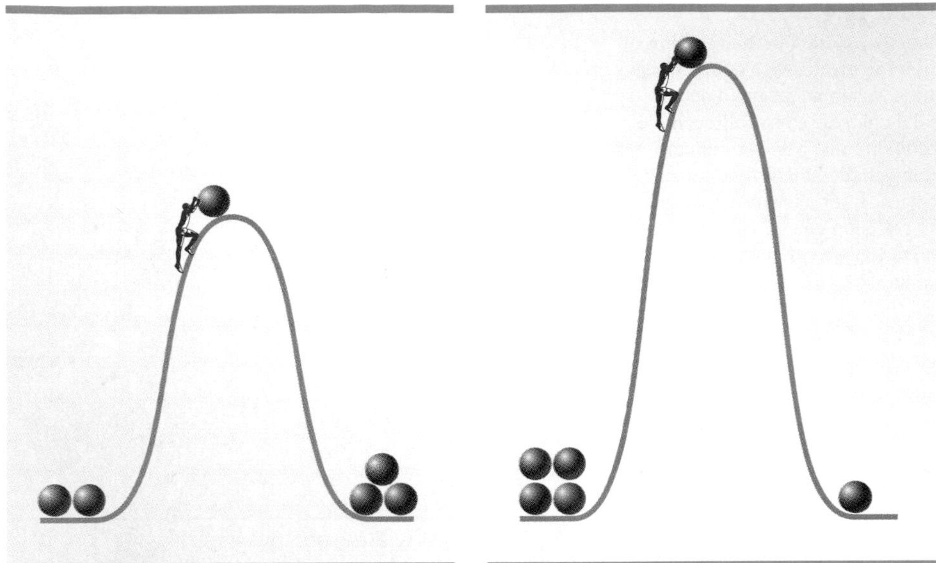

Exploration: Transition states and activation energy.

required. When a collision results in a reaction—unfortunately, in this case, an event that occurs all too often—ozone molecules are lost and holes appear in the ozone layer of the atmosphere.

How does a reaction's activation energy influence its rate? The higher the activation energy, the fewer the collisions with enough energy to cause bond breaking, and the slower the reaction. *A high activation energy means a slow reaction, and a lower activation energy means a faster reaction.*

Consider the person pushing a series of boulders over a hill in Figure 14.3. The height of that hill represents the magnitude of the activation energy in a chemical reaction. If the person does not supply enough energy, he will not be able to move a boulder over the hill. Moreover, more boulders can be pushed over a low hill in a given time than over a larger hill. In other words, moving the boulders over a low hill is faster than moving them over a larger hill. In like fashion, a fast chemical reaction has a low activation energy, and a slower reaction has a larger activation energy.

The relationship between the energies of the colliding reactant molecules and products shown in Figure 14.4 is similar to the analogy with the boulders. As you can see, the height of the hill in the diagram (E_a) represents the activation energy that colliding reactant molecules must overcome if they are to be converted to products. When an ozone molecule collides with a chlorine atom, the energy of the collision must allow the reaction to get over the hill to form products. If the energy of the collision is less than the activation energy, the ozone molecule and the chlorine atom will bounce apart without a reaction.

Video: Oscillating reaction.

14.2 Factors Affecting Reaction Rates

OBJECTIVE

- Using collision theory, explain why the rate of a chemical reaction depends on the identity of the reactants, their concentrations, the temperature, and the presence of a catalyst.

Why does a reaction's rate depend on the identity of the reactants? Why does the rate depend on concentration and temperature? What does a catalyst do? We are able to explain how these factors affect a reaction's rate in terms of collision theory.

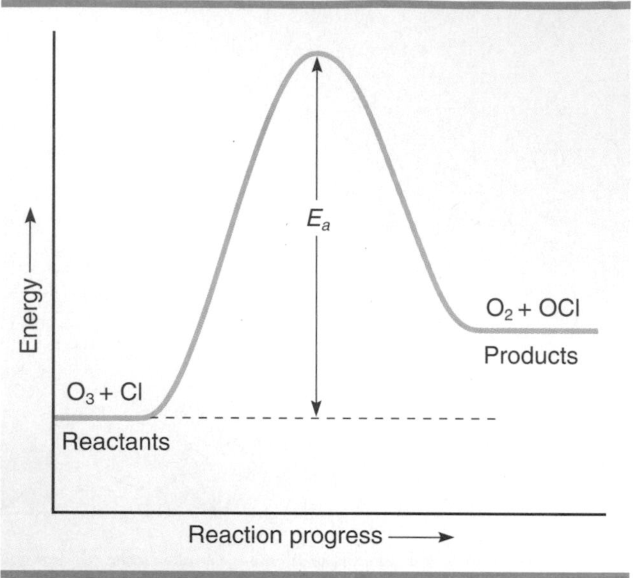

FIGURE 14.4
Energy profile for the reaction of chlorine atoms with ozone molecules. When an ozone molecule and a chlorine atom collide, the collision must have a minimum energy, called the activation energy (E_a), for a reaction to occur. If the collision energy is less than the activation energy, a reaction will not happen.

Identity of the Reactants

Wood burns quickly, but iron rusts slowly. *Each of these reactions and every other chemical reaction has its own characteristic rate,* a rate that is determined by which bonds must break and which new bonds must form. A reaction's activation energy, the energy that determines the inherent rate of the reaction, is the minimum energy required to bring about these events. Because some bonds are stronger than others, the activation energy for any given reaction is generally different from that of another reaction, so that each has its own inherent rate.

Concentration of the Reactants

Chemical reactions are usually faster when the concentrations of the reactants are increased (and slower when they are decreased) because more molecules exist in a given volume. More collisions will then occur and the rate of a reaction will increase. The chemical reactions shown in Figure 14.5 differ only in the concentration of one reactant. Zinc is reacting with sulfuric acid in both containers according to the chemical equation

$$Zn(s) + H_2SO_4(aq) \longrightarrow ZnSO_4(aq) + H_2(g)$$

The bubbles you see are hydrogen gas escaping from the solution, with the more vigorous evolution occurring in the reaction on the right. Each reaction began with the same amount of zinc, but the container on the left contained dilute acid and the one on the right contained more concentrated acid. The reaction is faster in the container with the more concentrated acid.

Temperature

Chemical reactions are faster when the temperature is increased (and slower when it is decreased). When we increase the temperature, kinetic energies and speeds increase

FIGURE 14.5
The reaction of zinc with sulfuric acid. Each container initially contained the same amount of zinc, but the container on the left contains dilute sulfuric acid, whereas the container on the right contains more concentrated acid. Clearly, the reaction is faster when the acid is more concentrated.

and, thus, the average energy of a collision also increases. As a result, the energy of any particular collision is more likely to exceed the activation energy. Thus, chemical reactions are faster at higher temperatures than at lower temperatures. At least one example can be taken from cooking. Some cooks use pressure cookers because water boils at a higher temperature when it is under pressure, so food will cook faster.

Video: Ammonia and oxygen.

Presence of a Catalyst

To get from New York to Baltimore, you could go by way of Chicago. However, this way is much slower than the direct route that would take you through New Jersey and Delaware. Similarly, a **catalyst,** *a substance that increases the rate of a chemical reaction without being consumed in the reaction,* provides an alternate but faster pathway for a reaction to occur. The reaction is faster because it has a lower activation energy than does an uncatalyzed reaction (Figure 14.6). The speeds of many chemical reactions in biological systems are increased by *enzymes* acting as catalysts. The following Chemical Perspective describes a catalyst that improves the quality of our lives.

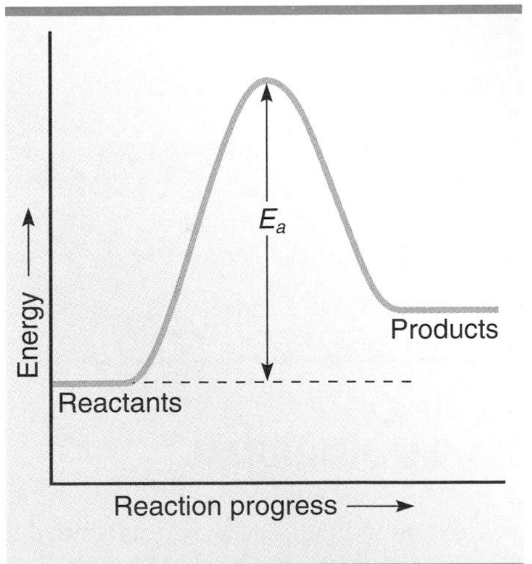

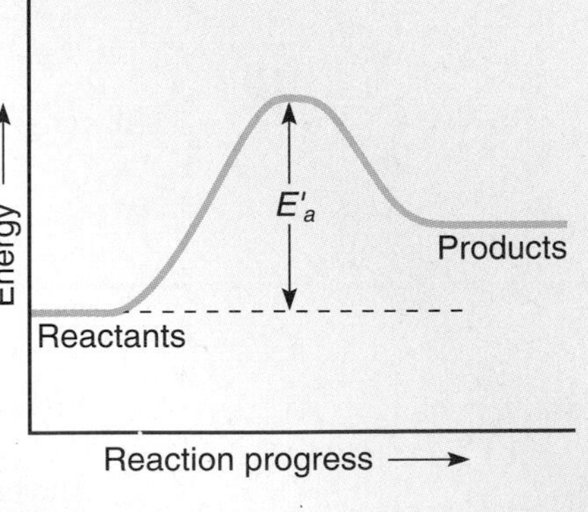

FIGURE 14.6
Activation energies of catalyzed and uncatalyzed reactions. *(A)* The activation energy (E_a) of an uncatalyzed reaction. *(B)* The same reaction in the presence of a catalyst. The activation energy is less, so the catalyzed reaction proceeds at a faster rate than the uncatalyzed reaction.

Chemical Perspective

A Car's Catalytic Converter

Although a catalyst is not consumed in a chemical reaction, it is not passive, as we show by tracing the events that take place in a car's catalytic converter. Combustion of gasoline in an automobile engine leads to substances (CO, NO, and others) that are pollutants if they are allowed to emerge from the tailpipe into the atmosphere. Although reactions that change these pollutants into relatively harmless compounds could occur, they are too slow; the pollutants are able to leave the exhaust system and pollute the environment. Much of the smog over Los Angeles and other cities occurs because of pollutants from automobile exhaust systems. The catalytic converter mounted in the exhaust systems of newer cars is designed to correct this problem (Figure 14.7). The catalysts in this device, typically platinum and rhodium, allow the reactions that destroy pollutants to occur rapidly so that most of the pollutants are used up before they can leave the car.

The pollutants form chemical bonds with the surfaces of the catalyst as the exhaust gases pass through the catalytic converter. Then the necessary reactions converting the pollutants to relatively harmless substances occur with lower activation energies than they would have otherwise. Next, the products are released from the surface so that the catalyst's surface is regenerated, ready to accept new molecules. Thus, the same surface can be used over and over. *The catalyst serves as a site to organize the reactants and hold them in proximity so that the reaction can occur.*

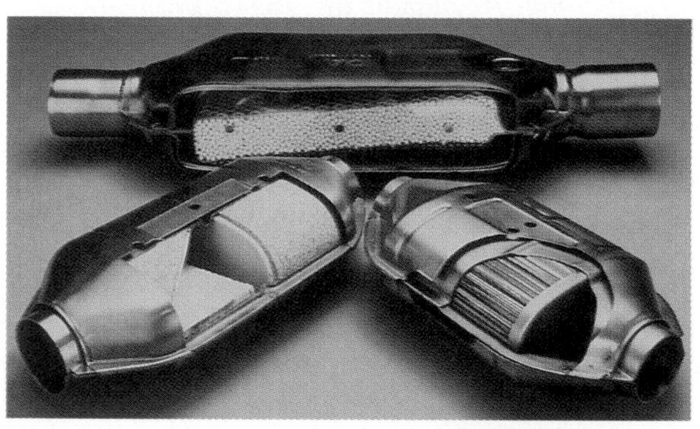

A

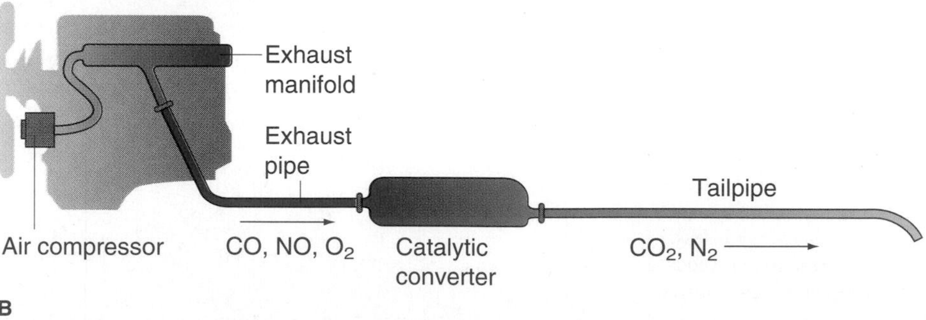

B

FIGURE 14.7

Catalytic converter. *(A)* Cutaway views of catalytic converters, showing beads on which the catalysts have been deposited. *(B)* Exhaust gases from the engine, including CO, NO, and O_2, pass through the catalytic converter, where the pollutants CO and NO are converted to harmless CO_2 and N_2 by catalyzed reactions.

CHEMICAL EQUILIBRIUM

The factors we examined in the last section control the rate of a chemical reaction from the time when it starts to the time at which it seems to stop. We assumed in Chapter 8 that a chemical reaction does not stop until *all* of the limiting reactant has been consumed in the reaction. Indeed, many reactions occurring in closed containers stop only when *essentially all* of the limiting reactant has been used. Other reactions under identical conditions seem to stop when a lot of the limiting reactant is unused, however.

Although we seem to be describing two different types of reactions, a common thread exists. If you could observe *any* reaction at the molecular level, you would see that the reaction had not stopped. Reactants are still being transformed into products, but the reverse reaction also occurs: that is, the products are being transformed back into the original reactants. Eventually the reaction reaches a point at which the amount of original reactants stops decreasing and remains constant, while the amount of products stops increasing and also remains constant. This is the point of *chemical equilibrium,* the point at which a reaction appears to stop, because we can no longer measure any increase in the products at the expense of the reactants. Intense activity at the molecular level still continues, however, and changing certain of the reaction conditions may alter the relative amount of reactants and products that are obtained at equilibrium. We explore and develop the concept of chemical equilibrium in the remainder of this chapter.

14.3 The Dynamic Nature of Chemical Equilibrium

OBJECTIVE

- Explain why chemical equilibrium is dynamic and not static.

In general, equilibrium can be defined as a balance between two opposing forces. One kind of equilibrium is shown with the children on the teeter-totter in Figure 14.8A. From time to time, they will stop their up-and-down movement and try to remain motionless with the teeter-totter in perfect balance. When this state is achieved, they are in equilibrium. It is a *static* equilibrium because there is no movement. Next, consider the small pool under the waterfall in Figure 14.8B. It, too, is at equilibrium even though water

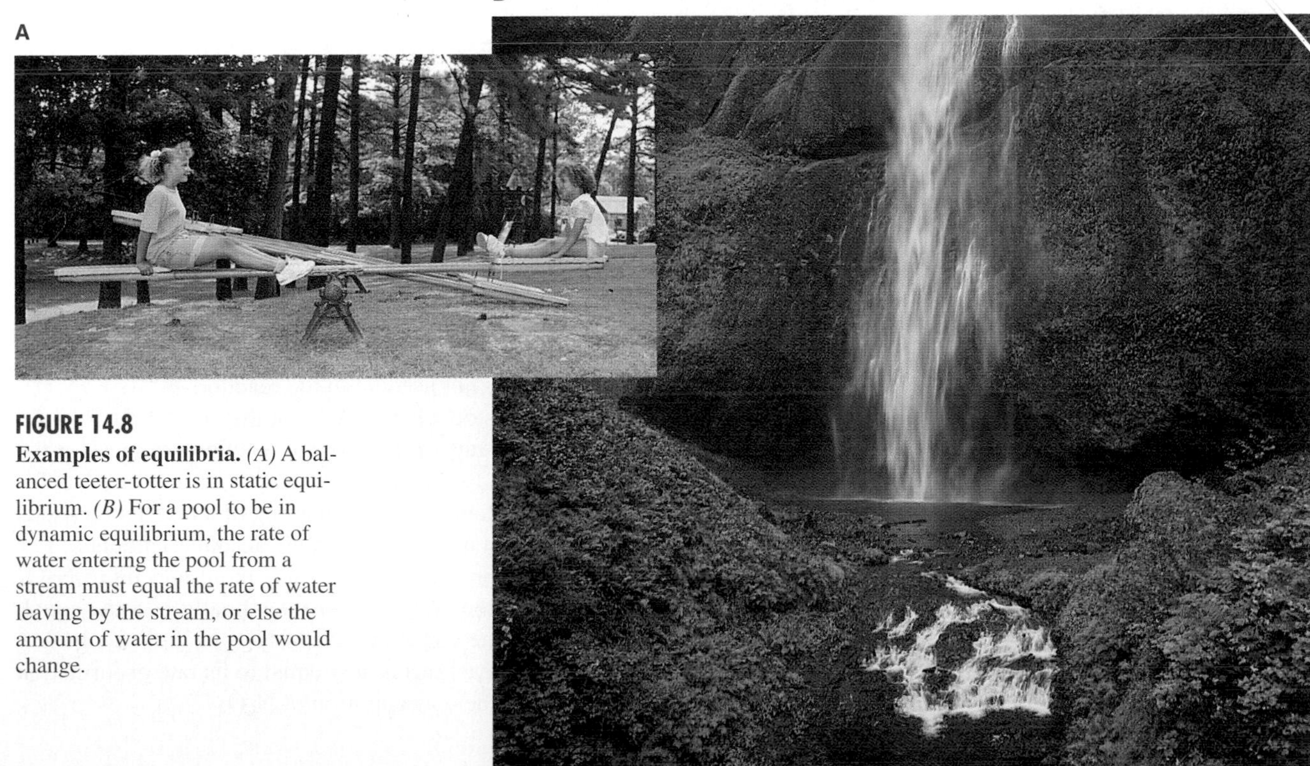

FIGURE 14.8
Examples of equilibria. *(A)* A balanced teeter-totter is in static equilibrium. *(B)* For a pool to be in dynamic equilibrium, the rate of water entering the pool from a stream must equal the rate of water leaving by the stream, or else the amount of water in the pool would change.

flows in and out of the pool. As long as the amount of water flowing into the pool is equal to the amount flowing out, the pool's level remains constant. This equilibrium is *dynamic* because movement of the water occurs.

You have encountered dynamic equilibria (the plural of *equilibrium*) before in this book. For example, recall that liquid water and ice coexist indefinitely if the temperature is maintained at 0°C. Melting occurs, but freezing also occurs at the same time and at a rate equal to that of melting. A dynamic balance is achieved: water freezing at the same rate at which ice is melting. This process is *reversible*. Liquid water becomes ice, but ice melts to give back the liquid. Although we could show these processes separately,

$$H_2O(l) \longrightarrow H_2O(s)$$
$$H_2O(s) \longrightarrow H_2O(l)$$

we do it in a more concise fashion by using two opposing arrows in a single equation,

$$H_2O(l) \rightleftharpoons H_2O(s)$$

We use the same method to show the reversible nature of chemical reactions. The usual way to show the reaction of hemoglobin with oxygen, the reaction we encountered in the opening of this chapter, is

$$Hb + 4O_2 \rightleftharpoons Hb(O_2)_4$$

to emphasize that the reaction proceeds in both directions.

We can now define **chemical equilibrium** as *a dynamic state in which the rates of the forward and the reverse reactions are equal*. Because these rates are identical, the quantities of the reactants and products must remain constant over time. The reaction may appear to stop, to the casual observer, but intense molecular activity continues. Collisions between molecules are occurring, such that reactants are still being transformed into products and products are reacting to become the original reactants.

We can observe chemical equilibrium in the reaction involving nitrogen dioxide (NO_2) and dinitrogen tetroxide (N_2O_4).

$$2NO_2(g) \rightleftharpoons N_2O_4(g)$$

In the forward reaction, one NO_2 molecule reacts with another NO_2 molecule to form N_2O_4, but N_2O_4 splits apart to give back two NO_2 molecules in the reverse reaction. Suppose we trace these events using the graph in Figure 14.9A. The results that you see apply to a known quantity of NO_2 after it is placed in a closed vessel. Only NO_2 exists at the beginning (time = zero); there is no N_2O_4 because the reaction has not begun. As time progresses, however, the concentration of NO_2 decreases because it begins to react with itself (the forward reaction). In the same period of time, the concentration of N_2O_4 increases even though some of the N_2O_4 molecules split apart (the reverse reaction). Eventually, the reaction appears to stop because the concentrations of NO_2 and N_2O_4 no longer change. Notice that all of the NO_2 is not used up in the reaction. ●

● Nitrogen dioxide (NO_2) is responsible for the brownish-yellow color of the smog found over major cities.

We can see why the reaction appears to stop when we look at the rates of the forward and reverse reactions in Figure 14.9B. The rate of the forward reaction decreases at the beginning because the concentration of NO_2 decreases, so that fewer and fewer collisions between NO_2 molecules can occur. Remember that the rate of a reaction depends on the *concentrations* of the reacting substances (*not* on their absolute amounts). For that reason, the rate of the reverse reaction increases because more and more N_2O_4 becomes available from the forward reaction. When these rates finally become equal, chemical equilibrium has been reached. The concentration of NO_2 cannot change any more because its rate of loss (the forward reaction) is now equal to its rate of formation (the reverse reaction). The same is true of the concentration of N_2O_4.

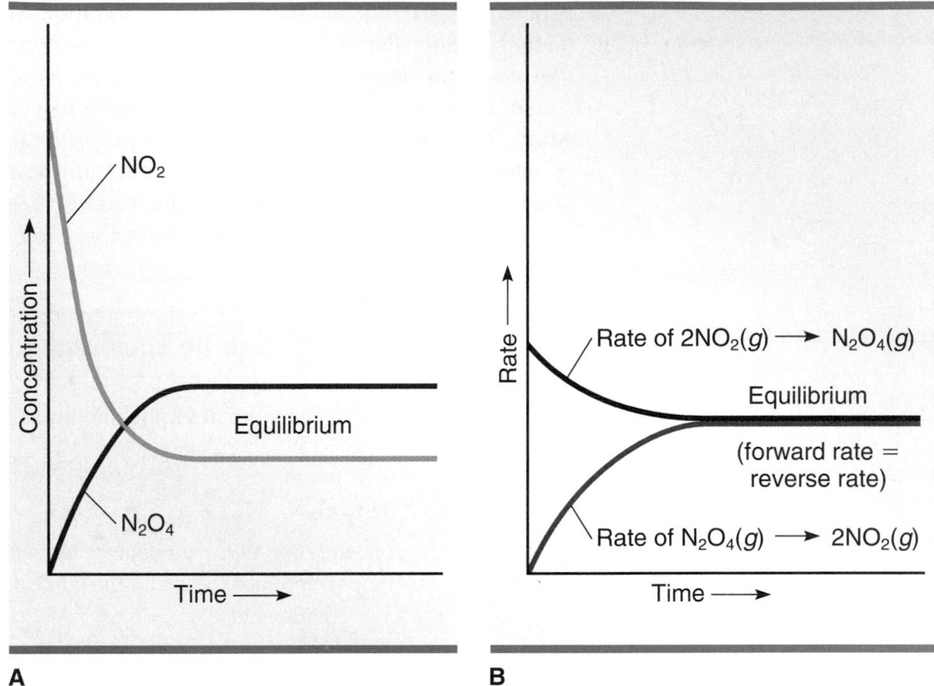

FIGURE 14.9
Time dependence in the reaction $2NO_2 \rightleftharpoons N_2O_4$. *(A)* Changes in the concentrations of NO_2 and N_2O_4 when the reaction starts with only NO_2 present. These changes no longer occur after equilibrium is established. *(B)* Changes in the rates of the forward and reverse reactions when the reaction begins with only NO_2 present. As time progresses, the rate of the forward reaction decreases because the concentration of NO_2 decreases. The rate of the reverse reaction is zero when the reaction begins because no N_2O_4 has been formed yet. As the concentration of that substance increases, the rate of the reverse reaction increases until it becomes equal to the rate of the forward reaction.

Now, we can see that the reaction of hemoglobin with oxygen seems to stop because the rates of the forward and reverse reactions become equal when equilibrium is reached. The synthesis of ammonia from nitrogen and hydrogen, a very important industrial process,

$$N_2(g) + 3H_2(g) \rightleftharpoons 2NH_3(g)$$

seems to stop for the same reason. Both equilibria are dynamic because collisions continue to happen. Collisions leading to reactions could not happen in a static system. ●

● Chemical equilibrium is *always* dynamic. It is never static.

14.4 The Equilibrium Expression

OBJECTIVE

- State the law of mass action and use it to write equilibrium expressions.

Because the concentrations of reactants and products become fixed and unchanging when equilibrium is reached, a mathematical relationship between these concentrations exists. Consider the general reaction

$$aA + bB \rightleftharpoons cC + dD$$

where the reactants are A and B and the products are C and D. We assume that this equation has been balanced with the coefficients *a*, *b*, *c*, and *d*. The **law of mass action** states that *each reaction has an equilibrium constant with its own characteristic value at a given temperature*. The equilibrium constant (K) is given by the following ratio of reactants to products.

$$K = \frac{[C]^c[D]^d}{[A]^a[B]^b}$$

● The brackets [ ] in the equation mean "concentration of" with units of mol/L. Molarities were discussed in Chapter 13. An early term for concentration was *active mass;* hence the words *mass action.*

where [A], [B], [C], and [D] are the concentrations (in mol/L) of the reactants and products *at equilibrium.* ● Note that *multiplying the concentrations of the products raised to the power of their coefficients, then dividing by the concentrations of the reactants raised to the power of their coefficients* results in a constant called the **equilibrium constant.** The entire equation is called the *equilibrium expression.* The equilibrium constant K does not depend on the initial concentration of any reactant or product, but it does depend on the temperature at which the reaction is carried out. The following example shows you how to set up an equilibrium expression.

■ EXAMPLE 14.1 Writing an Equilibrium Expression

Write the equilibrium expression for the important industrial reaction,

$$N_2(g) + 3H_2(g) \rightleftharpoons 2NH_3(g)$$

Problem Analysis

Multiply the concentration of the products, each raised to the power of the corresponding coefficient, and then divide by the concentrations of the reactants raised to the power of their coefficients.

Solution

Since the coefficient for N_2 is 1, we can write

$$1N_2(g) + 3H_2(g) \rightleftharpoons 2NH_3(g)$$

Coefficients

The coefficients become powers in the equilibrium expression. You must remember that

$$[N_2]^1 = [N_2]$$

With that in mind, the correct equilibrium expression is

$$K = \frac{[NH_3]^2}{[N_2]\,[H_2]^3}$$

Note that the symbol specifying the state of a substance—in this case (g)—is not included in the brackets.

Exercise 14.1

Balance the following equation and give the correct equilibrium expression.

$$NH_3(g) + O_2(g) \rightleftharpoons N_2(g) + H_2O(g)$$

(Try Problems 14.23, 14.24, 14.25, and 14.26.)

14.5 Calculating an Equilibrium Constant

OBJECTIVES

- Calculate the value of an equilibrium constant.
- Show that the same equilibrium constant is obtained with varying conditions as long as the temperature is constant.

If we write an equilibrium expression for

$$2NO_2(g) \rightleftharpoons N_2O_4(g)$$

we obtain

$$K = \frac{[N_2O_4]}{[NO_2]^2}$$

The equilibrium constant K is a characteristic constant for a reaction as long as the temperature remains unchanged. Let us calculate K for this reaction at 100°C. Suppose we add 1.00 mol of NO_2 to a 1.00-L flask. The initial concentrations are $[NO_2] = 1.00\ M$ and $[N_2O_4] = 0$. If we close the flask and let the reaction come to equilibrium at 100°C, we find from experiments that $[NO_2] = 0.36\ M$ and $[N_2O_4] = 0.32\ M$. ● Then we write

● Recall from Chapter 13 that M is the symbol for mol/L.

$$K = \frac{[N_2O_4]}{[NO_2]^2} = \frac{(0.32)}{(0.36)^2} = 2.5$$

These results are summarized in Table 14.1 (Experiment 1). Note that the units of K are omitted by custom.

Equilibrium can also be approached from the other direction, too. Suppose we imagine that all of the 1.00 mol of NO_2 is converted to N_2O_4. The coefficients in the chemical equation tell us that the quantity of N_2O_4 must now be 0.500 mol, or *one-half* of the original quantity of NO_2. Now, if we begin the reaction by adding this quantity of N_2O_4 to the 1.00-L flask, the initial concentrations will be $[NO_2] = 0$ and $[N_2O_4] = 0.500\ M$, and the reaction will proceed in the opposite direction until equilibrium is reached. Again, however, we find exactly the same equilibrium concentrations that we found before: $[NO_2] = 0.36\ M$ and $[N_2O_4] = 0.32\ M$. We also find $K = 2.5$ from these results. Once again, the results are summarized in Table 14.1 (Experiment 2).

These examples show that the concentrations of reactants and products at equilibrium are independent of the initial direction of the reaction. We started with only NO_2 present in the first example, and we started with only N_2O_4 in the second example. Yet the reaction reached the same equilibrium concentrations. These results show that the reaction is reversible.

TABLE 14.1

Results from Different Approaches to Equilibrium in the Reaction $2NO_2(g) \rightleftharpoons N_2O_4(g)$

	Initial (M)		Equilibrium (M)		
Experiment	$[NO_2]$	$[N_2O_4]$	$[NO_2]$	$[N_2O_4]$	$\frac{[N_2O_4]}{[NO_2]^2} = K$
1	1.00	0	0.36	0.32	$\frac{(0.32)}{(0.36)^2} = 2.5$
2	0	0.500	0.36	0.32	$\frac{(0.32)}{(0.36)^2} = 2.5$
3	2.00	0	0.54	0.73	$\frac{(0.73)}{(0.54)^2} = 2.5$
4	2.36	0.32	0.68	1.16	$\frac{(1.16)}{(0.68)^2} = 2.5$
5	0.72	0.64	0.54	0.73	$\frac{(0.73)}{(0.54)^2} = 2.5$

The equilibrium concentrations that we found in Experiments 1 and 2 are not the only ones corresponding to $K = 2.5$. As long as the temperature is constant, there are an *infinite* number of equilibrium concentrations, each set obtained from different initial concentrations of reactants and products, that correspond to the *same* equilibrium constant. Suppose we add 2.00 mol of NO_2 (rather than 1.00 mol as we did in Experiment 1) to a 1.00-L flask, close the flask, and let the reaction come to equilibrium at 100°C. Now, we find $[NO_2] = 0.54\ M$ and $[N_2O_4] = 0.73\ M$ after equilibrium is achieved, and

$$K = \frac{[N_2O_4]}{[NO_2]^2} = \frac{(0.73)}{(0.54)^2} = 2.5$$

This is the same value of K that we obtained in each of the other experiments (see Table 14.1, Experiment 3). We consider Experiments 4 and 5 in Section 14.8, but let's look at another example before we leave this topic.

EXAMPLE 14.2 Obtaining *K* from Equilibrium Concentrations

Ammonia, a fertilizer, is prepared in the Haber process by the reaction

$$N_2(g) + 3H_2(g) \rightleftharpoons 2NH_3(g)$$

When the concentrations of an equilibrium mixture of N_2, H_2, and NH_3 were measured at 470°C, the results were $[N_2] = 0.040\ M$, $[H_2] = 0.12\ M$, and $[NH_3] = 0.0027\ M$. Calculate the value of the equilibrium constant at this temperature.

Problem Analysis

Write the equilibrium expression. Substitute the known values of the concentrations of N_2, H_2, and NH_3 into this expression and do the calculation.

Solution

Because $[N_2] = 0.040\ M$, $[H_2] = 0.12\ M$, and $[NH_3] = 0.0027\ M$, you write

$$K = \frac{[NH_3]^2}{[N_2]\,[H_2]^3} = \frac{(0.0027)^2}{(0.040)\,(0.12)^3} = 0.10$$

The value of the equilibrium constant at 470°C is 0.10.

Exercise 14.2

Balance the following equation, write the equilibrium expression, and calculate the equilibrium constant if chemical analysis has shown that $[CH_4] = 1.10\ M$, $[H_2S] = 1.49\ M$, $[CS_2] = 1.10\ M$, and $[H_2] = 1.68\ M$ when equilibrium is reached at 900°C.

$$CH_4(g) + H_2S(g) \rightleftharpoons CS_2(g) + H_2(g)$$

(Try Problems 14.27, 14.28, 14.29, and 14.30.)

In the last three sections, we described the dynamic nature of chemical equilibrium, the equilibrium expression, and the equilibrium constant. We have calculated K for several reactions, but what do these equilibrium constants tell us? In the next section, we use equilibrium constants to make predictions about reactions.

14.6 Using an Equilibrium Constant

OBJECTIVES

- Use the equilibrium constant to predict the extent to which the reactants in a chemical reaction are converted to products.
- Use the equilibrium constant, the equilibrium expression, and all but one of the equilibrium concentrations of the reactants and products to calculate the missing equilibrium concentration.

Some chemical equilibria favor the right side of a chemical equation; that is, at equilibrium most of the reactants have been converted to products. Others favor the left side, so that very little of the reactants are converted to products. The magnitude of the equilibrium constant tells us which side of a chemical equation a particular reaction favors.

If the equilibrium constant is large (say, 10^2 or greater), the reaction significantly favors the right side of the chemical equation. (Most of the reactants are converted to products.) Consider the reaction in which ammonia is formed from its elements.

$$N_2(g) + 3H_2(g) \rightleftharpoons 2NH_3(g)$$

The equilibrium constant at 25°C is 4.1×10^8. This means that the numerator in the equilibrium expression must be 4.1×10^8 larger than the denominator, or

$$K = \frac{\overbrace{[NH_3]^2}^{\text{Numerator}}}{\underbrace{[N_2]\,[H_2]^3}_{\text{Denominator}}} = \frac{4.1 \times 10^8}{1}$$

Clearly, this reaction favors the right side of the equation (at least at this temperature) because most of the reactants are converted to ammonia. The reaction proceeds essentially to completion. Similarly, the stoichiometric reactions in Chapter 8 proceed essentially to completion because they have equilibrium constants that are very large.

When an equilibrium constant is very small (say, about 10^{-2} or less), the reaction significantly favors the left side of the chemical equation. (Very little of the reactants is converted to products.) If we look at the same reaction we just examined, but now at 600°C, we find that the equilibrium constant is only 1.2×10^{-2}. Thus, the denominator is considerably larger than the numerator. The reaction now favors the left side of the chemical equation, and very little of the reactants is converted to ammonia.

When the equilibrium constant is neither large nor small (around 1), neither the left side nor the right side of a chemical equation is favored. For example, the reaction leading to the formation of ammonia has an equilibrium constant of only 0.50 at 400°C. Neither the substances on the left side of the equation nor the substance on the right side is predominant at equilibrium. You will find an additional illustration of this use of an equilibrium constant in the following example.

EXAMPLE 14.3 Using an Equilibrium Constant to Predict the Extent of a Reaction

Phosgene ($COCl_2$), a substance used in manufacturing polyurethane plastics, is prepared from carbon monoxide and chlorine:

$$CO(g) + Cl_2(g) \rightleftharpoons COCl_2(g)$$

The equilibrium constant at 400°C is 1.2×10^3. Write the equilibrium expression. Do you expect a nearly complete reaction at this temperature? Explain.

Problem Analysis

Write the equilibrium expression. Use the magnitude of the equilibrium constant to judge the extent of the reaction. If the value of the equilibrium constant is 10^2 or greater, most of the reactants will be converted to the product. If the value of the equilibrium constant is 10^{-2} or smaller, very little of the reactants will be converted to the product. If the value of the equilibrium constant is neither large nor small, significant concentrations of the reactants and the product will be present at equilibrium.

Solution

The equilibrium expression for this reaction is

$$K = \frac{[COCl_2]}{[CO]\,[Cl_2]}$$

Because the equilibrium constant is greater than 10^2, most of the reactants will be converted to the product when the reaction reaches equilibrium.

Exercise 14.3

Suppose that the reaction for the manufacture of methanol (CH_3OH),

$$CO(g) + 2H_2(g) \rightleftharpoons CH_3OH(g)$$

has an equilibrium constant equal to 2.3 at a particular temperature. Write the equilibrium expression. Do you expect a nearly complete reaction at this temperature? Explain.

(Try Problems 14.31 and 14.32.)

Calculating an Unknown Concentration

If you know the equilibrium constant for a reaction and all but one of the concentrations of the reactants and products, you can use the equilibrium expression to calculate the missing concentration. The following example illustrates the procedure.

■ EXAMPLE 14.4 Calculating an Unknown Equilibrium Concentration

The equilibrium constant for the reaction represented by

$$N_2(g) + O_2(g) \rightleftharpoons 2NO(g)$$

is 4.00×10^{-4} at a particular temperature. During an experiment at that temperature, a chemist found that the equilibrium concentrations of nitrogen and nitrogen monoxide are 3.19×10^{-2} *M* and 1.05×10^{-3} *M*. Calculate the concentration of oxygen at equilibrium.

Problem Analysis

Write the equilibrium expression for the reaction from the chemical equation. Solve for $[O_2]$ by rearranging the expression so that it begins with $[O_2] =$. Insert the known values of the equilibrium constant and the equilibrium concentrations of N_2 and NO and do the calculation.

Solution

The equilibrium expression for this reaction is

$$K = \frac{[NO]^2}{[N_2]\,[O_2]}$$

After solving for $[O_2]$, you obtain

$$[O_2] = \frac{[NO]^2}{K \times [N_2]}$$

When you insert the known values of K, [NO], and $[N_2]$, you get

$$[O_2] = \frac{(1.05 \times 10^{-3})^2}{(4.00 \times 10^{-4})(3.19 \times 10^{-2})}$$

$$= 8.64 \times 10^{-2}\ M$$

Exercise 14.4

The reaction represented by the equation

$$CO(g) + 3H_2(g) \rightleftharpoons CH_4(g) + H_2O(g)$$

has an equilibrium constant of 3.9 at a particular temperature. What is the concentration of methane (CH_4) at that temperature if the mixture also consists of [CO] = 0.30 M, $[H_2]$ = 0.10 M, and $[H_2O]$ = 0.020 M? The reaction is at equilibrium.

(Try Problems 14.33 and 14.34.)

14.7 Heterogeneous Equilibria

OBJECTIVE

- Write equilibrium expressions for equilibria involving more than one state of matter.

Each equilibrium considered in the last three sections has been a **homogeneous equilibrium:** *one in which all substances are in a single state of matter.* Thus the reaction of elemental nitrogen and elemental hydrogen to form ammonia

$$N_2(g) + 3H_2(g) \rightleftharpoons 2NH_3(g)$$

leads to a homogeneous equilibrium because it involves only gaseous substances. On the other hand, a **heterogeneous equilibrium** is *one that involves more than one state of matter.* Consider an example. When an electric current is passed through liquid water, the water decomposes to give gaseous hydrogen and oxygen, as shown in Figure 14.10. The chemical equation for this reaction is

$$2H_2O(l) \rightleftharpoons 2H_2(g) + O_2(g)$$

It is heterogeneous because both liquid and gaseous substances are involved. More than one state of matter is present.

It might seem reasonable to write the equilibrium expression for this reaction as

$$K' = \frac{[H_2]^2\,[O_2]}{[H_2O]^2}$$

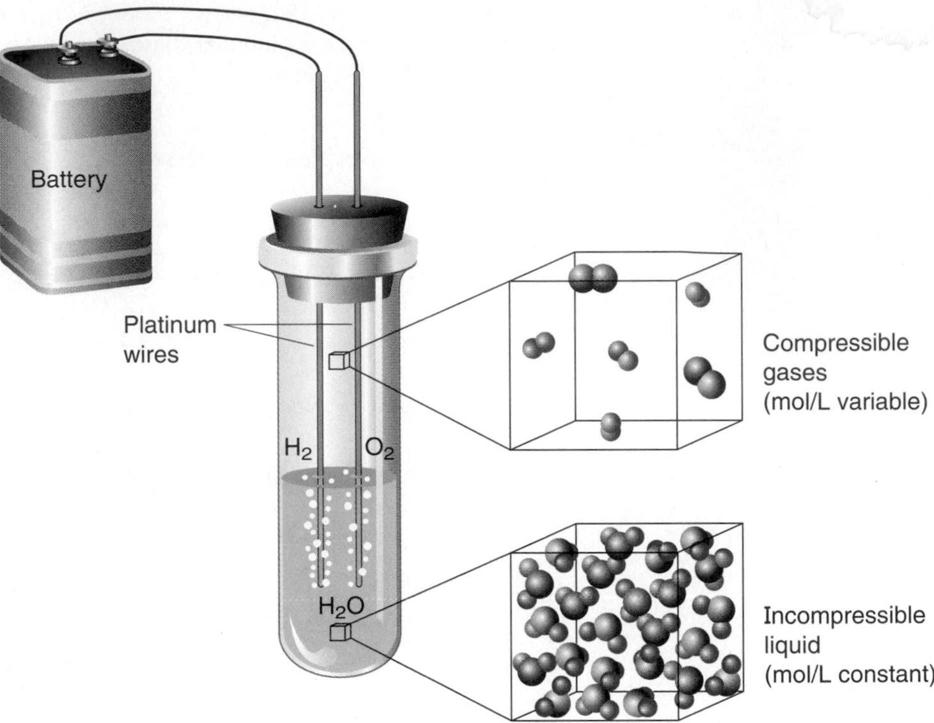

FIGURE 14.10
The decomposition of water. After this reaction comes to equilibrium, the concentrations of H_2 and O_2 can be altered by adding or removing one of them or simply by changing the pressure. However, the concentration of water is a constant because the number of water molecules per unit volume is incapable of change (as long as the temperature is constant).

We have written K' rather than K because K' is not the true equilibrium constant for this heterogeneous equilibrium, but we can modify K' to get the true equilibrium constant K. Let's consider the decomposition of water shown in Figure 14.10 more carefully to see why and how we make this modification. Notice that we could change the concentration of either H_2 or O_2 in the gas above the liquid by deliberately adding either one of these substances, or by merely compressing the gas that is already present. We could decrease $[H_2]$ and $[O_2]$ by doing the reverse: removing gas or causing it to expand. Both of these changes are possible because any gas is essentially empty space. Thus, $[H_2]$ and $[O_2]$ are *variables;* they can change. On the other hand, we cannot change the concentration of liquid water, even by adding it to or removing it from the container. Why? Unlike gases, pure liquids occupy a fixed volume and are therefore essentially incompressible (see Chapter 12). There are the same number of water molecules per unit volume at some fixed temperature, even if we add or remove water. Since concentration is expressed in moles per unit volume, we are saying that the concentration of water is always the same because the moles of water per unit volume cannot change. The concentration of water is not a variable; it is a *constant*. Adding or removing water or changing the pressure does not change the concentration of water.

With this in mind, we return to the equilibrium expression for the decomposition of H_2O, and recognize that K' and $[H_2O]^2$ are constants but each concentration in $[H_2]^2[O_2]$ is a variable.

Variables

$$K' = \frac{[H_2]^2\,[O_2]}{[H_2O]^2}$$

Constants

To write the true equilibrium expression, we rearrange the equation so that all the constants appear on the left side.

$$K' \times [H_2O]^2 = [H_2]^2[O_2]$$

Because all of the terms on the left side are constants, we can incorporate all of them into a single constant that we will call K.

$$K' \times [H_2O]^2 = K = [H_2]^2[O_2]$$

The net effect is to omit the concentrations of the pure liquid because it is now built into the equilibrium constant K. Thus, the equilibrium expression is

$$K = [H_2]^2[O_2]$$

A similar argument can be made for equilibria involving pure solids. Since the concentration of the solid is a constant, it is not included in the equilibrium expression. ● Thus, the equilibrium expression for the decomposition of $CaCO_3$,

● A pure solid, like a pure liquid, occupies a fixed volume and is incompressible, so its concentration in mol/L remains constant. Once the solid has been vaporized or has gone into solution, this is no longer true.

$$CaCO_3(s) \rightleftharpoons CaO(s) + CO_2(g)$$

is given by

$$K = [CO_2]$$

because neither $[CaCO_3]$ nor $[CaO]$ appears in the expression. In general, *we omit concentration terms for pure solids and pure liquids when we write the equilibrium expression for a heterogeneous equilibrium.*

Equilibria involving saturated solutions of certain sparingly soluble ionic substances are also handled in this manner. For example, consider the equilibrium occurring in a saturated solution of calcium carbonate in water (Figure 14.11). This process is described by

$$CaCO_3(s) \rightleftharpoons Ca^{2+}(aq) + CO_3^{2-}(aq)$$

The equilibrium constant for a reaction of this type is given a special name and a special symbol: the *solubility product constant*, K_{sp}. The equilibrium expression for this equation is

$$K_{sp} = [Ca^{2+}][CO_3^{2-}]$$
$$= 8.7 \times 10^{-9} \text{ at } 25°C$$

In this equilibrium expression, the concentration of calcium carbonate is not included because undissolved calcium carbonate is a pure solid. Notice that K_{sp} is very small, indicating that very little calcium carbonate dissolves to become the aqueous ions shown on the right side of the equation.

If we know the solubility constants for a group of sparingly soluble ionic substances, we can predict their relative solubilities as long as each formula unit of solute forms the same number of ions when it dissolves. For example, consider the following equilibria at 25°C.

$$CaCO_3(s) \rightleftharpoons Ca^{2+}(aq) + CO_3^{2-}(aq) \qquad K_{sp} = 8.7 \times 10^{-9}$$
$$AgBr(s) \rightleftharpoons Ag^{+}(aq) + Cl^{-}(aq) \qquad K_{sp} = 5.0 \times 10^{-13}$$

We can compare the solubilities of $CaCO_3$ and AgBr because each formula unit dissolves to produce two ions. The solubility product constant for $CaCO_3$ is considerably larger than that for AgBr. As a result, the equation with $CaCO_3$ proceeds further to the right (more of the reactant is transformed into products) than does the one with AgBr.

FIGURE 14.11

A saturated solution of calcium carbonate in water. Solid calcium carbonate is in equilibrium with Ca^{2+} and CO_3^{2-} ions in solution.

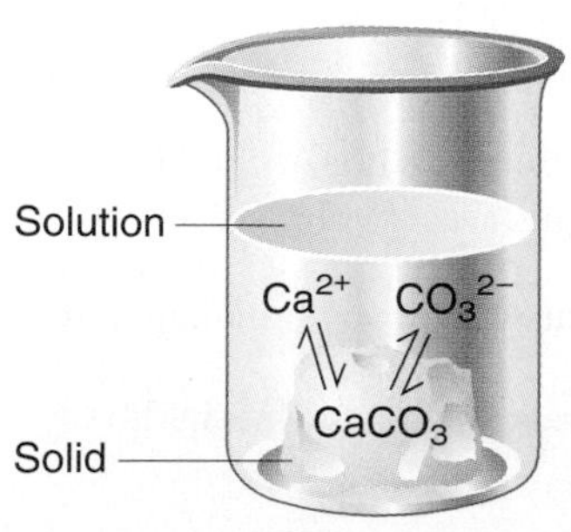

Since more $CaCO_3$ must dissolve than AgBr, $CaCO_3$ is more soluble than AgBr. You can obtain additional practice in predicting solubilities from the following example.

■ EXAMPLE 14.5 Determining Relative Solubilities by Comparing Equilibrium Constants

FIGURE 14.12

Two sparingly soluble minerals. Galena *(left)* is lead sulfide (PbS), and sphalerite *(right)* is zinc sulfide (ZnS).

Samples of the minerals galena (PbS) and sphalerite (ZnS) are shown in Figure 14.12. Determine the relative solubilities of these ionic substances from the following information.

$$PbS(s) \rightleftharpoons Pb^{2+}(aq) + S^{2-}(aq) \qquad K_{sp} = 7 \times 10^{-29} \quad (25°C)$$

$$ZnS(s) \rightleftharpoons Zn^{2+}(aq) + S^{2-}(aq) \qquad K_{sp} = 3 \times 10^{-22} \quad (25°C)$$

Problem Analysis

Notice that you can compare the solubilities of PbS and ZnS because each formula unit produces two ions in solution. The substance with the larger solubility product constant has the greater solubility.

Solution

Since K_{sp} for ZnS is larger than that for PbS, ZnS is more soluble in water than PbS.

Exercise 14.5

Determine the relative solubilities of $Cu(OH)_2$ and $Zn(OH)_2$ from the following information.

$$Cu(OH)_2(s) \rightleftharpoons Cu^{2+}(aq) + 2OH^-(aq) \qquad K_{sp} = 2.6 \times 10^{-19} \quad (25°C)$$

$$Zn(OH)_2(s) \rightleftharpoons Zn^{2+}(aq) + 2OH^-(aq) \qquad K_{sp} = 2.1 \times 10^{-16} \quad (25°C)$$

(Try Problems 14.39 and 14.40.)

Remember that relative solubilities can be predicted by comparing the values of the equilibrium constants only when the formula units of the substances produce the same number of ions. We cannot compare the solubilities of AgBr and Ag_2CrO_4 by this simple technique because AgBr produces two ions when a formula unit dissolves, as we have seen, but Ag_2CrO_4 produces three:

$$Ag_2CrO_4(s) \rightleftharpoons 2Ag^+(aq) + CrO_4^{2-}(aq)$$

The solubilities of substances producing a different number of ions when they dissolve can be compared by a mathematical method, but we do not describe it here.

Videos: Thermite reaction; Ammonium chloride and barium hydroxide.

14.8 Le Chatelier's Principle

OBJECTIVES

- Describe the general effects of disturbing a chemical reaction at equilibrium.
- Predict the result when the concentration of a substance involved in an equilibrium is changed.
- Predict the result when the volume or temperature of an equilibrium mixture of reactants and products is changed.
- Predict the result when a catalyst is added to a reaction mixture at equilibrium.

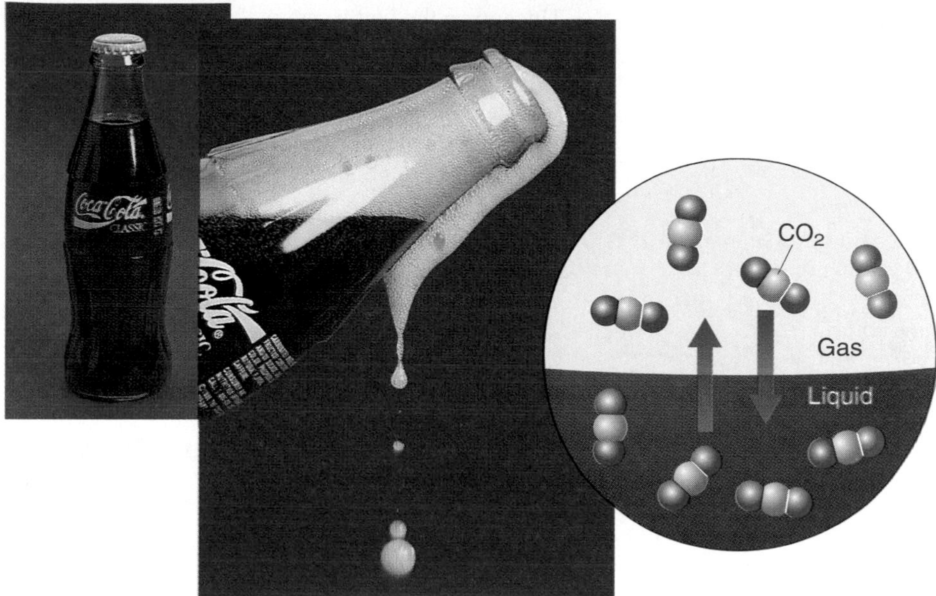

FIGURE 14.13
What happens when a bottle of carbonated drink is opened? Opening a bottle of a carbonated drink disturbs the equilibrium between gaseous carbon dioxide and carbon dioxide in solution. The drink fizzes and foams as equilibrium is reestablished. At equilibrium, the rate at which molecules of CO_2 enter the liquid is equal to the rate at which they leave the liquid, as shown by the double arrows on the right.

In previous sections, we explored dynamic equilibria in both homogeneous and heterogeneous reactions. In this section, we ask "What will happen to an equilibrium if one of the conditions affecting equilibrium is changed?" Let's consider the carbonated drink shown in Figure 14.13. Drinks like this one are sealed under a high pressure of carbon dioxide so that they have the characteristic effervescence and taste that most people like. Before carbonation, this drink contained no more CO_2 than the water we drink every day. (The air contains a small amount of CO_2, and some of this dissolves in any sample of water exposed to air.) The drink was carbonated by forcing carbon dioxide into it at high pressure. The amount of CO_2 in solution depends on the magnitude of the pressure applied by the gaseous CO_2, and that pressure remains after the bottle is capped. The reaction

$$CO_2(g) \rightleftharpoons CO_2(aq)$$

comes to equilibrium quickly, and then carbon dioxide enters the solution at the same rate that it leaves.

Uncapping the bottle disturbs the equilibrium. Because it was there under pressure, the gaseous carbon dioxide above the drink rushes out of the mouth of the uncapped bottle. The drink also begins to fizz and sometimes foam because CO_2 leaves the solution as it tries to reestablish the balance between carbon dioxide in the solution and carbon dioxide above it. Eventually, most of the CO_2 leaves the solution and the drink becomes "flat." Eventually, too, equilibrium is restored when the CO_2 left in the drink comes to equilibrium with atmospheric carbon dioxide.

We can use a diagram to show what happens to the carbonated drink after we uncap the bottle.

Decrease

$$\underbrace{CO_2(g)} \rightleftharpoons CO_2(aq)$$

← Shift

Uncapping the bottle causes the concentration of gaseous CO_2 to decrease, so the equilibrium shifts and CO_2 leaves the solution (producing fizz). *Notice that the reaction shifts in a way to counteract the change and to reestablish equilibrium.*

There are several ways to disturb a reaction at equilibrium: A substance involved in the equilibrium can be added or removed (just as CO_2 is removed when we uncap a bottle of a carbonated drink), or the volume or the temperature can be changed. The effects of changes such as these were studied systematically by Henri Louis Le Chatelier (1850–1936). ● His results have been generalized in what is known today as **Le Chatelier's principle:** *When a reaction at equilibrium is disturbed by a change in volume, temperature, or the concentration of one of the components, the reaction will shift in a way that tends to counteract the change and reestablish the equilibrium.* We examine applications of Le Chatelier's principle in this section.

● *Le Chatelier* is pronounced "le-sha-tuh-lee-a."

Exploration: Chemical equilibrium.

Adding or Removing a Substance Involved in a Reaction at Equilibrium

To study the effects of increasing or decreasing the concentration of any substance involved in a reaction at equilibrium, we work once again with the equilibrium between nitrogen dioxide and dinitrogen tetroxide:

$$2NO_2(g) \rightleftharpoons N_2O_4(g)$$

Let's look at the results of Experiment 1 in Table 14.1 again. What happens if we inject more NO_2 into the equilibrium mixture? The concentration of NO_2 increases suddenly, and the reaction is no longer at equilibrium. According to Le Chatelier's principle, the reaction should react to this increase in the concentration of NO_2 by shifting to counteract the change.

Increase

$$2NO_2(g) \rightleftharpoons N_2O_4(g)$$

Shift →

This shift causes the concentration of NO_2 to decrease; the shift also causes the production of more N_2O_4 and increases its concentration. Although a shift in concentrations occurs, the equilibrium constant does not change. ● We can demonstrate that K does not change by the following specific example. Suppose we add enough NO_2 to the reaction mixture in Experiment 1 (Table 14.1) to cause the concentration of NO_2 to increase instantaneously from its previous equilibrium value of 0.36 M to 2.36 M. At that instant, the concentration of N_2O_4 is still 0.32 M, the equilibrium value from Experiment 1. Now the reaction mixture has too much NO_2 to be at equilibrium. The reaction must shift from left to right so that some of the NO_2 is used up while more N_2O_4 is made. The equilibrium concentrations of these substances are shown in Experiment 4 in Table 14.1 as 0.68 M NO_2 and 1.16 M N_2O_4. When these concentrations are inserted into the equilibrium expression, the calculated value of K is identical to the one obtained in Experiment 1.

● The only change that alters the value of an equilibrium constant is a change in temperature.

On the other hand, what happens when we remove some NO_2 from the reaction mixture? Suddenly, the mixture does not contain enough NO_2 to be at equilibrium. Le Chatelier's principle indicates that the reaction should react to this decrease in the concentration of NO_2 by shifting to counteract the change. The equilibrium must shift from right to left to increase the concentration of NO_2. This shift causes the concentration of N_2O_4 to decrease.

Reduce

$$2NO_2(g) \rightleftharpoons N_2O_4(g)$$

Shift

We can make two generalizations based on our discussion about changing the concentration of NO_2 in an equilibrium mixture.

1. If a substance is added to a reaction at equilibrium, the reaction will shift away from that substance so that its concentration will *decrease*. Thus, we have

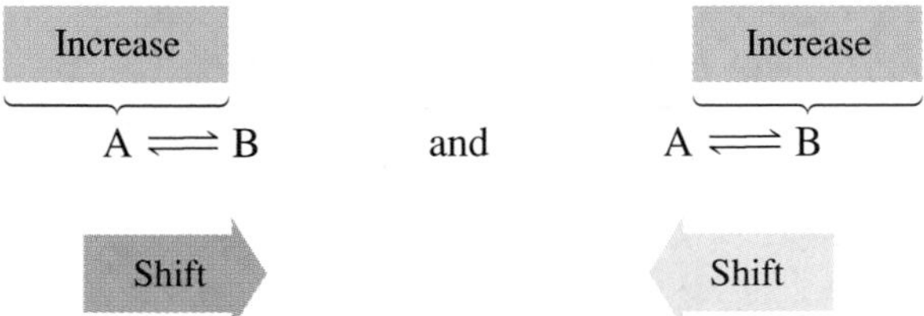

2. If a substance is removed from a reaction at equilibrium, the reaction will shift toward that substance so that its concentration will *increase*.

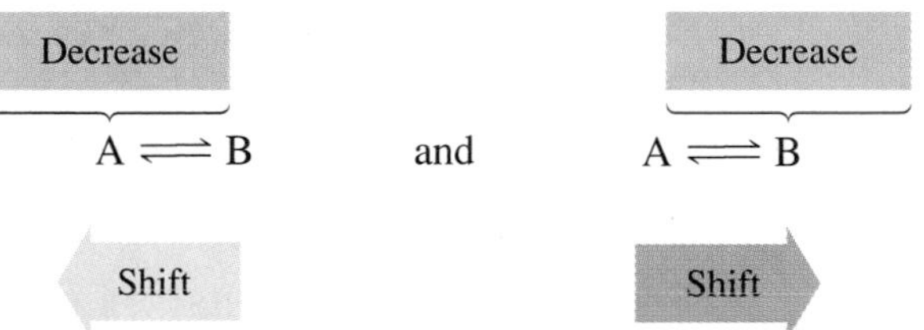

Let's use these rules in the following example, but keep in mind that the rules apply even when the reaction consists of more than one reactant or more than one product.

■ EXAMPLE 14.6 Using Le Chatelier's Principle to Predict the Outcome of Concentration Changes

Consider a reaction in which the following equilibrium has been established.

$$N_2(g) + 3H_2(g) \rightleftharpoons 2NH_3(g)$$

Predict the outcome of each of the following changes:
(a) Some hydrogen is removed.
(b) Some nitrogen is added.
(c) Some ammonia is added.

Problem Analysis

Remember that disturbing a reaction at equilibrium causes the reaction to shift to counteract the change and to establish equilibrium once again. Thus, if a substance is added to a reaction at equilibrium, the reaction will shift away from that substance so that its concentration will decrease. If a substance is removed, the reaction will shift toward that substance so that its concentration will increase.

Solution

(a) Since H_2 has been removed, the reaction will shift from right to left (toward H_2), replacing some of the hydrogen that has been removed.
(b) Since N_2 has been added, the reaction will shift from left to right (away from N_2) to remove some of the extra nitrogen.
(c) Because NH_3 has been added, the reaction will shift from right to left (away from NH_3) to remove some of the excess ammonia.

Exercise 14.6

If the equilibrium

$$H_2(g) + I_2(g) \rightleftharpoons 2HI(g)$$

is disturbed by increasing the concentration of hydrogen, what will happen?

(Try Problems 14.41 and 14.42.)

Changing the Volume

If one or more of the reactants or products of a reaction are gases, changing the volume of the reaction mixture already at equilibrium can cause the concentrations of the substances in the mixture to change in a way predicted by Le Chatelier's principle. For example, suppose we enclose some NO_2 in the syringe shown in Figure 14.14A. As you know, this brownish-yellow substance reacts with itself so that colorless N_2O_4 forms. Eventually, the familiar equilibrium shown in the following equation is reached.

$$2NO_2(g) \rightleftharpoons N_2O_4(g)$$

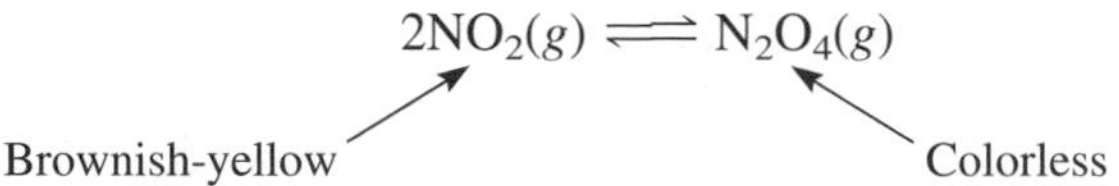

Next, we decrease the volume in the syringe so that the concentrations of both substances increase, as shown by the darker color in Figure 14.14B. As a result of the volume decrease, the reaction mixture experiences an increase in pressure. Pressure is the force per unit area exerted by the gaseous molecules (NO_2 and N_2O_4) on all of the walls of the syringe. According to Le Chatelier's principle, the reaction will shift in a way that counteracts this change. The only way it can do this is by decreasing the number of molecules in the syringe. Fewer molecules means a lower force per unit area and a lower pressure.

But how can this reaction shift toward fewer molecules? (Assume that there is no leak in the syringe, so no molecules can escape.) As the chemical equation shows, a single molecule of N_2O_4 results when two molecules of NO_2 combine.

$$\underset{\text{2 molecules}}{2NO_2(g)} \rightleftharpoons \underset{\text{1 molecule}}{N_2O_4(g)}$$

Thus, the number of molecules in the syringe will decrease if the reaction shifts from left to right, so we can predict that decreasing the volume of this mixture causes the reaction to generate colorless N_2O_4 at the expense of brownish-yellow NO_2. You can see this effect in Figure 14.14C, where the color of the gas mixture in the syringe has lightened after the new equilibrium is reached.

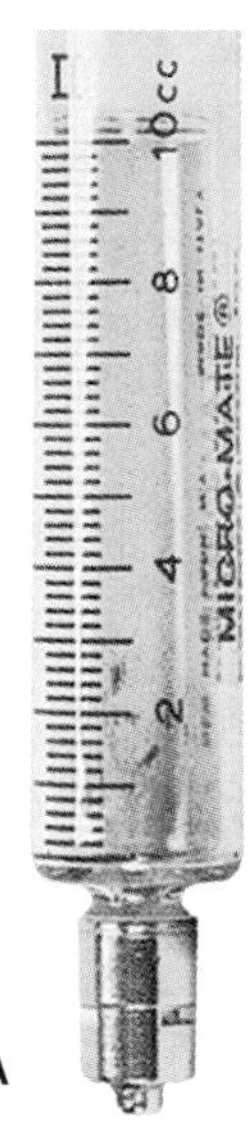

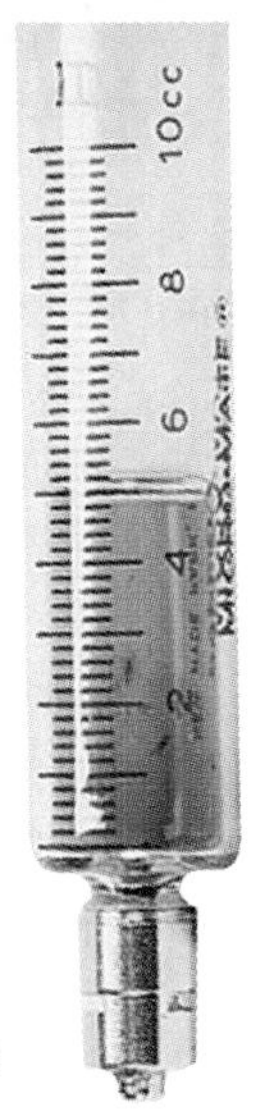

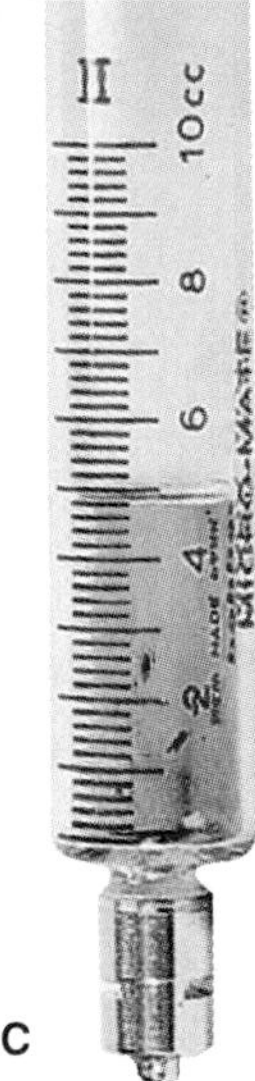

FIGURE 14.14
The effect of volume (and pressure) on the equilibrium between brownish-yellow NO_2 and colorless N_2O_4. *(A)* A syringe containing a mixture of the two gases at equilibrium. *(B)* Immediately after the volume is decreased, the concentrations of both substances increase. The darker color is caused by the increased concentration of NO_2. *(C)* The intensity of the color then diminishes as the reaction shifts so that N_2O_4 is formed at the expense of NO_2.

Similarly, if we increase the volume within the syringe, the pressure will decrease. To reduce the effect of this change, the reaction will shift from right to left to increase the number of molecules in the cylinder.

We can summarize our findings as follows.

1. If the volume decreases, the pressure will increase and the reaction will shift toward fewer molecules.
2. If the volume increases, the pressure will decrease and the reaction will shift toward more molecules.

Suppose, however, that we enclose the equilibrium reaction

$$H_2(g) + I_2(g) \rightleftharpoons 2HI(g)$$

in our syringe. Since the number of molecules on the left side of this equation equals the number on the right, volume changes will not affect this particular equilibrium.

Shifts due to volume changes do not affect the value of an equilibrium constant. We can demonstrate that K will not change from the following example. Suppose we enclose the equilibrium mixture from Experiment 1 in Table 14.1 in a syringe, and we cut the volume in half by pushing in on the plunger. The new concentrations are now twice those that were in the equilibrium mixture from Experiment 1. Since we have decreased the volume, the reaction will shift from left to right so that N_2O_4 is produced and the total number of molecules in the syringe decreases.

Animation: Equilibrium decomposition of N_2O_4.

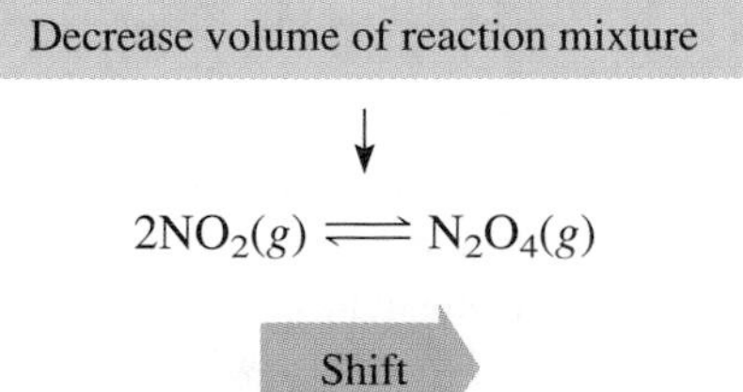

The new equilibrium concentrations of all substances in the reaction are shown in Experiment 5 in Table 14.1. When these concentrations are inserted into the equilibrium expression, the calculated value of K is identical to the one obtained in Experiment 1. Let's consider another example.

■ EXAMPLE 14.7 Using Le Chatelier's Principle to Predict the Outcome of Volume Changes

Recent research has been directed toward the economical conversion of coal to a gaseous fuel such as methane. Coal reacts with hot steam to form CO and H_2, and these substances react further in the presence of a catalyst to give methane and water vapor.

$$CO(g) + 3H_2(g) \rightleftharpoons CH_4(g) + H_2O(g)$$

When this reaction comes to equilibrium, predict the outcome of suddenly increasing the volume.

Problem Analysis

Remember that disturbing a reaction at equilibrium causes the reaction to shift, counteracting the change and establishing equilibrium once again. Thus, if the volume increases, the reaction shifts so that more molecules are formed (if possible). If the volume decreases, the reaction shifts so that fewer molecules are formed (if possible).

Solution

Because the volume increases, the pressure decreases and the reaction shifts toward more molecules.

$$\underbrace{CO(g) + 3H_2(g)}_{\text{4 molecules}} \rightleftharpoons \underbrace{CH_4(g) + H_2O(g)}_{\text{2 molecules}}$$

Thus, the reaction will shift from right to left until equilibrium is established again.

Exercise 14.7

If the equilibrium

$$N_2(g) + 3H_2(g) \rightleftharpoons 2NH_3(g)$$

is disturbed by decreasing the volume of the reaction mixture, what will happen to the concentrations of the substances in the equilibrium mixture?

(Try Problems 14.43 and 14.44.)

Changing the Temperature

Most chemical reactions are affected profoundly by temperature changes. Reaction rates increase when the temperature is increased so that equilibrium is reached sooner. In addition, equilibrium constants vary with the temperature. We explore this dependence next.

As you recall, energy changes are associated with chemical reactions. Many chemical reactions *liberate heat* and are said to be **exothermic.** An example is the reaction in which NO_2 is converted to N_2O_4.

$$2NO_2(g) \rightleftharpoons N_2O_4(g) + \text{heat}$$

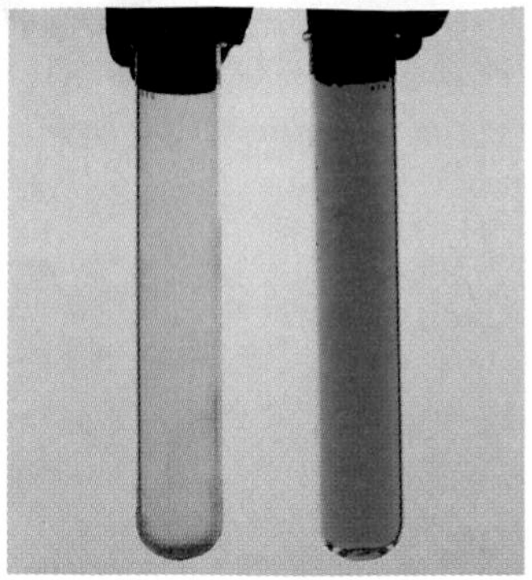

FIGURE 14.15
The effect of temperature on the equilibrium between brownish-yellow NO_2 and colorless N_2O_4. The concentration of NO_2 increases as the temperature increases from 0°C *(left)* to 100°C *(right).*

Heat is placed on the right side of the equation because it is evolved as N_2O_4 is formed. Other chemical reactions *absorb heat* from their surroundings and are called **endothermic.** An example is the decomposition of carbon dioxide to give carbon monoxide and oxygen.

$$2CO_2(g) + \text{heat} \rightleftharpoons 2CO(g) + O_2(g)$$

Heat is placed on the left side because it is used (absorbed) as molecules of CO_2 decompose.

Let's suppose that these reactions are at equilibrium. What will happen if we now increase the temperature of both of them by heating? According to Le Chatelier's principle, the reactions will shift in a way that will counteract the change. Consider the first of these reactions, an exothermic one:

$$2NO_2(g) \rightleftharpoons N_2O_4(g) + \text{heat}$$

As the temperature is increased, heat enters the reaction mixture, and the reaction shifts from right to left, absorbing the heat we added and attempting to counter the temperature increase (Figure 14.15).

$$2NO_2(g) \rightleftharpoons N_2O_4(g) + \text{heat}$$

Shift

Thus, the concentration of N_2O_4 will decrease, and the concentration of NO_2 will increase. A look at the equilibrium expression shows that the equilibrium constant will decrease in this case.

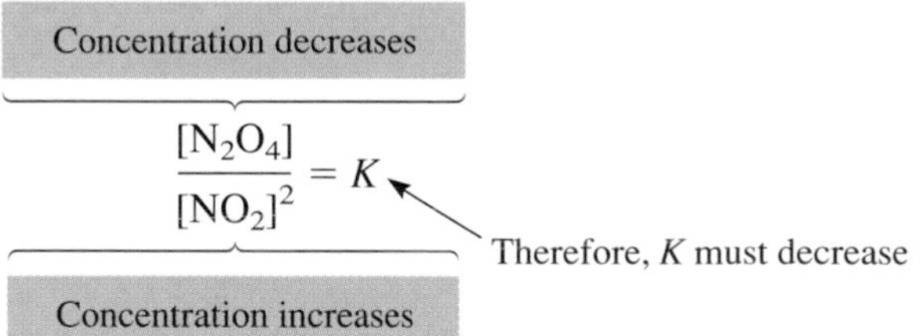

For this reason, K for this reaction is 1.3×10^3 at 0°C, but only 2.5 at 100°C.

Next, consider increasing the temperature of the endothermic reaction by heating the equilibrium mixture.

$$2CO_2(g) + \text{heat} \rightleftharpoons 2CO(g) + O_2(g)$$

The reaction shifts from left to right in order to absorb the added heat and to counteract the temperature increase. The concentration of CO_2 decreases, and the concentrations of the substances on the right side of the equation increase because of the shift. As a result, the equilibrium constant increases, as you can see from the equilibrium expression.

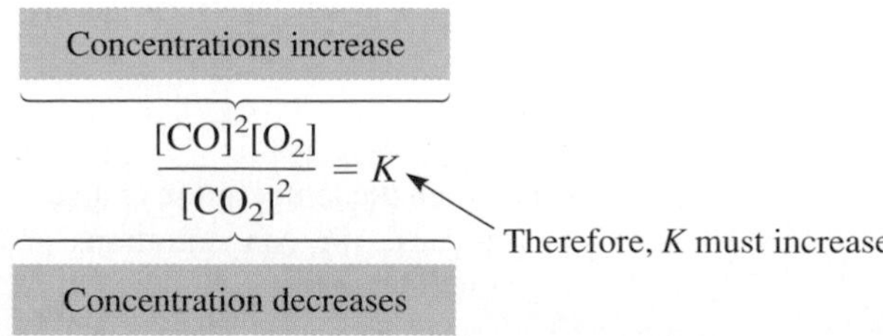

Notice that shifts in opposite directions would have occurred if the temperature had been decreased rather than increased. (You must remove heat to decrease the temperature.) Thus, *heat can be treated as if it were a component of the reaction.* Based on this example, we can make three generalizations:

1. If the temperature is increased, the reaction shifts away from the heat. For an exothermic reaction, we have

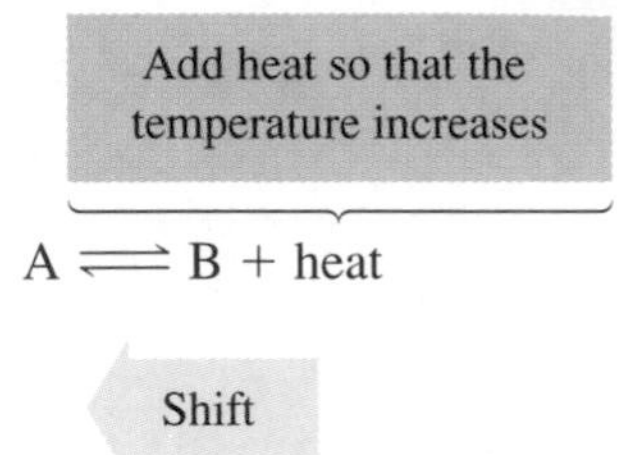

but for an endothermic reaction, we have

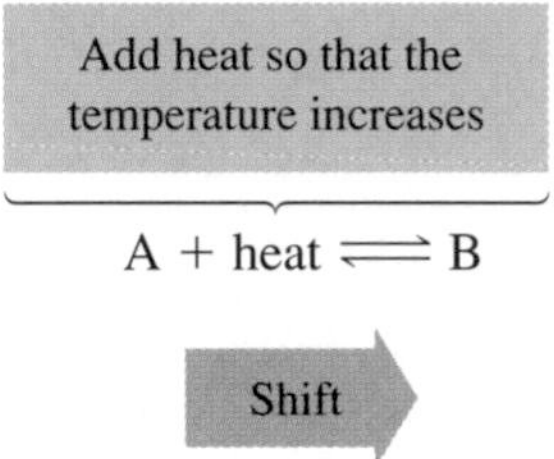

2. If the temperature is decreased, the reaction shifts toward the heat. If a reaction is exothermic, we obtain

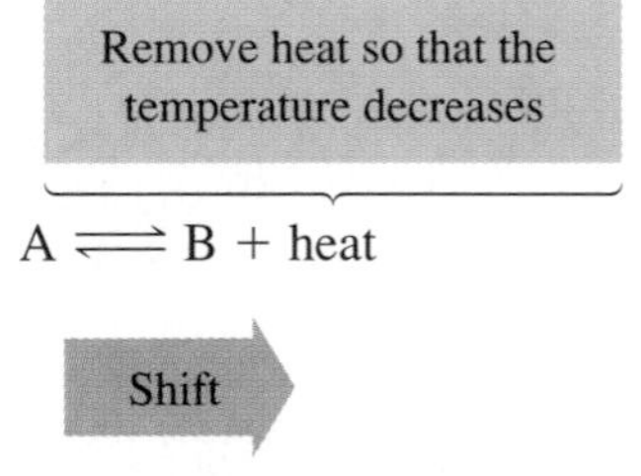

but an endothermic reaction leads to

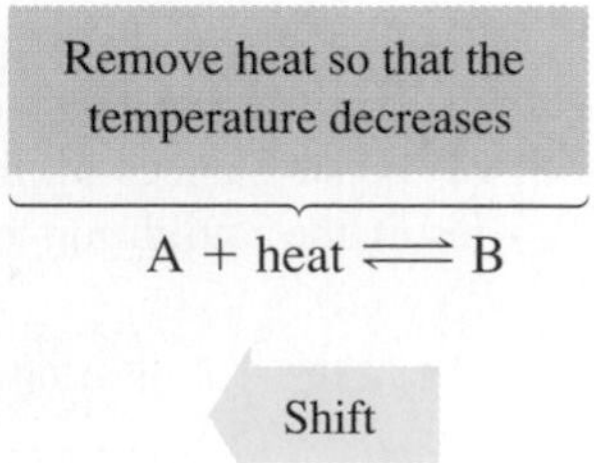

3. When the temperature changes, if the reaction shifts from left to right, K increases because the concentrations of the substances on the right side of the equation increase at the expense of those on the left. On the other hand, if a reaction shifts

from right to left, K decreases because concentrations of substances on the right side of the equation have decreased while those on the left have increased.

■ EXAMPLE 14.8 Using Le Chatelier's Principle to Predict the Outcome of Temperature Changes

Consider the exothermic reaction in the equation

$$N_2(g) + 3H_2(g) \rightleftharpoons 2NH_3(g)$$

Explain why the yield of ammonia increases as the temperature decreases.

Problem Analysis

Because the reaction is exothermic, you write

$$N_2(g) + 3H_2(g) \rightleftharpoons 2NH_3(g) + \text{heat}$$

Remember that heat can be treated as if it were a substance involved in the reaction. Also, remember that yield is the quantity of a substance that can be obtained from given quantities of reactants.

Solution

You can see that decreasing the temperature (removing heat) causes the reaction to shift from left to right. As a result, the concentration of NH_3 increases at the expense of the concentrations of N_2 and H_2. Because the concentration of ammonia increases, the yield must increase.

Exercise 14.8

Consider the endothermic reaction in the equation

$$2CO_2(g) + \text{heat} \rightleftharpoons 2CO(g) + O_2(g)$$

Will the concentration of carbon dioxide increase or decrease as the temperature is decreased?

(Try Problems 14.45 and 14.46.)

Adding a Catalyst

Since a catalyst can change the rate of a reaction, it is reasonable to wonder about the effect of adding a catalyst to a reaction mixture at equilibrium. Can the presence of a catalyst cause a reaction to shift so that an unfavorably small equilibrium constant changes into a larger, more favorable one? The answer is no because any substance that is a catalyst for the forward reaction of an equilibrium is also a catalyst for the reverse reaction. The reaction achieves equilibrium faster with a catalyst than it would without one, but *a catalyst has no effect on the equilibrium concentrations or the equilibrium constant of a reaction; a catalyst merely increases the rate at which equilibrium is reached.* For example, the Haber process for making ammonia,

$$N_2(g) + 3H_2(g) \rightleftharpoons 2NH_3(g)$$

relies on a catalyst to get the reaction to equilibrium faster, but the catalyst has no effect on the yield of ammonia at a given temperature.

Driving a Reaction to Completion

The chemical industry uses Le Chatelier's principle with many different chemical reactions to improve the yields of commercially important products, because higher yields translate into higher profits. As an example, let's consider the exothermic reaction leading to ammonia,

$$N_2(g) + 3H_2(g) \rightleftharpoons 2NH_3(g)$$

Because the reaction is exothermic, the yield of ammonia is improved by keeping the reaction mixture at a temperature that is as low as practical. If the temperature is too low, however, the rate of the reaction is too slow even with a catalyst. As a compromise, the reaction is usually conducted near 500°C. The yield is also improved by working at reduced volumes and high pressures because there are fewer molecules on the right side of the equation than on the left side. Typically, total pressures of 150 to 300 atm are used. Yet even with optimum conditions of temperature and pressure, the yield of ammonia does not exceed 20%. So another technique derived from Le Chatelier's principle is used to drive the reaction to completion. Using a cyclic process, the gases are forced to leave the reaction vessel and cooled. Ammonia liquefies because its boiling point is much higher than the boiling points of hydrogen and nitrogen. Liquid ammonia is drained from the reaction mixture before the hydrogen and nitrogen are recycled to the reaction vessel, where they can begin to react once again. The removal of ammonia shifts the reaction from left to right and drives the reaction to completion.

Chemical Perspective

Carbon Monoxide and Hemoglobin

A woman—let's call her Mildred—was admitted to a hospital in Philadelphia about two o'clock in the afternoon.* Her problem was a severe headache of three weeks' duration, extreme irritability, and a florid, flushed face. Because the hospital was very busy that day, her diagnostic interview was delayed until five o'clock. When a doctor asked how she felt, Mildred replied that she felt pretty good. It was a funny thing, she said, that her headache was gone. In the three hours since she had been admitted to the hospital, it had vanished completely along with her other symptoms. Extensive tests followed, but the doctor was unable to find any health problems or physical abnormalities.

Mildred was released, only to be admitted again within two days because the same symptoms had returned. This time the doctor realized that Mildred had carbon monoxide (CO) poisoning. When an inspector was sent to Mildred's house, he found that a partially blocked furnace flue was allowing smoke and gases from the furnace to escape into the house. Carbon monoxide is one of those gases.

Carbon monoxide is an asphyxiant. It deprives its victims of the oxygen they need by preferentially binding to hemoglobin in red blood cells. We discussed hemoglobin and its oxygenation briefly in the Chemical Perspective in Chapter 6 and again at the beginning of this chapter. Just as four molecules of oxygen will bind chemically to each hemoglobin molecule,

$$Hb + 4O_2 \rightleftharpoons Hb(O_2)_4$$

so too will four molecules of carbon monoxide bind to a single hemoglobin molecule,

$$Hb + 4CO \rightleftharpoons Hb(CO)_4$$

The carboxyhemoglobin molecule, $Hb(CO)_4$, is brilliant red in color, which accounts for Mildred's flushed and florid appearance.

Each carbon monoxide molecule binds about 300 times more strongly to hemoglobin than does an oxygen molecule. Stronger binding means that the equi-

*The story about Mildred was adapted from Berton Roueche, *Medical Detectives,* Vol. 2, Washington Square Press, New York, 1984.

librium is shifted more toward the right side of the equation. As a result, the equilibrium constant for the reaction of carbon monoxide is considerably larger than that for the oxygenation of hemoglobin, and hemoglobin will combine with carbon monoxide in preference to oxygen. Thus, inhaled carbon monoxide has a substantial impact on the transport of oxygen in the blood because hemoglobin that has combined with carbon monoxide is not available to carry oxygen.

Inhaling carbon monoxide can cause various adverse health effects. Relatively low levels will cause headaches or dizziness, and high levels can cause death. One of the most common sources of carbon monoxide is automobile exhaust. Police officers, parking garage attendants, and others who work for long hours around automobiles inhale substantial doses of carbon monoxide. Fortunately, most of the effects they experience are lost when they spend several hours breathing relatively pure air. Le Chatelier's principle tells why this recovery is possible.

Added from pure air — Dissipated when we exhale

$$Hb(CO)_4 + 4O_2 \rightleftharpoons Hb(O_2)_4 + 4CO$$

Shift →

Although the equilibrium normally favors the left side of this equation, adding oxygen as shown here causes a net shift from left to right. When Mildred left her home and entered the hospital, she was no longer breathing air polluted with carbon monoxide. So the reaction shifted and she lost the symptoms of carbon monoxide poisoning.

Cigarette smoke is another major source of carbon monoxide. Smokers inhale it directly through the burning cigarette, but even nonsmokers inhale carbon monoxide when they are exposed to cigarette smoke from others. Since hemoglobin that carries carbon monoxide is not used in transporting oxygen, the heart must pump more blood to get the needed oxygen. Moreover, heavy smokers do not give their hemoglobin much opportunity to recover. As a result, chronic exposure to carbon monoxide from smoking is believed to cause heart disease and heart attacks. Of course, the other substances in cigarette smoke can lead to lung cancer and respiratory problems. Clearly, smoking is a habit with many high risks (Figure 14.16).

FIGURE 14.16
A dangerous habit. Carbon monoxide from cigarette smoke is a poison to smokers and nearby people. Other substances resulting from the combustion in a lighted cigarette can lead to lung cancer and respiratory problems. Every pack of cigarettes manufactured in the United States carries a printed warning from the Surgeon General about these dangers.

CHAPTER REVIEW

Key Words

collision theory *(p. 438)*
activation energy *(p. 438)*
catalyst *(p. 441)*
chemical equilibrium *(p. 444)*
law of mass action *(p. 445)*
equilibrium constant *(p. 446)*
homogeneous equilibrium *(p. 451)*
heterogeneous equilibrium *(p. 451)*
Le Chatelier's principle *(p. 456)*
exothermic *(p. 460)*
endothermic *(p. 461)*

Key Equation

$$K = \frac{[C]^c[D]^d}{[A]^a[B]^b} \quad \text{for } aA + bB \rightleftharpoons cC + dD$$

Summary

The rate of a chemical reaction depends on collisions between the molecules of the reactants. We use *collision theory* to explain the effect of reactant concentrations and temperature on a reaction rate. According to this theory, molecules react after colliding but only when the collision energy is greater than the *activation energy.* The fraction of collisions having energy greater than the activation energy will increase as the concentrations of those molecules increase and as the temperature increases. *Catalysts,* substances that increase the speed of a reaction without being consumed by it, function by providing a new path with a smaller activation energy for the reaction.

Chemical equilibrium is the *dynamic* state achieved in a chemical reaction in which the forward and reverse reactions proceed at equal rates. The *law of mass action* states that the equilibrium expression is equal to a constant called the equilibrium constant. The equilibrium expression contains terms for the concentrations of the products in the numerator and terms for the concentrations of the reactants in the denominator. All terms appear for a *homogeneous equilibrium.* However, terms involving solids and pure liquids are omitted for a *heterogeneous equilibrium.* The equilibrium constant can be calculated from the equilibrium concentrations, and any one of the concentrations can be calculated from the equilibrium constant and the remainder of the equilibrium concentrations. According to *Le Chatelier's principle,* equilibrium concentrations can be altered by introducing or removing one of the substances involved in the reaction, by changing the volume, or by changing the temperature. However, the equilibrium constant cannot be changed by adding or removing a reactant or a product or by changing the volume; it is changed only when the temperature is changed. Temperature changes have opposite effects on *exothermic* reactions and *endothermic* reactions. A catalyst increases the rate at which equilibrium is attained, but it does not change the equilibrium concentrations.

Problem-Solving Skills

1. **Writing an Equilibrium Expression:** Given a chemical equation, write the equilibrium expression from the law of mass action (Example 14.1).
2. **Obtaining *K* from Equilibrium Concentrations:** Given the equilibrium concentrations of all substances taking part in a reaction, calculate the value of the equilibrium constant (Example 14.2).
3. **Using an Equilibrium Constant to Predict the Extent of a Reaction:** Given the equilibrium constant, predict the extent to which reactants are converted to products (Example 14.3).
4. **Calculating an Unknown Equilibrium Concentration:** Given the equilibrium constant, the equilibrium expression, and all but one of the equilibrium concentrations of the reactants and products, calculate the missing equilibrium concentration (Example 14.4).
5. **Determining Relative Solubilities by Comparing Equilibrium Constants:** Given the solubility product constants for two or more substances that produce the same number of ions when they dissolve, predict their relative solubilities (Example 14.5).
6. **Using Le Chatelier's Principle to Predict the Outcome of Concentration Changes:** Given a change in the concentration of one of the substances involved in a reaction at equilibrium, predict which direction the reaction will shift (Example 14.6).
7. **Using Le Chatelier's Principle to Predict the Outcome of Volume Changes:** Given a change in the volume of a reaction system at equilibrium, predict which direction, if any, the reaction will shift (Example 14.7).
8. **Using Le Chatelier's Principle to Predict the Outcome of Temperature Changes:** Given a change in the temperature of a reaction system at equilibrium and whether it is exothermic or endothermic, predict which direction the reaction will shift (Example 14.8).

Questions to Test Your Reading

14.1 What factors influence the rate of a reaction?

14.2 What must two reactant molecules do if they are to react with each other?

14.3 According to *collision theory,* will a collision between two reactant molecules always result in a reaction? Explain.

14.4 If the concentration of a reactant in a chemical reaction is increased, what will usually happen to the rate of the reaction?

14.5 In terms of collision theory, why do the concentrations of the reactants affect the rate of a reaction?

14.6 If the temperature of a reaction is increased, what will happen to the reaction rate?

14.7 How does a *catalyst* affect the rate of a reaction?

14.8 Compare the *activation energy* of an uncatalyzed reaction to its activation energy in the presence of a catalyst.

14.9 What is *chemical equilibrium*? Is it ever static? Explain.

14.10 What is an *equilibrium constant*? Is it always constant?

14.11 How many different sets of equilibrium concentrations of reactants and products may correspond to a given value of an equilibrium constant? Explain.

14.12 What qualitative information can you obtain from the magnitude of an equilibrium constant?

14.13 What is the difference between a *homogeneous equilibrium* and a *heterogeneous equilibrium*?

14.14 Which of the following reactions involves a homogeneous equilibrium and which involves a heterogeneous equilibrium?
(a) $AgCl(s) \rightleftharpoons Ag^+(aq) + Cl^-(aq)$
(b) $2N_2O(g) \rightleftharpoons 2N_2(g) + O_2(g)$

14.15 What is *Le Chatelier's principle*?

14.16 List three possible ways of altering the equilibrium concentrations of a reaction mixture.

14.17 How does an *exothermic* reaction differ from an *endothermic* reaction?

14.18 What will happen to the equilibrium concentrations of reactants and products during an exothermic reaction if the temperature is increased?

14.19 What will happen to the equilibrium concentrations of reactants and products during an endothermic reaction if the temperature is increased?

14.20 How will a catalyst affect the equilibrium concentrations of a reaction mixture?

14.21 Hydrogen will react with oxygen under certain conditions to form water. When these elements are mixed at 25°C, no reaction can be seen even though the equilibrium constant at this temperature is very large. However, the reaction takes place readily and rapidly if a piece of platinum is added to the mixture. Explain the role of platinum in this reaction. Will it affect the equilibrium concentrations of the reaction mixture?

14.22 Give four ways to increase the yield of ammonia in the exothermic reaction whose chemical equation is

$$N_2(g) + 3H_2(g) \rightleftharpoons 2NH_3(g)$$

if only a fixed amount of hydrogen is available.

Practice Problems

The Equilibrium Constant, the Equilibrium Expression, and Equilibrium Concentrations (Sections 14.4 14.6)

14.23 Balance the following chemical equations and write an equilibrium expression for each.
(a) $SO_2(g) + O_2(g) \rightleftharpoons SO_3(g)$
(b) $NOCl(g) \rightleftharpoons NO(g) + Cl_2(g)$
(c) $POCl_3(g) \rightleftharpoons POCl(g) + Cl_2(g)$
(d) $PCl_3(g) + Cl_2 \rightleftharpoons PCl_5(g)$

14.24 Balance the following chemical equations and write an equilibrium expression for each.
(a) $N_2(g) + Cl_2(g) \rightleftharpoons NCl_3(g)$
(b) $NO(g) + O_2(g) \rightleftharpoons NO_2(g)$
(c) $CO(g) + H_2(g) \rightleftharpoons CH_3OH(g)$
(d) $CO(g) + H_2O(g) \rightleftharpoons CO_2(g) + H_2(g)$

14.25 Balance the following chemical equations and write an equilibrium expression for each.
(a) $CS_2(g) + H_2(g) \rightleftharpoons CH_4(g) + H_2S(g)$
(b) $I_2(g) + Br_2(g) \rightleftharpoons IBr(g)$
(c) $COCl_2(g) \rightleftharpoons CO(g) + Cl_2(g)$
(d) $NH_4^+(aq) \rightleftharpoons NH_3(aq) + H^+(aq)$

14.26 Balance the following chemical equations and write an equilibrium expression for each.
(a) $NO_2(g) + H_2(g) \rightleftharpoons NH_3(g) + H_2O(g)$
(b) $IF(g) + F_2(g) \rightleftharpoons IF_3(g)$
(c) $HCl(g) + O_2(g) \rightleftharpoons Cl_2(g) + H_2O(g)$
(d) $HCN(aq) \rightleftharpoons H^+(aq) + CN^-(aq)$

14.27 Carbon monoxide can be converted to methane by the reaction shown in the unbalanced chemical equation

$$CO(g) + H_2(g) \rightleftharpoons CH_4(g) + H_2O(g)$$

Balance the equation and calculate the value of the equilibrium constant at 927°C if the following equilibrium concentrations are present: $[CO] = 0.0613\ M$, $[H_2] = 0.184\ M$, and $[CH_4] = [H_2O] = 0.0387\ M$.

14.28 Nitrogen monoxide reacts with hydrogen to give nitrogen and water. The unbalanced chemical equation is

$$NO(g) + H_2(g) \rightleftharpoons N_2(g) + H_2O(g)$$

Balance the equation and calculate the equilibrium constant at a particular temperature if the following equilibrium concentrations are present: $[NO] = 0.062\ M$, $[H_2] = 0.012\ M$, $[N_2] = 0.019\ M$, and $[H_2O] = 0.138\ M$.

14.29 Hydrogen iodide can be prepared from its elements according to the unbalanced chemical equation

$$H_2(g) + I_2(g) \rightleftharpoons HI(g)$$

Balance the equation and calculate the equilibrium constant at 425°C if the following concentrations are found at equilibrium: $[H_2] = [I_2] = 0.00213\ M$ and $[HI] = 0.0157\ M$.

14.30 Sulfur dioxide will react with oxygen to form sulfur trioxide. The unbalanced chemical equation is

$$SO_2(g) + O_2(g) \rightleftharpoons SO_3(g)$$

Balance the equation and calculate the equilibrium constant at 600°C if the following concentrations are found at equilibrium: $[SO_2] = 0.590\ M$, $[O_2] = 0.0450\ M$, and $[SO_3] = 0.260\ M$.

14.31 The value of the equilibrium constant for the formation of gaseous water from its elements is 3×10^{81} at a certain temperature. The chemical equation is

$$2H_2(g) + O_2(g) \rightleftharpoons 2H_2O(g)$$

Write the equilibrium expression. Do you expect a nearly complete reaction? Explain.

14.32 The value of the equilibrium constant for the formation of nitrogen dioxide from the elements is 3×10^{-17} at a certain temperature. The chemical equation is

$$N_2(g) + 2O_2(g) \rightleftharpoons 2NO_2(g)$$

Write the equilibrium expression. Do you expect a nearly complete reaction? Explain.

14.33 The reaction represented by the equation

$$CH_4(g) + 2H_2S(g) \rightleftharpoons CS_2(g) + 4H_2(g)$$

has an equilibrium constant of 3.59 at 900°C. What is the concentration of carbon disulfide (CS_2) at that temperature if the mixture also consists of $[CH_4] = 1.10\ M$, $[H_2S] = 1.49\ M$, and $[H_2] = 1.68\ M$? The reaction is at equilibrium.

14.34 The equilibrium constant for the reaction

$$H_2(g) + I_2(g) \rightleftharpoons 2HI(g)$$

is 51 at a particular temperature. What is the concentration of I_2 at that temperature if the mixture also consists of $[H_2] = 6.5 \times 10^{-5}\ M$ and $[HI] = 1.9 \times 10^{-3}\ M$. The reaction is at equilibrium.

Heterogeneous Equilibria (Section 14.7)

14.35 Balance the following equations and write an equilibrium expression for each.
(a) $C(s) + CO_2(g) \rightleftharpoons CO(g)$
(b) $FeO(s) + CO(g) \rightleftharpoons Fe(s) + CO_2(g)$
(c) $Na_2CO_3(s) + SO_2(g) + O_2(g) \rightleftharpoons Na_2SO_4(s) + CO_2(g)$

14.36 Balance the following equations and write an equilibrium expression for each.
(a) $NH_4Cl(s) \rightleftharpoons NH_3(g) + HCl(g)$
(b) $C(s) + N_2O(g) \rightleftharpoons CO_2(g) + N_2(g)$
(c) $NaHCO_3(s) \rightleftharpoons Na_2CO_3(s) + H_2O(g) + CO_2(g)$

14.37 Balance the following equations and write an equilibrium expression for each.
(a) $P_4(s) + O_2(g) \rightleftharpoons P_4O_{10}(s)$
(b) $H_2O_2(l) \rightleftharpoons H_2O(l) + O_2(g)$
(c) $PbI_2(s) + Cl_2(g) \rightleftharpoons PbCl_2(s) + I_2(g)$

14.38 Balance the following equations and write an equilibrium expression for each.
(a) $CuO(s) \rightleftharpoons Cu_2O(s) + O_2(g)$
(b) $NaBr(s) + Cl_2(g) \rightleftharpoons NaCl(s) + Br_2(l)$
(c) $Cr(s) + Cl_2(g) \rightleftharpoons CrCl_3(s)$

14.39 Determine the relative solubilities of PbI_2 and Ag_2CrO_4 from the following information.

$$PbI_2(s) \rightleftharpoons Pb^{2+}(aq) + 2I^-(aq) \quad K_{sp} = 6.5 \times 10^{-9}\ (25°C)$$

$$Ag_2CrO_4(s) \rightleftharpoons 2Ag^+(aq) + CrO_4^{2-}(aq) \quad K_{sp} = 1.1 \times 10^{-12}\ (25°C)$$

14.40 Determine the relative solubilities of CaF_2 and Ag_2S from the following information.

$$CaF_2(s) \rightleftharpoons Ca^{2+}(aq) + 2F^-(aq) \quad K_{sp} = 3.4 \times 10^{-11}\ (25°C)$$

$$Ag_2S(s) \rightleftharpoons 2Ag^+(aq) + S^{2-}(aq) \quad K_{sp} = 6 \times 10^{-50}\ (25°C)$$

Le Chatelier's Principle (Section 14.8)

14.41 Predict the direction of reaction when oxygen gas is added to an equilibrium mixture of SO_2, SO_3, and O_2. The chemical equation is

$$2SO_2(g) + O_2(g) \rightleftharpoons 2SO_3(g)$$

14.42 Predict the direction of reaction when methane, CH_4, is added to an equilibrium mixture of CO, H_2, CH_4, and H_2O. The chemical equation is

$$CO(g) + 3H_2(g) \rightleftharpoons CH_4(g) + H_2O(g)$$

14.43 What would be the effect of decreasing the volume of the reaction shown in the following chemical equation if it were at equilibrium before the change?

$$CO(g) + 2H_2(g) \rightleftharpoons CH_3OH(g)$$

14.44 What would be the effect of decreasing the volume of the reaction shown in the following chemical equation?

$$2HBr(g) \rightleftharpoons H_2(g) + Br_2(g)$$

14.45 What would be the effect of raising the temperature of the reaction shown in the following chemical equation? The reaction is exothermic.

$$CO(g) + 2H_2(g) \rightleftharpoons CH_3OH(g)$$

14.46 What would be the effect of raising the temperature of the reaction shown in the following chemical equation? The reaction is endothermic.

$$N_2(g) + O_2(g) \rightleftharpoons 2NO(g)$$

Additional Problems

14.47 Give the balanced equation that corresponds to the equilibrium expression

$$K = \frac{[CS_2]\,[H_2]^4}{[CH_4]\,[H_2S]^2}$$

14.48 Give the balanced equation that corresponds to the equilibrium expression

$$K = \frac{[NH_3]^4[O_2]^5}{[NO]^4[H_2O]^6}$$

14.49 Give the balanced equation that corresponds to the equilibrium expression

$$K = \frac{[N_2]\,[H_2O]^2}{[NO]^2[H_2]^2}$$

14.50 Give the balanced equation that corresponds to the equilibrium expression

$$K = \frac{[H_2O]^2[Cl_2]^2}{[HCl]^4[O_2]}$$

14.51 Give the equilibrium expression for

$$AgI(s) \rightleftharpoons Ag^+(aq) + I^-(aq)$$

and calculate the solubility product constant if $[Ag^+] = [I^-] = 9.1 \times 10^{-9}\ M$ at 25°C.

14.52 Give the equilibrium expression for

$$MnS(s) \rightleftharpoons Mn^{2+}(aq) + S^{2-}(aq)$$

and calculate the solubility product constant if $[Mn^{2+}] = [S^{2-}] = 1.6 \times 10^{-5}\ M$ at 25°C.

14.53 Give the equilibrium expression for

$$PbBr_2(s) \rightleftharpoons Pb^{2+}(aq) + 2Br^-(aq)$$

and calculate the solubility product constant if $[Pb^{2+}] = 0.010\ M$ and $[Br^-] = 0.020\ M$ at 25°C.

14.54 Give the equilibrium expression for

$$Ag_3PO_4(s) \rightleftharpoons 3Ag^+(aq) + PO_4^{3-}(aq)$$

and calculate the solubility product constant if $[Ag^+] = 4.8 \times 10^{-5} M$ and $[PO_4^{3-}] = 1.6 \times 10^{-5}\ M$.

14.55 Consider the exothermic reaction in the chemical equation

$$2NO(g) + 2H_2(g) \rightleftharpoons N_2(g) + 2H_2O(g)$$

If the reaction is at equilibrium, what changes will occur if
(a) some hydrogen is removed;
(b) some nitrogen is added;
(c) the volume of the reaction mixture is decreased;
(d) the temperature is decreased; or
(e) a catalyst is added?

14.56 Consider the endothermic reaction in the chemical equation

$$2SO_3(g) \rightleftharpoons 2SO_2(g) + O_2(g)$$

If the reaction is at equilibrium, what changes will occur if
(a) some sulfur trioxide is added;
(b) some sulfur dioxide is removed;
(c) the volume of the reaction mixture is decreased;
(d) the temperature is decreased; or
(e) a catalyst is added?

14.57 What will happen to the equilibrium constant during each of the changes in Problem 14.55?

14.58 What will happen to the equilibrium constant during each of the changes in Problem 14.56?

14.59 Consider the endothermic reaction in the chemical equation

$$PbCl_2(s) \rightleftharpoons Pb^{2+}(aq) + 2Cl^-(aq)$$

If the reaction is at equilibrium, what changes will occur if
(a) some Cl^- ion is removed;
(b) some $Pb(NO_3)_2$ (a source of Pb^{2+} ion) is added;
(c) the temperature is increased?

14.60 Consider the exothermic reaction in the chemical equation

$$BaO(s) + CO_2(g) \rightleftharpoons BaCO_3(s)$$

If the reaction is at equilibrium, what changes will occur if
(a) some CO_2 is added;
(b) the temperature is decreased;
(c) the volume is decreased?

14.61 What will happen to the equilibrium constant during each of the changes in Problem 14.59?

14.62 What will happen to the equilibrium constant during each of the changes in Problem 14.60?

Practice in Problem Analysis

For each problem, describe the thinking you would use (the problem analysis) before doing the actual solution, but do not solve the problem.

1. A chemist introduces 1.00 mol of N_2, 3.00 mol of H_2, and 2.00 mol of NH_3 into a 50.0-L reaction vessel, and then raises the temperature of the mixture to 400°C. Will more NH_3 form when the reaction comes to equilibrium, or will it decompose? The equilibrium constant for the reaction

$$N_2(g) + 3H_2(g) \rightleftharpoons 2NH_3(g)$$

is 0.500 at 400°C.

2. The amount of methanol (CH_3OH) formed in the reaction

$$CO(g) + 2H_2(g) \rightleftharpoons CH_3OH(g)$$

decreases as the temperature increases. Is this reaction exothermic or endothermic?

PRACTICE EXAM

1. Which statement is true?
(a) Some chemical reactions take years to reach completion.
(b) Decreasing the concentration of one of the reactants will increase the rate of chemical reaction.
(c) Decreasing the temperature of the reaction will increase the rate of chemical reaction.
(d) Removing a catalyst will increase the rate of chemical reaction.
(e) None of the above

2. Which statement is true?
(a) The collision of two reacting molecules is not always necessary for a chemical reaction to occur.
(b) A reaction with a very large activation energy will occur very rapidly.
(c) A chemical reaction has an activation energy because molecules attempt to avoid collisions.
(d) A reaction may occur even when the energy of the collision is less than the activation energy.
(e) None of the above

3. A chemical equilibrium is
(a) always dynamic.
(b) usually dynamic but sometimes static.
(c) always static.
(d) usually static but sometimes dynamic.
(e) None of the above

4. Which statement is true?
(a) Collisions between molecules cease when a reaction reaches chemical equilibrium.
(b) A reaction may reach chemical equilibrium before the rates of the forward and reverse reactions become equal.
(c) Reactants are still transformed into products when a reaction reaches chemical equilibrium.
(d) Products are no longer transformed into the original reactants when a reaction reaches chemical equilibrium.
(e) None of the above

5. What is the equilibrium expression for the reaction given in the following chemical equation?

$$2NO(g) + O_2(g) \rightleftharpoons 2NO_2(g)$$

(a) $K = [NO][O_2]/[NO_2]$
(b) $K = [NO]^2[O_2]/[NO_2]^2$
(c) $K = [NO_2]/[NO][O_2]$
(d) $K = [NO_2]^2/[NO]^2[O_2]$
(e) None of the above

6. Hydrogen iodide will decompose at elevated temperatures according to the equation

$$2HI(g) \rightleftharpoons H_2(g) + I_2(g)$$

A chemist found the following equilibrium concentrations at a particular temperature:
$[HI] = 0.196\ M$, $[H_2] = 0.0263\ M$, and $[I_2] = 0.0263\ M$. The value of the equilibrium constant at that temperature is
(a) 0.0180 (b) 55.5 (c) 0.00353 (d) 283
(e) None of the above

7. Hydrogen reacts with carbon monoxide to form methanol (CH_3OH) according to the equation

$$2H_2(g) + CO(g) \rightleftharpoons CH_3OH(g)$$

The value of the equilibrium constant for this reaction at 300°C is 1.3×10^{-4}. During an experiment at that temperature, a chemist finds that the equilibrium con-

centrations of hydrogen and carbon monoxide are 1.85 *M* and 3.33 *M*. The concentration of methanol at equilibrium is
(a) 0.0224 *M*
(b) 0.114 *M*
(c) 1.48×10^{-3} *M*
(d) 1.26×10^{-4} *M*
(e) None of the above

8. The reaction represented by the equation

$$H_2(g) + CO_2(g) \rightleftharpoons H_2O(g) + CO(g)$$

has an equilibrium constant of 0.26 at a certain temperature. If this reaction is at equilibrium at that temperature in an experiment and $[H_2] = 1.4\ M$, $[CO_2] = 1.8\ M$, and $[CO] = 0.26\ M$, then $[H_2O]$ must equal
(a) 0.66 *M* (b) 0.40 *M* (c) 2.5 *M*
(d) 3.9 *M* (e) None of the above

9. Consider the endothermic reaction

$$H_2(g) + CO_2(g) \rightleftharpoons H_2O(g) + CO(g)$$

If the reaction is at equilibrium, the addition of heat (by increasing the temperature)
(a) will not affect the equilibrium concentrations.
(b) will cause the reaction to shift to the right.
(c) will cause the reaction to shift to the left.
(d) will cause the reaction to shift unpredictably.
(e) None of the above

10. Consider the endothermic reaction

$$H_2(g) + CO_2(g) \rightleftharpoons H_2O(g) + CO(g)$$

If the reaction is at equilibrium, the addition of H_2
(a) will not affect the equilibrium concentrations.
(b) will cause the reaction to shift to the right.
(c) will cause the reaction to shift to the left.
(d) will cause the reaction to shift unpredictably.
(e) None of the above

11. Consider the endothermic reaction

$$H_2(g) + CO_2(g) \rightleftharpoons H_2O(g) + CO(g)$$

If the reaction is at equilibrium, the addition of some CO
(a) will not affect the equilibrium concentrations.
(b) will cause the reaction to shift to the right.
(c) will cause the reaction to shift to the left.
(d) will cause the reaction to shift unpredictably.
(e) None of the above

12. Consider the endothermic reaction

$$H_2(g) + CO_2(g) \rightleftharpoons H_2O(g) + CO(g)$$

If the reaction is at equilibrium, the removal of some CO_2
(a) will not affect the equilibrium concentrations.
(b) will cause the reaction to shift to the right.
(c) will cause the reaction to shift to the left.
(d) will cause the reaction to shift unpredictably.
(e) None of the above

13. Consider the endothermic reaction

$$H_2(g) + CO_2(g) \rightleftharpoons H_2O(g) + CO(g)$$

If the reaction is at equilibrium, the removal of some H_2O
(a) will not affect the equilibrium concentrations.
(b) will cause the reaction to shift to the right.
(c) will cause the reaction to shift to the left.
(d) will cause the reaction to shift unpredictably.
(e) None of the above

14. Consider the endothermic reaction

$$H_2(g) + CO_2(g) \rightleftharpoons H_2O(g) + CO(g)$$

If the reaction is at equilibrium, decreasing the volume
(a) will not affect the equilibrium concentrations.
(b) will cause the reaction to shift to the right.
(c) will cause the reaction to shift to the left.
(d) will cause the reaction to shift unpredictably.
(e) None of the above

15. Consider the endothermic reaction

$$H_2(g) + CO_2(g) \rightleftharpoons H_2O(g) + CO(g)$$

If the reaction is at equilibrium, the addition of a catalyst
(a) will not affect the equilibrium concentrations.
(b) will cause the reaction to shift to the right.
(c) will cause the reaction to shift to the left.
(d) will cause the reaction to shift unpredictably.
(e) None of the above

Chapter

15 Acids and Bases

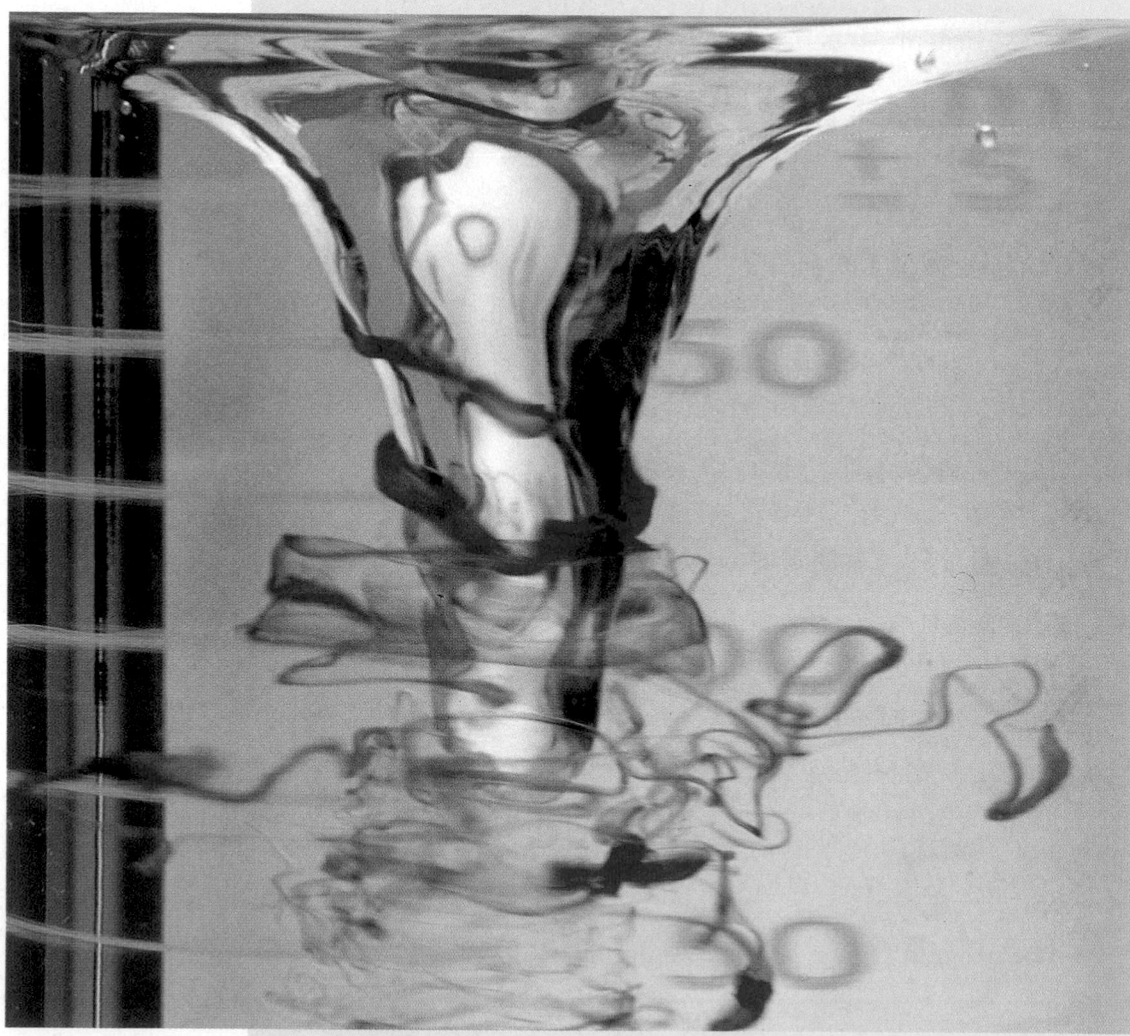

Acids and bases react with indicators to form highly colored substances.

Chapter Outline

Acids and bases are two interrelated types of chemicals that have been known since antiquity. The base that we call sodium hydroxide was manufactured and used to make soap from animal fats by the Phoenicians as early as the second century A.D. Medieval alchemists prepared many acids for the first time and used them during their investigations of metals (Figure 15.1).

Through the years, scientists and physicians have found that acids and bases play critical and complementary roles in the human body. For example, the stomach contains hydrochloric acid, which aids digestion. A liter of stomach fluid can contain as much as 0.1 mol of this acid, an amount that would be fatal anywhere else in the body. What protects the stomach lining from being dissolved by its own contents? Tissues surrounding the stomach secrete hydrogen carbonate ions (HCO_3^-), an important base in the body. These ions react with any acid that finds its way through the stomach's protective mucus coating. Blood in nearby capillaries cannot tolerate the acidity found in the stomach, but a combination of acidic and basic substances called a *buffer* keeps the acidity of the blood at a very low level.

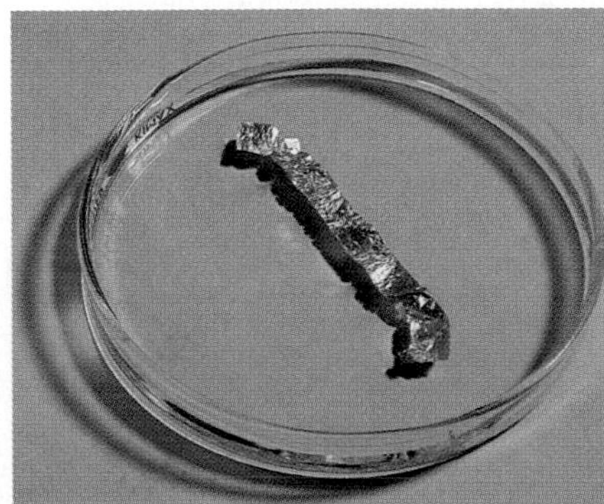
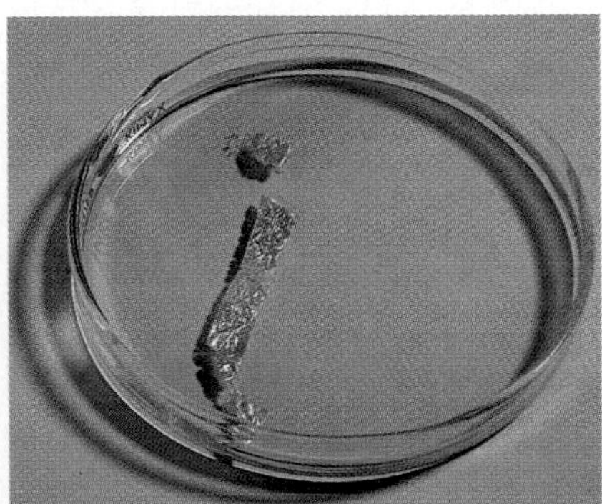
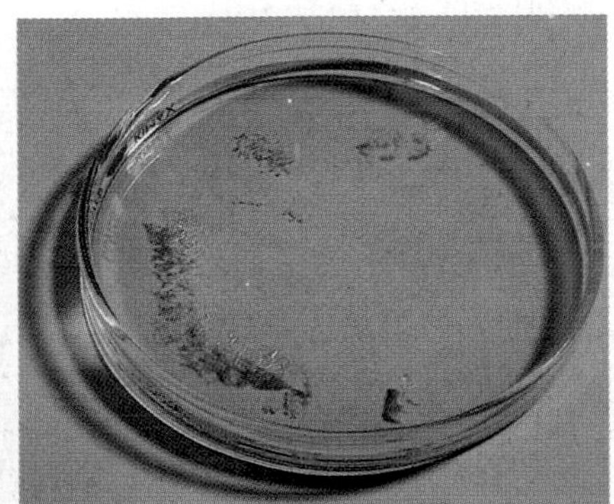

FIGURE 15.1
Acids can cause the toughest metals to disappear. Even gold, a metal so unreactive that it has been called the king of metals, will dissolve in a certain mixture of acids. This mixture still is known as *aqua regia* (royal water). Note that the gold foil disappears as it dissolves.

Many substances besides those in the fluids of the human body are acids or bases. All acids have some common properties, just as all of the bases have their own common properties. Moreover, acids share a common relationship with bases. In this chapter, we examine these properties and this relationship and show the chemical reasons for them. From this foundation, we move on to a discussion of the strengths of acids and bases, and the unique role of water as both an acid and a base. The remainder of the chapter describes techniques for expressing the acidity or basicity of solutions quantitatively.

ACID–BASE REACTIONS AND DEFINITIONS

Acids and bases have many practical applications. Industries throughout the world currently manufacture thousands of tons of them to be used as fertilizers, intermediates for other chemical processes, and household cleaners. As industrial capacity has grown, so too has the use of acids and bases in the home. Many people use antacids (a word coined by advertisers for certain bases) to overcome excess stomach acidity. Others purchase chemicals to control the acidity of water in swimming pools and aquariums. Acids and bases are routinely used for cooking, for drain cleaning, and for countless other purposes. But despite our everyday familiarity with these substances, chemists have not always known how to define exactly what they are.

This part of the chapter begins by looking at some of the physical and chemical properties of acids and bases. We discuss two theories that provide definitions for the terms *acid* and *base* and some explanations for their properties.

15.1 The Arrhenius Theory of Acids and Bases

OBJECTIVES

- List some of the physical properties of acids and bases.
- Use the Arrhenius theory to describe acids and bases.

Early chemists recognized that acids have a sour taste. The acetic acid in vinegar and the citric acid in lemon juice are examples (Figure 15.2). Indeed, the word *acid* is derived from the Latin word *acidus,* meaning "sour." Early chemists also knew that bases taste bitter and feel slippery to the touch, and that acids react with bases to form "salts" (a term that we will define soon). ●

● It is foolish and dangerous to taste an unknown substance to learn if it is sour or bitter. It is also very dangerous to feel an unknown material to see if it is slippery. No one knows how many of the ancient chemists died or were seriously injured after using these testing techniques.

These characteristics describe some of the properties of acids and bases, but they do not relate them (or any other properties) to the molecular composition and structure of acids and bases. As time passed, chemists wanted to know what structural component acids have in common, and what structural component bases have in common. A Swedish chemist, Svante Arrhenius (1859–1927), proposed the first answers to these questions. He defined an **acid** as *a substance that produces H^+ ions (protons) when it is dissolved in water,* and a **base** as *a substance that produces OH^- ions when it is dissolved in water.* These definitions should be familiar because we used them to identify acids and bases in Chapters 5 and 6. Substances that are Arrhenius acids in aqueous solutions include HCl, HNO_3, and H_2SO_4. Although each one of them is a molecular substance, they produce H^+ ions when they are dissolved in water. This process is called either *ionization* or *dissociation.* For example, the chemical equation that describes what happens when HNO_3 dissolves in water and ionizes is

$$HNO_3(l) \xrightarrow{H_2O} H^+(aq) + NO_3^-(aq)$$

Because of this common behavior, aqueous solutions of these substances are called hydrochloric acid, nitric acid, and sulfuric acid, respectively. ● On the other hand, Arrhenius bases include the ionic sodium hydroxide and calcium hydroxide, NaOH and $Ca(OH)_2$. Both of these substances produce OH^- ions when they are dissolved in water. For example, when solid $Ca(OH)_2$ dissolves in water, the reaction is described by

● These substances have other names when they are not dissolved in water. For example, HCl, a gas under ordinary conditions, is called hydrogen chloride.

$$Ca(OH)_2(s) \xrightarrow{H_2O} Ca^{2+}(aq) + 2OH^-(aq)$$

Arrhenius theory indicates that the physical properties of acids are due to H^+ ions and that the physical properties of bases arise because of OH^- ions. Arrhenius theory also provides an explanation of what happens when an acid reacts with a base, as we show in the next section.

Animation: Neutralization of a strong acid by a strong base.

15.2 Neutralization, Salts, and Net Ionic Equations

OBJECTIVES

- Describe neutralization of a base with an acid.
- Define a salt.
- Define a molecular equation, a total ionic equation, and a net ionic equation.
- Given a neutralization reaction, write the net ionic equation.

What happens when an acid "neutralizes" a base? Neutralization, a double-replacement reaction that you encountered in Chapter 6, leads to the formation of a salt. A **salt** is *an ionic compound containing the cation from a base and an anion from the acid.* Water is often another product. A typical neutralization occurs when hydrochloric acid reacts with aqueous sodium hydroxide according to the chemical equation,

$$HCl(aq) + NaOH(aq) \longrightarrow NaCl(aq) + H_2O(l)$$

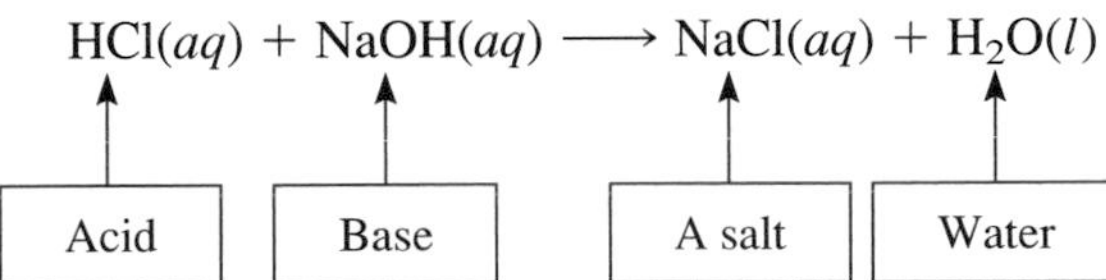

This is a **molecular equation,** *an equation in which each substance is written as if it were a molecular substance, even though it may actually exist as ions.* We can write this equation somewhat differently, however. When we write $HCl(aq)$, we really mean $H^+(aq)$ and $Cl^-(aq)$ because the Arrhenius theory tells us that these ions are formed when hydrogen chloride dissolves in water. Similarly, $NaOH(aq)$ represents $Na^+(aq)$ and $OH^-(aq)$, and $NaCl(aq)$ represents $Na^+(aq)$ and $Cl^-(aq)$. In contrast, H_2O is a molecular compound that must be written in molecular form. With this understanding of each reactant and product, we can replace the molecular equation with a **total ionic equation,** *an equation that shows all of the ions in solution.*

$$H^+(aq) + Cl^-(aq) + Na^+(aq) + OH^-(aq) \longrightarrow Na^+(aq) + Cl^-(aq) + H_2O(l)$$

Notice that the formation of water removes two of the four ions, leaving $Na^+(aq)$ and $Cl^-(aq)$. They are present in the solution before and after the reaction. These ions are **spectator ions,** *ions that do not take part in the reaction.* Since they have no role in the reaction, we can remove these ions from the total ionic equation.

$$H^+(aq) + \cancel{Cl^-(aq)} + \cancel{Na^+(aq)} + OH^-(aq) \longrightarrow \cancel{Na^+(aq)} + \cancel{Cl^-(aq)} + H_2O(l)$$

FIGURE 15.2
Vinegar and lemon juice are acidic. The sour tastes of vinegar and many citrus fruits are due to acids.

The result is a **net ionic equation,** *an equation that shows only the ions that take part in the reaction.*

$$H^+(aq) + OH^-(aq) \longrightarrow H_2O(l)$$

Because many neutralization reactions are described by this net ionic equation, *the same essential reaction occurs* in all of them. For example, this equation also characterizes the reaction between nitric acid and potassium hydroxide, as shown in Example 15.1.

EXAMPLE 15.1 Writing a Net Ionic Equation

Write a net ionic equation for the reaction described by the following molecular equation.

$$HNO_3(aq) + KOH(aq) \longrightarrow KNO_3(aq) + H_2O(l)$$

Problem Analysis

First, derive the total ionic equation by writing the formulas for the acid, the base, and the salt as separate ions. Water is a molecular substance, however, so you must write it in molecular form. Next, identify the spectator ions. After eliminating these ions, you are left with the net ionic equation.

Solution

Writing the formulas for the acid, the base, and the salt as separate ions, you obtain the total ionic equation.

$$H^+(aq) + NO_3^-(aq) + K^+(aq) + OH^-(aq) \longrightarrow K^+(aq) + NO_3^-(aq) + H_2O(l)$$

Potassium and nitrate ions are spectator ions because they have not changed—they appear on both sides of the equation. They can be eliminated.

$$H^+(aq) + \cancel{NO_3^-(aq)} + \cancel{K^+(aq)} + OH^-(aq) \longrightarrow \cancel{K^+(aq)} + \cancel{NO_3^-(aq)} + H_2O(l)$$

The ions that undergo a chemical change are $H^+(aq)$ and $OH^-(aq)$. They react to form water. Thus the net ionic equation is

$$H^+(aq) + OH^-(aq) \longrightarrow H_2O(l)$$

This equation is identical to the one obtained from the reaction of HCl with NaOH because the same essential reaction is occurring.

Exercise 15.1

When barium hydroxide neutralizes hydrochloric acid, the chemical equation is

$$Ba(OH)_2(aq) + 2HCl(aq) \longrightarrow BaCl_2(aq) + 2H_2O(l)$$

Write a total ionic equation and a net ionic equation for this reaction.

(Try Problems 15.29 and 15.30.)

Chemists often write a net ionic equation for *any* reaction involving ionic substances in an aqueous solution, including single-replacement reactions and double-replacement reactions other than neutralizations, in order to show the essential reactants and products. The procedure for obtaining the net ionic equation for any of these reactions is the same as the one we have used for neutralization reactions.

15.3 The Brønsted–Lowry Theory

OBJECTIVES

- Define acids and bases according to the Brønsted–Lowry theory.
- Describe the hydronium ion.
- Define a conjugate acid–base pair.

In the last section, you saw that the Arrhenius theory indicates that spontaneous ionization of an acid occurs when it dissolves in water. A Danish chemist, Johannes Brønsted (1879–1947), and an English chemist, Thomas Lowry (1874–1936), recognized independently that this process is not as simple as Arrhenius pictured it because a chemical reaction occurs. According to the Brønsted–Lowry theory, the spontaneous ionization of an acid in water is really an example of acid–base behavior. The theory states that an **acid** is *a proton (H^+) donor* and a **base** is *a proton acceptor.* ● When an acid dissolves, it transfers a proton to water, which functions as a base. Consider, for example, what happens when hydrofluoric acid dissolves in water.

● Chemists have invented other acid–base theories but the Brønsted–Lowry theory may be the most useful.

$$\underset{\text{Acid}}{HF(aq)} + \underset{\text{Base}}{H_2O(l)} \longrightarrow H_3O^+(aq) + F^-(aq)$$

As you can see, a proton is transferred from the acid (HF) to the base (H_2O) during this acid–base reaction. The result is the **hydronium ion** (H_3O^+) in which a covalent bond has been formed between the proton and the water molecule using one of water's lone pairs of electrons.

$$H^+ \quad + \quad :\ddot{O}:H \quad \longrightarrow \quad H:\ddot{O}:H^+$$
$$\qquad\qquad\quad \ddot{}\,H \qquad\qquad\qquad \ddot{}\,H$$

Recall, however, from Chapter 14 that reactions are reversible, so that the transfer of a proton from HF to H_2O must be accompanied by a reverse reaction. The reverse reaction,

$$HF(aq) + H_2O(l) \longleftarrow \underset{\text{Acid}}{H_3O^+(aq)} + \underset{\text{Base}}{F^-(aq)}$$

is also an acid–base reaction because a proton is transferred from an acid (H_3O^+) to a base (F^-). We emphasize that both reactions are important by using double arrows ($\rightleftharpoons$) in the equation. ●

● We saw in Chapter 14 that all reactions are reversible, even after they appear to stop, as long as one of the products is not removed.

$$\underset{\text{Acid}}{HF(aq)} + \underset{\text{Base}}{H_2O(l)} \rightleftharpoons \underset{\text{Acid}}{H_3O^+(aq)} + \underset{\text{Base}}{F^-(aq)}$$

As you can see, an acid and a base occur on each side of the chemical equation. According to the Brønsted–Lowry theory, any acid–base reaction involves a proton transfer between an acid and a base in the forward direction and another proton transfer between an acid and a base in the reverse direction. A general equation that describes what happens when an acid dissolves in water becomes

$$\underset{\text{Acid}}{HA(aq)} + \underset{\text{Base}}{H_2O(l)} \rightleftharpoons \underset{\text{Acid}}{H_3O^+(aq)} + \underset{\text{Base}}{A^-(aq)}$$

There are two pairs of substances, HA and A^- and H_2O and H_3O^+, in this equation that differ by a proton. HA becomes A^- when it loses a proton, whereas A^- becomes HA when it gains a proton. A similar statement applies to H_3O^+ and H_2O. Each of these pairs is called a *conjugate acid–base pair.* A **conjugate acid–base pair** consists of *two substances (one an acid and the other a base) in an acid–base reaction that differ by gain or loss of a proton.* The acid in such a pair is called the *conjugate acid* of the base, whereas the base is called the *conjugate base* of the acid. Thus, H_3O^+ is the conjugate acid of H_2O, and H_2O is the conjugate base of H_3O^+. The special relationship between the members of each pair of substances is emphasized when we show it in the chemical equation.

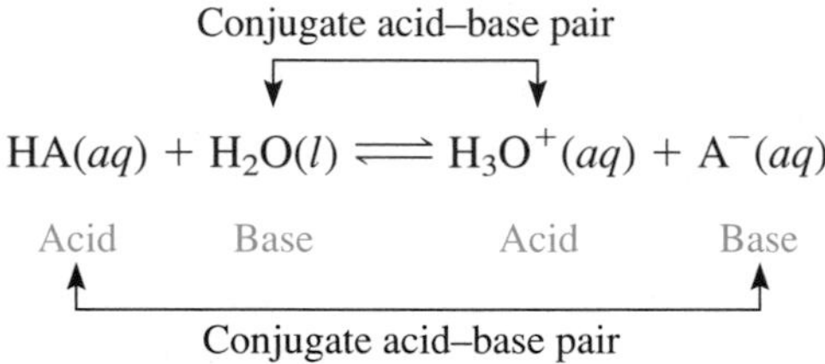

In the previous examples, water has been a base. However, water can also be an acid. Consider, for example, the reaction of ammonia with water.

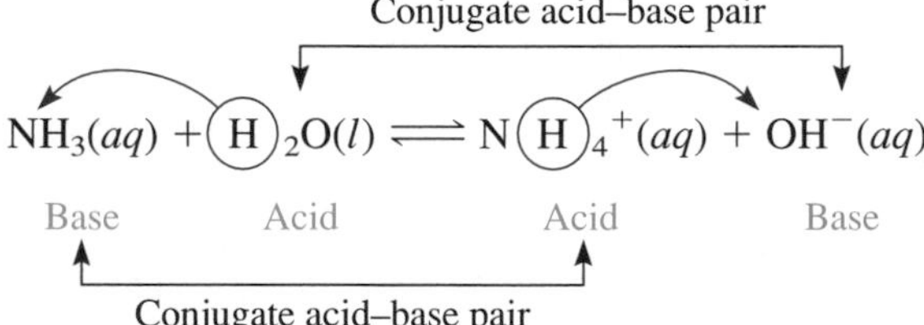

In this reaction, ammonia is a base because it accepts a proton from water. In the reverse reaction, the ammonium ion (NH_4^+) is the acid and the hydroxide ion is the base. The conjugate acid–base pairs are shown above and below the equation. You will get more practice with conjugate acid–base pairs by working the following examples.

EXAMPLE 15.2 Identifying Members of a Conjugate Acid–Base Pair

The hydrogen sulfate ion (HSO_4^-) can be a component of acid rain. What is this ion's conjugate base?

Problem Analysis

Remember that the substances in a conjugate acid–base pair differ by the gain or loss of a proton. The acid has the proton and the base does not.

Solution

Remove an H^+ ion from HSO_4^- to get the conjugate base. Therefore, SO_4^{2-} is the conjugate base of HSO_4^-.

Exercise 15.2

Give the conjugate acid of OH^-.

(Try Problems 15.33, 15.34, 15.35, and 15.36.)

■ EXAMPLE 15.3 Identifying Acids and Bases

The hydrogen carbonate ion (HCO_3^-), a component of human blood, reacts with acids and bases. Identify the acids and bases on both sides of the following equation and show the conjugate acid–base pairs.

$$HCO_3^-(aq) + OH^-(aq) \rightleftharpoons CO_3^{2-}(aq) + H_2O(l)$$

Problem Analysis

Identify the proton donors (acids) on each side of the equation, and then identify the corresponding proton acceptors (bases). Remember that the substances in a conjugate acid–base pair differ by the gain or loss of a proton.

Solution

Begin by examining the equation to determine the H^+ donor on each side. On the left, HCO_3^- is the donor, and on the right, water is the donor. Thus, these substances are the acids. The acceptors are OH^- and CO_3^{2-}; these are the bases. Now label the substances in the equation.

$$\underset{\text{Acid}}{HCO_3^-(aq)} + \underset{\text{Base}}{OH^-(aq)} \rightleftharpoons \underset{\text{Base}}{CO_3^{2-}(aq)} + \underset{\text{Acid}}{H_2O(l)}$$

Next, identify the pairs of substances that differ by the gain or loss of a proton.

Conjugate acid–base pair

$$\underset{\text{Acid}}{HCO_3^-(aq)} + \underset{\text{Base}}{OH^-(aq)} \rightleftharpoons \underset{\text{Base}}{CO_3^{2-}(aq)} + \underset{\text{Acid}}{H_2O(l)}$$

Conjugate acid–base pair

Thus, HCO_3^- and CO_3^{2-} are a conjugate acid–base pair, as are OH^- and H_2O.

Exercise 15.3

For the reaction

$$H_2CO_3(aq) + CN^-(aq) \rightleftharpoons HCN(aq) + HCO_3^-(aq)$$

label each substance as an acid or a base and show the conjugate acid–base pairs.

(Try Problems 15.37, 15.38, 15.39, and 15.40.)

Chemical Perspective

Acids, Bases, Baking Powder, and Dough

Although bread is often leavened with yeast, many cakes and cookies use baking powder. The term *leavened* refers to the addition of a chemical that makes the dough rise while it is baked. Baking powder is a mixture of $NaHCO_3$ (called sodium hydrogen carbonate by chemists, and baking soda by cooks and bakers), acidic salts such as $Ca(H_2PO_4)_2$ (calcium dihydrogen phosphate but often identified as calcium acid phosphate on food labels) and $KHC_4H_4O_6$ (potassium hydrogen tartrate but usually called cream of tartar on labels), and inert materials such as starch or flour to slow down chemical reactions. As the baking powder in this mixture dissolves in the water or milk the baker has added to make the dough, hydrogen ions coming from the acidic salt react with the baking soda to produce gaseous carbon dioxide. For example, the reaction with $Ca(H_2PO_4)_2$ begins with

$$\underset{\text{Base}}{HCO_3^-} + \underset{\text{Acid}}{H_2PO_4^-} \longrightarrow H_2CO_3 + HPO_4^{2-}$$

However, carbonic acid (H_2CO_3) is unstable, as you saw in Chapter 6; it decomposes to give carbon dioxide and water:

$$H_2CO_3 \longrightarrow CO_2 + H_2O$$

The dough rises because of the expanding gaseous carbon dioxide. The heat in the oven causes the carbon dioxide to escape from the dough, leaving behind a porous, light-textured cake or cookie.

Sodium aluminum sulfate is also used in baking powder, but the source of the acidity is not as obvious as the substances that we have already mentioned. Highly charged metal cations, such as Al^{3+} in sodium aluminum sulfate, react with water to produce $Al(OH)^{2+}$ ions and hydronium ions (H_3O^+),

$$Al^{3+} + 2H_2O \longrightarrow Al(OH)^{2+} + H_3O^+$$

The hydronium ions from this reaction react with baking soda to give carbon dioxide. Baking soda alone is another leavening agent, but its use is more limited than the others. Lactic acid from milk reacts with baking soda to release carbon dioxide. The reaction begins immediately, so the baker must work fast.

Although the action of yeast with sugar also produces carbon dioxide, Brønsted–Lowry acids and bases are not involved. Yeast is used often for bread dough.

Exploration: Weak and strong acids.

15.4 The Relative Strengths of Acids and Bases

OBJECTIVES

- Compare and contrast strong acids and weak acids in terms of their ability to transfer a proton to water.
- List the six strong acids.
- Compare the hydronium-ion concentrations of solutions containing equivalent quantities of acids but varying acid strengths.
- Describe how the strength of a base is related to the strength of its conjugate acid.

Acid Strengths

Some acids have a greater tendency to transfer their protons to water than others. The reaction that is described by

$$HA(aq) + H_2O(l) \rightleftharpoons H_3O^+(aq) + A^-(aq)$$

proceeds further from left to right with acids that have a stronger tendency to transfer their protons to water. Let's use an experiment to illustrate this idea.

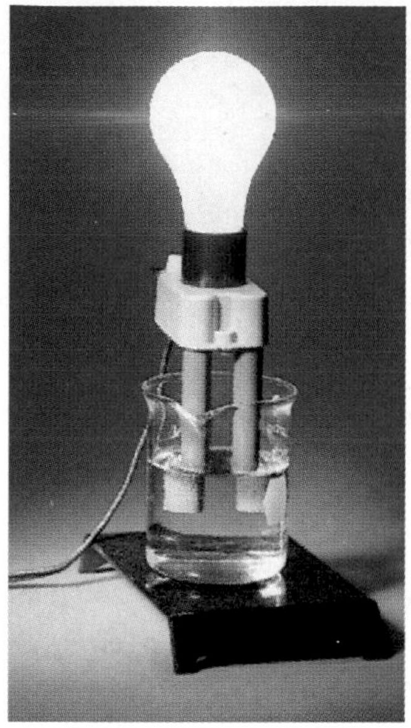

A

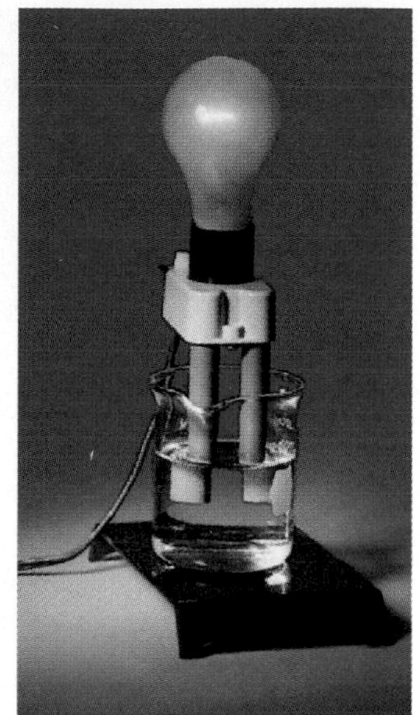

B

FIGURE 15.3
Conductivity of equimolar hydrochloric acid and acetic acid solutions. The light is brighter with the HCl solution *(A)* because more ions are in solution. This experiment shows that HCl must have a greater tendency to transfer protons to the base H_2O than does acetic acid *(B)*.

Aqueous solutions containing ions from ionic substances conduct electricity, as you saw in Chapter 4, because ions move through the solution, carrying the electric charge with them. Solutions that contain a lot of ions conduct electricity better than solutions that contain fewer ions. Let's compare the electrical conductivity of two solutions: one containing 0.1 *M* HCl and the other containing 0.1 *M* $HC_2H_3O_2$ (acetic acid). Wires from a source of electricity are connected to a light bulb, and the current through the bulb is also sent through each solution, as shown in Figure 15.3. If an acid readily dissociates in aqueous solution, transferring protons to water to produce ions, the solution will conduct electricity well, and the bulb will be brightly lit. If the acid has less tendency to dissociate and transfer protons, fewer ions will be produced and the bulb will be only dimly lit. The relevant equations are

$$HCl(aq) + H_2O(l) \rightleftharpoons H_3O^+(aq) + Cl^-(aq)$$
$$HC_2H_3O_2(aq) + H_2O(l) \rightleftharpoons H_3O^+(aq) + C_2H_3O_2^-(aq)$$

The number of ions and the intensity of the light will depend on how far these reactions proceed from left to right. The results are shown in Figure 15.3. The solution containing HCl is the better conductor, indicating that this equilibrium proceeds from left to right to a greater degree than the one involving acetic acid. Thus, aqueous HCl must have a greater tendency to transfer protons to the base (H_2O) than does aqueous acetic acid.

Imagine comparing many acids in this way. Because the base (water) remains the same no matter which acid is used, it is the nature of the acid and its ability to transfer a proton to water that determines *how far the equilibrium shifts from left to right* and how brightly the bulb is lit. Thus we can determine the **acid strength** of a series of acids. When the acid's ability to transfer a proton to water is very great, we say that the acid is *strong*. In contrast, a *weak* acid has a poor ability to transfer a proton to water. We only need to consider these two extremes because acids of intermediate strength are rare, and we can ignore them.

Six strong acids are listed in Table 15.1. You should memorize this list. The first five of these acids are **monoprotic acids,** *acids that are capable of transferring only one proton to a base*. With each of these monoprotic acids, the equilibrium proceeds so far

Acids		Bases*	
Formula	Name	Formula	Name
HCl	Hydrochloric acid	LiOH	Lithium hydroxide
HBr	Hydrobromic acid	NaOH	Sodium hydroxide
HI	Hydroiodic acid	$Mg(OH)_2$	Magnesium hydroxide
HNO_3	Nitric acid	$Ca(OH)_2$	Calcium hydroxide
$HClO_4$	Perchloric acid	$Sr(OH)_2$	Strontium hydroxide
H_2SO_4	Sulfuric acid	$Ba(OH)_2$	Barium hydroxide

*The strong bases are the Group IA and IIA metal hydroxides with the exception of beryllium hydroxide.

TABLE 15.1
The Six Strong Acids and Some of the Strong Bases

to the right that the reaction is essentially complete. Consider HCl, one of the strong acids, for example. Sometimes we replace the double arrow in the equation

$$HCl(aq) + H_2O(l) \rightleftharpoons H_3O^+(aq) + Cl^-(aq)$$

with a single arrow

$$HCl(aq) + H_2O(l) \longrightarrow H_3O^+(aq) + Cl^-(aq)$$

to emphasize that there are virtually no molecules of HCl in a hydrochloric acid solution. ● To a good approximation, all of the hydrogen in HCl has been transferred as protons to water, the base. A similar reaction can be written for each of the other four monoprotic acids. Thus, a **strong acid**—such as hydrochloric acid—is *an acid that is completely dissociated in water.*

● Although we have replaced $\rightleftharpoons$ with $\longrightarrow$, remember that all reactions, including acid–base reactions, are reversible.

$$HCl(aq) + H_2O(l) \xrightarrow{100\%} H_3O^+(aq) + Cl^-(aq)$$

Sulfuric acid, the last acid listed in Table 15.1, is a **diprotic acid,** *an acid capable of transferring two protons.* ● The dissociation of the first proton from this strong acid is complete.

● *Polyprotic* is the general term for an acid capable of transferring more than one proton.

$$H_2SO_4(aq) + H_2O(l) \xrightarrow{100\%} H_3O^+(aq) + HSO_4^-(aq)$$

However, the dissociation of the second proton is not complete so we write the chemical equation with double arrows.

$$HSO_4^-(aq) + H_2O(l) \rightleftharpoons H_3O^+(aq) + SO_4^{2-}(aq)$$

A weak acid, as we have said, is not able to transfer a proton to water as readily as a strong acid, so a **weak acid** is *an acid that is not completely dissociated in water.* We do not present a complete list of the weak acids because they are far too numerous. However, *any acid that is not one of the six strong acids is a weak acid.*

Table 15.2 compares the strengths of several acids. The first six entries are the strong acids, which have identical strengths in water. After the strong acids, the remainder are arranged according to decreasing strength. Notice that water is the last entry on the list. As you will see in the next section, a water molecule functions as a weak acid because it has a small tendency to transfer a proton to another water molecule, which acts as a base. Thus *water is both an acid and a base.*

Let's compare the acid strengths of HCO_3^-, HCl, and HF using the table. Remember that the six strong acids are at the top of the table and that acid strengths decrease as you pass down the table. Since HCl is a strong acid and the other two are not, HCl is the

TABLE 15.2
Strengths of Some Acids and Their Conjugate Bases

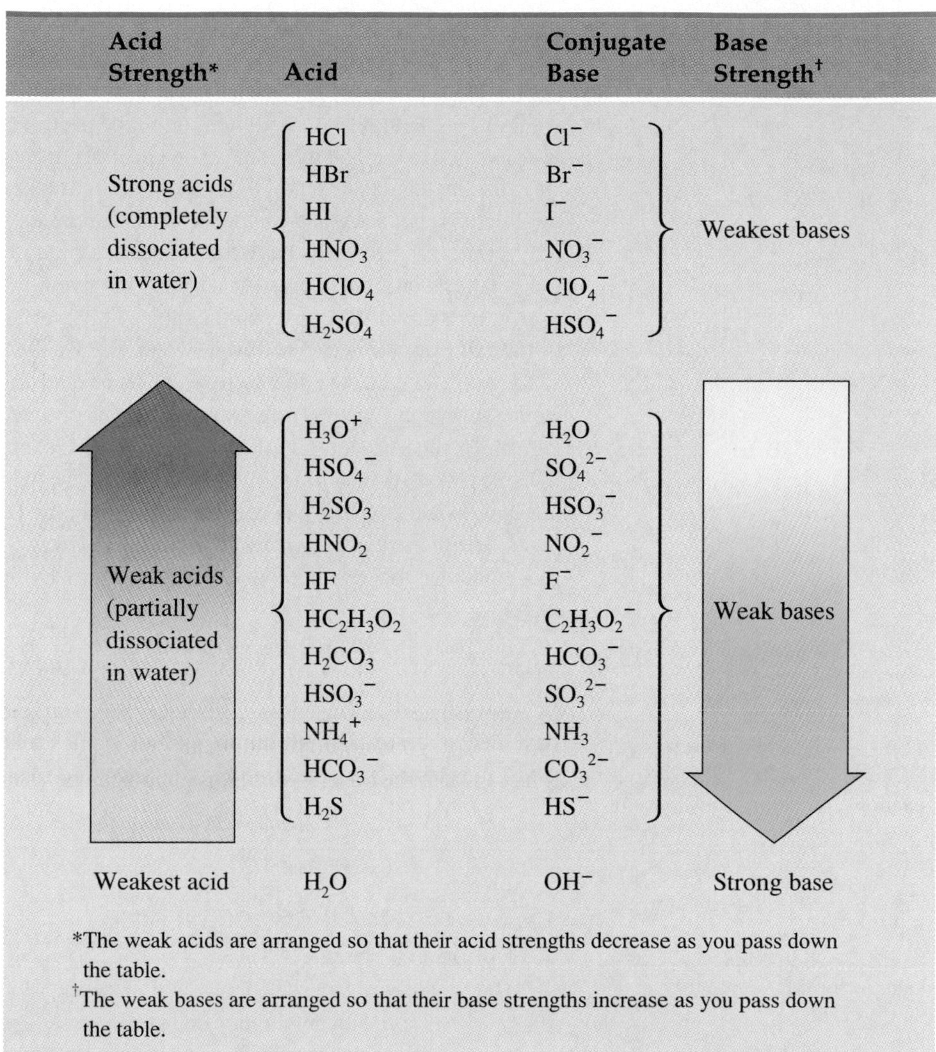

Acid Strength*	Acid	Conjugate Base	Base Strength†
Strong acids (completely dissociated in water)	HCl	Cl^-	Weakest bases
	HBr	Br^-	
	HI	I^-	
	HNO_3	NO_3^-	
	$HClO_4$	ClO_4^-	
	H_2SO_4	HSO_4^-	
Weak acids (partially dissociated in water)	H_3O^+	H_2O	Weak bases
	HSO_4^-	SO_4^{2-}	
	H_2SO_3	HSO_3^-	
	HNO_2	NO_2^-	
	HF	F^-	
	$HC_2H_3O_2$	$C_2H_3O_2^-$	
	H_2CO_3	HCO_3^-	
	HSO_3^-	SO_3^{2-}	
	NH_4^+	NH_3	
	HCO_3^-	CO_3^{2-}	
	H_2S	HS^-	
Weakest acid	H_2O	OH^-	Strong base

*The weak acids are arranged so that their acid strengths decrease as you pass down the table.

†The weak bases are arranged so that their base strengths increase as you pass down the table.

strongest acid. Next, you can see that HF is above HCO_3^- in the table, so HF is a stronger acid than HCO_3^-. Thus, the order of acid strength is given by HCl > HF > HCO_3^-. Consequently, a 1 *M* solution of HCl contains more H_3O^+ ions than a 1 *M* solution of HF, which in turn contains more of these ions than a 1 *M* solution of HCO_3^-.

Base Strengths

Table 15.2 also includes the conjugate base of each acid. Whereas the strengths of acids decrease passing down the table, the strengths of the conjugate bases increase. Thus the strength of a conjugate base is related to the strength of its acid. Let's explore the reason for this relationship by examining the following equation once again:

$$\underset{\text{Acid}}{HA(aq)} + \underset{\text{Base}}{H_2O(l)} \rightleftharpoons \underset{\text{Acid}}{H_3O^+(aq)} + \underset{\text{Base}}{A^-(aq)}$$

Each side of this equation represents an acid–base reaction—one between HA and H_2O and the other between H_3O^+ and A^-. These opposing reactions are competing for protons. If H_2O is a stronger base than A^-, then water will be able to hold on to the proton more strongly than A^-, so that water is converted to the hydronium ion, H_3O^+. As a result, the equilibrium will favor the right side of the equation, as it does with the strong acids. Thus, a strong acid must have a weak conjugate base. In contrast, if H_2O is a weaker base than A^-, A^- will hold on to the proton more strongly than water, so that A^- is converted to the HA molecule. The equilibrium will favor the left side of the equation. Because molecules of HA will tend to remain intact rather than dissociate, this situation applies to weak acids, so that a weak acid must have a strong conjugate base. In summary, the strongest acids have the weakest conjugate bases and the strongest bases have the weakest conjugate acids. As a result, the strengths of the bases in Table 15.2 increase as you pass down the entries in the table. The hydroxide ion (OH^-) is the strongest base in the table, and it is derived from water, the weakest acid.

If you wanted to use the hydroxide ion as a strong base, you would need to select an ionic compound containing this anion. Of course, these compounds also contain a cation. In general, you could choose from the Group IA and IIA metal hydroxides (with the exception of beryllium hydroxide) because these compounds are ionic. Examples are NaOH and $Ca(OH)_2$. A few more are given in Table 15.1.

A **strong base** is *a hydroxide compound that is totally ionized in water.* As an example, consider the result of dissolving sodium hydroxide, one of the strong bases, in water.

$$\text{NaOH}(s) \xrightarrow{100\%} \text{Na}^+(aq) + \text{OH}^-(aq)$$

In contrast, a **weak base** is *a base that is not totally ionized in water.* Ammonia is a weak base because only a small number of molecules react with water, leading to an equally small number of the ammonium and hydroxide ions.

$$\text{NH}_3(aq) + \text{H}_2\text{O}(l) \rightleftharpoons \text{NH}_4^+(aq) + \text{OH}^-(aq)$$

Animation: Self-ionization of water to form H^+ and OH^- in equilibrium.

15.5 Water: An Acid and a Base

OBJECTIVES

- Describe the self-ionization of water.
- Define an amphoteric substance.
- Write the equilibrium expression for the self-ionization of water, and define the ion-product constant from it.

You saw in Section 15.4 that water can act as an acid and as a base. When an acid dissolves in water, water acts as a base.

$$\underset{\text{Acid}}{\text{HF}(aq)} + \underset{\text{Base}}{\text{H}_2\text{O}(l)} \rightleftharpoons \text{H}_3\text{O}^+(aq) + \text{F}^-(aq)$$

When a base dissolves in water, however, water acts like an acid.

$$\underset{\text{Base}}{\text{NH}_3(aq)} + \underset{\text{Acid}}{\text{H}_2\text{O}(l)} \rightleftharpoons \text{NH}_4^+(aq) + \text{OH}^-(aq)$$

A substance (such as water) *that can behave as both an acid and a base* is called an **amphoteric substance.** The amphoteric nature of water is best seen in its **self-ionization,** *a process in which two identical molecules react to give ions.* A proton from one water molecule is transferred to another water molecule, leaving behind OH^- and forming H_3O^+.

$$\underset{\text{Acid}}{\text{H}_2\text{O}(l)} + \underset{\text{Base}}{\text{H}_2\text{O}(l)} \rightleftharpoons \text{H}_3\text{O}^+(aq) + \text{OH}^-(aq)$$

Writing the expression for the equilibrium constant for this reaction, we obtain

$$K_w = [H_3O^+][OH^-]$$

The term $[H_2O]^2$ does not appear in the denominator because the concentration of a pure liquid (or a pure solid) is not included in an equilibrium expression. We call the equilibrium constant K_w the **ion-product constant for water,** *the equilibrium value of the ion product* $[H_3O^+][OH^-]$.

At 25°C, experiments with *pure* water have shown that

$$[H_3O^+] = [OH^-] = 1.0 \times 10^{-7}\ M$$

where M means mol/L, as you saw in Chapter 13. The value of K_w at 25°C is

$$K_w = [H_3O^+]\,[OH^-] = (1.0 \times 10^{-7})(1.0 \times 10^{-7})$$
$$= 1.0 \times 10^{-14}$$

Although the concentrations of H_3O^+ and OH^- ions are equal in pure water, they will become unequal if we add an acid or a base. Nevertheless, the equilibrium expression

$$K_w = [H_3O^+][OH^-]$$

will still be valid and K_w will still equal 1.0×10^{-14} as long as the temperature is 25°C. We use this equation with both acidic and basic solutions in the next section.

QUANTIFYING ACID AND BASE SOLUTIONS

What does it mean when we say that vinegar and lemon juice are acidic? You already know that they taste sour. Given the right chemicals, you could show that the acids in vinegar and lemon juice transfer protons to bases, thus meeting the Brønsted–Lowry definition. But how acidic are vinegar and lemon juice? Is one solution more acidic than the other? The acidity of a solution depends on how many protons per liter are available for transfer to a base. That number, in turn, depends on the concentration (in mol/L) of each acid in an aqueous solution, and on the strength of each acid (the degree to which it dissociates to form ions). To measure and compare the acidity of two solutions, you need a measure of the proton concentration or, equivalently, the hydronium-ion concentration. Chemists have devised a scale, the pH scale, that quantitatively measures the hydronium-ion concentration—and thus the acidity of a solution—in conveniently sized numbers. On this scale, the lower the pH, the more acidic the solution. Lemon juice, with a pH of 2.3, is typically more acidic than household vinegar, with a pH of about 3.0. The use of pH, discussed in the remainder of this chapter, has many applications in everyday life.

Video: Ammonia fountain.

15.6 Acidic, Neutral, and Basic Solutions

OBJECTIVES

- Show how the equilibrium concentrations of $[H_3O^+]$ and $[OH^-]$ shift when an acid or a base is added to pure water.
- Given the hydronium-ion concentration of a solution, calculate the hydroxide-ion concentration, and vice versa.
- Define acidic, neutral, and basic solutions in terms of the concentration of hydronium ions in mol/L.

Remember from the last section that

$$K_w = [H_3O^+][OH^-]$$

and that K_w will always equal 1.0×10^{-14} as long as the temperature is 25°C. But what happens when an acid is added to pure water? Clearly, more H_3O^+ ions will form, so that $[H_3O^+]$ increases. At the same time, $[OH^-]$ must decrease so that the product $[H_3O^+][OH^-]$ will remain constant. Thus, if the addition of an acid causes $[H_3O^+]$ to increase from 1.0×10^{-7} *M* (the value of $[H_3O^+]$ in pure water at 25°C) to 1.0×10^{-3} *M*, then $[OH^-]$ must decrease simultaneously to 1.0×10^{-11} *M* so that the product $[H_3O^+][OH^-]$ is still equal to 1.0×10^{-14}.

$$K_w = [H_3O^+][OH^-] = (1.0 \times 10^{-3})(1.0 \times 10^{-11}) = 1.0 \times 10^{-14}$$

Similarly, the addition of a base to water will cause $[OH^-]$ to increase, causing $[H_3O^+]$ to decrease. However, the ion-product constant K_w and $[H_3O^+][OH^-]$ will still be equal to 1.0×10^{-14} (at 25°C).

The next example shows you how to calculate $[OH^-]$ if you know $[H_3O^+]$, and $[H_3O^+]$ if you know $[OH^-]$.

■ EXAMPLE 15.4 Calculating $[OH^-]$ from $[H_3O^+]$ and $[H_3O^+]$ from $[OH^-]$ in Solutions of Acids and Bases

Calculate (a) the concentration of hydroxide ions at 25°C in a drop of vinegar that has $[H_3O^+] = 3.3 \times 10^{-3}$ *M* and (b) the concentration of hydronium ions at 25°C in an egg white with $[OH^-] = 2.0 \times 10^{-7}$ *M*.

Problem Analysis

Remember that $K_w = [H_3O^+][OH^-] = 1.0 \times 10^{-14}$ at 25°C. If you know either $[H_3O^+]$ or $[OH^-]$, you can obtain $[OH^-]$ or $[H_3O^+]$ from the equilibrium expression.

Solution

(a) With $[H_3O^+] = 3.3 \times 10^{-3}$ *M*, obtain the concentration of hydroxide ions from the expression for K_w.

$$K_w = [H_3O^+][OH^-]$$
$$1.0 \times 10^{-14} = (3.3 \times 10^{-3}) \times [OH^-]$$

After rearranging the equation, you obtain

$$[OH^-] = \frac{1.0 \times 10^{-14}}{3.3 \times 10^{-3}} = 3.0 \times 10^{-12}\ M$$

(b) With $[OH^-] = 2.0 \times 10^{-7}$ *M*, calculate the concentration of hydronium ions from the expression for K_w.

$$K_w = [H_3O^+][OH^-]$$
$$1.0 \times 10^{-14} = [H_3O^+] \times (2.0 \times 10^{-7})$$

Rearranging terms, you find

$$[H_3O^+] = \frac{1.0 \times 10^{-14}}{2.0 \times 10^{-7}} = 5.0 \times 10^{-8}\ M$$

Exercise 15.4

A sample of impure water has a hydronium-ion concentration of 8.4×10^{-5} *M* at 25°C. What is the concentration of hydroxide ions?

(Try Problems 15.41, 15.42, 15.43, and 15.44.)

Next, we turn to three important definitions for any solution.

1. When the concentrations of H_3O^+ ions and OH^- ions in a solution are identical to those found in pure water (both are 1.0×10^{-7} *M* at 25°C), *the solution is neutral.*
2. In an *acidic solution,* the concentration of H^+ ions is greater than that of OH^- ions.
3. In a *basic solution,* the concentration of OH^- ions is greater than that of H_3O^+.

At 25°C, the following definitions apply:

In an acidic solution, $[H_3O^+] > 1.0 \times 10^{-7}\ M$
In a neutral solution, $[H_3O^+] = 1.0 \times 10^{-7}\ M$
In a basic solution, $[H_3O^+] < 1.0 \times 10^{-7}\ M$

EXAMPLE 15.5 Identifying Acidic, Basic, or Neutral Conditions from $[H_3O^+]$

Vinegar has a hydroxide-ion concentration of 3.0×10^{-12} *M* at 25°C. Is the solution acidic, basic, or neutral?

Problem Analysis

Calculate $[H_3O^+]$ from $K_w = [H_3O^+][OH^-]$. Then, remember that

In an acidic solution, $[H_3O^+] > 1.0 \times 10^{-7}\ M$
In a neutral solution, $[H_3O^+] = 1.0 \times 10^{-7}\ M$
In a basic solution, $[H_3O^+] < 1.0 \times 10^{-7}\ M$

Solution

Begin by calculating $[H_3O^+]$.

$$K_w = [H_3O^+]\,[OH^-]$$
$$1.0 \times 10^{-14} = [H_3O^+] \times (3.0 \times 10^{-12})$$
$$[H_3O^+] = \frac{1.0 \times 10^{-14}}{3.0 \times 10^{-12}} = 3.3 \times 10^{-3}\ M$$

Since 3.3×10^{-3} *M* is greater than 1.0×10^{-7} *M,* you see that the solution is acidic.

Exercise 15.5

If a solution has $[H_3O^+] = 4.7 \times 10^{-5}$ *M,* is the solution acidic, basic, or neutral?

(Try Problems 15.45, 15.46, 15.47, and 15.48.)

15.7 pH

OBJECTIVES

- Define pH.
- Calculate pH from $[H_3O^+]$ and $[H_3O^+]$ from pH.
- Define acidic, neutral, and basic solutions in terms of pH.

As you can see, the numbers we encounter for $[H_3O^+]$ and $[OH^-]$ are often very small, involving negative powers of ten. Since such numbers are unwieldy, chemists invented pH to provide a more concise and convenient way of expressing the hydronium-ion concentration. We define **pH** as *a unitless number obtained from the negative of the logarithm of the hydronium-ion concentration.* ●

● Chemists often use the shorthand notation, $pH = -\log[H^+]$.

$$pH = -\log [H_3O^+]$$

A **logarithm of a number** is *the power to which 10 must be raised to equal that number.* For example, 100 can also be written as 10^2, so the logarithm of 100 is 2. Similarly, 0.01 can also be written as 10^{-2}, and the logarithm of 0.01 is -2. These examples give us another way to think about the relationship between $[H_3O^+]$ and pH,

$$[H_3O^+] = 10^{-pH}$$

The logarithms of many other numbers are not so obvious. For example, the logarithm of 2 is 0.301029995 because 2 can also be written as $10^{0.301029995}$. We will rely on calculators to do operations like this one. Suppose you want to know the pH of a solution in which $[H_3O^+] = 5.4 \times 10^{-3}$ *M*. The pH can be obtained from

$$pH = -\log (5.4 \times 10^{-3})$$

On most scientific calculators, you would do the following operations. (Refer to your calculator's instruction manual for complete directions.)

Operation	Display
1. Press [5], [.], [4]	5.4
2. Press [EXP] (or [EE])	5.4 00
3. Press [3]	5.4 03
4. Press [+/−]	5.4 −03
5. Press [LOG]	−2.26760624
6. Press [+/−]	2.26760624

Although the calculator has given the correct mathematical answer, it does not take significant figures into account. *The number of digits after the decimal point in the pH must equal the number of significant figures in the concentration of H_3O^+ ions.* Because there are two significant figures in this problem's hydronium-ion concentration (5.4×10^{-3} *M*), we will retain two digits to the right of the decimal point. Thus, after rounding, the answer is pH = 2.27. The next example will give you another opportunity to practice.

EXAMPLE 15.6 Calculating pH from $[H_3O^+]$

The optimum pH range of a swimming pool is 7.2 to 7.6 because this range closely approximates the pH of human tears. The swimming pool in Figure 15.4 contains water with $[H_3O^+] = 3.9 \times 10^{-8}$ *M*. What is the pH, and is it within the optimum range?

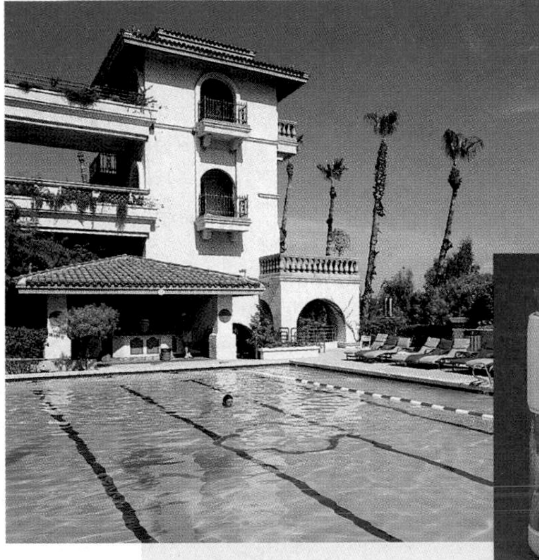

FIGURE 15.4
The pH of this water is important. The optimum pH range of a swimming pool is 7.2 to 7.6 because this range closely approximates the pH of human tears. When the pH is outside this range, swimmers may experience eye irritation. Chemicals such as those shown here are used to maintain the pH in the desired range.

Problem Analysis

Recall that $pH = -\log [H_3O^+]$ and that $[H_3O^+] = 10^{-pH}$.

Solution

Proceed by doing the following operations.

Operation	Display
1. Press [3], [.], [9]	3.9
2. Press [EXP] (or [EE])	3.9 00
3. Press [8]	3.9 08
4. Press [+/−]	3.9 −08
5. Press [LOG]	−7.408935393
6. Press [+/−]	7.408935393

After rounding, you obtain pH = 7.41. This pH is within the optimum range.

Exercise 15.6

What is the pH of a solution in which $[H_3O^+] = 2.2 \times 10^{-11}$ *M*?

(Try Problems 15.49 and 15.50.)

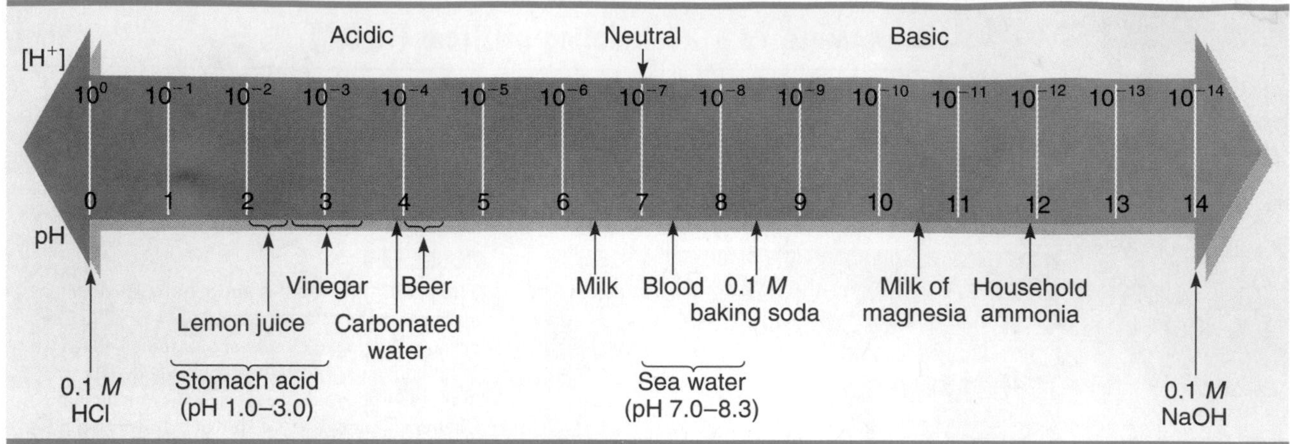

FIGURE 15.5

The pH scale. A solution having a pH of less than 7 is acidic; a solution with a pH of 7 is neutral; and a solution with a pH greater than 7 is basic. The temperature of each solution is 25°C.

The pH scale is shown in Figure 15.5. You can see that a pH of zero corresponds to $[H_3O^+] = 1\ M$ (or $10^0\ M$) and a pH of 14 corresponds to $[H_3O^+] = 10^{-14}\ M$. You need to know the following definitions at 25°C.

In an acidic solution, pH $<$ 7.00
In a neutral solution, pH = 7.00
In a basic solution, pH $>$ 7.00

Figure 15.5 also indicates approximate pH values of some common substances. For example, household ammonia with a pH of 11.9 is a basic solution. The following example gives you the opportunity to use pH to judge if a solution is acidic, neutral, or basic.

■ EXAMPLE 15.7 Calculating pH and Identifying Acidic, Neutral, or Basic Conditions

Some lemon juice has a concentration of hydroxide ions of $2 \times 10^{-12}\ M$ at 25°C. What is the pH? Is the solution acidic, neutral, or basic?

Problem Analysis

Calculate $[H_3O^+]$ from $K_w = [H_3O^+][OH^-]$. Remember that $pH = -\log [H_3O^+]$ and $[H_3O^+] = 10^{-pH}$. Also, remember that a solution at 25°C is acidic if pH $<$ 7.00, neutral if pH = 7.00, and basic if pH $>$ 7.00.

Solution

Proceed by calculating $[H_3O^+]$.

$$K_w = [H_3O^+]\,[OH^-]$$

$$1.0 \times 10^{-14} = [H_3O^+] \times (2 \times 10^{-12})$$

$$[H_3O^+] = \frac{1.0 \times 10^{-14}}{2 \times 10^{-12}} = 5 \times 10^{-3}\ M$$

Next, do the following operations with your calculator.

Operation	Display
1. Press [5]	5
2. Press [EXP] (or [EE])	5 00
3. Press [3]	5 03
4. Press [+/−]	5 −03
5. Press [LOG]	−2.301029996
6. Press [+/−]	2.301029996

After rounding, you obtain pH = 2.3. Only one digit after the decimal point is allowed here. Do you understand why? You can see that the solution is acidic.

Exercise 15.7

The concentration of hydronium ions in a solution is 5×10^{-8} *M*. What is the pH, and is the solution acidic, neutral, or basic?

(Try Problems 15.51, 15.52, 15.53, and 15.54.)

There are also times when you need to convert the pH of a solution to its hydronium-ion concentration. Remember that pH = −log $[H_3O^+]$. We can also write this equation as −pH = log $[H_3O^+]$. Next, we must take the antilog or inverse log of both sides of this equation because the inverse log of a logarithm will "cancel" the logarithm and allow us to solve for $[H_3O^+]$.

$$\text{Inverse log}\ (-\text{pH}) = \text{inverse log}\ (\log\ [H_3O^+])$$

$$\text{Inverse log}\ (-\text{pH}) = [H_3O^+]$$

This means the same thing as $[H_3O^+] = 10^{-\text{pH}}$, an equation we have previously given. We show how to do this type of problem with a calculator in the following example. (Again, refer to your particular calculator's instruction manual because the sequence of pressing keys might be different on your calculator.)

■ EXAMPLE 15.8 Calculating $[H_3O^+]$ from pH

The pH of a sample of arterial blood is 7.40. What is the concentration of H_3O^+ ions? Is the sample acidic, neutral, or basic?

Problem Analysis

Recall that pH = −log $[H_3O^+]$ and that $[H_3O^+] = 10^{-\text{pH}}$ = inverse log (−pH).

Solution

In this case,

$$[H_3O^+] = 10^{-7.40} = \text{inverse log}\ (-7.40).$$

The following operations on your calculator will lead to the answer.

Operation	Display
1. Press [7], [.], [4], [0]	7.40
2. Press [+/−]	−7.40
3. Press [INV]	−7.40
4. Press [LOG]	3.9810717 −08

With some calculators, you may need to replace the third and fourth steps of this operation with

3. Press [2nd] or [2nd FCN]	−7.40
4. Press [10^x]	3.9810717 −08

The result corresponds to 3.9810717×10^{-8}. After rounding to two significant figures, you can write

$$[H_3O^+] = 4.0 \times 10^{-8}\,M$$

The sample of blood is basic, as you can judge from either the pH or $[H_3O^+]$.

Exercise 15.8

If a certain carbonated drink has a pH of 3.16, what is the concentration of H_3O^+ ions? Is the drink acidic, neutral, or basic?

(Try Problems 15.57, 15.58, 15.59, and 15.60.)

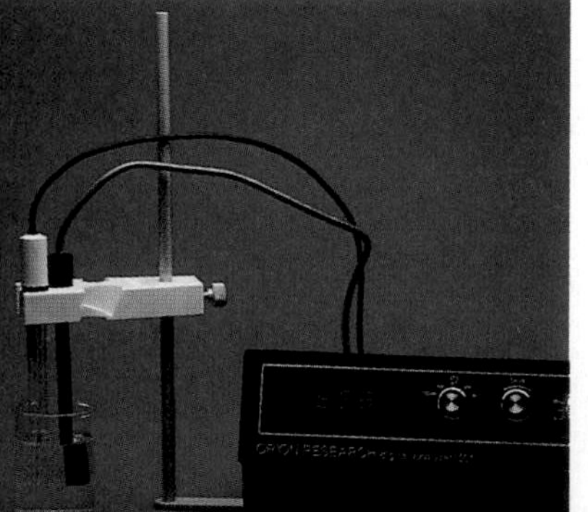

FIGURE 15.6
A pH meter can be used to measure pH. The electrodes are placed in a solution whose pH is to be measured. A voltage that depends on the pH is generated between the electrodes. This voltage is registered on the meter, which has been calibrated to read directly in pH.

15.8 Measuring pH

OBJECTIVE

- Describe two techniques for measuring pH.

A solution's pH can be measured accurately with a pH meter (Figure 15.6). This device has two specially designed electrodes that the user dips in a solution with an unknown pH. (Sometimes both electrodes are housed in the same glass container so that it appears that only one electrode is present.) A voltage that depends on the pH is generated between the electrodes. This voltage is registered on a meter that has been calibrated to read directly in pH. For example, the pH of the solution in Figure 15.6 is 6.58.

Acid–base indicators have also been used to estimate the pH of a solution. An **acid–base indicator** is *a weak acid whose solution will change color within a small pH range and indicate the pH of the solution by the color.* These intensely colored substances have rather exotic names, such as methyl red, phenolphthalein, or alizarin yellow R. In practice, we add a solution of an indicator in water to an aqueous solution with an unknown pH, and estimate the pH of the solution from the resulting color.

Let's see how an indicator works. We represent the acid form of the indicator by HIn and its conjugate base by In^-. The equilibrium between the acid form and the conjugate base is given by the same kind of chemical equation that we find with any acid,

$$HIn(aq) + H_2O(l) \rightleftharpoons H_3O^+(aq) + In^-(aq)$$

FIGURE 15.7

Color changes of some acid–base indicators. The color of the acid form of each indicator is shown on the left, and the color of its conjugate base is shown on the right. Because thymol blue is a diprotic acid, it has two different color changes.

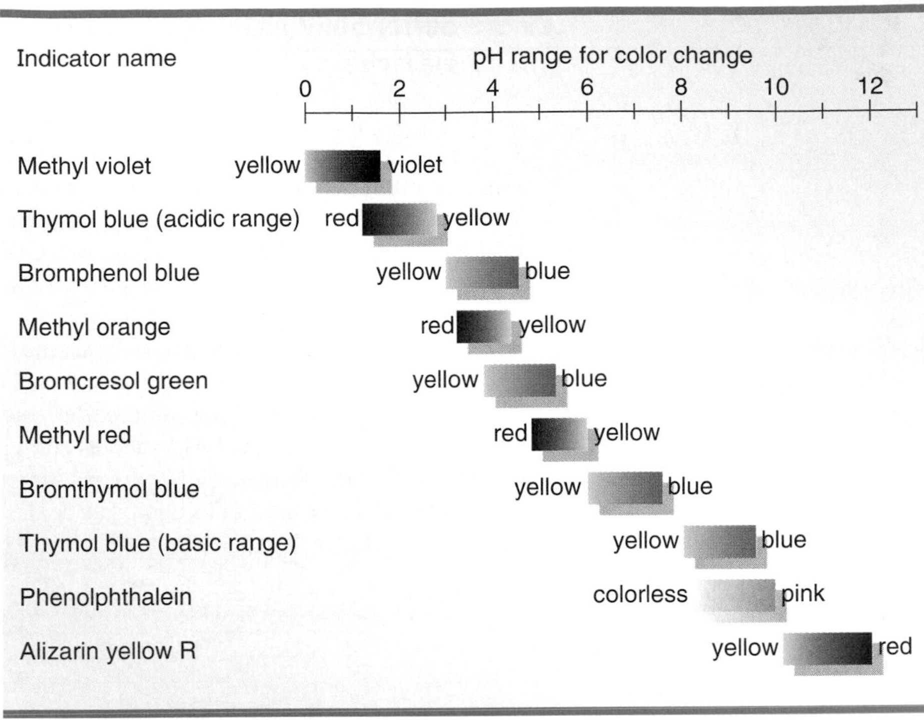

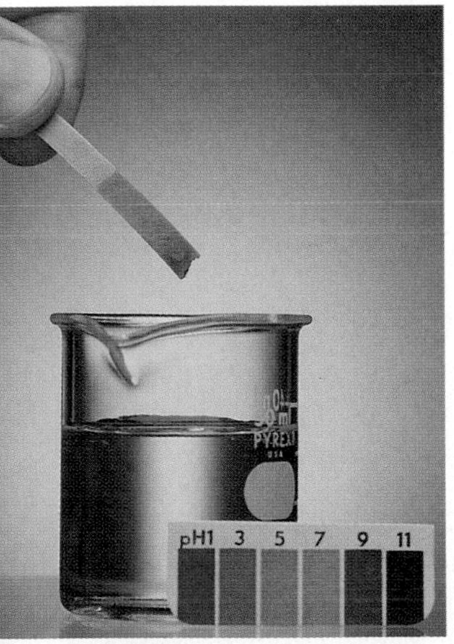

FIGURE 15.8

pH paper. Paper strips that are treated with several different indicators can be used to estimate pH. You can match the color appearing on the paper with a color corresponding to a known pH on a chart.

The indicator is useful because the acid form HIn has a very different color from the color of its conjugate base In^-. With methyl red, for example, the acid form is red and the conjugate base is yellow.

$$HIn(aq) + H_2O(l) \rightleftharpoons H_3O^+(aq) + In^-(aq)$$

HIn: Red with methyl red

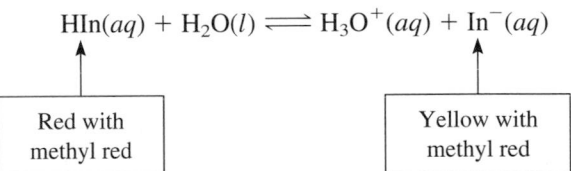

In^-: Yellow with methyl red

The colors of an indicator and its conjugate base are so intense that only a small quantity is needed, too small to influence the pH of the solution whose pH we wish to estimate (the test solution). Instead, the pH of the test solution or, more precisely, the test solution's hydronium-ion concentration will determine the concentrations of the acid and base forms of the indicator by causing the indicator reaction to shift until equilibrium is again attained. (Recall Le Chatelier's principle in Chapter 14.) Consequently, the pH of the test solution will dictate the color of the solution. For example, a solution containing the indicator methyl red will be red if the pH of the solution lies below 4.3, and yellow if the pH is above 6.2. Various hues of orange will occur when the pH of the solution lies between 4.3 and 6.2. Therefore, if you added this indicator to a solution and the color was orange, you would know that the pH must be within 4.3 and 6.2. Although this technique is not as precise as a direct measurement with a pH meter, it can be useful when a pH meter is not available.

Figure 15.7 shows color changes for various acid–base indicators. Sometimes paper strips that are treated with several indicators are used to estimate pH. These paper strips are called "pH paper" (Figure 15.8).

Video: Buffering.

15.9 Buffer Solutions

OBJECTIVES

- Define a buffer solution.
- Learn how a buffer solution works using chemical equations.

Exploration: Buffers.

A **buffer solution** is *a solution that can resist changes in pH when limited amounts of acid or base are added to it.* Pure water does not have this capacity. If 0.01 mol of hydrochloric acid is added to 1 liter of pure water, the pH will change from 7 (the pH of pure water) to 2—a pH change of 5 units. In contrast, adding 0.01 mol of HCl to 1 liter of a good buffer solution might change the pH by only 0.1 unit.

Buffer solutions resist pH changes because they contain a weak acid and its conjugate base. Examples are $HC_2H_3O_2$ (acetic acid) and $C_2H_3O_2^-$, or NH_4^+ and NH_3. Let's explore how a buffer works. Suppose a buffer solution contains approximately equal molar amounts of the weak acid (HA) and its conjugate base (A^-). When a strong acid is added to this buffer solution, the H_3O^+ ions from the strong acid react with the buffer solution's base and convert it to its conjugate acid.

$$H_3O^+(aq) + A^-(aq) \longrightarrow HA(aq) + H_2O(l)$$

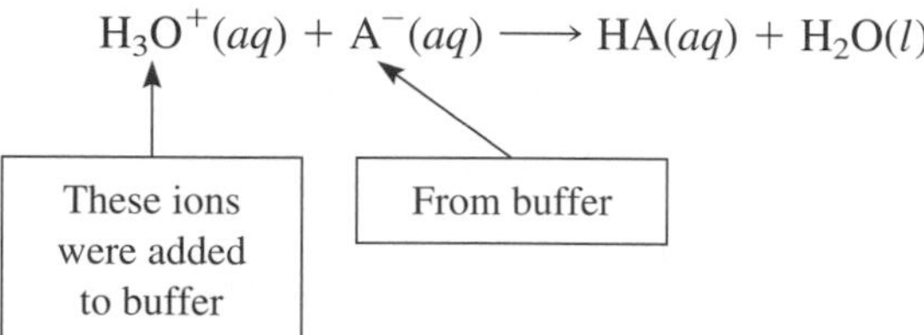

On the other hand, when a strong base is added to the buffer solution, the OH^- ions from the strong base react with the buffer solution's acid and convert it to its conjugate base.

$$OH^-(aq) + HA(aq) \longrightarrow A^-(aq) + H_2O(l)$$

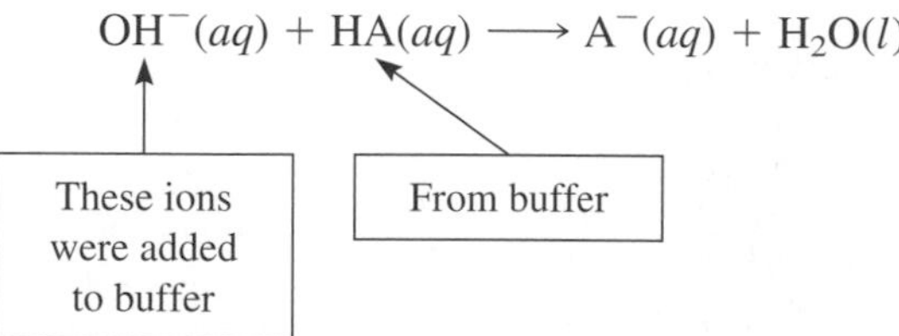

FIGURE 15.9
Buffers regulate pH. Drinks prepared from the contents of this package are buffered with citric acid and sodium citrate. The presence of the conjugate base (the citrate ion from sodium citrate) limits the tart, acid taste that would otherwise occur with citric acid alone.

A buffer solution resists changes in pH because of its ability to combine with both added H_3O^+ ions and added OH^- ions, ions that would otherwise cause a change in pH.

Blood is a buffered mixture whose pH is maintained at about 7.4. The pH of a healthy person's blood never changes by more than perhaps 0.2 pH unit from the average value. Death may result if the pH falls below 6.8 or rises above 7.8. Blood contains H_2CO_3 and HCO_3^- as well as other acid–conjugate base pairs to regulate the pH. Some artificial fruit drinks are also buffered. For example, the one in Figure 15.9 states that it contains both citric acid and citrate ion (provided by sodium citrate). The presence of the conjugate base regulates and limits the tart taste that would otherwise occur with citric acid alone.

Chemical Perspective

Acid Rain

Did you know that normal rain is acidic? The pH of normal rain is about 5.6. This acidity occurs because rain dissolves carbon dioxide from the air, and carbon dioxide reacts with water to form carbonic acid (H_2CO_3). This weak, diprotic acid undergoes ionization in water to produce H_3O^+ ions. Gases from volcanoes and other natural sources also contribute to the acidity of normal rain.

News reports of environmental damages often cite acid rain as the cause of the noticeable changes in natural waters, forests, crops, and even monuments and statues. Acid rain is rain that has a pH lower than that of normal rain, or a pH of less than 5.6. The pH of rain in eastern North America and northern and western Europe has been recorded routinely at 4 and sometimes lower. Rain with a pH of 1.5—over 10,000 times more acidic than normal rain—has fallen in Wheeling, West Virginia. Moreover, the West Coast of the United States is not immune to this problem. For example, Los Angeles has experienced fog with suspended water droplets having a pH as low as 2.5.

Sulfur oxides and nitrogen oxides produced from human activities are the principal substances that cause acid rain. These oxides eventually are converted to sulfuric acid and nitric acid, which constitute 95% of the acid components of acid rain. How do these oxides occur?

The combustion of coal in power plants is one of the primary sources of sulfur oxides. Coal typically contains 1% to 3% sulfur, which is converted to gaseous SO_2 when it is burned. In the presence of dust particles and other substances in polluted air, the sulfur dioxide is then converted to sulfur trioxide, SO_3, which reacts with rainwater to form liquid droplets containing sulfuric acid.

$$2SO_2(g) + O_2(g) \longrightarrow 2SO_3(g)$$
$$SO_3(g) + H_2O(l) \longrightarrow H_2SO_4(aq)$$

A small amount of sulfur dioxide also enters the atmosphere from natural causes. Scientists have estimated that the 1980 eruption of Mount St. Helens in the state of Washington spewed 400,000 tons of sulfur dioxide into the atmosphere. Although this is a huge amount, it was less than 2% of the sulfur dioxide contributed from human activities.

The other primary component of acid rain is nitric acid. This substance arises from natural causes when nitrogen monoxide is formed in lightning storms.

$$N_2(g) + O_2(g) \xrightarrow{\text{Lightning}} 2NO(g)$$

Nitrogen monoxide reacts with oxygen in the atmosphere to yield gaseous nitrogen dioxide, NO_2. This substance then reacts with rainwater to give nitric acid (as well as more nitrogen monoxide).

$$2NO(g) + O_2(g) \longrightarrow 2NO_2(g)$$
$$3NO_2(g) + H_2O(l) \longrightarrow 2HNO_3(aq) + NO(g)$$

However, much more nitric acid comes from the same reaction generated on the surfaces of automobile engines during the combustion of gasoline.

How does acid rain harm the environment? First, when acid rain falls in a lake, it decreases the pH of the water. The increased acidity can be detrimental to life in that water. Indeed, life has disappeared from several hundred lakes in New York State and Ontario, Canada, and from many others in northern Europe. The ultimate irony is that some of the lakes that have been destroyed by excess acidity are still *(continued)* very beautiful with water that is clear and sparkling. Second, low crop yields and deforestation have been linked to acid rain, though this observation has been disputed by some. Finally, marble and limestone monuments and statues, some of them centuries old, have begun to deteriorate rapidly (see Figure 15.10, for example). These materials are affected because they are different forms of the same substance, calcium carbonate. Hydronium ions from the acids in acid rain react with calcium carbonate and cause it to dissolve.

$$H_3O^+(aq) + CaCO_3(s) \longrightarrow Ca^{2+}(aq) + HCO_3^-(aq) + H_2O(l)$$

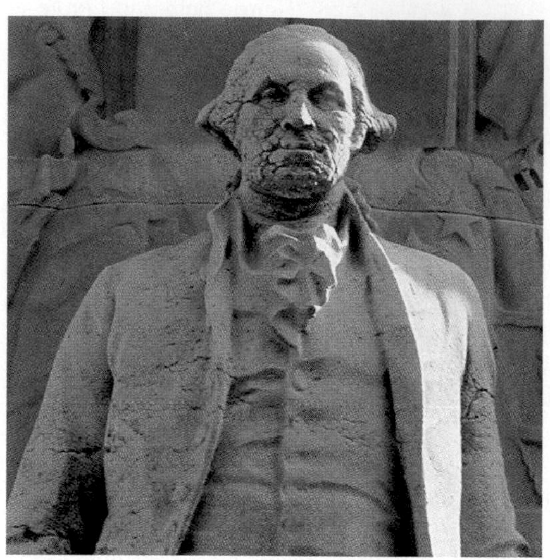

FIGURE 15.10

The effect of acid rain. Acid rain resulting from modern causes has polluted bodies of water, destroyed vegetation, and ruined treasured statues. Here, you see a statue located in New York City before and after 60 years of acid rain.

CHAPTER REVIEW

Key Words

acid *(pp. 474 and 477)*
base *(pp. 474 and 477)*
salt *(p. 475)*
molecular equation *(p. 475)*
total ionic equation *(p. 475)*
spectator ions *(p. 475)*
net ionic equation *(p. 476)*
hydronium ion *(p. 477)*
conjugate acid–base pair *(p. 478)*
acid strength *(p. 481)*
monoprotic acid *(p. 481)*
strong acid *(p. 482)*
diprotic acid *(p. 482)*
weak acid *(p. 482)*
strong base *(p. 484)*
weak base *(p. 484)*
amphoteric substance *(p. 484)*
self-ionization *(p. 484)*
ion-product constant for water *(p. 485)*
pH *(p. 488)*
logarithm of a number *(p. 488)*
acid–base indicator *(p. 492)*
buffer solution *(p. 494)*

Key Equations

$$K_w = [H_3O^+][OH^-]$$

$$pH = -\log [H_3O^+]$$

Summary

Acids and *bases* can be characterized by their tastes, the way that they feel on the skin, and the way that they behave with each other. Arrhenius proposed the first successful structural theory of these substances. According to the Arrhenius theory, an acid is a substance that produces H^+ ions when it is dissolved in water, whereas a base produces OH^- ions. When acids react with bases, neutralization occurs and a *salt* is formed. Many of these reactions can be described by a single net ionic equation, $H^+ + OH^- \longrightarrow H_2O$.

Another approach, the Brønsted–Lowry theory, defines an acid as an H^+ donor and a base as an H^+ acceptor. This theory recognizes that *hydronium ions* (H_3O^+) rather than H^+ ions are formed when an acid dissolves in water.

In this theory, an acid–base reaction can be viewed as an equilibrium competition for H^+ ions. The reaction consists of two *conjugate acid–base pairs,* and each pair consists of two substances that differ only by the gain or loss of a H^+ ion. The concentrations of these substances are determined by the strengths of the acids and bases. A *strong acid* loses an H^+ ion readily and is completely dissociated in water. There are six strong acids; five of them are *monoprotic,* and the other is *diprotic.* A *strong base* accepts an H^+ ion readily and is also completely dissociated in water. A *weak acid* or a *weak base* is not completely dissociated in water.

Water is *amphoteric* because it is both an acid and a base. It undergoes *self-ionization* to give small quantities of hydronium ions and hydroxide ions. The concentrations of these ions in pure water as well as in an aqueous solution are related by the *ion-product constant for water* (K_w). The acidity or basicity of a solution is determined by the hydronium-ion concentration. Often the acidity or basicity of a solution is measured in terms of the *pH,* the negative *logarithm* of the concentration of hydronium ions. The pH of a solution can be measured with a pH meter or estimated with an *acid–base indicator.*

A *buffer solution* is a solution that can resist changes in pH when small amounts of acid or base are added to it. A buffer contains a weak acid and its conjugate base.

Problem-Solving Skills

1. **Writing a Net Ionic Equation:** Given the molecular equation for a reaction, write the net ionic equation (Example 15.1).
2. **Identifying Members of a Conjugate Acid–Base Pair:** Given an acid (or a base), establish the identity of its conjugate base (or acid) (Example 15.2).
3. **Identifying Acids and Bases:** Given the chemical equation for an acid–base reaction, identify the conjugate acid–base pair on each side of the equation (Example 15.3).
4. **Calculating $[OH^-]$ from $[H_3O^+]$ and $[H_3O^+]$ from $[OH^-]$ in Solutions of Acids and Bases:** Given the concentration of hydronium ions (or hydroxide ions) in an aqueous solution at 25°C, calculate the concentration of hydroxide ions (or hydronium ions) (Example 15.4).
5. **Identifying Acidic, Basic, or Neutral Conditions from $[H_3O^+]$:** Given the hydronium-ion (or hydroxide-ion) concentration in a solution, tell whether the solution is acidic, basic, or neutral (Example 15.5).
6. **Calculating pH from $[H_3O^+]$:** Given the concentration of hydronium ions (or hydroxide ions), calculate the pH (Examples 15.6 and 15.7).
7. **Calculating $[H_3O^+]$ from pH:** Given the pH of a solution, calculate the concentration of hydronium ions (Example 15.8).

Questions to Test Your Reading

15.1 What are the definitions of an *acid* and a *base* according to Arrhenius? Give examples of an Arrhenius acid and an Arrhenius base.

15.2 Write the net ionic equation that describes the reaction between HBr and KOH. These substances are totally ionized in water.

15.3 What is a *salt*? Does ordinary table salt fit this definition?

15.4 What are the definitions of an acid and a base according to the Brønsted–Lowry theory? Give examples of a Brønsted–Lowry acid and a Brønsted–Lowry base.

15.5 Give the formula of the hydronium ion and draw its electron-dot structure.

15.6 What is a *conjugate acid–base pair*? Using the following equation, identify both of the acids, both of the bases, and the conjugate acid–base pairs in this acid–base reaction.

$$HSO_4^-(aq) + CN^-(aq) \rightleftharpoons SO_4^{2-}(aq) + HCN(aq)$$

15.7 Show with an example that a reaction of an acid with water is an acid–base reaction.

15.8 Using chemical equations, show that water is amphoteric.

15.9 Give an example of a monoprotic acid and a diprotic acid.

15.10 What is a *strong acid*? Name the strong acids and give their formulas.

15.11 What is a *weak acid*? Give the name and formula of an example.

15.12 What is a *strong base*? Name four strong bases and give their formulas.

15.13 What is a *weak base*? Give the name and formula of an example.

15.14 Give the equilibrium expression for the reaction of nitrous acid (HNO_2) with water.

15.15 Formic acid, $HCHO_2$, is a stronger acid than acetic acid, $HC_2H_3O_2$. Is the formate ion, CHO_2^-, a stronger base than the acetate ion, $C_2H_3O_2^-$? Explain.

15.16 Formic acid, $HCHO_2$, is a stronger acid than acetic acid, $HC_2H_3O_2$. Explain why a 0.1 *M* solution of formic acid would contain a greater concentration of hydronium ions than a 0.1 *M* solution of acetic acid.

15.17 Arrange the acids H_2SO_3 and HSO_3^- according to increasing acid strengths.

15.18 What is meant by *self-ionization*?

15.19 Define the ion-product constant for water. Write the expression for K_w. What is its value at 25°C?

15.20 Define acidic, neutral, and basic solutions in terms of hydronium-ion concentrations at 25°C.

15.21 What is *pH*? Define acidic, neutral, and basic solutions in terms of pH at 25°C.

15.22 If rainwater from a particular area has a pH of 4.3, is it acidic, neutral, or basic?

15.23 Describe two ways to measure pH.

15.24 Describe how an indicator works.

15.25 A solution containing the indicator bromcresol green is green. What can be said about the pH of this solution? Use Figure 15.7 during your consideration.

15.26 When a portion of a solution is treated with bromcresol green, the color turns yellow. When another portion is treated with methyl violet, the color is violet. Using Figure 15.7, estimate the pH of the solution.

15.27 What is a *buffer solution*? Give an example.

15.28 A buffer solution can be prepared by dissolving equimolar quantities of ammonia and ammonium chloride in water. Describe what happens when small quantities of either a strong acid or a strong base are added to this solution.

Practice Problems

Net Ionic Equations and Acid–Base Concepts (Sections 15.2 and 15.3)

15.29 Radium hydroxide reacts with hydrochloric acid according to the equation

$$Ra(OH)_2(aq) + 2HCl(aq) \longrightarrow RaCl_2(aq) + 2H_2O(l)$$

Write a total ionic equation and a net ionic equation. Assume that the acid, the base, and the salt dissociate completely in aqueous solution.

15.30 Lithium hydroxide reacts with nitric acid to give lithium nitrate and water.

$$LiOH(aq) + HNO_3(aq) \longrightarrow LiNO_3(aq) + H_2O(l)$$

Write a total ionic equation and a net ionic equation. Assume that the acid, the base, and the salt dissociate completely in water.

15.31 Using the hydronium ion, write a chemical equation describing the reaction that occurs when perchloric acid dissolves in water.

15.32 Using the hydronium ion, write the chemical equation describing the reaction that occurs when hydrogen iodide dissolves in water.

15.33 Give the conjugate base for each of the following acids.
(a) HF (b) H_2S (c) HS^- (d) H_2O

15.34 Give the conjugate base for each of the following acids.
(a) $HC_2H_3O_2$ (b) HCl (c) HClO
(d) HNO_2

15.35 Give the conjugate acid for each of the following bases.
(a) NH_3 (b) HS^- (c) S^{2-} (d) CN^-

15.36 Give the conjugate acid for each of the following bases.
(a) F^- (b) HSO_3^- (c) SO_3^{2-} (d) ClO_4^-

15.37 Identify the acids and bases on both sides of the following equations and show the conjugate acid–base pairs.
(a) $H_2SO_3(aq) + H_2O(l) \rightleftharpoons H_3O^+(aq) + HSO_3^-(aq)$
(b) $HSO_3^-(aq) + H_2O(l) \rightleftharpoons H_3O^+(aq) + SO_3^{2-}(aq)$
(c) $CH_3NH_2(aq) + H_2O(l) \rightleftharpoons CH_3NH_3^+(aq) + OH^-(aq)$
(d) $NH_4^+(aq) + CN^-(aq) \rightleftharpoons NH_3(aq) + HCN(aq)$

15.38 Identify the acids and bases on both sides of the following equations and show the conjugate acid–base pairs.
(a) $H_2CO_3(aq) + H_2O(l) \rightleftharpoons H_3O^+(aq) + HCO_3^-(aq)$
(b) $HCO_3^-(aq) + H_2O(l) \rightleftharpoons H_3O^+(aq) + CO_3^{2-}(aq)$
(c) $CH_3CH_2NH_2(aq) + H_2O(l) \rightleftharpoons CH_3CH_2NH_3^+(aq) + OH^-(aq)$
(d) $HPO_4^{2-}(aq) + NH_4^+(aq) \rightleftharpoons H_2PO_4^-(aq) + NH_3(aq)$

15.39 Identify the acids and bases on both sides of the following equations and show the conjugate acid–base pairs.
(a) $CN^-(aq) + H_2O(l) \rightleftharpoons HCN(aq) + OH^-(aq)$
(b) $HC_2H_3O_2(aq) + OH^-(aq) \rightleftharpoons C_2H_3O_2^-(aq) + H_2O(l)$
(c) $HPO_4^{2-}(aq) + H_3O^+(aq) \rightleftharpoons H_2PO_4^-(aq) + H_2O(aq)$
(d) $HNO_2(aq) + H_2O(l) \rightleftharpoons NO_2^-(aq) + H_3O^+(aq)$

15.40 Identify the acids and bases on both sides of the following equations and show the conjugate acid–base pairs.
(a) $F^-(aq) + H_2O(l) \rightleftharpoons HF(aq) + OH^-(aq)$
(b) $F^-(aq) + H_3O^+(aq) \rightleftharpoons HF(aq) + H_2O(l)$
(c) $H_2S(aq) + NH_3(aq) \rightleftharpoons NH_4^+(aq) + HS^-(aq)$
(d) $H_3O^+(aq) + HS^-(aq) \rightleftharpoons H_2S(aq) + H_2O(l)$

Concentrations of Hydronium and Hydroxide Ions (Section 15.6)

15.41 Calculate the concentration of OH^- ions in an aqueous solution at 25°C if the hydronium-ion concentration is
(a) 0.25 *M* (b) 0.0035 *M* (c) 1.2×10^{-2} *M*
(d) 4.2×10^{-11} *M*

15.42 Calculate the concentration of H_3O^+ ions in an aqueous solution at 25°C if the hydroxide-ion concentration is
(a) 0.15 *M* (b) 3×10^{-4} *M* (c) 5.3×10^{-9} *M*
(d) 0.00061 *M*

15.43 If a sample of acid rain has a hydronium-ion concentration of 3.7×10^{-4} *M* at 25°C, what is the concentration of hydroxide ions?

15.44 A cleaning solution containing ammonia at 25°C has a hydroxide-ion concentration of 8.8×10^{-4} *M*. Calculate the concentration of hydronium ions.

15.45 Consider the following solution concentrations at 25°C. Indicate whether each solution is acidic, neutral, or basic.
(a) 5×10^{-9} *M* H_3O^+ (b) 1×10^{-7} *M* OH^-
(c) 5×10^{-9} *M* OH^- (d) 2×10^{-7} *M* H_3O^+

15.46 Consider the following solution concentrations at 25°C. Indicate whether each solution is acidic, neutral, or basic.
(a) 2×10^{-4} *M* OH^- (b) 6×10^{-10} *M* OH^-
(c) 2×10^{-6} *M* H_3O^+ (d) 6×10^{-10} *M* H_3O^+

15.47 The concentration of hydroxide ions in a shampoo solution at 25°C is 1.5×10^{-7} *M*. Is the solution acidic, neutral, or basic?

15.48 The concentration of hydroxide ions in an antiseptic solution at 25°C is 8.4×10^{-5} *M*. Is the solution acidic, neutral, or basic?

Calculations Involving pH (Section 15.7)

15.49 Calculate the pH of each of the following solutions.
(a) 1.0×10^{-8} *M* H_3O^+
(b) 7.5×10^{-3} *M* H_3O^+
(c) 5.0×10^{-12} *M* H_3O^+
(d) 6.3×10^{-9} *M* H_3O^+

15.50 Calculate the pH of each of the following solutions.
(a) 1.0×10^{-4} *M* H_3O^+
(b) 2.3×10^{-5} *M* H_3O^+
(c) 3.2×10^{-10} *M* H_3O^+
(d) 2.9×10^{-11} *M* H_3O^+

15.51 If a quantity of vinegar has $[H_3O^+] = 7.5 \times 10^{-3}$ *M*, what is the pH? Is the solution acidic, neutral, or basic?

15.52 Some beer has a concentration of hydronium ions equal to 5.0×10^{-3} *M*. What is the pH? Is the solution acidic, neutral, or basic?

15.53 Calculate the pH of each of the following solutions.
(a) 1.0×10^{-8} *M* OH^-
(b) 7.5×10^{-3} *M* OH^-
(c) 5.0×10^{-12} *M* OH^-
(d) 6.3×10^{-9} *M* OH^-

15.54 Calculate the pH of each of the following solutions.
(a) 1.0×10^{-4} *M* OH^-
(b) 2.3×10^{-5} *M* OH^-
(c) 3.2×10^{-10} *M* OH^-
(d) 2.9×10^{-11} *M* OH^-

15.55 The concentration of hydroxide ions in a solution of washing soda (sodium carbonate, Na_2CO_3) is 0.0040 *M*. Calculate the pH. Is the solution acidic, neutral, or basic?

15.56 The concentration of hydroxide ions in a solution of lye (sodium hydroxide, NaOH) is 0.050 *M*. What is the pH? Is the solution acidic, neutral, or basic?

15.57 Convert each of the following pH values into the corresponding hydronium-ion concentration.
(a) 7.28 (b) 2.32 (c) 12.10 (d) 6.73

15.58 Convert each of the following pH values into the corresponding hydronium-ion concentration.
(a) 3.41 (b) 1.56 (c) 9.96 (d) 13.35

15.59 If a pH meter indicates that the pH of a solution is 8.74, what are the concentrations of hydronium ions and hydroxide ions in the solution?

15.60 The pH of a solution is 4.74. What are the concentrations of hydronium ions and hydroxide ions in the solution?

Additional Problems

15.61 What is the relative difference in the concentration of hydronium ions in solutions whose pH differs by 1 pH unit?

15.62 If one solution has a pH of 3 and another solution has a pH of 6, what is the relative difference in hydronium ions in these solutions?

15.63 A solution was prepared by dissolving 40. g of solid NaOH in 10. L of water. Remembering that NaOH is a strong base and dissociates completely in water, calculate the concentration of hydronium ions and the pH.

15.64 A solution was prepared by dissolving 18 g of gaseous HCl in 10. L of water. Remembering that HCl is a strong acid and dissociates completely in water, calculate the concentration of hydronium ions and the pH.

Practice in Problem Analysis

For each problem, describe the thinking you would use (the problem analysis) before doing the actual solution, but do not solve the problem.

1. What is the concentration of NH_4^+ ions in a 1.0 *M* solution of ammonia, NH_3, if the pH of the solution is 11.62?

2. A 1.00-L sample of aqueous NaOH was neutralized by adding 45.6 mL of 0.633 *M* HCl. How many grams of NaOH were present in the original sample?

PRACTICE EXAM

1. According to the Arrhenius theory of acids and bases, an acid is a substance that will
(a) increase the concentration of OH^- ions in a solution
(b) increase the concentration of H^+ ions in a solution
(c) donate H^+ ions
(d) accept H^+ ions
(e) None of the above

2. According to the Brønsted–Lowry theory of acids and bases, an acid is a substance that will
(a) increase the concentration of OH^- ions in a solution.
(b) increase the concentration of H^+ ions in a solution.
(c) donate H^+ ions.
(d) accept H^+ ions.
(e) None of the above

3. Which statement is true?
(a) You cannot write a molecular equation for a neutralization reaction that involves ionic acids and bases.
(b) If water is one of the products of a neutralization reaction, you cannot write a total ionic equation because water is a molecular substance.
(c) You can write the same net ionic equation for many neutralization reactions.
(d) You should not use a net ionic equation for a double-replacement reaction.
(e) None of the above

4. Which pair of substances is a conjugate acid–base pair?
(a) H_2O, OH^- (b) SO_3^{2-}, SO_4^{2-}
(c) HCl, OH^- (d) H_3O^+, Cl^-
(e) None of the above

5. What is the conjugate acid of H_2O in the following reaction?

$$HCN(aq) + H_2O(l) \leftrightarrows H_3O^+(aq) + CN^-(aq)$$

(a) HCN (b) CN^- (c) H_3O^+ (d) OH^-
(e) None of the above

6. Which of the following substances is amphoteric?
(a) H_2O (b) HF (c) NaOH (d) NaF
(e) None of the above

7. Which of the following substances is a strong acid?
(a) HCl (b) H_2O (c) NaOH (d) NaCl
(e) None of the above

8. Which substance is a weak acid?
(a) HCl (b) H_2O (c) NaOH (d) NaCl
(e) None of the above

9. Which of the following substances is a strong base?
(a) HCl (b) H_2O (c) NaOH (d) NaCl
(e) None of the above

10. If $[H_3O^+] = 1.0 \times 10^{-3}$ *M*, then $[OH^-] =$
(a) 1.0×10^{-11} *M* and the solution is basic.
(b) 1.0×10^{-11} *M* and the solution is acidic.
(c) 1.0×10^{-14} *M* and the solution is basic.
(d) Impossible to specify
(e) None of the above

11. If $[OH^-] = 1.0 \times 10^{-3}$ *M*, then $[H_3O^+] =$
(a) 1.0×10^{-11} *M* and the solution is basic.
(b) 1.0×10^{-11} *M* and the solution is acidic.
(c) 1.0×10^{-14} *M* and the solution is basic.
(d) Impossible to specify
(e) None of the above

12. If $[H_3O^+] = 4.2 \times 10^{-9}$ *M*, then the pH is
(a) 5.62 and the solution is basic.
(b) 5.62 and the solution is acidic.
(c) 8.38 and the solution is basic.
(d) 8.38 and the solution is acidic.
(e) None of the above

13. If the pH is 2.30, then $[H_3O^+] =$
(a) 2.0×10^{-11} *M* and the solution is basic.
(b) 2.0×10^{-11} *M* and the solution is acidic.
(c) 5.0×10^{-3} *M* and the solution is basic.
(d) 5.0×10^{-3} *M* and the solution is acidic.
(e) None of the above

14. If $[OH^-] = 4.0 \times 10^{-8}$ *M*, then the pH is
(a) 7.40 (b) 6.60 (c) 8.40 (d) 5.60
(e) None of the above

15. Which statement is true?
(a) A solution is buffered because it contains a strong acid and its conjugate base.
(b) When an acid is added to a buffered solution, it reacts with the acid in the buffer.
(c) When a base is added to a buffered solution, it reacts with the base in the buffer.
(d) A buffer solution resists changes in pH because its components react with any added acid or base so that H_3O^+ ions or OH^- ions do not accumulate.
(e) None of the above

Chapter

16 Oxidation–Reduction Reactions

Spectacular oxidation–reduction.

Chapter Outline

Hot, dry winds blew a fire across six southern California counties in October 1993. Only a short distance away, grass and trees continued to grow (Figure 16.1). Fire and plant growth—one a destructive process and the other a constructive process—share an important characteristic. In both of them electrons are transferred from one reactant to another. *Oxidation* and *reduction,* the subject of this chapter, are names for processes that occur when electrons are transferred. *Oxidation–reduction reactions* are found everywhere: in the human body, in the home, and throughout the environment. They also provide electricity in batteries so that you can start a car, make a watch work, or add, subtract, multiply, and divide with a calculator. Oxidation–reduction reactions are the subject of this chapter, and you will see how these reactions are able to generate electricity.

UNDERSTANDING OXIDATION–REDUCTION REACTIONS

Oxidation got its name from the chemical reactions of oxygen with other elements, and from the formation of oxides during these reactions. For example, oxygen reacts with magnesium to form magnesium oxide. The opposite reaction—obtaining an element from one of its compounds, such as an oxide—was called reduction. Chemists now recognize, however, that the reactions of chlorine, bromine, or sulfur with other elements are not very different from the reactions of oxygen. So they have adopted a broader view of oxidation and its counterpart, reduction. An **oxidation–reduction reaction** is *a reaction in which electrons are transferred from one reactant to another.* These reactions can occur with substances that are gases, liquids, or solids, or with substances that are in solution. You have already seen a few cases of oxidation and reduction in Chapter 6 when we discussed single-replacement reactions. In the next section, we examine the

FIGURE 16.1
Fires and plant growth. This fire swept part of southern California in October 1993. Plants continue to grow nearby. Fires and plant growth are both due to oxidation–reduction reactions.

common events that occur during every oxidation–reduction reaction. In subsequent sections, you learn about a bookkeeping method for keeping track of the electrons that are transferred. You also learn how to use this method to balance the chemical equations that describe oxidation–reduction reactions.

Exploration: Oxidation–reduction reactions.

16.1 Oxidation and Reduction

OBJECTIVES

- Define *oxidation, oxidizing agent, reduction,* and *reducing agent.*
- Given a reaction between a metal and a metal ion or a metal and a nonmetal, identify the oxidizing agent and the reducing agent.

When a strip of metallic zinc is placed in a blue solution of copper(II) sulfate, the zinc becomes coated with metallic copper (Figure 16.2). If we allow the reaction to proceed, the blue color will fade continuously until the solution becomes colorless. We can show that zinc ions have entered the solution while copper ions have left the solution. An oxidation–reduction reaction has occurred, and electrons have been transferred from one reactant to another. The chemical equation for this *single-replacement reaction* is

$$Zn(s) + CuSO_4(aq) \longrightarrow ZnSO_4(aq) + Cu(s)$$

How do we know that electrons were transferred? Recognizing that $CuSO_4$ and $ZnSO_4$ are ionic substances, we write the total ionic equation,

$$Zn(s) + Cu^{2+}(aq) + SO_4{}^{2-}(aq) \longrightarrow Zn^{2+}(aq) + SO_4{}^{2-}(aq) + Cu(s)$$

Videos: Zinc and iodine; Sugar, potassium chlorate, and sulfuric acid; Reactions of metals with acids; Dry ice and magnesium.

You can see that the sulfate ions do not take part in the reaction; they are spectator ions (see Section 15.2). So they can be canceled from both sides of the equation.

$$Zn(s) + Cu^{2+}(aq) + \cancel{SO_4{}^{2-}(aq)} \longrightarrow Zn^{2+}(aq) + \cancel{SO_4{}^{2-}(aq)} + Cu(s)$$

A

B

C

FIGURE 16.2
Reaction of zinc with aqueous copper(II) sulfate. *(A)* A strip of zinc and a blue solution of copper(II) sulfate. *(B)* Zinc reacts with $Cu^{2+}(aq)$ to give $Zn^{2+}(aq)$ and $Cu(s)$. *(C)* Some copper metal has plated out on the zinc strip.

The net ionic equation is

$$Zn(s) + Cu^{2+}(aq) \rightarrow Zn^{2+}(aq) + Cu(s)$$

Now you can see that zinc and copper are active participants because their charges change.

$$Zn \longrightarrow Zn^{2+}$$
$$Cu^{2+} \longrightarrow Cu$$

A zinc atom must have lost two electrons during the reaction to go from Zn to Zn^{2+}.

$$Zn(s) \longrightarrow Zn^{2+}(aq) + 2e^-$$

Similarly, each Cu^{2+} ion must have gained two electrons in order to become a copper atom with no charge.

$$Cu^{2+}(aq) + 2e^- \longrightarrow Cu(s)$$

We define **oxidation** as *the loss of electrons* and **reduction** as *the gain of electrons.* In our reaction, zinc was oxidized because it lost electrons, and the Cu^{2+} ion in copper(II) sulfate was reduced because it gained electrons.

$$Zn(s) \longrightarrow Zn^{2+}(aq) + 2e^- \qquad \text{(Oxidation)}$$
$$Cu^{2+}(aq) + 2e^- \longrightarrow Cu(s) \qquad \text{(Reduction)}$$

Each reaction is called a *half-reaction.* One half-reaction describes the oxidation portion of the reaction and the other describes the reduction portion. The two electrons that each zinc atom lost were gained by each Cu^{2+} ion. Thus, the electrons were transferred from zinc to the copper ions. ●

● The sum of the two half-reactions is the original net ionic equation after the electrons are canceled.

Zn 2e⁻ → Cu^{2+}

An oxidation–reduction reaction requires two substances: an *oxidizing agent* and a *reducing agent.* The **oxidizing agent** *causes the oxidation of the other substance (the reducing agent) in an oxidation–reduction reaction.* The oxidizing agent causes the other substance to lose electrons by accepting them. In this process, the oxidizing agent is itself reduced (gains electrons). The **reducing agent** *causes the reduction of the other* *substance (the oxidizing agent) in an oxidation–reduction reaction.* It causes the other substance to gain electrons by supplying (losing) those electrons. In the process, the reducing agent is itself oxidized. Notice that *an oxidizing agent requires a reducing agent, and a reducing agent cannot function without an oxidizing agent.*

Returning to the net ionic equation for the reaction of zinc with Cu^{2+} ions, Cu^{2+} is the oxidizing agent because it causes the oxidation of Zn. The oxidizing agent is reduced during this process ($Cu^{2+} \longrightarrow Cu$). Zinc is the reducing agent because it causes the reduction of Cu^{2+} ions. The reducing agent is oxidized during the reaction ($Zn \longrightarrow Zn^{2+}$). The relationships between these terms are shown in the following diagram.

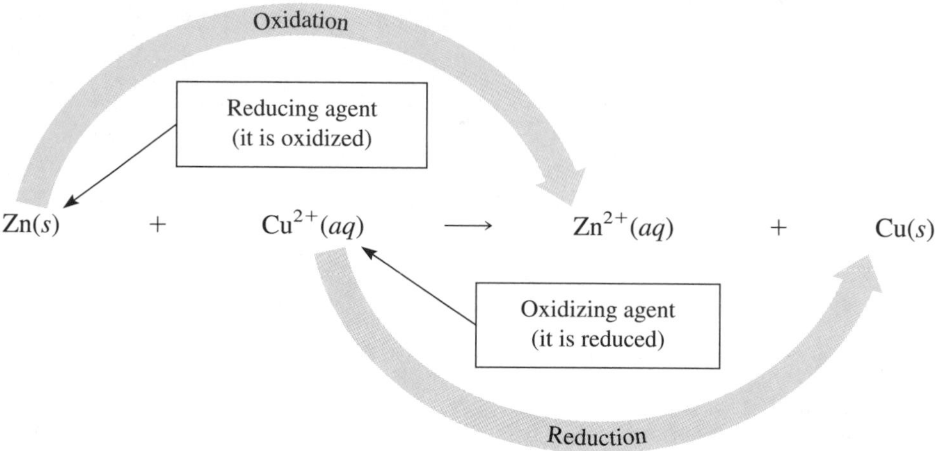

We can summarize these ideas in another way:

Reactant	What Happens to the Reactant	The Reactant Is a
Zn	Oxidized (loses e^-)	Reducing agent
Cu^{2+}	Reduced (gains e^-)	Oxidizing agent

In reactions between a metal and a metal ion or a metal and a nonmetal, it is usually easy to decide that the reactions are oxidation–reduction reactions and to pinpoint the oxidizing agents and the reducing agents. Here are two rules for these types of reactions.

Rules for Identifying Simple Oxidizing and Reducing Agents

1. If the charge on a monatomic substance (such as Pb or Pb^{2+}) increases (becomes more positive) as a result of a reaction without any other changes, the substance has been oxidized and it is the reducing agent.
2. If the charge on a monatomic substance decreases (becomes more negative) as a result of a reaction without any other changes, the substance has been reduced and it is the oxidizing agent.

Chemical Perspective

Vitamin C, the Antioxidant

Why should we include vitamin C (ascorbic acid) in our diets? We know that it prevents scurvy, but it also serves other purposes. It has an essential role in the formation of the structural protein collagen, a fibrous protein that is a major component of bone, teeth, cartilage, tendons, and ligaments. It is also one of three vitamins known as *antioxidant vitamins.* (The other two antioxidant vitamins are vitamins A and E.) An antioxidant is a reducing agent that is so easily oxidized that it protects other more valued substances from oxidation. Let's examine why nutritionists believe that we need vitamin C as an antioxidant.

Our cells are continually exposed to harmful oxidizing agents that come from normal metabolic processes, as well as external sources such as air pollution and tobacco smoke. If cellular components are oxidized, a cell is either altered or destroyed. The vitamin is like a dedicated bodyguard: It is ready to sacrifice itself to save the life of the cell. When vitamin C intervenes, it reacts with these oxidizing agents, and the vitamin is oxidized. Using AH_2 to represent its formula, the half-reaction for the oxidation of vitamin C is

$$AH_2 + 2H_2O \longrightarrow A + 2H_3O^+ + 2e^-$$

The oxidation product of vitamin C is dehydroascorbic acid, which is represented by A in this half-reaction. The harmful oxidizing agents are reduced during the reaction, so they are no longer capable of oxidizing and damaging cellular components.

Processed food manufacturers sometimes add vitamin C to foods to protect them from oxidation by oxygen in the air. You can test the effectiveness of the vitamin as an antioxidant in foods with a simple experiment. Sliced apples will darken over a period of time when they are exposed to air because of oxidation. Place each half of a sliced apple in separate bowls. Coat the cut surface of one of the halves with two spoonfuls of tap water; coat the surface of the other half with two spoonfuls of water containing vitamin C (from a crushed tablet). After about an hour, you should be able to see that the half that was treated with water alone has begun to darken because of oxidation. It will be more noticeable after two hours. Some people keep sliced fruit from darkening by applying a little lemon juice, a source of vitamin C. You will find another example of sacrificial oxidation in the Chemical Perspective at the end of this chapter.

EXAMPLE 16.1 Identifying Oxidizing and Reducing Agents in a Net Ionic Equation

A battery is able to discharge electricity because of an oxidation–reduction reaction. For example, the nickel–cadmium or NiCad battery in many cordless appliances and tools delivers electricity because of the reaction (in simplified form)

$$Cd + 2Ni^{3+} \longrightarrow Cd^{2+} + 2Ni^{2+}$$

Identify the atom or ion that is oxidized and the atom or ion that is reduced. What is the oxidizing agent and what is the reducing agent?

Problem Analysis

Identify the monatomic substance whose charge increases; it has been oxidized and is the reducing agent. The other monatomic substance should show a charge decrease; it has been reduced and is the oxidizing agent.

Solution

The charge on cadmium increases ($Cd \longrightarrow Cd^{2+}$) so Cd is oxidized (lost electrons), and it is the reducing agent. The charge on nickel decreases ($Ni^{3+} \longrightarrow Ni^{2+}$) so Ni^{3+} is reduced (gained electrons); it is the oxidizing agent.

Exercise 16.1

When a NiCad battery is recharged, the reaction causing the discharge of electricity is reversed so that it becomes

$$Cd^{2+} + 2Ni^{2+} \longrightarrow Cd + 2Ni^{3+}$$

Identify the atom or ion that is oxidized and the atom or ion that is reduced. What is the oxidizing agent and what is the reducing agent?

(Try Problems 16.27 and 16.28.)

Some common oxidizing and reducing agents are given in Table 16.1. Oxidation–reduction reactions with these chemicals and others are pervasive. The following Chemical Perspective discusses the behavior of vitamin C toward certain oxidizing agents.

16.2 Activity Series

OBJECTIVES

- Use the activity series to determine the relative ease of oxidation of some of the elements.
- Predict whether an element in the activity series will react with a monatomic ion of another element in the series.
- Predict which metals will dissolve in acids.

As we have seen, zinc reacts with copper(II) ion according to the equation

$$Zn(s) + Cu^{2+}(aq) \longrightarrow Zn^{2+}(aq) + Cu(s)$$

Zinc will also react with certain acids, as shown in the equation

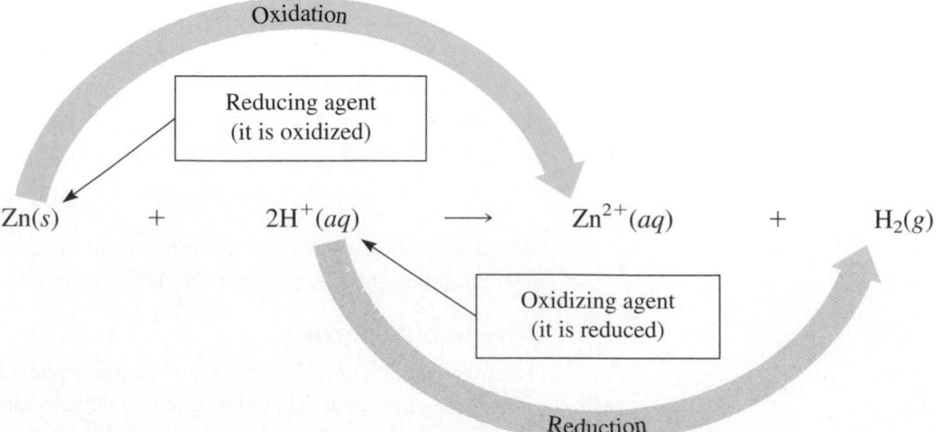

The relative ease with which the substances on the left side of these equations gain or lose electrons determines whether or not these oxidation–reduction reactions will

TABLE 16.1
Some Common Oxidizing and Reducing Agents

Oxidizing Agents	Reducing Agents
Hydrogen peroxide, H_2O_2*	Hydrogen peroxide, H_2O_2*
Sodium hypochlorite, NaClO	Iron, Fe
Chlorine, Cl_2	Sulfur dioxide, SO_2
Oxygen, O_2	Hydrogen, H_2

*Hydrogen peroxide can be used as either an oxidizing agent or a reducing agent, depending on what other substances are present.

● Recall from Chapter 15 that hydrogen ions do not exist as bare protons in aqueous solution; instead they react with water to form hydronium ions. We write them as H^+ here in order to simplify the notation of oxidation–reduction equations.

occur. Although zinc readily gives up electrons to hydrogen ions, no reaction occurs in the seemingly similar case of copper and hydrogen ions. ●

$$Cu(s) + 2H^+(aq) \longrightarrow \text{no reaction}$$

We can say that zinc is more reactive than copper because of this difference.

You have seen in Chapter 6 that the outcome of a potential single-replacement reaction can be predicted by using the *activity series* (Table 16.2). We can now say that this series is an arrangement of the elements in order of their ease of losing electrons during reactions in aqueous solutions. The ease of losing electrons decreases as you pass down the table. Elements near the top of the table lose electrons most easily (are oxidized readily), so they are the strongest reducing agents. Elements near the bottom of the table are the weakest reducing agents.

TABLE 16.2
Activity Series

Metal	Ion Formed
Lithium Li	Li^+
Potassium K	K^+
Barium Ba	Ba^{2+}
Calcium Ca	Ca^{2+}
Sodium Na	Na^+
Magnesium Mg	Mg^{2+}
Aluminum Al	Al^{3+}
Manganese Mn	Mn^{2+}
Zinc Zn	Zn^{2+}
Chromium Cr	Cr^{3+}
Iron Fe	Fe^{2+}
Cadmium Cd	Cd^{2+}
Cobalt Co	Co^{2+}
Nickel Ni	Ni^{2+}
Tin Sn	Sn^{2+}
Lead Pb	Pb^{2+}
Hydrogen H_2	$2H^+$
Copper Cu	Cu^{2+}
Silver Ag	Ag^+
Mercury Hg	Hg^{2+}
Platinum Pt	Pt^{2+}
Gold Au	Au^{3+}

Under ordinary circumstances, *an element in the table will react with a monatomic ion from another element in the table if the element is above the other element in the activity series.* For example, you can see from the table that cobalt is above nickel. Therefore, cobalt reacts with Ni^{2+} ions.

$$Co(s) + Ni^{2+}(aq) \longrightarrow Co^{2+}(aq) + Ni(s)$$

Cobalt is oxidized and aqueous Ni^{2+} ions are reduced. The table also indicates that nickel will not react with aqueous Co^{2+} ions, because nickel is below cobalt.

$$Ni(s) + Co^{2+}(aq) \longrightarrow \text{no reaction}$$

● Copper will dissolve in nitric acid, but the oxidizing agent is the nitrate ion, rather than hydrogen ions.

The activity series also shows which elements will dissolve in acids. The metals above hydrogen (H_2) in the activity series will dissolve in acids with the liberation of hydrogen gas. For example, you can see that the activity series predicts our previous observation that zinc will dissolve in an acid but copper will not. ●

■ EXAMPLE 16.2 Using the Activity Series

If you swallowed a silver coin, would it dissolve in stomach acid (hydrochloric acid) to produce hydrogen gas?

Problem Analysis

Recall that hydrochloric acid is an aqueous solution containing H^+ and Cl^- ions. If an element is to dissolve in this acid to produce hydrogen gas, it must lie above hydrogen in the activity series.

Solution

Silver is below hydrogen in the activity series. Therefore, a reaction between Ag and hydrogen ions will not occur, and silver will not dissolve in hydrochloric acid.

$$Ag(s) + H^+(aq) \longrightarrow \text{no reaction}$$

Exercise 16.2

Will magnesium react with aqueous silver ions (Ag^+) to give magnesium ions and silver?

(Try Problems 16.29 and 16.30.)

When we discuss atoms or monatomic ions, such as Zn atoms or Cu^{2+} ions, we know how many electrons belong to each species. As a result, it is an easy matter to decide when electrons have been transferred from one substance to another. When we deal with other types of substances, however, such as molecules or polyatomic ions, we need an easy and quick bookkeeping device that will allow us to keep track of electrons in oxidation–reduction reactions without resorting to an actual count. We give that method in the next section.

16.3 Oxidation Numbers

OBJECTIVES

- Given a molecule, formula unit, polyatomic ion, or even a monatomic ion, assign an oxidation number to each atom.
- Use oxidation numbers to identify oxidizing and reducing agents.

The chemical equation

$$2H_2(g) + O_2(g) \longrightarrow 2H_2O(g)$$

describes an oxidation–reduction reaction, but how would you know it and how would you identify the oxidizing and reducing agents? To answer those questions, we introduce oxidation numbers.

Assigning Oxidation Numbers

Our bookkeeping device for keeping track of electrons is the **oxidation number** (also called oxidation state). It is defined as either

1. *the charge on an atom or a monatomic ion,* or
2. *the charge that an atom in a substance would have if the shared pair of electrons belonged to the more electronegative atom in the bond.* ●

● Recall that electronegativity is a measure of the ability of an atom in a covalent bond to draw bonding electrons to itself.

Using the first definition, we can assign oxidation numbers to the ions in a binary ionic compound. Consider sodium chloride (NaCl). The oxidation numbers of Na^+ and Cl^- are $+1$ and -1, respectively, because these are the charges on these ions. Thus, we have

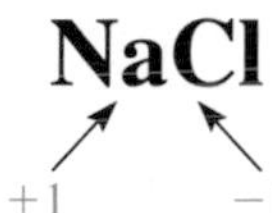

for NaCl. Notice that the sum of the oxidation numbers is zero, as it must be because a compound such as NaCl does not have a net charge.

The second definition applies to covalent molecules and polyatomic ions. First recall Lewis's definition of a covalent bond: a chemical bond formed by sharing a pair of electrons between two atoms. If the two atoms are not identical, the pair of electrons is not shared equally, so that the more electronegative atom draws these electrons to itself more strongly than the less electronegative atom. By assigning both electrons of the shared pair to the more electronegative atom, we give that atom a *hypothetical charge* rather than a real charge, and that hypothetical charge is the oxidation number.

Consider, for example, the covalent compound hydrogen fluoride (HF). What are the oxidation numbers of the atoms in a molecule of this substance? You can begin by writing the electron-dot formula.

$$\mathrm{H} : \ddot{\underset{\cdot\cdot}{\mathrm{F}}} :$$

Even though hydrogen and fluorine share a pair of bonding electrons, we assign both of them to fluorine because fluorine is more electronegative than hydrogen (see Figure 10.5). When we do this, the fluorine atom has one more valence electron than the neutral atom (eight rather than seven). Thus, the hypothetical charge is -1, so its oxidation number is -1. Now the hydrogen atom has no electrons, so its hypothetical charge and oxidation number are both $+1$. Again, the sum of the oxidation numbers is zero for this neutral compound.

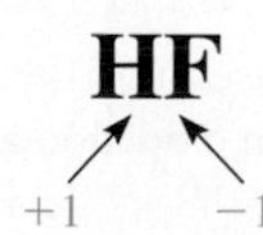

Fortunately, you will not need to examine a Lewis electron-dot formula each time you wish to assign oxidation numbers. Instead, you can short-cut that procedure by applying the following rules, which incorporate both definitions for oxidation numbers.

Rules for Assigning Oxidation Numbers

1. The oxidation number of an atom in an element is zero. For example, the oxidation number of a Zn atom is zero, as is the oxidation number of each fluorine atom in F_2 or each phosphorus atom in P_4.
2. The oxidation number of a monatomic ion is equal to the charge on the ion. As an example, the oxidation numbers of a Cu^{2+} ion or a O^{2-} ion are $+2$ and -2, respectively.
3. The oxidation number of fluorine is -1 in all of its compounds. Therefore, the oxidation number of fluorine is -1 in HF, NF_3, and SF_6 even though these compounds are not ionic.
4. The oxidation number of chlorine, bromine, and iodine in binary compounds containing only one type of Group VIIA element is -1. For example, the oxidation number of each iodine in NI_3 is -1.
5. The oxidation number of hydrogen is $+1$ in its compounds with nonmetals. As a result, the oxidation number of hydrogen in HCl, H_2O, and NH_3 is $+1$.
6. The oxidation number of oxygen is usually -2. The major exception occurs with peroxides, such as H_2O_2 and Na_2O_2, in which the oxidation number of each oxygen is -1. The peroxides contain the polyatomic ion O_2^{2-}.
7. The oxidation number of a Group IA atom in any compound is $+1$; the oxidation number of a Group IIA atom in any compound is $+2$. Thus, the oxidation numbers of sodium and calcium in Na_2CO_3 and $Ca_3(PO_4)_2$ are $+1$ and $+2$, respectively.
8. The sum of the oxidation numbers of the atoms in a compound will always equal zero. In a polyatomic ion, however, the sum of the oxidation numbers of the constituent atoms will add up to the charge on the ion.

Rule 8 can be used to obtain an oxidation number that is not otherwise specified by the seven preceding rules, as demonstrated by the following examples.

■ EXAMPLE 16.3 Assigning Oxidation Numbers to Atoms in a Compound

Photographs of Io, one of Jupiter's moons, from the *Voyager I* space probe showed recently active volcanoes and a landscape composed of sulfur and sulfur dioxide (SO_2) frost (Figure 16.3). Assign oxidation numbers to each atom in a molecule of sulfur dioxide.

Problem Analysis

Use rules 1 through 7 to get as many oxidation numbers as you can. Use rule 8 to get an oxidation number that has not yet been assigned.

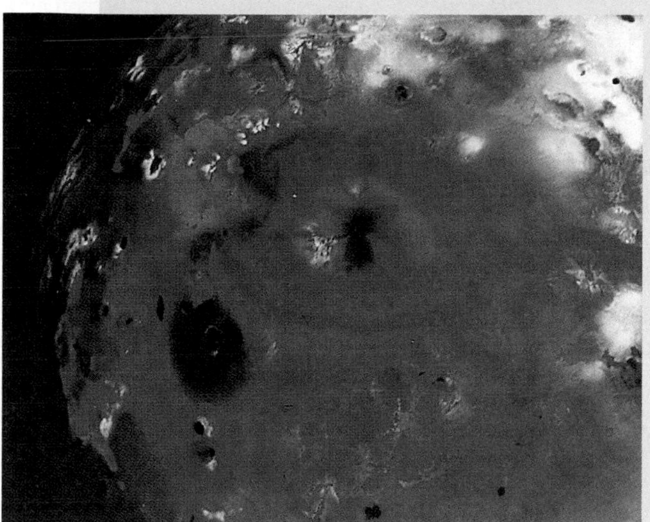

FIGURE 16.3
A photograph of Io taken from the *Voyager I* space probe. The dark circular features are recently active volcanoes. Astronomers believe that the red, orange, and yellow colors are due to sulfur at various temperatures. The white patches are sulfur dioxide frost.

Solution

First, use rule 6 to get the oxidation number for oxygen.

$$SO_2$$

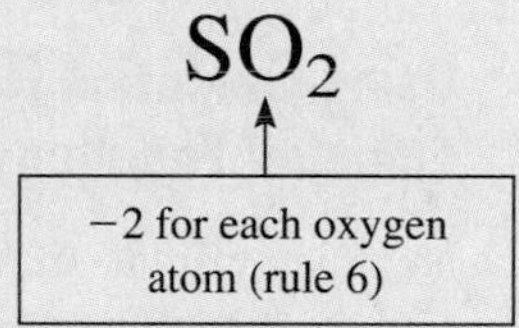

Next, use rule 8 to obtain the oxidation number for sulfur. Because the sum of the oxidation numbers for the atoms in a compound is zero, write

$$\text{Oxidation number for S} + (2 \times \text{oxidation number for oxygen}) = 0$$

The oxidation number for oxygen must be multiplied by 2 because there are two oxygen atoms in this molecule. Substituting the known oxidation number for oxygen, you get

$$\text{Oxidation number for sulfur} + 2 \times (-2) = 0$$
$$\text{Oxidation number for sulfur} - 4 = 0$$
$$\text{Oxidation number for sulfur} = +4$$

Thus, the oxidation number of sulfur in SO_2 is +4.

Exercise 16.3

Determine the oxidation number of chromium in Na_2CrO_4 and $Na_2Cr_2O_7$.

(Try Problems 16.31, 16.32, 16.33, and 16.34.)

■ EXAMPLE 16.4 Assigning Oxidation Numbers in a Polyatomic Ion

Limestone is principally calcium carbonate. What is the oxidation number of carbon in the carbonate ion (CO_3^{2-})?

Problem Analysis

Use rules 1 through 7 to get as many oxidation numbers as you can. Use rule 8 to get an oxidation number that has not yet been assigned.

Solution

The oxidation number of oxygen is -2 (rule 6). Obtain the oxidation number for carbon from rule 8, which states that the sum of the oxidation numbers of the atoms in a polyatomic ion will equal the charge on the ion.

$$\text{Oxidation number for C} + (3 \times \text{oxidation number for O}) = -2$$
$$\text{Oxidation number for C} + 3 \times (-2) = -2$$
$$\text{Oxidation number for C} - 6 = -2$$
$$\text{Oxidation number for C} = +4$$

Exercise 16.4

What is the oxidation number of each sulfur atom in the thiosulfate ion, $S_2O_3^{2-}$?

(Try Problems 16.35, 16.36, 16.37, and 16.38.)

Using Oxidation Numbers in Chemical Equations

Next, we want to show how to use oxidation numbers to identify a reaction as an oxidation–reduction reaction. In terms of oxidation numbers, oxidation is an increase in oxidation number (because electrons are lost, making the hypothetical or real charge more positive), and reduction is a decrease in oxidation number (because electrons are gained, making the hypothetical or real charge more negative). Let's consider, once again, the combination of elemental hydrogen and oxygen to form water.

$$2H_2(g) + O_2(g) \longrightarrow 2H_2O(g)$$

and let's assign oxidation numbers to the atoms in each substance.

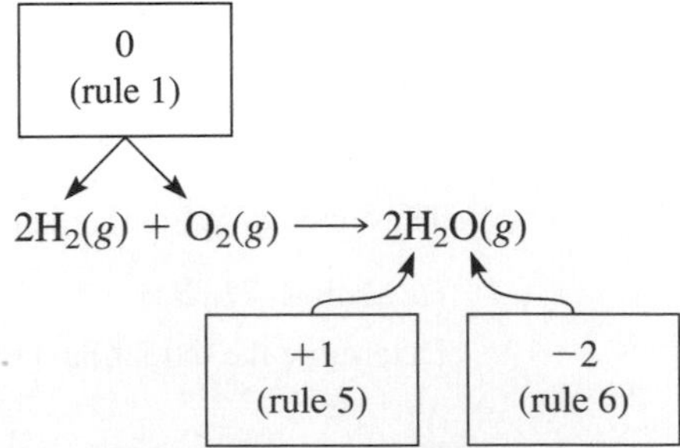

Note that the oxidation number for hydrogen changes from 0 to $+1$, indicating that hydrogen has been oxidized; therefore, H_2 is the reducing agent. The oxidation number for oxygen changes from 0 to -2, indicating that oxygen has been reduced; therefore, O_2 is the oxidizing agent.

Chemists designate a whole molecule or an entire formula unit as the oxidizing or reducing agent, not just the atom that has undergone a change in oxidation number, as shown in the following example.

■ EXAMPLE 16.5 Identifying Oxidizing and Reducing Agents Using Oxidation Numbers

Although you might think that tungsten is a rare element, you probably use it every time you turn on an incandescent light. Tungsten is used to make filaments for incandescent bulbs because it has the highest melting point (3410°C) and highest boiling point (5900°C) of any metal (Figure 16.4). The metal is obtained from tungsten(VI) oxide by heating it in a stream of hydrogen.

$$WO_3(s) + 3H_2(g) \longrightarrow W(s) + 3H_2O(g)$$

Identify the oxidizing and reducing agents, if any.

FIGURE 16.4
Tungsten filament in an incandescent light bulb. The tungsten wire becomes white-hot when an electric current is passed through it.

Problem Analysis

Assign oxidation numbers to each atom using the rules given in this section. Remember that oxidation causes an increase in oxidation number, and reduction causes a decrease in oxidation number. Also, remember that a substance is an oxidizing agent if it is reduced, and a reducing agent if it is oxidized.

Solution

First, you need to assign oxidation numbers to each atom in every substance.

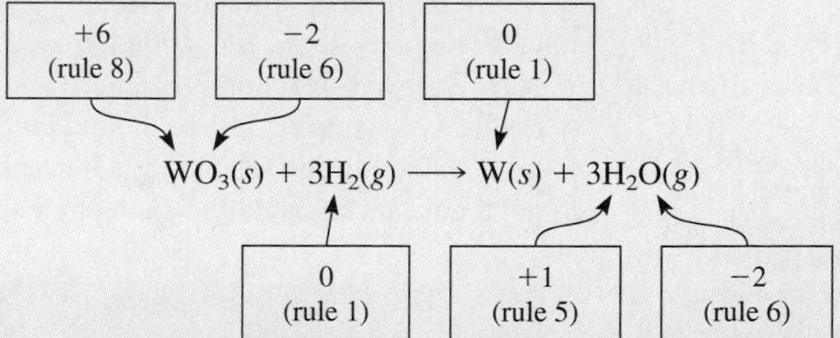

You can see that the oxidation number of tungsten decreases, so it belongs to the compound that is reduced. Similarly, the oxidation number of hydrogen increases, indicating oxidation. Therefore, WO_3 (and not just W) is the oxidizing agent and H_2 is the reducing agent.

Exercise 16.5

Identify the oxidizing agent and reducing agents, if any, in the following reaction.

$$Na_2O(s) + H_2O(l) \longrightarrow 2NaOH(aq)$$

(Try Problems 16.39 and 16.40.)

In this section, we have shown that oxidation numbers can be used to identify the oxidizing and reducing agents in an oxidation–reduction reaction. Next, we show that oxidation numbers can be used to balance oxidation–reduction equations.

16.4 Balancing Oxidation–Reduction Equations by the Half-Reaction Method

OBJECTIVE

- Balance oxidation–reduction equations by the half-reaction method.

Many simple oxidation–reduction equations can be balanced by inspection. Consider, for example, the single-replacement reaction describing the reduction of *pyrolusite* (Figure 16.5), an ore containing manganese dioxide, with aluminum. This reaction has been used at high temperatures to obtain metallic manganese. The unbalanced equation for this reaction is

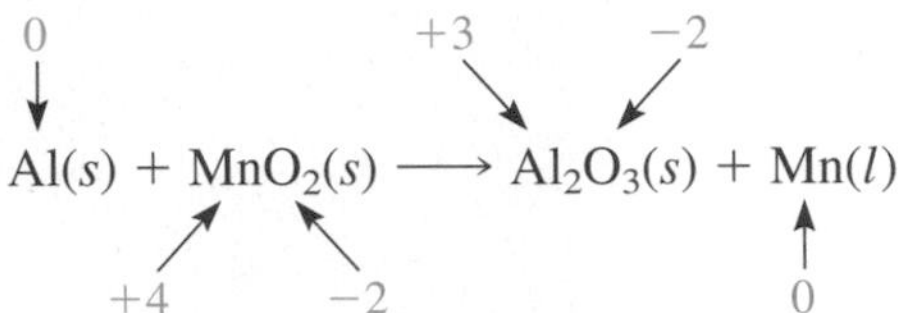

$$\overset{0}{\text{Al}}(s) + \underset{+4\ \ -2}{\text{MnO}_2}(s) \longrightarrow \overset{+3\ \ -2}{\text{Al}_2\text{O}_3}(s) + \underset{0}{\text{Mn}}(l)$$

The oxidation numbers that are written above and below the equation clearly show that MnO_2 has been reduced and Al has been oxidized. It is easy to balance this equation using the systematic approach that we outlined in Chapter 6. The results are

$$4\text{Al}(s) + 3\text{MnO}_2(s) \longrightarrow 2\text{Al}_2\text{O}_3(s) + 3\text{Mn}(l)$$

When aqueous ions are in the equation, however, we must do more than balance each kind of atom; we must also balance the net ionic charge on each side of the equation. The half-reaction method is the most popular way to accomplish both of these tasks. Moreover, it leads to an understanding of the reactions in electrochemical cells and batteries, the topics in Sections 16.5, 16.6, and 16.7.

The half-reaction method separates the equation that we wish to balance into two half-reactions—*one for oxidation and one for reduction*—just as we did in Section 16.1. Each half-reaction is balanced and multiplied by an appropriate number so that the number of electrons lost in the oxidation half-reaction is equal to the number of electrons gained in the reduction half-reaction. The half-reactions are then recombined to get the balanced oxidation–reduction equation.

FIGURE 16.5

Pyrolusite. This mineral is found in Brazil (whose second largest city, Rio de Janeiro, is shown here), India, Cuba, and South Africa. Because it is principally manganese dioxide (MnO_2), pyrolusite is the chief ore of manganese, an element that has been used with iron to make very tough steel.

The starting point for using the half-reaction method is a *skeletal equation,* an unbalanced net ionic equation containing only those substances having atoms that change oxidation number during the reaction. Because the reaction occurs in aqueous solution, you will need to add water (H_2O) and either H^+ or OH^- to balance the equation. Which one you add depends on whether the solution is acidic or basic. A procedure that allows for reactions under acidic or basic conditions is summarized by the following steps.

Rules for Using the Half-Reaction Method

1. Assign oxidation numbers to each atom in the skeletal equation and decide which atoms are oxidized and which atoms are reduced.
2. Split the skeletal equation into a half-reaction for oxidation and a half-reaction for reduction.
3. Balance all of the atoms in each half-reaction except hydrogen and oxygen.
4. Balance oxygen by adding H_2O where needed.
5. In acidic solutions, balance hydrogen by adding H^+ where needed. In basic solutions, first balance hydrogen by adding H^+ where needed, and then change each half-reaction from acidic conditions to basic conditions by adding as many OH^- ions to both sides of the equation as there are H^+ ions. (For example, if you added $2H^+$ to one side of the equation to balance hydrogen, you would now add $2OH^-$ to both sides.) Both H^+ and OH^- ions will appear on one side of the equation. Combine them to give water. (For example, $3H^+$ and $3OH^-$ give $3H_2O$.) Then cancel water molecules, if necessary, so that water appears only on one side of the equation or perhaps not at all.
6. Balance the charge by adding electrons where needed.
7. Multiply the half-reactions by factors that will lead to the same number of electrons in each half-reaction.
8. Add the two half-reactions. Cancel equal amounts of any substance (including electrons) on both sides of the overall equation.
9. Check the equation to make sure it is balanced.

Let's apply the half-reaction method to the reaction between intensely purple potassium permanganate and potassium iodide in an acidic aqueous solution (Figure 16.6). After we omit the spectator ions (K^+), the skeletal net ionic equation for this reaction is

$$MnO_4^-(aq) + I^-(aq) \longrightarrow Mn^{2+}(aq) + I_2(aq)$$

The step-by-step procedure is as follows.

1. Assign oxidation numbers to each atom in the equation and decide which atoms are oxidized and which atoms are reduced.

$$\overset{+7}{Mn}\overset{-2}{O_4}{}^- + \underset{-1}{I^-} \longrightarrow \overset{+2}{Mn^{2+}} + \underset{0}{I_2}$$

   Clearly, iodine is oxidized because its oxidation number increases, and MnO_4^- is reduced because the oxidation number of manganese decreases. The MnO_4^- ion (not just Mn) is the oxidizing agent and the I^- ion is the reducing agent.

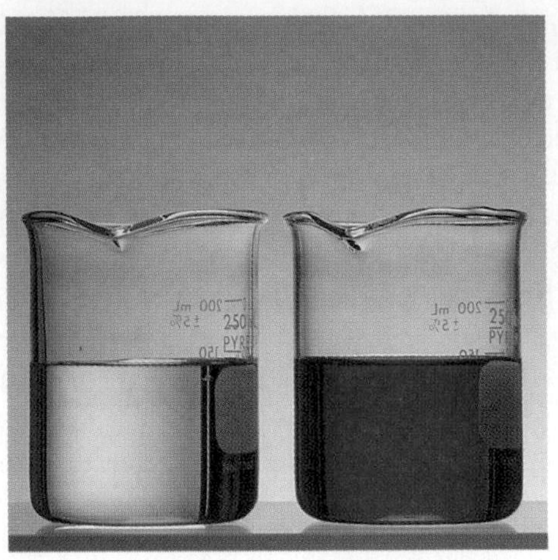

A B

FIGURE 16.6

The reaction of aqueous solutions of potassium permanganate and potassium iodide. *(A)* The solution of potassium permanganate ($KMnO_4$), a strong oxidizing agent, is intensely purple. The solution of potassium iodide (KI), a moderately strong reducing agent, is colorless. *(B)* When the solutions are mixed, the color of potassium permanganate disappears, and the color of aqueous iodine (I_2) appears.

2. Split the skeletal equation into a half-reaction for oxidation and a half-reaction for reduction.

$$I^- \longrightarrow I_2 \qquad \text{(Oxidation)}$$
$$MnO_4^- \longrightarrow Mn^{2+} \qquad \text{(Reduction)}$$

3. Balance all of the atoms in each half-reaction except hydrogen and oxygen.

$$2I^- \longrightarrow I_2 \qquad \text{(Oxidation)}$$
$$MnO_4^- \longrightarrow Mn^{2+} \qquad \text{(Reduction)}$$

4. Balance oxygen by adding H_2O where needed.

$$2I^- \longrightarrow I_2 \qquad (H_2O \text{ is not needed})$$
$$MnO_4^- \longrightarrow Mn^{2+} + 4H_2O$$

5. Balance hydrogen by adding H^+ where needed.

$$2I^- \longrightarrow I_2 \qquad (H^+ \text{ is not needed})$$
$$MnO_4^- + 8H^+ \longrightarrow Mn^{2+} + 4H_2O$$

   Because the reaction occurs in acidic solution, you do not need to introduce OH^- ions.

6. Balance the charge by adding electrons where needed.

$$2I^- \longrightarrow I_2 + 2e^- \qquad \text{(Net charge on each side} = -2)$$
$$MnO_4^- + 8H^+ + 5e^- \longrightarrow Mn^{2+} + 4H_2O \qquad \text{(Net charge on each side} = +2)$$

7. Multiply the half-reactions by factors that will lead to the same number of electrons in each half-reaction.

$$5 \times [2I^- \longrightarrow I_2 + 2e^-] = 10I^- \longrightarrow 5I_2 + 10e^-$$
$$2 \times [MnO_4^- + 8H^+ + 5e^- \longrightarrow Mn^{2+} + 4H_2O] =$$
$$2MnO_4^- + 16H^+ + 10e^- \longrightarrow 2Mn^{2+} + 8H_2O$$

8. Add the two half-reactions. Cancel equal amounts of any substance (including electrons) on both sides of the overall equation.

$$10I^- \longrightarrow 5I_2 + 10e^-$$
$$2MnO_4^- + 16H^+ + 10e^- \longrightarrow 2Mn^{2+} + 8H_2O$$
$$2MnO_4^- + 10I^- + 16H^+ + \cancel{10e^-} \longrightarrow 2Mn^{2+} + 5I_2 + 8H_2O + \cancel{10e^-}$$

The electrons cancel, as you can see. The result is

$$2MnO_4^-(aq) + 10I^-(aq) + 16H^+(aq) \longrightarrow 2Mn^{2+}(aq) + 5I_2(aq) + 8H_2O(l)$$

9. Check the equation to make sure it is balanced.

	Left	Right
Mn	2	2
O	8	8
I	10	10
H	16	16
Charge	+4	+4

We use this procedure again in the next example.

EXAMPLE 16.6 Balancing Equations by the Half-Reaction Method in Acidic Solution

Police use a device called a Breathalyzer (Figure 16.7) to test a person's breath for ethanol, an alcohol whose formula is C_2H_6O. The Breathalyzer contains an acidic solution of potassium dichromate ($K_2Cr_2O_7$). Reaction with ethanol causes the solution to change from the characteristic orange color of the $Cr_2O_7^{2-}$ ion to a green color due to Cr^{3+} ions. The change in color is proportional to the amount of ethanol in the sample of exhaled air, which in turn is directly related to the alcohol content of the person's blood. The skeletal equation for this oxidation–reduction reaction is

$$Cr_2O_7^{2-} + C_2H_6O \longrightarrow Cr^{3+} + CO_2$$

Balance this equation by the half-reaction method.

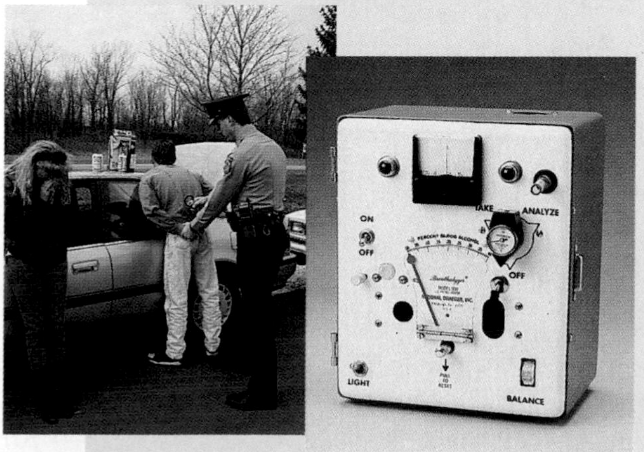

FIGURE 16.7
Chemistry in action with the Breathalyzer. Police use this device to test the breath of a driver suspected of drinking too much alcohol. It works because of the oxidation–reduction reaction between the alcohol and an acidic, aqueous solution of potassium dichromate ($K_2Cr_2O_7$), a strong oxidizing agent.

Problem Analysis

Use the nine rules for balancing an oxidation–reduction equation by the half-reaction method. Note that the reaction occurs in an acidic solution.

Solution

1. Assign oxidation numbers to each atom in the equation and decide which atoms are oxidized and which atoms are reduced.

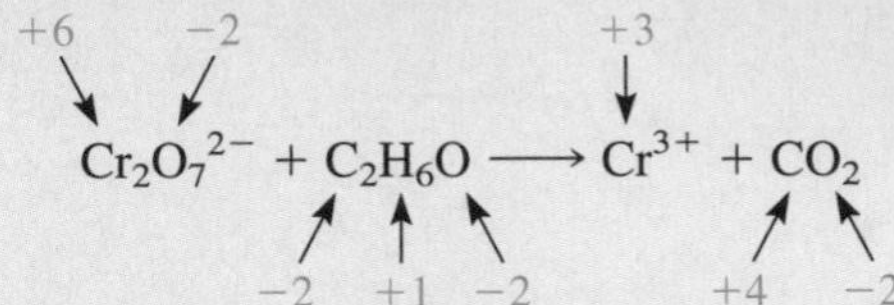

The anion $Cr_2O_7^{2-}$ is reduced because the oxidation number of chromium decreases; ethanol is oxidized because the oxidation number of carbon increases. The $Cr_2O_7^{2-}$ ion is the oxidizing agent and ethanol is the reducing agent.

2. Split the skeletal equation into a half-reaction for oxidation and a half-reaction for reduction.

$$C_2H_6O \longrightarrow CO_2 \quad \text{(Oxidation)}$$
$$Cr_2O_7^{2-} \longrightarrow Cr^{3+} \quad \text{(Reduction)}$$

3. Balance all of the atoms in each half-reaction except hydrogen and oxygen.

$$C_2H_6O \longrightarrow 2CO_2$$
$$Cr_2O_7^{2-} \longrightarrow 2Cr^{3+}$$

4. Balance oxygen by adding H_2O where needed.

$$C_2H_6O + 3H_2O \longrightarrow 2CO_2$$
$$Cr_2O_7^{2-} \longrightarrow 2Cr^{3+} + 7H_2O$$

5. Balance hydrogen by adding H^+ where needed.

$$C_2H_6O + 3H_2O \longrightarrow 2CO_2 + 12H^+$$
$$Cr_2O_7^{2-} + 14H^+ \longrightarrow 2Cr^{3+} + 7H_2O$$

Since the reaction occurs in acidic solution, you do not need to introduce OH^- ions.

6. Balance the charge by adding electrons where needed.

$$C_2H_6O + 3H_2O \longrightarrow 2CO_2 + 12H^+ + 12e^- \quad \text{(Net charge on each side = 0)}$$
$$Cr_2O_7^{2-} + 14H^+ + 6e^- \longrightarrow 2Cr^{3+} + 7H_2O \quad \text{(Net charge on each side = +6)}$$

7. Multiply the half-reactions by factors that will lead to the same number of electrons in each half-reaction.

$$1 \times [C_2H_6O + 3H_2O \longrightarrow 2CO_2 + 12H^+ + 12e^-] =$$
$$C_2H_6O + 3H_2O \longrightarrow 2CO_2 + 12H^+ + 12e^-$$
$$2 \times [Cr_2O_7^{2-} + 14H^+ + 6e^- \longrightarrow 2Cr^{3+} + 7H_2O] =$$
$$2Cr_2O_7^{2-} + 28H^+ + 12e^- \longrightarrow 4Cr^{3+} + 14H_2O$$

8. Add the two half-reactions.

$$C_2H_6O + 3H_2O \longrightarrow 2CO_2 + 12H^+ + 12e^-$$
$$2Cr_2O_7^{2-} + 28H^+ + 12e^- \longrightarrow 4Cr^{3+} + 14H_2O$$

$$2Cr_2O_7^{2-} + C_2H_6O + \underset{16}{28H^+} + \cancel{3H_2O} + \cancel{12e^-} \longrightarrow 4Cr^{3+} + 2CO_2 + \cancel{12H^+} + \underset{11}{\cancel{14}H_2O} + \cancel{12e^-}$$

Notice that 12 electrons on each side have been cancelled. In addition, the $12H^+$ ions on the right cancel the same number on the left, causing a decrease from $28H^+$ to $16H^+$. Similarly, the $3H_2O$ molecules on the left cancel an identical number on the right, leaving $11H_2O$. The result is

$$2Cr_2O_7^{2-} + C_2H_6O + 16H^+ \longrightarrow 4Cr^{3+} + 2CO_2 + 11H_2O$$

9. Check the equation to make sure it is balanced.

	Left	Right
Cr	4	4
O	15	15
C	2	2
H	22	22
Charge	+12	+12

The balanced equation is

$$2Cr_2O_7^{2-} + C_2H_6O + 16H^+ \longrightarrow 4Cr^{3+} + 2CO_2 + 11H_2O$$

Exercise 16.6

Hydrogen sulfide is oxidized by the nitrate ion in acidic solution. The skeletal net ionic equation is

$$H_2S(aq) + NO_3^-(aq) \longrightarrow S_8(s) + NO(g)$$

Balance this equation by the half-reaction method.

(Try Problems 16.45, 16.46, 16.47, and 16.48.)

Sometimes oxidation–reduction reactions occur under basic conditions. For example, neither manganese(III) oxide (Mn_2O_3) nor manganese(IV) oxide (MnO_2) is stable under acidic or neutral conditions. However, these compounds can be prepared and used under basic conditions, as you can see in the next example.

■ EXAMPLE 16.7 Balancing Equations by the Half-Reaction Method in Basic Solution

An alkaline battery (Figure 16.8) discharges electricity because of an oxidation–reduction reaction involving zinc and manganese(IV) oxide under basic conditions.

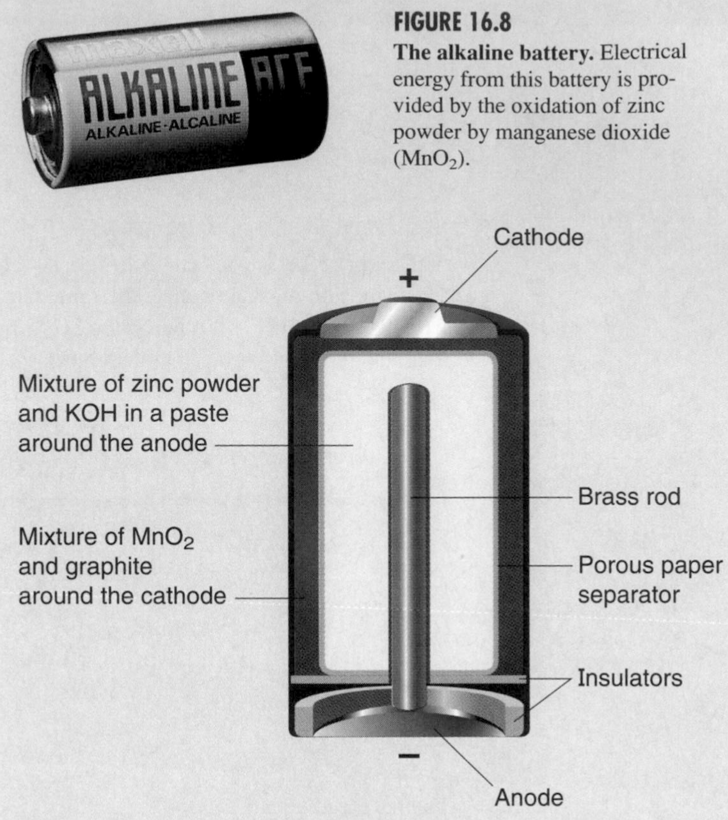

FIGURE 16.8
The alkaline battery. Electrical energy from this battery is provided by the oxidation of zinc powder by manganese dioxide (MnO_2).

The skeletal unbalanced equation for the reaction is

$$Zn(s) + MnO_2(s) \longrightarrow Zn(OH)_2(s) + Mn_2O_3(s)$$

Balance this equation by the half-reaction method.

Problem Analysis

Use the nine rules for balancing an oxidation–reduction equation by the half-reaction method. Note that the reaction occurs under basic conditions.

Solution

1. Assign oxidation numbers to each atom in the equation and decide which atoms are oxidized and which atoms are reduced.

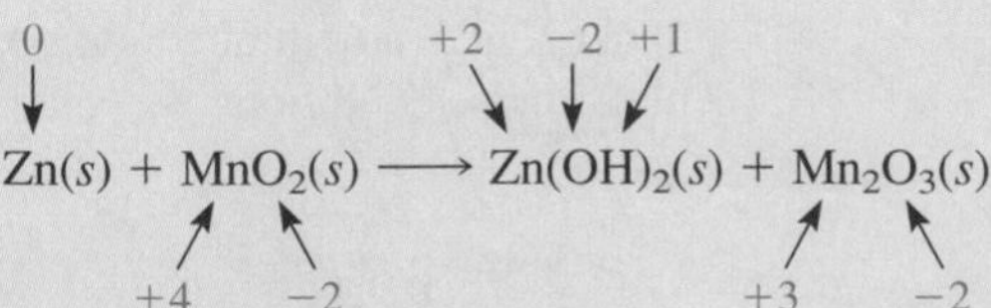

   Zinc is oxidized and it is the reducing agent; MnO_2 is reduced (because the oxidation number of manganese decreases) and it is the oxidizing agent.

2. Split the skeletal equation into a half-reaction for oxidation and a half-reaction for reduction.

$$Zn \longrightarrow Zn(OH)_2 \qquad \text{(Oxidation)}$$
$$MnO_2 \longrightarrow Mn_2O_3 \qquad \text{(Reduction)}$$

3. Balance all of the atoms in each half-reaction except hydrogen and oxygen.

$$Zn \longrightarrow Zn(OH)_2$$
$$2MnO_2 \longrightarrow Mn_2O_3$$

4. Balance oxygen by adding H_2O where needed.

$$Zn + 2H_2O \longrightarrow Zn(OH)_2$$
$$2MnO_2 \longrightarrow Mn_2O_3 + H_2O$$

5. Balance hydrogen by adding H^+ where needed.

$$Zn + 2H_2O \longrightarrow Zn(OH)_2 + 2H^+$$
$$2MnO_2 + 2H^+ \longrightarrow Mn_2O_3 + H_2O$$

Notice that $2H^+$ occurs in both half-reactions. Because this reaction occurs under basic conditions, change the half-reactions from acidic conditions to basic conditions by adding as many OH^- ions to both sides of each half-reaction as there are H^+ ions. (It is a coincidence here that both half-reactions require the same procedure.)

$$Zn + 2H_2O + 2OH^- \longrightarrow Zn(OH)_2 + \underbrace{2H^+ + 2OH^-}_{2H_2O}$$
$$2MnO_2 + \underbrace{2H^+ + 2OH^-}_{2H_2O} \longrightarrow Mn_2O_3 + H_2O + 2OH^-$$

Combine the $2H^+ + 2OH^-$ on the right side of the first half-reaction to give two H_2O molecules. Repeat with the left side of the second half-reaction.

$$Zn + 2H_2O + 2OH^- \longrightarrow Zn(OH)_2 + 2H_2O$$
$$2MnO_2 + 2H_2O \longrightarrow Mn_2O_3 + H_2O + 2OH^-$$

Cancel equal numbers of water molecules.

$$Zn + \cancel{2H_2O} + 2OH^- \longrightarrow Zn(OH)_2 + \cancel{2H_2O}$$
$$2MnO_2 + \overset{1}{\cancel{2}}H_2O \longrightarrow Mn_2O_3 + \cancel{H_2O} + 2OH^-$$

The results are

$$Zn + 2OH^- \longrightarrow Zn(OH)_2$$
$$2MnO_2 + H_2O \longrightarrow Mn_2O_3 + 2OH^-$$

6. Balance the charge by adding electrons where needed.

$$Zn + 2OH^- \longrightarrow Zn(OH)_2 + 2e^-$$
$$2MnO_2 + H_2O + 2e^- \longrightarrow Mn_2O_3 + 2OH^-$$

7. Multiply the half-reactions by factors that will lead to the same number of electrons in each half-reaction. In this case, the half-reactions already have the same number of electrons.

8. Add the two half-reactions. Cancel equal amounts of any substance (including electrons) on both sides of the overall equation.

$$Zn + 2OH^- \longrightarrow Zn(OH)_2 + 2e^-$$
$$2MnO_2 + H_2O + 2e^- \longrightarrow Mn_2O_3 + 2OH^-$$
$$Zn + 2MnO_2 + \cancel{2OH^-} + H_2O + \cancel{2e^-} \longrightarrow Zn(OH)_2 + Mn_2O_3 + \cancel{2OH^-} + \cancel{2e^-}$$

The result is

$$Zn + 2MnO_2 + H_2O \longrightarrow Zn(OH)_2 + Mn_2O_3$$

9. Check the equation to make sure it is balanced.

	Left	Right
Zn	1	1
Mn	2	2
O	5	5
H	2	2
Charge	0	0

The balanced equation is

$$Zn(s) + 2MnO_2(s) + H_2O(l) \longrightarrow Zn(OH)_2(s) + Mn_2O_3(s)$$

Exercise 16.7

Balance the skeletal equation

$$MnO_4^-(aq) + I^-(aq) \longrightarrow MnO_2(s) + I_2(aq)$$

assuming that the reaction occurs in a basic solution.

(Try Problems 16.49 and 16.50.)

ELECTROCHEMISTRY

In the first section of this chapter, we showed that metallic zinc reduces the copper(II) ions in a solution of copper(II) sulfate, according to the net ionic equation

$$Zn(s) + Cu^{2+}(aq) \longrightarrow Zn^{2+}(aq) + Cu(s)$$

Recall that electrons flow from zinc to copper(II) ions as the zinc is oxidized and the Cu^{2+} ions are reduced. This is a **spontaneous process,** *a physical or chemical change that occurs by itself.* Spontaneous processes do not require an outside agent (such as a source of energy) to make them happen. ●

● If a rock is placed on a hill, it will roll downhill spontaneously. The rolling of the rock uphill by itself is a nonspontaneous process.

Oxidation–reduction reactions such as this one always involve electrons moving spontaneously from the reducing agent to the oxidizing agent. In the next two sections, we explore how **electricity**—*the flow of electrons through a conductor*—can be obtained from a spontaneous oxidation–reduction reaction. This is how batteries work. We also look at using electricity to make a nonspontaneous reaction occur, which is the basis for recharging a battery. Spontaneous oxidation–reduction reactions that produce electricity and nonspontaneous oxidation–reduction reactions that are caused by electrical currents are the subjects of **electrochemistry,** *the study of the conversion of stored* *chemical energy into electrical energy and vice versa*. An electrochemical reaction usually occurs in an **electrochemical cell,** *an apparatus that either generates or uses an electric current*. There are two types of these cells: *voltaic cells* and *electrolytic cells*. We study these in the following two sections and then look at a few practical uses of voltaic cells.

Animation: Electrochemical half-reactions in a galvanic cell.

16.5 Voltaic Cells

OBJECTIVES

- Define a *voltaic (galvanic) cell*.
- Describe how to construct a voltaic cell.
- Use the activity series to predict the direction of electron flow in a voltaic cell.

Electricity from batteries allows us to start cars, to tell time from watches that do not require winding, and to use cordless appliances and tools. When you look at the battery in one of these devices, it may not be clear that the battery case is a container for an oxidation–reduction reaction—one in which chemical energy is converted to electrical energy. How does this process work? Let's begin by returning to a familiar reaction: the spontaneous reaction of zinc with Cu^{2+} ions.

Consider Figure 16.2 again. Because electrons flow directly from zinc to the copper ions in solution, we are not able to harness the electrons so they can be used to light a bulb or to power a cordless tool. However, suppose we dip a strip of zinc in a beaker containing a solution of zinc(II) sulfate and, separately in another beaker, a strip of copper in a solution of copper(II) sulfate, as shown in Figure 16.9A. The direct oxidation–reduction reaction cannot occur because we have physically isolated Zn from the Cu^{2+} ions. Nevertheless, the tendency for the transfer of electrons between these substances remains. We can take advantage of that tendency by connecting the zinc and copper strips by a wire. Now we have a path for electrons to flow, but the circuit is not complete until we add a means for ions to flow from one beaker to the other—the inverted U-shaped tube (salt bridge) in Figure 16.9B. The voltmeter connected to the wire between the strips shows that we have been successful—electricity is flowing through the wire.

Video: Galvanic (voltaic) cells.

The device in Figure 16.9 is a **voltaic cell,** *a cell in which a spontaneous chemical reaction generates an electric current*. Voltaic cells are named in honor of Alessandro Volta (1745–1827), who discovered the first primitive cells of this type around 1800. It is also called a **galvanic cell** after Luigi Galvani (1737–1798), who is generally believed to have discovered electricity. The voltaic cell consists of two *half-cells* that are connected electrically. A **half-cell** is *that portion of an electrochemical cell in which a half-reaction takes place*. The half-reactions in this case are

$$Zn(s) \longrightarrow Zn^{2+}(aq) + 2e^- \qquad \text{(Oxidation)}$$
$$Cu^{2+}(aq) + 2e^- \longrightarrow Cu(s) \qquad \text{(Reduction)}$$

Exploration: Electrochemistry.

Each metal strip is called an *electrode*. Oxidation occurs at the zinc electrode and reduction occurs at the copper electrode. *The electrode at which oxidation occurs* is called the **anode.** The **cathode** is *the electrode at which reduction occurs*. These definitions of anode and cathode also apply to the electrolytic cells we examine in the next section.

Because oxidation occurs at the zinc electrode (the anode), electrons leave that electrode and flow through the external wire to the copper electrode (the cathode), where reduction occurs. This cell can produce about 1 *volt,* but the exact voltage depends on

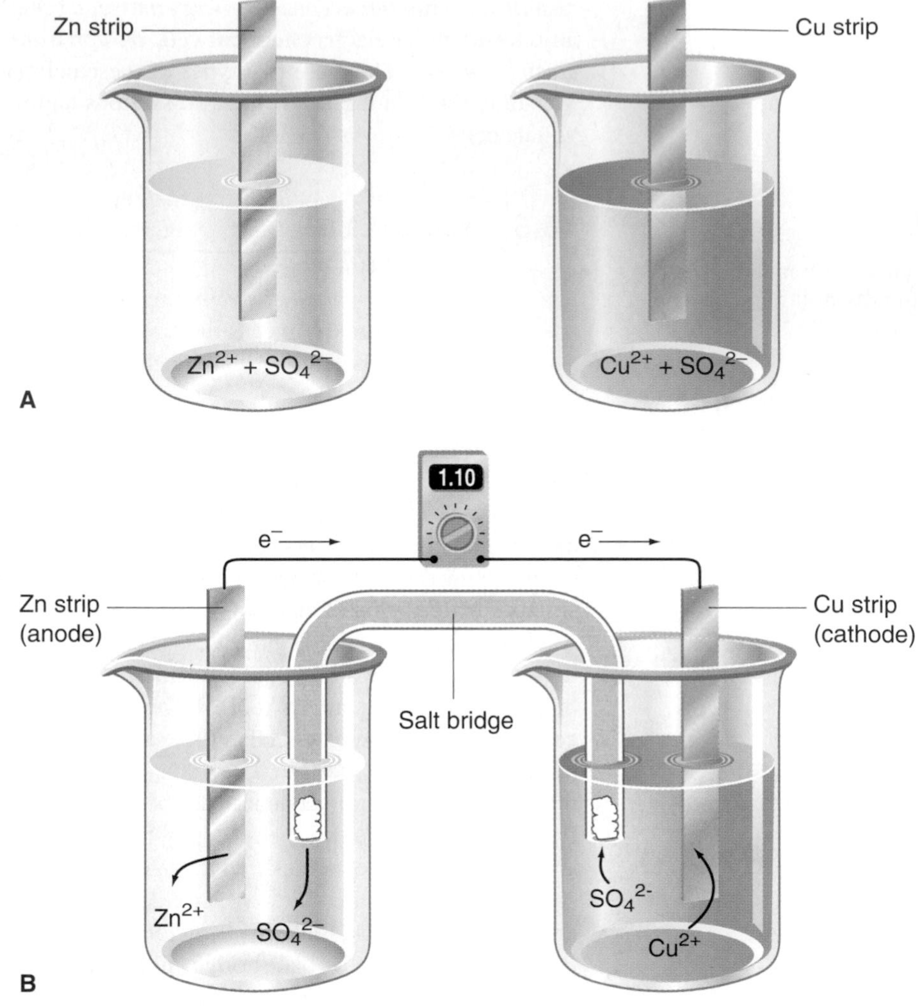

FIGURE 16.9
Constructing a voltaic cell. *(A)* A zinc strip in a solution of zinc sulfate and a copper strip in a solution of copper sulfate are the chemical components of a well-known voltaic cell. There is a natural tendency for oxidation–reduction to occur and for electrons to flow from the zinc strip to the copper strip, but the reaction cannot occur because the components are physically separated. *(B)* When the components are connected by the wire and the salt bridge, an electric current flows. The voltage is 1.10 V when the concentrations of zinc sulfate and copper sulfate are identical.

the concentrations of the Zn^{2+} and Cu^{2+} ions. A **volt** (V) is *an SI unit describing the energy per unit charge.* The higher the voltage, the more energy the electrons carry. Other voltaic cells may produce just a few hundredths of a volt or up to 2 or 3 V. If we use a voltaic cell continuously for an extended period of time, the voltage will decrease very slowly until eventually it is zero. When the voltage is zero, the spontaneous chemical reaction generating the flow of electrons has reached equilibrium. Reactions at equilibrium cannot provide any usable energy.

There is more to be said about the need for the inverted U-tube in Figure 16.9B. Electrons leave the anode as zinc metal is oxidized to Zn^{2+} ions. The electrons travel through the external circuit to the cathode, where they are captured by Cu^{2+} ions as these ions are reduced to copper metal. As this process occurs, Zn^{2+} ions enter the solution surrounding the zinc electrode, while Cu^{2+} ions simultaneously leave the solution around the copper electrode. If nothing else happened, there would be too few sulfate ions at the anode and too many at the cathode because of these reactions. Because ionic systems must be electrically neutral (with the number of positive charges equal to the number of negative charges), a voltaic cell must have some means for ions to pass from one side of the cell to the other. The inverted U-tube in Figure 16.9B permits the passage of ions through a gelatinous material containing an ionic salt, such as sodium sulfate (Na_2SO_4). The U-tube can be replaced with a porous glass plate. In either case, the passage of ions allows the solutions at the anode and cathode to remain electrically neutral.

A voltaic cell can be constructed with any one of many different anodes and any one of many different cathodes. How can you tell which way the electrons will flow and which are the oxidizing and reducing agents without doing an experiment? You can use the activity series in Table 16.2 to predict the behavior of an electrochemical cell whose electrodes are constructed from two of the substances in the table. For each electrode, you need a strip of the metal, and it must be placed in a solution of some salt of the metal. If you make the appropriate electrical connections, electrons will flow spontaneously from the electrode made of the more active metal (the one higher in the activity series) to the electrode made of the less active metal. Consider the following example.

■ EXAMPLE 16.8 Predicting the Behavior of a Voltaic Cell

An amateur chemist is experimenting with voltaic cells. She dips a magnesium strip in a solution of $Mg(NO_3)_2$ and a silver strip in a solution of $AgNO_3$. She uses beakers that are connected by a porous glass plate so that ions can pass. Then she connects the metal strips with a wire. A sketch of her apparatus is shown in Figure 16.10. Use the activity series in Table 16.2 to make some predictions about her cell. Which substance is oxidized and which one is reduced? What is the anode and what is the cathode? In which direction will electrons flow during the reaction? What are the half-cell reactions, and what is the overall chemical reaction?

Problem Analysis

Recall that oxidation occurs at the anode and reduction occurs at the cathode. Electrons flow from the anode, which is the more active metal (the one higher in the activity series), to the cathode, the less active metal.

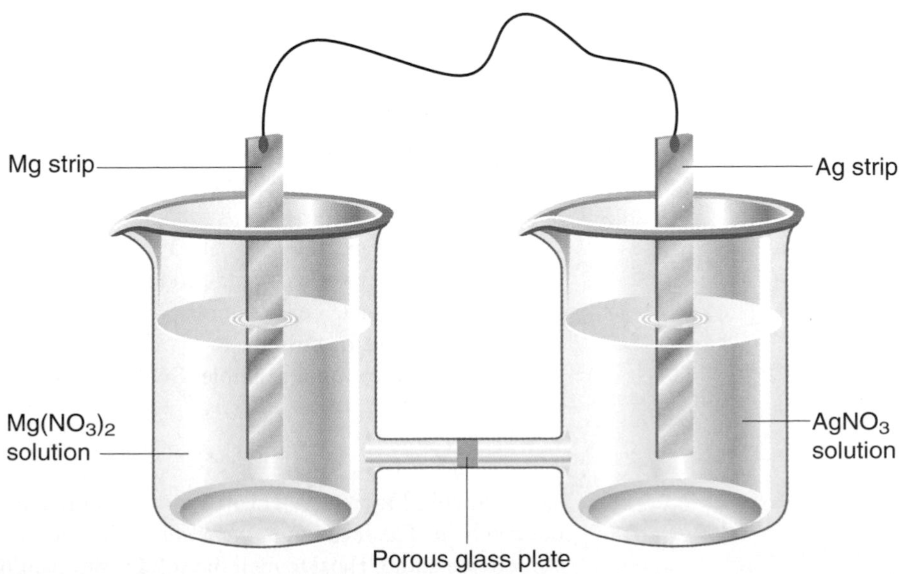

FIGURE 16.10

A voltaic cell. A voltaic cell can be made from almost any pair of different metals and solutions of their salts. In this case, the cell has a magnesium electrode dipped in a solution of $Mg(NO_3)_2$ and a silver electrode in a solution of $AgNO_3$. Electrons flow through the external wire. The circuit is completed by ions passing through the porous glass plate.

Solution

Because magnesium is above silver in Table 16.2, magnesium metal (Mg) will be oxidized and silver ions (Ag^+) will be reduced. Thus, the magnesium strip is the anode and the silver strip is the cathode. Electrons flow from the magnesium half-cell through the wire to the silver half-cell. These statements give you the half-cell reactions.

$$Mg(s) \longrightarrow Mg^{2+}(aq) + 2e^- \quad \text{(Oxidation)}$$
$$Ag^+(aq) + e^- \longrightarrow Ag(s) \quad \text{(Reduction)}$$

After we multiply the half-reaction for reduction by 2, addition of the half-reactions leads to cancellation of the electrons and the cell's overall chemical reaction.

$$\begin{array}{r}
Mg(s) \longrightarrow Mg^{2+}(aq) + 2e^- \\
2Ag^+(aq) + 2e^- \longrightarrow 2Ag(s) \\
\hline
Mg(s) + 2Ag^+(aq) + \cancel{2e^-} \longrightarrow Mg^{2+}(aq) + 2Ag(s) + \cancel{2e^-}
\end{array}$$

Thus, you predict that the overall chemical reaction (written as a net ionic equation) is

$$Mg(s) + 2Ag^+(aq) \longrightarrow Mg^{2+}(aq) + 2Ag(s)$$

Exercise 16.8

Consider a voltaic cell containing a copper electrode in a solution containing copper(II) ions and a lead electrode in a solution containing lead(II) ions. Which substance is oxidized and which one is reduced? What is the anode and what is the cathode? In which direction do electrons flow during the reaction? What are the half-cell reactions, and what is the overall chemical reaction?

(Try Problems 16.51, 16.52, 16.53, and 16.54.)

16.6 Electrolytic Cells

OBJECTIVES

- Define an *electrolytic cell.*
- Show how an electrolytic cell works.

A voltaic cell produces electricity because an oxidation–reduction reaction occurs spontaneously. In contrast, an **electrolytic cell** is *an electrochemical cell in which an external electric current drives a nonspontaneous oxidation–reduction reaction.* The process occurring in this type of cell is called *electrolysis,* and it has great practical importance. Charging the battery of a car and producing aluminum for soft-drink cans are examples of electrolysis. In each case, a nonspontaneous oxidation–reduction reaction is driven by an external electric current. ●

● Production of aluminum metal uses a little more than 4% of all the electricity in the United States.

Let's look at a simple example. Sodium reacts violently with chlorine to give ionic sodium chloride,

$$2Na(s) + Cl_2(g) \longrightarrow 2NaCl(s)$$

as you saw in Figure 6.9. This reaction is spontaneous since it occurs by itself with no outside help. The reaction in the reverse direction is not spontaneous, however. Even if we provide enough heat to melt the solid sodium chloride, it will not decompose into the elements.

$$NaCl(l) \xrightarrow{\Delta} \text{no reaction}$$

Nevertheless, we can force this reaction by passing an external electric current through the melted salt. Sodium metal and gaseous chlorine have been produced commercially by this process since 1921. When a salt such as sodium chloride melts, the positive and negative ions are free to move about. The positive ions move to the cathode, where they are reduced, and the negative ions move to the anode, where they are oxidized. The half-reactions for this electrochemical reaction are simple.

$$\text{Cathode:} \quad Na^+ + e^- \longrightarrow Na \qquad \text{(Reduction)}$$

$$\text{Anode:} \quad 2Cl^- \longrightarrow Cl_2 + 2e^- \qquad \text{(Oxidation)}$$

The site of oxidation is the anode, and the site of reduction is the cathode, just as it is with a voltaic cell. When the first half-reaction is multiplied by 2 and when the electrons are canceled, the result is

$$2NaCl(l) \longrightarrow 2Na(l) + Cl_2(g)$$

● Electroplating of one metal on another is done by electrolysis. Old automobiles have steel bumpers with chromium plating to protect them from corrosion.

Electrolysis is also an important commercial technique for producing aluminum from minerals containing this element, and hydrogen from water. In the next section, we show how electrolysis can be used to recharge a battery. ●

16.7 Three Important Batteries

OBJECTIVES

- Describe a battery.
- Show how a lead storage battery, a nickel–cadmium battery, and a mercury battery work.

A **battery** is *a single voltaic cell or a series of voltaic cells.* If more than one cell is used, they are arranged so that the anode of one is connected to the cathode of another. The voltage from such a series of cells is equal to the voltage from one cell in the series multiplied by the number of cells. Some batteries are rechargeable and others are not. Let's look at three batteries that may be familiar to you.

Lead Storage Battery

Because of its use in cars, the best known battery is probably the lead storage battery (Figure 16.11A). It consists of a series of lead grids alternately packed with spongy elemental lead (to form the anode) and lead dioxide (to form the cathode). The grids are immersed in an aqueous solution of sulfuric acid. The spontaneous half-reactions during discharge of electricity are

$$\text{Anode:} \quad Pb(s) + H_2SO_4(aq) \longrightarrow PbSO_4(s) + 2H^+(aq) + 2e^- \qquad \text{(Oxidation)}$$

$$\text{Cathode:} \quad PbO_2(s) + H_2SO_4(aq) + 2H^+(aq) + 2e^- \longrightarrow PbSO_4(s) + 4H_2O(l) \qquad \text{(Reduction)}$$

Both reactions produce insoluble $PbSO_4$. The net reaction is the sum of these reactions, or

$$Pb(s) + PbO_2(s) + 2H_2SO_4(aq) \longrightarrow 2PbSO_4(s) + 2H_2O(l)$$

Each adjacent pair of grids in Figure 16.11A is a voltaic cell, and each pair furnishes about 2 V. The familiar battery found in most cars consists of six cells wired one after the other. As a result, the battery will provide about 2 V × 6 or about 12 V.

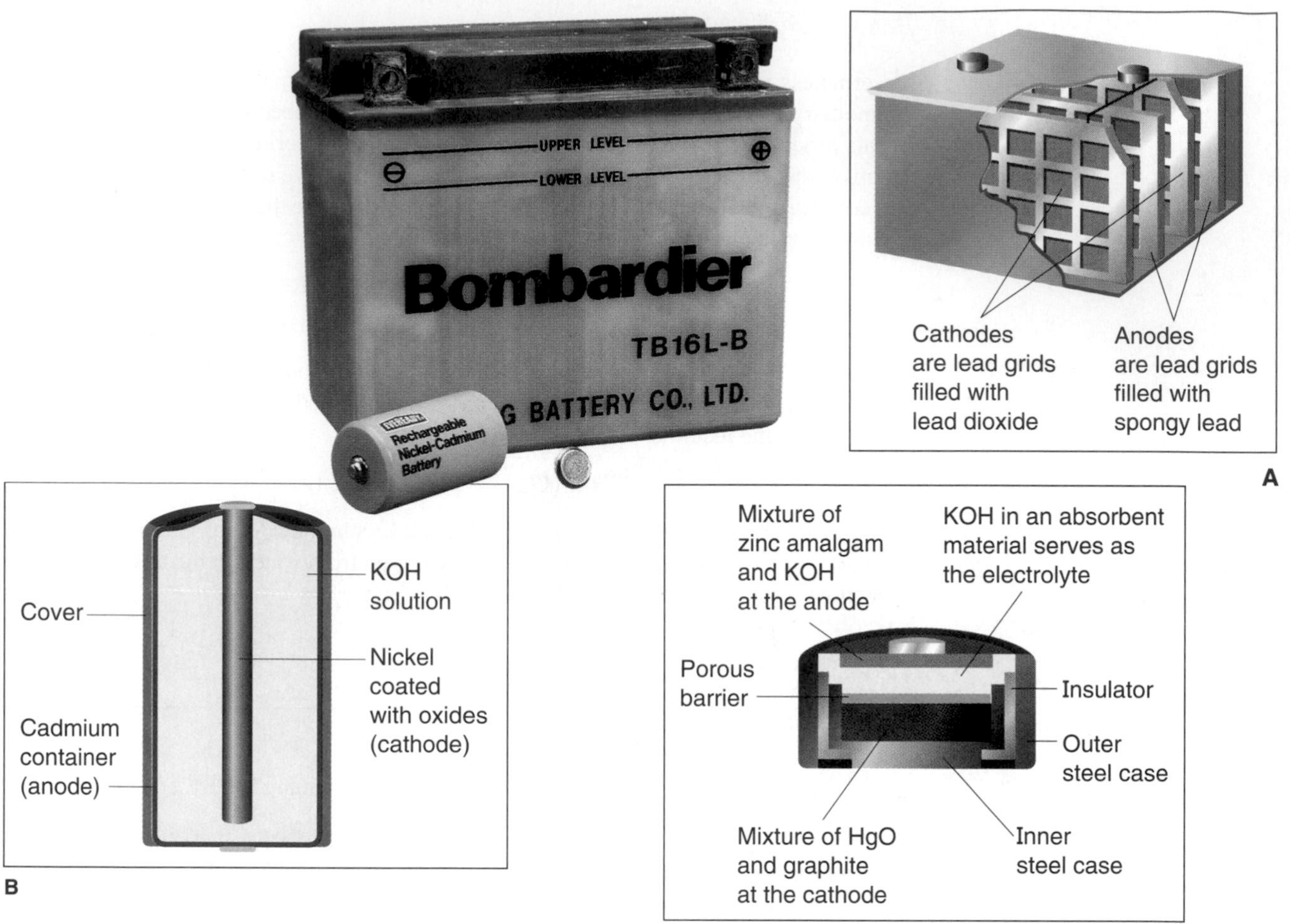

FIGURE 16.11
Three batteries. *(A)* A lead storage battery, *(B)* a nickel–cadmium battery, and *(C)* a mercury battery.

The lead storage battery can be recharged. The reaction during recharging is just the reverse of the spontaneous reaction, or

$$2PbSO_4(s) + 2H_2O(l) \rightarrow Pb(s) + PbO_2(s) + 2H_2SO_4(aq)$$

Because this reaction is not spontaneous, it must be driven by an external electric current from some other source. Each cell in the lead storage battery becomes an electrolytic cell during recharging. The external electric current is supplied by the car's alternator under normal running conditions, or by a battery recharger if the battery is "dead." After the battery is recharged, it is ready to operate spontaneously once again.

Nickel–Cadmium Battery

The nickel–cadmium or NiCad battery (Figure 16.11B) is used extensively with portable rechargeable appliances and tools. Unlike the lead storage battery, only a single voltaic cell is used in many cases. The electrode reactions are

Anode:	$Cd + 2OH^- \longrightarrow Cd(OH)_2 + 2e^-$	(Oxidation)
Cathode:	$NiO(OH) + H_2O + e^- \longrightarrow Ni(OH)_2 + OH^-$	(Reduction)

This battery produces about 1.3 V and has a long life. It can be recharged.

Mercury Battery

The mercury battery is used extensively in watches and pacemakers because of its reliability (Figure 16.11C). It also produces about 1.3 V. The electrode reactions are

$$\text{Anode:} \quad Zn + 2OH^- \longrightarrow Zn(OH)_2 + 2e^- \quad \text{(Oxidation)}$$
$$\text{Cathode:} \quad HgO + H_2O + 2e^- \longrightarrow Hg + 2OH^- \quad \text{(Reduction)}$$

This battery cannot be recharged. ●

● We also discussed another kind of battery, the alkaline battery, in Example 16.7.

Chemical Perspective

Electrochemistry and Rusting

The 172-ft U.S.S. *Monitor* lies bottom up about 220 ft below the surface of the Atlantic Ocean near Cape Hatteras, North Carolina. This Civil War ironclad sank during a storm ten months after its historic 1862 battle with the Confederate C.S.S. *Virginia,* a ship formerly known as the U.S.S. *Merrimack* (Figure 16.12). Since its discovery in 1973, the *Monitor* has been designated as this country's first undersea national historic landmark. Officials of the National Oceanic and Atmospheric Administration want to recover the ship, but the rusting hull is so badly corroded that salvage would be difficult if not impossible. Marine archaeologists may attach pieces of zinc to divert corrosion away from the aging metal of the ship, while the officials ponder the tricky question of recovery (*Time,* June 22, 1987). The corrosion destroying the *Monitor* is an electrochemical process. So, too, is the remedy offered by zinc.

Let's look at rusting first. Figure 16.13 shows a drop of water on a piece of iron. The water contains an ionic substance (such as salt used on wintery, slippery roads). When this combination is exposed to oxygen in the air, the edge of the drop becomes the cathode of a voltaic cell. Oxygen is reduced to hydroxide ions around the circumference of the drop. (*continued*)

FIGURE 16.12
A historic battle in the Civil War. The ironclad U.S.S. *Monitor* squared off against the armored frigate, C.S.S. *Virginia* (formerly the U.S.S. *Merrimack*), on March 9, 1862, off the coast of Norfolk, Virginia.

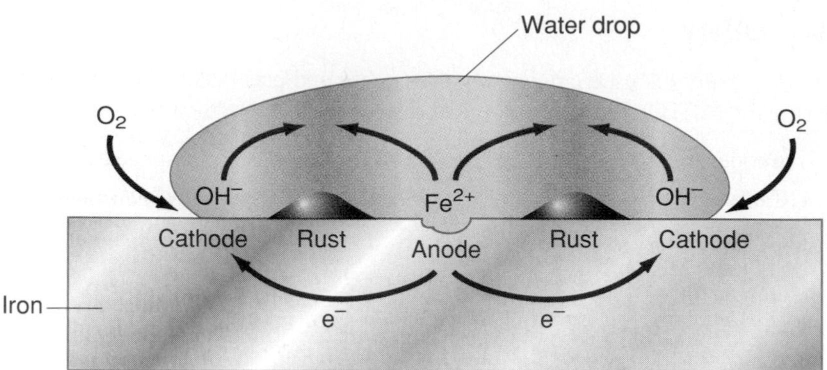

FIGURE 16.13
Rusting, an electrochemical process. As shown here, rust can be caused by a single drop of water, provided it contains ions. The anode is at the center of the drop, and iron is oxidized to Fe^{2+} ions. The edges of the drop serve as the cathode, where oxygen from air is reduced to OH^- ions. The Fe^{2+} and OH^- ions migrate toward each other and combine so that insoluble $Fe(OH)_2$ is formed. Further oxidation by more oxygen results in the formation of the iron(III) compound called rust.

The half-reaction is

$$O_2(g) + 2H_2O(l) + 4e^- \longrightarrow 4OH^-(aq)$$

The electrons for this half-reaction are provided by the oxidation of iron at the center of the drop. This site is the anode.

$$Fe(s) \longrightarrow Fe^{2+}(aq) + 2e^-$$

Electrons flow from the center of the drop to the drop's edge through the metallic iron. This conductor serves the same purpose as the external wire in the zinc-copper cell in Figure 16.9. The circuit is completed by ions moving within the drop. When the Fe^{2+} ions encounter the OH^- ions in this movement, insoluble iron(II) hydroxide precipitates.

$$Fe^{2+}(aq) + 2OH^-(aq) \longrightarrow Fe(OH)_2(s)$$

This precipitate is quickly oxidized by more oxygen to the insoluble iron(III) compound that is known as rust. These are the processes responsible for the corrosion of cars, tools, and other iron and steel objects above the sea. Below the sea, the iron of the *Monitor,* the salt-water, and dissolved oxygen provide a similar kind of voltaic cell, so that rusting occurs as easily below the surface as above it.

Rusting can be prevented by connecting iron to a metal more active than it is, one that will react with oxygen more readily. This metal replaces iron as the anode of the electrochemical reaction. Because the metal is used up in preference to iron, it is called a *sacrificial anode*. It forces the iron to become part of the cathode, where the reduction of oxygen occurs. The term *cathodic protection* is often used to describe this process. In principle, any metal higher than iron in the activity series (Table 16.2) can be used for this purpose. In practice, however, a very active metal cannot be used because it would be consumed by other reactions. For example, sodium would react violently with water, and would not be available for the electrochemical process.

Zinc is the sacrificial anode under consideration for the *Monitor.* The half-reactions that will protect her are

$$Zn(s) \longrightarrow Zn^{2+}(aq) + 2e^-$$

$$O_2(aq) + 2H_2O(l) + 4e^- \longrightarrow 4OH^-(aq)$$

Notice that the pieces of zinc are consumed, so they need to be replaced periodically. Sacrificial anodes and cathodic protection are not new ideas. Magnesium has been used successfully in the past to protect buried steel pipes and fuel tanks. (Steel is a homogeneous mixture of iron, carbon, and other metals.) And aluminum nails have been used to protect steel rain gutters on houses.

CHAPTER REVIEW

Key Words

oxidation–reduction reaction *(p. 503)*
oxidation *(p. 505)*
reduction *(p. 505)*
oxidizing agent *(p. 505)*
reducing agent *(p. 505)*
oxidation number *(p. 511)*
spontaneous process *(p. 524)*
electricity *(p. 524)*
electrochemistry *(p. 524)*
electrochemical cell *(p. 525)*
voltaic (galvanic) cell *(p. 525)*
half-cell *(p. 525)*
anode *(p. 525)*
cathode *(p. 525)*
volt *(p. 526)*
electrolytic cell *(p. 528)*
battery *(p. 529)*

Summary

A transfer of electrons from one atom to another occurs in an *oxidation–reduction reaction*. The atom that loses electrons is oxidized, and this atom belongs to the *reducing agent*. The atom that gains electrons is reduced, and it belongs to the *oxidizing agent*. Oxidation cannot occur without reduction. The activity series arranges the metals so that the atoms of the ones uppermost in the series are most easily oxidized and the ions of the ones lower in the series are most easily reduced. Many oxidation–reduction reactions can be balanced by inspection, but those involving aqueous ions in acidic or basic solution probably require the *half-reaction method*.

Oxidation–reduction reactions are found in *electrochemical cells*. A *voltaic (galvanic) cell* is an electrochemical cell that generates an electric current from a *spontaneous* oxidation–reduction reaction. The oxidation and reduction *half-reactions* are usually separated into *half-cells*. The half-cell in which oxidation takes place contains the *anode*. The half-cell in which reduction takes place contains the *cathode*. The other type of electrochemical cell, *an electrolytic cell,* requires an electric current from an outside source to drive a nonspontaneous reaction.

A *battery* is either a single voltaic cell or several of them wired so that the anode of one is connected to the cathode of another. When a *lead storage battery,* the battery found in an automobile, discharges electricity, it can continue to do so until the chemical reaction in each cell comes to *equilibrium*. When the battery is recharged, each cell becomes an electrolytic cell. The *nickel–cadmium battery* can also be recharged, but a *mercury battery* cannot be recharged.

Problem-Solving Skills

1. **Identifying Oxidizing and Reducing Agents in a Net Ionic or Molecular Equation:** Given a net ionic or molecular equation, identify the oxidizing and reducing agents, if any (Examples 16.1 and 16.5).
2. **Using the Activity Series:** Given an element in the activity series and an ion of another element in the activity series, determine whether an oxidation–reduction reaction involving these substances will occur (Example 16.2).
3. **Assigning Oxidation Numbers:** Given the formula of a compound or polyatomic ion, assign oxidation numbers to every atom in the formula (Examples 16.3 and 16.4).
4. **Balancing Oxidation–Reduction Equations:** Given the unbalanced or skeletal equation for an oxidation–reduction reaction, complete and balance the equation (Examples 16.6 and 16.7).
5. **Predicting the Behavior of a Voltaic Cell:** Given electrodes of two different metals dipped in solutions of their salts, predict which is the anode and which is the cathode, name the oxidized and reduced substances and the direction of electron flow, and specify the half-cell reactions and the overall chemical reaction (Example 16.8).

Questions to Test Your Reading

16.1 What is an oxidation–reduction reaction? What is oxidation? What is reduction? Can oxidation occur without reduction? Explain.

16.2 What is an oxidizing agent? What is a reducing agent?

16.3 What is the activity series?

16.4 Will lead react with aqueous Cu^{2+} ions? Explain.

16.5 Will tin dissolve in hydrochloric acid? Explain.

16.6 Would you expect powdered aluminum to be an oxidizing agent or a reducing agent? Explain.

16.7 Is copper a better reducing agent than cadmium? Explain.

16.8 What is an oxidation number? Why do we use oxidation numbers?

16.9 What is the oxidation number of a Cr atom in the free element? What are the oxidation numbers of the atoms in O_2, P_4, and S_8?

16.10 What are the oxidation numbers of each atom in $MgCl_2$ and HI?

16.11 What are the oxidation numbers of each atom in H_2O and H_2O_2?

16.12 How will the oxidation number of an atom change during oxidation? during reduction?

16.13 Is the reaction of hydrochloric acid with sodium hydroxide, $HCl(aq) + NaOH(aq) \longrightarrow NaCl(aq) + H_2O(l)$, an oxidation–reduction reaction? If so, identify the oxidizing and reducing agents.

16.14 When coal (which is essentially carbon) is burned (that is, allowed to react with elemental oxygen in the air), the product is carbon dioxide. Is this reaction an oxidation–reduction reaction? If so, identify the oxidizing and reducing agents.

16.15 What is a half-reaction? Give an example of a half-reaction for oxidation and one for reduction.

16.16 Name the method that is used to balance oxidation–reduction equations.

16.17 What is an electrochemical cell? a voltaic cell? a galvanic cell? an electrolytic cell?

16.18 What is a spontaneous process? Give an example from everyday life.

16.19 What is a half-cell?

16.20 Explain the difference between an anode and a cathode in a voltaic cell and in an electrolytic cell.

16.21 What is a battery?

16.22 What substance is oxidized in the lead storage battery? What substance is reduced? What has occurred when this battery is "dead"? What occurs when it is recharged?

16.23 Overcharging a lead storage battery can produce hydrogen gas. What reaction do you think is taking place?

16.24 What substance is oxidized in a nickel–cadmium battery? What substance is reduced?

16.25 If you fasten a wire across the terminals of a nickel–cadmium battery, what will happen?

16.26 What substance is oxidized in a mercury battery? What substance is reduced?

Practice Problems

Oxidation–Reduction Reactions (Sections 16.1 and 16.2)

16.27 In each of the following reactions, identify the oxidizing and reducing agents.
(a) $Sn^{2+}(aq) + 2Ce^{4+}(aq) \longrightarrow Sn^{4+}(aq) + 2Ce^{3+}(aq)$
(b) $Zn(s) + 2H^{+}(aq) \longrightarrow Zn^{2+}(aq) + H_2(g)$
(c) $Pb^{2+}(aq) + Cd(s) \longrightarrow Pb(s) + Cd^{2+}(aq)$
(d) $2Al(s) + 3Cl_2(g) \longrightarrow 2Al^{3+}(aq) + 6Cl^{-}(aq)$

16.28 In each of the following reactions, identify the oxidizing and reducing agents.
(a) $Cr^{2+}(aq) + Fe^{3+}(aq) \longrightarrow Cr^{3+}(aq) + Fe^{2+}(aq)$
(b) $I_2(aq) + Fe(s) \longrightarrow 2I^{-}(aq) + Fe^{2+}(aq)$
(c) $2H^{+}(aq) + Mg(s) \longrightarrow H_2(g) + Mg^{2+}(aq)$
(d) $Pb(s) + Hg^{2+}(aq) \longrightarrow Pb^{2+}(aq) + Hg(l)$

16.29 Which of the following reactions, if any, actually occur? Use Table 16.2 to make your predictions.
(a) $Pb(s) + Mg^{2+}(aq) \longrightarrow Pb^{2+}(aq) + Mg(s)$
(b) $Zn(s) + 2Ag^{+}(aq) \longrightarrow Zn^{2+}(aq) + 2Ag(s)$
(c) $Cr(s) + 2H^{+}(aq) \longrightarrow Cr^{2+}(aq) + H_2(g)$
(d) $2Au(s) + 6H^{+}(aq) \longrightarrow 2Au^{3+}(aq) + 3H_2(g)$

16.30 Which of the following reactions, if any, actually occur? Use Table 16.2 to make your predictions.
(a) $Sn(s) + Zn^{2+}(aq) \longrightarrow Sn^{2+}(aq) + Zn(s)$
(b) $Hg(l) + 2H^{+}(aq) \longrightarrow Hg^{2+}(aq) + H_2(g)$
(c) $H_2(g) + Ni^{2+}(aq) \longrightarrow 2H^{+}(aq) + Ni(s)$
(d) $3Ag^{+}(aq) + Al(s) \longrightarrow 3Ag(s) + Al^{3+}(aq)$

Oxidation Numbers (Section 16.3)

16.31 Assign oxidation numbers to the atoms in the following substances.
(a) HBr (b) Na_2O (c) CH_4 (d) C_2H_6
(e) O_2 (f) O_3

16.32 Assign oxidation numbers to the atoms in the following substances.
(a) NH_3 (b) H_2O (c) Al_2O_3
(d) $Al(OH)_3$ (e) CrO (f) CrO_3

16.33 Assign oxidation numbers to the atoms in the following substances.
(a) $NaClO_4$ (b) $NaClO_3$ (c) $NaClO_2$
(d) $NaClO$ (e) $NaCl$ (f) Cl_2

16.34 Assign oxidation numbers to the atoms in the following substances.
(a) $BaCrO_4$ (b) $Sr(MnO_4)_2$ (c) $Al(NO_3)_3$
(d) H_3PO_4 (e) H_2SO_4 (f) P_4

16.35 Assign oxidation numbers to the atoms in the following ions.
(a) Br^{-} (b) N^{3-} (c) NH_2^{-} (d) OH^{-}
(e) MnO_4^{-} (f) HS^{-}

16.36 Assign oxidation numbers to the atoms in the following ions.
(a) S^{2-} (b) SO_3^{2-} (c) HSO_3^{-}
(d) HSO_4^{-} (e) CO_3^{2-} (f) HCO_3^{-}

16.37 Assign oxidation numbers to the atoms in the following ions.
(a) $H_2PO_4^{-}$ (b) HPO_4^{2-} (c) PO_4^{3-}
(d) NH_4^{+} (e) NO_3^{-} (f) NO_2^{-}

16.38 Assign oxidation numbers to the atoms in the following ions.
(a) CrO_4^{2-} (b) $HCrO_4^{-}$ (c) $Cr_2O_7^{2-}$
(d) VO^{2+} (e) VO_2^{+} (f) IO_3^{-}

16.39 Identify the oxidizing and reducing agents, if any, in the following reactions.
(a) $N_2(g) + O_2(g) \longrightarrow 2NO(g)$
(b) $P_4(s) + 5O_2(g) \longrightarrow P_4O_{10}(s)$
(c) $P_4O_{10}(s) + 6H_2O(l) \longrightarrow 4H_3PO_4(aq)$
(d) $2CO(g) + O_2(g) \longrightarrow 2CO_2(g)$

16.40 Identify the oxidizing and reducing agents, if any, in the following reactions.
(a) $N_2(g) + 3H_2(g) \longrightarrow 2NH_3(g)$
(b) $K_2O(s) + H_2O(l) \longrightarrow 2KOH(aq)$
(c) $2Na(s) + 2H_2O(l) \longrightarrow 2NaOH(aq) + H_2(g)$
(d) $C_2H_4(g) + H_2(g) \longrightarrow C_2H_6(g)$

Balancing Oxidation–Reduction Equations (Section 16.4)

16.41 Balance the following equations by inspection. Identify the oxidizing and reducing agents.
(a) $Li(s) + O_2(g) \longrightarrow Li_2O(s)$
(b) $Al(s) + O_2(g) \longrightarrow Al_2O_3(s)$
(c) $Cr(s) + HCl(aq) \longrightarrow CrCl_3(aq) + H_2(g)$
(d) $CH_4(g) + O_2(g) \longrightarrow CO_2(g) + H_2O(l)$

16.42 Balance the following equations by inspection. Identify the oxidizing and reducing agents.
(a) $Na(s) + O_2(g) \longrightarrow Na_2O_2(s)$
(b) $Fe(s) + O_2(g) \longrightarrow Fe_2O_3(s)$
(c) $B(s) + H_2(g) \longrightarrow B_2H_6(g)$
(d) $C_2H_4(g) + O_2(g) \longrightarrow CO_2(g) + H_2O(l)$

16.43 Balance the following equations by inspection. Identify the oxidizing and reducing agents.
(a) $Br_2(aq) + SO_2(aq) + H_2O(l) \longrightarrow HBr(aq) + H_2SO_4(aq)$
(b) $HI(aq) + HNO_3(aq) \longrightarrow I_2(aq) + NO(g) + H_2O(l)$
(c) $MnO_2(s) + HBr(aq) \longrightarrow MnBr_2(aq) + Br_2(aq) + H_2O(l)$
(d) $CuO(s) + NH_3(g) \longrightarrow Cu(aq) + N_2(g) + H_2O(g)$

16.44 Balance the following equations by inspection. Identify the oxidizing and reducing agents.
(a) $Fe_2O_3(s) + CO(g) \longrightarrow Fe(s) + CO_2(g)$
(b) $H_2SO_4(l) + HBr(g) \longrightarrow SO_2(g) + Br_2(g) + H_2O(g)$
(c) $CuCl(s) \longrightarrow CuCl_2(aq) + Cu(s)$
(d) $HI(aq) + O_2(g) \longrightarrow I_2(aq) + H_2O(l)$

16.45 Balance the following equations by the half-reaction method. The reactions occur in acidic aqueous solutions. Identify the oxidizing and reducing agents.
(a) $Cr_2O_7^{2-}(aq) + Fe^{2+}(aq) \longrightarrow Cr^{3+}(aq) + Fe^{3+}(aq)$
(b) $VO_2^{+}(aq) + Zn(s) \longrightarrow VO^{2+}(aq) + Zn^{2+}(aq)$
(c) $VO_2^{+}(aq) + Zn(s) \longrightarrow V^{3+}(aq) + Zn^{2+}(aq)$
(d) $VO_2^{+}(aq) + Zn(s) \longrightarrow V^{2+}(aq) + Zn^{2+}(aq)$

16.46 Balance the following equations by the half-reaction method. The reactions occur in acidic aqueous solutions. Identify the oxidizing and reducing agents.
(a) $MnO_4^{-}(aq) + Sn^{2+}(aq) \longrightarrow Mn^{2+}(aq) + Sn^{4+}(aq)$
(b) $Cl_2(g) + IO_3^{-}(aq) \longrightarrow Cl^{-}(aq) + IO_4^{-}(aq)$
(c) $Cu_2O(s) + Cl_2(g) \longrightarrow Cu^{2+}(aq) + Cl^{-}(aq)$
(d) $I_2(aq) + Cl_2(g) \longrightarrow HIO_3(aq) + Cl^{-}(aq)$

16.47 Balance the following equations by the half-reaction method. The reactions occur in acidic aqueous solutions. Identify the oxidizing and reducing agents.
(a) $NO_3^{-}(aq) + S_8(s) \longrightarrow NO_2(g) + SO_4^{2-}(aq)$
(b) $MnO_4^{-}(aq) + SO_2(g) \longrightarrow Mn^{2+}(aq) + SO_4^{2-}(aq)$
(c) $Cr_2O_7^{2-}(aq) + HNO_2(aq) \longrightarrow Cr^{3+}(aq) + NO_3^{-}(aq)$
(d) $MnO_4^{-}(aq) + HNO_2(aq) \longrightarrow Mn^{2+}(aq) + NO_3^{-}(aq)$

16.48 Balance the following equations by the half-reaction method. The reactions occur in acidic aqueous solutions. Identify the oxidizing and reducing agents.
(a) $MnO_4^{-}(aq) + V^{2+}(aq) \longrightarrow Mn^{2+}(aq) + VO_2^{+}(aq)$
(b) $MnO_4^{-}(aq) + V^{3+}(aq) \longrightarrow Mn^{2+}(aq) + VO_2^{+}(aq)$
(c) $MnO_4^{-}(aq) + VO^{2+}(aq) \longrightarrow Mn^{2+}(aq) + VO_2^{+}(aq)$
(d) $MnO_4^{-}(aq) + C_2O_4^{2-}(aq) \longrightarrow Mn^{2+}(aq) + CO_2(aq)$

16.49 Balance the following equations by the half-reaction method. The reactions occur in basic aqueous solutions. Identify the oxidizing and reducing agents.
(a) $Al(s) + NO_3^{-}(aq) \longrightarrow Al(OH)_4^{-}(aq) + NH_3(aq)$
(b) $S^{2-}(aq) + I_2(aq) \longrightarrow SO_4^{2-}(aq) + I^{-}(aq)$
(c) $Mn(OH)_2(s) + H_2O_2(aq) \longrightarrow MnO_2(s) + H_2O(l)$
(d) $Cl_2(g) + IO_3^{-}(aq) \longrightarrow Cl^{-}(aq) + IO_4^{-}(aq)$

16.50 Balance the following equations by the half-reaction method. The reactions occur in basic aqueous solutions. Identify the oxidizing and reducing agents.
(a) $Pb(OH)_4^{2-}(aq) + ClO^{-}(aq) \longrightarrow PbO_2(s) + Cl^{-}(aq)$
(b) $IO_3^{-}(aq) + I^{-}(aq) \longrightarrow I_2(aq)$
(c) $Cr(OH)_2(s) + ClO^{-}(aq) \longrightarrow Cr(OH)_3(s) + Cl^{-}(aq)$
(d) $CrO_4^{2-}(aq) + Fe(OH)_2(s) \longrightarrow Cr(OH)_3(s) + Fe(OH)_3(s)$

Electrochemistry (Sections 16.5, 16.6, and 16.7)

16.51 A voltaic cell consists of a tin–tin(II) half-cell and a copper–copper(II) half-cell. What is the anode and what is the cathode? What is the overall chemical reaction?

16.52 A voltaic cell consists of an iron–iron(II) half-cell and a zinc–zinc(II) half-cell. What is the anode and what is the cathode? What is the overall chemical reaction?

16.53 A voltaic cell consists of a lead–lead(II) half-cell and a chromium–chromium(II) half-cell. What is the anode and what is the cathode? What is the overall chemical reaction?

16.54 A voltaic cell consists of a cadmium–cadmium(II) half-cell and a nickel–nickel(II) half-cell. What is the anode and what is the cathode? What is the overall chemical reaction?

16.55 The electrolysis of molten $AlBr_3$ produces aluminum metal and Br_2. What are the half-reactions?

16.56 The electrolysis of water produces H_2 and O_2. What are the half-reactions?

Additional Problems

16.57 A mixture of solid chromium(III) hydroxide and iron(III) hydroxide can be separated because $Cr(OH)_3$ is easily oxidized by hydrogen peroxide in basic solution but $Fe(OH)_3$ is not. The unbalanced skeletal ionic equation describing the oxidation is

$$Cr(OH)_3(s) + H_2O_2(aq) \longrightarrow CrO_4^{2-}(aq)$$

Balance this equation by the half-reaction method.

16.58 Hydrogen peroxide can be oxidized by potassium permanganate in basic solution according to the unbalanced skeletal ionic equation

$$H_2O_2(aq) + MnO_4^-(aq) \longrightarrow MnO_2(s) + O_2(g)$$

Balance this equation by the half-reaction method.

16.59 Iron(II) hydroxide is a greenish precipitate that is formed from iron(II) ion by the addition of a base such as sodium hydroxide. This precipitate is gradually oxidized by O_2 in the air to the reddish-brown iron(III) hydroxide. Write a balanced equation for this oxidation–reduction reaction.

16.60 A sensitive test for bismuth(III) ion consists of shaking a solution that may contain the ion with a basic solution of sodium stannite, Na_2SnO_2. If the ion is present, it is reduced under these conditions to a black precipitate of bismuth metal. Stannite ion (SnO_2^{2-}) is oxidized to stannate ion (SnO_3^{2-}). Write a balanced equation for this reaction.

16.61 Molecular chlorine is simultaneously oxidized and reduced in a basic aqueous solution. The skeletal equation is

$$Cl_2(aq) \longrightarrow Cl^-(aq) + ClO^-(aq)$$

Balance the equation.

16.62 In a process called photosynthesis, green plants convert carbon dioxide and water to carbohydrates (sugars and starches) with the release of oxygen. The energy required comes from sunlight. The unbalanced equation for the production of the sugar (glucose) is

$$CO_2(g) + H_2O(l) \longrightarrow C_6H_{12}O_6(s) + O_2(g)$$

Is this an oxidation–reduction reaction? Balance the equation.

Practice in Problem Analysis

For each problem, describe the thinking you would use (the problem analysis) before doing the actual solution, but do not solve the problem.

1. In 1800 Alessandro Volta assembled a pile of alternating zinc and silver disks with each disk separated from its neighbor by a paper disk soaked in saltwater. When he touched both ends of this pile, he experienced an electric shock. What do you think happened?

2. Although you can recharge a lead storage battery and a nickel–cadmium battery, you cannot recharge a mercury battery. Why?

PRACTICE EXAM

1. Which statement is true?
(a) Oxidation is a gain of electrons.
(b) An oxidizing agent is oxidized during an oxidation–reduction reaction.
(c) Electrons flow from the reducing agent to the oxidizing agent during an oxidation–reduction reaction.
(d) Oxidation can occur without reduction.
(e) None of the above

2. Identify the reaction in which Pb^{2+} ions are reduced.
(a) $Pb(s) + 2Ag^+(aq) \longrightarrow Pb^{2+}(aq) + 2Ag(s)$
(b) $Pb^{2+}(aq) + 2Cl^-(aq) \longrightarrow PbCl_2(s)$
(c) $Mg(s) + Pb^{2+}(aq) \longrightarrow Mg^{2+}(aq) + Pb(s)$
(d) $2Pb(s) + O_2(g) \longrightarrow 2PbO(s)$
(e) None of the above

3. The oxidation number of carbon in CO_2 is
(a) +1 (b) +2 (c) +4 (d) −4
(e) None of the above

4. The oxidation number of chlorine in the ClO_2^- ion is
(a) +3 (b) −3 (c) +1 (d) −1
(e) None of the above

5. The oxidation number of oxygen in the ClO_2^- ion is
(a) +4 (b) −4 (c) +2 (d) −2
(e) None of the above

6. The oxidation number of phosphorus in K_3PO_4 is
(a) +1 (b) +3 (c) +5 (d) +6
(e) None of the above

7. The oxidation number of phosphorus in $K_4P_2O_7$ is
(a) +1 (b) +3 (c) +5 (d) +6
(e) None of the above

8. Which substance is reduced in the reaction

$$Br_2(l) + 2I^-(aq) \rightarrow 2Br^-(aq) + I_2(aq)$$

(a) Br_2 (b) I^- (c) Br^- (d) I_2
(e) Reduction is not occurring

9. Which substance is the reducing agent in the reaction given by the equation

$$Br_2(l) + 2I^-(aq) \longrightarrow 2Br^-(aq) + I_2(aq)$$

(a) Br_2 (b) I^- (c) Br^- (d) I_2
(e) Reduction is not occurring

10. Which reaction is an oxidation–reduction reaction?
(a) $N_2(g) + 3H_2(g) \longrightarrow 2NH_3(g)$
(b) $Na_2O(s) + H_2O(l) \longrightarrow 2NaOH(aq)$
(c) $FeO(s) + 2HCl(aq) \longrightarrow FeCl_2(aq) + H_2O(l)$
(d) $Cr^{3+}(aq) + 3OH^-(aq) \longrightarrow Cr(OH)_3(s)$
(e) None of the above

11. Which reaction is an oxidation–reduction reaction?
(a) $CH_4(g) + 3O_2(g) \longrightarrow CO_2(g) + 2H_2O(g)$
(b) $C_3H_8(g) + 5O_2(g) \longrightarrow 3CO_2(g) + 4H_2O(g)$
(c) $C(s) + O_2(g) \longrightarrow CO_2(g)$
(d) $2CH_4(g) + 3O_2(g) \longrightarrow 2CO(g) + 4H_2O(g)$
(e) All of the above

12. Balance the half-reaction

$$ClO_3^-(aq) \longrightarrow Cl^-(aq)$$

for an acidic aqueous solution. The balanced half-reaction includes
(a) six electrons on the left side
(b) six electrons on the right side
(c) one electron on the left side
(d) one electron on the right side
(e) None of the above

13. Balance the half-reaction

$$ClO_3^-(aq) \longrightarrow Cl^-(aq)$$

for an acidic aqueous solution. The balanced half-reaction includes
(a) six H^+ ions on the left side
(b) six H^+ ions on the right side
(c) one H^+ ion on the left side
(d) one H^+ ion on the right side
(e) None of the above

14. Balance the reaction

$$Cr_2O_7^{2-}(aq) + I^-(aq) \longrightarrow Cr^{3+}(aq) + I_2(aq)$$

The reaction occurs in acidic aqueous solution. The coefficient of H^+ in the balanced equation is
(a) 1 (b) 4 (c) 8 (d) 14
(e) None of the above

15. Which single-replacement reaction could occur only in an electrolytic cell? Remember a reaction in an electrolytic cell will not be spontaneous.
(a) $Al(s) + Cr^{2+}(aq) \longrightarrow Al^{3+}(aq) + Cr(s)$
(b) $Co(s) + Fe^{2+}(aq) \longrightarrow Co^{2+}(aq) + Fe(s)$
(c) $Cu(s) + 2Ag^+(aq) \longrightarrow Cu^{2+}(aq) + Ag(s)$
(d) $Pb(s) + 2H^+(aq) \longrightarrow Pb^{2+}(aq) + H_2(g)$
(e) None of the above

Chapter

17 Nuclear Chemistry

Dating with uranium-238 indicates that the oldest rocks on earth have existed for more than a billion years.

Chapter Outline

A burst of fire and smoke that looks like a giant mushroom, cooling towers shaped a bit like old-fashioned milk bottles, or an eerie skeleton suitable for a Halloween night of trick-or-treating are pictures you might have in your mind when radioactivity and atomic energy are mentioned (Figure 17.1). Indeed, atomic bombs, nuclear power plants, and medical imaging are uses for the power held within the nucleus of an atom. In this chapter, we look at these nuclear processes and many others. You learn how to

FIGURE 17.1
Are these pictures in your mind when you think of radioactivity? The well-known symbol for nuclear energy is accompanied by photographs of the explosion of an atomic bomb, a nuclear power plant, and an image of a person's skeleton that is made possible through modern medical technology.

write nuclear symbols and nuclear reactions, how to express the rate of certain reactions, and how to date important historical artifacts. You also learn about the medical applications and biological effects of radioactivity, the origins of the energy unleashed in the explosion of an atomic bomb, and why a nuclear power plant is not an atomic bomb waiting to explode.

RADIOACTIVITY

Radioactivity is the result of a nuclear reaction; it is *radiation in the form of particles or energy coming from the nucleus of an atom undergoing spontaneous disintegration.* The atom can lose its identity and become some other type of atom. For example, a radioactive carbon atom can become a nitrogen atom. Up to this point, we have dealt solely with chemical reactions, reactions in which the electron structures of atoms and molecules have changed without changing the nuclei (the plural of *nucleus*) of the atoms. A nuclear reaction is very different from a chemical reaction because the nuclear composition of an atom is altered, and a different atom results, at least in many cases. Let's begin this chapter by reviewing some aspects of nuclear structure.

17.1 The Nuclear Model Revisited

OBJECTIVES

- Define *nucleon* and *nuclide*.
- Given the number of protons and neutrons in the nucleus of an atom, write the correct isotope symbol.

In Section 4.2, we described the nuclear model of the atom. The incredibly tiny, dense nucleus is made up of particles called **nucleons,** a term applied to both *neutrons and protons*. Recall that the *atomic number* of an atom is the number of protons in the atom's nucleus, and the *mass number* is the total number of protons and neutrons in the nucleus. Usually we represent the atomic number by the symbol Z and the mass number by the symbol A. The number of neutrons in a nucleus is equal to $A - Z$. *Isotopes* are atoms whose nuclei have the same atomic number but different mass numbers. Isotope symbols consisting of the atomic symbol, the atomic number, and the mass number were introduced in Chapter 4.

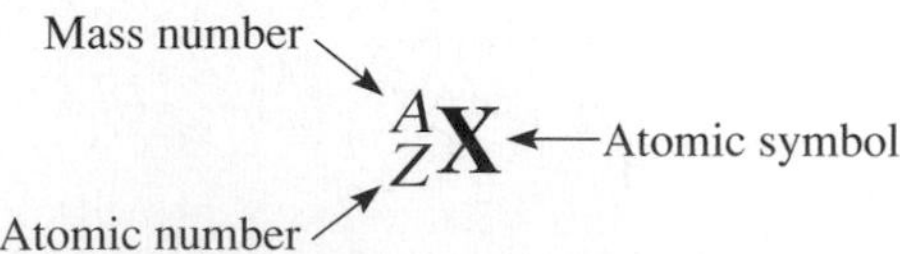

For example, the symbols for oxygen-16 and uranium-238 are

$$^{16}_{8}\mathrm{O} \qquad ^{238}_{92}\mathrm{U}$$

The nucleus of a specific isotope is called a **nuclide.** As an example, the nucleus of $^{16}_{8}\mathrm{O}$ is called the oxygen-16 nuclide.

Isotopes are usually of little concern in ordinary chemical reactions, but they are very important in nuclear chemistry. Let's consider the three isotopes of hydrogen.

^1_1H	^2_1H	^3_1H
Protium (ordinary hydrogen)	Deuterium	Tritium

Each one of these isotopes has one proton in its nucleus ($Z = 1$). The nucleus of the most common isotope, protium or ordinary hydrogen, contains a proton and nothing else; there are no neutrons. This isotope constitutes 99.984% of naturally occurring hydrogen. The remainder is almost totally deuterium, whose nucleus contains a neutron in addition to the proton. Neither ordinary hydrogen nor deuterium is radioactive. Trace quantities of the radioactive isotope, tritium, are also present because of nuclear reactions in the upper atmosphere. The nucleus of this isotope consists of two neutrons in addition to the single proton. All three isotopes are chemically identical (because each one has only one electron). For example, each one occurs as a diatomic molecule, and each one reacts with oxygen to form water. In general, some of the isotopes of any element are radioactive, but all of them are chemically identical because they have the same number of electrons.

17.2 Radioactive Decay

OBJECTIVES

- List and describe four types of radioactive decay.
- Write nuclear equations describing radioactive decay.
- Given the nucleus that decays and all but one of the products, write a nuclear equation for that reaction.
- Explain what happens in a radioactive decay series.

Radioactivity was discovered accidentally by Antoine Henri Becquerel in 1896 when he found that a mineral containing uranium produced its image on a photographic plate in the absence of light. Since then, about 2000 nuclides have been discovered, and most of them are radioactive. Only about 280 are stable, meaning they are not radioactive. The ratio of neutrons to protons in a nucleus determines its stability. Up to an atomic number of 20, nearly all of the stable nuclides have a neutron-to-proton ratio of 1:1, or equal numbers of neutrons and protons. Above $Z = 20$, this ratio gradually increases in the stable nuclides until it is about 1.5:1 at $Z = 83$ (bismuth). This means about three neutrons for every two protons. Stable nuclides are not known for elements with atomic numbers greater than 83. When the neutron-to-proton ratio of a nuclide is too large or too small (too many neutrons or too few), the nuclide is radioactive, and the nucleus disintegrates. This event, *the spontaneous disintegration of a nucleus,* is called **radioactive decay.**

Types of Radioactive Decay

We look at four types of radioactive decay in the following discussion. Each type is also listed in Table 17.1.

1. **Alpha emission:** *emission of a ^4_2He nucleus from the decaying nucleus. A ^4_2He nucleus* is called an **alpha particle.** Nuclides with atomic numbers greater than 83 often decay by alpha emission. The radioactive decay of uranium-238 by this process

TABLE 17.1
Four Types of Radiation

Type	Symbol	Example
Alpha emission	${}^{4}_{2}He$	${}^{238}_{92}U \longrightarrow {}^{234}_{90}Th + {}^{4}_{2}He$
Beta emission	${}_{-1}e$	${}^{14}_{6}C \longrightarrow {}^{14}_{7}N + {}^{0}_{-1}e$
Positron emission	${}^{0}_{1}e$	${}^{22}_{11}Na \longrightarrow {}^{22}_{10}Ne + {}^{0}_{1}e$
Gamma emission	${}^{0}_{0}\gamma$	${}^{99m}_{43}Tc \longrightarrow {}^{99}_{43}Tc + {}^{0}_{0}\gamma$

is shown in Figure 17.2. Generally, we use a **nuclear equation** (*a symbolic representation of a nuclear reaction*) to show this process.

$$ {}^{238}_{92}U \longrightarrow {}^{234}_{90}Th + {}^{4}_{2}He $$

The products of this nuclear reaction are thorium-234 and an alpha particle. Nuclear equations must be balanced. To be balanced, both A and Z must be conserved; that is, the values of A on both sides of the equation must give the same sum, as must the values of Z.

	Left Side	Right Side
A:	238	= 234 + 4
Z:	92	= 90 + 2

In practice, conservation of A means that the total number of nucleons does not change. The decaying nucleus has 238 nucleons and so do the combined product nuclei. Conservation of Z means that we must account for nuclear charge: The charge on the left side must equal the total charge on the right side.

2. **Beta emission:** *emission of a high-speed electron from the decaying nucleus. An electron ejected from a nucleus* is called a **beta particle.** In beta emission, the product nuclide has an atomic number that is one greater than that of the decaying nuclide. Beta emission is equivalent to the conversion of a neutron (${}^{1}_{0}n$) to a proton and an electron.

$$ {}^{1}_{0}n \longrightarrow {}^{1}_{1}H + {}^{0}_{-1}e $$

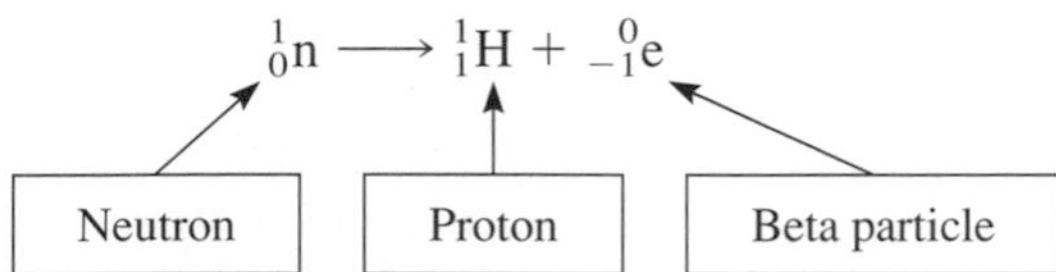

Here ${}^{0}_{-1}e$ is the symbol for a beta particle. An example is found in the radioactive decay of carbon-14,

$$ {}^{14}_{6}C \longrightarrow {}^{14}_{7}N + {}^{0}_{-1}e $$

where a carbon-14 nucleus changes to a nitrogen-14 nucleus by emitting a beta particle. Note that the mass number of this electron is zero even though the actual mass is not zero (but is very small in comparison with even the lightest nuclides). As you can see, A and Z are again conserved.

3. **Positron emission:** *emission of a positron from the decaying nucleus.* A **positron** is *a particle with the same mass as an electron, but with a positive charge rather than a*

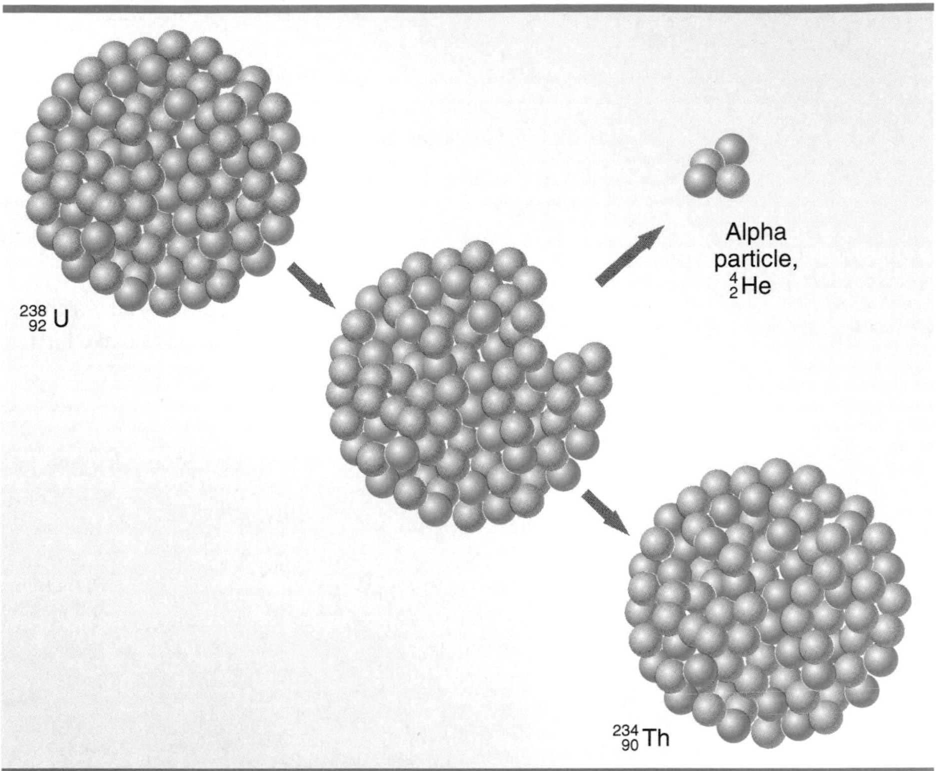

FIGURE 17.2

Alpha emission. Radioactive decay of uranium-238 leads to thorium-234 and an alpha particle ($^{4}_{2}He$).

negative charge. Positron emission is equivalent to the conversion of a proton to a neutron and a positron.

$$^{1}_{1}H \longrightarrow {}^{1}_{0}n + {}^{0}_{1}e$$

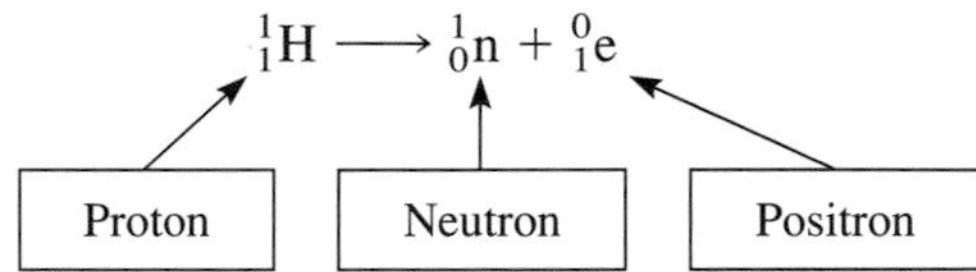

Here $^{0}_{1}e$ is the symbol for a positron. An example occurs in the radioactive decay of sodium-22.

$$^{22}_{11}Na \longrightarrow {}^{22}_{10}Ne + {}^{0}_{1}e$$

Here a sodium-22 nucleus decays, and the products are neon-22 and a positron.

4. **Gamma emission:** *emission of pure energy (a photon) with a single wavelength from an excited nucleus.* In many cases, radioactive decay by alpha, beta, or positron emission results in a new nucleus that is highly excited. It is said to be in a metastable state. This nucleus relaxes by emitting a gamma ray, which has neither mass nor charge. An example occurs with metastable technetium-99, denoted by $^{99m}_{43}Tc$, an isotope that is used in medical diagnosis.

$$^{99m}_{43}\text{Tc} \longrightarrow ^{99}_{43}\text{Tc} + ^{0}_{0}\gamma$$

Here $^{0}_{0}\gamma$ is the symbol for a gamma-ray photon. Gamma rays are a form of electromagnetic radiation (see Figure 9.6).

Consider the following example that deals with writing a nuclear equation.

EXAMPLE 17.1 Writing a Nuclear Equation

In the past, glow-in-the-dark numerals on watches were manufactured by including radioactive radium in the paint used to make the numerals (Figure 17.3). When radium-226 decays, the product is radon-222 and an alpha particle. Write the equation that describes this nuclear reaction.

FIGURE 17.3

A radium-painted watch dial. These glow-in-the-dark numerals were obtained from paint containing radium. Watches like this are no longer sold in the United States.

Problem Analysis

Write the nuclear equation with radium-226 to the left of the arrow and radon-222 and the alpha particle on the right. Use the periodic table or the table on the inside front cover of this book to find the atomic numbers of radium and radon. Recall that the symbol for an alpha particle is $^{4}_{2}\text{He}$.

Solution

The atomic numbers for radium and radon are 88 and 86, respectively, so their symbols are $^{226}_{88}\text{Ra}$ and $^{222}_{86}\text{Rn}$. Because the symbol for an alpha particle is $^{4}_{2}\text{He}$, the correct equation is

$$^{226}_{88}\text{Ra} \longrightarrow ^{222}_{86}\text{Rn} + ^{4}_{2}\text{He}$$

Notice that A and Z are conserved.

Exercise 17.1

Potassium-40 is a naturally occurring radioactive isotope. It decays to calcium-40 by emission of a beta particle. Write the nuclear equation for this decay.

(Try Problems 17.27, 17.28, 17.29, and 17.30.)

If all but one of the reactants and products of a nuclear reaction are known, you can find the identity of the unknown nuclide by remembering that A and Z must be conserved, as shown in the following example.

EXAMPLE 17.2 Finding an Unknown Product in a Nuclear Equation

Technetium-99, a nuclide that has been used in medical imaging, decays by the emission of a single beta particle. Identify the product nucleus and write the complete nuclear equation.

Problem Analysis

Because *A* and *Z* must be conserved, the values of *A* on both sides of the equation must give the same sum, as must the values of *Z*. You know or can readily find *A* and *Z* for two of the three particles involved, so you can solve for the missing *A* and *Z* to reveal the unknown product nucleus.

Solution

The table on the inside front cover of this book shows that the atomic number of technetium is 43. Thus, the symbol for technetium-99 is ${}^{99}_{43}\text{Tc}$. The symbol for the beta particle is ${}^{0}_{-1}\text{e}$. If you write ${}^{A}_{Z}\text{X}$ for the unknown product nucleus, the nuclear equation becomes

$${}^{99}_{43}\text{Tc} \longrightarrow {}^{A}_{Z}\text{X} + {}^{0}_{-1}\text{e}$$

Because *A* and *Z* must be conserved, the values of *A* on both sides of the equation must give the same sum, as must the values of *Z*. Use these relationships to identify ${}^{A}_{Z}\text{X}$ by writing

	Left Side	Right Side	
A:	99	$= A + 0$	or $A = 99$
Z:	43	$= Z - 1$	or $Z = 44$

You can see that *A* for the product nucleus must be 99 and *Z* must be 44. Using the periodic table, you find that the element whose atomic number is 44 is ruthenium. The symbol for the product nucleus is ${}^{99}_{44}\text{Ru}$, and the complete nuclear equation is

$${}^{99}_{43}\text{Tc} \longrightarrow {}^{99}_{44}\text{Ru} + {}^{0}_{-1}\text{e}$$

Exercise 17.2

A nucleus of plutonium-239 decays by emission of an alpha particle. What nucleus is the product of this decay? Write the nuclear equation.

(Try Problems 17.31, 17.32, 17.33, and 17.34.)

Radioactive Decay Series

In all of the nuclear reactions that you have seen so far, an unstable nucleus decays in one step to a stable one. Sometimes, however, a single radioactive decay will lead to a product that is also unstable and so must also decay. The result is a **radioactive decay series,** *a sequence of decay steps that continues until a stable nucleus is reached.*

Most of the naturally occurring radioactive nuclides are found in one of three radioactive decay series. ● One of these begins with uranium-238, as shown in Figure 17.4. Each long diagonal line represents a step involving alpha emission, and each short

● Carbon-14 and potassium-40, two radioactive nuclides that are not in these series, also occur naturally but in small quantities.

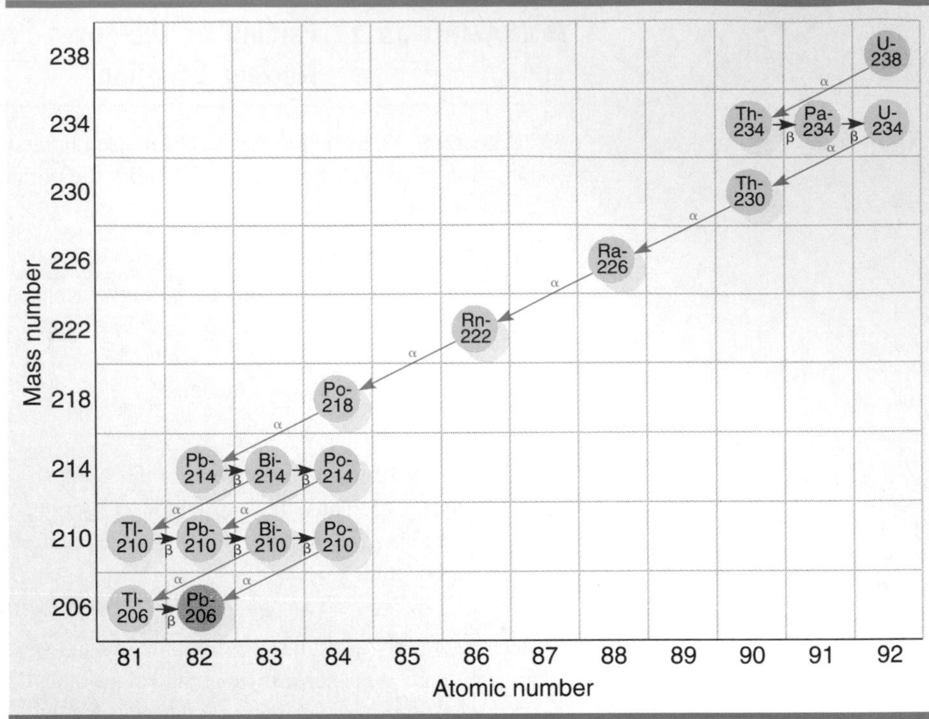

FIGURE 17.4
Uranium-238 radioactive decay series. This series starts with uranium-238 and ends with lead-206, a nuclide that is not radioactive. Alpha decay is shown by a diagonal line, and beta decay is shown by a short horizontal line.

horizontal line is a step involving beta emission. The final product of this series is stable lead-206. A second series starts with uranium-235 and ends with stable lead-207, and a third begins with thorium-232 and ends with stable lead-208.

17.3 Nuclear Transmutation

OBJECTIVE

- Describe how one element can be changed into another by bombardment.

For about 16 centuries, alchemists attempted unsuccessfully to cause the **transmutation** *(the change of one element into another)* of cheap metals, such as iron and lead, into gold by artificial means. They failed because they tried to use chemical methods to achieve nuclear reactions. Natural transmutation has happened since the dawn of time because it occurs spontaneously when a radioactive element decays. Each radioactive decay in the radioactive decay series shown in Figure 17.4 involves spontaneous transmutation. Successful artificial transmutation in a laboratory awaited Ernest Rutherford's studies of nuclear reactions in 1919. He let alpha particles coming from a radioactive element collide with the nuclei of nitrogen atoms. In other words, each nitrogen nucleus acted as a target for projectiles that were alpha particles. The products were oxygen-17 atoms and protons. The nuclear equation is

$$^{14}_{7}\text{N} + ^{4}_{2}\text{He} \longrightarrow ^{17}_{8}\text{O} + ^{1}_{1}\text{H}$$

If the target nucleus has a large atomic number, it will contain a large number of protons. When an alpha particle (or any other positively charged particle) is directed toward this nucleus, it will be deflected because of the repulsion between the positively charged target and the positively charged projectile. But if we accelerate the alpha particle so that it has a great deal of energy, we can overcome this repulsion and shoot the alpha

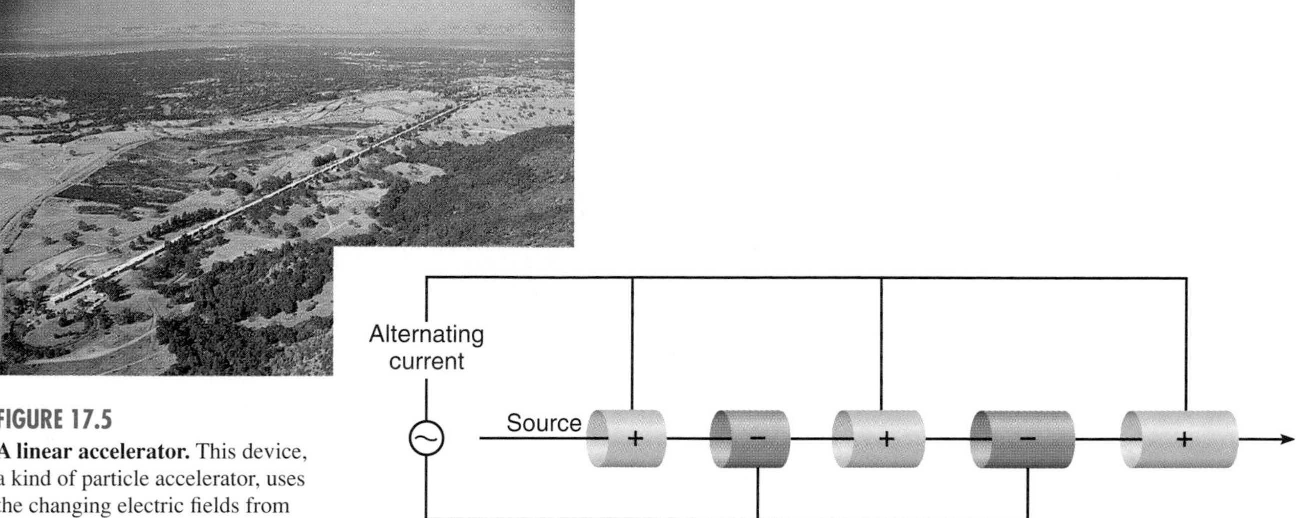

FIGURE 17.5
A linear accelerator. This device, a kind of particle accelerator, uses the changing electric fields from an alternating current to impart high velocities to particles on a very long linear path.

particle into the target nucleus. A **particle accelerator,** *a device used to accelerate electrons, protons, and nuclei (such as alpha particles) to very high speeds,* is used to impart the required energy (Figure 17.5).

In 1940, element 92 (uranium) was the heaviest known element. Since then, transmutation using particle accelerators has extended the periodic table up to element 112. All of these new elements are radioactive. They are called **transuranium elements** because they are *elements with atomic numbers greater than that of uranium ($Z = 92$),* the naturally occurring element with the largest atomic number.

Transmutations involving target nuclei and nuclear projectiles are not confined to laboratories; they have occurred in the earth's atmosphere for many millennia. **Cosmic rays,** *radiation that originates from nuclear reactions in the sun and other stars,* cause transmutations when they bombard certain nuclei in the upper atmosphere, as in the following example.

■ EXAMPLE 17.3 Finding an Unknown Nuclide in a Nuclear Transmutation Reaction

When a nitrogen-14 nucleus in the upper atmosphere is bombarded by a neutron from cosmic rays, the result is the nucleus of another element and a proton ($^{1}_{1}H$). Identify the other element by writing the nuclear equation.

Problem Analysis

Because A and Z must be conserved, the values of A on both sides of the equation must give the same sum, as must the values of Z. You know or can readily find Z for three of the four particles involved, so you can solve for the missing Z to reveal the unknown product element.

Solution

The atomic number of nitrogen is 7, so the symbol for nitrogen-14 is $^{14}_{7}N$. The symbol for a neutron is $^{1}_{0}n$. If you write $^{A}_{Z}X$ for the unknown product nucleus, the nuclear equation becomes

$$^{14}_{7}N + ^{1}_{0}n \longrightarrow ^{A}_{Z}X + ^{1}_{1}H$$

Because *A* and *Z* must be conserved, the values of *A* on both sides of the equation must give the same sum, as must the values of *Z*. Thus, you can write

	Left Side	Right Side	
A:	14 + 1	= *A* + 1	or *A* = 14
Z:	7 + 0	= *Z* + 1	or *Z* = 6

You see that *A* for the product nucleus must be 14 and *Z* must be 6. The element whose atomic number is 6 is carbon. The symbol for the product nucleus is $^{14}_{6}C$. The nuclear equation is

$$^{14}_{7}N + ^{1}_{0}n \longrightarrow ^{14}_{6}C + ^{1}_{1}H$$

Exercise 17.3

When an element is bombarded by alpha particles, the result is phosphorus-30 and a neutron. What is the target nucleus?

(Try Problems 17.35, 17.36, 17.37, and 17.38.)

17.4 Rate of Radioactive Decay and Half-Life

OBJECTIVES

- Define *half-life* and show how it can be used to calculate the amount of a radioactive sample remaining after a given number of half-lives have elapsed.
- Describe the method for determining the age of archaeological artifacts.

Some radioactive elements that you have encountered in this chapter, such as uranium-238, are found in nature, although they are constantly undergoing radioactive decay, because the earth is not yet old enough for these long-lived nuclei to have disappeared by decay. Others, such as iodine-131, have short lives so they disappeared from earth a long time ago. They must be prepared in the laboratory if they are needed, since they do not occur in nature. Each radioactive nuclide decays at its own characteristic rate, and iodine-131 decays much more rapidly than uranium-238.

The rate of radioactive decay can be expressed in terms of half-lives. A **half-life** is *the time required for half of any amount of a radioactive nuclide to decay.* Each radioactive nuclide has its own characteristic half-life. Let's look at the decay of iodine-131, a radioactive nuclide used in cancer therapy, as an example. This nuclide decays by beta emission,

$$^{131}_{53}I \longrightarrow ^{131}_{54}Xe + ^{0}_{-1}e$$

with a half-life of 8.1 days. Consequently, a 1.0-g sample of iodine-131 decays to 0.50 g in 8.1 days. Because the half-life is independent of the amount of the radioactive nuclide, the remaining 0.50 g will decay to 0.25 g in another 8.1 days, and so forth. The pattern for this decay is shown in Figure 17.6.

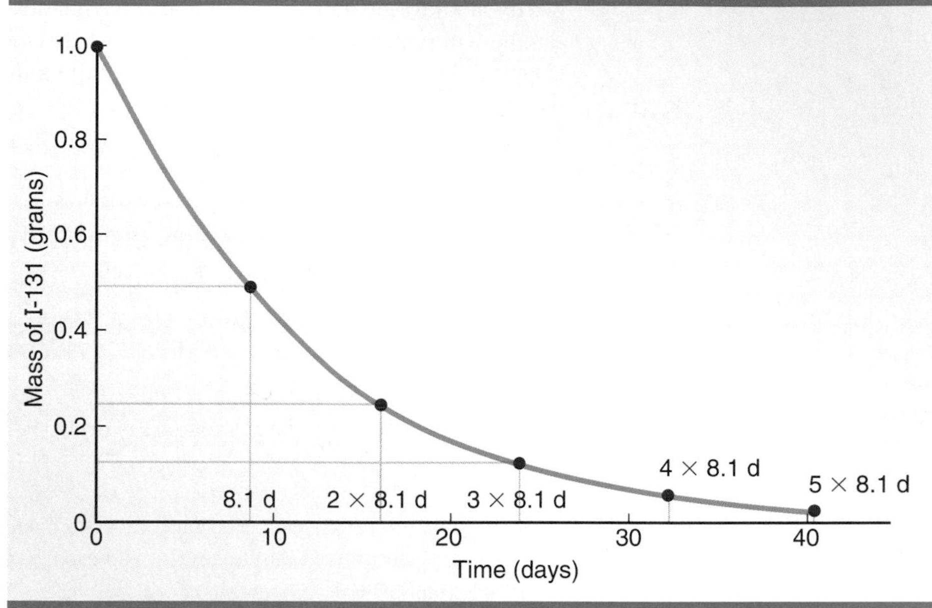

FIGURE 17.6
Radioactive decay of 1.0 g of iodine-131. The half-life of this nuclide is 8.1 days. After 8.1 days, only 0.50 g of this nuclide remains. After another half-life, only 0.25 g is left, and so forth. In general, the sample decays by one-half each time a half-life of 8.1 days has elapsed.

Some radioactive nuclides have half-lives as short as a fraction of a second, and others have half-lives as long as a few billion years. The half-lives of the nuclides in the radioactive decay series from Figure 17.4 are given in Table 17.2 as examples. Notice that the half-life of polonium-214 is only 1.64×10^{-4} second but the half-life of uranium-238 is 4.51×10^{9} years. Notice also that the half-lives of different radioactive isotopes of the same element are not equal. For example, the half-life of thorium-234 is 24.1 days, but it is 8.0×10^{4} years for thorium-230.

TABLE 17.2
Half-Lives of the Nuclides in the Uranium-238 Radioactive Decay Series

Nuclide	Half-Life
U-238	4.51×10^{9} years
Th-234	24.1 days
Pa-234	6.75 hours
U-234	2.47×10^{5} years
Th-230	8.0×10^{4} years
Ra-226	1.6×10^{3} years
Rn-222	3.82 days
Po-218	3.05 minutes
Pb-214	26.8 minutes
Bi-214	19.7 minutes
Po-214	1.64×10^{-4} second
Tl-210	1.3 minutes
Pb-210	20.4 years
Bi-210	5.01 days
Po-210	138 days
Tl-206	4.19 minutes
Pb-206	Stable

If you know the half-life of a radioactive nuclide, you can predict how much of this sample will remain after a given number of half-lives have elapsed, as in the following example. This kind of information is very important to people dealing with radioactive wastes from hospitals and to people concerned about the longevity of the fallout from tests of nuclear bombs.

EXAMPLE 17.4 Determining the Quantity of a Radioactive Sample Remaining After a Given Number of Half-Lives Have Elapsed

Strontium-90, ${}^{90}_{38}\text{Sr}$, is a nuclide in radioactive fallout from tests of atomic bombs. This nuclide decays by beta emission to yttrium-90,

$${}^{90}_{38}\text{Sr} \longrightarrow {}^{90}_{39}\text{Y} + {}^{0}_{-1}\text{e}$$

with a half-life of 28 years. Because of its chemical similarity to calcium, strontium-90 can be incorporated into bones if it is present in food. What fraction of a sample of strontium-90 will remain 84 years after a test?

Problem Analysis

Recall that half-life is the time required for half of any amount of a radioactive nuclide to decay. Half of any amount of strontium-90 will decay in 28 years. You need to determine how many half-lives have elapsed in 84 years, and from that you can calculate the fraction of the strontium-90 sample remaining after that time.

Solution

First, calculate the number of half-lives in 84 years.

$$84\,\cancel{\text{yr}} \times \frac{1\text{ half-life}}{28\,\cancel{\text{yr}}} = 3.0\text{ half-lives}$$

Because 84 years is equivalent to 3.0 half-lives, you need to find what fraction of a sample of this nuclide will remain after 3.0 half-lives. After one half-life, $\frac{1}{2}$ of the original atoms remain. After two half-lives, $\frac{1}{2} \times \frac{1}{2}$ ($\frac{1}{2}$ of $\frac{1}{2}$) $= \frac{1}{4}$ of the original atoms are left. After three half-lives, $\frac{1}{2} \times \frac{1}{4} = \frac{1}{8}$ of the original atoms remain.

Exercise 17.4

If a sample contains 32,000,000 radioactive atoms of cobalt-60, a radioactive nuclide used in cancer therapy, how many atoms will remain after four half-lives have elapsed?

(Try Problems 17.39, 17.40, 17.41, and 17.42.)

Dating Old Objects

Next, we describe a very important use of radioactive decay rates: determining the age of old objects. This process, commonly called **radioactive dating,** is *a technique for determining the age of certain old objects that relies on the known decay rate of radioactive nuclides in the object.* For example, dating old pieces of wood, charcoal, leather, paper, cloth, and other carbon-containing objects can be done with radioactive carbon-14, a nuclide with a half-life of 5730 years. Carbon-14 is present in the atmosphere and the earth as a result of cosmic ray bombardment (see Example 17.3). After its formation, carbon-14 decays by beta emission.

$$^{14}_{6}C \longrightarrow\ ^{14}_{7}N + \ ^{0}_{-1}e$$

Because of a constant rate of production of carbon-14 and a constant rate of decay, a small, constant quantity of this nuclide is always present on and above the earth. Atoms of this nuclide, along with those of carbon-12 and carbon-13 (the stable isotopes of carbon), react with oxygen to form carbon dioxide. As a result, when living plants incorporate atmospheric carbon dioxide into their tissues, they have a constant amount of carbon-14, which matches that in the atmosphere. Living animals also have the same amount of this nuclide because they eat plants or other animals who have eaten plants.

When a plant or an animal dies, it cannot take in carbon-14 any longer, but the carbon-14 that is already present continues to decay. The rate of decay can be used as a clock to determine how long ago the plant or animal died. For example, a dead plant that has one-half of the carbon-14 found in a living plant must have died one half-life or 5730 years ago. Similarly, a wooden artifact that has one-fourth of the carbon-14 that is found in a living tree must be two half-lives or 11,460 years old. This method is limited to objects up to about 50,000 years old, because older objects have too little carbon-14 to measure. ● The next example uses carbon-14 to date a volcanic eruption.

● One of the drawbacks to carbon-14 dating is that a portion of the artifact must be destroyed during the analysis.

EXAMPLE 17.5 Dating an Old Object

Carbon-14 decays by beta emission. Freshly cut wood gives 16 beta particles per minute per gram of total carbon. If a piece of charcoal from a tree killed by volcanic eruption gives 8 beta particles per minute per total gram of carbon, when did the volcano erupt?

Problem Analysis

One beta particle is produced each time a carbon-14 nucleus decays. Since the rate of beta particle emission has been halved, the number of carbon-14 nuclei must be down by half also, and you can relate this directly to the half-life of carbon-14 to determine how much time has elapsed.

Solution

Because the number of beta particles has been halved, one half-life must have elapsed, so the age of the charcoal is one half-life or about 5700 years old. The volcano must have erupted in approximately

$$2000 - 5700 = -3700 \text{ A.D. or } 3700 \text{ B.C.}$$

where 2000 is being used as today's approximate date.

Exercise 17.5

A jawbone from an archaeological site was dated by analysis of its radioactive carbon. The activity of the carbon from the jawbone was 4 disintegrations per minute per gram of total carbon. Carbon from living material gives 16 disintegrations per minute per gram of total carbon. When did the animal die?

(Try Problems 17.43, 17.44, 17.45, and 17.46.)

Chemical Perspective

The Shroud of Turin

The shroud of Turin is a piece of woven linen cloth about 4.3 meters (14 feet) long and 1.1 meters (3.5 feet) wide with the image of a man on it (Figure 17.7). Many people believe that it is the burial shroud of Jesus Christ, and that the image belongs to the crucified Christ. The first historical record of the shroud comes from the late sixteenth century, when it was brought to the cathedral in Turin, Italy. It is still there today.

If the cloth is Christ's shroud, it must be at least 2000 years old. Because linen is woven from thread derived from a plant called flax, the age of the shroud could be determined from its carbon-14 content. In 1988, three independent laboratories used carbon-14 dating to show that the cloth is only about 700 years old, so it cannot be the burial shroud of Christ. The origin of the image on the shroud remains a mystery.

FIGURE 17.7
The Shroud of Turin. Carbon-14 dating has shown that this piece of cloth cannot be the burial shroud of Jesus Christ because it is only about 700 years old. The origin of the image on the shroud remains a mystery.

17.5 Radioactivity Detection and Measurement

OBJECTIVES

- Name and describe two instruments that detect and measure radioactivity.
- Define *curie*.

Scientists and others who work around radioactive materials wear film badges containing photographic film. The film is developed at regular intervals to determine how much radiation has been directed at the wearer of the badge. Because this method is not a very accurate way to detect and measure radioactivity, other ways are also used. We describe two of them in this section.

A Geiger-Müller counter consists of a metal tube fitted with a thin window through which the radiation enters. The tube contains argon. A positively charged wire runs down the center of the tube, and it is insulated from the negatively charged walls where the wire is in contact with them at the end of the tube. Current cannot flow normally because the argon is an electrical insulator. But when radiation—alpha particles, beta particles, or gamma rays—enters the tube, it can collide with an argon atom and, if the radiation has sufficient energy, cause its ionization.

$$Ar(g) \longrightarrow Ar^{+}(g) + e^{-}$$

FIGURE 17.8
A Geiger-Müller counter. The metal tube in the foreground contains argon. When this substance is ionized by radiation, a pulse of electricity flows through the wire to the meter.

Because the ionized gas conducts electricity, there is a momentary surge of electric current as argon ions migrate to the walls of the tube and the electrons are attracted to the wire. This pulse of electricity activates a digital counter or causes an audible "click." The rate of clicking indicates how much radiation is entering the tube (Figure 17.8).

A scintillation counter is another common device for detecting alpha particles, beta particles, or gamma rays. It contains a substance such as sodium iodide or zinc sulfide that emits a flash of light each time it is struck with radiation. Each flash is detected and counted.

The unit describing the amount of radioactivity is the *curie*. One **curie** corresponds to *37 billion disintegrations per second.* This unit was chosen because it is the number of disintegrations in 1 gram of pure radium. Because 1 curie is an enormous amount of radioactivity, the millicurie and the microcurie, corresponding to one-thousandth and one-millionth of a curie, are used more commonly.

17.6 Medical Applications

OBJECTIVE

- Give three examples of radioactive nuclides in medicine.

Radioactive nuclides have had a profound effect on both treatment and diagnosis in medicine. They were first used in the treatment of cancer a few years after the discovery of radioactivity. Radiation adversely affects rapidly dividing cells, such as cancer cells, more than it does those that divide more slowly. Today, gamma radiation from cobalt-60 is commonly used to irradiate cancer cells in the hope of killing or shrinking the tumors.

The use of radioactive nuclides is not limited to cancer therapy, however, as shown in Figure 17.9. A nuclide may be used as a **radioactive tracer,** *a radioactive isotope that is added to a chemical or biological system to trace the path of a nonradioactive isotope that is normally used by the system.* For example, when a solution of sodium iodide, NaI, containing radioactive iodine is ingested, a portion of the radioactive iodine is taken up by the patient's thyroid gland. A scanner that detects the emission of radiation

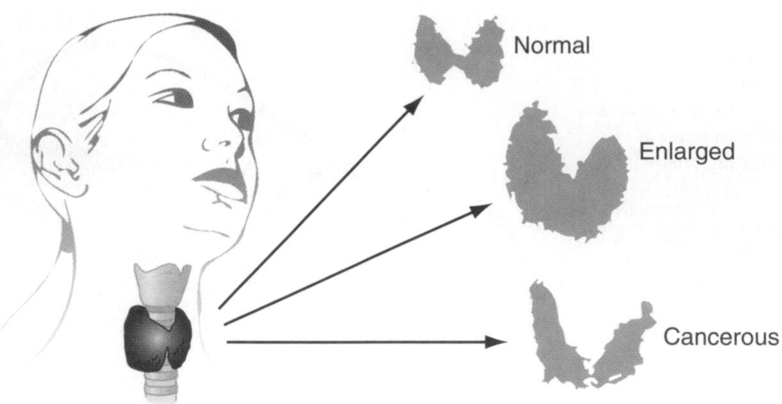

FIGURE 17.9

Diagnosing thyroid problems. The thyroid gland normally takes up iodine (including radioactive iodine-123 or iodine-131) and uses it in making hormones such as thyroxine. After a person ingests sodium iodide containing radioactive iodine, the distribution of radiation from the patient's thyroid gland can be scanned to locate abnormal growths.

from the radioactive nucleus provides an image of the thyroid gland. The gland's size, shape, activity, and pathology can be assessed with this technique.

Positron emission tomography (PET) is a modern technique for scanning the brain, the heart, and other organs. First, a positron-emitting nuclide with a short half-life is injected into a patient's body. Examples are carbon-11 with a half-life of 20 minutes and oxygen-15 with a half-life of 2 minutes. After a positron is emitted, it travels only a few millimeters until it encounters an electron and disappears with the formation of a gamma ray. The gamma rays are detected by a bank of scintillation counters placed around the patient. You can see the circular bank of detectors in Figure 17.10. The detectors record the distribution of gamma radiation, and a computer constructs an image from the distribution. By comparing the PET scan to that of a healthy patient, a physician can diagnose the presence or absence of disease.

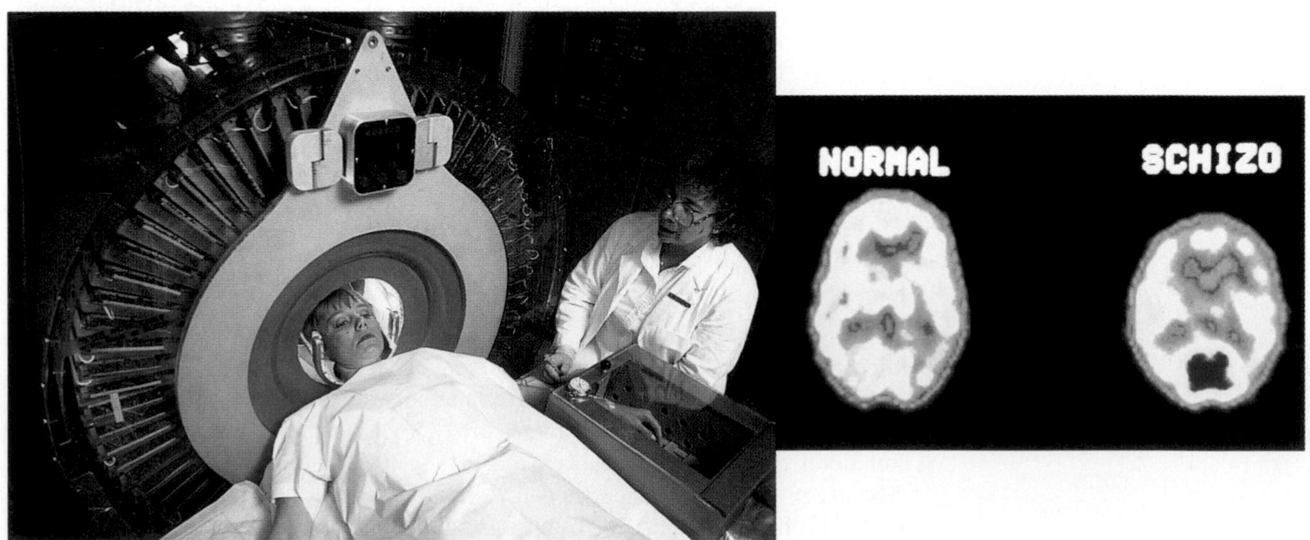

FIGURE 17.10

PET scanning. The patient is undergoing a PET scan of the brain that may reveal abnormalities. For example, the scan of a normal patient and the scan of a schizophrenic patient reveal clear differences in brain activities (as indicated by the different colors).

17.7 Everyday Sources and Biological Effects of Radiation

OBJECTIVES

- Describe some of the effects of radiation on living tissue in terms of ionizing radiation and the formation of free radicals.
- Give sources and doses of radiation experienced by the average person over a year's time.
- Describe the dangers of radon.

Researchers have known about the damaging effects of radioactivity for a long time. Early experimenters found that glass containers used to store radium would turn purple and then crack. They also found that radiation from radium caused slow-healing wounds. These effects are due to the ionizing nature of radiation. Alpha particles, beta particles, and gamma rays (in addition to x rays) can have energies that greatly exceed ordinary bond energies and ionization energies. Consequently, these forms of radiation are able to fragment and ionize the molecules of matter as they pass through it.

Ionizing radiation damages living tissue, which is composed of 70% to 90% water. When the tissue is irradiated, the water can become ionized.

$$H_2O \longrightarrow H_2O^+ + e^-$$

An H_2O^+ ion can then react with another water molecule according to the chemical equation

$$H_2O^+ + H_2O \longrightarrow H_3O^+ + OH$$

One of the products is the hydronium ion, H_3O^+, which we encountered in Chapter 15. The other product is not the negatively charged hydroxide ion, OH^-, but rather the unstable and highly reactive OH molecule, an example of a **free radical,** *a molecule with an unpaired valence electron.* Free radicals are among the strongest oxidizing agents that can exist in an aqueous media. Through oxidation, they destroy vital biological molecules by either removing electrons or removing hydrogen atoms. ●

● Some theories view aging as the cumulative effects of free radicals running amuck. Antioxidants, such as vitamin E, are supposed to react with free radicals, detoxify them, and slow down the aging process.

If the ionizing radiation is *outside* the body, its energy and penetrating power determine how much damage will be done. Like x rays, gamma rays are very dangerous because they can penetrate so deeply. Beta particles have less penetrating power than gamma rays but more penetrating power than alpha particles, which can be stopped by the skin. If a radioactive nuclide is *inside* the body, however, the situation is reversed. The body must then absorb all of the energy released by the radiation. Alpha particles cannot travel very far, but they leave a wake of damaged biological molecules in a very small area. Beta particles travel farther within the body and distribute a somewhat lesser amount of damage over a larger area.

The unit for a radiation dose is the *rem* (from *r*oentgen *e*quivalent for *m*an). ● This unit was invented to indicate the danger posed to humans by radiation. A dental x ray gives an exposure of about 0.005 rem. Doses of 25 to 200 rems can cause a decrease in the number of white blood cells and increasing incidents of fatigue, nausea, and diarrhea. Doses of 500 rems or greater can cause death when they are delivered quickly. Most scientists believe that the effects of radiation are in proportion to the exposure (the number of rems) and that any amount of radiation may cause some damage even if it cannot be detected.

● A roentgen (pronounced rentgen), a unit named after Wilhelm Roentgen, the discoverer of x rays, is a measure of the intensity of radiation.

During any year, the average person in the United States is exposed to about 0.360 rem. The distribution of this radiation among the various sources is shown in Figure 17.11. As you can see, more than 80% is from natural sources. These sources include radon in the air, uranium and other radioactive elements in soil and rocks, and cosmic rays (radiation from the sun and other stars), as well as potassium-40 and carbon-14 in

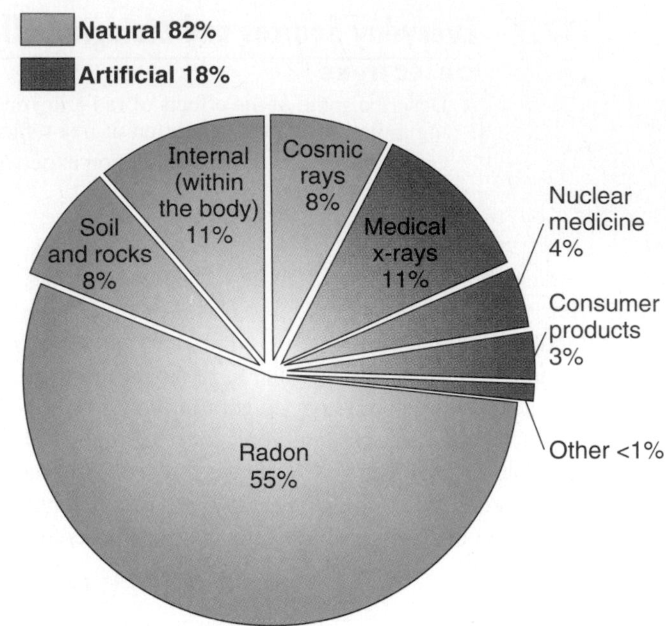

FIGURE 17.11
Yearly average exposures to radiation. More than 80% of the radiation to which we are exposed comes from natural, unavoidable sources. Less than 20% comes from artificial sources. The category listed as "Other" includes radiation from nuclear power plants. This information was obtained from the U.S. Nuclear Regulatory Commission.

the body. Lifestyle and daily activities can cause the annual dose to vary somewhat, however. For example, someone living in Denver, Colorado, rather than in a city at sea level will receive an additional average annual dose of 0.070 rem because the dose from cosmic rays is larger at higher elevations and Rocky Mountain soil contains larger concentrations of natural radioactive materials than lowland soils.

About 55% of the average annual radiation dose comes from radon in the air. Radon-222 is a decay product in the uranium-238 radioactive decay series (Figure 17.4), so it is generated continually as the uranium in the soil and rocks decays. Because radon is one of the noble gases, it is unreactive and can be inhaled without any chemical effects. However, radon-222 decays by emission of an alpha particle with a half-life of 3.82 days (see Table 17.2). If a nuclide of radon-222 happens to decay before it is exhaled, then the following reaction occurs.

$$^{222}_{86}\text{Rn} \longrightarrow {}^{218}_{84}\text{Po} + {}^{4}_{2}\text{He}$$

We said earlier that alpha particles are very dangerous when they are emitted inside the body. Indeed, many scientists believe that inhaled radon can produce lung cancer.

According to Figure 17.11, less than 20% of our average annual radiation dose comes from human (artificial) sources. It is noteworthy that less than 1% is derived from nuclear power plants.

How do the dose levels in Figure 17.11 translate into risks? The U.S. Nuclear Regulatory Commission believes that a dose of 0.001 rem corresponds to an annual risk of death from radiation-induced cancer of about 5 in 10 million. Thus, the annual, unavoidable dose of 0.360 rem from natural sources corresponds to a risk of about 1800 in 10 million (if a direct relationship between dose and the risk of death is assumed).

NUCLEAR ENERGY

When chemical reactions occur, they are accompanied by changes in energy. Nuclear reactions also involve energy changes, and these changes are enormous in comparison

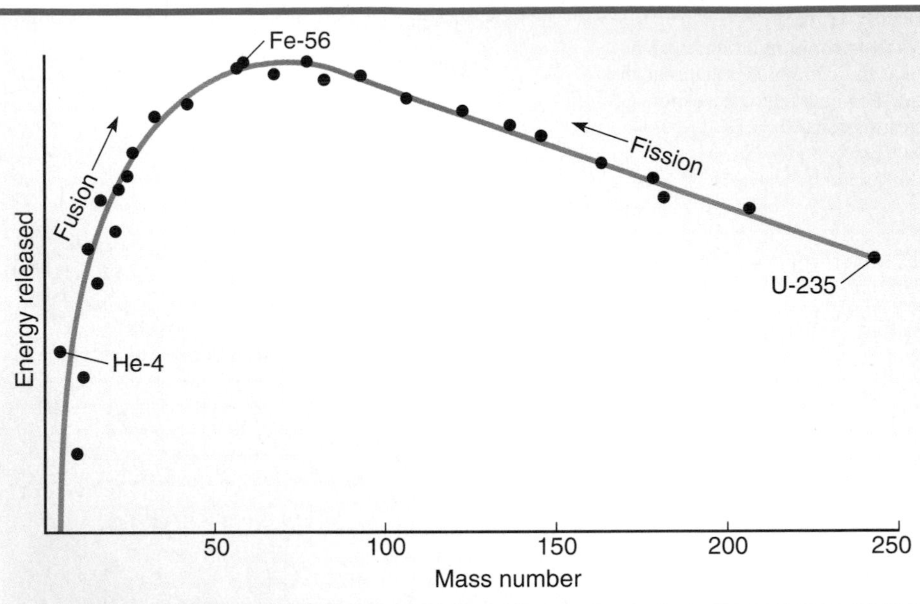

FIGURE 17.12

Nuclear stability as a function of mass number. This graph shows the energy released when a nuclide is formed from the required number and type of nucleons. The most stable nuclides—those with the greatest amount of released energy—occur near mass number 50. Thus, lighter nuclei might be expected to undergo fusion to approach this mass number and gain this stability, whereas heavier nuclei might undergo fission to gain stability.

to the energy involved with chemical reactions. There are two ways to obtain this energy.

Nuclides whose mass numbers are near 50 appear to have the most stable nuclei. As a result, nuclides with mass numbers greater than 50 might be expected to undergo **nuclear fission,** *a process in which the heavy nucleus splits to form two lighter and more stable nuclei.* In contrast, nuclides with mass numbers of less than 50 undergo **nuclear fusion,** *a process in which light nuclei combine to give heavier, more stable nuclei.* Energy is released during both processes (Figure 17.12). We consider each process in turn.

17.8 Nuclear Fission

OBJECTIVE

- Describe nuclear fission and nuclear reactors.

Some nuclides will undergo fission spontaneously—that is, without any outside agent inducing the nucleus to break apart. For example, out of every 100 radioactive californium-252 nuclei, 97 will decay by alpha emission, but 3 will decay by spontaneous fission. The fission can occur in many different ways and result in many different pairs of lighter nuclei. Neutrons can also be released during the process. One possible way for the spontaneous fission to happen can be seen in the nuclear equation

$$^{252}_{98}\text{Cf} \longrightarrow {}^{142}_{56}\text{Ba} + {}^{106}_{42}\text{Mo} + 4\,^{1}_{0}\text{n}$$

The products in this case are a barium-142 nucleus, a molybdenum-106 nucleus, and four neutrons.

FIGURE 17.13
Nuclear chain reaction. Each nuclear fission produces neutrons that can then be used to cause more nuclear fissions.

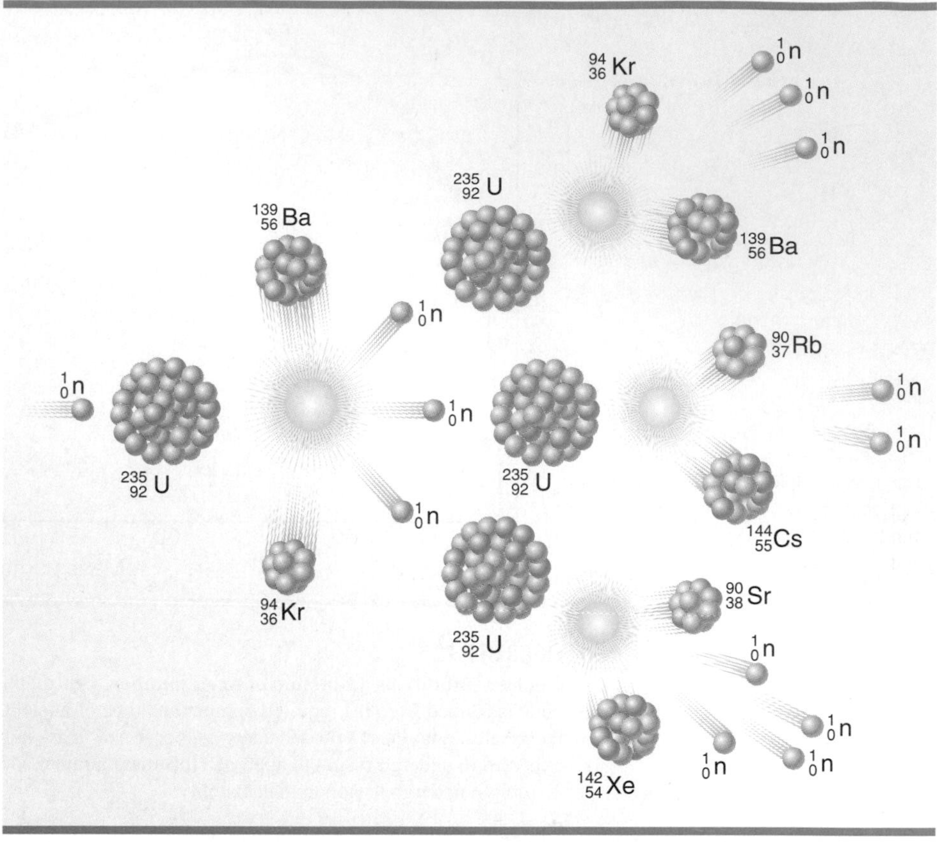

Fission can be induced in other heavy nuclei by bombarding them with neutrons. A very important example occurs with uranium-235, which also splits in many different ways when it undergoes fission. Three of these are shown in the following equations.

$$^{1}_{0}n + ^{235}_{92}U \longrightarrow \begin{cases} ^{142}_{54}Xe + ^{90}_{38}Sr + 4^{1}_{0}n \\ ^{139}_{56}Ba + ^{94}_{36}Kr + 3^{1}_{0}n \\ ^{144}_{55}Cs + ^{90}_{37}Rb + 2^{1}_{0}n \end{cases}$$

In each case, neutrons are also ejected. These neutrons are then absorbed by other uranium-235 nuclei, causing these nuclei to split also. When they do, even more neutrons are released, only to be absorbed by even more uranium-235 nuclei (Figure 17.13). This process can be self-sustaining under the right conditions. *A self-sustaining series of nuclear fissions caused by the absorption of neutrons from previous nuclear fissions* is called a **chain reaction.**

To sustain a chain reaction, there must be enough uranium-235 present to absorb the released neutrons. If the sample is too small, neutrons escape before being absorbed by the fissionable nuclei. A chain reaction can be sustained if the sample has the **critical mass,** *the minimum mass that will allow fission to become a chain reaction.* If a supercritical mass is used, the amount of fissionable material will be larger than the critical mass, very few of the neutrons will be able to escape, and the chain reaction will then continue rapidly. Such a rapid chain reaction in uranium-235 releases a prodigious amount of energy, as is evident in the explosion of a uranium atomic bomb (Figure 17.1). The United States used two of these bombs in World War II.

Only about 0.7% of naturally occurring uranium is uranium-235. The more abundant isotope, uranium-238, will absorb neutrons liberated in fission but it will not undergo fission itself. As a result, the atomic bomb required uranium that had been enriched to about 90% uranium-235.

Uranium enriched to only about 3% uranium-235 is currently used in nuclear power plants, in which controlled nuclear fission occurs (Figure 17.14). Fission energy in the

FIGURE 17.14

A nuclear power plant. Heat generated in the reactor core converts water to steam, which is used in a turbine-driven generator that produces electricity. Cooling water that passes through a massive cooling tower condenses the steam after it comes from the turbine. The photograph shows a containment building and cooling tower from the Trojan nuclear power plant in Oregon.

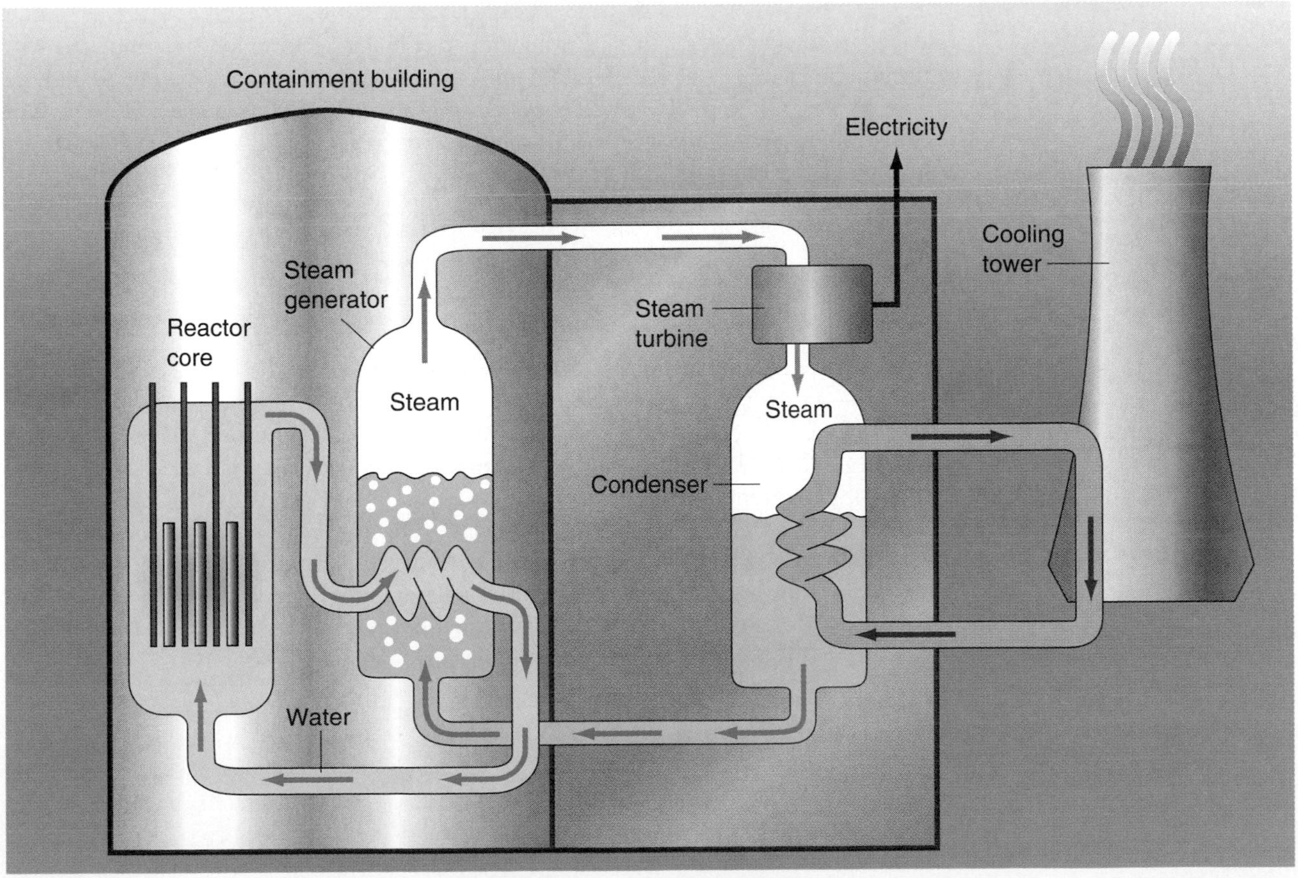

form of heat produces steam, which drives a turbine, which in turn drives an electric generator. Fuel rods in the reactor core contain enriched uranium dioxide, UO_2, and control rods contain substances such as boron or cadmium that absorb neutrons and slow the chain reaction.

The opposition to nuclear power plants has been intense for at least four reasons. First, plutonium-239, a nuclide that comes from a series of nuclear reactions occurring after the absorption of a neutron by nonfissionable uranium-238, is itself fissionable. Thus, it can be used to construct atomic bombs. Many people fear that the availability of this substance (from spent fuel rods) will increase the possibility for small countries and terrorist groups to build atomic bombs. Second, since the spent fuel rods are highly radioactive, their safe disposal is a severe problem for which there is no current satisfactory solution. Third, many people fear that a nuclear power plant is itself an atomic bomb waiting to explode. This belief is without foundation because the fuel rods are not enriched enough for an uncontrolled chain reaction leading to a nuclear explosion. Fourth, many fear that a malfunction will cause the release of dangerous radioactive materials over a populated area. This type of accident has already occurred on a small scale at Three Mile Island near Harrisburg, Pennsylvania, in 1979 and on a much larger scale with far greater consequences at Chernobyl in the Ukraine in 1986.

17.9 Nuclear Fusion

OBJECTIVE

- Describe nuclear fusion and some of the difficulties involved in its control.

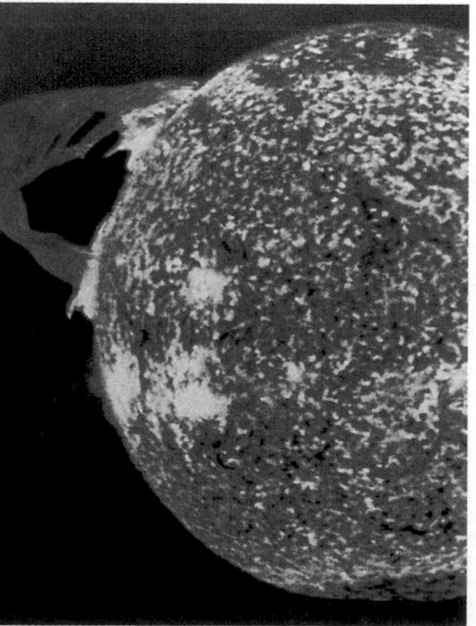

FIGURE 17.15
The sun during a major eruption. This star, like others, is a giant nuclear fusion reactor.

We have already noted that lighter nuclei can combine to form a heavier nucleus with the release of energy, a process called *nuclear fusion*. Stars produce their energy in this way. The sun—the star in our solar system—is about 75% hydrogen and 25% helium, and the helium has come from the fusion of hydrogen nuclei (Figure 17.15). Some of the fusion reactions that are believed to occur are

$$^{1}_{1}H + ^{1}_{1}H \longrightarrow ^{2}_{1}H + ^{0}_{1}e$$

$$^{1}_{1}H + ^{2}_{1}H \longrightarrow ^{3}_{2}He$$

$$^{1}_{1}H + ^{3}_{2}He \longrightarrow ^{4}_{2}He + ^{0}_{1}e$$

The sun uses fusion reactions such as these to produce about 4×10^{26} J of energy every second—a truly prodigious quantity!

The hydrogen bomb is the only successful human use of nuclear fusion as a source of energy. Its power is awesome and, when unleashed, beyond human control. Though technology currently exists to generate and sustain controlled fission reactions, controlled fusion reactions have not yet been achieved. Current research is aimed at finding the means to overcome the repulsive forces between two positively charged nuclei so that they can fuse in a controlled way. But temperatures as great as 40 million °C may be required to give the nuclei enough kinetic energy to overcome the repulsions. Achieving this temperature and maintaining it are the main problems.

Chemical Perspective

Like Mother, Like Daughter

This tale is a short story about two hard-working, highly intelligent women—a mother and her daughter—and their pioneering research in radioactivity and nuclear reactions.

Radioactivity was discovered accidentally by Antoine Henri Becquerel in 1896. The name for this phenomenon, however, was provided a few years later by Marie Curie (Figure 17.16A), who did her doctoral work with Becquerel. Her thesis has been described as one of the most important ever presented for a doctoral degree. It included her account of using radioactivity to discover two new elements, polonium and radium. In 1903, Becquerel, Marie Curie, and her husband Pierre, who worked with her, were awarded the Nobel Prize in physics. After her husband's death in a street accident, Marie Curie continued to study radioactivity and was awarded another Nobel Prize—this time in chemistry—in 1911. She was the first scientist to receive two Nobel awards. Marie Curie died in 1934 of a disease that may have developed because of her long exposure to radioactivity.

Marie and Pierre had two daughters, Eve and Irene. Irene followed in her parents' footsteps by becoming a scientist (Figure 17.16B). When she married Frederic Joliot, the couple adopted the combined surname of Joliot-Curie, since the Curies had no sons to carry on their name. After their marriage, Irene and Frederic Joliot-Curie became interested in nuclear transmutation, a process that had been discovered a few years earlier by Ernest Rutherford (Section 17.3). Together, the Joliot-Curies discovered induced radioactivity (radioactivity from an artificially produced nuclide as opposed to the spontaneous radioactivity of naturally occurring substances such as uranium and radium). The discovery occurred in 1934 when they bombarded an aluminum target with alpha particles. The unexpected products were neutrons and a previously unknown nuclide, phosphorus-30.

$$^{27}_{13}\text{Al} + ^{4}_{2}\text{He} \longrightarrow ^{30}_{15}\text{P} + ^{1}_{0}\text{n}$$

The unexpected discoveries did not stop there. Phosphorus-30, the first radioactive nucleus produced in the laboratory, decays to silicon-30 by ejecting a positron.

$$^{30}_{15}\text{P} \longrightarrow ^{30}_{14}\text{Si} + ^{0}_{1}\text{e}$$

In 1935, one year after the death of Marie Curie, Irene and Frederic Joliot-Curie received the Nobel Prize for their discoveries. Like her mother, Irene Joliot-Curie died of a disease that may have been caused by years of exposure to radioactivity.

A

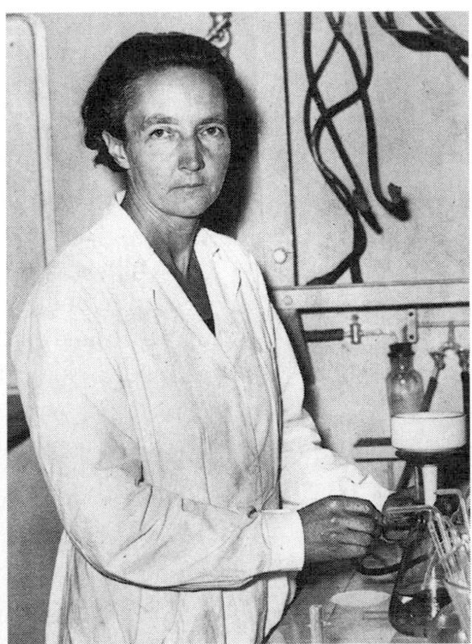

B

FIGURE 17.16

Mother and daughter Curie. (*A*) Marie Curie (1867–1934) and (*B*) her daughter, Irene Joliot-Curie (1897–1956), were pioneers in the discovery of radioactivity and nuclear reactions.

CHAPTER REVIEW

Key Words

radioactivity *(p. 540)*
nucleons *(p. 540)*
nuclide *(p. 540)*
radioactive decay *(p. 541)*
alpha emission *(p. 541)*
alpha particle *(p. 541)*
nuclear equation *(p. 542)*
beta emission *(p. 542)*
beta particle *(p. 542)*
positron emission *(p. 542)*
positron *(p. 542)*
gamma emission *(p. 543)*
radioactive decay series *(p. 545)*
transmutation *(p. 546)*
particle accelerator *(p. 547)*
transuranium elements *(p. 547)*
cosmic rays *(p. 547)*
half-life *(p. 548)*
radioactive dating *(p. 550)*
curie *(p. 553)*
radioactive tracer *(p. 553)*
free radical *(p. 555)*
nuclear fission *(p. 557)*
nuclear fusion *(p. 557)*
chain reaction *(p. 558)*
critical mass *(p. 558)*

Summary

Radioactivity is the result of nuclear reactions. These reactions are represented by nuclear equations, with each nucleus being denoted by a symbol whose form is $^{A}_{Z}X$. These equations must be balanced in charge (subscripts) and in *nucleons* (superscripts).

The ratio of neutrons to protons in an atom's nucleus determines the stability of that nucleus, that is, its tendency to undergo *radioactive decay*. Four types of radiation arising from the decay of an unstable nucleus are *alpha emission* (emission of a helium-4 nucleus), *beta emission* (emission of a high-speed electron), *positron emission* (emission of a positively charged particle whose mass is equal to that of an electron), and *gamma emission* (emission of the energy of a single wavelength).

In many nuclear reactions, an unstable nucleus decays in one step to a stable one. Sometimes, however, a single radioactive decay will lead to an immediate product that is also unstable, so that it must decay, too. The result is a *radioactive decay series*. Decay continues until a stable nuclide is reached.

Transmutation of elements (the changing of one element into another) has been carried out in the laboratory by bombarding nuclei with various atomic particles. A *particle accelerator* is often used to provide the necessary energy. Many of the *transuranium elements* were prepared in this manner.

Each radioactive nuclide decays with its own characteristic rate that can be expressed in terms of its *half-life*. Knowing the half-life allows us to calculate how long it will take for a given radioactive sample to decay by a certain fraction. Methods of *radioactive dating* depend on determining the fraction of a radioactive nuclide that has decayed and, from this, the object's age.

Radiation can be detected by either Geiger-Müller counters or scintillation counters. Both of these counters give the number of disintegrations per unit time for radioactive material.

Radioactive nuclides are used extensively in medicine. Radiation therapy for the treatment of cancer is effective because rapidly dividing cancer cells are more adversely affected by radiation than those that divide more slowly. Radioactive nuclides are also used extensively as *radioactive tracers*.

Biological molecules can become fragmented and ionized by radiation. The unit for a radiation dose is the rem. The average annual dose for a person in the United States is about 0.360 rem, and more than 80% of that dose comes from natural sources.

Nuclides whose mass numbers are near 50 appear to have the most stable nuclei. As a result, nuclides with mass numbers greater than 50 tend to split, a process called *nuclear fission,* and nuclides with mass numbers less than 50 tend to combine, a process called *nuclear fusion*. Energy is released during both processes. Nuclear fission was used in the atomic bomb and is currently used in conventional nuclear power plants. Nuclear fusion, the basis of the sun's energy, is difficult to control. Reactors based on nuclear fusion are still in the experimental stage.

Problem-Solving Skills

1. **Writing a Nuclear Equation:** Given a word description of a radioactive decay process, write the nuclear equation (Example 17.1).
2. **Finding an Unknown Product in a Nuclear Equation:** Given the reactant and all but one of the products, deduce the missing nuclide or particle (Example 17.2).
3. **Finding an Unknown Nuclide in a Nuclear Transmutation Reaction:** Given all but one species in a nuclear transmutation reaction, predict the missing nuclide or particle (Example 17.3).

4. **Determining the Quantity of a Radioactive Sample Remaining After a Given Number of Half-Lives Have Elapsed:** Given the half-life of a radioactive nuclide, and the time elapsed, calculate the fraction of the nuclide that remains (Example 17.4).

5. **Dating an Old Object:** Given the disintegrations of carbon-14 nuclei per minute per gram of carbon in a dead organic object, calculate the approximate age of the object—that is, the date when it died (Example 17.5).

Questions to Test Your Reading

17.1 What is radioactivity?

17.2 What is the atomic number of nitrogen-14? What is its mass number? How many protons, neutrons, and nucleons are in the nucleus? What is the correct symbol?

17.3 Define *isotope* and *nuclide*.

17.4 What is radioactive decay? Name four types of radiation that are emitted during radioactive decay.

17.5 What is a radioactive decay series?

17.6 Give the nuclides that begin each of the three major naturally occurring radioactive decay series.

17.7 Write a nuclear equation that describes the formation of the last member of the series shown in Figure 17.4.

17.8 Name two naturally occurring radioactive nuclides that are not in any of the naturally occurring radioactive decay series.

17.9 What is a particle accelerator? Why is it required in certain nuclear reactions?

17.10 In what major way has the discovery of the transuranium elements affected the form of the periodic table?

17.11 The half-life of cesium-137 is 30.2 years. Explain the meaning of that statement.

17.12 All of the isotopes of technetium are radioactive, as are those of uranium. Uranium occurs naturally, but any isotope of technetium must be prepared in a laboratory. Why?

17.13 Consider the nuclides in the radioactive decay series beginning with uranium-238 (see Table 17.2). If you initially had 10,000 atoms of each member of this series, which of them would be present in the largest quantity after one billion years had elapsed? Which of them would be present in the smallest quantity after one day had elapsed?

17.14 Define *radioactive dating*.

17.15 If a piece of wood contains half as much carbon-14 as an uncut tree, what can be said about the age of the piece of wood?

17.16 If a piece of wood contains no detectable quantities of carbon-14, what can be said about the age of the wood?

17.17 What is a Geiger-Müller counter, and which forms of radiation does it count? What is a scintillation counter, and which forms of radiation does it count?

17.18 Phosphorus is an essential nutrient in the human body. If a compound containing radioactive phosphorus were administered to a living organism, could it function as a radioactive tracer? If so, what kind of information might we obtain?

17.19 What is the principal source of the radiation that an average person receives over a year's time?

17.20 What is a rem? If you were told that you had been exposed to 0.10 rem of radioactivity, what would be the implications to you? What questions might you ask?

17.21 Why is the inhalation of radon, a noble gas, dangerous?

17.22 How does nuclear fission differ from nuclear fusion?

17.23 Would you expect nuclei of carbon-12 to undergo fission under appropriate conditions or undergo fusion with other carbon-12 nuclei? Give a reason for your choice.

17.24 Describe the events that occur during the fission of uranium-235.

17.25 What is a critical mass?

17.26 Some people believe that a nuclear power plant is a potential atomic bomb waiting to explode. Do you agree or disagree? Explain your answer.

Practice Problems

Radioactivity (Sections 17.1 and 17.2)

17.27 Rubidium-87, which forms about 28% of natural rubidium, is radioactive. It decays by emitting a beta particle to strontium-87. Write the nuclear equation for this decay.

17.28 Write the nuclear equation for the decay of phosphorus-32 to sulfur-32 by beta emission

17.29 Thorium is a naturally occurring radioactive element. Thorium-232 decays by emitting an alpha particle to radium-228. Write the nuclear equation that describes this decay.

17.30 The naturally occurring isotopes of carbon are carbon-12, carbon-13, and carbon-14. Another isotope, carbon-11, is produced from nuclear reactions in the laboratory. It decays by emitting a positron to boron-11. Write the nuclear equation for this decay.

17.31 Polonium-210 decays by emitting a single alpha particle. Write the nuclear equation that describes this decay.

17.32 Actinium-227 decays by emitting a single alpha particle. Write the nuclear equation that describes this decay.

17.33 Oxygen-18 is formed when fluorine-18, an artificially produced radioactive nuclide, decays. Write the nuclear equation for this decay.

17.34 Helium-3 is produced when tritium, the artificially produced radioactive isotope of hydrogen, decays. Write the nuclear equation for this decay.

Nuclear Transmutation (Section 17.3)

17.35 When lithium-6 is bombarded with neutrons, hydrogen-3 and another nuclide are formed. Identify the unknown nuclide.

17.36 When aluminum-27 is bombarded by hydrogen-3, magnesium-27 and another nuclide are formed. Identify the unknown nuclide.

17.37 Curium was first synthesized by bombarding an element with alpha particles. The products were curium-242 and a neutron. What was the target element?

17.38 Californium was first synthesized by bombarding an element with alpha particles. The products were californium-245 and a neutron. What was the target element?

Half-Lives and Radioactive Dating (Section 17.4)

17.39 Molybdenum-99 decays by beta emission with a half-life of 67 hours. What fraction of a sample of this nuclide will remain after three half-lives have elapsed?

17.40 Hydrogen-3 (tritium) decays by emission of a beta particle with a half-life of 12.3 years. What fraction of a sample of this nuclide will remain after six half-lives have elapsed?

17.41 Gold-198 has a half-life of 2.69 days. For a sample containing 0.00100 g of this nuclide, approximately how much would remain after 8 days?

17.42 Cesium-134 has a half-life of 2.05 years. For a sample containing 1,000,000,000 atoms of cesium-134, approximately how many atoms will remain after 200 years?

17.43 Carbon from a cypress beam obtained from the tomb of an ancient Egyptian king gives 8 disintegrations per minute per gram of total carbon. Carbon from living material gives 16 disintegrations per minute per gram of total carbon. How old is the cypress beam? Give the approximate date that the tree was felled.

17.44 Carbon in some mammoth bones gives 2 disintegrations per minute per gram of total carbon. Carbon from living material gives 16 disintegrations per minute per gram of total carbon. How old are the bones? Give the approximate date that the animal died.

17.45 A sample of a wooden artifact gives 4 disintegrations per minute per gram of total carbon. Carbon from living material gives 16 disintegrations per minute per gram of total carbon. What is the age of the artifact?

17.46 Carbon from an ancient object gives about 1 disintegration per minute per gram of total carbon. Carbon from living material gives 16 disintegrations per minute per gram of total carbon. What is the age of the artifact?

Additional Problems

17.47 A bismuth-209 nucleus reacts with an alpha particle to produce an astatine nucleus and two neutrons. Write the nuclear equation for this reaction.

17.48 A bismuth-209 nucleus reacts with a deuterium nucleus to produce polonium and a neutron. Write the nuclear equation for this reaction.

17.49 Complete the following equation by filling in the blank.

$$^{238}_{92}\text{U} + ^{12}_{6}\text{C} \longrightarrow \underline{\qquad\qquad} + 4\,^{1}_{0}\text{n}$$

17.50 Complete the following equation by filling in the blank.

$$^{246}_{96}\text{Cm} + ^{12}_{6}\text{C} \longrightarrow \underline{\qquad\qquad} + 4\,^{1}_{0}\text{n}$$

Practice in Problem Analysis

For each problem, describe the thinking you would use (the problem analysis) before doing the actual solution, but do not solve the problem.

1. If a meteorite contains roughly equal quantities of uranium-238 and lead-206, how old is the meteorite according to your best estimate?

2. Alpha particles are eventually converted to helium. What volume of helium would be collected at 0°C and 760 mmHg during the decay of 1.00 g of polonium-210 after one half-life has elapsed? This isotope decays by alpha emission.

PRACTICE EXAM

1. How many neutrons are in an atom of carbon-11?
(a) 5 (b) 6 (c) 11 (d) 12
(e) None of the above

2. How many protons are in an atom of carbon-11?
(a) 5 (b) 6 (c) 11 (d) 12
(e) None of the above

3. How many electrons are in an atom of carbon-11?
(a) 5 (b) 6 (c) 11 (d) 12
(e) None of the above

4. A particle emitted from the nucleus of an atom contains two protons and has a mass of 4 amu. The particle is
(a) a neutron (b) an electron
(c) a positron (d) an alpha particle
(e) None of the above

5. Beta particles are
(a) neutrons (b) electrons
(c) positrons (d) helium nuclei
(e) None of the above

6. Gamma radiation is composed of
(a) neutrons (b) electrons
(c) positrons (d) helium nuclei
(e) None of the above

7. When a thorium-234 atom decays by emitting a beta particle, the other product is
(a) actinium-234 (b) protactinium-234
(c) thorium-233 (d) thorium-234
(e) None of the above

8. When $^{230}_{90}Th$ decays, the products are $^{226}_{88}Ra$ and
(a) a neutron (b) an electron
(c) a positron (d) an alpha particle
(e) None of the above

9. Uranium-238 has a half-life of 4.51×10^9 yr. If experiments show that about half of the uranium-238 in a rock has decayed, the approximate age of the rock is
(a) 1 yr (b) $\frac{1}{2}$ yr (c) 1×10^9 yr
(d) 4.51×10^9 yr (e) None of the above

10. The half-life of tritium is 12 yr. How many years are required for a sample of tritium to be $\frac{1}{16}$ as radioactive as it was originally?
(a) 12 yr (b) 24 yr (c) 36 yr (d) 48 yr
(e) None of the above

11. After 32 s, a sample of a radioactive element is $\frac{1}{16}$ as radioactive as it was originally. The half-life of this nuclide is
(a) 4 s (b) 32 s (c) 64 s (d) 128 s
(e) None of the above

12. An old piece of cotton cloth could be dated using radioactivity from naturally occurring
(a) nitrogen-14 (b) carbon-14 (c) carbon-13
(d) oxygen-16 (e) None of the above

13. An old piece of wood could be dated using radioactivity from naturally occurring
(a) nitrogen-14 (b) carbon-14 (c) carbon-13
(d) oxygen-16 (e) None of the above

14. Which statement is true?
(a) A nuclear power plant is an atomic bomb waiting to explode.
(b) A nuclear power plant uses fusion to provide the energy for converting water to steam.
(c) A nuclear power plant captures beta particles before converting them to electricity.
(d) A nuclear power plant depends on nuclear fission.
(e) None of the above

15. Which of the following would be susceptible to nuclear fission?
(a) hydrogen-2 (b) helium-3
(c) plutonium-239 (d) carbon-11
(e) None of the above

Chapter

18 Organic Chemistry

An oxyacetylene torch.

Chapter Outline

Suppose you made a list of substances or materials around your home that have originated from manufactured chemicals. Perhaps you have aspirin in the medicine cabinet. You might also find rubbing alcohol (an aqueous solution of isopropyl alcohol). Elsewhere in your home, there might be nail polish remover, a solution of acetone (Figure 18.1), or antifreeze, a solution of ethylene glycol. Your clothes are likely to be made, at least in part, of textile fibers manufactured from chemicals, such as polyesters and nylons. All these substances are *organic compounds*. In fact, most of the materials you see around a home are organic compounds or mixtures containing organic compounds, including drugs, cosmetics, dyes, plastics, textiles, and rubber. The vast majority of the millions of known compounds are organic compounds.

Originally, organic compounds were defined as substances obtained from plant and animal materials. These compounds were thought to be fundamentally different from mineral substances. Organic compounds, early chemists thought, were produced by a "vital force" present only in living material. The concept of a vital force was later abandoned as more and more plant and animal substances were reproduced from nonliving materials in the laboratory. Today we define organic compounds as substances that contain the element carbon. Carbon is unique in the variety and complexity of molecular

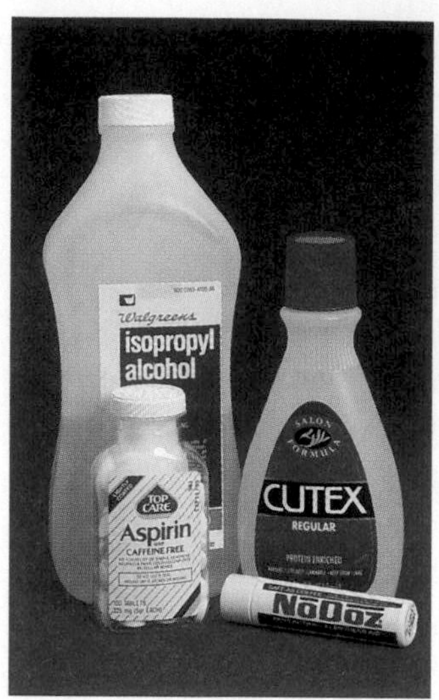

FIGURE 18.1
Some common organic compounds. *Left to right:* aspirin, rubbing alcohol (isopropyl alcohol), nail polish remover (acetone), and NoDoz (caffeine).

substances that it forms. In this chapter, we explore some of the most common types of organic compounds, describing their molecular structures and some of their chemical reactions.

BONDING AND STRUCTURE IN ORGANIC COMPOUNDS

Organic chemistry is *the chemistry of carbon compounds.* The formal study of organic chemistry began more than 150 years ago and has developed into a subject of enormous breadth, describing the chemistry of molecules that range from a few atoms to millions of atoms. What gives the carbon atom its ability to form so many different compounds and in such wide variety and complexity? In the next two sections, we look at some general bonding and structural features of the carbon atom that help explain this variety and complexity.

18.1 Carbon-Atom Bonding

OBJECTIVES

- Identify the four bonds to carbon in a variety of compounds.
- Identify chains of carbon atoms in organic compounds.

In almost all organic compounds, each carbon atom forms four bonds (counting double bonds as two bonds and triple bonds as three bonds). Thus, carbon is *tetravalent.* The carbon atom can have several different bonding arrangements.

$$-\overset{|}{\underset{|}{C}}- \qquad \gt C= \qquad =C= \qquad -C\equiv$$

These different possible arrangements are in part responsible for the variety of organic compounds.

Another reason for the variety of organic compounds is that carbon forms strong bonds to many elements. For example, when a carbon atom bonds to four hydrogen atoms, it forms the methane molecule, CH_4. (Methane is the major constituent of natural gas.)

$$\begin{array}{c} H \\ | \\ H-C-H \\ | \\ H \end{array}$$

As you know, the dash represents a shared electron pair, or covalent bond. Carbon–hydrogen bonds are relatively strong, and they exist in most organic compounds.

If, instead of bonding to hydrogen atoms, the carbon atom bonds to four chlorine atoms, it forms the carbon tetrachloride molecule, CCl_4. (Carbon tetrachloride is a liquid, formerly used as a dry-cleaning solvent.)

$$\begin{array}{c} Cl \\ | \\ Cl-C-Cl \\ | \\ Cl \end{array}$$

This molecule, like methane, has a *tetrahedral shape* (Figure 18.2).

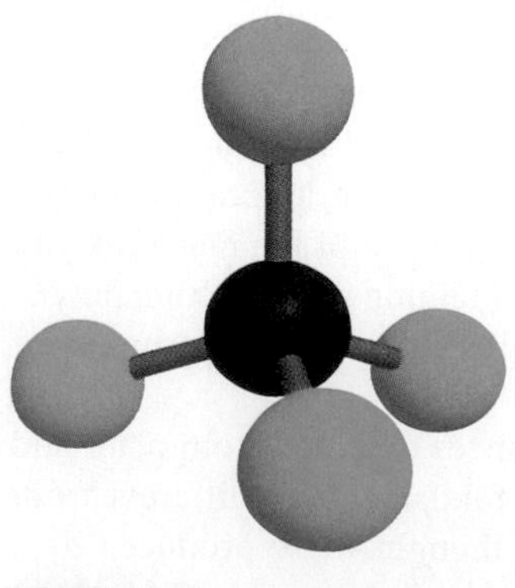

FIGURE 18.2
Molecular model of carbon tetrachloride. A ball-and-stick model showing the tetrahedral shape of the carbon tetrachloride molecule.

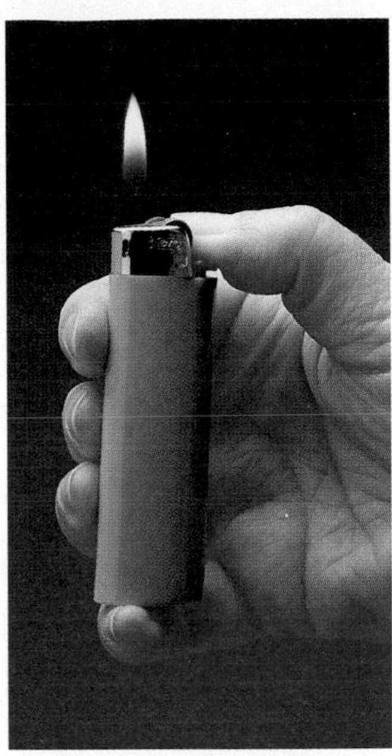

FIGURE 18.3
Butane lighter. The lighter contains butane liquid under pressure. When you press a valve, you release the pressure, and butane gas escapes. The flame results from butane gas burning in air.

Carbon also bonds strongly to itself to form molecules with carbon-atom chains, as in these examples.

```
     H  H  H  H            H    H  H  H
     |  |  |  |             \   |  |  |
  H—C—C—C—C—H               C=C—C—C—H
     |  |  |  |             /      |  |
     H  H  H  H            H       H  H

       Butane                 1-Butene
```

Butane is a fuel used in lighters (Figure 18.3). Both butane and 1-butene have chains of four carbon-atoms. Other molecules have thousands of carbon atoms bonded together to form long carbon-atom chains. Carbon is unique in its ability to form long chains of atoms, as well as branched chains and rings, and it is this ability that accounts for much of the variety and complexity of organic compounds.

18.2 Structural Formulas and Isomers

OBJECTIVES

- Differentiate between a Lewis formula and a structural formula.
- Write the structural formulas of isomers, given their molecular formulas.
- Write a condensed structural formula, and vice versa, given a full structural formula.

The formulas we used in the previous section are *structural formulas;* we discussed them briefly in Section 4.7.

Structural Formulas

Structural formulas are similar to Lewis electron-dot formulas except that they generally do not indicate the lone pairs of electrons. For example, dimethyl ether has the following Lewis formula and structural formula.

```
     H     H                H     H
     |  ·· |                |     |
  H—C—O—C—H             H—C—O—C—H
     |  ·· |                |     |
     H     H                H     H

  Lewis formula          Structural formula
```

In a structural formula, we are merely indicating how atoms are bonded to one another; we are not trying to describe the molecular shape. Consequently, we have some latitude in writing such formulas. For example, the structural formula of dimethyl ether might also be written

```
           H
           |
        H—C—H
     H     |
     |     |
  H—C————O
     |
     H
```

Isomers

Structural formulas are necessary in organic chemistry because often the same molecular formula represents more than one chemical substance. For example, both ethanol

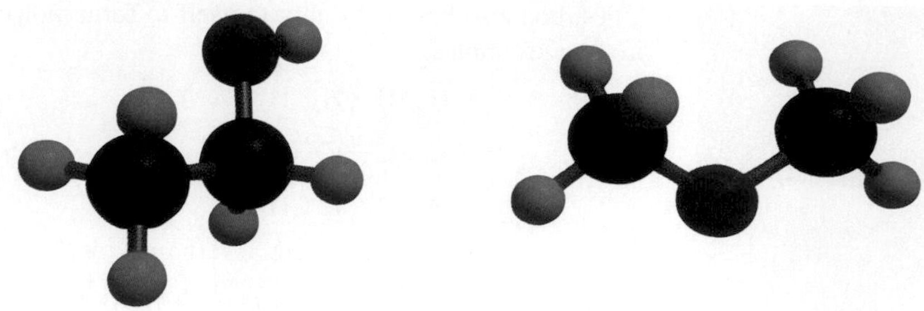

FIGURE 18.4
Isomers of C_2H_6O. Ball-and-stick models of ethanol *(left)* and dimethyl ether *(right)*. Note that each molecule contains the same atoms, but in a different bonding arrangement.

(a liquid compound, commonly called alcohol) and dimethyl ether (a gas) have the molecular formula C_2H_6O. But the compounds have different structural formulas.

```
    H  H                     H     H
    |  |                     |     |
 H—C—C—O—H               H—C—O—C—H
    |  |                     |     |
    H  H                     H     H

   Ethanol                Dimethyl ether
```

These formulas clearly show that the atoms bond to one another differently. In ethanol, the oxygen atom bonds to only one carbon atom and to a hydrogen atom, whereas in dimethyl ether the oxygen atom bonds to the two carbon atoms.

Compounds that have the same molecular formula but different structural formulas (that is, molecules with the same number of each kind of atom but with the atoms bonded in different ways) are **isomers.** Thus, ethanol and dimethyl ether are isomers. Each has the molecular formula C_2H_6O; they differ by having distinct structural formulas. Figure 18.4 shows ball-and-stick models of these isomers.

As the number of carbon atoms in a formula grows, the number of isomers increases enormously. There are three isomers with the molecular formula C_5H_{12}; here are their structural formulas:

```
                                   H                    H
                                   |                    |
                                 H—C—H                H—C—H
                                   |                    |
    H  H  H  H  H              H   |  H  H          H   |   H
    |  |  |  |  |              |   |  |  |          |   |   |
 H—C—C—C—C—C—H            H—C—C—C—C—H       H—C—C—C—H
    |  |  |  |  |              |   |  |  |          |   |   |
    H  H  H  H  H              H   H  H  H          H   |   H
                                                        |
                                                      H—C—H
                                                        |
                                                        H
```

There are 75 isomers with the molecular formula $C_{10}H_{22}$.

Condensed Structural Formulas

In the preceding section, you saw how important structural formulas are for specifying an organic compound. Because such formulas can be cumbersome, organic chemists have devised condensed structural formulas. A **condensed structural formula** is *a structural formula that uses established abbreviations for various groups of atoms*. The following are some of the abbreviations used.

$$\text{CH}_3 \text{ means } \begin{array}{c} \text{H} \\ | \\ \text{H—C—} \\ | \\ \text{H} \end{array} \qquad \text{CH}_2 \text{ means } \begin{array}{c} \text{H} \\ | \\ \text{—C—} \\ | \\ \text{H} \end{array} \qquad \text{CH means } \begin{array}{c} | \\ \text{—C—} \\ | \\ \text{H} \end{array}$$

$$\text{OH means } \text{—O—H} \qquad \text{O means } \text{—O—}$$

Earlier we wrote full structural formulas for the isomers of C_5H_{12}. Using condensed structural formulas, you would write these as follows.

$$CH_3CH_2CH_2CH_2CH_3 \qquad \begin{array}{l} \quad CH_3 \\ \quad | \\ CH_3CHCH_2CH_3 \end{array} \qquad \begin{array}{c} CH_3 \\ | \\ CH_3\text{—}C\text{—}CH_3 \\ | \\ CH_3 \end{array}$$

HYDROCARBONS

For classification purposes, all organic compounds can be considered hydrocarbons or compounds derived structurally from hydrocarbons. A **hydrocarbon** is *a compound containing only carbon and hydrogen.* For example, methane (CH_4) is a hydrocarbon. Carbon tetrachloride (CCl_4) is a structural derivative of methane obtained by replacing the H atoms of methane with Cl atoms. In the next four sections, we look at the principal kinds of hydrocarbons. In the last part of this chapter, we look at some derivatives of hydrocarbons. ●

● We are using the term *derivative* in a structural sense. A particular hydrocarbon derivative may or may not be prepared commercially from the hydrocarbon.

18.3 Alkanes

OBJECTIVES

- Define *alkane.*
- Write the structural formulas and names of the first ten straight-chain alkanes.
- Write the structures of the methyl and ethyl groups.
- Name a branched-chain alkane given the structural formula; write the structural formula given the name.
- Describe the commercial sources and the uses of alkanes.
- Write a balanced equation for the complete burning of an alkane.
- Write a chemical equation for the substitution reaction of an alkane with chlorine.

Methane is the first member of the alkane (pronounced "al′-kane") series of hydrocarbons. An **alkane** is *a hydrocarbon containing only single bonds and having the general molecular formula* C_nH_{2n+2}. The simplest alkanes are straight-chain hydrocarbons, whose first four members have the following full structural formulas. ●

● The so-called straight-chain hydrocarbons do not actually have carbon atoms in a straight line, as you can see from Figure 18.5.

$$\begin{array}{c} \text{H} \\ | \\ \text{H—C—H} \\ | \\ \text{H} \\ \\ \text{Methane} \end{array} \qquad \begin{array}{c} \text{H H} \\ |\ \ | \\ \text{H—C—C—H} \\ |\ \ | \\ \text{H H} \\ \\ \text{Ethane} \end{array} \qquad \begin{array}{c} \text{H H H} \\ |\ \ |\ \ | \\ \text{H—C—C—C—H} \\ |\ \ |\ \ | \\ \text{H H H} \\ \\ \text{Propane} \end{array} \qquad \begin{array}{c} \text{H H H H} \\ |\ \ |\ \ |\ \ | \\ \text{H—C—C—C—C—H} \\ |\ \ |\ \ |\ \ | \\ \text{H H H H} \\ \\ \text{Butane} \end{array}$$

These *straight-chain alkanes* are also referred to as *normal alkanes.* Figure 18.5 shows a ball-and-stick model of butane. The condensed structural formulas of the first four

FIGURE 18.5

Molecular model of butane. Ball-and-stick models of butane, C_4H_{10}, show different atomic arrangements (conformations) of the same molecule. Note that the bond angles do not allow a truly straight carbon chain.

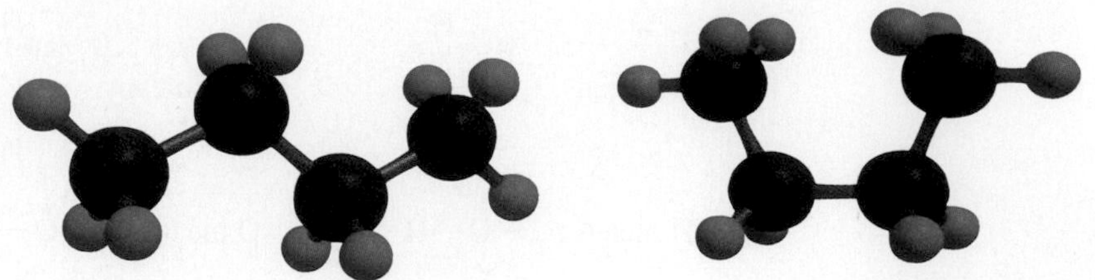

normal alkanes are

CH_4	CH_3CH_3	$CH_3CH_2CH_3$	$CH_3CH_2CH_2CH_3$
Methane	Ethane	Propane	Butane

The corresponding molecular formulas are CH_4, C_2H_6, C_3H_8, and C_4H_{10}. Check these formulas to verify that they follow the general formula C_nH_{2n+2}, where $n = 1, 2, 3$, and 4, respectively.

Alkanes may also have branched chains. Here is an example of a *branched-chain alkane*.

```
                  H
                  |
                H—C—H
                  |
                H—C—H
                  |
      H       H   |   H   H   H
      |       |   |   |   |   |
    H—C———C———C———C———C———C———C—H
      |   |   |   |   |   |   |
      H   |   H   H   H   H   H
          |
        H—C—H
          |
          H
```

These branched-chain alkanes are isomers of the straight-chain alkanes with the same number of carbon atoms.

Naming Alkanes

More than ten million organic compounds exist. To understand and classify these compounds, organic chemists began late in the nineteenth century to devise a system of naming organic compounds that depends on their structure. An international body, the International Union of Pure and Applied Chemistry (IUPAC, pronounced "eye-you-pac"), constantly reviews the rules for naming organic compounds (as well as considering other issues of importance to chemists).

The IUPAC rules start with the names of the straight-chain alkanes, which have names denoting the number of carbon atoms in the chain (except for the first four members of the series, methane, ethane, propane, and butane). The name of a straight-chain alkane of five or more carbon atoms consists of a stem from the Greek that denotes the number of carbon atoms and a suffix *-ane*. For example, we call the straight-chain alkane with five carbon atoms *pentane*. The stem *pent-* indicates five carbon atoms; the suffix *-ane* indicates an alkane. Table 18.1 lists the first ten members of the straight-chain alkanes.

The name of a branched-chain alkane specifies both the longest carbon-atom chain (the parent chain) in the molecule and its branch chains. We give the parent chain the

TABLE 18.1
The First Ten Straight-Chain Alkanes

Name	Molecular Formula	Condensed Structural Formula
Methane	CH_4	CH_4
Ethane	C_2H_6	CH_3CH_3
Propane	C_3H_8	$CH_3CH_2CH_3$
Butane	C_4H_{10}	$CH_3CH_2CH_2CH_3$
Pentane	C_5H_{12}	$CH_3CH_2CH_2CH_2CH_3$
Hexane	C_6H_{14}	$CH_3CH_2CH_2CH_2CH_2CH_3$
Heptane	C_7H_{16}	$CH_3CH_2CH_2CH_2CH_2CH_2CH_3$
Octane	C_8H_{18}	$CH_3CH_2CH_2CH_2CH_2CH_2CH_2CH_3$
Nonane	C_9H_{20}	$CH_3CH_2CH_2CH_2CH_2CH_2CH_2CH_2CH_3$
Decane	$C_{10}H_{22}$	$CH_3CH_2CH_2CH_2CH_2CH_2CH_2CH_2CH_2CH_3$

name of the corresponding straight-chain alkane, and we precede this with the names of the branch chains, each named as an alkyl group.

An **alkyl group** is *a group of atoms obtained by removing one hydrogen atom from an alkane*. When we remove a hydrogen atom from methane, CH_4, we obtain the methyl group, $CH_3—$; when we remove a hydrogen atom from ethane, CH_3CH_3, we obtain the ethyl group, $CH_3CH_2—$. Table 18.2 lists some common alkyl groups. The branched-chain alkane given earlier has a methyl branch and an ethyl branch.

```
                    CH2CH3        Ethyl group
                    |
CH3CHCH2CHCH2CH2CH3
     |
     CH3     Methyl group
```

(Here and in the following text, we will use condensed structural formulas.)

TABLE 18.2
Some Common Alkyl Groups

Original Alkane	Structure of Alkyl Group	Name of Alkyl Group
Methane, CH_4	$CH_3—$	Methyl
Ethane, CH_3CH_3	$CH_3CH_2—$	Ethyl
Propane, $CH_3CH_2CH_3$	$CH_3CH_2CH_2—$	Propyl
Propane, $CH_3CH_2CH_3$	CH_3CHCH_3 (bond down from central C)	Isopropyl
Butane, $CH_3CH_2CH_2CH_3$	$CH_3CH_2CH_2CH_2—$	Butyl
Isobutane, CH_3CHCH_3 with CH_3 on central C	CH_3CCH_3 with bond up and CH_3 down from central C	*Tertiary*-butyl (*t*-butyl)

(From Darrell D. Ebbing, *General Chemistry,* 5th ed. [Boston: Houghton Mifflin, 1996], p. 1028.)

Rules for Naming a Branched-Chain Alkane

1. **Find the parent chain.** Determine the longest carbon-atom chain in the structural formula of the molecule; this then is the parent chain, from which you begin the name of the branched-chain alkane. The name of the parent chain is the same as the name of the straight-chain alkane of the same number of carbon atoms. For example, the longest chain, or parent chain, in the following alkane has four carbon atoms, so the name of the parent chain is butane. (See Table 18.1 for the names of the straight-chain alkanes.)

$$\begin{array}{l} \quad CH_3 \\ \quad\ | \\ CH_3CHCH_2CH_3 \end{array}$$

The parent chain in any given structural formula may be twisted or bent (since in drawing a structural formula you are merely indicating the connections, or bonding, between atoms). The following alkane has a parent chain of seven carbon atoms, so the parent name of the alkane is heptane (see Table 18.1).

$$\begin{array}{rcl} & H & \\ & | & \\ CH_3CH_2CH_2CH_2- & C & -CH_3 \\ & | & \\ & CH_2 & \\ & | & \\ & CH_3 & \end{array}$$

2. **Find the branch chains.** Any branch off the parent chain is named as an alkyl group. The preceding formula has a *methyl* group as a branch off the parent *heptane* chain. (See Table 18.2 for the names of some common alkyl groups.)

$$\begin{array}{rcl} & H & \\ & | & \\ CH_3CH_2CH_2CH_2- & C & -CH_3 \leftarrow \text{Methyl group} \\ & | & \\ & CH_2 & \\ & | & \\ & CH_3 & \end{array}$$

3. **Add location numbers for each branch chain.** The complete name of a branch requires a number that locates that branch on the parent chain. To obtain it, number the carbon atoms of the parent chain beginning with the end of the chain that is closest to the branch. You number the carbon atoms in the preceding formula as follows.

$$\begin{array}{rcl} & H & \\ 7\ \ 6\ \ 5\ \ 4\ \ \ 3 & | & \\ CH_3CH_2CH_2CH_2- & C & -CH_3 \\ & | & \\ 2 & CH_2 & \\ & | & \\ 1 & CH_3 & \end{array}$$

Note that the methyl group is attached to carbon atom 3. Therefore, the branch name is *3-methyl,* where the number indicates the position of the branch on the parent chain. (Note that a hyphen separates the number and the name of the alkyl group.)

4. **Write the complete name of the alkane.**

   a. *An alkane with one branch.* The name of an alkane with a single branch is given the name of the parent chain with the alkyl branch name written as a prefix, without a space between the parent name and the branch name. Thus, the name of the preceding alkane is

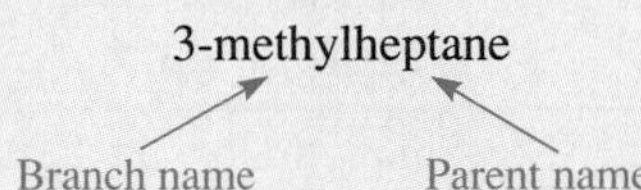

   b. *An alkane with two or more similar branches.* When the same alkyl branch occurs more than once on a parent chain (say, two methyl groups are attached to a pentane chain), you indicate the number of these same alkyl groups by a Greek prefix *di-* (for two), *tri-* (for three), *tetra-* (for four), and so forth. Consider the following alkane.

$$\begin{array}{c} CH_3 \\ \overset{1}{C}H_3\overset{2}{C}H\overset{3}{C}H\overset{4}{C}H_2\overset{5}{C}H_3 \\ CH_3 \end{array}$$

   The longest chain has five carbon atoms, so the parent name is pentane. Now you number the parent chain from the end closest to a branch. There are two methyl branches. You indicate these branches, including their locations on the parent chain, by the prefix 2,3-dimethyl. Note that you use location numbers for each methyl group and separate the numbers by a comma. The complete name of the alkane is 2,3-dimethylpentane.

   c. *An alkane with two or more different branches.* When the parent chain has two or more different branches, the branch names are listed before the parent name in alphabetical order, with hyphens between the branch names. Consider the alkane

Methyl group

$$\begin{array}{c} CH_3 \\ \overset{1}{C}H_3\overset{2}{C}H\overset{3}{C}H\overset{4}{C}H_2\overset{5}{C}H_3 \\ CH_2CH_3 \end{array}$$

Ethyl group

   The branches are 2-methyl and 3-ethyl. These branches are listed in alphabetical order: ethyl, then methyl. Therefore, the name of the alkane is 3-ethyl-2-methylpentane.

   Note that numbers such as 2 and 3 are not alphabetized (as "two" and "three"). Similarly, when you have an alkane with two similar branches and a different branch, you ignore the prefixes *di-*, *tri-*, and so forth. For example, if the previous alkane had two methyl groups on the number 2 carbon, you would name it 3-ethyl-2,2-dimethylpentane (ignoring the prefix *di-* and the numbers in alphabetizing the alkyl branch names).

■ EXAMPLE 18.1 Naming Alkanes

Write the IUPAC names of the following alkanes.

(a) 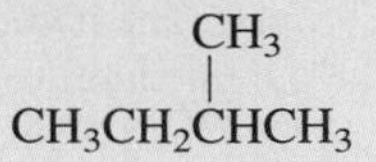
$$\begin{array}{c} \quad\quad\quad CH_3 \\ \quad\quad\quad | \\ CH_3CH_2CHCH_3 \end{array}$$

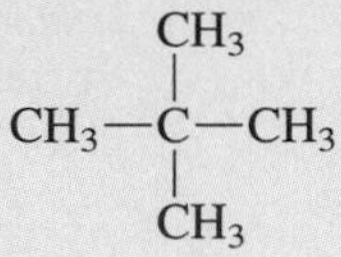

(b)
$$\begin{array}{c} CH_3 \\ | \\ CH_3{-}C{-}CH_3 \\ | \\ CH_3 \end{array}$$

(c)
$$\begin{array}{c} CH_3 \\ | \\ CH_2 \\ | \\ CH_3CHCH_3 \end{array}$$

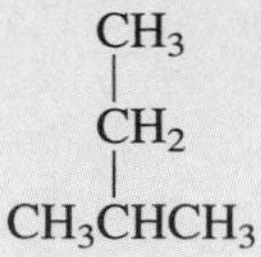

Problem Analysis

Simply follow the IUPAC rules for naming an alkane: First find the longest carbon chain in each alkane (the parent chain); find the branch chains attached to the parent chain; number the carbon atoms in the parent chain, writing the full name of the branches; and, finally, name the alkane by placing the branch names before the name of the parent chain.

Solution

(a) The longest carbon chain contains four atoms, so the parent name of the alkane is butane (see Table 18.1). Note that the butane chain has one methyl branch (see the alkyl names in Table 18.2).

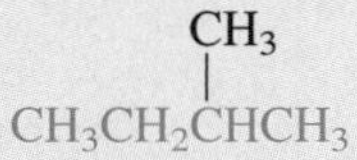

$$\begin{array}{c} \quad\quad\quad CH_3 \\ \quad\quad\quad | \\ CH_3CH_2CHCH_3 \end{array}$$

Number the parent chain, starting from the right (because this carbon is closest to the methyl group).

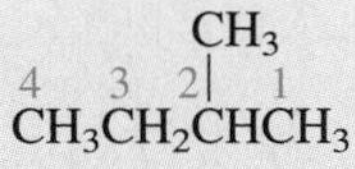

$$\begin{array}{c} \quad\quad\quad CH_3 \\ \overset{4}{C}H_3\overset{3}{C}H_2\overset{2}{C}H\overset{1}{C}H_3 \end{array}$$

The name of the alkane is 2-methylbutane.

(b) The longest carbon chain has three atoms, so the parent name of the alkane is propane. This propane chain has two branches, both methyl groups.

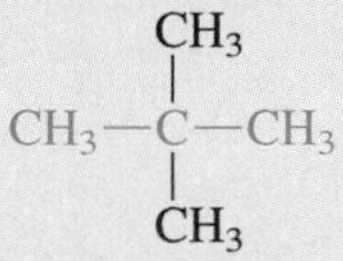

$$\begin{array}{c} CH_3 \\ | \\ CH_3{-}C{-}CH_3 \\ | \\ CH_3 \end{array}$$

You can number the carbon atoms in either direction.

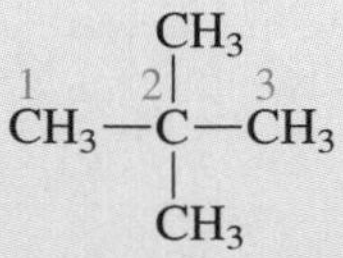

$$\begin{array}{c} CH_3 \\ | \\ \overset{1}{C}H_3{-}\overset{2}{C}{-}\overset{3}{C}H_3 \\ | \\ CH_3 \end{array}$$

The name of the alkane is 2,2-dimethylpropane.

(c) The longest chain has four carbon atoms, so the parent name of the alkane is butane. This chain has a methyl branch.

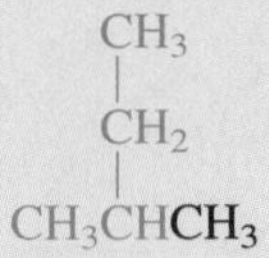

$$\begin{array}{c} CH_3 \\ | \\ CH_2 \\ | \\ CH_3CHCH_3 \end{array}$$

Begin numbering the carbon atoms from the lower left or lower right.

$$\begin{array}{c} {}^{4}\,CH_3 \\ | \\ {}^{3}\,CH_2 \\ {}^{1}\quad{}^{2}| \\ CH_3CHCH_3 \end{array}$$

The name of the alkane is 2-methylbutane. It is not 2-ethylpropane, which is the name you would have written had you taken the parent chain to be the one along the bottom of the structural formula. Notice that this is the same alkane as in (a). The structural formulas given in (a) and (c) are simply different representations of the same compound.

Exercise 18.1

What are the names of the following alkanes?

(a)
$$\begin{array}{c} CH_3 \\ | \\ CH_3CHCHCH_3 \\ | \\ CH_2CH_3 \end{array}$$

(b)
$$\begin{array}{c} CH_3 \\ | \\ CH_3CHCHCH_3 \\ | \\ CH_3 \end{array}$$

(Try Problems 18.43 and 18.44.)

When you see the name of an alkane in print, you can obtain the structural formula by simply taking the parts of the name (the parent name and the branch names) and writing down the corresponding features of the structural formula.

1. Split the name of the alkane into the parent name and the names of the branch chains.
2. Write down the carbon chain corresponding to the parent chain.
3. Number the carbon atoms in this chain from left to right.
4. Attach the alkyl branches at the numbered locations.
5. Fill out the carbon chain with hydrogen atoms to satisfy the tetravalence of carbon.

The following example illustrates the method.

■ EXAMPLE 18.2 Writing the Structural Formula of an Alkane

Write the structural formulas for (a) 3-ethylhexane; (b) 2,2,3-trimethylhexane.

Problem Analysis

Split the name of the alkane into the parent name and the names of the branch chains. Then write down the parent chain and number the carbon atoms. Finally, attach the branch alkyl groups to the appropriate carbon atoms of the chain. Then fill out the bonds to the carbon atoms with hydrogen atoms.

Solution

(a) The parent name of the alkane is hexane, and there is an ethyl branch at carbon 3. The numbered parent chain is

$$\overset{1}{C}-\overset{2}{C}-\overset{3}{C}-\overset{4}{C}-\overset{5}{C}-\overset{6}{C}$$

You attach the ethyl group at carbon 3.

$$\overset{1}{C}-\overset{2}{C}-\overset{3}{\underset{}{C}}\overset{\displaystyle CH_2CH_3}{\big|}\!\!\!-\overset{4}{C}-\overset{5}{C}-\overset{6}{C}$$

Now you add hydrogen atoms to satisfy the tetravalence of carbon. For example, carbon atom 1 has only one bond in the preceding formula, so you add three hydrogen atoms (giving CH_3—). Carbon atom 2 has two bonds in the preceding formula, so you add two hydrogen atoms (giving —CH_2—). Proceeding this way, you obtain the final structural formula.

$$CH_3CH_2\overset{\displaystyle CH_2CH_3 \atop |}{C}HCH_2CH_2CH_3$$

(b) The parent name is hexane, as in (a), and there are three methyl branches (two methyls at carbon 2 and one methyl at carbon 3).

$$\overset{1}{C}-\underset{\displaystyle | \atop CH_3}{\overset{\displaystyle CH_3 \atop 2|}{C}}-\overset{\displaystyle CH_3 \atop 3|}{C}-\overset{4}{C}-\overset{5}{C}-\overset{6}{C}$$

After filling in with hydrogen atoms, you obtain this structural formula.

$$CH_3-\underset{\displaystyle | \atop CH_3}{\overset{\displaystyle CH_3 \atop |}{C}}-\underset{\displaystyle | \atop H}{\overset{\displaystyle CH_3 \atop |}{C}}-CH_2CH_2CH_3$$

Exercise 18.2

Write the structural formula for 3-methylpentane.

(Try Problems 18.45 and 18.46.)

Sources and Uses of Alkanes

The major commercial sources of alkanes are natural gas and petroleum. These materials are the end products of the decay of organic material such as marine plants that have been buried under sediments. Chemical changes have occurred in these materials over millions of years as the result of heat and pressure from being buried deep in the earth. ●

● Coal, which is a complex material containing a high percentage of carbon, can also be chemically processed to produce alkanes.

Natural gas is mostly methane, CH_4 (about 80%), with smaller amounts of ethane, propane, butane, and 2-methylpropane (isobutane). Petroleum contains various alkanes, including those having very long chains. On distillation, petroleum separates into fractions with various hydrocarbon components, including those used to produce gasoline, diesel fuel, lubricating oils, and asphalt (Table 18.3).

Natural gas and petroleum alkanes are used primarily as fuels and as starting materials for the production of organic chemicals. Natural gas is used for home and industrial heating; and propane and butane, which are gases obtained from natural gas, are widely available as liquids in fuel cylinders (Figure 18.6). Gasoline is produced in enormous quantities from petroleum as an automotive fuel.

TABLE 18.3
Petroleum Fractions

Boiling Range, °C	Name	Range of Carbon Atoms per Molecule	Use
Below 20	Gases	C_1 to C_4	Heating, cooking, and petrochemical raw material
20–200	Naphtha; straight-run gasoline	C_5 to C_{12}	Fuel; lighter fractions (such as petroleum ether, b.p. 30°C–60°C) are also used as laboratory solvents
200–300	Kerosene	C_{12} to C_{15}	Fuel
300–400	Fuel oil	C_{15} to C_{18}	Heating homes, diesel fuel
Over 400		Over C_{18}	Lubricating oil, greases, paraffin waxes, asphalt

(From Harold Hart, *Organic Chemistry: A Short Course,* 8th ed. [Boston: Houghton Mifflin, 1991], p. 102.)

The petroleum fraction that is destined for gasoline requires chemical processing (known as *petroleum refining*) to produce a fuel suitable for modern automobiles (Figure 18.7). The fuel mixture (hydrocarbon plus air) in a gasoline engine should begin to burn in the engine cylinder at the precise moment when the piston head starts downward, after the fuel mixture has been compressed. The burning is initiated as the result of a spark from the spark plug. The fuel mixture should not "preignite," or begin to burn before this spark, because then the piston is moving upward as the gases are being compressed. When that happens, the engine "knocks" as a result of irregular stress on the piston head and crankshaft. The straight-chain alkanes present in petroleum, however, do tend to preignite.

The *octane rating* of a gasoline is a measure of the antiknock performance of a gasoline compared with that of the branched-chain alkane 2,2,4-trimethylpentane (an isomer of octane), which has an assigned octane rating of 100. Straight-chain hydrocarbons give gasolines of low octane rating, whereas branched-chain alkanes give gasolines of higher octane rating. In petroleum refining, straight-chain hydrocarbons are heated with certain catalysts to change them to branched-chain alkanes, resulting in fuels of higher octane rating.

FIGURE 18.6
Propane cylinder. Propane, normally a gas, is sold as a liquid under pressure in steel cylinders.

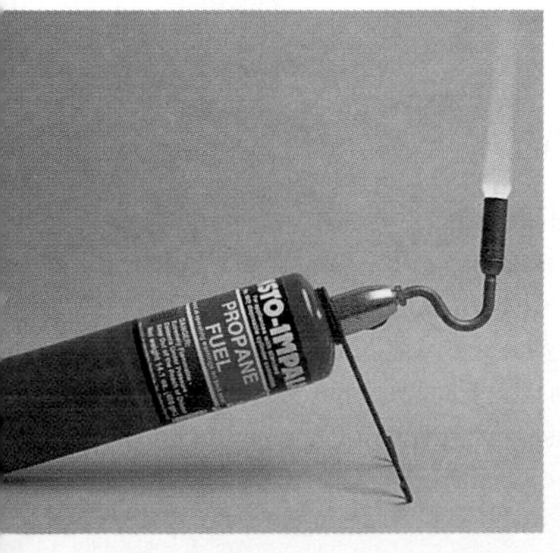

Heating petroleum alkanes in the presence of suitable catalysts also breaks ("cracks") carbon–hydrogen and carbon–carbon bonds, giving alkanes of shorter chain length and hydrocarbons with multiple bonds. Ethylene ($CH_2{=}CH_2$) is produced this way from natural gas and low-boiling petroleum fractions. Ethylene is the major starting material for the manufacture of organic chemicals and products such as polyethylene plastic. We look at polyethylene plastics in more detail in Section 18.6.

Reactions of Alkanes

The alkanes were at one time known as *paraffin* hydrocarbons (from the Latin *parum affinis* for "little affinity"). The name arose as a result of their relative unreactivity (little affinity for reaction). However, alkanes do undergo two important kinds of reactions: reaction with oxygen (burning) and substitution of hydrogen atoms on the hydrocarbon with chlorine and other halogens.

FIGURE 18.7
Petroleum refining. Shown here is a catalytic cracking unit. Hydrocarbon cracking is the breaking of carbon–carbon and carbon–hydrogen bonds to form new hydrocarbons.

All of the alkanes burn in air. The lower molecular weight members burn vigorously (even explosively) in air, releasing large quantities of heat. We use this explosive reaction in the gasoline engine. In burning, the alkanes react with the oxygen in air to produce carbon dioxide and water. (Carbon monoxide is also a product, if there is an insufficient supply of oxygen.) For example, methane burns in an excess of oxygen as in the following reaction.

$$CH_4(g) + 2O_2(g) \longrightarrow CO_2(g) + 2H_2O(g)$$

Methane and other alkanes react with chlorine in sunlight or at high temperature in a substitution reaction. A **substitution reaction** is *a reaction in which one atom (or atom group) substitutes for another atom (or group) on a molecule*. The reaction of methane with chlorine is a reaction in which one or more chlorine atoms substitute for hydrogen atoms of the methane molecule.

$$\begin{array}{c} \text{H} \\ | \\ \text{H—C—H} \\ | \\ \text{H} \end{array} \quad + \quad \text{Cl—Cl} \quad \longrightarrow \quad \begin{array}{c} \text{H} \\ | \\ \text{H—C—Cl} \\ | \\ \text{H} \end{array} \quad + \quad \text{H—Cl}$$

Methane — Chloromethane (methyl chloride)

(We give the IUPAC name directly under the structural formula of a compound, and an alternative name—usually the common name—in parentheses under this.) Substitution of one hydrogen atom yields chloromethane, or methyl chloride. In an excess of chlorine (Cl_2) more atoms of hydrogen are replaced by chlorine atoms. The final product is tetrachloromethane, CCl_4 (which is also named by the rules of binary compounds as carbon tetrachloride). In general, the reaction of an alkane with chlorine gives a mixture of substitution products called *alkyl chlorides*. ●

● These alkyl chlorides are examples of chlorinated hydrocarbons, which are used in pesticides, plastics, and other products. Many are being phased out because of environmental concerns.

■ EXAMPLE 18.3 Writing Equations for Alkane Reactions

Write equations for (a) the burning of butane gas in an excess of oxygen; (b) the formation of one possible product in the reaction of propane with chlorine.

Problem Analysis

(a) Simply write the balanced equation for the reaction of the alkane with oxygen to give carbon dioxide and water. (b) A number of products are possible; you are asked to write only one equation, yielding one product. The reaction will be a substitution of one chlorine atom for one hydrogen atom on the hydrocarbon.

Solution

(a) Butane is $CH_3CH_2CH_2CH_3$; the molecular formula is C_4H_{10}. The equation (not balanced) is

$$C_4H_{10}(g) + O_2(g) \longrightarrow CO_2(g) + H_2O(g)$$

The balanced equation is

$$2C_4H_{10}(g) + 13O_2(g) \longrightarrow 8CO_2(g) + 10H_2O(g)$$

(b) You obtain one possibility when chlorine substitutes for an end hydrogen atom on propane; the equation is

$$CH_3CH_2CH_3 + Cl_2 \longrightarrow \underset{\text{1-Chloropropane}}{ClCH_2CH_2CH_3} + HCl$$

Chlorine can also substitute for a hydrogen atom in the middle of the propane molecule.

$$CH_3CH_2CH_3 + Cl_2 \longrightarrow \underset{\text{2-Chloropropane}}{CH_3CHClCH_3} + HCl$$

Reactions involving the substitution of more than one hydrogen atom also occur, so that the result is a mixture of alkyl chlorides.

Exercise 18.3

(a) Write the balanced equation for the complete burning of propane in oxygen.
(b) Write the equation for one reaction of chlorine with butane.

(Try Problems 18.47, 18.48, 18.49, and 18.50.)

18.4 Alkenes and Alkynes

OBJECTIVES

- Define *alkene* and *alkyne*.
- Write the name of an alkene or alkyne, given the structural formula.
- Describe sources and uses of the alkenes and alkynes.
- Write the chemical equation for the addition of a halogen, hydrogen halide, or water to an alkene or alkyne.

Ethylene and propylene (whose IUPAC names are ethene and propene) are at the top of the list of key industrial organic chemicals.

Ethene (ethylene)

$$\begin{matrix} H & & H \\ & C=C & \\ H & & H \end{matrix}$$

(condensed formula $CH_2{=}CH_2$)

Propene (propylene)

$$\begin{matrix} H & & H & H & \\ & & | & | & \\ & C= & C- & C- & H \\ & & & | & \\ H & & & H & \end{matrix}$$

(condensed formula $CH_2{=}CHCH_3$)

Ethylene is the simplest alkene (pronounced "al′-keen"). An **alkene** is *a hydrocarbon containing a carbon–carbon double bond and having the general molecular formula* C_nH_{2n}.

Acetylene, H—C≡C—H, is the simplest alkyne (pronounced "al′-kine"). An **alkyne** is *a hydrocarbon containing a carbon–carbon triple bond and having the general molecular formula* C_nH_{2n-2}.

Naming Alkenes and Alkynes

The IUPAC rules for naming alkenes and alkynes are similar to those for naming alkanes. The major differences are these: *You take the parent name for the hydrocarbon from the longest carbon-atom chain that contains the double or triple bond, and you number this chain beginning with the end closest to the multiple bond.* You also need to indicate the position of the multiple bond along the parent chain.

Rules for Naming Alkenes and Alkynes

1. **Find the parent chain.** Determine the longest carbon-atom chain in the structural formula of the molecule that contains the double (or triple) bond; this then is the parent chain, from which you obtain the parent name of the alkene or alkyne. The name of the parent chain is given the same name as the alkane with the same number of carbon atoms, except that you replace the suffix *-ane* by *-ene* for the alkene and *-yne* for the alkyne.
2. **Find the branch chains.** Branch chains off the parent chain are named as alkyl groups.
3. **Add location numbers for the multiple bond and for each branch.** To obtain the location numbers, number the carbon atoms of the parent chain beginning with the end of the chain that is closest to the double (or triple) bond. The position of the multiple bond is denoted by the position of the first carbon atom in that bond. For example, the alkene $CH_3CH{=}CHCH_3$ is named 2-butene.
4. **Write the complete name of the alkene or alkyne.** Write the parent name of the hydrocarbon preceded by the names of any branches.

Consider the hydrocarbon

$$\begin{matrix} & CH_3 \\ & | \\ CH_3CH{=}CHCH_2 & CHCH_3 \end{matrix}$$

This compound has a double bond, so it is an alkene. The longest carbon-atom chain containing the double bond is shown in blue. It has six carbon atoms, so you name the hydrocarbon as a hexene. You number this hexene chain beginning with the end closest to the double bond (not the end closest to the branch, as you did for alkanes). Thus, the complete name for the parent chain is 2-hexene, and the name for the compound is 5 methyl-2-hexene. ●

● Two isomers of 5-methyl-2-hexene exist. One isomer has two hydrogen atoms on the same side of the carbon–carbon double bond, and the other has a hydrogen atom on one side of the carbon–carbon double bond and another hydrogen atom on the other side.

$$\overset{1}{C}H_3\overset{2}{C}H{=}\overset{3}{C}H\overset{4}{C}H_2\overset{\overset{CH_3}{|}}{\overset{5}{C}}H\overset{6}{C}H_3$$

The following example further illustrates the naming of alkenes and alkynes.

EXAMPLE 18.4 Naming Alkenes and Alkynes

Write the IUPAC names of the following compounds.

(a) $CH{\equiv}CCH_2CH_3$ (b) $CH_3CH_2CH_2\overset{\overset{CH_2CH_3}{|}}{C}{=}CH_2$

Problem Analysis

Follow the rules for naming alkenes and alkynes: Find the longest chain containing the multiple bond (the parent chain), naming it with the appropriate suffix, either *-ene* or *-yne*; find the branch chains; number the parent chain beginning with the carbon atom nearest the multiple bond, adding the location number for the multiple bond to the name of the parent chain; and, finally, name the compound by writing the parent name preceded by the names of any branches.

Solution

(a) Find the parent chain. In this case, there are no branch chains; the parent chain contains four carbon atoms and one triple bond. Therefore, you name this as a butyne. You start numbering the carbon atoms from the end closest to the triple bond.

$$\overset{1}{C}H{\equiv}\overset{2}{C}\overset{3}{C}H_2\overset{4}{C}H_3$$

You locate the triple bond by giving the number of the first carbon atom in that bond; the name of the compound is 1-butyne.

(b) Find the parent chain, which is the longest chain that includes the multiple bond. It contains five carbon atoms and a double bond, so you name it pentene. Note the ethyl branch near the double-bond end of the pentene chain. You number the pentene chain from the end nearest the double bond.

$$\overset{5}{C}H_3\overset{4}{C}H_2\overset{3}{C}H_2\overset{\overset{CH_2CH_3}{|}}{\overset{2}{C}}{=}\overset{1}{C}H_2$$

The name of the alkene is 2-ethyl-1-pentene.

Exercise 18.4

Give the IUPAC names of the following hydrocarbons.

(a) $CH_3CH_2CH_2C{\equiv}CCH_3$ (b) $CH_3\overset{\overset{CH_3}{|}}{C}HCH{=}CHCH_2CH_2CH_3$

(Try Problems 18.51 and 18.52.)

FIGURE 18.8
Preparation of acetylene. Water reacts with pieces of calcium carbide, releasing acetylene gas. The gas burns in air with a sooty flame.

Sources and Uses of Alkenes and Alkynes

Ethylene (ethene) and propylene (propene) are present in small quantities in natural gas and petroleum, but enormous quantities of these compounds are needed by the chemical industry. They are obtained by heating ethane and propane, and other alkanes.

$$\underset{\text{Ethane}}{CH_3CH_3} \xrightarrow{\Delta} \underset{\substack{\text{Ethene}\\\text{(ethylene)}}}{CH_2{=}CH_2} + H_2$$

$$\underset{\text{Propane}}{CH_3CH_2CH_3} \xrightarrow{\Delta} \underset{\substack{\text{Propene}\\\text{(propylene)}}}{CH_2{=}CHCH_3} + H_2$$

In these reactions, carbon–hydrogen bonds are broken in the alkanes to produce the corresponding alkenes. Heating larger alkanes breaks carbon–carbon bonds as well and produces ethylene and propylene as part of the products. Ethylene and propylene are used to prepare plastics (polyethylene and polypropylene) and various organic chemicals, including ethanol (alcohol), acetic acid, and ethylene glycol (antifreeze).

Acetylene, $CH{\equiv}CH$, is prepared by heating methane. In this process the carbon–hydrogen bonds are broken to give reactive molecular fragments that form carbon–carbon bonds.

$$2CH_4 \xrightarrow{1250°C} CH{\equiv}CH + 3H_2$$

Acetylene is also prepared from calcium carbide (CaC_2), which is obtained by heating lime (calcium oxide) with coke (carbon).

$$CaO(s) + 3C(s) \xrightarrow{2000°C} CaC_2(l) + CO(g)$$

The calcium carbide is then cooled and solidified. It reacts with water to produce acetylene (Figure 18.8). ●

● Calcium carbide was at one time used in coal miners' lamps; water dripping on the calcium carbide produced acetylene, which fueled the lamp (see Figure 7.8).

$$CaC_2(s) + 2H_2O(l) \longrightarrow Ca(OH)_2(aq) + CH{\equiv}CH(g)$$

Acetylene was at one time the starting material for many industrial organic chemicals, most of which are now prepared from ethylene. Acetylene is widely used in the oxyacetylene welding torch because it burns in oxygen with a very hot flame (Figure 18.9).

FIGURE 18.9
Oxyacetylene torch. In this welding torch, acetylene burns in oxygen with a very hot flame.

Reactions of Alkenes and Alkynes

In contrast to the alkanes, alkenes and alkynes are very reactive hydrocarbons. The reactions they undergo are mostly addition reactions. An **addition reaction** is *a reaction in which parts of a reactant molecule are added to each carbon atom of a carbon–carbon multiple bond; a carbon–carbon double bond becomes a single bond and a carbon–carbon triple bond becomes a double bond, then a single bond.* The reaction of propylene (propene) with bromine is a simple example of an addition reaction.

$$\underset{\text{Propene (propylene)}}{CH_2{=}CHCH_3} + Br_2 \longrightarrow CH_3\underset{\displaystyle Br}{\underset{|}{C}}H-\underset{\displaystyle Br}{\underset{|}{C}}H_2$$

The product is 1,2-dibromopropane. (The numbers locate the bromine atoms on the propane chain.) This type of reaction with bromine is often used as a simple test for multiple bonds because the red-brown color of bromine disappears as it reacts with the multiple bond.

Ethanol is prepared from ethylene by reacting the ethylene with steam in the presence of an acid catalyst (such as sulfuric acid).

$$CH_2{=}CH_2 + HOH \longrightarrow \underset{\displaystyle H}{\underset{|}{C}}H_2-\underset{\displaystyle OH}{\underset{|}{C}}H_2 \qquad (CH_3CH_2OH)$$

The two parts of the water molecule (H and OH) add to different carbon atoms of the double bond.

Alkynes also undergo similar addition reactions. For example, acetylene adds hydrobromic acid, HBr.

$$CH{\equiv}CH + HBr \longrightarrow \underset{\displaystyle H}{\underset{|}{C}}H{=}\underset{\displaystyle Br}{\underset{|}{C}}H$$

The product is 1-bromoethene, which has a double bond. In the presence of an excess of hydrobromic acid, 1-bromoethene also adds HBr. ●

● This second addition of HBr produces two different products. We have shown one; can you draw the structure of the other?

$$\underset{\displaystyle H}{\underset{|}{C}}H{=}\underset{\displaystyle Br}{\underset{|}{C}}H + HBr \longrightarrow \underset{\displaystyle H}{\underset{|}{\overset{\displaystyle H}{\overset{|}{C}}}}H-\underset{\displaystyle Br}{\underset{|}{\overset{\displaystyle Br}{\overset{|}{C}}}}H \qquad (CH_3CHBr_2)$$

18.5 Polyalkene Polymers

OBJECTIVES

- Define *polymer, monomer,* and *addition polymer.*
- Write equations, using structural formulas, for the preparation of polyethylene and polypropylene.

Plastics, rubber, and textile fibers are materials made of polymers (Figure 18.10). A **polymer** is *a very large molecule consisting of many repeating units of low molecular weight.* Many polymers are of natural origin; we look at some of these in Chapter 19.

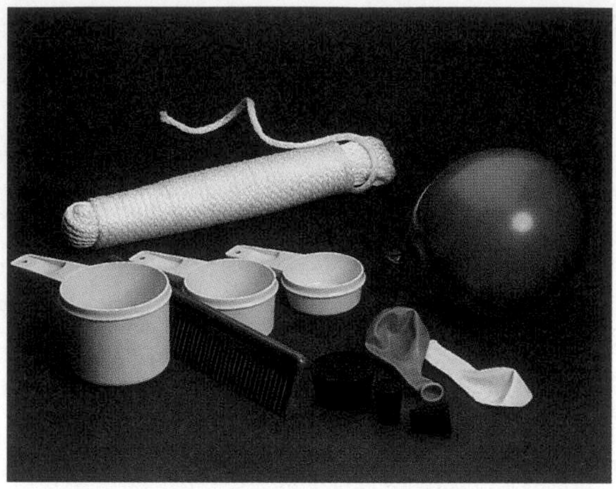

FIGURE 18.10
Polymer products. Molded plastic (measuring cups and comb), nylon (rope), and synthetic rubber (stoppers and balloons).

Others are synthetic, manufactured from substances of low molecular weight. Polyethylene is an example of a polymer used to make plastic bottles and packaging film. The following is the structural formula of a small piece of a polyethylene molecule.

$$\begin{array}{cccccccccccccc} & \mathrm{H} & \mathrm{H} & \mathrm{H} & \mathrm{H} & \mathrm{H} & \mathrm{H} & \mathrm{H} & \mathrm{H} & \mathrm{H} & \mathrm{H} & \mathrm{H} & \mathrm{H} & \\ & | & | & | & | & | & | & | & | & | & | & | & | & \\ - & \mathrm{C}- & \mathrm{C}- & \mathrm{C}- & \mathrm{C}- & \mathrm{C}- & \mathrm{C}- & \mathrm{C}- & \mathrm{C}- & \mathrm{C}- & \mathrm{C}- & \mathrm{C}- & \mathrm{C}- & \\ & | & | & | & | & | & | & | & | & | & | & | & | & \\ & \mathrm{H} & \mathrm{H} & \mathrm{H} & \mathrm{H} & \mathrm{H} & \mathrm{H} & \mathrm{H} & \mathrm{H} & \mathrm{H} & \mathrm{H} & \mathrm{H} & \mathrm{H} & \end{array}$$

The repeating unit in polyethylene, shown here in blue, is an ethylene unit, C_2H_4. The polymer chain varies in length but consists of thousands of repeating units.

A **monomer** is *a compound used to prepare a polymer; the monomer gives rise to the polymer's repeating unit.* As you have just seen, the monomer in polyethylene is ethylene. The prefix *poly-* means "many." Thus, polyethylene means "many ethylene monomer units."

Polyethylene is an example of an addition polymer. An **addition polymer** is *a polymer formed by linking many monomer molecules through addition reactions.* When ethylene is heated at high pressures in the presence of oxygen or peroxides (compounds with an oxygen–oxygen bond), it forms polyethylene.

$$\begin{array}{c} \mathrm{H}\;\;\mathrm{H} \\ |\;\;\;\;| \\ \mathrm{C}{=}\mathrm{C} \\ |\;\;\;\;| \\ \mathrm{H}\;\;\mathrm{H} \end{array} + \begin{array}{c} \mathrm{H}\;\;\mathrm{H} \\ |\;\;\;\;| \\ \mathrm{C}{=}\mathrm{C} \\ |\;\;\;\;| \\ \mathrm{H}\;\;\mathrm{H} \end{array} + \begin{array}{c} \mathrm{H}\;\;\mathrm{H} \\ |\;\;\;\;| \\ \mathrm{C}{=}\mathrm{C} \\ |\;\;\;\;| \\ \mathrm{H}\;\;\mathrm{H} \end{array} + \cdots \longrightarrow \begin{array}{cccccccc} & \mathrm{H} & \mathrm{H} & \mathrm{H} & \mathrm{H} & \mathrm{H} & \mathrm{H} & \\ & | & | & | & | & | & | & \\ - & \mathrm{C}- & \mathrm{C}- & \mathrm{C}- & \mathrm{C}- & \mathrm{C}- & \mathrm{C}- & \\ & | & | & | & | & | & | & \\ & \mathrm{H} & \mathrm{H} & \mathrm{H} & \mathrm{H} & \mathrm{H} & \mathrm{H} & \end{array}$$

One ethylene molecule adds to the double bond of another ethylene molecule, which in turn adds to another ethylene molecule, and so forth, until a polymer chain forms. The result is a long alkane chain.

Polypropylene is another example of an addition polymer. As the name implies, the monomer is propylene.

$$\begin{array}{c} \mathrm{CH_3}\;\mathrm{H} \\ |\;\;\;\;\;\;| \\ \mathrm{C}{=\!=}\mathrm{C} \\ |\;\;\;\;\;\;| \\ \mathrm{H}\;\;\;\;\mathrm{H} \end{array} + \begin{array}{c} \mathrm{CH_3}\;\mathrm{H} \\ |\;\;\;\;\;\;| \\ \mathrm{C}{=\!=}\mathrm{C} \\ |\;\;\;\;\;\;| \\ \mathrm{H}\;\;\;\;\mathrm{H} \end{array} + \begin{array}{c} \mathrm{CH_3}\;\mathrm{H} \\ |\;\;\;\;\;\;| \\ \mathrm{C}{=\!=}\mathrm{C} \\ |\;\;\;\;\;\;| \\ \mathrm{H}\;\;\;\;\mathrm{H} \end{array} + \cdots \longrightarrow \begin{array}{cccccccc} & \mathrm{CH_3} & \mathrm{H} & \mathrm{CH_3} & \mathrm{H} & \mathrm{CH_3} & \mathrm{H} & \\ & | & | & | & | & | & | & \\ - & \mathrm{C}- & \mathrm{C}- & \mathrm{C}- & \mathrm{C}- & \mathrm{C}- & \mathrm{C}- & \\ & | & | & | & | & | & | & \\ & \mathrm{H} & \mathrm{H} & \mathrm{H} & \mathrm{H} & \mathrm{H} & \mathrm{H} & \end{array}$$

In this case, the polymer is a long branched-chain alkane. Polypropylene is used to make carpet fibers, molded plastic parts, and tubing.

18.6 Aromatic Hydrocarbons

OBJECTIVES

- Define *aromatic hydrocarbon.*
- Describe the delocalized bonding in benzene, and write the resonance description of the bonding.
- Write the structural formulas for the derivatives of benzene, given their names.
- Draw the structural formula of naphthalene.
- Describe, with a chemical equation, the substitution reaction of benzene with a halogen.

Benzene (C_6H_6) consists of six carbon atoms bonded to each other in a ring, with hydrogen atoms bonded to each carbon atom. It is the simplest aromatic hydrocarbon. An **aromatic hydrocarbon** is *a hydrocarbon that has a structure based on the benzene ring.* These hydrocarbons were named aromatic because many derivatives of benzene have an aroma, or aromatic odor. One of these derivatives is benzaldehyde, C_6H_5CHO, which has the odor of almond oil. (The condensed formula group C_6H_5— is a benzene group in the compound.) Others are cinnamaldehyde, $C_6H_5CH{=}CHCHO$, which has the odor of cinnamon oil, and vanillin, $C_6H_3(OH)(CH_3O)CHO$, which has the odor of vanilla (Figure 18.11).

Structure of Benzene

The simplest structural formula that you can write for the benzene ring is

```
            H
            |
    H     //C\      H
      \  //   \   /
        C       C
        |       ||
        C       C
      /  \\   /   \
    H     \\C/      H
            |
            H
```

Note, however, that you could also write this formula with the single bonds drawn as double bonds and vice versa.

```
            H
            |
    H      /C\\     H
      \  /    \\  /
        C       C
        ||      |
        C       C
      /  \    //  \
    H     \C//      H
            |
            H
```

On the basis of these formulas, you might expect benzene to react similarly to an alkene, easily undergoing addition reactions. In fact, this is not the case. The Lewis structural formula suffers from a limitation that prevents it from fully representing the actual chemical bonding in the benzene molecule. A Lewis formula represents each bonding electron pair as fixed in position, that is, as localized between two atoms. Thus, in each of the preceding formulas, a bond has one or two pairs of electrons between two carbon atoms. In fact, one pair of electrons in each double bond occupies a molecular orbital encompassing all six carbon atoms. These electron pairs in benzene are *delocalized.*

FIGURE 18.11
Some benzene derivatives used as flavors. *Left to right:* cinnamon flavor (cinnamaldehyde), vanilla flavor (vanillin), and almond flavor (benzaldehyde).

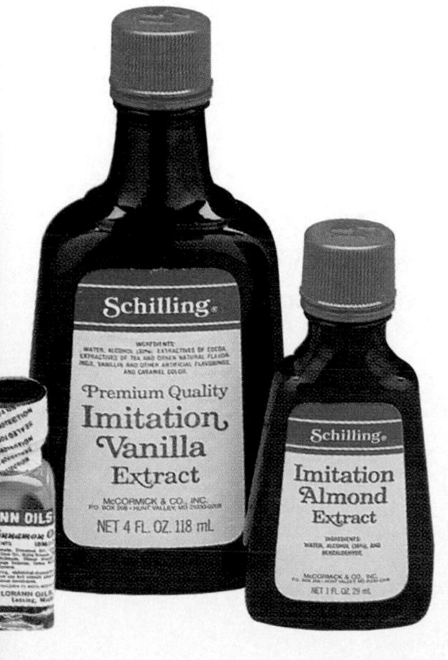

To retain the simplicity of the Lewis representation, we write the two preceding formulas connected by a double arrow as described in Section 10.5.

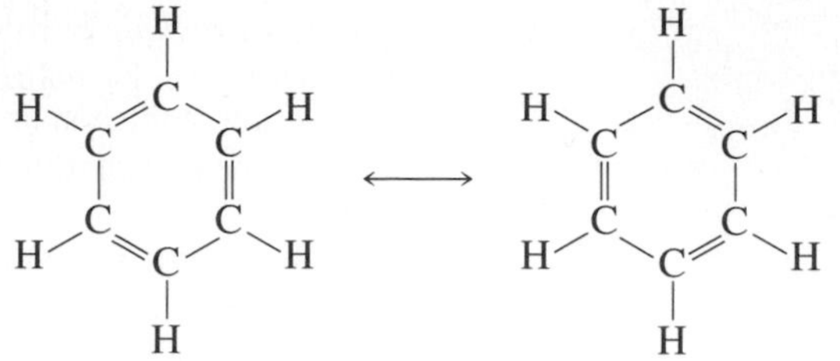

This is the *resonance description* of the benzene molecule. The meaning of the double arrow is simply that the extra pairs of electrons represented by double bonds are actually delocalized over the benzene ring, rather than localized between two carbon atoms as shown by any single formula. All six carbon atoms share the six electrons of the three double bonds, giving the benzene molecule great stability.

Often a simpler notation is used in which benzene is represented by a hexagon with a circle representing the delocalized electron pairs.

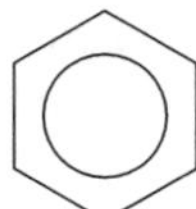

Each corner of the hexagon represents a carbon atom with its attached hydrogen atom. However, when this representation is used for a derivative of benzene in which the hydrogen atom has been substituted by another atom or group, that atom or group must be explicitly symbolized. For example, when a chlorine atom is substituted for a hydrogen atom of benzene, the structure is written as follows.

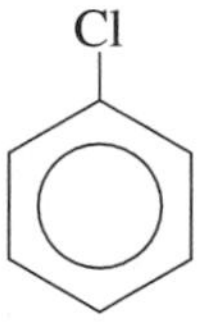

Cl

The hydrocarbon group obtained by removing a hydrogen atom from the benzene ring is called a *phenyl* group. Its formula, C_6H_5—, is used in condensed formulas involving benzene derivatives, as we noted at the beginning of this section. Thus, the condensed formula for the preceding structural formula is written C_6H_5Cl.

Derivatives of Benzene

The simplest aromatic hydrocarbons, other than benzene, are the monosubstituted alkyl derivatives of benzene. Their structures are obtained by substituting one hydrogen atom on the benzene ring by an alkyl group.

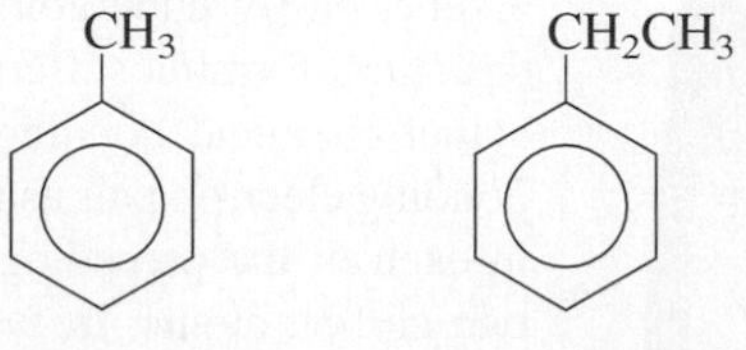

CH_3 CH_2CH_3

Methylbenzene (toluene) Ethylbenzene

These are condensed formulas, in which the benzene ring is written as a hexagon with a circle inside. Occasionally, we may also write these as $C_6H_5CH_3$ and $C_6H_5CH_2CH_3$.

These hydrocarbons are generally named as alkyl derivatives of benzene (for example, ethylbenzene). However, methylbenzene is better known as toluene, a name accepted by the IUPAC rules. This naming system of benzene derivatives extends to compounds in which the *substituent* (the group that substitutes for a hydrogen atom) is not an alkyl. Examples are

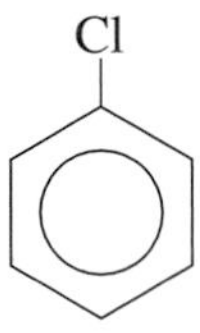

Cl Br NO_2

Chlorobenzene 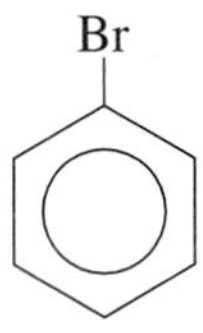 Bromobenzene 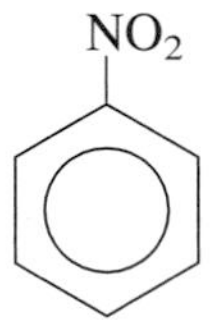 Nitrobenzene

When there are two alkyl substituents on a benzene ring, three isomers are possible. These isomers are distinguished by the prefixes *ortho- (o-), meta- (m-),* and *para- (p-).* For example,

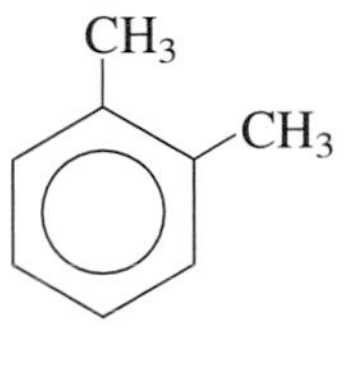

CH_3 CH_3 CH_3 CH_3 CH_3 CH_3

o-Dimethylbenzene (*o*-xylene) *m*-Dimethylbenzene (*m*-xylene) *p*-Dimethylbenzene (*p*-xylene)

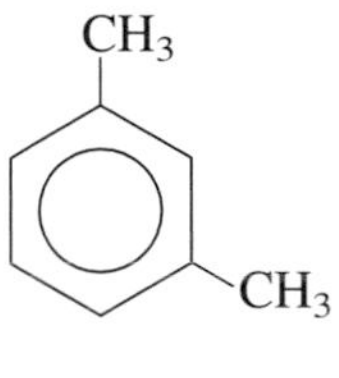

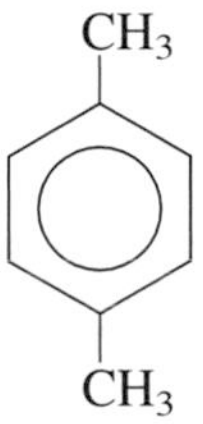

The prefix *ortho-* means that the two substituents are on adjacent carbon atoms; the prefix *meta-* means that the substituents are on carbon atoms separated by one carbon atom; and the prefix *para-* means that the substituents are on carbon atoms on opposite sides of the ring. The condensed formula of *o*-xylene may also be written *o*-$C_6H_4(CH_3)_2$.

Generally, the disubstituted benzenes (those with two substituents) are named as benzene derivatives. However, the dimethylbenzenes are better known as xylenes, which are accepted IUPAC names. The prefixes *ortho-, meta-,* and *para-* are also applied to derivatives with substituents other than alkyls. Thus, the dichloro derivative of benzene in which the chlorine atoms are on opposite sides of the ring is named *p*-dichlorobenzene. ●

● *Para*-dichlorobenzene is used as a moth repellent. It has a penetrating, sweetish odor.

To name derivatives of benzene having more than two substituents, you number the carbon atoms of the benzene ring, starting at a carbon with a substituent.

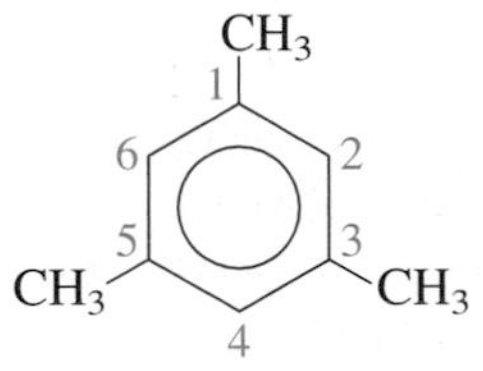

CH_3 1 2 3 CH_3 4 5 CH_3 6

CH_3 1 2 NO_2 3 4 NO_2 5 6 NO_2

1,3,5-Trimethylbenzene 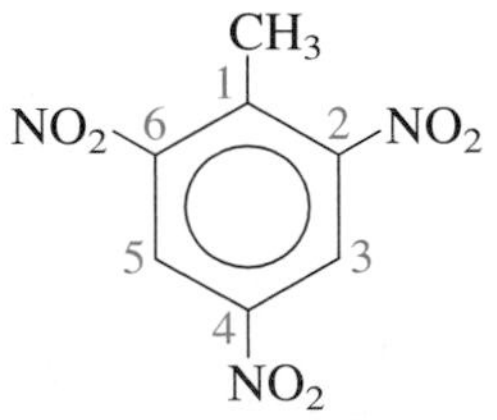 2,4,6-Trinitrotoluene (TNT)

Substituent names include the position numbers (for example, 1,3,5-trimethylbenzene). The second compound (the explosive TNT) is named as a derivative of toluene (the accepted IUPAC name of methylbenzene), rather than as a derivative of benzene.

■ EXAMPLE 18.5 Writing the Structural Formula of a Derivative of Benzene

Write structural formulas for the following compounds.
(a) propylbenzene
(b) 1,2,3-trinitrobenzene
(c) *p*-dichlorobenzene
(d) *o*-chlorotoluene

Problem Analysis

Look at the name of the benzene derivative. Monosubstituted derivatives are named using the substituent name in the form of a prefix. Disubstituted derivatives also use one of the prefixes *o*-, *m*-, or *p*-. With more substituents, a numbering system is used.

Solution

(a) Draw a benzene ring with a propyl substituent.

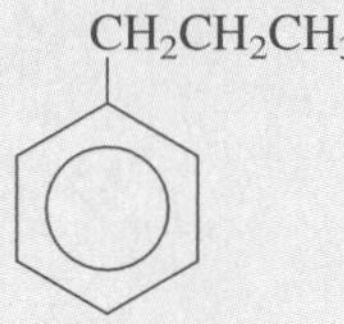

$CH_2CH_2CH_3$

(b) Draw a benzene ring, numbering the carbon atoms. Then add nitro groups (—NO_2) to the positions 1, 2, and 3.

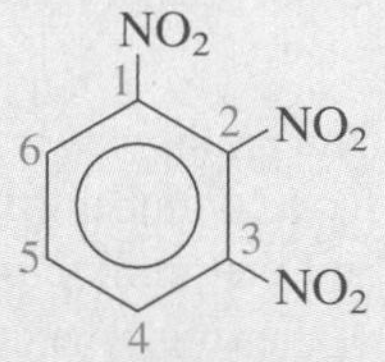

NO_2 1, 2 NO_2, 3 NO_2, 4, 5, 6

(c) The prefix *p*- (for *para*-) means that the ring substituents are opposite one another. Draw a benzene ring, adding chloro groups (—Cl) opposite one another.

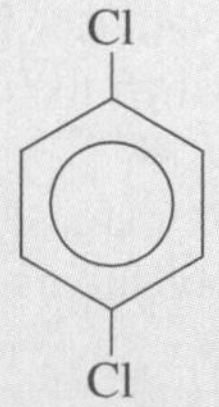

Cl

Cl

(d) Toluene is the accepted IUPAC name of methylbenzene. A chlorotoluene has two substituents on the benzene ring, a methyl group and a chloro group. The prefix *o*- (for *ortho*-) means that the substituents are adjacent to one another. The structural formula is

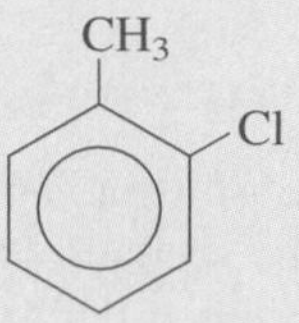

CH_3

Cl

Exercise 18.5

Write the structural formulas of the following.

(a) *m*-chloronitrobenzene
(b) 1,2,4-tribromobenzene
(c) *p*-bromotoluene
(d) nitrobenzene

(Try Problems 18.59 and 18.60.)

Polycyclic Aromatic Hydrocarbons

Naphthalene ($C_{10}H_8$), a compound commonly used as a moth repellent, is the simplest example of a polycyclic aromatic hydrocarbon. The structural formula is represented as

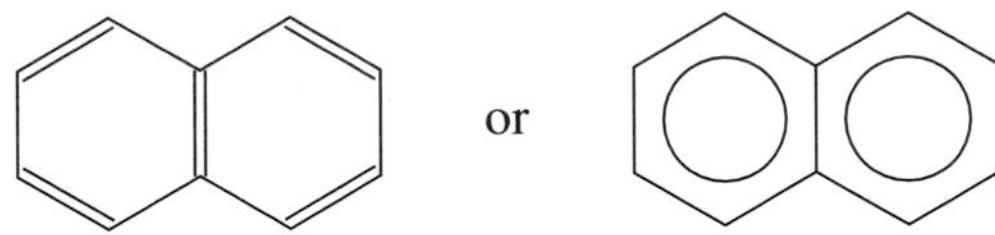

The polycyclic aromatic hydrocarbons are also called *fused-ring aromatic hydrocarbons.*

Reactions of Benzene

As we noted earlier, the electrons in the double bonds in benzene are delocalized and the molecule does not normally undergo addition reactions, as you might expect if you were to write a single Lewis formula for the compound. When you shake an alkene with bromine, the red-brown color of bromine vanishes as the bromine adds across the double bond. Benzene is normally unreactive with bromine unless a catalyst such as iron(III) bromide is present. In that case, the bromine reacts in a *substitution reaction;* the bromine substitutes for hydrogen atoms on the benzene ring.

$$C_6H_6 + Br_2 \xrightarrow{FeBr_3} C_6H_5Br + HBr$$

Chlorine reacts similarly with benzene, using iron(III) chloride as a catalyst.

OXYGEN DERIVATIVES OF HYDROCARBONS

In the preceding sections, we discussed the main series of hydrocarbons. We can consider all other organic compounds as derivatives of hydrocarbons. In these compounds, one or more hydrogen atoms of a hydrocarbon are replaced by other atoms to give a *functional group.* In the next five sections, we look at various functional groups.

A **functional group** is *a reactive portion of a molecule that undergoes predictable reactions.* You have already encountered two such functional groups, the carbon–carbon double bond (C=C) and the carbon–carbon triple bond (C≡C). A compound

Functional Group	Example Compound	Structural Formula*
Alkene	Ethene (ethylene)	$H_2C{=}CH_2$ (H, H on each C; C=C)
Alkyne	Ethyne (acetylene)	$H{-}C{\equiv}C{-}H$
Alcohol	Ethanol	$CH_3CH_2{-}O{-}H$
Ether	Diethyl ether	$CH_3CH_2{-}O{-}CH_2CH_3$
Aldehyde	Ethanal (acetaldehyde)	$CH_3{-}C({=}O){-}H$
Ketone	Propanone (acetone)	$CH_3{-}C({=}O){-}CH_3$
Carboxylic acid	Ethanoic acid (acetic acid)	$CH_3{-}C({=}O){-}O{-}H$
Ester	Methyl ethanate (methyl acetate)	$CH_3{-}C({=}O){-}O{-}CH_3$
Amine	Methylamine	$CH_3{-}NH_2$

*The functional group is shown in red.

TABLE 18.4
Some Organic Functional Groups

having such a functional group would be expected to undergo addition reactions; for example, you would expect it to add bromine across the double bond. Table 18.4 lists some functional groups. Note that many contain an atom other than carbon. In the three sections that follow, we look at functional groups containing an oxygen atom.

18.7 Alcohols

OBJECTIVES

- Define *alcohol.*
- Describe the commercial preparation and uses of methanol and ethanol.
- Name an alcohol, given its structural formula.
- Write the structural formula of the product (if any) of the controlled oxidation of an alcohol.

Alcohols and ethers are classes of organic compounds that may be considered as structurally related to water. If we take a model of the water molecule and replace one hydrogen atom by an alkyl group (represented by R), we obtain an alcohol. If we replace the remaining hydrogen atom by another alkyl group (R′), we obtain an ether.

$H{-}O{-}H$	$R{-}O{-}H$	$R{-}O{-}R'$
Water	An alcohol	An ether

We look at the characteristics of alcohols in this section.

Definition of Alcohols

An **alcohol** is *a compound of formula ROH that contains an —OH group bonded to a tetrahedral carbon atom* (that is, to an alkyl group). The simplest examples are methanol (wood alcohol) and ethanol (grain alcohol).

CH_3OH	CH_3CH_2OH
Methanol	Ethanol

Note that the —OH group is covalently bonded to the carbon atom. Therefore, alcohols are quite different chemically from the metal hydroxides, such as NaOH, in which the hydroxide is present as the ion. ●

● A phenol is a compound in which one or more —OH groups are attached to a benzene ring. Their properties are different from those of alcohols. The simplest member, phenol, C_6H_5OH, is an antiseptic.

Alcohols are classified by the number of carbon atoms attached to the carbon atom to which the —OH group is bonded. A primary alcohol has one such carbon atom, a secondary alcohol has two such carbon atoms, and a tertiary alcohol has three such carbon atoms.

```
     H  H               H  H  H               CH3
     |  |               |  |  |                |
  H—C—C—O—H          H—C—C—C—H          CH3—C—O—H
     |  |               |  |  |                |
     H  H               H  O  H               CH3
                           |
                           H
```

Ethanol (a primary alcohol) — 2-Propanol (a secondary alcohol) — 2-Methyl-2-propanol (a tertiary alcohol)

Ethanol has one methyl group (one carbon atom) attached to the carbon atom bearing the hydroxyl group; 2-propanol (also called isopropyl alcohol) has two methyl groups (two carbon atoms) attached to the carbon atom bearing the hydroxyl group; and 2-methyl-2-propanol has three methyl groups (three carbon atoms) attached to the carbon atom bearing the hydroxyl group.

Methanol and Ethanol

Methanol or methyl alcohol, CH_3OH, is a volatile liquid boiling at 65°C. It is poisonous and can cause blindness or death if taken internally. The compound is also known as wood alcohol because at one time it was obtained by heating wood chips or sawdust; the liquid distilled from the decomposing wood contains methanol. Today the alcohol is produced from the reaction of carbon monoxide and hydrogen using a metal oxide catalyst.

$$CO(g) + 2H_2(g) \xrightarrow[400°C,\ 150\ atm]{ZnO—Cr_2O_3} CH_3OH(g)$$

Methanol is a common solvent for organic materials, such as shellac and varnish. It is the active ingredient in automobile windshield washer compound, where it functions as both a solvent and an antifreeze. Methanol is also used as a fuel, and it can be mixed with gasoline as an automotive fuel. The chemical industry is the principal market for methanol, however. About one-half of the methanol produced is used in the manufacture of formaldehyde, HCHO, which is used to make plastics and adhesive resins. Methanol is also used to produce acetic acid, CH_3COOH. ●

● Methanol is mixed with ethanol to "denature" it, that is, to make ethanol undrinkable. Methanol is a poison, because the body changes it to formaldehyde (embalming fluid).

Ethanol or ethyl alcohol, CH_3CH_2OH, is the most important alcohol and likely to be the compound meant when someone refers simply to "alcohol." It is a volatile liquid,

FIGURE 18.12
Gasohol. Gasohol is a mixture of ethanol in gasoline.

boiling at 78°C. Large quantities of ethanol are produced as alcoholic beverages by the yeast fermentation of glucose, a sugar present in fruit juices and sprouted grains.

$$\underset{\text{Glucose}}{C_6H_{12}O_6} \xrightarrow{\text{Yeast}} \underset{\text{Ethanol}}{2CH_3CH_2OH} + 2CO_2(g)$$

Most industrial ethanol is produced by the addition of water, as steam, to ethylene (see earlier discussion under Reactions of Alkenes and Alkynes).

Ethanol, like methanol, is used as a solvent and as a reactant in the preparation of many organic substances. Some of the materials produced from ethanol are pharmaceutical drugs, plastics, lacquers, and cosmetics. Also, like methanol, ethanol can be used as a fuel. Gasohol, an automotive fuel, is a mixture of 10% ethanol in gasoline (Figure 18.12).

Naming Alcohols

The IUPAC rules for naming alcohols are similar to those for naming alkanes and alkenes.

Rules for Naming Alcohols

1. **Find the parent chain.** Determine the longest carbon–carbon chain that contains the —OH group. The name of the parent chain is the same as that of the alkane with the same number of carbon atoms, except that the final *-e* is changed to *-ol*.
2. **Find the branch chains.** The branch chains are named as alkyls.
3. **Add the location number for the —OH group and branch chains.** Number the carbon atoms of the parent chain beginning with the carbon atom closest to the —OH group. The position of the —OH group is the number of the carbon atom to which it is attached. This number precedes the name of the parent chain. Also add numbers to the branch chains.
4. **Write the complete name of the alcohol.** Write the parent name of the alcohol preceded by the names of any branches.

Consider the example of the tertiary alcohol given earlier.

$$\begin{array}{c} \quad\;\; CH_3 \\ \quad\;\; | \\ CH_3—C—O—H \\ \quad\;\; | \\ \quad\;\; CH_3 \end{array}$$

The longest chain containing the —OH group is shown in blue. Number the chain, in this case starting with the carbon atom of either methyl group.

$$\begin{array}{c} \quad\;\; {}^{1}\,CH_3 \\ \quad\;\; {}^{2}\;| \\ CH_3—C—O—H \\ \quad\;\; | \\ \quad\;\; {}^{3}\,CH_3 \end{array}$$

The parent alcohol is 2-propanol, and the complete name of the alcohol is 2-methyl-2-

propanol, which is the name given earlier. The following example further illustrates the naming of alcohols.

EXAMPLE 18.6 Naming Alcohols

Give the IUPAC name of the compound with the following structural formula.

```
OH       CH3
|        |
CH3CHCH2—C—CH3
         |
         CH3
```

Problem Analysis

Apply the rules for naming alcohols: Find the parent chain and branch chains, and then number the position of the —OH group and branch chains. Finally, write the complete name, using the parent name preceded by the branch names.

Solution

The longest chain containing the —OH group has five carbon atoms, so the parent name is pentanol. There are two methyl branch groups at the other end of the parent chain. Begin numbering at the end of the parent chain nearest the —OH group (not the end nearest the methyl branch groups, as you would if this compound were an alkane).

```
   OH       CH3
1 2|  3    4|  5
CH3CHCH2—C—CH3
            |
            CH3
```

The parent chain is 2-pentanol; the name of the alcohol is 4,4-dimethyl-2-pentanol.

Exercise 18.6

What is the IUPAC name of the following alcohol?

```
     CH2CH3  CH3
     |       |
CH3—C——CH2CHCH3
     |
     OH
```

(Try Problems 18.65 and 18.66.)

Reactions of Alcohols

Alcohols burn in oxygen to produce carbon dioxide and water.

$$CH_3CH_2OH(l) + 3O_2(g) \longrightarrow 2CO_2(g) + 3H_2O(g)$$

● To write the balanced equation for the complete oxidation of an alcohol, you write the equation with alcohol plus O_2 on the left and CO_2 and H_2O on the right; then balance.

This *complete oxidation* is the basis of ethanol as a fuel. ● However, *controlled oxidation* of alcohols is most important for the production of organic compounds. In controlled oxidation, one uses more moderate conditions than those in burning. For example, when you oxidize ethanol with an oxidizing agent such as potassium dichromate in acidic solution, you obtain acetaldehyde, CH_3CHO, which further oxidizes to acetic

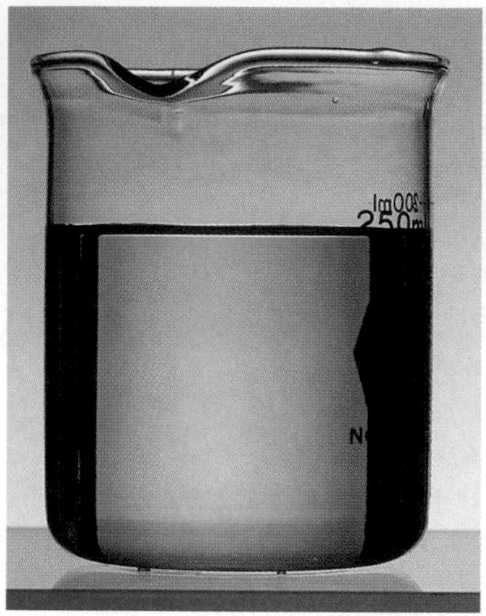

FIGURE 18.13
Controlled oxidation of ethanol. An acidic solution of potassium dichromate oxidizes ethanol to acetic acid. During the reaction, dichromate ion (orange) changes to a chromium(III) species (green).

acid, CH_3COOH (Figure 18.13). For simplicity, we can represent these oxidations as follows.

$$\underset{\text{Ethanol}}{CH_3CH_2OH} + \underset{\text{An oxidizing agent}}{(O)} \longrightarrow \underset{\text{Acetaldehyde}}{CH_3\overset{\overset{\displaystyle O}{\|}}{C}-H} + H_2O$$

$$\underset{\text{Acetaldehyde}}{CH_3\overset{\overset{\displaystyle O}{\|}}{C}-H} + \underset{\text{An oxidizing agent}}{(O)} \longrightarrow \underset{\text{Acetic acid}}{CH_3\overset{\overset{\displaystyle O}{\|}}{C}-OH}$$

(Note that CH_3CHO and CH_3COOH are the condensed formulas for acetaldehyde and acetic acid, respectively.) The net ionic equation for the oxidation of ethanol to acetaldehyde using dichromatic ion in acidic solution is

$$3CH_3CH_2OH + Cr_2O_7^{2-} + 8H^+ \longrightarrow 3CH_3CHO + 2Cr^{3+} + 7H_2O$$

We generally use the simpler representation, using (O) for the oxidizing agent, since our emphasis here is on the organic reactants and products.

The compound 2-propanol (whose common name is isopropyl alcohol) reacts with similar oxidizing agents to give propanone (whose common name is acetone).

$$\underset{\substack{\text{2-Propanol}\\\text{(isopropyl alcohol)}}}{CH_3\overset{\overset{\displaystyle OH}{|}}{C}HCH_3} + (O) \longrightarrow \underset{\substack{\text{Propanone}\\\text{(acetone)}}}{CH_3\overset{\overset{\displaystyle O}{\|}}{C}CH_3} + H_2O$$

The condensed formula of acetone is CH_3COCH_3.

We have just covered three new series of organic compounds as reaction products: aldehydes, ketones, and carboxylic acids. Aldehydes are compounds with the general formula RCHO; ketones are compounds with the general formula RCOR′. We look at aldehydes and ketones in detail in the next section. Carboxylic acids are organic acids with a —COOH group, and we look at them in Section 18.9.

In general, primary alcohols (such as ethanol) and secondary alcohols (such as isopropyl alcohol) are easily oxidized. Both types of alcohols have a hydrogen atom bonded to the carbon atom to which the —OH group is attached. This hydrogen atom is removed during the oxidation to produce the C=O group. Primary alcohols are oxidized with oxidizing agents such as acidic dichromate ion to aldehydes, then to carboxylic acids. Secondary alcohols are oxidized with moderate oxidizing agents to ketones. Tertiary alcohols are unreactive with these oxidizing agents. These alcohols do not have a hydrogen atom on the carbon to which the —OH group is attached.

18.8 Aldehydes and Ketones

OBJECTIVES

- Define *carbonyl compound, aldehyde,* and *ketone.*
- Describe the industrial preparation and uses of formaldehyde and acetone.
- Name an aldehyde or ketone, given the compound's structural formula.

Aldehydes and ketones are carbonyl compounds. A **carbonyl compound** is *a compound containing the carbonyl, or C=O, group.* (*Carbonyl* is pronounced "carbo-neel.") An **aldehyde** is *a compound having the formula RCHO that contains a carbonyl group bonded to a hydrocarbon group R and a hydrogen atom.* A **ketone** is *a compound having the formula RCOR′ that contains a carbonyl group bonded to two hydrocarbon groups.*

$\diagdown$C=O (with bonds to two unspecified groups)	R(H)C=O	R(R′)C=O
A carbonyl group	An aldehyde	A ketone

Formaldehyde and Acetone

The simplest aldehyde is formaldehyde, HCHO, which has two hydrogen atoms bonded to the carbonyl group.

$$\begin{matrix} \text{H} & \\ & \diagdown \\ & & \text{C}=\text{O} \\ & \diagup \\ \text{H} & \end{matrix}$$

Methanal
(formaldehyde)

The IUPAC name of formaldehyde is methanal. Formaldehyde, HCHO, is produced in large quantities from methanol, CH_3OH.

$$CH_3OH(g) \xrightarrow[650°C]{\text{Ag catalyst}} HCHO(g) + H_2(g)$$

Formaldehyde is a gas with a characteristic irritating odor. The compound is normally shipped as an aqueous solution containing methanol (to prevent the formation of a polymer). The solution, called formalin, is used as a preservative and disinfectant. The principal use of formaldehyde, however, is in the manufacture of polymers (Figure 18.14).

A

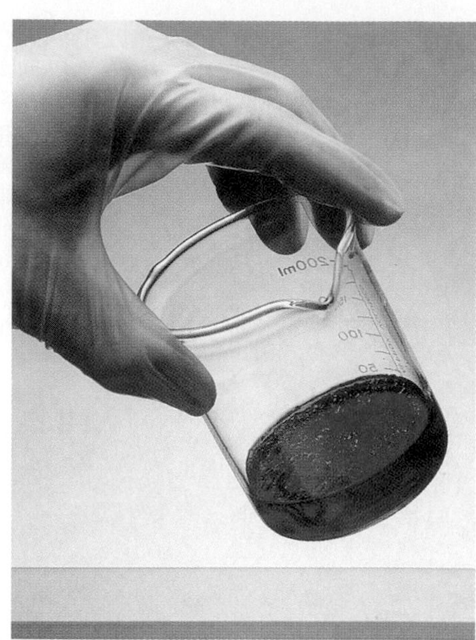

B

FIGURE 18.14

A formaldehyde plastic. *(A)* The clear solution contains formaldehyde, HCHO, and another compound, resorcinol, $m\text{-}C_6H_4(OH)_2$. An experimenter adds a few drops of potassium hydroxide solution to initiate the polymer reaction. *(B)* Later, the red polymer fills the bottom of the beaker.

Urea–formaldehyde polymer is an example. It is formed by *condensation reactions,* in which monomer units are chemically linked by eliminating small molecules (in this case H_2O).

$$H_2N{-}\overset{\overset{\displaystyle O}{\|}}{C}{-}NH_2 \;+\; H{-}\overset{H}{\underset{}{C}}{=}O \;+\; H_2N{-}\overset{\overset{\displaystyle O}{\|}}{C}{-}NH_2 \;+\; \cdots \longrightarrow$$

Urea Formaldehyde Urea

$$-\underset{}{\overset{H}{N}}-\overset{\overset{\displaystyle O}{\|}}{C}-\overset{H}{N}-\underset{H}{\overset{H}{C}}-\overset{H}{N}-\overset{\overset{\displaystyle O}{\|}}{C}-\overset{H}{N}- \;+\; nH_2O$$

Urea–formaldehyde polymer

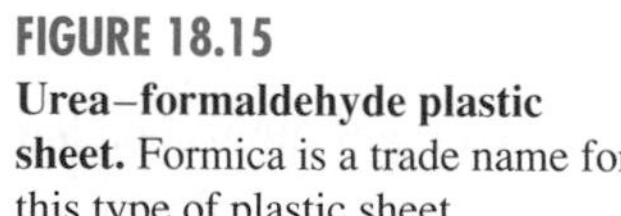

FIGURE 18.15

Urea–formaldehyde plastic sheet. Formica is a trade name for this type of plastic sheet.

Polymer chains can bond to one another to form a cross-linked polymer. Depending on the amount of cross-linking (or connections) between polymer chains, the result is either a resin or a hard plastic. The resin is used as an adhesive for plywood. Formica is a brand name of a hard plastic sheet made of urea–formaldehyde polymer (Figure 18.15).

Acetone, CH_3COCH_3, is the simplest ketone. In one industrial process, it is produced from isopropyl alcohol.

$$CH_3\overset{\overset{\displaystyle OH}{|}}{C}HCH_3 \xrightarrow[500°C]{\text{Cu catalyst}} CH_3\overset{\overset{\displaystyle O}{\|}}{C}CH_3 + H_2$$

2-propanol (isopropyl alcohol) → Propanone (acetone)

Acetone is a flammable, volatile liquid with a slightly aromatic odor. It is an excellent solvent for certain resins and acetate plastics. Some cements contain cellulose acetate plastic dissolved in acetone. Nail polish and nail polish remover also contain acetone. The major use of this ketone is in the manufacture of methyl methacrylate, used to make the addition polymer polymethyl methacrylate (trade names Lucite and Plexiglas).

$$CH_2{=}\underset{\underset{\displaystyle OCH_3}{|}}{\underset{|}{\overset{\overset{\displaystyle CH_3}{|}}{C}}}{-}\!\!\!\!\!\!\!\!\!\quad C{=}O$$

Methyl methacrylate

$$-CH_2-\overset{CH_3}{\underset{C(=O)OCH_3}{C}}-CH_2-\overset{CH_3}{\underset{C(=O)OCH_3}{C}}-CH_2-\overset{CH_3}{\underset{C(=O)OCH_3}{C}}-$$

Polymethyl methacrylate

Aldehydes can be distinguished from ketones using *Tollen's test.* In this test, aldehydes (but not ketones) reduce silver ion, Ag^+, to silver metal. ("Reduced" means that the charge, or oxidation number, of silver is reduced, in this case from 1 to zero; see Chapter 16.) For example, if formaldehyde is put into a glass flask containing a solution of silver nitrate, silver metal deposits on the glass, coating the inside of the flask with a silver mirror (Figure 18.16). ●

● High-quality mirrors can be produced by this reaction.

Naming Aldehydes and Ketones

Aldehydes are named by IUPAC rules as derivatives of the longest chain that contains the carbonyl function, C=O (the aldehyde group —CHO). The parent chain is named after the corresponding alkane, but the final *-e* of the alkane is changed to *-al* for the aldehyde. To obtain location numbers for branch chains, we number the parent chain starting from the carbon atom of the aldehyde function. Here are some examples (with common names in parentheses).

$$H-\overset{\overset{\displaystyle O}{\|}}{C}-H \qquad CH_3-\overset{\overset{\displaystyle O}{\|}}{C}-H \qquad CH_3\overset{\overset{\displaystyle CH_3}{|}}{C}HCH_2-\overset{\overset{\displaystyle O}{\|}}{C}-H$$

Methanal (formaldehyde) — Ethanal (acetaldehyde) — 3-Methylbutanal

FIGURE 18.16

Tollen's test. Aldehydes (such as formaldehyde) react with silver ion to produce silver metal, which plates the wall of the flask.

We name ketones by IUPAC rules in a similar manner; that is, we look for the longest carbon-atom chain that contains the carbonyl function, C=O, numbering the chain beginning with the carbon atom nearest this carbonyl group. We name the parent chain after the name of the corresponding alkane, changing the final *-e* of the alkane to *-one* for the ketone. We precede the name of the parent chain by the location number of the carbonyl function; and, as with aldehydes, we add the names of any branch chains before the name of the parent chain. Here are some examples (with common names in parentheses).

$$CH_3-\overset{\overset{\displaystyle O}{\|}}{C}-CH_3 \qquad CH_3-\overset{\overset{\displaystyle O}{\|}}{C}-CH_2CH_3 \qquad CH_3-\overset{\overset{\displaystyle O}{\|}}{C}-\overset{\overset{\displaystyle CH_3}{|}}{C}HCH_3$$

Propanone (acetone) — 2-Butanone (methyl ethyl ketone) — 3-Methyl-2-butanone (methyl isopropyl ketone)

We do not need a number for propanone, because the carbonyl group of a ketone can only be on the number 2 carbon atom; otherwise, the compound would be an aldehyde.

EXAMPLE 18.7 Naming Aldehydes and Ketones

Give the IUPAC names of the following.

(a) $\underset{\displaystyle CH_3CHCH_2CHO}{\overset{\displaystyle CH_3}{|}}$ (b) $CH_3\overset{\overset{\displaystyle O}{\|}}{C}-\overset{\overset{\displaystyle CH_3}{|}}{C}HCH_2CH_3$

Problem Analysis

Name aldehydes and ketones by locating the longest chain containing the carbonyl group, numbering the chain from the end nearest this group. Add the suffix *-al* or *-one,* depending on whether the compound is an aldehyde or ketone. (For the ketone, you need to add a number for the position of the carbonyl group.) Precede the name of the parent chain with the name of any branch chains.

Solution

(a) The oxygen is bonded to a terminal carbon, so the compound is an aldehyde; the numbering of the parent chain is

$$\overset{4}{C}H_3\overset{\overset{\displaystyle CH_3}{3|}}{C}H\overset{2}{C}H_2\overset{1}{C}HO$$

The IUPAC name is 3-methylbutanal.

(b) The oxygen is not bonded to a terminal carbon, so the compound is a ketone; the numbering of the parent chain is

$$\overset{1}{C}H_3\overset{\overset{\displaystyle O}{2\|}}{C}-\overset{\overset{\displaystyle CH_3}{3|}}{C}H\overset{4}{C}H_2\overset{5}{C}H_3$$

The IUPAC name is 3-methyl-2-pentanone.

Exercise 18.7

Give the IUPAC names of the following.

(a) $H\overset{\overset{\displaystyle O}{\|}}{C}CH_2CH_2\overset{\overset{\displaystyle CH_2CH_3}{|}}{C}HCH_3$ (b) $CH_3CH_2CH_2COCH_3$

(Try Problems 18.69 and 18.70.)

18.9 Carboxylic Acids and Esters

OBJECTIVES

- Define *carboxylic acid.*
- Describe the change in properties of the carboxylic acids as the length of the hydrocarbon chain increases.
- Describe some uses of salts and esters of carboxylic acids.
- Describe the difference between a salt and an ester of a carboxylic acid.

Carboxylic acids, like aldehydes and ketones, contain a carbonyl group; however, the carbon atom of the carbonyl group also has an —OH group attached to it. A **carboxylic acid** is *a compound containing the carboxyl group, —COOH.*

$$\underset{\text{Carboxyl group}}{-\overset{\overset{\large O}{\|}}{C}-O-H} \qquad \underset{\text{A carboxylic acid}}{R-\overset{\overset{\large O}{\|}}{C}-O-H} \qquad \underset{\substack{\text{Ethanoic acid}\\ \text{(acetic acid)}}}{CH_3-\overset{\overset{\large O}{\|}}{C}-O-H}$$

The acidity of such an acid is due to the hydrogen atom of the carboxyl group (—COOH), which is readily lost.

The simplest members of this series of acids are formic acid (HCOOH) and acetic acid (CH_3COOH). Formic acid has a sharp, irritating odor. Acetic acid, a constituent of vinegar, has a sharp, vinegary odor. As the chain length of the R group becomes longer, the odors of the acids become more disagreeable. Butyric acid, $CH_3CH_2CH_2COOH$, has an intensely disagreeable odor when pure. The acid is present in small concentrations in rancid butter (from which it was first isolated), in cheeses along with other odorous carboxylic acids, and in perspiration. In small concentrations, odorous carboxylic acids lend their "charm" to cheeses, such as Limburger and blue cheese.

Once the chain length becomes long enough, the carboxylic acid is no longer volatile and so has no odor. The so-called fatty acids are long-chain carboxylic acids isolated from animal fats and plant oils. Stearic acid, with 18 carbon atoms, is an example of a saturated fatty acid. "Saturated" means that the compound has no carbon–carbon multiple bonds. ●

● Fats are compounds formed from glycerol (glycerin, an alcohol) and fatty acids. When you boil a fat with aqueous sodium hydroxide, you free the glycerol and obtain the sodium salts of the fatty acids, which are soaps.

$$CH_3CH_2CH_2CH_2CH_2CH_2CH_2CH_2CH_2CH_2CH_2CH_2CH_2CH_2CH_2CH_2CH_2COOH$$

Octadecanoic acid
(stearic acid)

Naming Carboxylic Acids

The IUPAC rules for naming the carboxylic acids are the same as those for naming aldehydes, except that the final *-e* of the alkane name is replaced by *-oic acid.* The following are some examples.

$$\underset{\substack{\text{Butanoic acid}\\ \text{(butyric acid)}}}{CH_3CH_2CH_2COOH} \qquad \underset{\text{3-Methylbutanoic acid}}{\overset{4}{C}H_3\overset{3}{\underset{}{C}}H\overset{\overset{CH_3}{|}}{}\overset{2}{C}H_2\overset{1}{C}OOH}$$

Both compounds consist of parent chains of four carbon atoms, so they are named as butanoic acids (pronounced "byu′-tan-oh′-ic").

Since many carboxylic acids were known long before systematic naming was established, some of their common names are more frequently used than their IUPAC names. These common names usually stem from the original source of the acid. For example, the substance responsible for the irritating bite of the red ant is formic acid, which was first isolated from this insect, whose biological name is *Formica rufus*. Table 18.5 lists the common name, origin of name, and condensed structural formula of several carboxylic acids.

Acid Properties and Ester Formation

The carboxylic acids are weak acids that can lose the hydrogen atom of the carboxyl group.

Common Name	Origin of Name	Structural Formula
Formic acid	Latin *formica* (ant)	$HCOOH$
Acetic acid	Latin *acetum* (vinegar)	CH_3COOH
Propionic acid	Greek *pro-* and *pion* (first fat)	CH_3CH_2COOH
Butyric acid	Latin *butyrum* (butter)	$CH_3CH_2CH_2COOH$ or $CH_3(CH_2)_2COOH$
Caproic acid	Latin *caper* (goat)	$CH_3CH_2CH_2CH_2CH_2COOH$ or $CH_3(CH_2)_4COOH$
Caprylic acid	Latin *caper* (goat)	$CH_3CH_2CH_2CH_2CH_2CH_2CH_2COOH$ or $CH_3(CH_2)_6COOH$
Capric acid	Latin *caper* (goat)	$CH_3CH_2CH_2CH_2CH_2CH_2CH_2CH_2CH_2COOH$ or $CH_3(CH_2)_8COOH$
Palmitic acid	Latin *palma* (palm tree)	$CH_3(CH_2)_{14}COOH$
Stearic acid	Greek *stear* (tallow)	$CH_3(CH_2)_{16}COOH$

TABLE 18.5
Some Carboxylic Acids

$$RCOOH + H_2O \rightleftharpoons RCOO^- + H_3O^+$$

For example, acetic acid ionizes in water solution by losing the hydrogen atom (as H^+) from its carboxyl group to a water molecule.

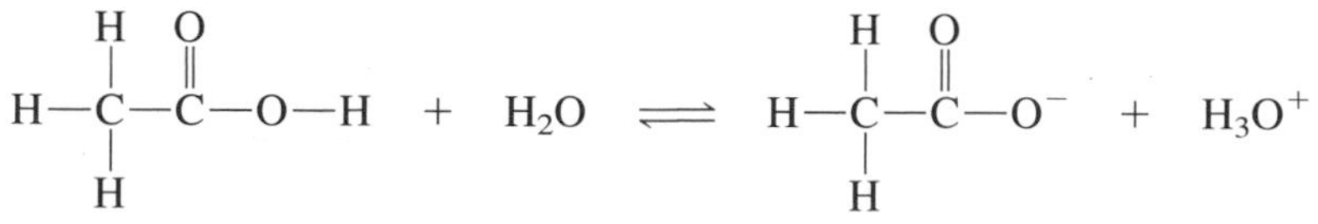

$$\mathrm{H{-}\underset{\displaystyle H}{\overset{\displaystyle H}{\underset{|}{\overset{|}{C}}}}{-}\overset{\displaystyle O}{\overset{\|}{C}}{-}O{-}H} + H_2O \rightleftharpoons \mathrm{H{-}\underset{\displaystyle H}{\overset{\displaystyle H}{\underset{|}{\overset{|}{C}}}}{-}\overset{\displaystyle O}{\overset{\|}{C}}{-}O^-} + H_3O^+$$

Acetic acid — Acetate ion

The hydrogen atoms on the methyl group are not easily lost and therefore do not make a substance acidic.

Like any acid, a carboxylic acid reacts with a base to form a salt. Thus, if sodium hydroxide (a base) is added to a solution of acetic acid, the product is sodium acetate.

$$CH_3COOH(aq) + NaOH(aq) \longrightarrow CH_3COONa(aq) + H_2O(l)$$

Acetic acid — Sodium hydroxide — Sodium acetate

Sodium acetate (like sodium hydroxide) is an ionic compound composed of Na^+ and CH_3COO^- ions. Soaps are mixtures of fatty acid salts. Bar soap is a mixture of sodium salts; liquid soap is a mixture of potassium salts. Lubricating grease is made by adding a lithium soap as a thickening agent to liquid petroleum hydrocarbons.

When a carboxylic acid is heated with an alcohol in acidic solution, it forms an ester. An **ester** is *a compound formed from a carboxylic acid, RCOOH, and an alcohol, R′OH*. The general structure of an ester is

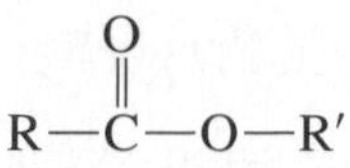

$$\mathrm{R{-}\overset{\displaystyle O}{\overset{\|}{C}}{-}O{-}R'}$$

An ester

Name	Formula	Odor*
Ethyl formate	$HCOOCH_2CH_3$	Rum
Pentyl acetate	$CH_3COOCH_2CH_2CH_2CH_2CH_3$	Banana
Octyl acetate	$CH_3COOCH_2CH_2CH_2CH_2CH_2CH_2CH_2CH_3$	Orange
Methyl butyrate	$CH_3CH_2CH_2COOCH_3$	Apple
Ethyl butyrate	$CH_3CH_2CH_2COOCH_2CH_3$	Pineapple
Pentyl butyrate	$CH_3CH_2CH_2COOCH_2CH_2CH_2CH_2CH_3$	Apricot
Methyl salicylate	*o*-$C_6H_4(OH)COOCH_3$	Wintergreen

*Natural flavors are generally complex mixtures of esters and other constituents.
(From Darrell D. Ebbing, *General Chemistry,* 5th ed. [Boston: Houghton Mifflin, 1996], p. 1050.)

TABLE 18.6
Some Esters and Their Odors

An ester is named from the alcohol and the acid in the manner of a salt, but an ester is a covalent molecular compound, not an ionic compound like a salt. Thus, the ester made from methanol and acetic acid is named methyl (from the alcohol) acetate (from the acid). Esters often have very pleasant odors (see Table 18.6). Many fruit odors are due to esters.

As an example of ester formation, consider the reaction of butyric acid with ethanol.

$$\underset{\text{Butyric acid}}{CH_3CH_2CH_2\text{—}\overset{\overset{\displaystyle O}{\|}}{C}\text{—}OH} + \underset{\text{Ethanol}}{HOCH_2CH_3} \xrightarrow{H^+}$$

$$\underset{\text{Ethyl butyrate}}{CH_3CH_2CH_2\text{—}\overset{\overset{\displaystyle O}{\|}}{C}\text{—}O\text{—}CH_2CH_3} + H_2O$$

The reaction involves the elimination of a small molecule (in this case H_2O); such a reaction is called a *condensation reaction,* discussed earlier for the formation of the urea–formaldehyde polymer. The ester is ethyl butyrate, which is used to make pineapple flavor. Its odor is quite unlike that of the corresponding carboxylic acid, butyric acid, which has a very disagreeable odor, as we mentioned earlier.

NITROGEN DERIVATIVES OF HYDROCARBONS

Nitrogen derivatives of hydrocarbons are essential compounds in living organisms. Proteins, which are present in muscle and connective tissue and enzymes (biological catalysts), are organic nitrogen compounds. Deoxyribonucleic acid (DNA), which is present in genetic material, is also an organic nitrogen compound. Many commercially important organic compounds contain nitrogen. Some of these are amines (pronounced "uh meens′"), which we discuss in the next section.

18.10 Amines

OBJECTIVES

- Define *amine*.
- Write a chemical equation in which an amine acts as a base.

An **amine** is *a compound that is structurally derived by replacing one or more hydrogen atoms of ammonia (NH_3) with hydrocarbon groups.*

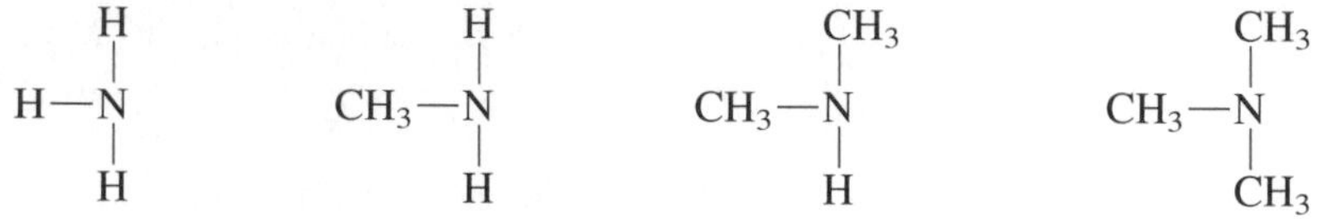

Ammonia	Methylamine	Dimethylamine	Trimethylamine
$H-N(H)-H$	$CH_3-N(H)-H$	$CH_3-N(CH_3)-H$	$CH_3-N(CH_3)-CH_3$

Simple alkyl amines can be named by preceding *-amine* with the name of the alkyl groups, as in methylamine and dimethylamine. Amines of low molecular weight have fishy odors.

Amines are bases, just as ammonia is a base.

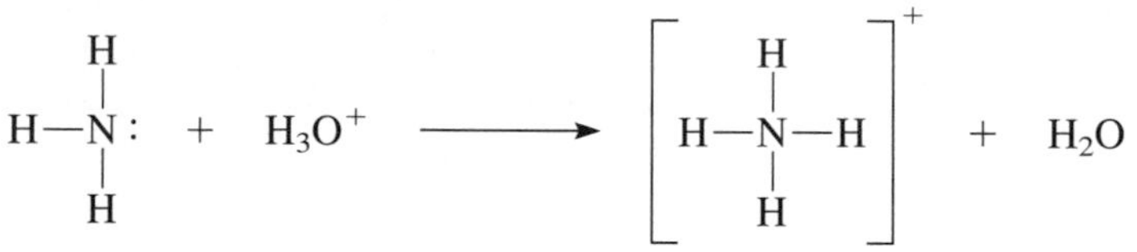

$$H-\ddot{N}(H)-H\ + \ H_3O^+ \longrightarrow [H-N(H)(H)-H]^+ \ + \ H_2O$$

Here, ammonia accepts a proton (H^+) from the hydronium ion (H_3O^+), present in acid solution, to form the ammonium ion and water. In the same way, an amine reacts with H_3O^+ to form a substituted ammonium ion.

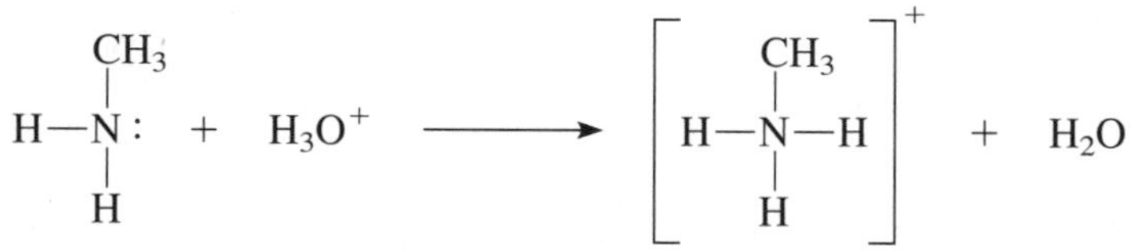

$$\underset{\text{Methylamine}}{CH_3-\ddot{N}(H)-H}\ + \ H_3O^+ \longrightarrow \underset{\text{Methylammonium ion}}{[CH_3-N(H)(H)-H]^+} \ + \ H_2O$$

Aniline is an aromatic amine that is used in the manufacture of various polymers, such as polyurethanes, and in the preparation of dyes.

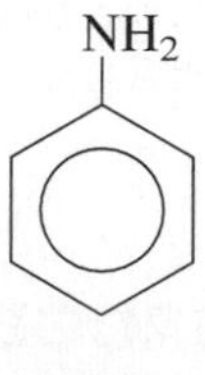

$$C_6H_5NH_2$$

Aniline

18.11 Amides and Polyamides

OBJECTIVES

- Define *amide*.
- Write the reaction describing the formation of an amide from ammonia (or an amine) and a carboxylic acid.
- Describe how a compound with two $-NH_2$ groups and another compound with two $-COOH$ or two $-COCl$ groups react to form a polyamide.

An **amide** is *a compound derived from the reaction of ammonia or an amine with a carboxylic acid.* For example, when ammonia is strongly heated with acetic acid, the reaction produces the amide acetamide, CH_3CONH_2.

$$CH_3{-}\overset{\displaystyle O}{\overset{\|}{C}}{-}OH + H{-}\underset{\displaystyle H}{\underset{|}{\overset{\displaystyle H}{\overset{|}{N}}}} \xrightarrow{\Delta} \underset{\text{Acetamide}}{CH_3{-}\overset{\displaystyle O}{\overset{\|}{C}}{-}NH_2} + H_2O$$

Like ester formation, this is also a condensation reaction. During the reaction, an H_2O molecule is removed (using an —OH group from the acid and an H atom from the ammonia), and a bond forms between the C atom of the carboxyl group and the N atom. This same amide can be produced at room temperature if instead of the carboxylic acid we use the acid chloride derived from the acid. The acid chloride of acetic acid is called acetyl chloride. Acetyl chloride and ammonia react to give acetamide.

$$\underset{\text{Acetyl chloride}}{CH_3{-}\overset{\displaystyle O}{\overset{\|}{C}}{-}Cl} + H{-}\underset{\displaystyle H}{\underset{|}{\overset{\displaystyle H}{\overset{|}{N}}}} \longrightarrow \underset{\text{Acetamide}}{CH_3{-}\overset{\displaystyle O}{\overset{\|}{C}}{-}NH_2} + HCl$$

In this case, an HCl molecule is removed. A similar reaction occurs with an amine in place of ammonia.

$$\underset{\text{Acetyl chloride}}{CH_3{-}\overset{\displaystyle O}{\overset{\|}{C}}{-}Cl} + \underset{\text{Methylamine}}{H{-}\underset{\displaystyle H}{\underset{|}{\overset{\displaystyle CH_3}{\overset{|}{N}}}}} \longrightarrow \underset{N\text{-Methylacetamide}}{CH_3{-}\overset{\displaystyle O}{\overset{\|}{C}}{-}\underset{\displaystyle H}{\underset{|}{\overset{\displaystyle CH_3}{\overset{|}{N}}}}} + HCl$$

Nylon is a *polyamide,* a group of polymers consisting of many amide monomers. One such polymer, nylon-6,6 is prepared commercially by heating adipic acid, $HOOC(CH_2)_4COOH$, with hexamethylene diamine, $NH_2(CH_2)_6NH_2$, under pressure to eliminate water. The polymer is recovered as a molten mass, which after purification is extruded through small holes to produce a fine fiber. In the laboratory, it is convenient to react adipoyl chloride (the acid chloride corresponding to adipic acid) with hexamethylene diamine at room temperature (Figure 18.17).

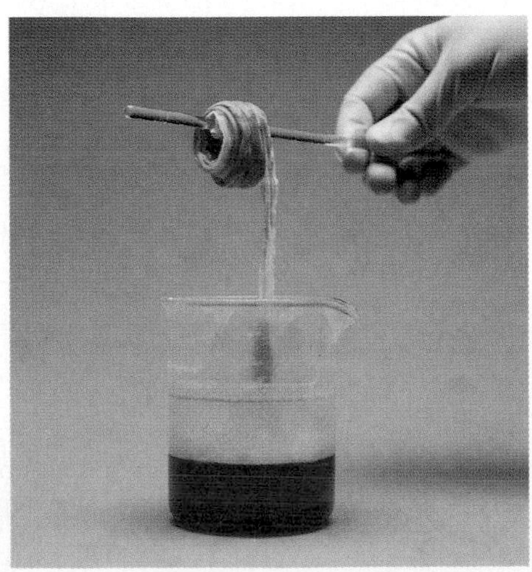

FIGURE 18.17
Preparation of nylon. In this laboratory experiment, hexamethylene diamine, in the bottom water layer, reacts with adipoyl chloride, in the upper hexane layer, to produce nylon-6,6 at the interface of the two layers. The nylon is being pulled out and rolled onto the stirring rod.

$$\underset{\text{Hexamethylene diamine}}{\text{H}-\overset{\text{H}}{\overset{|}{\text{N}}}(\text{CH}_2)_6\overset{\text{H}}{\overset{|}{\text{N}}}-\text{H}} + \underset{\text{Adipoyl chloride}}{\text{Cl}-\underset{\text{O}}{\underset{\|}{\text{C}}}(\text{CH}_2)_4\underset{\text{O}}{\underset{\|}{\text{C}}}-\text{Cl}} + \underset{\text{Hexamethylene diamine}}{\text{H}-\overset{\text{H}}{\overset{|}{\text{N}}}(\text{CH}_2)_6\overset{\text{H}}{\overset{|}{\text{N}}}-\text{H}} + \cdots \longrightarrow$$

$$\underset{\text{Nylon-6,6}}{-\overset{\text{H}}{\overset{|}{\text{N}}}(\text{CH}_2)_6\overset{\text{H}}{\overset{|}{\text{N}}}-\underset{\text{O}}{\underset{\|}{\text{C}}}(\text{CH}_2)_4\underset{\text{O}}{\underset{\|}{\text{C}}}-\overset{\text{H}}{\overset{|}{\text{N}}}(\text{CH}_2)_6\overset{\text{H}}{\overset{|}{\text{N}}}-} \quad + \quad n\text{HCl}$$

The monomer molecules bond like links in a chain, eliminating HCl molecules. Nylon is an example of a **condensation polymer,** *a polymer formed from monomer molecules by condensation reactions.*

Video: Synthesis of nylon

Chemical Perspective

The Discovery of Nylon

The announcement in October 1938 of the discovery of nylon, the first completely synthetic fiber, and the imminent production of nylon stockings was a momentous occasion. The press was fascinated by a fiber described as "strong as steel" and made from nothing more than coal, air, and water. And the public was eager to see stockings made from a fiber that it thought, being as strong as steel, must last forever. From our perspective, the discovery of nylon was important because it was the beginning of the synthetic polymers industry. Its discovery also established the study of polymers as a basic science.

Wallace Hume Carothers, the discoverer of nylon, was 31 years old in 1928 when he left Harvard University for the DuPont Company. (He had obtained his Ph.D. four years earlier in organic chemistry at the University of Illinois.) DuPont had decided to establish a group devoted to fundamental research, an idea that was rather novel at the time. The plan was that Carothers would work in an area that, while potentially useful to the company, did not have to have immediate commercial application.

Carothers proposed to work on the structure of materials like rubber, silk, and wool. The German chemist Hermann Staudinger had earlier suggested that these materials were actually macromolecules (that is, huge molecules) consisting of long chains of similar groups of atoms (in other words, polymers). At the time, most chemists thought that these materials were composed of aggregates of many small molecules held together by some unknown force that was different from the normal forces of chemical bonding.

Carothers felt that he could determine the truth of Staudinger's view by using the methods of organic chemistry to synthesize (or build up) macromolecules in such a way as to establish their structure. He had a simple but brilliant idea.

To understand Carothers's idea, consider two functional groups, such as a carboxylic acid group (—COOH) and an alcohol group (HO—), that chemically link together. We can write a reaction between two compounds containing these groups this way.

$$\sim\!\sim\overset{\text{O}}{\overset{\|}{\text{C}}}-\text{OH} + \text{HO}\sim\!\sim \longrightarrow \sim\!\sim\overset{\text{O}}{\overset{\|}{\text{C}}}-\text{O}\sim\!\sim + \text{H}_2\text{O}$$

(where ∼∼ stands for the rest of the molecule)

Even more simply, we can represent the linking of functional groups, such as a carboxyl group and an alcohol (or amine) group, as the linking of a hook and eye (the functional groups).

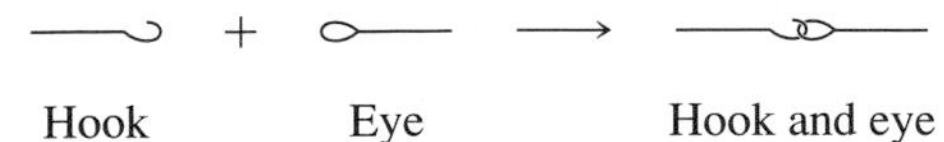

Now suppose we find a molecule having two of the same functional groups (such as two carboxyl groups) at the ends of the molecule; this would be the equivalent of a molecule with hooks at both ends. Then the reaction of several such molecules would result in a larger molecule.

With many such molecules, the product would be a long-chain molecule—a macromolecule. Carothers felt that if he could synthesize such a macromolecule with the properties of a textile fiber, he could establish the basis for further study of natural polymers, and he could go on to develop synthetic polymers (such as nylon) with improved properties.

Carothers's first macromolecule of this type was a polyester prepared from two different molecules, a dicarboxylic acid (a molecule with two carboxylic acid groups) and a dialcohol (two alcohol groups). The product, although interesting, tended to decompose in hot water and had a low melting point, so as a fabric it would hardly withstand washing or ironing. For his next experiments he prepared a series of polyamides, each from a dicarboxylic acid and a diamine. One polyamide from this series, called nylon-6,6, had properties that were promising. (The 6,6 refers to the number of carbon atoms in the dicarboxylic acid and the diamine, six in each case.) At this point, other teams at DuPont developed nylon-6,6 as a fiber that could be spun into hosiery.

Carothers might well have been awarded the Nobel Prize for his outstanding contributions to chemistry. In less than ten years, he had established synthetic polymer science and had discovered neoprene rubber and polyesters, as well as nylon. He had numerous interests, including politics, music, and sports, and he also prided himself on his writing skills. Unfortunately, Carothers had been troubled since youth with more and more frequent bouts of depression. After the discovery of nylon, he went into a severe depression, feeling a failure as a scientist. On April 29, 1937, he committed suicide by drinking lemonade containing potassium cyanide. His death was a tremendous loss.

CHAPTER REVIEW

Key Words

organic chemistry *(p. 568)*
isomers *(p. 570)*
condensed structural formula *(p. 570)*
hydrocarbon *(p. 571)*
alkane *(p. 571)*
alkyl group *(p. 573)*
substitution reaction *(p. 580)*
alkene *(p. 582)*
alkyne *(p. 582)*
addition reaction *(p. 585)*
polymer *(p. 585)*
monomer *(p. 586)*
addition polymer *(p. 586)*
aromatic hydrocarbon *(p. 587)*
functional group *(p. 591)*
alcohol *(p. 593)*
carbonyl compound *(p. 597)*
aldehyde *(p. 597)*
ketone *(p. 597)*
carboxylic acid *(p. 600)*
ester *(p. 602)*
amine *(p. 604)*
amide *(p. 605)*
condensation polymer *(p. 606)*

Summary

Organic chemistry is the chemistry of carbon compounds. The variety and complexity of organic compounds result from the strong bonds carbon forms with many elements, including itself. In bonding to itself, it forms chains and rings. *Isomers* are common in organic chemistry. Because of the possibility of isomers, a molecular formula may represent more than one compound. To describe an organic compound, we need to give its structural formula, which we frequently abbreviate using a *condensed structural formula.*

For classification purposes, organic compounds are considered to be *hydrocarbons* or derivatives of hydrocarbons containing *functional groups.* The simplest hydrocarbons are the *alkanes.* The straight-chain alkanes are named using a Greek stem denoting the number of carbon atoms, followed by the suffix *-ane* (except for the first four alkanes, which have special names). Branched-chain alkanes are named as derivatives of the straight-chain alkanes, preceding the alkane name by the names of the *alkyl* branches. The position of a branch is given by a number. The alkanes frequently react by *substitution,* for example, by substituting a chlorine atom for a hydrogen atom.

Alkenes and *alkynes* are hydrocarbons containing multiple bonds. They are named similarly to alkanes, except that the position of the multiple bond is given as a prefix, and the suffix is *-ene* for alkenes and *-yne* for alkynes. These hydrocarbons tend to react by addition of parts of a reactant molecule across the multiple bond. Commercially, the most important of the alkenes are ethene (ethylene), $CH_2{=}CH_2$, and propene (propylene), $CH_2{=}CHCH_3$. These compounds are used in the preparation of polymers.

Polymers are very large molecules made of many repeating units, or *monomers*. Polyethylene, for example, is an *addition polymer* consisting of many ethylene units. Polypropylene is a similar addition polymer.

Aromatic hydrocarbons have structures based on the benzene ring. The simplest representation of benzene, C_6H_6, is as a ring of carbon atoms connected by alternating single and double bonds. In reality, the three extra pairs of electrons representing the double bonds are delocalized over the benzene ring. For simplicity, we retain the Lewis representation by writing a resonance description of the molecule. The simplest aromatic hydrocarbons are named as derivatives of benzene, for example, ethylbenzene. Disubstituted benzenes are named as *ortho-*, *meta-*, and *para-* compounds. Numbers are also used to locate groups on the benzene ring. Polycyclic aromatic hydrocarbons are fused-ring hydrocarbons. Benzene generally reacts with substances such as chlorine by a substitution reaction.

In the last sections of the chapter, we looked at oxygen derivatives and nitrogen derivatives of the hydrocarbons. Methanol (CH_3OH) and ethanol (CH_3CH_2OH) are alcohols, compounds of the form ROH, where R is a hydrocarbon group. Primary alcohols are oxidized by an acidic dichromate ion to aldehydes, then to carboxylic acids. Secondary alcohols are oxidized under similar circumstances to ketones, and tertiary alcohols are not reactive.

Aldehydes and *ketones* are carbonyl compounds containing a carbonyl group, C=O. The best-known examples are formaldehyde, HCHO, and acetone, CH_3COCH_3. Formaldehyde is used to manufacture polymers, such as urea–formaldehyde polymer. Acetone is used in the manufacture of the polymer polymethyl methacrylate. *Carboxylic acids* also contain a carbonyl group, but it is incorporated into the carboxyl group, —COOH, in which the H atom is acidic. Acetic acid, CH_3COOH, is the most common example. Carboxylic acids react with alcohols in a condensation reaction to produce *esters,* many of which have pleasant odors.

Amines are compounds that have hydrocarbon groups in place of H atoms of ammonia, NH_3. Like ammonia, they are bases. An *amide* is a compound prepared from the condensation reaction of a carboxylic acid with ammonia or an amine. Polyamides (nylons) are *condensation polymers* containing many amide groups.

Problem-Solving Skills

1. **Naming Alkanes:** Given the structural formula of an alkane, use the IUPAC rules to name it (Example 18.1).
2. **Writing the Structural Formula of an Alkane:** Given the IUPAC name of an alkane, write its structural formula (Example 18.2).
3. **Writing Equations for Alkane Reactions:** Given a verbal description of a reaction of an alkane, involving either burning or a substitution reaction, write the chemical equation (Example 18.3).
4. **Naming Alkenes and Alkynes:** Given the structural formula of an alkene or alkyne, use IUPAC rules to name it (Example 18.4).
5. **Writing the Structural Formula of a Derivative of Benzene:** Given the name of a benzene derivative, write the structural formula (Example 18.5).
6. **Naming Alcohols:** Given the structural formula of an alcohol, give the IUPAC name (Example 18.6).
7. **Naming Aldehydes and Ketones:** Given the structural formula of an aldehyde or ketone, use IUPAC rules to name it (Example 18.7).

Questions to Test Your Reading

18.1 Make a list of organic chemicals or mixtures containing organic chemicals that you might find around a house.

18.2 Give the present definition of *organic chemistry.* How does this differ from its original definition?

18.3 Draw some structural formulas of molecules that display the carbon atom in different bonding situations.

18.4 What is an isomer? Give an example.

18.5 Give the condensed structural formula of the following.

```
    H   H   H   H
    |   |   |   |
H—C—C—C—C—O—H
    |   |   |   |
    H   H   H   H
```

18.6 The first four members of the alkane series of hydrocarbons have long-established names. What are they?

18.7 What is the name of the straight-chain alkane having seven carbon atoms?

18.8 What is the name of the alkyl group obtained by removing an end hydrogen atom from propane?

18.9 What are the major commercial sources of alkanes? What are the principal uses of these alkanes?

18.10 What is the purpose of petroleum refining?

18.11 Why do straight-chain alkanes tend to be unsuitable as gasoline in automobile combustion engines?

18.12 Alkanes react with oxygen in burning and with chlorine in a substitution reaction. Give examples of both reactions.

18.13 Give the structural formulas of two simple alkenes and one alkyne.

18.14 What is the commercial source of ethylene? Give one use of ethylene.

18.15 Bromine adds to ethylene. What is the chemical reaction? How can bromine be used as a test for multiple bonds?

18.16 What is a *polymer*? a *monomer*? Give an example illustrating these terms.

18.17 The major use of propylene is in the production of polypropylene. Draw a portion of this molecule.

18.18 Describe the structure of benzene in terms of its resonance description. What is the meaning of resonance in benzene?

18.19 What does one mean when one refers to a *para*- derivative of benzene?

18.20 What is meant by the term *functional group?* Give some examples of functional groups.

18.21 What is an *alcohol*? Give an example.

18.22 What are the reactants in the commercial preparation of methanol? What are the principal uses of methanol?

18.23 Describe, by means of a chemical equation, the commercial production of ethanol from ethylene.

18.24 Give the structural formula of the final product of the oxidation of 1-propanol by an oxidizing agent such as acidic dichromate ion.

18.25 Which of the following compounds is expected to react with acidic dichromate ion? Which one does not react?
(a) 3-methyl-3-heptanol (b) 2-methyl-3-heptanol

18.26 For each of the following, identify the type of compound by the functional group (that is, as an alcohol, aldehyde, or ketone).

(a) $CH_3CH(CH_3)CH_2C(=O)-H$ (b) $CH_3CH_2CH(OH)CH_2CH_3$

(c) $CH_3CH_2C(=O)CH_2CH_3$

18.27 Formaldehyde and acetone are important carbonyl compounds. Describe the physical state of each of these compounds. Give one commercial use of each.

18.28 To write the IUPAC name of most ketones, you need the position number of the carbonyl group on the parent chain. No number is needed, however, for propanone. Why not?

18.29 Draw a structural formula for a carboxylic acid. Circle the carboxyl group. Now write the chemical equation indicating the ionization in water of the acidic hydrogen of the carboxylic acid.

18.30 Write the chemical equation for the reaction of methanol with acetic acid to form the ester methyl acetate.

18.31 Write the chemical equation for the reaction of aniline, $C_6H_5NH_2$, with hydronium ion, H_3O^+.

18.32 Write the chemical equation for the reaction in which you heat acetic acid with methylamine.

Practice Problems

Bonding, Structure, and Isomers (Sections 18.1 and 18.2)

18.33 The following are some organic compounds. What is the shape of the molecule? If you need help, review Section 10.7.

(a) $H-C(Cl)_2-Cl$ (central C bonded to H and three Cl)
Chloroform

(b) $H_2C=O$
Formaldehyde

(c) $H-C\equiv N$
Hydrogen cyanide

18.34 The following are some organic compounds. What is the shape of the molecule? If you need help, review Section 10.7.

(a) $Cl_2C=O$
Phosgene

(b) $H-C(H)_2-Br$ (central C bonded to three H and Br)
Methylbromide

(c) $Cl-C\equiv N$
Cyanogen chloride

18.35 Write structural formulas for all of the isomers of the following molecular formulas.
(a) C_4H_{10} (b) C_3H_8O

18.36 Write structural formulas for all of the isomers of the following molecular formulas.
(a) C_6H_{14} (b) $C_4H_{10}O$

18.37 Write structural formulas for all of the isomers in each of the following (chlorine is monovalent).
(a) $C_2H_4Cl_2$ (b) C_3H_7Cl

18.38 Write structural formulas for all of the isomers in each of the following (chlorine and fluorine are monovalent).
(a) $C_2H_3ClF_2$ (b) $C_3H_6Cl_2$

18.39 Draw full structural formulas for each of the following condensed formulas.

$$\text{(a)}\ \begin{array}{c} CH_3 \\ | \\ CH_3CHCH_2OH \end{array} \qquad \text{(b)}\ \begin{array}{c} CH_3 \\ | \\ CH_3OCHCH_2CH_3 \end{array}$$

18.40 Draw full structural formulas for each of the following condensed formulas.

$$\text{(a)}\ HOCH_2CH_2OH \qquad \text{(b)}\ \begin{array}{c} CH_3 \\ | \\ CH_3OCHCH_2CH_2OH \end{array}$$

Alkanes (Section 18.3)

18.41 What is the molecular formula of the alkane with eight carbon atoms? Draw the full structural formula and condensed structural formula of the corresponding straight-chain alkane.

18.42 What is the molecular formula of the alkane with nine carbon atoms? Draw the full structural formula and condensed structural formula of the corresponding straight-chain alkane.

18.43 Give the IUPAC names of the following alkanes.

$$\text{(a)}\ \begin{array}{c} CH_3 \\ | \\ CH_3CH_2CH_2CHCH_2CH_3 \end{array}$$

$$\text{(b)}\ \begin{array}{c} CH_3 \\ | \\ CH_3-C-CH_2CH_3 \\ | \\ CH_3 \end{array}$$

$$\text{(c)}\ \begin{array}{c} CH_2CH_2CH_3 \\ | \\ CH_3CH_2CHCH_2CH_3 \end{array}$$

18.44 Give the IUPAC names of the following alkanes.

$$\text{(a)}\ \begin{array}{c} CH_3 \\ | \\ CH_3CH_2CH_2CHCHCH_3 \\ | \\ CH_3 \end{array}$$

$$\text{(b)}\ \begin{array}{c} CH_2CH_3 \\ | \\ CH_3CH_2CH_2-C-CH_2CH_2CH_3 \\ | \\ CH_3 \end{array}$$

$$\text{(c)}\ \begin{array}{c} CH_2CH_2CH_3 \\ | \\ CH_3CH_2CCH_2CH_3 \\ | \\ CH_3 \end{array}$$

18.45 Draw the structural formulas of the following.
(a) 2-methylbutane
(b) 3-ethyl-2-methylpentane

18.46 Write structural formulas for the following molecules.
(a) 2,3-dimethylhexane
(b) 4-propyloctane

18.47 Write the balanced equation for the complete burning of heptane in oxygen.

18.48 Write the balanced equation for the complete burning of 2,2,4-trimethylpentane in oxygen.

18.49 Write the chemical equation for one possible substitution reaction of pentane by chlorine.

18.50 Write the chemical equation for one possible substitution reaction of 2-methylbutane by bromine.

Alkenes and Alkynes (Sections 18.4 and 18.5)

18.51 What are the IUPAC names of the following compounds?

(a) 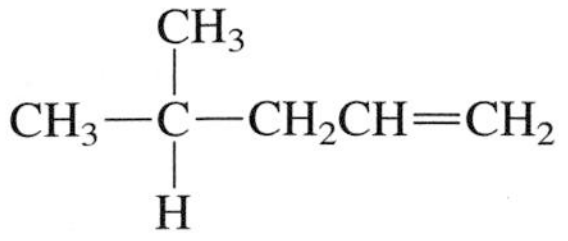 $CH_3-\underset{H}{\overset{CH_3}{C}}-CH_2CH{=}CH_2$

(b) 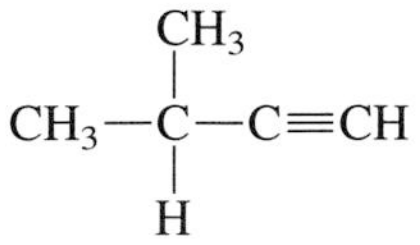 $CH_3-\underset{H}{\overset{CH_3}{C}}-C{\equiv}CH$

18.52 What are the IUPAC names of the following compounds?

(a) 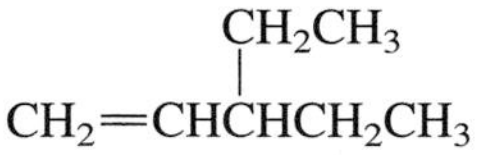 $CH_2{=}CH\overset{CH_2CH_3}{C}HCH_2CH_3$

(b) 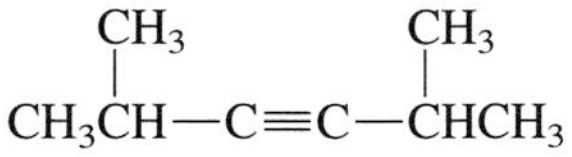 $CH_3\overset{CH_3}{C}H-C{\equiv}C-\overset{CH_3}{C}HCH_3$

18.53 Complete the following equation.

$$CH_3CH{=}CH_2 + Br_2 \longrightarrow$$

18.54 Complete the following equation.

$$CH_2{=}CHCH_2CH_3 + Cl_2 \longrightarrow$$

18.55 Polytetrafluoroethylene is a polymer (with Teflon as a trade name) obtained from the monomer tetrafluoroethylene.

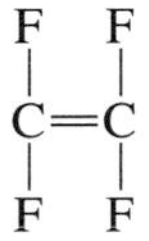

$$\begin{matrix} F & & F \\ | & & | \\ C & = & C \\ | & & | \\ F & & F \end{matrix}$$

Use structural formulas to show the reaction of the monomer to give the polymer. (The reaction is similar to the formation of polyethylene.)

18.56 Polyvinyl chloride (PVC) is a polymer obtained from the monomer vinyl chloride.

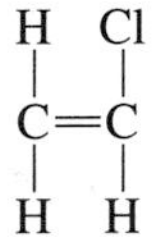

$$\begin{matrix} H & & Cl \\ | & & | \\ C & = & C \\ | & & | \\ H & & H \end{matrix}$$

Use structural formulas to show the reaction of the monomer to give the polymer. (The reaction is similar to the formation of polyethylene.)

Aromatic Hydrocarbons (Section 18.6)

18.57 Draw the resonance description of the toluene molecule, $C_6H_5CH_3$.

18.58 Draw the resonance description of the *p*-xylene molecule, *p*-$C_6H_4(CH_3)_2$.

18.59 Write structural formulas, using hexagons with circles for benzene rings, for the following.
(a) *m*-diethylbenzene
(b) *o*-bromotoluene
(c) 1,3,5-tribromobenzene

18.60 Write structural formulas, using hexagons with circles for benzene rings, for the following.
(a) *p*-chloronitrobenzene
(b) 1,2,4-trichlorobenzene
(c) *p*-nitrotoluene

18.61 Give the name of each of the following compounds.

(a) 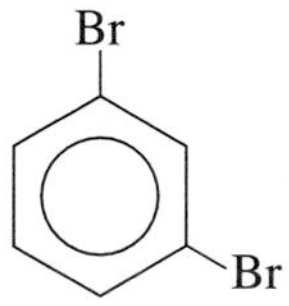 Br, Br (b) 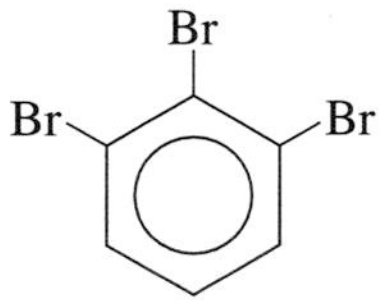 Br, Br, Br

18.62 Give the name of each of the following compounds.

(a) 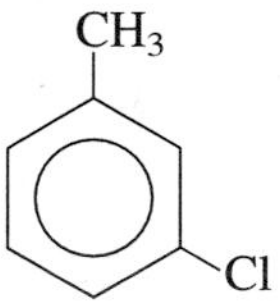 CH_3, Cl (b) 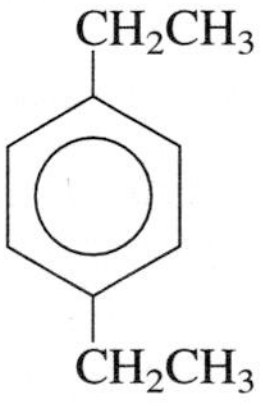 CH_2CH_3, CH_2CH_3

Alcohols (Section 18.7)

18.63 Classify each of the following alcohols as primary, secondary, or tertiary.

(a) $$\begin{array}{l} \quad\ \ OH \\ \quad\ \ \ | \\ CH_3CCH_2CH_3 \\ \quad\ \ \ | \\ \quad\ \ CH_3 \end{array}$$

(b) $CH_3CH_2CH_2CH_2OH$

(c) $$\begin{array}{l} \quad\ \ OH \\ \quad\ \ \ | \\ CH_3CHCH_2CH_3 \end{array}$$

18.64 Classify each of the following alcohols as primary, secondary, or tertiary.

(a) $$\begin{array}{l} CH_3CHOH \\ \quad\ \ \ | \\ CH_3CHCH_2CH_3 \end{array}$$

(b) $$\begin{array}{l} \quad\ \ CH_3 \\ \quad\ \ \ | \\ CH_3C{-}OH \\ \quad\ \ \ | \\ \quad CH_3CHCH_2CH_3 \end{array}$$

(c) $$\begin{array}{l} \quad\ \ CH_2OH \\ \quad\ \ \ | \\ CH_3CHCH_2CH_3 \end{array}$$

18.65 Give the IUPAC name of the following compound.

$$\begin{array}{l} CH_3CHOH \\ \quad\ \ \ | \\ CH_3CHCH_2CH_3 \end{array}$$

18.66 Give the IUPAC name of the following compound.

$$\begin{array}{l} \quad\ \ CH_3 \\ \quad\ \ \ | \\ CH_3C{-}OH \\ \quad\ \ \ | \\ CH_3CHCH_2CH_3 \end{array}$$

18.67 Write the chemical equation, if a reaction occurs, for each of the following.

(a) complete burning of 1-butanol in oxygen
(b) oxidation with acidic dichromate ion of 1-butanol
(c) oxidation with acidic dichromate ion of 2-butanol
(d) oxidation with acidic dichromate ion of 2-methyl-2-butanol

18.68 Write the chemical equation, if a reaction occurs, for each of the following.

(a) complete burning of 2-methyl-2-butanol
(b) oxidation with acidic dichromate ion of 3-methyl-2-butanol
(c) oxidation with acidic dichromate ion of 3-ethyl-3-pentanol
(d) oxidation with acidic dichromate ion of 1-pentanol

Aldehydes and Ketones (Section 18.8)

18.69 Give the IUPAC name for each of the following.

(a) $$\begin{array}{l} \ O \\ \ \| \\ HCCHCH_2CH_3 \\ \quad\ | \\ \quad CH_3 \end{array}$$

(b) $$\begin{array}{l} \qquad\ \ O \\ \qquad\ \ \| \\ CH_3CCHCH_3 \\ \qquad\quad | \\ \qquad\ \ CH_3 \end{array}$$

18.70 Give the IUPAC name for each of the following.

(a) $$\begin{array}{l} \quad\ \ CH_3\ \ O \\ \quad\ \ \ |\qquad \| \\ CH_3CCH_2CH \\ \quad\ \ \ | \\ \quad\ \ CH_3 \end{array}$$

(b) $$\begin{array}{l} \qquad\qquad O \\ \qquad\qquad \| \\ CH_3CHCCHCH_3 \\ \qquad\ \ | \\ \qquad CH_3 \end{array}$$

18.71 Draw structural formulas for each of the following.

(a) 2,3-dimethylpentanal
(b) 3,3-dimethyl-2-pentanone

18.72 Draw structural formulas for each of the following.

(a) 3-ethylhexanal
(b) 3-ethyl-2-hexanone

Carboxylic Acids and Esters (Section 18.9)

18.73 Give the IUPAC name for the following compound.

$$\begin{array}{l} \qquad\quad CH_3\ \ O \\ \qquad\qquad |\quad\ \ \| \\ CH_3{-}C{-}COH \\ \qquad\qquad | \\ \qquad\quad CH_3 \end{array}$$

18.74 Give the IUPAC name for the following compound.

$$\begin{array}{l} \quad CH_3\qquad\quad O \\ \qquad |\qquad\qquad \| \\ CH_3CHCHCH_2COH \\ \qquad\qquad | \\ \qquad\quad CH_3 \end{array}$$

18.75 Draw the structural formula of 2-methylpropanoic acid.

18.76 Draw the structural formula of 2,2-dimethylbutanoic acid.

18.77 Using structural formulas, give the reaction of 2-propanol and butanoic acid in acid solution.

18.78 Using structural formulas, give the reaction of 2-butanol and propanoic acid in acid solution.

Amines and Amides (Sections 18.10 and 18.11)

18.79 Give the structural formula of butylamine. Show the reaction of butylamine with hydronium ion.

18.80 Give the structural formula of diethylamine. Show the reaction of diethylamine with hydronium ion.

18.81 Using structural formulas, show the reaction of ammonia and propanoic acid when heated.

18.82 Using structural formulas, show the reaction of methylamine and propanoic acid when heated.

Additional Problems

18.83 For each of the following, draw the structural formula of a two-carbon compound containing the indicated functional group.
(a) alcohol (b) aldehyde (c) carboxylic acid
(d) alkene

18.84 For each of the following, draw the structural formula of a two-carbon compound containing the indicated functional group.
(a) alkene (b) alkyne (c) amine (d) ester

18.85 Circle and name the oxygen-containing or nitrogen-containing functional groups in the following compounds.

(a) 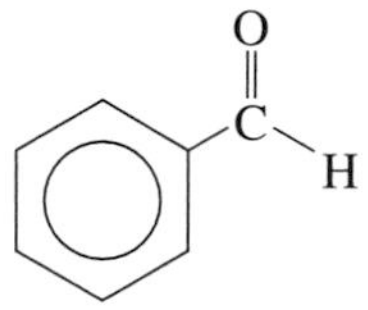

(b) 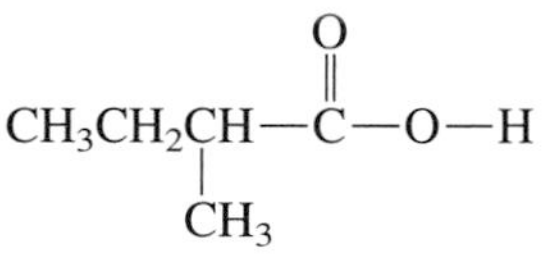

(c) 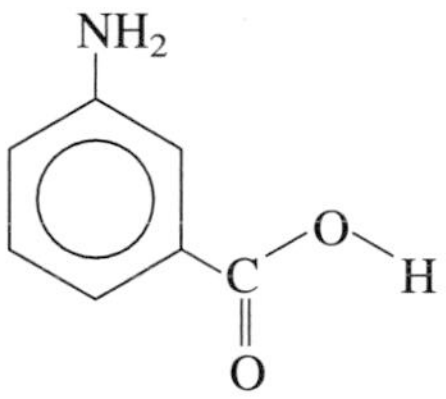

18.86 Circle and name the oxygen-containing or nitrogen-containing functional groups in the following compounds.

(a) 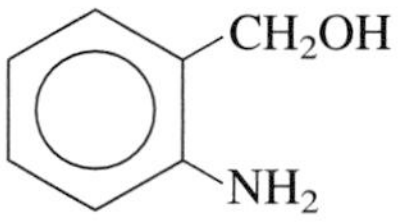

(b) 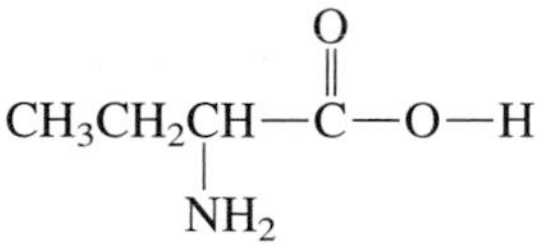

(c) 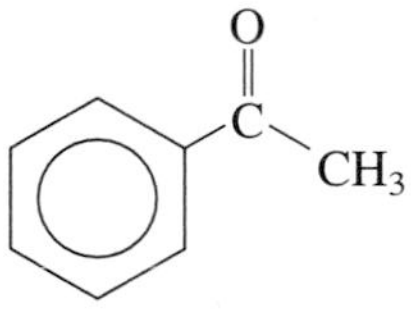

18.87 Give the IUPAC names for each of the following hydrocarbons.

(a) 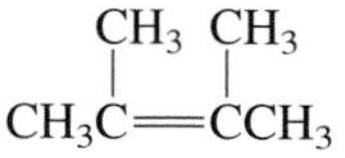

(b) 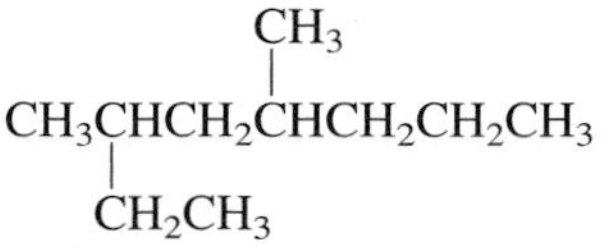

(c) 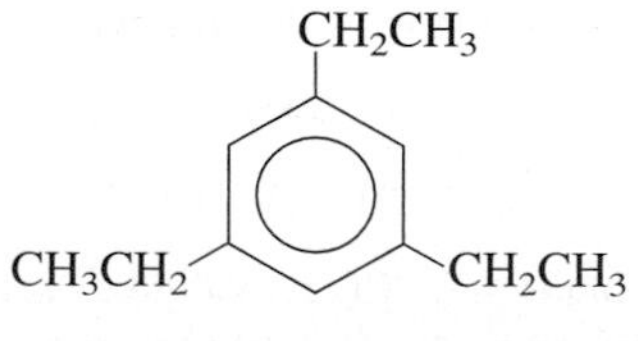

18.88 Give the IUPAC names for each of the following hydrocarbons.

(a) 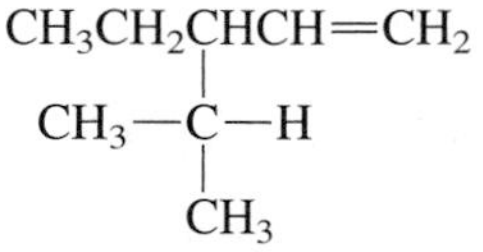

(b) 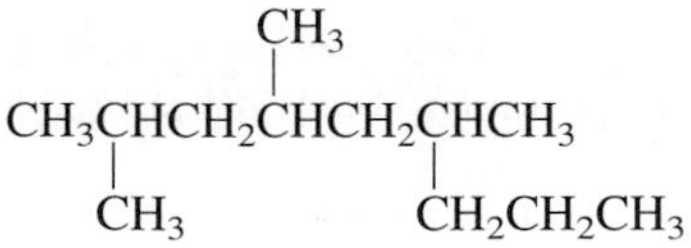

(c) 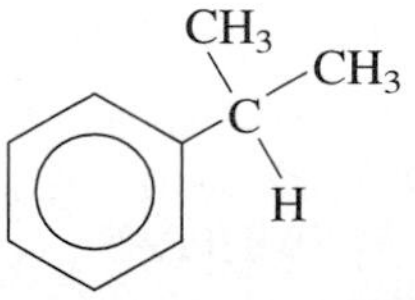

18.89 Draw structural formulas for each of the following hydrocarbons.
(a) 4-ethyl-2,2-dimethyloctane
(b) 2,3-dimethyl-2-heptene
(c) 1,2,3-triethylbenzene

18.90 Draw structural formulas for each of the following hydrocarbons.
(a) 3-ethyl-4,4-dimethyloctane
(b) 2,3-dimethyl-1-heptene
(c) 1,3,5-triethylbenzene

18.91 Draw structural formulas of all isomers having one chlorine atom substituted on pentane.

18.92 Draw structural formulas of all isomers having one chlorine atom substituted on 2-methylbutane.

18.93 Start from an alkene and show by means of a chemical equation how you could prepare 2,3-dibromobutane.

18.94 Start from an alkene and show by means of a chemical equation how you could prepare 2-bromopentane.

18.95 Start from an alcohol and show by means of a chemical equation how you could prepare 2,2-dimethyl-butanoic acid.

18.96 Start from an alcohol and show by means of a chemical equation how you could prepare 2,4-dimethyl-3-pentanone.

18.97 The following names, though not correctly written according to IUPAC rules, do lead to the correct structures. Draw each structural formula; then rename each compound using the IUPAC rules.
(a) 2,2-dimethyl-3-butene
(b) 2-methyl-4-pentanone

18.98 The following names, though not correctly written according to IUPAC rules, do lead to the correct structures. Draw each structural formula; then rename each compound using the IUPAC rules.
(a) 2,3-dimethyl-5-hexanone
(b) 2-propylbutane

18.99 Draw the structural formula of the polyamide that would be formed by heating $HOOC(CH_2)_2COOH$ (succinic acid) with $H_2N(CH_2)_4NH_2$ (tetramethylene diamine).

18.100 Draw the structural formula of the polyester that would be formed by heating $HOOC(CH_2)_2COOH$ (succinic acid) with $HOCH_2CH_2OH$ (ethylene glycol).

Practice in Problem Analysis

For each problem, describe the thinking you would use (the problem analysis) before doing the actual solution, but do not solve the problem.

1. Write the IUPAC name of the following compound:

```
            H
            |
          H-C-H
            |
          H-C-H
            |
    H  H    |  H  H
    |  |    |  |  |
  H-C--C----C--C--C-H
    |  |    |  |  |
    H  H    H  H  H
```

2. Write the structural formula for 2,3-dimethylpentane.

PRACTICE EXAM

1. What are the IUPAC names of the two isomers of butane, C_4H_{10}?
(a) butane, isobutane
(b) 1-methylpropane, 2-methylpropane
(c) butane, 2-methylpropane
(d) 3-methylpropane, 2-methylpropane
(e) butane, 2-methylbutane

2. What is the IUPAC name of the following compound?

```
            CH3
            |
CH3CH2CH2CHCH2CH3
```

(a) 2-propylbutane (b) 3-methylhexane
(c) 4-methylhexane (d) 2-ethylpentane
(e) 4-ethylpentane

3. Which is the structural formula of 2,4-dimethylhexane?

(a) 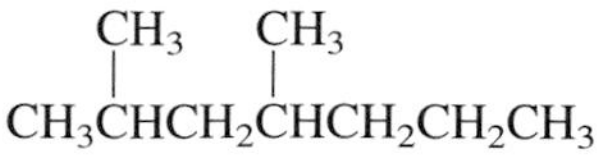
$$\begin{array}{l}\quad\ \ CH_3\quad CH_3\\ \quad\ \ |\qquad\ \ \ |\\ CH_3CHCH_2CHCH_2CH_2CH_3\end{array}$$

(b) 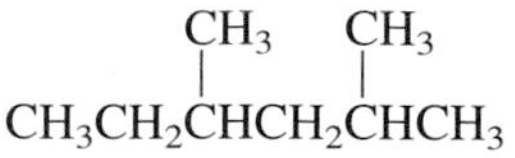
$$\begin{array}{l}\qquad\quad\ \ CH_3\quad CH_3\\ \qquad\quad\ \ |\qquad\ \ \ |\\ CH_3CH_2CHCH_2CHCH_3\end{array}$$

(c) 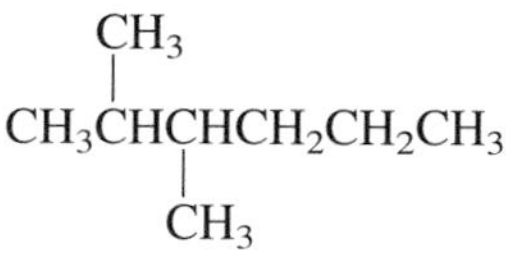
$$\begin{array}{l}\quad\ \ CH_3\\ \quad\ \ |\\ CH_3CHCHCH_2CH_2CH_3\\ \qquad\quad |\\ \qquad\quad CH_3\end{array}$$

(d) 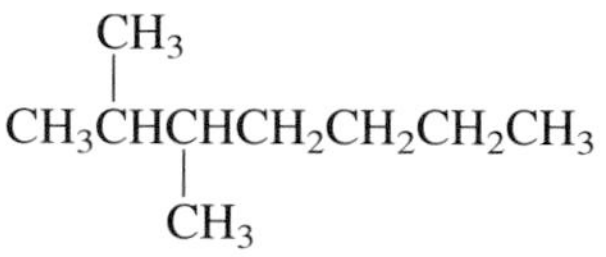
$$\begin{array}{l}\quad\ \ CH_3\\ \quad\ \ |\\ CH_3CHCHCH_2CH_2CH_2CH_3\\ \qquad\quad |\\ \qquad\quad CH_3\end{array}$$

(e) 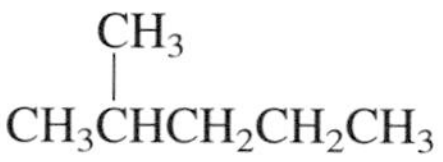
$$\begin{array}{l}\quad\ \ CH_3\\ \quad\ \ |\\ CH_3CHCH_2CH_2CH_3\end{array}$$

4. The total of all coefficients in the balanced equation for the complete burning of propane, C_3H_8, in an excess of oxygen is

(a) 9 (b) 10 (c) 11 (d) 12 (e) 13

5. What is the IUPAC name of the following compound?

$$\begin{array}{l}\qquad\qquad\quad\ \ CH_3\\ \qquad\qquad\quad\ \ |\\ CH_3CH_2CH_2CHCH{=}CH_2\end{array}$$

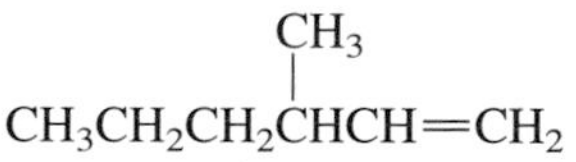

(a) 4-methyl-5-hexene (b) 3-propyl-1-butene
(c) 3-methyl-2-hexene (d) 3-methyl-1-hexene
(e) 3-propyl-2-butene

6. Which is the structural formula of *m*-dimethylbenzene?

(a) 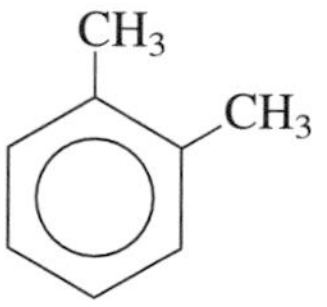 benzene ring with CH_3 at top and CH_3 at upper right
(b) 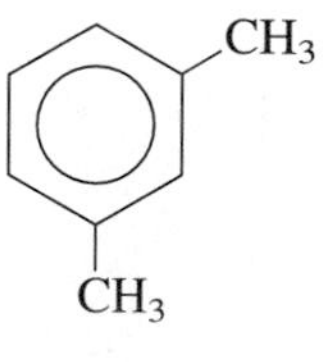 benzene ring with CH_3 at upper right and CH_3 at bottom
(c) 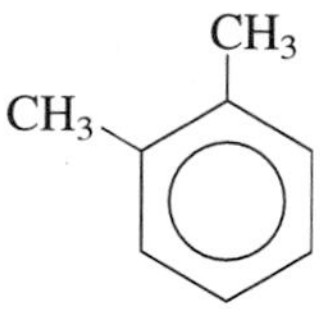 benzene ring with CH_3 at top and CH_3 at upper left
(d) 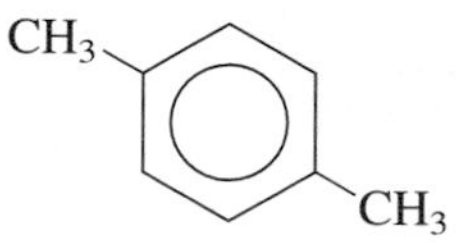 benzene ring with CH_3 at upper left and CH_3 at lower right
(e) 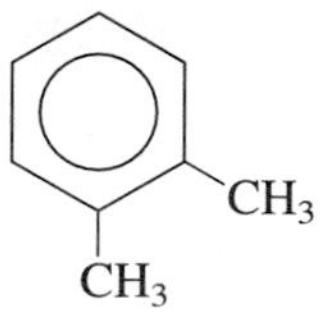 benzene ring with CH_3 at lower right and CH_3 at bottom

7. What is the name of the following compound?

Benzene ring with Cl at top, Cl at upper right, and Cl at bottom

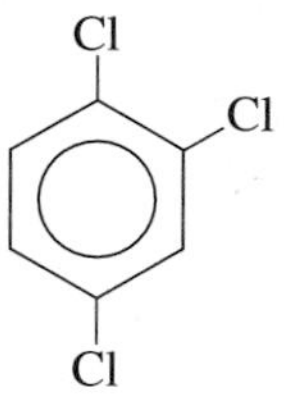

(a) *m*-trichlorobenzene
(b) 1,2,4-phenyltrichloride
(c) 1,2,4-trichlorobenzene
(d) *o,p*-trichlorobenzene
(e) 1,2,4-trichlorophenylene

8. What is the IUPAC name of the following compound?

$$\begin{array}{l}\quad\ \ CH_3\\ \quad\ \ |\\ CH_3CHCH_2CH_2OH\end{array}$$

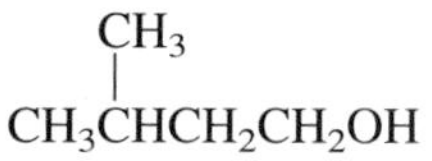

(a) 1-hydroxy-3-pentane (b) 3-methyl-1-butanone
(c) 1,3-pentanol (d) 3-butyl-1-methanol
(e) 3-methyl-1-butanol

9. What is the IUPAC name of the following compound?

$$\begin{array}{l}\quad\ \ CH_2CH_3\\ \quad\ \ |\\ CH_3CHCH_2CH_2CHO\end{array}$$

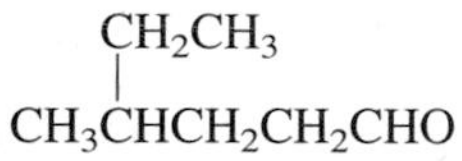

(a) 4-ethyl-1-pentanal (b) 4-ethylpentanal
(c) 4-methyl-1-hexanal (d) 4-methylhexanal
(e) 4-ethylhexanol

10. What is the IUPAC name of the following compound?

$$\begin{array}{l}CH_3\quad\ \ O\\ |\qquad\quad\ \ \|\\ CH_2CH_2CCH_3\end{array}$$

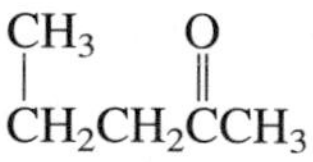

(a) 2-pentanone (b) 4-methyl-2-butanone
(c) 1-methyl-3-butanone (d) 4-pentanone
(e) 1,4-dimethylpropanone

Chapter

19 Biochemistry

The structure of DNA.

Chapter Outline

In recent years, scientists have made enormous strides in biochemistry, the study of the chemical basis of living organisms. Every discovery, it seems, leads to ever more fascinating areas of investigation. What may be the most fascinating result of this research is the knowledge that all living things, whether bacteria or human beings, consist of the same basic molecules and use essentially the same chemical processes. In every living organism, the fundamental information for that organism, which is passed from one generation to the next, is stored in molecules of deoxyribonucleic acid, or DNA, a particular kind of nucleic acid. A DNA molecule is an enormously long polymer chain; the cell uses the positions of different monomer units on the chain as a code with which to construct protein molecules.

Protein molecules are crucial to the chemical processes of a cell. They catalyze biochemical reactions, transport biologically important molecules to different parts of an organism, and play specific physiological roles. These molecules, too, are similar in bacteria and in human beings, consisting of chains whose units are amino acids, each having an amine group and a carboxylic acid group. Of all the possible amino acids, only 20 of them occur in the majority of proteins in living organisms.

In this chapter we present an overview of cell structure and a short discussion of the types of molecules found in biological organisms, followed by a description of the main classes of biological molecules and some of their functions.

THE CELL

The fundamental unit of most living organisms is the cell. The simplest forms of life consist of a single cell; more complex forms of life are multicellular organisms. In the following two sections, we look first at the basic structure and functions of the cell and then briefly describe the molecules used to carry out those functions.

19.1 Cell Structure

OBJECTIVES

- State the major difference between prokaryotic and eukaryotic cells.
- List the parts of a eukaryotic cell and their functions.

The simplest cells are *prokaryotes*. All prokaryotes are single-celled organisms, such as bacteria and some algae. The machinery of the cell is contained within a cell membrane, and the various functions of the cell are distributed throughout its interior. The workings of the cell are directed by its *nuclear region,* which contains a DNA (deoxyribonucleic acid) molecule that is attached to the cell membrane.

All multicelled organisms, such as plants and animals, are *eukaryotes,* as are some single-celled organisms such as yeasts and many algae. Eukaryotic cells are generally

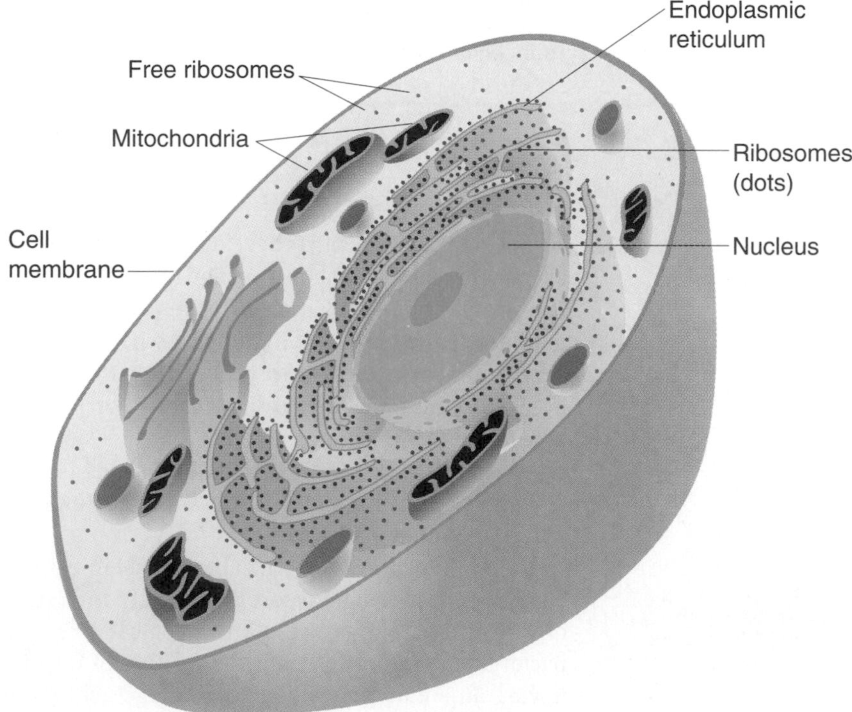

FIGURE 19.1

A eukaryotic cell. This diagram of the cell shows some of the major organelles of the cell, including the cell nucleus and mitochondria.

larger and more organized than prokaryotic cells, the main difference being that they have *organelles,* distinct parts with distinct functions; each organelle is separated from the rest of the cell by a membrane (Figure 19.1). Thus, a eukaryotic cell has a distinct cell *nucleus,* which, like the nuclear region in a prokaryote, directs the functions of the cell. The information for the preparation of the cell's proteins is contained within this nucleus. This information is passed from one generation of organism to the next. ●

● A current hypothesis states that the organelles of a eukaryotic cell originated as prokaryotic cells absorbed by another cell.

The region of the cell between the nucleus and the cell membrane is called the *cytoplasm.* The cytoplasm consists of a semiliquid medium (called the *cytosol*) in which various organelles, such as the mitochondria, are suspended. *Mitochondria* are organelles that produce the energy of the cell from food molecules such as glucose (a kind of sugar). Cells require energy to synthesize, or build, protein and other molecules and to perform cell functions. This protein synthesis occurs on cell particles called *ribosomes.* Many of these ribosomes are suspended in the cytosol; others are attached to the *endoplasmic reticulum,* a membrane material within the cell. Other chemical processes besides protein synthesis occur on the endoplasmic reticulum, including processes involving the production of fats.

19.2 Biological Molecules: An Overview

OBJECTIVE

- List four classes of biological molecules and identify their general functions.

The realm of biochemistry includes a few relatively simple substances and many very complex ones. Most of the substances of biological importance are organic compounds, usually containing several functional groups, but some inorganic substances are also of importance in lesser amounts. Three important classes of biomolecules are considered essential food constituents. These are *proteins, carbohydrates,* and *lipids.* In addition, *nucleic acids* carry genetic information and regulate the formation of proteins. In the chapter introduction, we briefly mentioned proteins and nucleic acids. Also, in the preceding section, we mentioned the sugar glucose, which is a kind of carbohydrate, and we mentioned fats, which are one type of lipid. In this section, we look briefly at each of these classes of compounds and then we look more closely at each type of compound in later sections of this chapter.

Proteins

Protein molecules are of fundamental importance to living organisms. Here are some of their major functions.

1. **Catalysis.** Many protein molecules act as *enzymes,* organic catalysts for the reactions that occur in living organisms. ●
2. **Transport.** Some proteins function as molecular transporters. Cholesterol and fats, for example, are carried in the blood by *lipoproteins,* a certain type of protein.
3. **Structural support.** Many proteins, such as those in skin and bone, add structural support to an organism.
4. **Movement of the organism.** Muscle tissue in animals is composed of proteins.

● Until recently, biochemists thought that all enzymes were proteins. However, Thomas R. Cech at the University of Colorado and Sidney Altman at Yale University have shown that some ribonucleic acids behave as enzymes. They shared the Nobel Prize in chemistry in 1989 for their discovery.

Protein molecules are essentially chain polymers composed of amino acids, which as we noted are organic molecules containing an amine group and a carboxylic acid group. The properties of a protein are determined by the number of amino acids it contains, the precise order in which the different amino acids are linked in the protein chain, and the geometry it assumes in the cell environment. We look further at the structure of amino acids and proteins in Sections 19.3 to 19.5.

Carbohydrates

Carbohydrates are either simple sugars or else compounds consisting of two or more simple sugar units. Simple sugars have a carbonyl (C=O) group and two or more hydroxyl (—OH) groups.

Carbohydrates occur widely in plants and animals. Glucose, or dextrose, a simple sugar whose molecular formula is $C_6H_{12}O_6$, occurs in many fruits, such as grapes. (Hence one common name of glucose is *grape sugar.*) Glucose is also present in blood (and so is also called blood sugar); blood carries the glucose to body cells as an energy source. Ribose, $C_5H_{10}O_5$, is another simple sugar. It and a derivative, deoxyribose, are constituents of nucleic acids.

Sucrose, $C_{12}H_{22}O_{11}$, a sugar consisting of two simple sugar units (glucose and fructose) linked together, is present in fruits, beets, and sugar cane. Sucrose is common table sugar, sometimes called beet sugar or cane sugar, depending on the commercial source of the sugar. ●

● Whether pure beet sugar or pure cane sugar, the chemical identity of the substance is the same: sucrose.

Starch and cellulose are glucose polymers with the general formula $(C_6H_{10}O_5)_n$. Starch is used by plants as a way to store glucose compactly. Glycogen is a similar polymer used by animals. Cellulose is the material that lends rigidity to plants. Plant cells, like animal cells, are enclosed by a membrane, but are also surrounded by a cell wall composed of cellulose. Wood is composed mostly of cellulose.

Nucleic Acids

Nucleic acids are polymers composed of *nucleotide* units. In turn, each nucleotide unit consists of either deoxyribose or ribose to which a phosphate unit and a nitrogen-containing base are attached. If the nucleic acid contains deoxyribose, it is called deoxyribonucleic acid (DNA). If the nucleic acid contains ribose, it is called ribonucleic acid (RNA).

Nucleic acids have two related functions in a cell. First, deoxyribonucleic acid (DNA) stores the cell's genetic information. It contains the information needed to produce the sequence of amino acids in each of the cell's proteins. This information is encoded in the cell's DNA by the order of the different bases in the polymer chain. During cell division, the DNA is duplicated, with one copy of the DNA going to each daughter cell.

The second function of nucleic acids is to provide the information for the cell's protein synthesis. This process begins when a code for a protein is copied from a piece of a DNA molecule to a smaller nucleic acid molecule, called *messenger RNA* (a ribonucleic acid). Each messenger RNA molecule forms a template for the synthesis of a given protein molecule.

We look at the structure of nucleotides, the building blocks of nucleic acids, in Section 19.8. Then we look at DNA in Section 19.9 and RNA in Section 19.10.

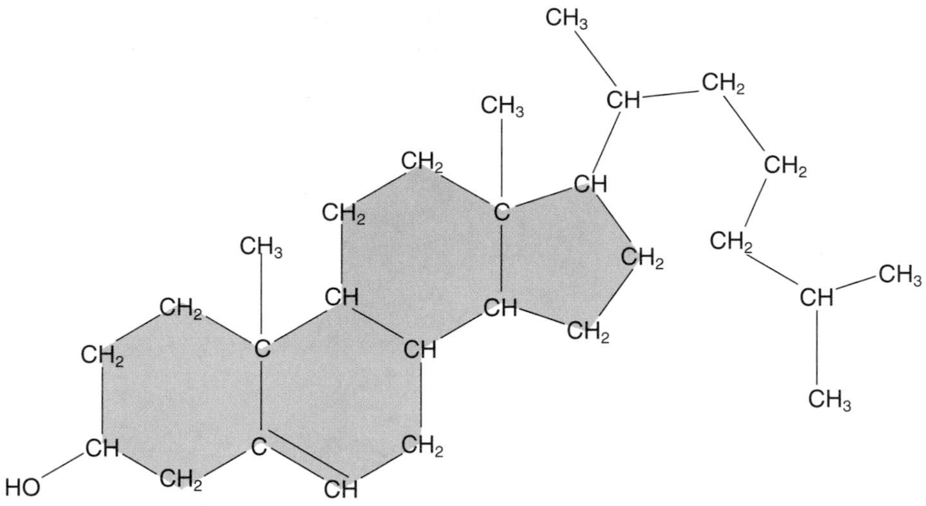

FIGURE 19.2

The steroid cholesterol. Cholesterol is an essential constituent of eukaryotic cell membranes. However, high levels of cholesterol in the blood play a role in the development of atherosclerosis, a disease condition in which lipid deposits clog arteries.

Lipids

Lipids are biological substances that are soluble in organic solvents. There are two main classes of lipids.

1. **Steroids.** These are fused-ring compounds that include sex hormones and cholesterol (see Figure 19.2). Cholesterol is an important constituent of biological membranes.
2. **Molecules having ester linkages.** An important example is a fat, known chemically as a *triacylglycerol* or a triglyceride (older name). A fat molecule consists of glycerol, a compound with three hydroxy (—OH) groups, bonded through ester linkages to three fatty acids. Fats are important energy storage molecules. A *phospholipid* is a similar compound containing a phosphate group in place of one of the fatty acids. Phospholipids are constituents of cell membranes.

We look at triacylglycerols (fats) in Section 19.11 and phospholipids in Section 19.12.

PROTEINS

Proteins, in some form, make up most of body tissue. Enzymes, viruses, hormones, antibodies, muscle, tendons, hair, and fingernails are all composed at least in part of proteins. Proteins are composed largely of carbon, hydrogen, oxygen, and nitrogen, with lesser amounts of other elements such as sulfur, iodine, phosphorus, iron, and copper. They are very complex substances of high molecular weight (about 15,000 to several million amu). Earlier we noted that a protein is essentially a chain polymer of amino acids. We also noted that the sequence of these amino acids is responsible for the properties of the protein. In Section 19.3, we look at the structure of amino acids. Then, in Sections 19.4 and 19.5, we look at the structure of proteins.

19.3 Amino Acids

OBJECTIVES

- Define *amino acid.*
- Identify the structural features held in common by the amino acids.

An **amino acid** is *an organic molecule that contains an amine group and a carboxylic acid group*. The amino acids that are the constituents of most proteins have a similar structure. Glycine and alanine are two simple examples.

$$\begin{array}{c} COOH \\ | \\ H_2N-C-H \\ | \\ R \end{array} \qquad \begin{array}{c} COOH \\ | \\ H_2N-C-H \\ | \\ H \end{array} \qquad \begin{array}{c} COOH \\ | \\ H_2N-C-H \\ | \\ CH_3 \end{array}$$

An amino acid — Glycine — Alanine

Other amino acids have a different hydrocarbon group or hydrocarbon derivative, R, in place of the methyl group, CH_3, in alanine. This R group, which distinguishes one amino acid from another, is referred to as the amino acid *side chain,* because in a protein molecule these groups hang off the side of the main chain of atoms.

Only 20 amino acids are commonly encountered in protein molecules (Table 19.1). Each of these has a structure similar to that of glycine and alanine. Except for one of these amino acids (proline), each consists of a carbon atom to which a carboxylic acid group (—COOH), an amine group (—NH_2), a hydrogen atom, and a side chain (R) are attached. The side chains of the amino acids in Table 19.1 are shown in orange. In the case of proline, one hydrogen atom of the simple amine group NH_2 has been substituted by the hydrocarbon group R, resulting in a cyclic structure. (Note the structure in Table 19.1.)

The amino acids in Table 19.1 are *α-amino acids*, in which the amine group is attached to the carbon atom (the α-carbon) immediately next to the carboxylic acid group, rather than further away. A *β-amino acid* has the amine group on the β-carbon atom that is one more atom away from the —COOH group.

$$\begin{array}{cccccccc} & H & H & H & & O & & \\ & \gamma| & \beta| & \alpha| & & // & & \\ H- & C- & C- & C- & C & & & \\ & | & | & | & & \backslash & & \\ & H & H & NH_2 & & O-H & & \end{array} \qquad \begin{array}{cccccccc} & H & H & H & & O & & \\ & \gamma| & \beta| & \alpha| & & // & & \\ H- & C- & C- & C- & C & & & \\ & | & | & | & & \backslash & & \\ & H & NH_2 & H & & O-H & & \end{array}$$

An α-amino acid — A β-amino acid

19.4 Primary Structure of a Protein

OBJECTIVES

- Define *protein.*
- Identify the peptide bond in a peptide.
- Write the structural formulas for the peptide formed by two or three selected amino acids.
- Identify the amino acids from which a peptide is formed.
- Define the *primary structure* of a protein.

A **protein** is *a biological polymer whose monomer units are amino acids linked by peptide (amide) bonds.* Each amino acid in a protein is linked to another amino acid by

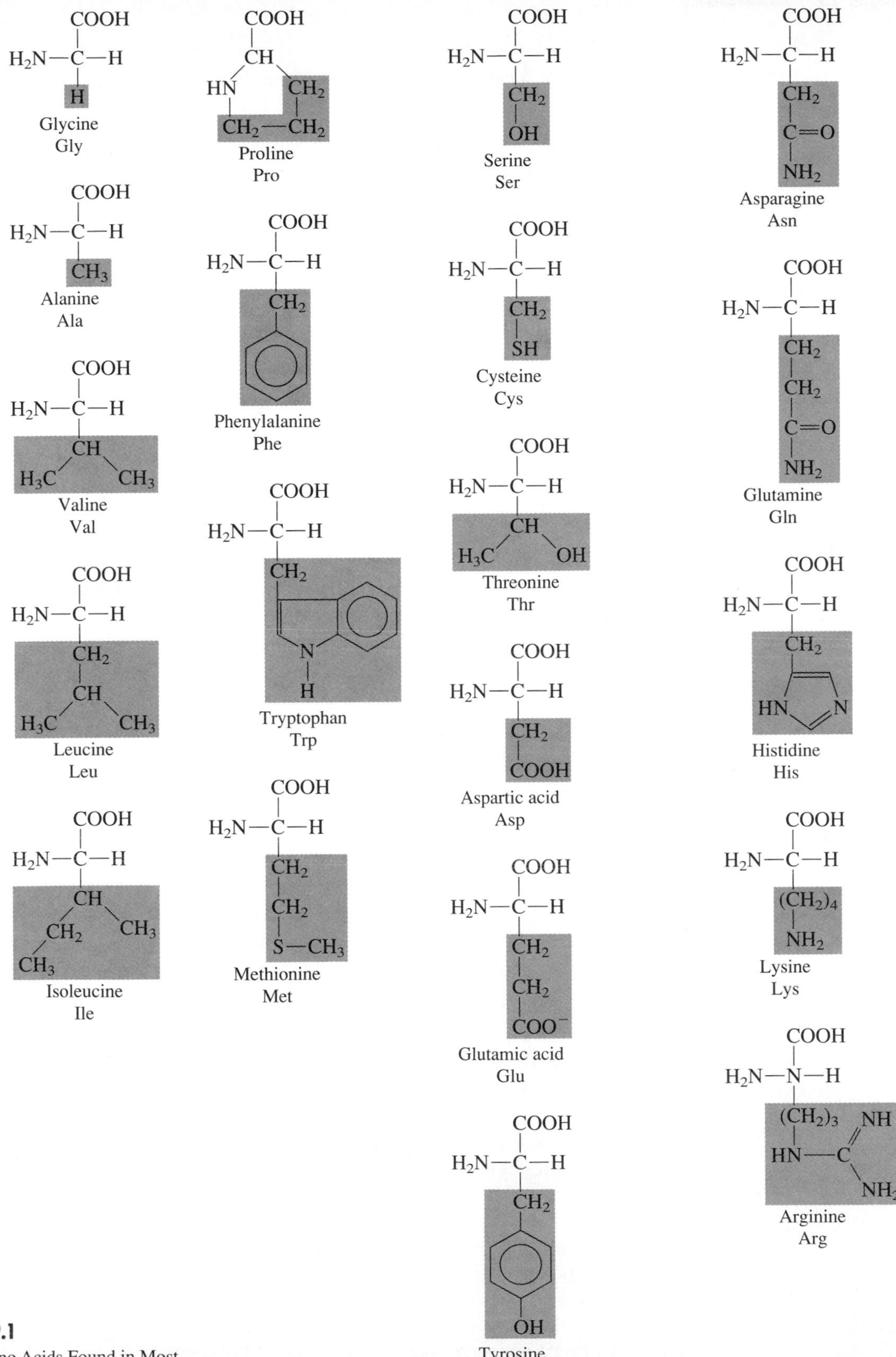

TABLE 19.1

The Amino Acids Found in Most Proteins

forming an amide functional group between them. For example, glycine and alanine can link as follows.

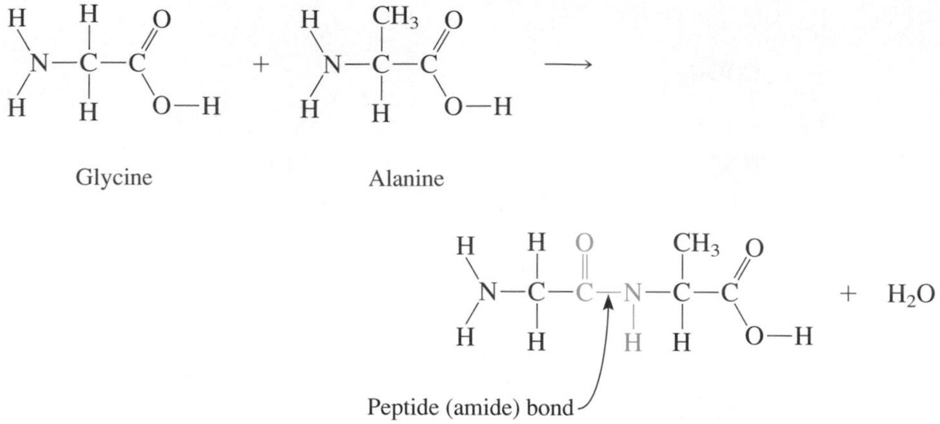

$$\underset{\text{Glycine}}{H_2N{-}CH_2{-}C({=}O){-}O{-}H} + \underset{\text{Alanine}}{H_2N{-}CH(CH_3){-}C({=}O){-}O{-}H} \longrightarrow H_2N{-}CH_2{-}C({=}O){-}N(H){-}CH(CH_3){-}C({=}O){-}O{-}H + H_2O$$

Peptide (amide) bond

An amide group (shown here in blue) forms by a condensation reaction of a carboxylic acid group of one amino acid (glycine in the example) with the amine group of the other amino acid (alanine). *The bond linking the carbon atom and the nitrogen atom in the amide group of a protein molecule* is called a **peptide bond,** or **amide bond** (see Section 18.11).

Amino Acid Sequence

A short sequence of amino acids (from 2 to less than 50) linked by peptide bonds is referred to as a *peptide.* The molecule formed by glycine and alanine, which we just described, is a dipeptide, a peptide of two amino acid units. Here is the general structural formula of a tripeptide, in which the amino acid side chains are R_1, R_2, and R_3.

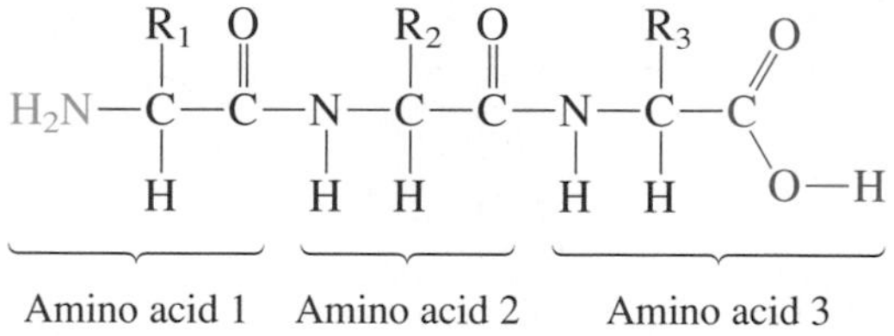

$$\underbrace{H_2N{-}CH(R_1){-}C({=}O)}_{\text{Amino acid 1}}{-}\underbrace{N(H){-}CH(R_2){-}C({=}O)}_{\text{Amino acid 2}}{-}\underbrace{N(H){-}CH(R_3){-}C({=}O){-}O{-}H}_{\text{Amino acid 3}}$$

Polypeptides are similar to simple peptides but are longer chains of amino acids linked by peptide bonds. Proteins are polypeptides having biological activity. A protein may have as few as 50 or more than 1000 amino acid units. Note that one end of the peptide has an amine group (shown in blue) and the other end has a carboxylic acid group (shown in red). This will be true for any peptide or protein molecule: One end has an amine group; the other has a carboxylic acid group.

What distinguishes one peptide or protein from another is the identity of the side chains and their order; that is, the sequence of the amino acids. We can show this sequence conveniently by using standard three-letter abbreviations for the amino acids, which are given in Table 19.1. Thus, glycine is abbreviated Gly, and alanine is abbreviated Ala. The dipeptide of glycine and alanine discussed earlier is written

Gly–Ala

In this notation, it is understood that the amine end of the peptide is at the left and the carboxylic acid end is at the right. The following dipeptide, however, is a different molecule. If you write its structural formula you will see that this is so.

Ala–Gly

■ EXAMPLE 19.1 Writing Structural Formulas for Peptides

Write structural formulas for the two dipeptides that could be formed between alanine and leucine.

Problem Analysis

Two amino acids can form a peptide bond in two ways, depending on which amino acid reacts at the $—NH_2$ position and which reacts at the —COOH position.

Solution

The two possible reactions to form a dipeptide bond result in Ala–Leu and Leu–Ala. By convention, we place the $—NH_2$ group on the left. The structural formulas are

Alanine + leucine

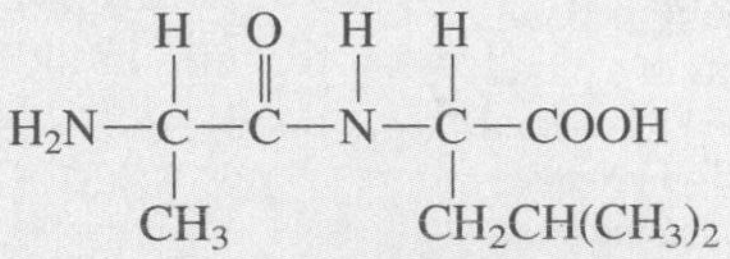

$$H_2N-\underset{\underset{CH_3}{|}}{\overset{\overset{H}{|}}{C}}-\overset{\overset{O}{\|}}{C}-\overset{\overset{H}{|}}{N}-\underset{\underset{CH_2CH(CH_3)_2}{|}}{\overset{\overset{H}{|}}{C}}-COOH$$

Leucine + alanine

$$H_2N-\underset{\underset{CH_2CH(CH_3)_2}{|}}{\overset{\overset{H}{|}}{C}}-\overset{\overset{O}{\|}}{C}-\overset{\overset{H}{|}}{N}-\underset{\underset{CH_3}{|}}{\overset{\overset{H}{|}}{C}}-COOH$$

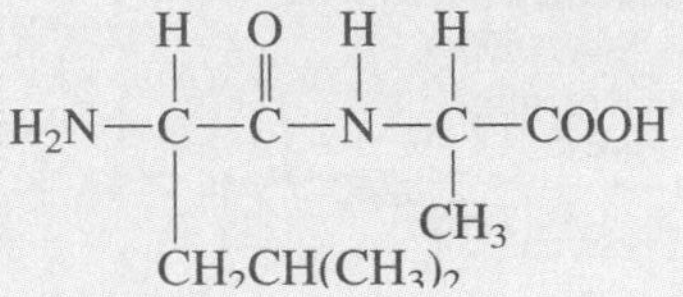

Exercise 19.1

Write structural formulas for the two dipeptides that could be formed between glycine and phenylalanine.

(Try Problems 19.37 and 19.38.)

In addition to peptide bonds, another type of covalent bond occurs in some proteins in which two cysteine (Cys) amino acid units on the protein chain link different parts of the polymer chain. Each of these cross-links is a disulfide bond formed between the two cysteine units.

Disulfide Bonds

The amino acid cysteine has the following structure.

Cysteine

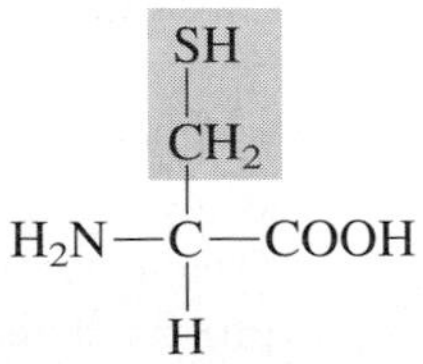

$$H_2N-\underset{\underset{H}{|}}{\overset{\overset{\overset{SH}{|}}{CH_2}}{\overset{|}{C}}}-COOH$$

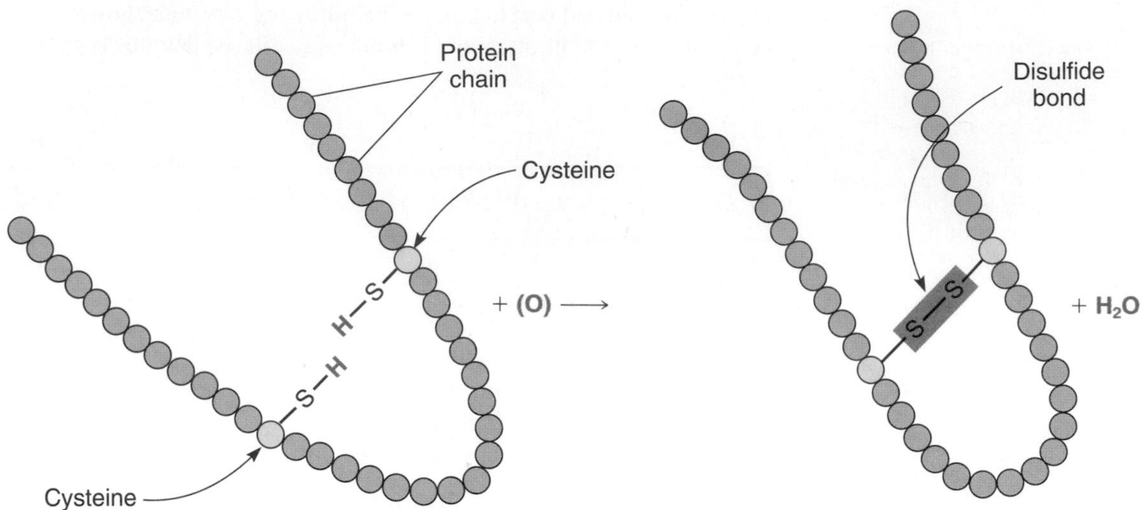

FIGURE 19.3
Formation of a disulfide bond. The amino acid units of a protein chain are represented by colored circles. The sulfhydryl groups (—S—H) in two cysteine units (yellow circles) react to form a disulfide group (—S—S—) linking two parts of the same protein chain.

The sulfhydryl (—S—H) groups on two cysteine units of a protein chain can link by reacting in the presence of an oxidizing agent, shown here as (O), to form a disulfide bond (—S—S—).

$$\begin{array}{ccccccc} & \text{H} & & \text{H} & & \text{O} & \\ & | & & | & & \| & \\ - & \text{N} & - & \text{C} & - & \text{C} & - \\ & & & | & & & \\ & & & \text{CH}_2 & & & \\ & & & | & & & \\ & & & \text{SH} & & & \\ & & & \text{SH} & & & \\ & & & | & & & \\ & & & \text{CH}_2 & & & \\ & & & | & & & \\ - & \text{N} & - & \text{C} & - & \text{C} & - \\ & | & & | & & \| & \\ & \text{H} & & \text{H} & & \text{O} & \end{array} \quad + \quad (\text{O}) \quad \longrightarrow \quad \begin{array}{ccccccc} & \text{H} & & \text{H} & & \text{O} & \\ & | & & | & & \| & \\ - & \text{N} & - & \text{C} & - & \text{C} & - \\ & & & | & & & \\ & & & \text{CH}_2 & & & \\ & & & | & & & \\ & & & \text{S} & & & \\ & & & | & & & \\ & & & \text{S} & & & \\ & & & | & & & \\ & & & \text{CH}_2 & & & \\ & & & | & & & \\ - & \text{N} & - & \text{C} & - & \text{C} & - \\ & | & & | & & \| & \\ & \text{H} & & \text{H} & & \text{O} & \end{array} \quad + \quad \text{H}_2\text{O}$$

Figure 19.3 shows the cross-linking of a protein chain through the formation of a disulfide bond.

Primary Structure

The **primary structure** of a protein or a peptide refers to *the order, or sequence, of the amino acid units in the protein or peptide molecule*. Thus, the primary structure of a protein describes all of the covalent bonds in the molecule. Here is the primary structure of the pentapeptide leucine enkephalin (a brain opiate, a naturally occurring compound present in the brain that acts as a painkiller).

Try–Gly–Gly–Phe–Leu

Figure 19.4 shows the primary structure of human proinsulin protein. (Insulin, which is formed from proinsulin, is a hormone that regulates glucose levels in the blood.) The protein consists of 86 amino acid units. Note the three disulfide bonds that cross-link the protein chain.

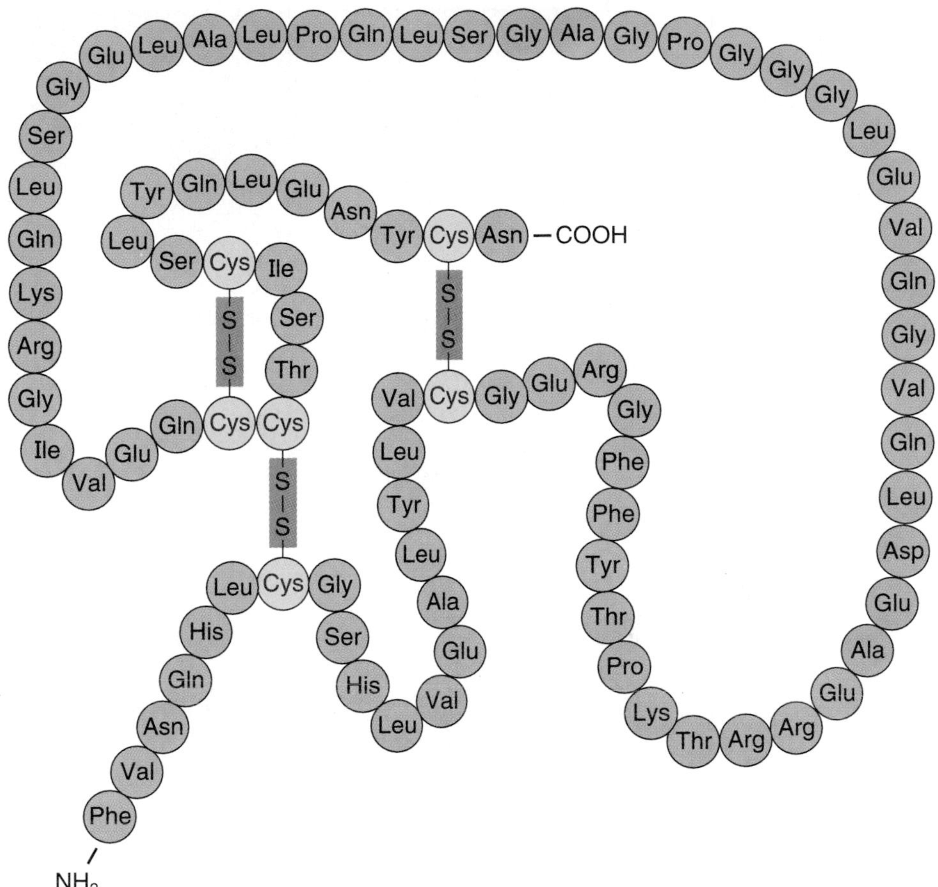

FIGURE 19.4
Primary structure of human proinsulin. Note the three disulfide bonds. The diagram shows only how the amino acids are linked in the protein. It does not show the actual three-dimensional shape of the molecule.

Human Digestion of Proteins

Human digestion breaks down proteins into their component amino acids. Protein digestion begins in the stomach and ends in the small intestine. In both the stomach and the small intestine, enzymes present in digestive fluids catalyze the reaction of amide groups in the protein with water to yield first smaller proteins, or peptides, and finally individual amino acids. (The general name for a chemical reaction with water is *hydrolysis;* thus, the digestion of proteins involves their hydrolysis.) The amino acids released by the hydrolysis of proteins are absorbed primarily from the small intestine into the blood, where they provide the body with the raw materials to synthesize new proteins. If the amino acids are not needed for protein synthesis, they may be used as an energy source or they may be eliminated from the body.

The human body can synthesize some amino acids from others. However, certain amino acids, called *essential amino acids,* must be included in the diet since they cannot be synthesized by the body. Table 19.2 lists the ten essential amino acids. Foods that contain proteins having all of the essential amino acids in about the same proportions are called *complete protein foods.* Some examples of complete protein foods are meat, milk, and eggs. Other protein foods, though lacking in some of the essential amino acids, can be combined with other protein foods to provide a complete protein diet. For

TABLE 19.2
The Essential Amino Acids

Name	Abbreviation
Isoleucine	Ile
Leucine	Leu
Lysine	Lys
Methionine	Met
Phenylalanine	Phe
Threonine	Thr
Tryptophan	Trp
Valine	Val
Arginine*	Arg
Histidine†	His

*Essential in growing children, but not in adults.
†Essential in infants.

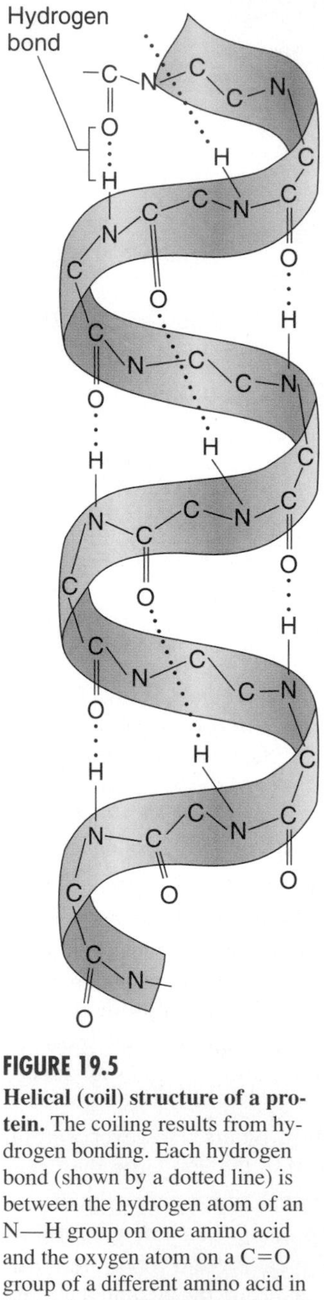

example, beans and rice, a staple food combination in many countries, provide all of the essential amino acids, even though beans and rice separately do not.

19.5 Three-Dimensional Structure of a Protein

OBJECTIVE

- Describe how the primary structure of a protein influences its three-dimensional shape.

A protein with a specific primary structure spontaneously coils, and these coils also spontaneously fold into a characteristic three-dimensional shape in water solution. Disulfide bonds between different parts of the protein chain, of course, force some folding of the chain. However, noncovalent forces are responsible for much of the three-dimensional structure of a protein. Large portions of many proteins consist of helixes, or coils, that result from hydrogen bonding between atoms along the main chain of the protein. Each hydrogen bond is between the hydrogen atom on an amine group of one amino acid unit and the oxygen atom on a carbonyl group of another amino acid unit (Figure 19.5).

In globular proteins, the coils fold themselves into compact spherical shapes (or globules) in which the nonpolar side chains of the protein, which are not attracted to water, are inside the globule. The polar groups are on the outside of the globule, where they are attracted to water. Figure 19.6 shows the three-dimensional structure of myoglobin, a globular protein that binds oxygen in muscle tissue and functions as an oxygen storage depot for the muscles.

Except for the disulfide bonds, the forces responsible for the coiling and folding of a protein chain are weak. A protein loses its three-dimensional structure when heated or treated with certain chemical agents, because these processes disrupt the weak noncovalent forces in the protein. The **denaturation** of a protein is *the loss of a protein's three-dimensional shape through the unfolding and uncoiling of the protein as a result of the breaking of the weak forces (such as hydrogen bonding) that hold the protein in its normal three-dimensional shape.* Heating a protein causes it to denature because the protein molecule vibrates more vigorously at higher temperatures, and eventually this

FIGURE 19.5
Helical (coil) structure of a protein. The coiling results from hydrogen bonding. Each hydrogen bond (shown by a dotted line) is between the hydrogen atom of an N—H group on one amino acid and the oxygen atom on a C=O group of a different amino acid in the protein chain. R groups have been omitted for clarity.

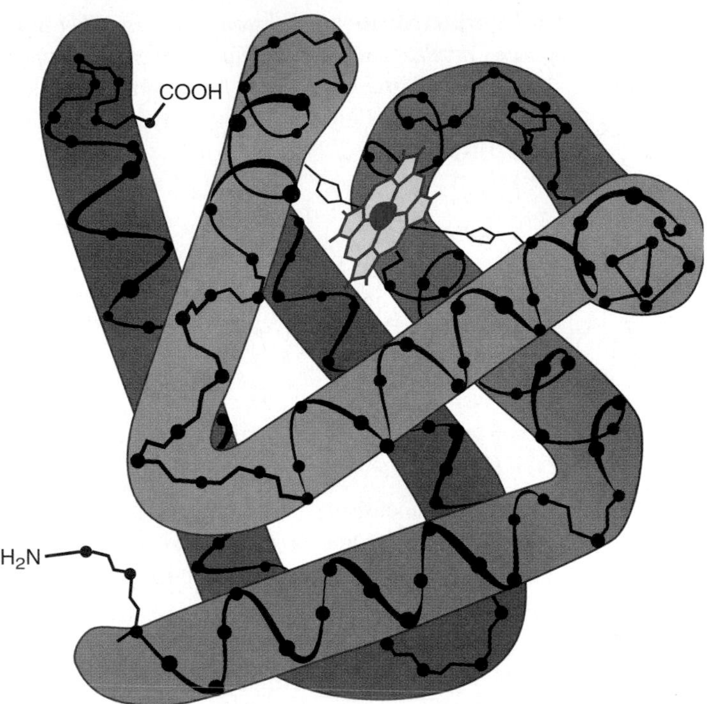

FIGURE 19.6
The three-dimensional structure of myoglobin. Note that many sections of the protein have a helical, or coiled, structure, which results from hydrogen bonding. The helical sections fold so that the nonpolar side chains are inside the globular molecule and the polar side chains point outward. The dots throughout the structure represent α-carbon atoms. The planar structure in the upper right is a heme group, which contains an iron ion that can bind to an O_2 molecule in respiration. After R. E. Dickerson, *The Proteins,* H. Neurath, ed., 2nd ed., Academic Press, 1964.

vibration disrupts the noncovalent forces holding the protein in its three-dimensional shape. When you heat an egg, the proteins denature. The uncooked, clear egg white contains globular proteins, such as albumin, suspended in water. These globular proteins unravel when heated, and the long protein chains entangle, resulting in an opaque material (cooked egg white).

The three-dimensional shape of a protein is essential to its function. In a classic experiment, the enzyme ribonuclease, which catalyzes the hydrolysis of ribonucleic acids, was denatured chemically in a way that also broke the disulfide bonds (—S—S—), reducing them to sulfhydryl groups (—S—H). The free chains, with the forces that held them in their globular shape disrupted by chemical action, lost their enzymatic activity. When the denatured protein was purified from the chemical denaturing agent and exposed to oxygen in air, it slowly recovered its enzymatic activity. The oxygen in the air oxidized the sulfhydryl groups to disulfide bonds, and the protein chains spontaneously recoiled and refolded as the result of the reforming of hydrogen bonds and other weak forces. The experiment not only shows the importance of the three-dimensional structure of a protein to its function, but also shows that the primary structure of a protein contains all the information needed to give the protein's three-dimensional shape in water solution.

CARBOHYDRATES

The term *carbohydrate* arises from the molecular formulas of these substances, which can be written in the general form $C_m(H_2O)_n$. From this formula, it appears that the substances are hydrates of carbon, that is, carbon associated with water. The actual structure of these compounds, however, is far different from this. ●

● If you heat sucrose, it decomposes by releasing water vapor, leaving a residue of carbon behind.

A **carbohydrate** is *a substance that is either a polyhydroxy aldehyde or polyhydroxy ketone, or else a substance that yields such compounds if the carbohydrate hydrolyzes (reacts with water).* The term *polyhydroxy* refers to a compound having two or more hydroxyl groups (—OH groups). Glyceraldehyde and dihydroxyacetone are the simplest carbohydrates.

```
H     O                 H
 \\  //                 |
   C                 H—C—OH
   |                    |
H—C—OH                 C=O
   |                    |
H—C—OH               H—C—OH
   |                    |
   H                    H
```

Glyceraldehyde   Dihydroxyacetone

In both substances, one carbon atom has a carbonyl functional group (C=O), whereas the other carbon atoms have hydroxyl groups attached. Glyceraldehyde has an aldehyde functional group; dihydroxyacetone has a ketone functional group.

19.6 Monosaccharides

OBJECTIVES

- List examples of some simple monosaccharides.
- Distinguish between the structures of aldoses and ketoses.

Carbohydrates are of three types: monosaccharides, oligosaccharides, and polysaccharides. A **monosaccharide,** also called a simple sugar, is *a carbohydrate that cannot be broken down by hydrolysis into simpler carbohydrates.* Glyceraldehyde and dihydroxyacetone are examples of monosaccharides. Monosaccharides are further classified as *aldoses,* sugars that contain an aldehyde group (such as glyceraldehyde), and *ketoses,* sugars that contain a ketone group (such as dihydroxyacetone). Each carbon atom in either an aldose or a ketose is bonded to an oxygen atom. All but one of these oxygen atoms is in the form of a hydroxyl group (—OH); the other oxygen atom is present as a carbonyl group (C=O). Figure 19.7 shows the structures of several important monosaccharides: glucose, fructose, ribose, and 2-deoxyribose. (We number the carbon atoms of a monosaccharide starting from the end closest to the carbonyl group. Thus, the 2-deoxy in 2-deoxyribose refers to the loss of an oxygen atom on carbon atom 2 of ribose.) Note that glucose, ribose, and 2-deoxyribose are aldoses; fructose is a ketose.

Fructose is the sweetest monosaccharide (about twice as sweet as sucrose, common table sugar), and glucose is about half as sweet as sucrose. Glucose is prepared by the acid hydrolysis of starch. (The acid acts as a catalyst.) Starting from cornstarch and dilute hydrochloric acid, the hydrolysis yields corn syrup.

$$\text{Cornstarch} + \text{water} \xrightarrow{\text{HCl}(aq)} \text{glucose} \quad \text{(corn syrup)}$$

Fructose is prepared industrially from glucose (in the form of corn syrup) with the aid of the enzyme glucose isomerase to yield a high-fructose syrup, now commonly used as a sweetener in processed foods. Ribose and 2-deoxyribose are sugars that occur in nucleic acids, which we discuss later in Sections 19.8 to 19.10.

FIGURE 19.7

Some monosaccharides. The structural formulas at the right show open-chain forms of glucose, fructose, ribose, and 2-deoxyribose. Note the aldehyde groups in red and the ketone group in blue. (A) Glucose solutions are sometimes given to patients intravenously (that is, through a tube into a vein). (B) Honey is a concentrated solution of glucose and fructose. (C) DNA, shown here as a molecular model, contains 2-deoxyribose units, as well as base units and phosphate groups.

Glucose	Fructose	Ribose	2-Deoxyribose
$H-C=O$ (aldehyde)	CH_2OH	$H-C=O$ (aldehyde)	$H-C=O$ (aldehyde)
$H-C-OH$	$C=O$	$H-C-OH$	$H-C-H$
$HO-C-H$	$HO-C-H$	$H-C-OH$	$H-C-OH$
$H-C-OH$	$H-C-OH$	$H-C-OH$	$H-C-OH$
$H-C-OH$	$H-C-OH$	CH_2OH	CH_2OH
CH_2OH	CH_2OH		

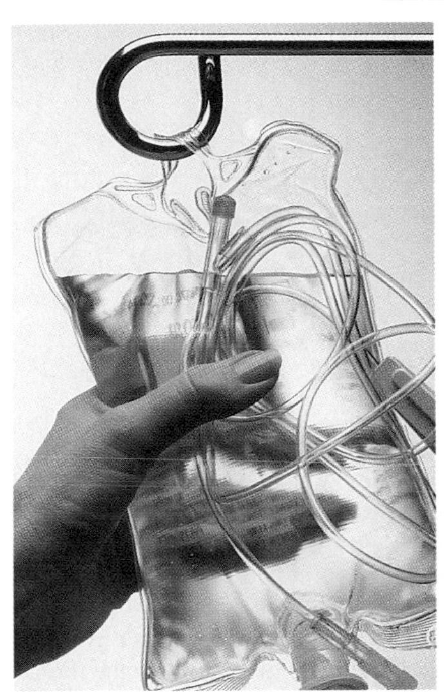

A

B

C

■ EXAMPLE 19.2 Writing Structural Formulas for Aldoses and Ketoses

Draw the structural formula of an aldose and a ketose, each with four carbon atoms.

Problem Analysis

Draw the four carbon atoms in each chain, then decide where the carbonyl group goes. In an aldose the carbonyl group is located at a terminal carbon atom, whereas in a ketose the carbonyl group is located at one of the inner carbon atoms. Fill in hydroxyl groups and H atoms on the remaining carbon atoms.

Solution

In the aldose, the first carbon atom is double bonded to an oxygen atom to form a carbonyl group and the other carbon atoms are bonded to —OH groups. In the ketose, either the second or the third carbon atom forms the carbonyl group. The two structures are

Aldose

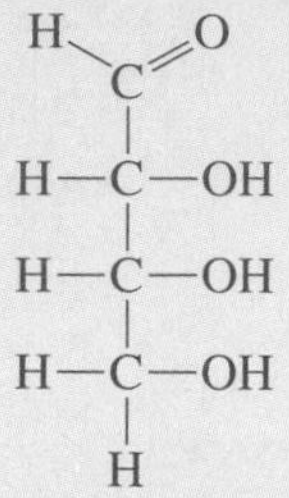

```
H   O
 \ //
  C
  |
H—C—OH
  |
H—C—OH
  |
H—C—OH
  |
  H
```

Ketose

```
  H
  |
H—C—OH
  |
  C=O
  |
H—C—OH
  |
H—C—OH
  |
  H
```

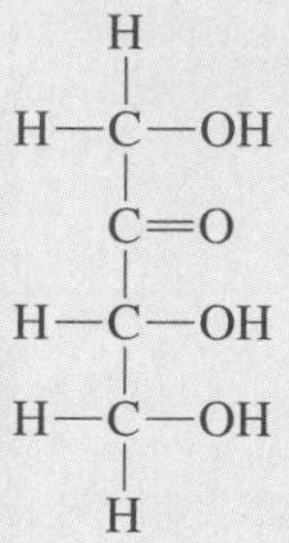

Exercise 19.2

Draw the structural formula of an aldose and a ketose, each with seven carbon atoms.

(Try Problems 19.41 and 19.42.)

Although we have represented the simple sugars in Figure 19.7 as open-chain, or straight-chain, molecules, the predominant species in a solution of a five-carbon-atom or six-carbon-atom sugar is cyclic. There are two reasons for this. First, the open-chain species is a flexible molecule that can easily bend to form a ring. Second, a hydroxyl group can react with a carbonyl group. In the case of glucose, we can show the formation of the cyclic form as follows.

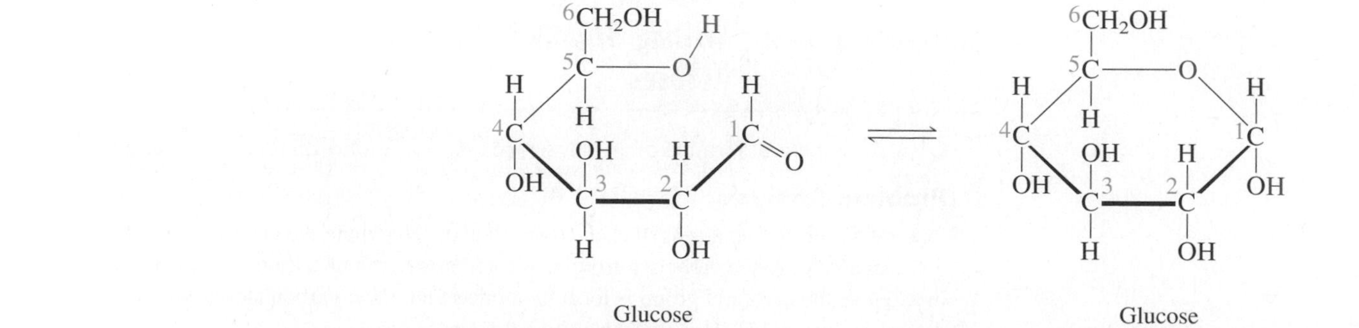

```
     6CH2OH    H                           6CH2OH
       |      /                              |
      5C-----O                              5C-----O
 H   /  |           H               H     /  |      \    H
 |  /   H           |               |    /   H       \   |
4C                 1C  O    ⇌      4C                  1C
 |  \   OH    H   /  \\             |  \   OH    H    /  |
 OH  \  |3   2|  /                  OH  \  |3   2|   /   OH
      \ C-----C /                        \ C-----C  /
        |     |                            |     |
        H     OH                           H     OH
```

Glucose (open-chain form)

Glucose (cyclic form)

Here we have drawn the open-chain form of glucose bent so that carbon atom 1 is near

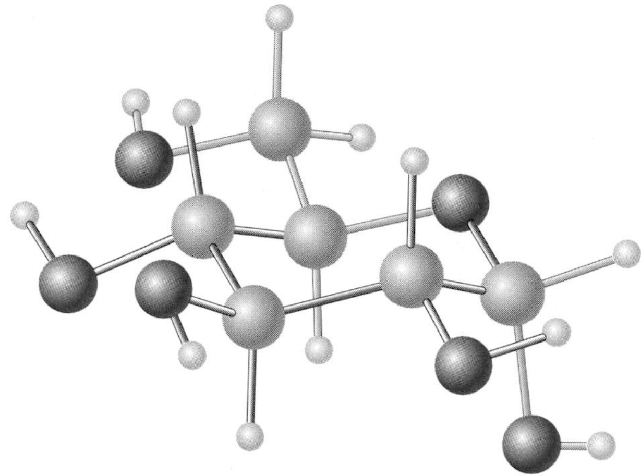

FIGURE 19.8
Cyclic form of the glucose molecule. The ball-and-stick molecular model shows that the cyclic form is not flat, but slightly puckered.

the oxygen atom attached to carbon atom 5. After bond formation between carbon atom 1 and this oxygen atom, a ring forms, and the carbonyl oxygen becomes an —OH group. (For simplicity, the ring structures have been drawn flat, which is a common convention. However, as Figure 19.8 shows, the cyclic form of glucose is a puckered ring.)

19.7 Oligosaccharides and Polysaccharides

OBJECTIVES

- Distinguish between oligosaccharides and polysaccharides.
- Describe how oligosaccharides and polysaccharides are converted to monosaccharides in human digestion.

Oligosaccharides and polysaccharides are polymers of monosaccharides. An **oligosaccharide** is *an oligomer, or short polymer, containing from two to ten monosaccharide (simple sugar) units.* Sucrose, which contains a glucose unit and a fructose unit, is an example of a two-unit oligosaccharide. If you heat an aqueous solution of sucrose, made slightly acidic, the sucrose hydrolyzes to give a mixture of glucose and fructose.

$$\underset{\text{Sucrose}}{C_{12}H_{22}O_{11}} + H_2O \xrightarrow{H_3O^+} \underset{\text{Glucose}}{C_6H_{12}O_6} + \underset{\text{Fructose}}{C_6H_{12}O_6}$$

In human digestion, the enzyme sucrase in the small intestine catalyzes the hydrolysis of sucrose to glucose and fructose, and these simple sugars then enter the bloodstream.

A **polysaccharide** is *a polymer containing many monosaccharide (simple sugar) units.* Starch and cellulose are examples. Both substances hydrolyze in the presence of a catalyst such as H_3O^+ to give glucose. The human digestion of starch begins in the mouth, where the enzyme salivary amylase catalyzes the partial hydrolysis of the starch into oligosaccharides. The digestion of starch continues in the small intestine with the aid of pancreatic amylase and other enzymes, and the final product is glucose, which is absorbed into the blood. Cellulose, although also a polymer of glucose, is not hydrolyzed by human digestive processes and therefore does not function as a food source. ●

● In animals such as rabbits and cattle that use high-cellulose food sources, such as grasses, bacteria in the digestive tract digest the cellulose.

NUCLEIC ACIDS

Nucleic acids, as we noted earlier, are the molecular carriers of genetic information. These little bits of matter contain all the information needed for plants and animals to grow and reproduce. They also contain the biological information needed to characterize and distinguish one species from another. Nucleic acids are polymers of nucleotides. Apart from being the monomeric units of nucleic acids, some nucleotides have direct chemical importance. In the following section, we discuss nucleotides and their structures. In two succeeding sections, we explore the biological role of nucleic acids.

19.8 Nucleotides

OBJECTIVES

- Describe the structure of nucleotides.
- List the bases that are found in nucleotides.
- Describe how AMP, ADP, and ATP are involved in the transfer of energy in biochemical processes.

A **nucleotide** is *a molecule consisting of a sugar, either ribose or 2-deoxyribose, to which is attached a phosphate group and a nitrogen-containing base.* We can draw the general structure as follows.

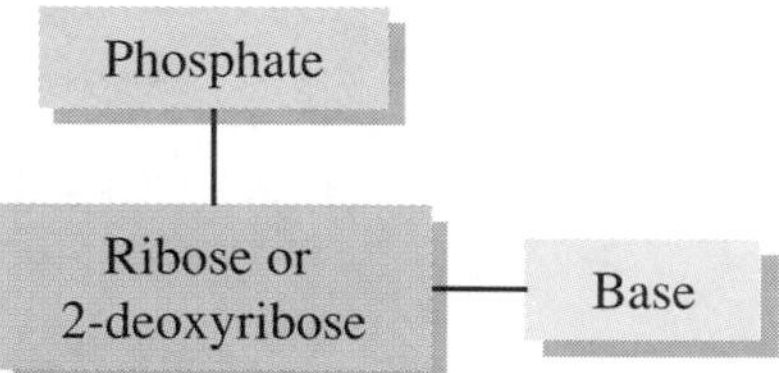

A nucleotide composed of ribose is a *ribonucleotide,* and a nucleotide composed of 2-deoxyribose is a *deoxyribonucleotide*. Deoxyribonucleic acid (DNA) is composed of deoxyribonucleotide monomers, whereas ribonucleic acid (RNA) is composed of ribonucleotide monomers.

The structural formulas of the nitrogen-containing bases that occur in DNA and RNA are shown in Figure 19.9. They are essentially amines (compounds containing $—NH_2$, —NHR, or $—NR_2$ groups), which is why they are bases. The bases occurring in nucleotides are *heterocyclic* compounds, that is, compounds having rings composed of carbon atoms and another kind of atom, in this case nitrogen.

The bases that occur in DNA (and their one-letter abbreviations) are cytosine (C), thymine (T), adenine (A), and guanine (G). The bases that occur in RNA are the same, except that uracil (U) replaces thymine.

Adenine (A) Guanine (G) Cytosine (C) Uracil (U) Thymine (T)

FIGURE 19.9

The nitrogen-containing bases in DNA and RNA. The names and one-letter abbreviations of the bases accompany their structures. Adenine and guanine are similar; each consists of two fused rings containing four nitrogen atoms. Cytosine, uracil, and thymine are similar; each consists of a single ring containing two nitrogen atoms. In nucleotides and nucleic acids, each base connects from the nitrogen atom shown in red to carbon atom 1 of the sugar molecule.

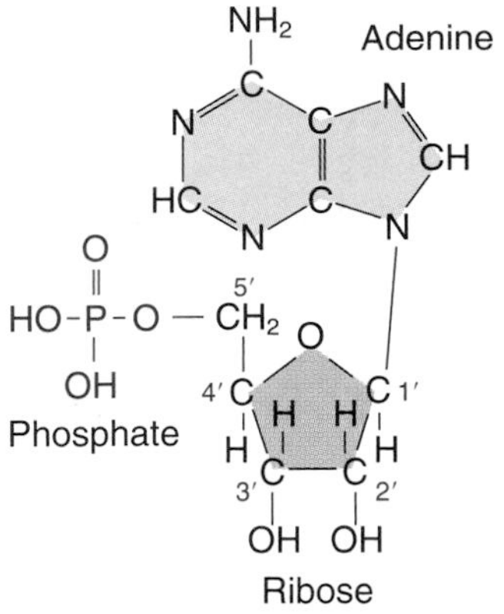

FIGURE 19.10
Adenosine-5′-monophosphate (AMP). This nucleotide consists of ribose (purple) linked to phosphate (red type) and adenine (green). The monophosphate group is shown in the acid form; in the environment of a cell, the monophosphate group is in the ionized form, —PO_4^{2-}.

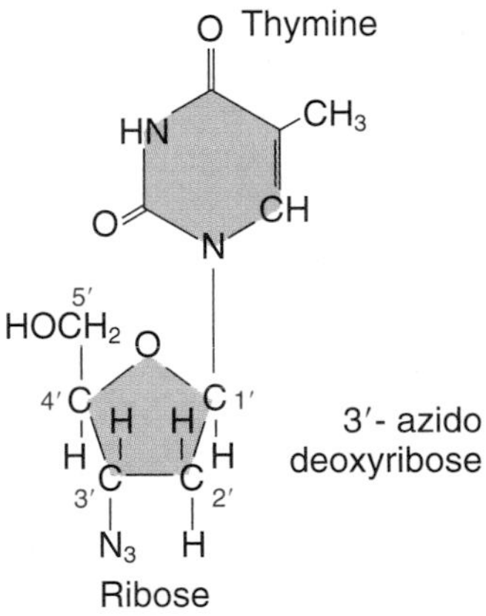

FIGURE 19.11
3′-Azido-3′-deoxythymidine (AZT). AZT is a drug used to treat AIDS. It consists of the base thymine linked to a derivative of 2-deoxyribose (in which the —OH group on carbon atom 3′ has been replaced by an azide group, —N_3). J. P. Horowitz, then at the Michigan Cancer Foundation, first prepared AZT in the early 1960s, hoping to find an anticancer drug. AZT did not prove successful as an anticancer drug, but investigators later showed that it inhibits the process in which the AIDS virus inserts itself into a cell. However, many patients suffer anemia and other serious side effects from using AZT.

Figure 19.10 shows the structural formula of a typical nucleotide, adenosine-5′-monophosphate (AMP), which is a monophosphate derivative of adenosine. Adenosine is composed of two parts: adenine (in green) and ribose (in purple). By monophosphate, we mean the following group.

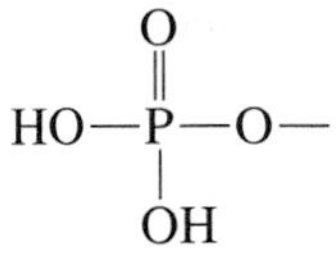

$$\mathrm{HO{-}P({=}O)(OH){-}O{-}}$$

Monophosphate (acid form)

The monophosphate group (in red in Figure 19.10) is shown here in the protonated form (with H atoms attached). The monophosphate group is attached to the 5′ carbon atom of ribose. (The carbon atoms of ribose in a nucleotide have numbers followed by a prime; the atoms of the base are given unprimed numbers.)

Corresponding diphosphate and triphosphate derivatives also are possible. The diphosphate and triphosphate groups have the following structures.

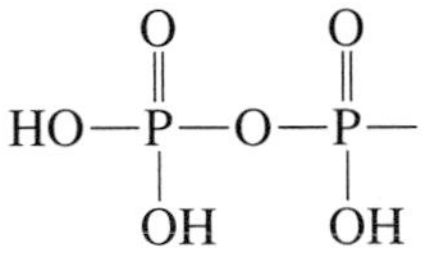

$$\mathrm{HO{-}P({=}O)(OH){-}O{-}P({=}O)(OH){-}} \qquad \mathrm{HO{-}P({=}O)(OH){-}O{-}P({=}O)(OH){-}O{-}P({=}O)(OH){-}}$$

Diphosphate (acid form) 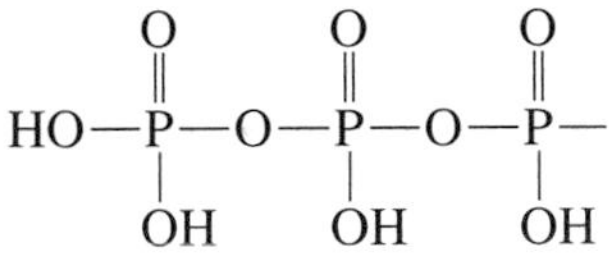 Triphosphate (acid form)

If the monophosphate group in AMP is replaced by a diphosphate group, the resulting compound is adenosine-5′-diphosphate (ADP); if replaced by a triphosphate group, the compound is adenosine-5′-triphosphate (ATP).

ATP is used by living organisms as an immediate source of energy. For example, when glucose is broken down in a cell, eventually to carbon dioxide and water, the energy obtained is stored in ATP molecules by coupling the breakdown of glucose with the reaction of ADP with phosphate ion, which absorbs the energy of the reaction.

$$\text{ADP} + \text{phosphate ion} + \text{energy} \longrightarrow \text{ATP} + H_2O$$

During biochemical processes that require energy, the reverse process occurs, in which ATP breaks down into ADP and phosphate ion, with the release of energy. Many biochemical processes use other nucleotides. Figure 19.11 shows the structure of a related compound, 3′-azido-3′-deoxythymidine (AZT), which is a drug used in the treatment of AIDS (acquired immune deficiency syndrome).

19.9 Deoxyribonucleic Acid and the Double Helix

OBJECTIVES

- Describe how complementary base pairing determines the structure of DNA.
- Explain why the only base pairing found in DNA involves adenine with thymine, or guanine with cytosine.
- Describe the process of DNA replication.

Deoxyribonucleic acid (DNA), as we mentioned, is *a polymer of deoxyribonucleotides, a polynucleotide*. Actually, DNA consists of two strands of polynucleotide chains twisted together (a *double helix*). Figure 19.12 shows a diagrammatic representation of a short length of one polymer chain, a polynucleotide with six nucleotides. The sequence of different bases (base 1, base 2, base 3, and so forth) on a DNA chain is a code representing the sequence of amino acids in a series of protein molecules. A DNA chain has many millions of such bases.

As mentioned, DNA forms a double-helix (double-coil) structure; see Figure 19.13. A length of DNA consists of two related polynucleotide chains, or strands, intertwined. The relationship between two strands of the double helix depends on **complementary base pairing,** which is *the pairing (through hydrogen bonding) of certain bases in DNA (or RNA) with other bases.* DNA has only four bases; the base adenine (A) pairs only with thymine (T), and guanine (G) pairs only with cytosine (C). We say that adenine and thymine are complementary bases; similarly, guanine and cytosine are complementary bases. ●

● James D. Watson and Francis H. C. Crick discovered the structure of DNA in 1953. For a first-hand account of their work, see Watson's book *The Double Helix* (Signet Books, New York, 1967).

Complementary base pairing occurs because several hydrogen bonds form between complementary bases. Figure 19.14 shows the hydrogen bonding between these base pairs. Note that there are two hydrogen bonds between adenine and thymine, and three hydrogen bonds between guanine and cytosine.

One polynucleotide strand in a length of DNA is the complement of the other strand. When a cell separates into two daughter cells during cell division, the DNA helix begins to unravel. As it does so, new, complementary deoxyribonucleotides attach to the original, unraveled strands. The result is the formation of two DNA double helixes where there had been only one. This process of forming a new DNA double helix from an original one is called *DNA replication* (Figure 19.15). During cell division, one strand of each double helix goes with each daughter cell. Thus, the double-helix structure provides a mechanism for the duplication of the genetic code of the cell. As we will see in the next section, it also provides a way for synthesizing cell proteins while protecting the original genetic code.

■ EXAMPLE 19.3 Determining the Base Sequence in Complementary Chains of Nucleic Acids

A segment of DNA chain has the base sequence A–T–G. What is the base sequence of the complementary chain?

Problem Analysis

Complementary chains of DNA form by the pairing of bases in the combinations of A with T, or of G with C. If the first chain has base A, the corresponding position in the second chain must have base T. Similarly, T corresponds to A, G corresponds to C, and C corresponds to G.

Solution

The first base is A, so the complementary chain will have T in this position. The second base is T, so the complementary chain will have A in the corresponding position. Finally, the third base is G, so the complementary chain will have C in this position. The base sequence in the complementary chain is thus T–A–C.

Exercise 19.3

A segment of DNA chain has the base sequence T–A–C–G. What is the base sequence of the complementary chain?

(Try Problems 19.45 and 19.46.)

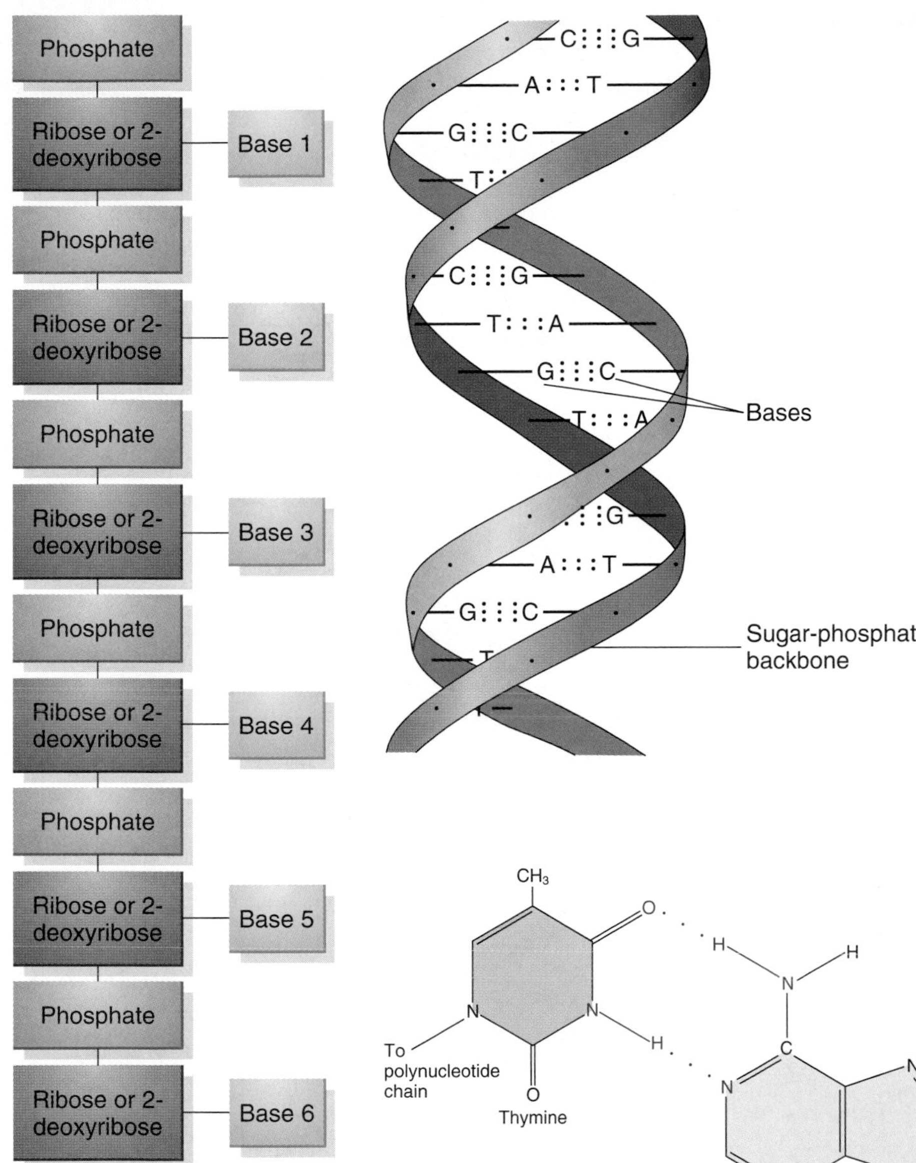

FIGURE 19.12

Short length of a polynucleotide chain. This diagrammatic representation of a polynucleotide chain (a small piece of a single chain from DNA or RNA) shows how the bases protrude from the main chain. The order of these bases is a code for the order of amino acids in a protein chain. In DNA, the sugar is 2-deoxyribose; in RNA the sugar is ribose.

FIGURE 19.13

The double-helix structure of DNA. DNA consists of two polynucleotide chains coiled together. The main chains are represented by colored ribbons, one purple, the other red. The DNA bases extend from the main chains (represented here by horizontal lines labeled by single letters denoting the bases), and each base pairs with only one other base (A pairs with T, and G pairs with C). Thus, the two chains in the double helix must be related to one another; one chain is the complement of the other.

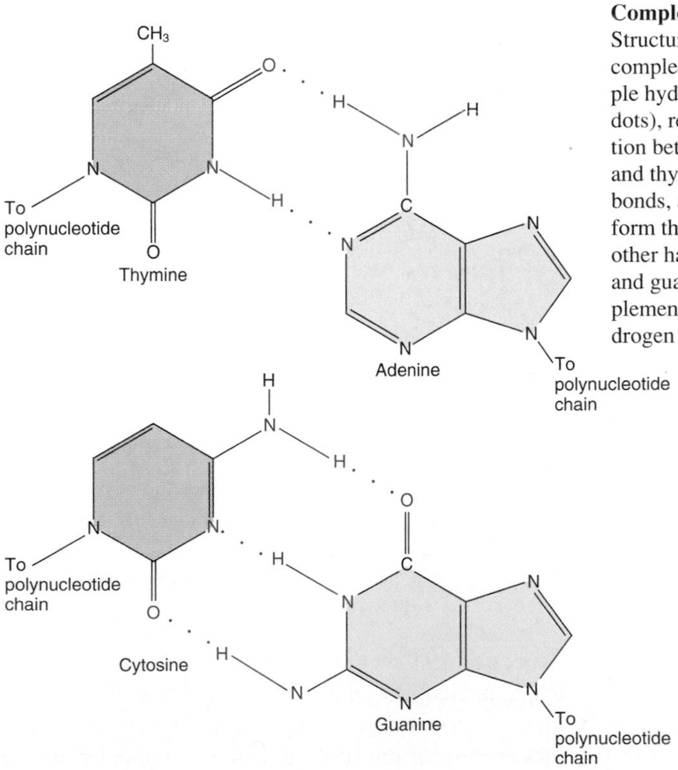

FIGURE 19.14

Complementary base pairing. Structural formulas show how two complementary bases form multiple hydrogen bonds (shown by dots), resulting in a strong attraction between the bases. Adenine and thymine form two hydrogen bonds, and guanine and cytosine form three hydrogen bonds. On the other hand, bases such as adenine and guanine, which are not complementary, do not form such hydrogen bonds.

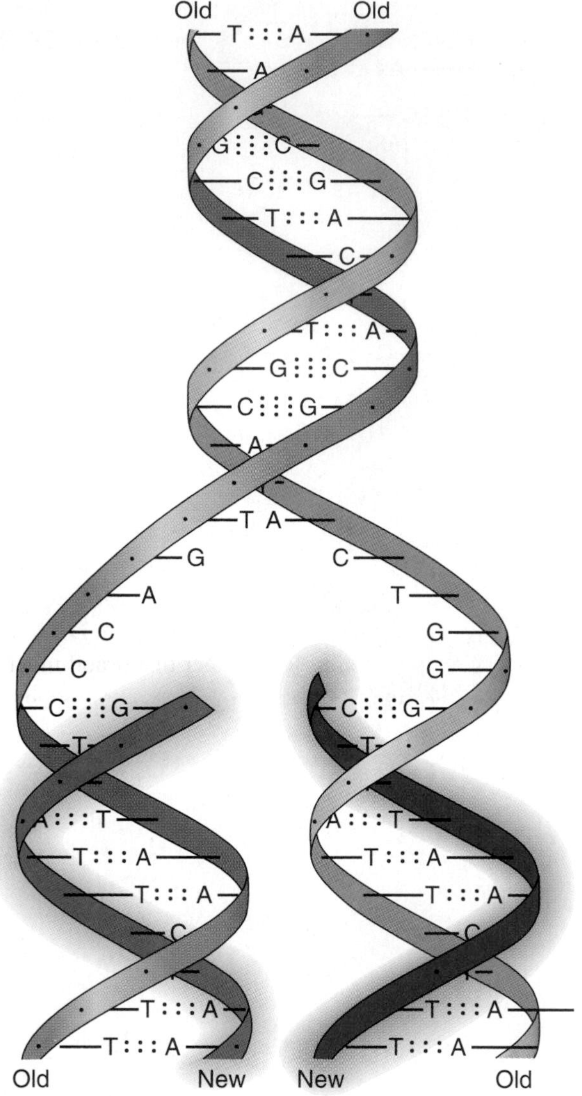

FIGURE 19.15
DNA replication. The first stage of replication involves the unraveling of the two chains of the DNA double helix. As the chains begin to unravel, the exposed bases on each chain attract complementary bases to themselves. The process is finished when there are two DNA double helixes where before there was only one.

19.10 Ribonucleic Acid and Protein Biosynthesis

OBJECTIVES

- List the different types of RNA and identify their functions.
- Describe the processes that occur during protein biosynthesis.
- Explain the significance of the base sequence of mRNA in protein biosynthesis.
- Use Table 19.3 to match base sequences in mRNA with amino acid sequences in proteins.

In Section 19.9 we said that DNA contains in its sequence of bases a code that represents the sequence of amino acids in a series of cell proteins. To construct a protein, a cell must decode (translate) a portion of this base sequence. Before the decoding can occur, this portion of the base sequence is copied from DNA to **ribonucleic acid (RNA),** *a* *polymer of ribonucleotides.* Ribonucleic acids are much smaller molecules than deoxyribonucleic acids, because they consist of far fewer monomer units. There are several types of RNA.

1. *Messenger RNA* (mRNA) is a complementary copy of a segment of DNA that contains the code for a particular protein.
2. *Ribosomal RNA* (rRNA), along with certain proteins, forms the ribosomes, the cell particles on which proteins are made. Recall that ribosomes can be either suspended in the cytosol or attached to the endoplasmic reticulum (see Figure 19.1). ●
3. *Transfer RNA* (tRNA) is the molecule that actually decodes the nucleic acid base sequence into the amino acid sequence in a protein.

● See Section 19.1 for a discussion of the parts of a eukaryotic cell.

Protein biosynthesis, *the building of protein molecules in a cell,* begins with the process of *transcription,* in which the base sequence information from a DNA chain in the cell nucleus is transferred to a messenger RNA molecule. Figure 19.16 shows the process by which transcription occurs: DNA unravels and a segment of its base sequence is copied into an RNA base sequence before the DNA can recombine into its usual double helix. After transcription is complete, the messenger RNA molecule has a sequence complementary to that of the corresponding DNA segment. Thymine in DNA pairs with adenine in RNA, and adenine in DNA pairs with uracil in RNA; guanine and cystosine always pair. (Remember that uracil in RNA has the role of thymine in DNA.) For example, the DNA segment ATG (denoting just the base sequence, with letter abbreviations for the bases) is transcribed to messenger RNA as UAC.

The messenger RNA molecule migrates from the cell nucleus to the cytoplasm, where ribosomes bind to it. The next step in protein biosynthesis is *translation,* in which the messenger RNA code is translated into an amino acid sequence. Messenger RNA provides the template on which the amino acids are assembled with the aid of a ribosome and transfer RNA. The flow of information is as follows:

A ribosome begins at one end of the messenger RNA, and transfer RNA brings the amino acids to the ribosome and attaches them, following the template. By the time the ribosome makes its way to the other end of the messenger RNA chain, a protein molecule has been synthesized; the protein is then released. In the meantime, other

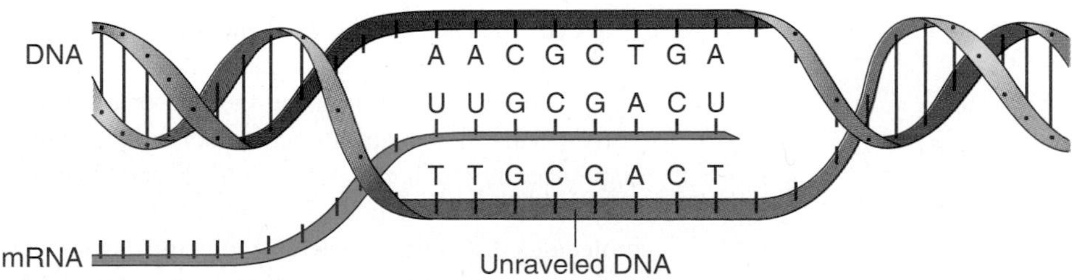

FIGURE 19.16

Formation of messenger RNA (transcription). A segment of a DNA double helix corresponding to a single protein unravels, and complementary ribonucleotides line up, attracted by hydrogen bonding to one of the DNA chains. The ribonucleotides are then linked to form messenger RNA. After being released from the DNA, the messenger RNA migrates from the cell nucleus into the cytoplasm, where protein biosynthesis occurs.

Code	Amino Acid
UUU or UUC	Phe
UUA, UUG, CUU, CUC, CUA, or CUG	Leu
AUU, AUC, or AUA	Ile
AUG	Met (or Begin codon)
GUU, GUC, GUA, or GUG	Val
UCU, UCC, UCA, UCG, AGU, or AGC	Ser
CCU, CCC, CCA, or CCG	Pro
ACU, ACC, ACA, or ACG	Thr
GCU, GCC, GCA, or GCG	Ala
UAU or UAC	Tyr
CAU or CAC	His
CAA or CAG	Gln
AAU or AAC	Asn
AAA or AAG	Lys
GAU or GAC	Asp
GAA or GAG	Glu
UGU or UGC	Cys
UGG	Trp
CGU, CGC, CGA, CGG, AGA, or AGG	Arg
GGU, GGC, GGA, or GGG	Gly
UAA, UAG, or UGA	Stop codon

TABLE 19.3
Genetic Codes for Amino Acids

ribosomes have begun the trip along the messenger RNA, so that eventually a single RNA template results in many protein molecules.

The base sequence on the messenger RNA molecule is decoded in groups of three, or triplets, reading from the beginning of the sequence to the end. Each triplet of bases, called a *codon* (pronounced "coh′-don"), corresponds to a particular amino acid. For example, the codon UCA (uracil–cytosine–adenine) corresponds to the amino acid serine. The amino acid assignments of the different codons are given in Table 19.3. There are 64 different combinations of RNA base triplets composed of the four bases uracil (U), cytosine (C), adenine (A), and guanine (G). Because there are only 20 amino acids, several codons sometimes correspond to the same amino acid. Three of the codons are "stop" codons, which indicate the end of the mRNA base sequence. The codon AUG corresponds to the amino acid methionine, unless it is at the beginning of the base sequence, in which case it functions as the "begin" code.

The RNA base sequence is actually translated or decoded by a set of transfer RNA molecules. Each transfer RNA molecule carries a particular amino acid attached to one end and an anticodon, consisting of a triplet of RNA bases, at the other end. The anticodon attaches to its complementary codon on the messenger RNA chain. Thus, each three-base codon on the messenger RNA chain links to the proper transfer RNA molecule having the matching (complementary) anticodon. That transfer RNA molecule in turn carries a particular amino acid. As the ribosome "reads" the messenger RNA from one end of the chain to the other, transfer RNA brings in specific amino acids, which are then linked to one another by peptide bonds, forming the protein (Figure 19.17).

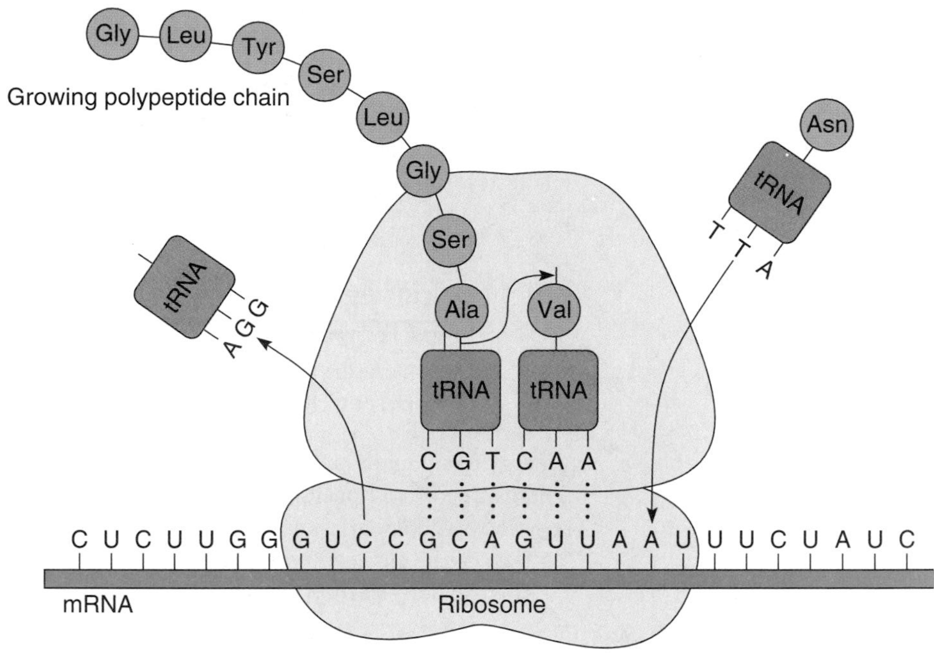

FIGURE 19.17
Synthesis of protein molecules (translation). Messenger RNA provides the code needed to determine the sequence of amino acids to be added to the protein. Transfer RNA brings the necessary amino acid to the mRNA in the ribosome. The tRNA bonds to the mRNA, and the amino acid is transferred to the growing protein chain. After transfer of the amino acid, the tRNA is released from the mRNA.

EXAMPLE 19.4 Writing the Base Sequence in mRNA That Corresponds to a Specific Amino Acid Sequence in a Protein

Write a possible base sequence for a segment of messenger RNA that codes for the amino acid sequence Phe–Cys–Trp.

Problem Analysis

Find the codon that supplies the genetic code for each amino acid in Table 19.3. (In most cases, there is more than one possible codon; select any one of these.) Write down the abbreviations for the nucleic acids in the order given and repeat this procedure for each amino acid in the sequence.

Solution

The code for the amino acid Phe is UUU or UUC. You may arbitrarily select the first of these. The code for Cys is UGU or UGC; again you may select the first of these. The only code for Trp is UGG. The base sequence is thus U–U–U–U–G–U–U–G–G. Other selections of codons would give other base sequences for the same amino acid sequence. The other possibilities are:

U–U–C–U–G–U–U–G–G

U–U–U–U–G–C–U–G–G

U–U–C–U–G–C–U–G–G

Exercise 19.4

Write a possible base sequence for a segment of messenger RNA that codes for the amino acid sequence Glu–Met–Tyr.

(Try Problems 19.47 and 19.48.)

LIPIDS

A **lipid** is *a biological substance belonging to one of several structurally different classes of substances that are soluble in organic solvents such as chloroform, $CHCl_3$.* As mentioned earlier, lipids include substances needed in a well-balanced diet: fats and cholesterol (as well as other steroids). In the remaining sections of this chapter, we look at two classes of lipids: triacylglycerols (triglycerides) and phospholipids.

19.11 Triacylglycerols

OBJECTIVES

- Describe the structure of fats.
- Describe how soaps remove oil or grease spots.

Energy compounds, such as glucose, that are not required by the body for its immediate energy needs are transformed to triacylglycerols, or fats. Triacylglycerols ("tri-ay′-sil-glis′-e-rols′") are also known as *triglycerides*. A **triacylglycerol** is *an ester of glycerol (a trihydroxy alcohol) and three fatty acids (long-chain carboxylic acids).* The acyl groups are derived from carboxylic acid groups.

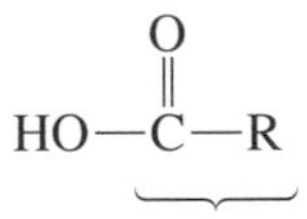

$$HO\!-\!\underbrace{\overset{\displaystyle O}{\overset{\|}{C}}\!-\!R}_{\text{Acyl group}}$$

Glycerol and a typical triacylglycerol have the following structures.

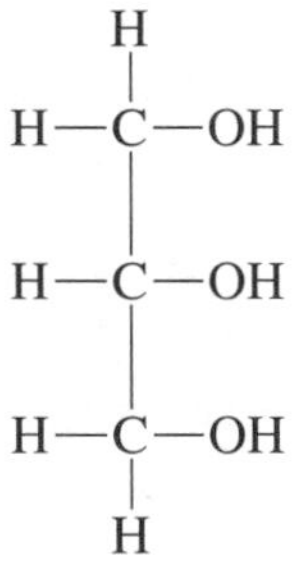

$$\begin{array}{c} H \\ | \\ H-C-OH \\ | \\ H-C-OH \\ | \\ H-C-OH \\ | \\ H \end{array} \qquad \begin{array}{l} \quad\ H \qquad\ \ O \\ \quad\ | \qquad\ \ \| \\ H-C-O-C(CH_2)_{16}CH_3 \\ \quad\ | \qquad\ \ O \\ \quad\ | \qquad\ \ \| \\ H-C-O-C(CH_2)_{14}CH_3 \\ \quad\ | \qquad\ \ O \\ \quad\ | \qquad\ \ \| \\ H-C-O-C(CH_2)_7CH{=}CH(CH_2)_7CH_3 \\ \quad\ | \\ \quad\ H \end{array}$$

Glycerol 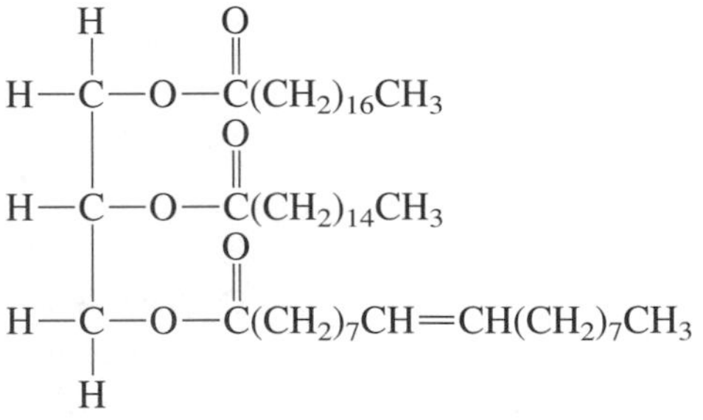 — A triacylglycerol (with stearic, palmitic, and oleic acids)

When a fat (a triacylglycerol) is heated with a solution of sodium hydroxide (lye), it is hydrolyzed, yielding glycerol and the sodium salts of the fatty acids. This is the well-known reaction for making soap. **Saponification** (from the Latin *sapon* meaning "soap") is *the general term for the base-catalyzed hydrolysis of an ester, especially a fat.*

A soap is therefore the sodium salt of a fatty acid. A fatty-acid ion, such as the stearate ion, has a carboxylate ion end, which is polar (and therefore attracted to water and to other polar molecules or ions), and a long hydrocarbon end, which is nonpolar (and not attracted to water but attracted to nonpolar substances).

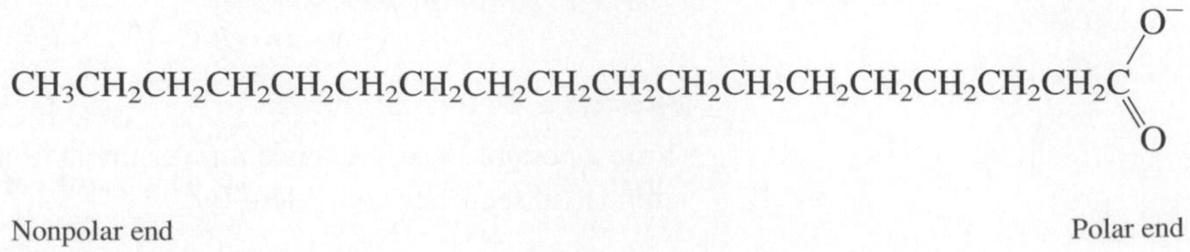

$$CH_3CH_2CH_2CH_2CH_2CH_2CH_2CH_2CH_2CH_2CH_2CH_2CH_2CH_2CH_2CH_2CH_2C\begin{array}{l} \diagup O^- \\ \\ \diagdown\!\diagdown O \end{array}$$

Nonpolar end — Polar end

Stearate ion

Long ions or molecules of this type tend to aggregate in water solution as *micelles* (see Figure 13.13). Such a micelle in a stearate (soap) solution consists of many stearate ions arranged with their hydrocarbon ends pointing inward and their carboxylate ion ends pointing outward toward the water solution. The cleansing action of soap occurs because oil and grease, which are hydrocarbons themselves, are absorbed into the hydrocarbon centers of the soap micelles and washed away (see Figure 13.14).

19.12 Phospholipids

OBJECTIVES

- Compare and contrast phospholipids and soap ions.
- Describe the structure of a cell membrane.

A **phospholipid** is *a lipid compound that contains a phosphate group*. Most of these are *phosphoacylglycerols*, which are similar to triacylglycerols but have one fatty acid of the triacylglycerol replaced by phosphoric acid; the phosphoric acid is itself bonded to another alcohol. In *lecithins* ("les′-i-thins"), phospholipids that are commonly present in cell membranes, this alcohol is choline, which also is a nitrogen-containing base ion. Free choline is present as the hydroxide salt.

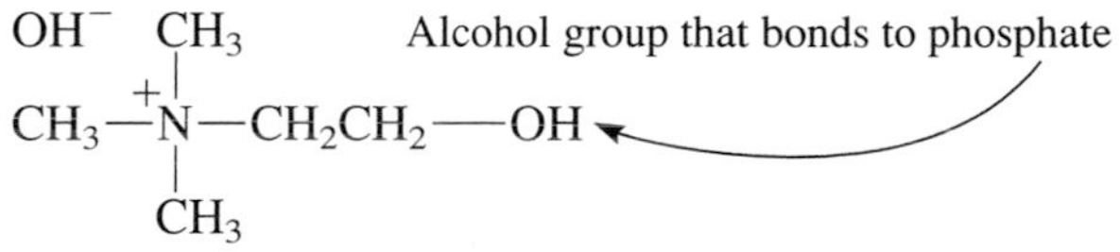

$$OH^- \quad CH_3-\overset{+}{N}(CH_3)_2-CH_2CH_2-OH$$

Alcohol group that bonds to phosphate

Choline hydroxide

● Lecithins are used in food processing as emulsifiers, agents that are used to suspend oil droplets in water. You make mayonnaise by beating a vegetable oil with an egg (which contains lecithins) into a vinegar solution.

Figure 19.18 shows the structure of a phospholipid (a lecithin). ● The phospholipid shown in this figure has a polar end (choline) and a nonpolar end (the fatty-acid groups), similar to a soap ion. Soap ions, as we noted earlier, form micelles in water solution. Phospholipid molecules spontaneously form similar structures in water. However, instead of micelles, the phospholipids form *lipid bilayers,* sheets that are two molecules thick in which the nonpolar ends of the phospholipid molecules are inside the sheet and

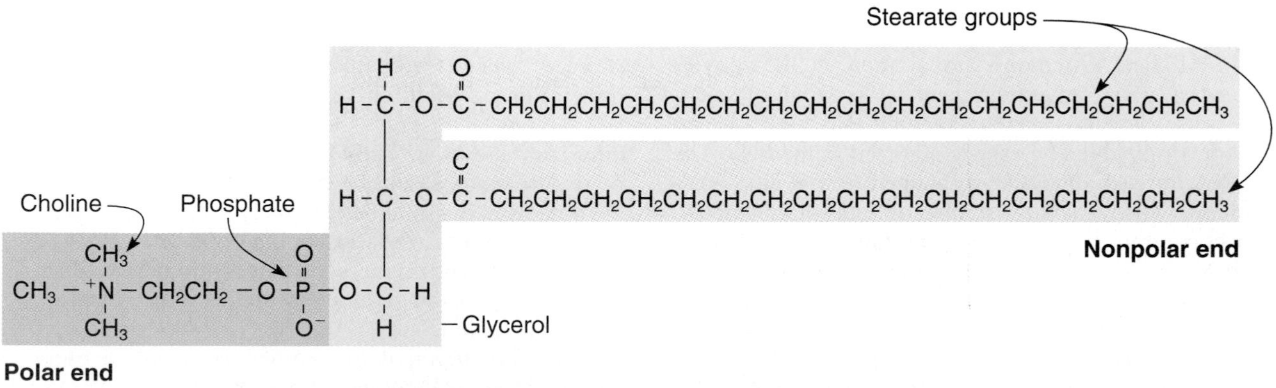

FIGURE 19.18

A phospholipid. This phospholipid (a lecithin) consists of glycerol to which is attached by ester linkages two stearic acid molecules (stearate groups) and a phosphoric acid molecule (the phosphate group); choline is then attached to the other end of the phosphoric acid. The resulting molecule has a polar end and a nonpolar end.

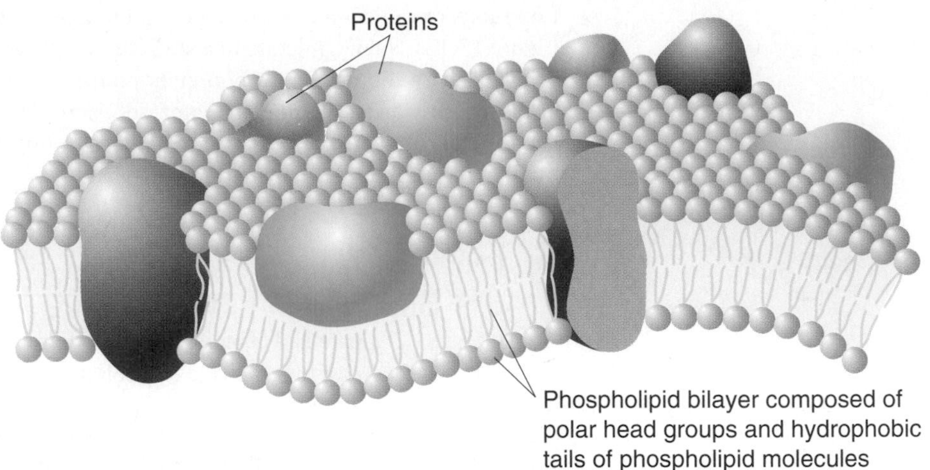

FIGURE 19.19

Cell membrane (diagram of a model). The membrane consists of a lipid bilayer with protein molecules embedded in it. One function of these membrane proteins is to transport substances into or out of the cell.

the polar ends are outside. They form such sheets because the two fatty-acid (hydrocarbon) groups are bulky and prevent the formation of more compact micelles.

Figure 19.19 shows a model of a cell membrane in which the main structure of the membrane is such a lipid bilayer; protein molecules are embedded in the lipid bilayer. The lipid bilayer presents an effective barrier to ions and molecules having polar groups, preventing their entrance to the interior of the cell. However, some of the protein molecules in the membrane do transport certain ions or polar molecules needed by the cell through the membrane.

Chemical Perspective

Dinosaurs, Human Origins, and Ancient Molecules

In Michael Crichton's novel (and in the movie) *Jurassic Park,* an entrepreneur has a group of scientists use ancient DNA to clone dinosaurs for his island park, from which he expects to profit immensely. The DNA for each dinosaur is obtained from a dinosaur's blood, which was the last meal of a mosquito that had become trapped in tree resin and preserved in the fossil amber. This is fiction, of course, but George O. Poinar and Raúl J. Cano (at California Polytechnic State University in San Luis Obispo) have reproduced part of the DNA of a wood beetle dated at 120 million years old (in the Jurassic period, or the age of the dinosaurs), which they obtained from a specimen of amber (Figure 19.20). They have also nearly reconstructed a gene (the complete DNA sequence for one protein) of a 40-million-year-old bacterium. Molecular biologists dispute the possibility of cloning dinosaurs, however. The DNA would be expected to be quite damaged, and because of the complexity of the organism, it would be extremely difficult to reconstruct the genome (the total of the organism's DNA). On the other hand, some investigators do believe that it would be possible to clone an ancient virus, or possibly a bacterium.

Molecular archaeology is a new science that uses ancient biological molecules to obtain information about past human populations, information that bears on the question of human origins. The requirements

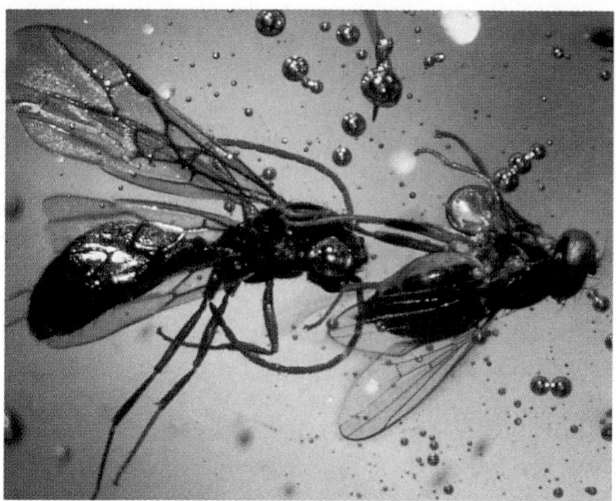

FIGURE 19.20

Insect trapped in amber. Insects trapped in tree resin are preserved in the fossil amber. Such insects are potential sources of ancient genetic material.

are much less demanding than trying to clone an organism. It is only necessary to reconstruct certain pieces of DNA, pieces whose base sequences vary distinctively among human populations and can be used to deduce human migration paths and population intermixings. The first DNA sequences from an ancient human being were obtained in 1985 from the skin of a 4400-year-old mummy. Since then, the biochemical techniques for dealing with minute quantities of DNA have improved tremendously.

The *polymerase chain reaction* (PCR) is one of the most exciting developments in the analysis of minute quantities of DNA. PCR is a laboratory technique used to amplify a quantity of DNA in a sample. Polymerase is an enzyme involved in the replication of DNA (the biological process of copying DNA). Using PCR, a biochemist heats a sample of DNA with polymerase. The heat separates a DNA double-helix molecule into its single strands, and the polymerase catalyzes the formation of two new pieces of double-stranded DNA. Each step, carried out automatically, doubles the quantity of DNA, and after 25 to 40 steps, the quantity of DNA in a sample is sufficient to analyze its base sequence.

In general, PCR works only with undamaged DNA. A specimen of ancient DNA consists of DNA that is mostly damaged over time by its environment. But, as long as a specimen contains a few intact DNA molecules, a molecular archaeologist can use PCR to amplify this undamaged DNA, from which the scientist can obtain an analyzable sample. Using this technique, DNA has been isolated from the brain tissue of an 8000-year-old human skull obtained from an ancient burial site in Florida.

The application of the techniques of biochemistry is transforming archaeology. By adding ancient molecules to human tools and pot shards, archaeologists have immensely broadened the evidence at their disposal. We should eventually have a much better picture of our human ancestry.

CHAPTER REVIEW

Key Words

amino acid *(p. 622)*
protein *(p. 622)*
peptide (amide) bond *(p. 624)*
primary structure *(p. 626)*
denaturation *(p. 628)*
carbohydrate *(p. 630)*
monosaccharide *(p. 630)*
oligosaccharide *(p. 633)*
polysaccharide *(p. 633)*
nucleotide *(p. 634)*
deoxyribonucleic acid (DNA) *(p. 635)*
complementary base pairing *(p. 636)*
ribonucleic acid (RNA) *(p. 638)*
protein biosynthesis *(p. 639)*
lipid *(p. 642)*
triacylglycerol *(p. 642)*
saponification *(p. 642)*
phospholipid *(p. 643)*

Summary

Biochemistry is the study of the chemical basis of living organisms. The fundamental unit of living organisms is the *cell*. Eukaryotic cells contain organelles such as the nucleus, which is the region that directs the workings of the cell. The region of the cell outside the nucleus is the cytoplasm. Protein synthesis occurs on ribosomes that are in the cytoplasm. Cells are composed mainly of four classes of molecules: proteins, carbohydrates, nucleic acids, and lipids.

Proteins are crucial molecules in a cell. The building blocks of proteins are *amino acids*. Twenty amino acids are commonly encountered in protein molecules. They are α-amino acids, consisting of an amino group, a carboxyl group, a hydrogen atom, and a side chain (R) attached to a carbon atom. The amino acids in a protein link together through *peptide bonds* to form a peptide chain. The *primary structure* of a protein refers to the protein's amino acid sequence. During human digestion, proteins are hydrolyzed back to their constituent amino acids. A protein with a specific primary structure spontaneously coils and folds into a characteristic three-dimensional shape in water solution. This shape is crucial to the function of the protein. When the protein is *denatured,* it loses its biological function.

Carbohydrates are either *monosaccharides* (polyhydroxy aldehydes or ketones), *oligosaccharides,* or *polysaccharides.* Glucose and 2-deoxyribose are examples of monosaccharides, or simple sugars. Sucrose is an oligosaccharide consisting of the simple sugars glucose and fructose linked together. Starch is a polysaccharide, a polymer of glucose.

Nucleic acids, the molecular carriers of genetic information, are polynucleotides. *Deoxyribonucleic acid (DNA)* consists of a chain of phosphate and ribose groups with nitrogen-containing bases attached to the ribose units. A DNA molecule has many millions of base units; the sequence of these bases is a code representing the sequence of amino acids in a series of protein molecules. The DNA actually consists of a double helix of polynucleotide chains. Each strand of the double helix is the *base-pair complement* of the other. During DNA replication, these strands unravel, and the deoxyribonucleotides attach to the unraveled strands to produce two double helixes where there had been only one.

Protein biosynthesis begins with transcription, in which the information for a protein is copied from a DNA molecule in the cell nucleus to a messenger RNA molecule. The messenger RNA, which migrates to the cytoplasm, becomes the template for the synthesis of the protein molecule. The synthesis of protein takes place in ribosomes that attach to the messenger RNA. Each triplet of bases (called a codon) on the messenger RNA chain codes for a particular amino acid.

Lipids, the last class of biological molecules that we discuss, are biological substances that dissolve in organic solvents. *Triacylglycerols,* or triglycerides (fats), are esters of glycerol and three fatty acids. These compounds serve as the body's energy storage. When a triacylglycerol is *saponified* (hydrolyzed in the presence of a base), it yields glycerol and the salts of fatty acids (soaps). In water, the fatty-acid ions of soap form micelles, which absorb oil particles. *Phospholipids* are lipids, similar to the acylglycerols, but containing a phosphate group to which an alcohol and a nitrogen-containing base ion are attached. This gives the phospholipid a polar and a nonpolar end, similar to soap ions. Phospholipid molecules spontaneously assemble in water solution to form lipid bilayers, which are similar to soap micelles, although they are flat structures rather than spherical. The phospholipids are important constituents of cell membranes.

Problem-Solving Skills

1. **Writing Structural Formulas for Peptides:** Given any two amino acids, write the structural formulas for the two dipeptides that could be formed from them (Example 19.1).
2. **Writing Structural Formulas for Aldoses and Ketoses:** Write the structural formulas for an aldose and a ketose for carbon chains containing several carbon atoms (Example 19.2).
3. **Determining the Base Sequence in Complementary Chains of Nucleic Acids:** Construct the base sequence of a chain of nucleic acids that is complementary to another nucleic acid chain (Example 19.3).
4. **Writing the Base Sequence in mRNA That Corresponds to a Specific Amino Acid Sequence in a Protein:** Given the amino acid sequence in a protein, construct the corresponding base sequence in mRNA (Example 19.4).

Questions to Test Your Reading

19.1 What are the different kinds of cells? Describe the difference between these two kinds of cells. Give an example of each kind.

19.2 Name some of the organelles in a eukaryotic cell. Describe the function of each of these organelles.

19.3 Name the four major classes of biological molecules. Describe the distinctive feature of each class.

19.4 What are some of the major functions of proteins?

19.5 Give some examples of carbohydrates.

19.6 What are the two kinds of nucleic acids? Describe their different functions in the cell.

19.7 Give examples of the two different kinds of lipids.

19.8 Draw the structural formula of an α-amino acid in which the side chain is —CH_2OH. What is the name of this amino acid (see Table 19.1)?

19.9 Draw the structure of a β-amino acid.

19.10 Draw the structural formula of the dipeptide Ala–Gly. How does it differ from the dipeptide Gly–Ala?

19.11 Show the different possible tripeptides that you could form from the three amino acids Phe, Ala, and Gly.

19.12 What do we mean by the primary structure of a protein?

19.13 Write the chemical equation, using structural formulas, showing the reaction between two cysteine molecules with an oxidizing agent (O) to form a disulfide bond.

19.14 What happens to proteins during digestion?

19.15 Certain protein foods, such as kidney beans, are said to be incomplete protein foods. When mixed with some other incomplete protein foods, the food mixture is said to provide a complete protein diet. Explain.

19.16 When the linear chain of a certain protein is placed in water, it spontaneously coils and folds into a globular shape. Explain the nature of the forces responsible.

19.17 Explain what happens when a protein is denatured. What happens to its normal biochemical function?

19.18 What is the origin of the term *carbohydrate?* Why is this term, taken literally, inappropriate for these compounds?

19.19 Draw structures of a ketose and an aldose.

19.20 Give examples of a monosaccharide, an oligosaccharide, and a polysaccharide.

19.21 What is the final compound obtained in the complete digestion of starch? What is the chemical name of the type of reaction that occurs in the digestion of starch?

19.22 What is the name of the substance that carries genetic information in a cell? What is the name of the sugar constituent in that substance?

19.23 What are the names of the four bases contained in DNA? What are the names of the bases contained in RNA?

19.24 What is the function of ATP in biological cells? What is the reaction that occurs when ATP functions this way?

19.25 Describe in a general way the structure of a single strand of DNA. How does this structure carry genetic information?

19.26 Describe how two strands of DNA give the double-helix structure of DNA. In your description, explain what forces hold the two strands together. Also, explain the relationship between these two strands.

19.27 Compare the relative sizes of DNA and messenger RNA molecules. What accounts for the difference in size?

19.28 Describe the flow of protein information during the biosynthesis of a protein molecule.

19.29 What is one possible codon for the amino acid phenylalanine? How many codons for this amino acid are there?

19.30 What is the difference in function between messenger RNA and transfer RNA?

19.31 Define the term *lipid.* Give two examples of lipids.

19.32 Describe the process of preparing a sodium soap from fat.

19.33 Describe how a soap cleans an oily fabric.

19.34 Explain why a phospholipid forms bilayers in water, whereas a triacylglycerol does not.

Practice Problems

Proteins (Sections 19.3 and 19.4)

19.35 Draw structural formulas of alanine, an α-amino acid, and an isomer that is a β-amino acid.

19.36 Draw structural formulas of valine, an α-amino acid, and an isomer that is a β-amino acid.

19.37 Write chemical equations, using structural formulas, to show the formation of a peptide bond between alanine and phenylalanine. There are two ways to do this; show both reactions and compare them.

19.38 Write chemical equations, using structural formulas, to show the formation of a peptide bond between valine and methionine. There are two ways to do this; show both reactions and compare them.

19.39 Using the conventional three-letter abbreviations for amino acids, draw all possible tetrapeptides of the amino acids glycine, alanine, methionine, and leucine. Note the amine and carboxyl ends of the peptides.

19.40 Using the conventional three-letter abbreviations for amino acids, draw all possible tetrapeptides of the amino acids tyrosine, glycine, phenylalanine, and leucine. Note the amine and carboxyl ends of the peptides.

Carbohydrates (Sections 19.6 and 19.7)

19.41 Draw the structural formula of two aldoses, one with five carbon atoms, the other with six carbon atoms.

19.42 Draw the structural formula of two ketoses, one with five carbon atoms, the other with six carbon atoms.

19.43 Lactose (milk sugar) is a disaccharide (two-unit oligosaccharide) composed of a glucose unit and a galactose unit (galactose is an isomer of glucose). Using molecular formulas and name labels, write the chemical equation for the hydrolysis of lactose.

19.44 Maltose (a sugar obtained in the partial digestion of starch) is a disaccharide (two-unit oligosaccharide) composed of two glucose units. Using molecular formulas and name labels, write the chemical equation for the hydrolysis of maltose.

Nucleic Acids (Sections 19.8, 19.9, and 19.10)

19.45 A segment of DNA chain has the base sequence A–T–G. What is the base sequence of the complementary chain?

19.46 A segment of DNA chain has the base sequence G–A–T–A. What is the base sequence of the complementary chain?

19.47 Write a possible base sequence for a segment of messenger RNA that codes for the amino acid sequence Ala–Gly–Phe.

19.48 Write a possible base sequence for a segment of messenger RNA that codes for the amino acid sequence Leu–Met–Gly.

19.49 Write the base sequence for the segment of DNA that corresponds to the segment of messenger RNA that you gave in Problem 19.47.

19.50 Write the base sequence for the segment of DNA that corresponds to the segment of messenger RNA that you gave in Problem 19.48.

Lipids (Sections 19.11 and 19.12)

19.51 Draw a structural formula of a triacylglycerol (fat) containing the following fatty acids.

$CH_3(CH_2)_{10}COOH$ (lauric acid)

$CH_3(CH_2)_{16}COOH$ (stearic acid)

$CH_3(CH_2)_7CH{=}CH(CH_2)_7COOH$ (oleic acid)

19.52 Draw a structural formula of a triacylglycerol (fat) containing the following fatty acids.

$CH_3(CH_2)_{16}COOH$ (stearic acid)

$CH_3(CH_2)_4CH{=}CHCH_2CH{=}CH(CH_2)_7COOH$ (linoleic acid)

$CH_3(CH_2)_{12}COOH$ (myristic acid)

19.53 Draw the structural formula of a lecithin (a phospholipid) using a selection of the fatty acids listed in Problem 19.51.

19.54 Draw the structural formula of a lecithin (a phospholipid) using a selection of the fatty acids listed in Problem 19.52.

Additional Problems

19.55 Write the structural formulas of a tripeptide having the amino acid units alanine, glycine, and valine. How many tripeptides are possible?

19.56 Write the structural formulas of a tripeptide having the amino acid units leucine, glycine, and methionine. How many tripeptides are possible?

19.57 A peptide has the following amino acid sequence:

His–Leu–Cys–Gly–Ser–His–Thr–Cys–Gln–Glu

Draw this sequence again, but bent backward with a disulfide bond (—S—S—) between two parts of the sequence.

19.58 A peptide has the following amino acid sequence:

Leu–Val–Cys–Gly–Glu–Arg–Glu–Asn–Tyr–Cys–Asn

Draw this sequence again, but bent backward with a disulfide bond (—S—S—) between two parts of the sequence.

19.59 The base sequence in a segment of DNA is copied to give a messenger RNA segment, which in turn is used as the template for the amino acid segment of a protein. The DNA base sequence is T–T–G–C–T–T–C–A–A. What is the messenger RNA base sequence? What is the sequence of amino acids in the protein segment?

19.60 The base sequence in a segment of DNA is copied to give a messenger RNA segment, which in turn is used as the template for the amino acid segment of a protein. The DNA base sequence is C–T–A–T–G–C–T–A–A. What is the messenger RNA base sequence? What is the sequence of amino acids in the protein segment?

Practice in Problem Analysis

For each problem, describe the thinking you would use (the problem analysis) before doing the actual solution, but do not solve the problem.

1. Write the structural formula of Gly–Ala–Ser.

2. Write a possible base sequence for the portion of DNA corresponding to the amino acid sequence in the previous problem.

PRACTICE EXAM

1. Which of the following is *not* a function of proteins?
(a) genetic code
(b) catalysis
(c) transport of biological molecules
(d) structural support
(e) movement of organism

2. Identify the α-amino acid.
(a) $CH_3CH_2C(=O)-NH_2$ (b) $CH_3C(NH_2)-C(=O)OH$
(c) $NH_2CH_2CH_2C(=O)OH$ (d) CH_3CH_2CN
(e) $HOCH_2CH_2C(=O)OH$

3. Which of the following is a dipeptide?
(a) $H_2C{=}C(H)-C(H){=}C(H)-CHCH_3$
(b) $H_2N-CH_2CH_2-C(=O)-NH_2$
(c) $H_2N-CH_2CH_2C(=O)OH$
(d) $H_2N-CH_2C(=O)-NHCH_2C(=O)OH$
(e) $H_2N-CH_2CH_2-NH_2$

4. Which of the following molecular structures is a ketose?
(a) $H-C(H)(OH)-C(H)(OH)-CH_3$ (b) $H-C(H)(OH)-C{=}O-C(H)(OH)-H$
(c) $H-C(=O)-C(H)(OH)-C(H)(OH)-H$ (d) $CH_3-C{=}O-CH_3$ (e) $H-C(=O)-CH_2-CH_3$

5. Which of the molecular structures given in Question 4 is an aldose?

6. A segment of DNA has the base sequence T–A–G. Which of the following is the complementary sequence?
(a) Ala (b) Ser (c) Glu (d) C–T–A
(e) A–T–C

7. DNA consists of two strands of
(a) polypeptides (b) polynucleotides
(c) polyethylenes (d) polysaccharides
(e) polylipids

8. The type of force holding the two strands in DNA is
(a) carbon bonding (b) hydrogen bonding
(c) ionic bonding (d) covalent bonding
(e) disulfide bonding

Mathematical Skills

Appendix A

The study of chemistry requires a firm grasp of a few basic mathematical skills. In this appendix, we review the use of scientific notation in arithmetic, simple algebraic operations, and basic components of graphs.

SCIENTIFIC NOTATION

Chapter 2 shows that any number can be written in scientific notation as $A \times 10^n$, where A is a number greater than or equal to 1 and less than 10, and the exponent n is a negative or positive whole number (integer). Sometimes you may want to add or subtract two numbers written in scientific notation, or multiply or divide them. Before doing this, you must express both numbers to the same power of 10. For example, let's do the arithmetic in

$$(9.42 \times 10^{-2}) + (7.6 \times 10^{-3})$$

You can shift the decimal point in either of these numbers so that both are expressed to the same power of 10. For example, to get both numbers expressed in terms of 10^{-2}, you shift the decimal point in 7.6×10^{-3} one place to the left and add 1 to the exponent: 0.76×10^{-2}. Note that adding 1 to a negative exponent makes that exponent less negative or closer to 0 ($-3 + 1 = -2$). These two operations are the same as dividing 7.6 by 10 and multiplying 10^{-3} by 10. The result of the operation $7.6/10 \times (10^{-3} \times 10)$ is 0.76×10^{-2}. *Since you have both multiplied by 10 and divided by 10, the value of the number has not changed.* Now you can add the two numbers.

$$\begin{array}{r} 9.42 \times 10^{-2} \\ +\ 0.76 \times 10^{-2} \\ \hline 10.18 \times 10^{-2} \end{array}$$

Since 10.18 is not between 1 and 10, you will need to shift the decimal point and adjust the exponent to express the final result properly.

$$10.18 \times 10^{-2} = 1.018 \times 10^{-1}$$

Here again, you shift the decimal point one place to the left and increase the exponent by 1.

Exercise 1

Complete the following arithmetic.

(a) $(5.2 \times 10^9) + (6.3 \times 10^9)$

(b) $(3.142 \times 10^6) + (2.8 \times 10^4)$

(c) $(3.142 \times 10^6) - (2.8 \times 10^4)$

(d) $(4.1 \times 10^8) + (4 \times 10^7)$

(e) $(4.1 \times 10^8) - (4 \times 10^7)$

Multiplication and division are handled very differently. To multiply two numbers expressed in scientific notation, you multiply the two powers of 10 by adding their exponents and then multiplying the leading factors. For example, let's find the answer to

$$(6.3 \times 10^2) \times (2.4 \times 10^5)$$

You obtain

$$(6.3 \times 2.4) \times 10^7 = 15.12 \times 10^7 = 1.512 \times 10^8$$

and you report 1.5×10^8 because the answer should have two significant figures.

Division is handled in a similar way. First, you move any power of 10 in the denominator to the numerator, changing the sign of the exponent. After multiplying the two powers of 10 by adding their exponents, you carry out the division of the leading factors. For example, suppose you are confronted with

$$\frac{6.4 \times 10^2}{2.0 \times 10^5}$$

You obtain

$$\frac{6.4 \times 10^2 \times 10^{-5}}{2.0} = \frac{6.4}{2.0} \times 10^{-3} = 3.2 \times 10^{-3}$$

Exercise 2

Do the following arithmetic and express your answers in scientific notation. Report your answer with the correct number of significant figures.

(a) $(5.4 \times 10^{-7}) \times (1.8 \times 10^8)$

(b) $\dfrac{3.0 \times 10^{-6}}{6.0 \times 10^3}$

(c) $\dfrac{3.0 \times 10^6}{6.0 \times 10^{-3}}$

(d) $(6.1 \times 10^9) \times (2.3 \times 10^2)$

(e) $\dfrac{2.5 \times 10^4}{5.0 \times 10^6}$

ALGEBRA: REARRANGING AN EQUATION

Sometimes you are given an equation that needs to be solved for an unknown (the thing you want to calculate). Suppose, for example, you want to calculate a Celsius temperature from a Kelvin temperature using the formula

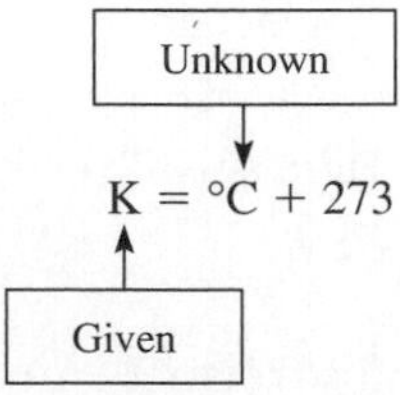

The basic rule is that whatever is on the left side of an equation must equal whatever is on the right side. Therefore, *whatever you do to one side, you must also do to the other side.* In this problem, you want °C on one side all by itself. You can isolate °C by subtracting 273 from the right side of the equation. This means that you must also subtract 273 from the left side, or

$$K - 273 = °C + 273 - 273$$

After canceling, $K - 273 = °C + \cancel{273} - \cancel{273}$

You obtain $K - 273 = °C$

This equation says that the Kelvin temperature minus 273 equals the Celsius temperature. You can also say that the Celsius temperature equals the Kelvin temperature minus 273. That is, you can turn the equation around so that it reads

$$°C = K - 273$$

Now you have the unknown all by itself on the left side, and you are in a position to begin the problem. For example, the Celsius temperature corresponding to 12 K is given by

$$°C = 12 - 273 = -261°C$$

Let's do another problem. The ideal gas law is

$$PV = nRT$$

This equation means that P times V is equal to n times R times T. If you are given values for P, n, R, and T, you can solve for the unknown V.

Unknown

↓

$$PV = nRT$$

↑ ↑↑↑

Given

In other words, you would like to get V on one side so that it is all by itself. You can get rid of P on the left side by dividing that side by P, but the rule tells you that you must also divide the right side by P, or

$$\frac{PV}{P} = \frac{nRT}{P}$$

Cancellation gives $\frac{\cancel{P}V}{\cancel{P}} = \frac{nRT}{P}$

You obtain $V = \frac{nRT}{P}$

Now you can substitute the known values of n, R, T, and P and calculate the unknown V.

Exercise 3

Solve the following equations for the indicated unknown.

(a) $V = bT$; solve for b

(b) $PV = nRT$; solve for T

(c) $\frac{V_1}{T_1} = \frac{V_2}{T_2}$; solve for T_2

(d) Density $= \frac{\text{mass}}{\text{volume}}$; solve for volume

(e) $°C = (5/9)(°F - 32)$; solve for °F

READING A GRAPH

Graphs are often used in newspapers and magazines to show how some feature of the economy, such as the gross national product, has changed from year to year. In chemistry, we use graphs in a similar way. For example, we can use a graph to show how the volume of a gas changes when we change its pressure systematically. The graph in Figure A.1 has a vertical axis corresponding to the volume: as you read up the axis, the volume increases. It also has a horizontal axis corresponding to pressure. The pressure increases as you read from left to right. The solid curved line gives you the volume of the gas between 0 and about 6 L as the pressure increases from 0 to about 6 atm. (Atm is an abbreviation for atmosphere, a unit of pressure.)

Let's see how you can find the volume and pressure of this gas sample at points A and B.

1. At point A, a horizontal line has been drawn to intersect the vertical axis at 4 L. A vertical line has also been drawn until it intersects the horizontal axis at 1 atm. Thus, point A occurs at volume = 4 L and pressure = 1 atm. Point A means that the volume of the gas is 4 L when the pressure is 1 atm.
2. At point B, you follow the same procedure and find that it occurs at volume = 2 L and pressure = 2 atm. Thus, the volume of the gas is 2 L when the pressure is 2 atm.

FIGURE A.1
Sample graph.

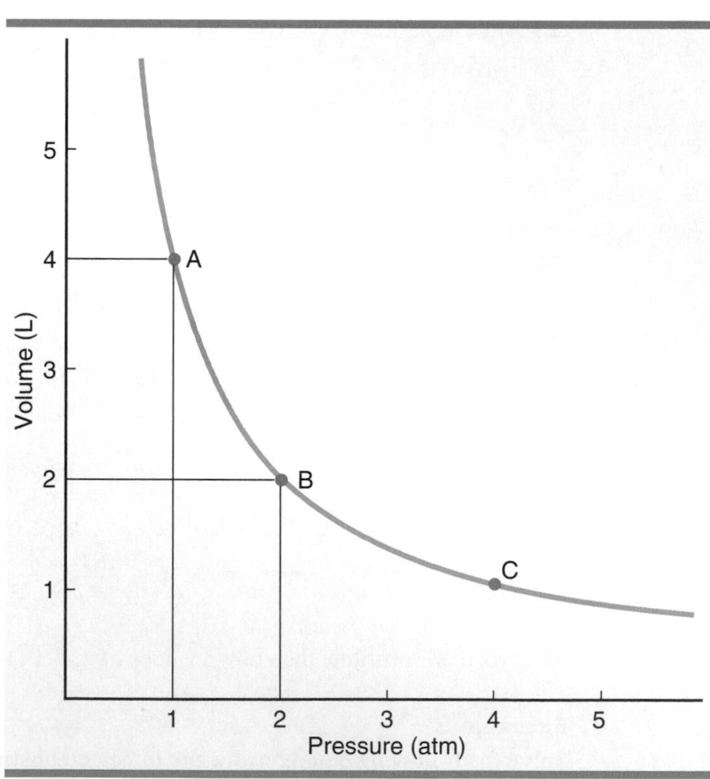

Exercise 4

Using the graph in Figure A.1, find the volume and pressure of the gas at point C.

A graph can have almost any shape—it can be a straight line or it can be an irregular, saw-toothed line. Whatever the shape, you read the graph using the same procedure that you used with the example just given.

Using a Scientific Calculator

Appendix B

Some of the chapters in this book contain calculations involving addition, subtraction, multiplication, and division. This appendix explains how to do these operations with a scientific calculator.

The instructions given here are general because one brand of calculator may differ from another in many different ways. Therefore, be sure to read the instructions that came with your calculator before you begin using this appendix. We will begin by showing you how to enter a number into a calculator.

ENTERING A NUMBER

Suppose you want to enter 3.84. You do it by keying in 3, a decimal point, and then 8 followed by 4, or

Operation	Display
1. Press [3], [.], [8], [4]	3.84

The key that changes a positive number to a negative number (or changes a negative number to a positive number) is

So, if you wanted to enter −3.84 instead of 3.84, you would do one extra step.

Operation	Display
1. Press [3], [.], [8], [4]	3.84
2. Press [+/−]	−3.84

SCIENTIFIC NOTATION

Scientific notation is a way to express very large or very small numbers quickly and easily. For example, Chapter 2 explains that

$$9{,}900{,}000{,}000{,}000{,}000{,}000{,}000 = 9.9 \times 10^{21}$$
$$0.000{,}000{,}000{,}13 = 1.3 \times 10^{-10}$$

You cannot enter the numbers on the left into your calculator because they have too many zeros for the calculator display to handle. However, once you put a very large or very small number in scientific notation (the numbers on the right), you can then enter it into your calculator. You first key in the number to the left of the multiplication sign and then the exponent along with its sign. For example, to enter 9.9×10^{21}, you key in the 9.9 and the 21 separately. The display on your calculator will look a little different than the number you are actually using. It will not show "× 10." Before you key in the exponent, you will need to press a special key that is often labeled

[EXP] or [EE]

We will now show how to enter 9.9×10^{21} and 1.3×10^{-10} into your calculator. To enter 9.9×10^{21}, you do the following.

Operation	Display
1. Press [9], [.], [9]	9.9
2. Press [EXP] (or [EE])	9.9 00
3. Press [2], [1]	9.9 21

This display represents 9.9×10^{21}

To enter 1.3×10^{-10}, you do the following.

Operation	Display
1. Press [1], [.], [3]	1.3
2. Press [EXP] (or [EE])	1.3 00
3. Press [1], [0]	1.3 10
4. Press [+/−]	1.3 −10

This display represents 1.3×10^{-10}

UNITS, SIGNIFICANT FIGURES, AND ROUNDING

The following sections discuss how to do a few mathematical operations with a calculator. However, bear in mind that calculators cannot do three things.

1. Calculators do not take *units* into account; you will need to add whatever units are appropriate to the answer that you get.
2. Calculators do not take *significant figures* into account. You will need to consider

each answer in terms of the proper number of significant figures; following the rules on pages 27 and 29.

3. If *rounding* is necessary, your calculator will not do it for you. The rules for rounding are on page 31.

ADDITION AND SUBTRACTION

Suppose you want to convert 78°C to a Kelvin temperature. You use the formula K = °C + 273. To obtain the temperature in kelvins, you must add 78 and 273, so you could say

78 plus 273 equals ?

You key in 78, press the key for "plus,"

then key in 273, and then press the key for "equals,"

These steps will appear on your calculator as

Operation	Display
1. Press [7], and [8]	78
2. Press [+]	78
3. Press [2], [7], [3]	273
4. Press [=]	351

After you add the proper unit, you see that the temperature is 351 K.

If you were asked to convert a Kelvin temperature such as 90 K to the corresponding Celsius temperature, you would use the formula °C = K − 273. Thus, you would need to subtract 273 from 90., and you can set this problem up by saying

90. minus 273 equals

The key for "minus" is

The operations on your calculator become

Operation	Display
1. Press [9], and [0]	90
2. Press [−]	90
3. Press [2], [7], [3]	273
4. Press [=]	−183

After you add units, your answer is −183°C.

Appendix A explains that whenever you add or subtract numbers with exponents, the exponents of the numbers must be the same. However, this holds true only if you are not using a calculator. When you use a calculator for addition or subtraction, you do not need to have identical exponents. For example, consider adding 2.34×10^{-2} and 0.8918.

Operation	Display	
1. Press [2], [.], [3], [4]	2.34	
2. Press [EXP] (or [EE])	2.34	00
3. Press [2]	2.34	2
4. Press [+/−]	2.34	−2
5. Press [+]	2.34	−2
6. Press [.], [8], [9], [1], [8]	0.8918	
7. Press [=]	0.9152	

The answer is 0.9152, or you could change it to 9.152×10^{-1} if you wished to use scientific notation.

MULTIPLICATION AND DIVISION

The keys for multiplication and division look like

[×] and [÷]

and they are used in a manner similar to the addition and subtraction keys.

If you want to convert 5.614 inches to centimeters, you need to set up the problem using a conversion factor that will change inches to centimeters. There are 2.54 cm to each inch. Therefore,

$$5.614\ \cancel{\text{in}} \times \frac{2.54\ \text{cm}}{1\ \cancel{\text{in}}}$$

You need to multiply 5.614 by 2.54 to get the answer, so you say

5.614 times 2.54 equals ?

and the operations on your calculator are the following.

Operation	Display
1. Press [5], [.], [6], [1], [4]	5.614
2. Press [×]	5.614
3. Press [2], [.], [5], [4]	2.54
4. Press [=]	14.259956

After rounding to four significant figures (2.54 is an exact number), you obtain 14.26. With units, the answer is 14.26 cm.

Turning to division, suppose you know that the mass of a sample is 5.5 g and its volume is 7.7 cm^3, and you want to calculate the density.

$$\text{Density} = \frac{\text{mass}}{\text{volume}} = \frac{5.5\ \text{g}}{7.7\ \text{cm}^3}$$

You say 5.5 divided by 7.7 equals ?

and the operations on your calculator are

Operation	Display
1. Press [5], [.], [5]	5.5
2. Press [÷]	5.5
3. Press [7], [.], [7]	7.7
4. Press [=]	0.714285714

When you round to two significant figures and add units, the answer becomes 0.71 g/cm^3.

MULTIPLE OPERATIONS

Sometimes you may be confronted with problems that require a series of steps. For example, let's calculate the answer for

$$\frac{2.3 \times 4.7}{6.1 \times 0.52}$$

Both multiplication and division are required, and you will need to divide by both 6.1 and 0.52. In words, you would say

2.3 multiplied by 4.7, divided by 6.1, divided by 0.52 equals ?

Following the words faithfully, the operations are

Operation	Display
1. Press [2], [.], [3]	2.3
2. Press [×]	2.3
3. Press [4], [.], [7]	4.7
4. Press [÷]	10.81
5. Press [6], [.], [1]	6.1
6. Press [÷]	1.772131148
7. Press [.], [5], [2]	0.52
8. Press [=]	3.407944515

After rounding to two significant figures, the answer is 3.4.

When a multiple operation involves both addition (or subtraction) and multiplication (or division), it may be necessary to carry out the operations in a specific order. For example, the following two calculations will give different answers (4 and 6.5).

$$\frac{5 + 3}{2} \qquad 5 + \frac{3}{2}$$

The order is determined by the following rules.

1. First carry out any operations that are enclosed inside parentheses. A series of addition or subtraction operations located together above or below a division line are considered to be enclosed in parentheses.
2. Carry out multiplication or division operations in individual terms before addition or subtraction operations.

Exercise 1

Complete the following arithmetic with your calculator.

(a) $(5.67 \times 10^{-3}) + (2.1 \times 10^{-4})$

(b) $(3.0 \times 10^{7}) - (3 \times 10^{6})$

(c) $(4.3 \times 10^{-2}) \times (9 \times 10^{10})$

(d) $\dfrac{7.2 \times 10^{-5}}{3.6 \times 10^{-7}}$

(e) $\dfrac{(4.4 \times 10^{3}) \times (1.7 \times 10^{-2})}{(5.1 \times 10^{-3}) \times (4.4 \times 10^{-5})}$

(f) $\dfrac{(2.5 \times 10^{-2}) + (1.0 \times 10^{-1})}{(5.5 \times 10^{-2}) - (5.0 \times 10^{-3})}$

(g) $(1.1 \times 10^{2}) + (2.0 \times 10^{3}) \times (1.5 \times 10^{-2}) - \dfrac{3.6 \times 10^{9}}{2.4 \times 10^{7}}$

LOGARITHMS

Using your calculator to obtain logarithms is covered in Chapter 15 (p. 488).

Answers to Exercises

Note: Your answers may differ from those given here in the last significant figure.

Chapter 2 **2.1** (a) 6.9252×10^{-3} (b) 6.300×10^{-2} (c) 8.0200×10^{2} (d) 9.002×10^{3} (e) 7.07×10^{-1} **2.2** (a) 0.0004834 (b) 6250. (c) 4.89 **2.3** (a) 4 sig. figs. (b) 4 sig. figs. (c) 4 sig. figs. (d) at least 2 sig. figs. (e) 5 sig. figs. **2.4** (a) 9.71×10^{1} (b) 2.83×10^{1} (c) 3.08×10^{1} (d) 3.2 **2.5** (a) 126 (b) 5.46 **2.6** 23°C **2.7** (a) 95°F (b) 39°C **2.8** (a) 2.58×10^{-3} kg (b) 5.54×10^{-1} m (c) 7.11×10^{3} mL **2.9** (a) 2.45×10^{-6} ms (b) 7.38×10^{6} mg (c) 5.02×10^{-3} mL (d) 8.97×10^{-4} km (e) 6.11×10^{3} μs **2.10** 3.10 g/cm^3 or 3.10 g/mL **2.11** 13 g **2.12** 1.46×10^{3} cm^3

Chapter 3 **3.1** solid **3.2** Its physical properties are that it is a colorless liquid that freezes at −114°C and boils at 78°C. Its chemical property is that it reacts with oxygen to produce carbon dioxide and water. **3.3** 11.6 cal **3.4** 4.0×10^{2} J **3.5** 1.33×10^{-1} J/(g·°C)

Chapter 4 **4.1** $^{35}_{17}Cl$ **4.2** 9 protons, 10 neutrons **4.3** 5 protons, 5 electrons, 6 neutrons. The protons and neutrons are located in the nucleus; the electrons are located around and far away from the nucleus. **4.4** 63.55 amu **4.5** magnesium (Mg) **4.6** nitrogen (N) and phosphorus (P) **4.7** Molecular. Because it is a gas at room temperature, its melting point is below room temperature, which is expected of a molecular substance. **4.8** P_4O_{10}

Chapter 5 **5.1** +2 **5.2** (a) LiF (b) Li_2S (c) Li_3N **5.3** (a) potassium sulfide (b) lithium nitride (c) aluminum oxide **5.4** BaS, LiCl **5.5** Tin(II) oxide, stannous oxide, tin(IV) oxide, stannic oxide **5.6** (a) K_2SO_3, potassium sulfite (b) $Ba(ClO)_2$, barium hypochlorite (c) NH_4CN, ammonium cyanide **5.7** $Ba_3(PO_4)_2$ **5.8** (a) P_2O_3, diphosphorus trioxide (b) ClF_3, chlorine trifluoride (c) CCl_4, carbon tetrachloride **5.9** hydrobromic acid **5.10** H_2SO_3, sulfurous acid

Chapter 6 **6.1** (a) Two molecules of hydrogen gas and one molecule of oxygen gas react in the presence of a platinum catalyst to form two molecules of liquid water. (b) Two formula units of aqueous sodium iodide react with one molecule of chlorine gas to form two formula units of aqueous sodium chloride and one molecule of aqueous iodine. (c) One molecule of carbon monoxide gas reacts with three molecules of hydrogen gas to form one molecule of methane gas and one molecule of water gas. (d) Two molecules of aqueous hydrogen peroxide react when heated to form two molecules of liquid water and one molecule of oxygen gas. **6.2** (a) $NH_4NO_3(s) \xrightarrow{\Delta} N_2O(g) + 2H_2O(g)$ (b) $2N_2O_5(g) \longrightarrow 4NO_2(g) + O_2(g)$ (c) $Ca(s) + 2H_2O(l) \longrightarrow Ca(OH)_2(aq) + H_2(g)$ (d) $2As_2S_3(s) + 9O_2(g) \xrightarrow{\Delta} 2As_2O_3(s) + 6SO_2(g)$ **6.3** $4Fe(s) + 3O_2(g) \longrightarrow 2Fe_2O_3(s)$ **6.4** (a) decomposition reaction (b) combination reaction (c) neither a combination reaction nor a decomposition reaction **6.5** No. The equation $2P(s) + 5Cl_2(g) \longrightarrow 2PCl_5(s)$ is a combination reaction because it can be represented by $A + B \longrightarrow AB$. **6.6** Yes. $Mg(s) + 2AgNO_3(aq) \longrightarrow 2Ag(s) + Mg(NO_3)_2(aq)$ **6.7** Yes. $(NH_4)_2CO_3(aq) + CaBr_2(aq) \longrightarrow CaCO_3(s) + 2NH_4Br(aq)$. **6.8** $CaCO_3(s) + 2HNO_3(aq) \longrightarrow Ca(NO_3)_2(aq) + CO_2(g) + H_2O(l)$ **6.9** $Ca(OH)_2(aq) + H_2SO_4(aq) \longrightarrow CaSO_4(s) + 2H_2O(l)$, calcium sulfate **6.10** (a) $C_3H_8(g) + 5O_2(g) \longrightarrow 3CO_2(g) + 4H_2O(g)$, combustion reaction (b) $Mg(s) + 2HBr(aq) \longrightarrow MgBr_2(aq) + H_2(g)$, single replacement reaction (c) $CS_2(l) + 3O_2(g) \longrightarrow CO_2(g) + 2SO_2(g)$, combustion reaction

Chapter 7 **7.1** 102.04 amu **7.2** 58.33 amu, 208.22 amu **7.3** 1.35×10^{23} $C_6H_{12}O_6$ molecules; 8.13×10^{23} C atoms, 1.63×10^{24} H atoms, 8.13×10^{23} O atoms **7.4** 253.80 g, 226.0 g **7.5** 163.94 g **7.6** 0.193 mol **7.7** 0.0501 g **7.9** NH_2CONH_2 **7.10** CH_4O **7.11** $C_6H_{12}O_2$ **7.12** C_6H_6

Chapter 8 **8.1** 7.76 mol **8.2** 25 mol **8.3** 0.247 mol **8.4** 7.49 g **8.5** 33.6 g **8.6** 49.0 g **8.7** (a) no limiting reactant (b) H_2 (c) N_2 **8.8** 0.50 mol $CaCl_2$, 0.5 mol $CaCO_3$ (remaining) **8.9** $NaHCO_3$ is the limiting reactant, 0.524 g CO_2 **8.10** Theoretical yield: 44.69 g H_2O, percentage yield: 0.72%

Chapter 9 **9.1** 9 **9.2** n = 5 shell and *d*-type subshell **9.3** 3 subshells: 3*s*, 3*p*, and 3*d*. 1, 3, and 5, respectively. **9.4** 2 electrons in the 1*s* subshell, 2 electrons in the 2*s* subshell, 6 electrons in the 2*p* subshell, 2 electrons in the 3*s* subshell, and 5 electrons in the 3*p* subshell. **9.5** $1s^22s^22p^63s^23p^5$ **9.6** $1s^22s^22p^63s^23p^63d^{10}4s^24p^2$ **9.7** $4s^24p^2$ **9.8** F **9.9** F^- **9.10** O

Chapter 10

10.1 (a) $\cdot\dot{Al}\cdot + 3\,\cdot\ddot{F}: \longrightarrow Al^{3+} + 3\,[:\ddot{F}:]^-$

(b) $\dot{Ca}\cdot + \cdot\ddot{O}: \longrightarrow Ca^{2+} + [:\ddot{O}:]^{2-}$

10.2 (a) H : Ö : H or H—O—H

(b) :O: above Cl in :O : Cl : O : H or :O: bonded to Cl in :O—Cl—O—H

10.3 (a) : N : : : N : or : N≡N :

(b) S : : C : : S or S=C=S

10.4 $[$:F : B : F: with :F: above and below B$]^-$ or $[$:F—B—F: with :F: bonded above and below B$]^-$

10.5 (a) tetrahedral (b) trigonal pyramidal

Chapter 11

11.1 541 mmHg **11.2** 9.52 L **11.3** 520. mmHg **11.4** 447 mL **11.5** 373°C **11.6** 44 g/mol **11.7** 70.8 g/mol **11.8** 3.89 atm **11.9** 0.0164 mol, 0.721 g **11.10** $P_{\text{Carbon dioxide}} = 0.245$ atm, $P_{\text{Oxygen}} = 0.489$ atm, $P_{\text{Total}} = 0.734$ atm **11.11** 3.76 L

Chapter 12

12.1 The formation of frost is deposition. **12.2** 1.1×10^{-2} kJ **12.3** 3.0×10^{-1} kJ **12.4** 1.8 kJ **12.5** 8.16×10^{-3} mol **12.6** Hydrogen (H_2) and bromine (Br_2): dispersion forces only; hydrogen bromide (HBr): dispersion forces and dipole–dipole forces. **12.7** Vapor pressure increases as molecular weight decreases. In order of increasing vapor pressure, $C_3H_8 < C_2H_6 < CH_4$. **12.8** Both have only dispersion forces, so SiH_4 will have the higher boiling point because it has the larger molecular weight. **12.9** Zinc (Zn) is a metallic solid; sodium bromide (NaBr) is an ionic solid; and methane (CH_4) is a molecular solid. **12.10** Magnesium chloride ($MgCl_2$) is an ionic solid; methanol (CH_3OH) is a molecular solid; and argon (Ar) is an atomic solid. In order of increasing melting point, Ar < CH_3OH < $MgCl_2$.

Chapter 13

13.1 6.5% **13.2** 6.1% **13.3** 2.1 g potassium dichromate, 40.4 g water **13.4** 0.167 *M* **13.5** 6.37 g **13.6** 0.427 g **13.7** 0.427 g **13.8** 40. mL **13.9** 1.0×10^2 mL **13.10** 2.20 m **13.11** −1.02°C **13.12** 100.28°C

Chapter 14

14.1 $4NH_3(g) + 3O_2(g) \rightleftharpoons 2N_2(g) + 6H_2O(g)$; $K = \dfrac{[N_2]^2[H_2O]^6}{[NH_3]^4[O_2]^3}$ **14.2** $CH_4(g) + 2H_2S(g) \rightleftharpoons CS_2(g) + 4H_2(g)$; $K = \dfrac{[CS_2][H_2]^4}{[CH_4][H_2S]^2} = \dfrac{(1.10)(1.68)^4}{(1.10)(1.49)^2} = 3.5\underline{8}8 = 3.59$ **14.3** $K = \dfrac{[CH_3OH]}{[CO][H_2]^2}$. Significant concentrations of reactants and products will be present at equilibrium because $K = 2.3$. **14.4** $[CH_4] = 0.059\ M$. **14.5** $Zn(OH)_2$ is more soluble. **14.6** The reaction will shift from left to right (away from H_2). **14.7** The concentrations of N_2 and H_2 will decrease and the concentration of NH_3 will increase. **14.8** $[CO_2]$ will increase.

Chapter 15

15.1 Total ionic equation: $Ba^{2+}(aq) + 2OH^-(aq) + 2H^+(aq) + 2Cl^-(aq) \rightarrow Ba^{2+}(aq) + 2Cl^-(aq) + 2H_2O(l)$. Net ionic equation: $2OH^-(aq) + 2H^+(aq) \rightarrow 2H_2O(l)$. **15.2** H_2O.
15.3 $H_2CO_3(aq) + CN^-(aq) \rightarrow HCN(aq) + HCO_3^-(aq)$;
Acid Base Acid Base
H_2CO_3 and HCO_3^- are a conjugate acid–base pair, as are CN^- and HCN. **15.4** $1.2 \times 10^{-10}\ M$ **15.5** Acidic **15.6** pH = 10.66 **15.7** 7.3, basic **15.8** $6.9 \times 10^{-4}\ M$, acidic

Chapter 16

16.1 Ni^{2+} is oxidized and is the reducing agent. Cd^{2+} is reduced and is the oxidizing agent. **16.2** yes **16.3** +6, +6 **16.4** +2 **16.5** not an oxidation–reduction reaction **16.6** $16H^+(aq) + 16NO_3^-(aq) + 24H_2S(aq) \longrightarrow 16NO(g) + 32H_2O(l) + 3S_8(s)$ **16.7** $2MnO_4^-(aq) + 6I^-(aq) + 4H_2O(l) \longrightarrow 2MnO_2(s) + 3I_2(aq) + 8OH^-(aq)$ **16.8** Pb is oxidized and Cu^{2+} is reduced. The anode is Pb; the cathode is Cu. Electrons flow from Pb (the anode) to Cu (the cathode). $Pb \longrightarrow Pb^{2+} + 2e^-$; $Cu^{2+} + 2e^- \longrightarrow Cu$; $Pb(s) + Cu^{2+}(aq) \longrightarrow Pb^{2+}(aq) + Cu(s)$.

Chapter 17

17.1 ${}^{40}_{19}K \longrightarrow {}^{40}_{20}Ca + {}^{0}_{-1}e$ **17.2** ${}^{239}_{94}Pu \longrightarrow {}^{235}_{92}U + {}^{4}_{2}He$ **17.3** ${}^{27}_{13}Al$ **17.4** 2,000,000 atoms **17.5** 9400 B.C.

Chapter 18

18.1 (a) 2,3-dimethylpentane (b) 2,3-dimethylbutane

18.2 $H_3C-CH_2-C(CH_3)H-CH_2-CH_3$ (H—C—C—C—C—C—H, with H atoms on each carbon and CH_3 on the central carbon)

18.3 (a) $C_3H_8(g) + 5O_2(g) \longrightarrow 3CO_2(g) + 4H_2O(g)$
(b) $CH_3CH_2CH_2CH_3 + Cl_2 \longrightarrow CH_3CH_2CH_2CH_2Cl + HCl$,
$CH_3CH_2CH_2CH_3 + Cl_2 \longrightarrow CH_3CH_2CHClCH_3 + HCl$
18.4 (a) 2-hexyne (b) 2-methyl-3-heptene

18.5

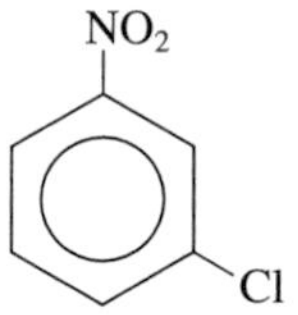

(a) *m*-chloronitrobenzene 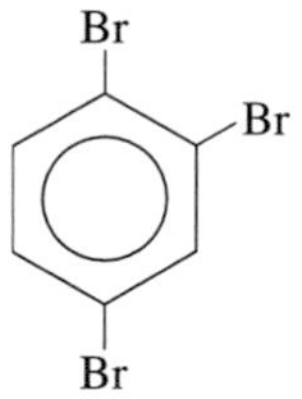 (b) 1,2,4-tribromobenzene

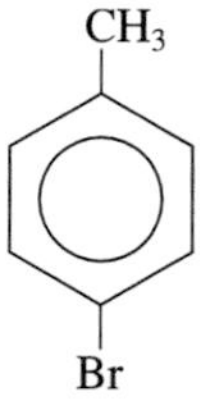

(c) *p*-bromotoluene 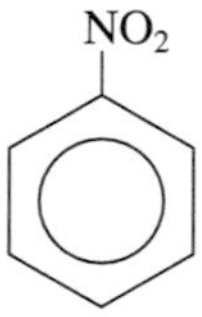 (d) nitrobenzene

18.6 3,5-dimethyl-3-hexanol **18.7** (a) 4-methylhexanal
(b) 2-pentanone

Chapter 19

19.1

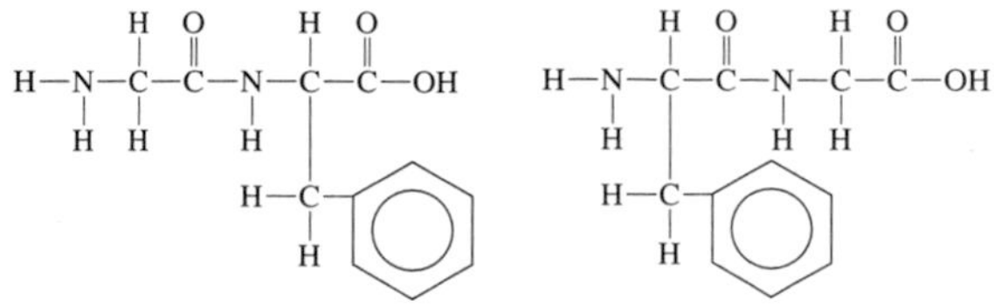

19.2 aldose ketose

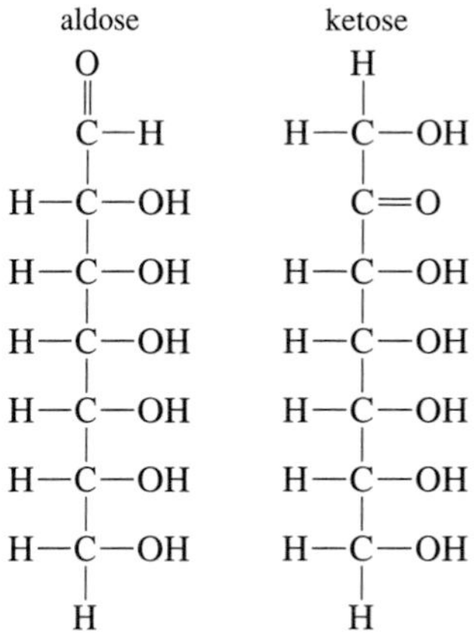

19.3 A–T–G–C **19.4** G–A–A–A–U–G–U–A–C,
G–A–A–A–U–G–U–A–U, G–A–G–A–U–G–U–A–C,
G–A–G–A–U–G–U–A–U

Answers to Odd-Numbered Problems

Chapter 2 **2.21** (a) 3.0402×10^4 (b) 6.58×10^{-3} (c) 2.0900×10^2 (d) 3.5×10^{-6} (e) 1.046×10^3 **2.23** (a) 890.45 (b) 0.005126 (c) 6.13 (d) 0.00007302 (e) 20,375.9 **2.25** (a) 26.8 (b) 0.004598 (c) 0.00783 (d) 8,536. (e) 3,189.1 **2.27** (a) 3.69×10^{-3} m (b) 3.489×10^4 m (c) 9.2×10^{-11} s (d) 5.67×10^{-8} g **2.29** (a) 4.93 km (b) 2.3 μs (c) 5.68 mg (d) 1.568 cm (e) 9.3 ng **2.31** (a) 4 sig. figs. (b) 3 sig. figs. (c) 4 sig. figs. (d) at least 2 sig. figs. (e) 2 sig. figs. **2.33** 0.00800 **2.35** (a) 308.0 (b) 45.15 (c) 126.6 (d) 50.5092 (e) 1.29379 **2.37** (a) 6.2975×10^2 (b) 4.78 (c) 2.31×10^2 (d) 9.4×10^1 (e) 7.6×10^{-1} **2.39** (a) 14.19 (b) 201.9 (c) 1.013 (d) 2×10^{-2} (e) 29 **2.41** (a) 329 K (b) 49°C (c) 308 K (d) 185°C (e) 294 K **2.43** 303 K **2.45** (a) 29°C (b) −40.°F (c) −23°C (d) 131°F (e) 257°F **2.47** −108°F **2.49** (a) 1.65×10^5 mL (b) 4.8×10^{-8} g (c) 4.7×10^3 ms **2.51** (a) 1.91×10^1 cm (b) 1.2×10^2 g **2.53** (a) 5.69×10^{-3} km (b) 1.25 dm **2.55** 7.7×10^{-8} mm **2.57** 1.59 g/cm^3 **2.59** 2.6 g/cm^3 **2.61** 19.36 g **2.63** 23.4 mL **2.65** 10.05 mL **2.67** (a) 9 (b) 3 (c) 2 (d) 3 (e) 3 **2.69** (a) 6.41 (b) 15 (c) 8.30×10^{-1} (d) 6.07 (e) 2.479×10^2 **2.71** 2×10^{12} **2.73** (a) 4.37×10^{-5} (b) 5.720×10^5 (c) 1.0×10^{-5} (d) 1.5×10^{-5} (e) 1.61×10^{-2} **2.75** (a) 5.489 km (b) 723 ng (c) 21.64 km (d) 6.50×10^2 μL (e) 2.3 ms **2.77** 235°F, 386 K **2.79** 34°C, 307 K **2.81** 3.70×10^{-2} cm^3 or 3.70×10^{-2} mL **2.83** 2.50 g **2.85** 1.92×10^3 g **2.87** 5×10^2 μm **2.89** 3 kg **2.91** 2.4×10^{-3} in **2.93** 79.8 kg **2.95** 25 mi/hr

Chapter 3 **3.25** liquid **3.27** gas **3.29** Its physical properties are: colorless liquid, 69°C boiling point, and −95°C freezing point. Its chemical property is that it burns to form CO_2 and H_2O. **3.31** Its physical properties are: colorless, oily liquid, and density of 1.59 g/cm^3. Its chemical properties are: dilates blood vessels, and explodes, forming N_2, CO_2, H_2O, and O_2. **3.33** (a) mixture (b) element (c) compound (d) mixture **3.35** 2.89×10^4 cal **3.37** 5.02×10^4 J **3.39** 3.18×10^3 J **3.41** 8.97×10^2 J **3.43** 4.80×10^{-1} J/(g·°C) **3.45** 12.2°C **3.47** 45.2°C **3.49** (a) gas (b) gas (c) gas (d) liquid (e) solid **3.51** (a) physical (b) chemical (c) physical (d) physical (e) chemical **3.53** (a) solid (b) solid (c) solid (d) solid (e) gas **3.55** (a) solid (b) liquid (c) liquid (d) gas (e) gas **3.57** (a) substance (b) het. mixt. (c) solution (d) substance **3.59** (a) physical (b) chemical (c) physical (d) chemical **3.61** 1.14×10^5 cal, 4.78×10^5 J **3.63** 2.3×10^3 g **3.65** 551 J

Chapter 4 **4.23** (a) $^{20}_{10}Ne$ (b) $^{28}_{14}Si$ (c) $^{31}_{15}P$ (d) $^{39}_{19}K$ (e) $^{55}_{25}Mn$ **4.25** (a) 7p + 7n (b) 15p + 17n (c) 16p + 18n (d) 17p + 18n (e) 24p + 28n **4.27** 30 protons, 34 neutrons, and 30 electrons. The protons and neutrons are in the nucleus; the electrons are far outside the nucleus. **4.29** 34 electrons **4.31** Arsenic-75 has 33 protons and 42 neutrons in the nucleus and 33 electrons outside the nucleus. **4.33** nitrogen-14 **4.35** 10.81 amu **4.37** 20.17 amu (The element is neon.) **4.39** (a) In, indium (b) Si, silicon (c) Mn, manganese (d) Ti, titanium (e) As, arsenic **4.41** Bismuth (Bi) **4.43** Phosphorus: atomic no. = 15; atomic wt. = 30.97 amu, group VA, period 3, nonmetal. **4.45** (a) molecular (b) molecular (c) ionic (d) ionic (e) molecular **4.47** Ionic. This prediction agrees with the high melting point. **4.49** (a) H_2SO_4 (b) Cl_2O_7 (c) N_2O_4 (d) C_3H_8 (e) N_2F_4 **4.51** 11 protons, 11 electrons, and 13 neutrons. **4.53** $^{47}_{20}Ca$ **4.55** They differ by two neutrons; they are alike in that both have 28 protons and 28 electrons. **4.57** The proton is 1.8×10^3 times heavier than the electron. **4.59** 52.00 amu **4.61** Group IVA: C (nonmetal); Si (metalloid); Ge (metalloid); Sn (metal); and Pb (metal). The trend in this group does agree with the general rule. **4.63** $CHCl_3$. **4.65** The $SO_4{}^{2-}$ ion formula means that there is one sulfur atom and four oxygen atoms and an excess of two electrons in the ion.

Chapter 5 **5.11** (a) +1 (b) −1 (c) −3 (d) +2 (e) +2 **5.13** Two Na^+ ions and one $CO_3{}^{2-}$ ion **5.15** Al_2O_3 **5.17** (a) MgO (b) $CaCl_2$ (c) Na_2O (d) AlF_3 (e) CsBr (f) Zn_3N_2 **5.19** (a) Li^+ and O^{2-} (b) Ba^{2+} and O^{2-} (c) Zn^{2+} and Cl^- (d) K^+ and P^{3-} (e) Na^+ and Br^- (f) Ca^{2+} and I^- **5.21** potassium iodide **5.23** (a) lithium bromide (b) calcium bromide (c) rubidium oxide (d) barium nitride (e) radium sulfide (f) aluminum fluoride **5.25** (a) BaO (b) $SrCl_2$ (c) Na_2S (d) NaBr (e) Al_2S_3 (f) K_3N **5.27** (a) cobalt(III) oxide, cobaltic oxide (b) cobalt(II) oxide, cobaltous oxide (c) lead(IV) oxide, plumbic oxide (d) lead(II) oxide, plumbous oxide (e) copper(II) oxide, cupric oxide (f) copper(I) oxide, cuprous oxide **5.29** (a) $SnCl_4$ is tin(IV) chloride, or stannic chloride (b) correct (c) $CuBr_2$ is copper(II) bromide, or cupric bromide (d) Fe_2O_3 is iron(III) oxide, or ferric oxide (e) correct (f) $CaCl_2$ is calcium chloride **5.31** $CaCO_3$, calcium carbonate **5.33** (a) $Ca(HCO_3)_2$, calcium hydrogen carbonate (b) NH_4Cl, ammonium chloride (c) $Mg(ClO)_2$, magnesium hypochlorite (d) AgCN, silver cyanide (e) NH_4NO_3, ammonium nitrate (f) $Al(ClO_4)_3$, aluminum perchlorate (g) $Ba_3(PO_4)_2$, barium phosphate (h) KNO_2, potassium nitrite (i) $NaC_2H_3O_2$, sodium acetate

5.35 (a) $Al(OH)_3$ (b) $BaSO_3$ (c) $BaSO_4$ (d) Li_2HPO_4 (e) LiH_2PO_4 (f) Li_3PO_4 **5.37** nitrogen trihydride **5.39** (a) N_2O, dinitrogen monoxide (b) CCl_4, carbon tetrachloride (c) SiF_4, silicon tetrafluoride (d) ClF_3, chlorine trifluoride (e) Cl_2O_7, dichlorine heptoxide **5.41** (a) NBr_3 (b) XeO_4 (c) OF_2 (d) Cl_2O_7 (e) SF_6 (f) PI_3 **5.43** hydrogen sulfide (binary compound), hydrosulfuric acid (acid) **5.45** (a) phosphoric acid (b) nitrous acid (c) carbonic acid (d) nitric acid (e) hypochlorous acid (f) perchloric acid **5.47** (a) HNO_2 (b) HNO_3 (c) HI (d) H_2SO_4 (e) H_2SO_3 (f) HBr **5.49** $NaHCO_3$, sodium hydrogen carbonate **5.51** $CaCO_3$, calcium carbonate **5.53** (a) $FeSO_4$ (b) $Fe_2(SO_4)_3$ (c) $CuC_2H_3O_2$ (d) $Cu(C_2H_3O_2)_2$ **5.55** (a) $NiCl_2$ (b) TiF_4 (c) MnO_2 (d) Cr_2S_3 **5.57** (a) $Pb(NO_3)_2$ (b) $Co(OH)_2$ (c) $Cu_3(PO_4)_2$ (d) $CuNO_2$ (e) $Fe(ClO_2)_2$ (f) $Sn(ClO_4)_4$ **5.59** (a) tin(II) phosphate (b) iron(II) nitrate (c) chromium(II) sulfate (d) aluminum chlorate **5.61** (a) BrO_3^-, bromate ion (b) $N_2O_2^{2-}$, hyponitrite ion (c) $S_2O_3^{2-}$, thiosulfate ion (d) AsO_4^{3-}, arsenate ion

Chapter 6 **6.11** (a) Two atoms of solid potassium and one molecule of liquid bromine react to form two formula units of solid potassium bromide. (b) One molecule of solid phosphorus and six molecules of chlorine gas react to form four molecules of liquid phosphorus trichloride. **6.13** (a) One formula unit of solid barium carbonate reacts when heated to produce one formula unit of solid barium oxide and one molecule of carbon dioxide gas. (b) Two molecules of aqueous hydrogen peroxide react in the presence of a potassium iodide catalyst to produce two molecules of liquid water and one molecule of oxygen gas. **6.15** (a) $2CH_3OH + 3O_2 \longrightarrow 2CO_2 + 4H_2O$ (b) $2Mg + SiO_2 \longrightarrow 2MgO + Si$ (c) $Cl_2O_7 + H_2O \longrightarrow 2HClO_4$ (d) $TiCl_4 + 2H_2O \longrightarrow TiO_2 + 4HCl$ **6.17** (a) $CO + 3H_2 \longrightarrow CH_4 + H_2O$ (b) $Ba + 2H_2O \longrightarrow Ba(OH)_2 + H_2$ (c) $2H_2S + 3O_2 \longrightarrow 2H_2O + 2SO_2$ (d) $4H_3PO_3 \longrightarrow 3H_3PO_4 + PH_3$ **6.19** $N_2(g) + O_2(g) \longrightarrow 2NO(g)$ **6.21** $4NO_2(g) + O_2(g) \longrightarrow 2N_2O_5(g)$ **6.23** (a) Neither a combination nor a decomposition reaction. (b) Combination reaction. (c) Decomposition reaction. (d) Combination reaction. **6.25** (a) $2CH_4O(l) + 3O_2(g) \longrightarrow 2CO_2(g) + 4H_2O(g)$ (b) $2NO_2(g) \longrightarrow 2NO(g) + O_2(g)$. Decomposition reaction. (c) $N_2(g) + 3H_2(g) \longrightarrow 2NH_3(g)$. Combination reaction. (d) $4Li(s) + O_2(g) \longrightarrow 2Li_2O(g)$. Combination reaction. **6.27** (a) $Cd(s) + 2AgNO_3(aq) \longrightarrow Cd(NO_3)_2(aq) + 2Ag(s)$. Single replacement reaction. (b) $C_3H_6(g) + H_2(g) \longrightarrow C_3H_8(g)$. **6.29** (a) no (b) yes **6.31** $Pb(NO_3)_2(aq) + Na_2CO_3(aq) \longrightarrow PbCO_3(s) + 2NaNO_3(aq)$. **6.33** (a) $MgSO_4(aq) + 2NaOH(aq) \longrightarrow Mg(OH)_2(s) + Na_2SO_4(aq)$. (b) No precipitate will form. (c) No precipitate will form. (d) No precipitate will form. **6.35** $Na_2CO_3(aq) + 2HC_2H_3O_2(aq) \longrightarrow 2NaC_2H_3O_2(aq) + CO_2(g) + H_2O(l)$. **6.37** (a) No gas will form. (b) No gas will form. (c) $(NH_4)_2CO_3(aq) + 2HCl(aq) \longrightarrow 2NH_4Cl(aq) + CO_2(g) + H_2O(l)$. (d) No gas will form. **6.39** (a) base (b) acid (c) base (d) acid (e) neither (f) neither (g) neither (h) base (i) neither **6.41** (a) $HNO_3(aq) + NaOH(aq) \longrightarrow NaNO_3(aq) + H_2O(l)$ (b) $Ba(OH)_2(aq) + 2HCl(aq) \longrightarrow BaCl_2(aq) + 2H_2O(l)$ (c) $LiOH(aq) + HBr(aq) \longrightarrow LiBr(aq) + H_2O(l)$ (d) $HC_2H_3O_2(aq) + NaOH(aq) \longrightarrow NaC_2H_3O_2(aq) + H_2O(l)$ **6.43** The reaction of octane (C_8H_{18}) with oxygen is a combustion reaction. $2C_8H_{18}(l) + 25O_2(g) \longrightarrow 16CO_2(g) + 18H_2O(g)$, $2C_8H_{18}(l) + 17O_2(g) \longrightarrow 16CO(g) + 18H_2O(g)$ **6.45** $NH_4Cl(s) \xrightarrow{\Delta} NH_3(g) + HCl(g)$ **6.47** $6CO_2 + 6H_2O \longrightarrow C_6H_{12}O_6 + 6O_2$ **6.49** $2NaC_{18}H_{36}O_2(aq) + CaCl_2(aq) \longrightarrow Ca(C_{18}H_{36}O_2)_2(s) + 2NaCl(aq)$ **6.51** $6Li(s) + N_2(g) \longrightarrow 2Li_3N(s)$ **6.53** (a) $HNO_3(aq) + NaOH(aq) \longrightarrow NaNO_3(aq) + H_2O(l)$. The reaction is driven by the formation of water. (b) $HNO_3(aq) + NaCN(aq) \longrightarrow NaNO_3(aq) + HCN(g)$. The reaction is driven by the formation of a gas. (c) $Pb(NO_3)_2(aq) + 2NaCl(aq) \longrightarrow 2NaNO_3(aq) + PbCl_2(s)$. The reaction is driven by the formation of a precipitate. **6.55** (a) soluble (b) soluble (c) insoluble (d) insoluble (e) insoluble (f) soluble (g) insoluble (h) soluble (i) insoluble **6.57** (a) Single replacement reaction: $2Na(s) + 2H_2O(l) \longrightarrow 2NaOH(aq) + H_2(g)$ (b) Double replacement reaction: $CrCl_3(aq) + 3NaOH(aq) \longrightarrow Cr(OH)_3(s) + 3NaCl(aq)$ (c) Single replacement reaction: $Zn(s) + CuSO_4(aq) \longrightarrow ZnSO_4(aq) + Cu(s)$ (d) Combustion reaction: $C_6H_{12}(l) + 9O_2(g) \longrightarrow 6CO_2(g) + 6H_2O(g)$ **6.59** (a) $BaCO_3(s) + 2HNO_3(aq) \longrightarrow Ba(NO_3)_2(aq) + CO_2(g) + H_2O(l)$ (b) $BaCl_2(aq) + H_2SO_4(aq) \longrightarrow BaSO_4(s) + 2HCl(aq)$ (c) $Ca(OH)_2(s) + 2HC_2H_3O_2(aq) \longrightarrow Ca(C_2H_3O_2)_2(aq) + 2H_2O(l)$ (d) $2Al(s) + 3NiSO_4(aq) \longrightarrow Al_2(SO_4)_3(aq) + 3Ni(s)$ **6.61** (a) $2Li(s) + Cl_2(g) \longrightarrow 2LiCl(s)$ (b) $2C_{10}H_{22}(s) + 31O_2(g) \longrightarrow 20CO_2(g) + 22H_2O(g)$ (c) There is no one answer. (d) $2Au_2O_3(s) \xrightarrow{\Delta} 4Au(s) + 3O_2(g)$ **6.63** $2PbS(s) + 3O_2(g) \xrightarrow{\Delta} 2PbO(s) + 2SO_2(g)$ **6.65** $Fe_2O_3(s) + 3CO(g) \xrightarrow{\Delta} 2Fe(l) + 3CO_2(g)$

Chapter 7 **7.9** (a) 38.00 amu (b) 125.97 amu (c) 80.07 amu (d) 60.05 amu (e) 174.11 amu (f) 342.30 amu **7.11** 86.09 amu, 180.16 amu **7.13** (a) 319.18 amu (b) 266.68 amu (c) 30.07 amu (d) 80.05 amu (e) 159.70 amu (f) 161.98 amu **7.15** (d) **7.17** 2.81×10^{24} C atoms, 8.42×10^{24} H atoms, 1.40×10^{24} S atoms **7.19** 4.23×10^{23} Mg^{2+} ions **7.21** 2.016 g, 4.003 g **7.23** 55.85 g, 256.56 g **7.25** (a) 32.04 g (b) 125.97 g (c) 80.07 g (d) 60.05 g (e) 180.16 g (f) 342.30 g **7.27** (a) 7.949 g (b) 223.21 g (c) 173.29 g (d) 601.93 g (e) 122.99 g (f) 158.27 g **7.29** (a) 0.496 mol (b) 0.144 mol (c) 0.0435 mol (d) 0.0256 mol (e) 0.0117 mol (f) 7.52×10^{-3} mol **7.31** (a) 0.156 mol (b) 0.0397 mol (c) 0.0624 mol (d) 0.0833 mol (e) 0.0278 mol (f) 0.0146 mol **7.33** 0.378 g **7.35** 1.3×10^{-3} g **7.37** 2.239% H, 26.68% C, 71.08% O **7.39** 49.02% C, 2.743% H, 48.23% Cl **7.41** $MgCl_2$ **7.43** P_2O_5 **7.45** C_3H_3O **7.47** $C_6H_6Cl_6$

7.49 P_4O_{10} **7.51** $C_2H_2O_4$ **7.53** $C_6H_4Cl_2$ **7.55** 0.53 mol, 3.2×10^{23} atoms **7.57** 48.4 g **7.59** 756 mg **7.61** 23.3 mol Cl, 1.40×10^{25} Cl ions **7.63** 7.528×10^{22} S_8 molecules **7.65** 2.0×10^{22} Ca^{2+} ions **7.67** 1.6×10^{23} H atoms **7.69** $BaSO_4$ **7.71** K_2MnO_4; $KMnO_4$ **7.73** MgO **7.75** CO_2

Chapter 8 **8.13** 1.40 mol **8.15** 24.7 mol **8.17** 3.7 mol **8.19** 1.50 mol **8.21** 5.34 mol **8.23** 0.800 mol Na_2CO_3; 0.800 mol CO_2 **8.25** 0.649 g **8.27** 1.40 g **8.29** 0.908 g **8.31** 9.54 g **8.33** 1.61 g HCl **8.35** 80. g **8.37** (a) H_2O (b) H_2O (c) H_2O **8.39** (a) Cl_2 (b) Cl_2 (c) Cl_2 **8.41** 3.0 mol NaCl; 1.5 mol Cl_2; 0 mol Na **8.43** 1.3 mol Al_2O_3; 0.62 mol O_2; 0 mol Al **8.45** 0.53 g NH_3 **8.47** 34.3 g CH_3OH; 25.7 g H_2 (remaining) **8.49** 96.9% **8.51** 54.8 g; 80.7% **8.53** $CaCO_3 + 2HCl \longrightarrow CaCl_2 + CO_2 + H_2O$; 44.8 mol HCl; 22.4 mol $CaCl_2$; 22.4 mol CO_2; 22.4 mol H_2O **8.55** $Fe_2O_3 + 6HCl \longrightarrow 2FeCl_3 + 3H_2O$; 7.38 mol HCl; 2.46 mol $FeCl_3$; 3.69 mol H_2O **8.57** $3NaOH + H_3PO_4 \longrightarrow Na_3PO_4 + 3H_2O$; 6.83 g Na_3PO_4; 2.92 g H_3PO_4 (remaining) **8.59** $Ca(HCO_3)_2 + Ca(OH)_2 \longrightarrow 2CaCO_3 + 2H_2O$; 20.1 g **8.61** $3Ba(OH)_2 + 2H_3PO_4 \longrightarrow Ba_3(PO_4)_2 + 6H_2O$; 2.56 g $Ba(OH)_2$; 0.977 g H_3PO_4 **8.63** $2C + O_2 \longrightarrow 2CO$; $ZnO + CO \longrightarrow Zn + CO_2$; 60.3 g **8.65** 2.61 g

Chapter 9 **9.27** violet light **9.29** green light **9.31** $n = 4$ **9.33** $n = 5$ level to the $n = 2$ level **9.35** (a) 4 (b) 25 **9.37** (a) The notation $3p$ refers to the $n = 3$ shell and a p-type subshell. (b) The notation $4d$ refers to the $n = 4$ shell and a d-type subshell. (c) The notation $2s$ refers to the $n = 2$ shell and an s-type subshell. **9.39** $2s$ and $2p$; 1 and 3, respectively **9.41** $3s < 3p < 3d$ **9.43** There are 2 electrons in the $1s$ subshell, 2 electrons in the $2s$ subshell, 6 electrons in the $2p$ subshell, 2 electrons in the $3s$ subshell, and 4 electrons in the $3p$ subshell; 16 **9.45** 8 **9.47** $1s^22s^22p^63s^23p^2$ **9.49** $1s^22s^22p^63s^23p^63d^{10}4s^24p^3$ **9.51** $5s^2$ **9.53** Period 4 and Group VIA (or 6); selenium **9.55** (a) Br (b) K **9.57** (a) O^{2-} (b) Cl^- **9.59** (a) F (b) Br **9.61** (a) 13 (b) 5 (c) 18 (d) 35 **9.63** (a) allowed (b) Not allowed. The $n = 2$ shell has only 2 subshells, $2s$ and $2p$. (c) Not allowed. The $n = 3$ shell has only 3 subshells, $3s$, $3p$, and $3d$. (d) allowed
9.65 (a), (b), and (d) are allowed

9.67

Element	Electron configuration
carbon (C)	$1s^22s^22p^2$
silicon (Si)	$1s^22s^22p^63s^23p^2$
germanium (Ge)	$1s^22s^22p^63s^23p^63d^{10}4s^24p^2$
tin (Sn)	$1s^22s^22p^63s^23p^63d^{10}4s^24p^64d^{10}5s^25p^2$

9.69 titanium (Ti) **9.71** (a) 1 (b) 2 (c) 4 (d) 6

9.73

Element	Valence-shell configuration
boron (B)	$2s^22p^1$
aluminum (Al)	$3s^23p^1$
gallium (Ga)	$4s^24p^1$
indium	$5s^25p^1$

9.75 Be **9.77** Li **9.79** F < Cl < S **9.81** S > Mg > Sr **9.83** $F^- < Cl^- < Br^- < I^-$ **9.85** $Se^{2-} > Br^- > Rb^+ > Sr^{2+}$ **9.87** Na **9.89** Cl

Chapter 10 **10.31** (a) 4.0 − 0.9 = 3.1 (b) 3.0 − 2.5 = 0.5 (c) 4.0 − 2.5 = 1.5 (d) 2.5 − 2.5 = 0 Order of bonds (pure covalent to ionic): C—C, C—Cl, F—C, F—Na.

10.33 (a) 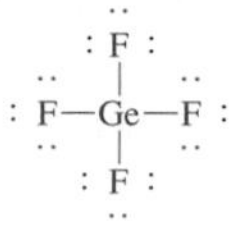 GeF_4: Ge central, single-bonded to four F, each F with three lone pairs (b) 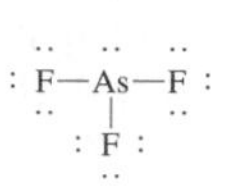 AsF_3: F—As—F with F below As; lone pair on As, three lone pairs on each F (c) 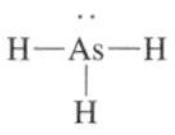 H—As—H with H below As; lone pair on As (d) 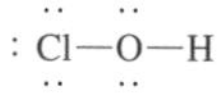 :Cl—O—H with three lone pairs on Cl, two on O **10.35** (a) 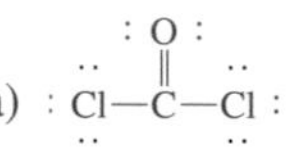 Cl—C—Cl with C=O; two lone pairs on O, three on each Cl (b) :C≡O: (c) :Br—C≡N: with three lone pairs on Br **10.37** (a) 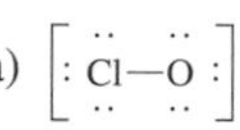 [:Cl—O:]$^-$ with three lone pairs on Cl and on O (b) 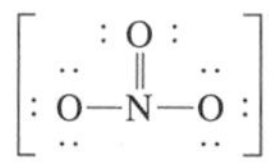 [O—N—O with N=O above; lone pairs on O atoms and N]$^-$

10.39 (a) tetrahedral (b) trigonal pyramidal (c) trigonal planar **10.41** LiCl is ionic (metal + nonmetal) and Cl_2 is covalent (two nonmetals). The formation of the ionic bond in LiCl can be described as follows:

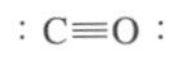

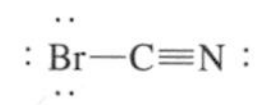

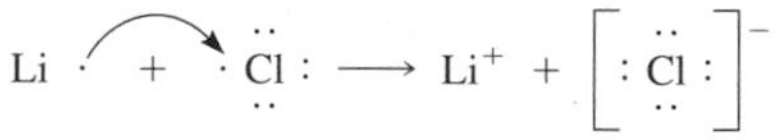

Li· + ·Cl: (seven dots) ⟶ Li^+ + [:Cl:]$^-$ (eight dots)

The formation of the covalent bond in Cl_2 is as follows:

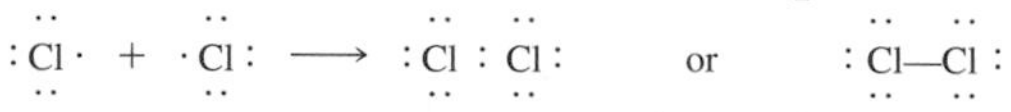

:Cl· + ·Cl: ⟶ :Cl:Cl: or :Cl—Cl:

10.43 [Xe]$4f^{14}5d^{10}6s^2$ **10.45** (a) Na—O (b) H—O (c) F—H
10.47 (a) 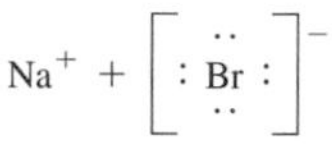 Na^+ + [:Br:]$^-$ (b) 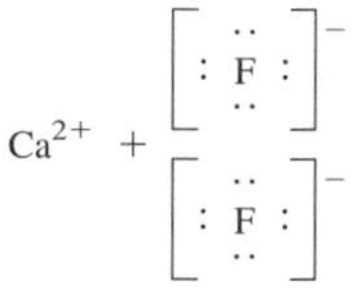 Ca^{2+} + [:F:]$^-$ + [:F:]$^-$ (c) 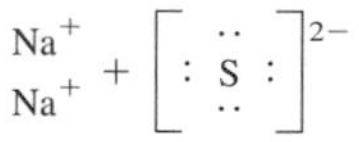 Na^+ Na^+ + [:S:]$^{2-}$

10.49 (a) 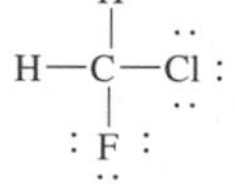 H—C—Cl: with H above C and :F: below C (b) O=S—O: with lone pairs (c) :O: double-bonded above N; :O—N—O—H (d) 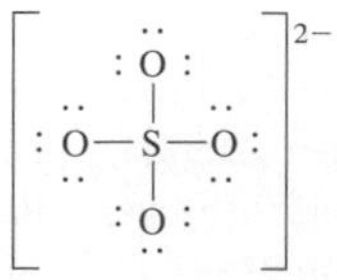 [O—S—O with O above and below S, all O with lone pairs]$^{2-}$

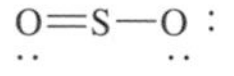

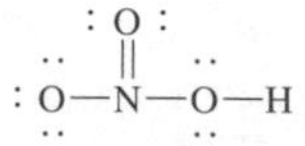

10.51 (a) In SO_2, 3 electron groups surround the central atom so the arrangement is trigonal planar. (b) In $SeCl_2$, 4 electron groups surround the central atom so the arrangement is tetrahedral. (c) In ClO_4^-, 4 electron groups surround the central atom so the arrangement is tetrahedral. (d) In PH_3, 4 electron groups surround the central atoms so the arrangement is tetrahedral. **10.53** (a) tetrahedral (b) bent or angular (c) trigonal planar (d) tetrahedral **10.55** N—Cl **10.57** (a) ionic (b) covalent (c) ionic (d) covalent **10.59** (a) HF (b) NF_3 (c) H_2O

Chapter 11 **11.21** 1.484 atm **11.23** 3.25 L **11.25** 39.3 L **11.27** 1.00×10^3 mmHg **11.29** 0.879 atm **11.31** 1.22×10^3 L **11.33** 1.20×10^2 mL **11.35** 123°C **11.37** 333°C **11.39** 28.1 g/mol **11.41** 39.9 g/mol **11.43** 7.63 atm **11.45** 1.22 atm **11.47** 44.9 L **11.49** 3.50 L **11.51** 4.99×10^{-3} mol **11.53** 0.0227 mol **11.55** 143°C **11.57** −152°C **11.59** P_{Oxygen} = 0.0148 atm, P_{Helium} = 0.0384 atm, P_{Total} = 0.0532 atm **11.61** 0.419 L **11.63** 27.0 L **11.65** 21.4 g **11.67** 0.255 g **11.69** $BaCO_3(s) \xrightarrow{\Delta} BaO(s) + CO_2(g)$; 41.4 g **11.71** 3.12×10^{20} atoms **11.73** 0.115 mol

Chapter 12 **12.27** (a) condensation (b) freezing **12.29** 25.0 kJ **12.31** 8.76 kJ **12.33** 169 kJ **12.35** 41.5 kJ **12.37** 36.8 kJ **12.39** 84.0 kJ **12.41** 720. mmHg, 8.84×10^{-3} mol **12.43** (a) dispersion forces (b) dispersion forces, hydrogen bonding (c) dispersion forces, dipole–dipole forces **12.45** $CCl_4 < SiCl_4 < GeCl_4$; dispersion forces increase as molecular weight increases. **12.47** CCl_4; the greater the molecular weight, the lower the vapor pressure. **12.49** $CCl_4 < SiCl_4 < GeCl_4$; the greater the molecular weight, the greater the boiling point. **12.51** water because of hydrogen bonding **12.53** molecular solid **12.55** ionic solid **12.57** water because of hydrogen bonding **12.59** phosphorus because of greater dispersion forces **12.61** 1.22×10^3 g **12.63** Yes. Hydrogen has smaller dispersion forces because of its lower molecular weight.

Chapter 13 **13.33** (a) 33% (b) 10.9% **13.35** 0.909% **13.37** 21 g copper(II) sulfate; 404 g H_2O **13.39** 68 g **13.41** 2.82% **13.43** (a) 0.180 *M* (b) 0.549 *M* **13.45** (a) 0.0640 *M* (b) 0.00390 *M* **13.47** 5.61 g **13.49** 566 mL **13.51** 0.224 g **13.53** 0.167 g sodium carbonate; 0.267 g sodium nitrate **13.55** 0.829 *M* **13.57** 0.329 g **13.59** 1.4 mL **13.61** 1.2×10^2 mL **13.63** 0.570 *M* **13.65** 1.44 m **13.67** 3.0×10^2 g **13.69** −0.47°C **13.71** 100.18°C **13.73** −0.056°C, 100.016°C **13.75** 387 amu **13.77** 1.92 m, 1.81 *M* **13.79** $BaCl_2$ **13.81** 0.747 m, 5.27% **13.83** 101.2°C **13.85** 0.141 *M*, 0.106 g **13.87** 11.6 mL **13.89** 5.05 *M* **13.91** 41°C/m

Chapter 14 **14.23** (a) $2SO_2(g) + O_2(g) \rightleftharpoons 2SO_3(g)$; $K = \dfrac{[SO_3]^2}{[SO_2]^2[O_2]}$ (b) $2NOCl(g) \rightleftharpoons 2NO(g) + Cl_2(g)$; $K = \dfrac{[NO]^2[Cl_2]}{[NOCl]^2}$ (c) $POCl_3(g) \rightleftharpoons POCl(g) + Cl_2(g)$; $K = \dfrac{[POCl][Cl_2]}{[POCl_3]}$ (d) $PCl_3(g) + Cl_2(g) \rightleftharpoons PCl_5(g)$; $K = \dfrac{[PCl_5]}{[PCl_3][Cl_2]}$

14.25 (a) $CS_2(g) + 4H_2(g) \rightleftharpoons CH_4(g) + 2H_2S(g)$; $K = \dfrac{[CH_4][H_2S]^2}{[CS_2][H_2]^4}$ (b) $I_2(g) + Br_2(g) \rightleftharpoons 2IBr(g)$; $K = \dfrac{[IBr]^2}{[I_2][Br_2]}$ (c) $COCl_2(g) \rightleftharpoons CO(g) + Cl_2(g)$; $K = \dfrac{[CO][Cl_2]}{[COCl_2]}$ (d) $NH_4^+(aq) \rightleftharpoons NH_3(aq) + H^+(aq)$; $K = \dfrac{[NH_3][H^+]}{[NH_4^+]}$ **14.27** $CO(g) + 3H_2(g) \rightleftharpoons CH_4(g) + H_2O(g)$; $K = 3.92$ **14.29** $H_2(g) + I_2(g) \rightleftharpoons 2HI(g)$; $K = 54.3$

14.31 $K = \dfrac{[H_2O]^2}{[H_2]^2[O_2]} = 3 \times 10^{81}$. The reaction will go to completion. **14.33** $[CS_2] = 1.10\ M$

14.35 (a) $C(s) + CO_2(g) \rightleftharpoons 2CO(g)$; $K = \dfrac{[CO]^2}{[CO_2]}$

(b) $FeO(s) + CO(g) \rightleftharpoons Fe(s) + CO_2(g)$; $K = \dfrac{[CO_2]}{[CO]}$

(c) $2Na_2CO_3(s) + 2SO_2(g) + O_2(g) \rightleftharpoons 2Na_2SO_4(s) + 2CO_2(g)$; $K = \dfrac{[CO_2]^2}{[SO_2]^2[O_2]}$

14.37 (a) $P_4(s) + 5O_2(g) \rightleftharpoons P_4O_{10}(s)$; $K = \dfrac{1}{[O_2]^5}$

(b) $2H_2O_2(l) \rightleftharpoons 2H_2O(l) + O_2(g)$; $K = [O_2]$

(c) $PbI_2(s) + Cl_2(g) \rightleftharpoons PbCl_2(s) + I_2(g)$; $K = \dfrac{[I_2]}{[Cl_2]}$

14.39 PbI_2 is more soluble. **14.41** The reaction will shift from left to right (away from O_2). **14.43** The equilibrium shifts from left to right. **14.45** The reaction shifts from right to left. **14.47** $CH_4 + 2H_2S \rightleftharpoons CS_2 + 4H_2$ **14.49** $2NO + 2H_2 \rightleftharpoons N_2 + 2H_2O$ **14.51** $K_{sp} = [Ag^+][I^-] = 8.3 \times 10^{-17}$ **14.53** $K_{sp} = [Pb^{2+}][Br^-]^2 = 4.0 \times 10^{-6}$ **14.55** (a) Shift from right to left. (b) Shift from right to left. (c) Shift from left to right. (d) Shift from left to right. (e) No change. **14.57** For parts (a), (b), (c), and (e), there is no effect on *K*. For part (d), *K* increases. **14.59** (a) Shift from left to right. (b) Shift from right to left. (c) Shift from left to right. **14.61** For parts (a) and (b), there is no effect on *K*. For part (c), *K* increases.

Chapter 15 **15.29** Total ionic equation: $Ra^{2+}(aq) + 2OH^-(aq) + 2H^+(aq) + 2Cl^-(aq) \longrightarrow Ra^{2+}(aq) +$

$2Cl^-(aq) + 2H_2O(l)$. Net ionic equation: $2OH^-(aq) + 2H^+(aq) \longrightarrow 2H_2O(l)$. **15.31** $HClO_4(l) + H_2O(l) \longrightarrow H_3O^+(aq) + ClO_4^-(aq)$ **15.33** (a) F^- (b) HS^- (c) S^{2-} (d) OH^- **15.35** (a) NH_4^+ (b) H_2S (c) HS^- (d) HCN **15.37** (a) $H_2SO_3(aq) + H_2O(l) \rightleftharpoons H_3O^+(aq) + HSO_3^-(aq)$

Acid Base Acid Base

The conjugate acid–base pairs are H_2SO_3 with HSO_3^-, and H_2O with H_3O^+.

(b) $HSO_3^-(aq) + H_2O(l) \rightleftharpoons H_3O^+(aq) + SO_3^{2-}(aq)$

Acid Base Acid Base

The conjugate acid–base pairs are HSO_3^- with SO_3^{2-}, and H_2O with H_3O^+.

(c) $CH_3NH_2(aq) + H_2O(l) \rightleftharpoons CH_3NH_3^+(aq) + OH^-(aq)$

Base Acid Acid Base

The conjugate acid–base pairs are CH_3NH_2 with $CH_3NH_3^+$, and H_2O with OH^-.

(d) $NH_4^+(aq) + CN^-(aq) \rightleftharpoons NH_3(aq) + HCN(aq)$

Acid Base Base Acid

The conjugate acid–base pairs are NH_4^+ with NH_3, and CN^- with HCN.

15.39 (a) $CN^-(aq) + H_2O(l) \rightleftharpoons HCN(aq) + OH^-(aq)$

Base Acid Acid Base

The conjugate acid–base pairs are CN^- with HCN, and H_2O with OH^-.

(b) $HC_2H_3O_2(aq) + OH^-(aq) \rightleftharpoons C_2H_3O_2^-(aq) + H_2O(l)$

Acid Base Base Acid

The conjugate acid–base pairs are $HC_2H_3O_2$ with $C_2H_3O_2^-$, and OH^- with H_2O.

(c) $HPO_4^{2-}(aq) + H_3O^+(aq) \rightleftharpoons H_2PO_4^-(aq) + H_2O(l)$

Base Acid Acid Base

The conjugate acid–base pairs are HPO_4^{2-} with $H_2PO_4^-$, and H_3O^+ with H_2O.

(d) $HNO_2(aq) + H_2O(l) \rightleftharpoons NO_2^-(aq) + H_3O^+(aq)$

Acid Base Base Acid

The conjugate acid–base pairs are HNO_2 with NO_2^-, and H_2O with H_3O^+.

15.41 (a) 4.0×10^{-14} *M* (b) 2.9×10^{-12} *M* (c) 8.3×10^{-13} *M* (d) 2.4×10^{-4} *M* **15.43** 2.7×10^{-11} *M* **15.45** (a) Basic (b) Neutral (c) Acidic (d) Acidic **15.47** Basic **15.49** (a) 8.00 (b) 2.12 (c) 11.30 (d) 8.20 **15.51** 2.12, acidic **15.53** (a) pH = 6.00 (b) pH = 11.88 (c) pH = 2.70 (d) pH = 5.80 **15.55** 11.60, basic **15.57** (a) 5.2×10^{-8} *M* (b) 4.8×10^{-3} *M* (c) 7.9×10^{-13} *M* (d) 1.9×10^{-7} *M* **15.59** $[H_3O^+] = 1.8 \times 10^{-9}$ *M*, $[OH^-] = 5.5 \times 10^{-6}$ *M* **15.61** The concentrations of H_3O^+ differ by a factor of 10. **15.63** $[H_3O^+] = 1.0 \times 10^{-13}$; pH = 13.00

Chapter 16

16.27 (a) Ce^{4+} is the oxidizing agent and Sn^{2+} is the reducing agent. (b) H^+ is the oxidizing agent and Zn is the reducing agent. (c) Pb^{2+} is the oxidizing agent and Cd is the reducing agent. (d) Cl_2 is the oxidizing agent and Al is the reducing agent. **16.29** (a) will not occur (b) will occur (c) will occur (d) will not occur **16.31** (a) H: +1, Br: −1 (b) Na: +1, O: −2 (c) C: −4, H: +1 (d) C: −3, H: +1 (e) O: 0 (f) O: 0 **16.33** (a) Na: +1, Cl: +7, O: −2 (b) Na: +1, Cl: +5, O: −2 (c) Na: +1, Cl: +3, O: −2 (d) Na: +1, Cl: +1, O: −2 (e) Na: +1, Cl: −1 (f) Cl: 0 **16.35** (a) Br: −1 (b) N: −3 (c) N: −3, H: +1 (d) O: −2, H: +1 (e) Mn: +7, O: −2 (f) H: +1, S: −2 **16.37** (a) H: +1, P: +5, O: −2 (b) H: +1, P: +5, O: −2 (c) P: +5, O: −2 (d) N: −3, H: +1 (e) N: +5, O: −2 (f) N: +3, O: −2 **16.39** (a) N_2 is the reducing agent and O_2 is the oxidizing agent. (b) P_4 is the reducing agent and O_2 is the oxidizing agent. (c) This equation does not represent an oxidation–reduction reaction. (d) CO is the reducing agent and O_2 is the oxidizing agent. **16.41** (a) $4Li(s) + O_2(g) \longrightarrow 2Li_2O(s)$. O_2 is the oxidizing agent and Li is the reducing agent. (b) $4Al(s) + 3O_2(g) \longrightarrow 2Al_2O_3(s)$. O_2 is the oxidizing agent and Al is the reducing agent. (c) $2Cr(s) + 6HCl(aq) \longrightarrow 2CrCl_3(aq) + 3H_2(g)$. HCl is the oxidizing agent and Cr is the reducing agent. (d) $CH_4(g) + 2O_2(g) \longrightarrow CO_2(g) + 2H_2O(l)$. O_2 is the oxidizing agent and CH_4 is the reducing agent. **16.43** (a) $Br_2(aq) + SO_2(aq) + 2H_2O(l) \longrightarrow 2HBr(aq) + H_2SO_4(aq)$. Br_2 is the oxidizing agent and SO_2 is the reducing agent. (b) $6HI(aq) + 2HNO_3(aq) \longrightarrow 3I_2(aq) + 2NO(g) + 4H_2O(l)$. HNO_3 is the oxidizing agent and HI is the reducing agent. (c) $MnO_2(s) + 4HBr(aq) \longrightarrow MnBr_2(aq) + Br_2(aq) + 2H_2O(l)$. MnO_2 is the oxidizing agent and HBr is the reducing agent. (d) $3CuO(s) + 2NH_3(g) \longrightarrow 3Cu(s) + N_2(g) + 3H_2O(g)$. CuO is the oxidizing agent and NH_3 is the reducing agent. **16.45** (a) $Cr_2O_7^{2-}(aq) + 6Fe^{2+}(aq) + 14H^+(aq) \longrightarrow 2Cr^{3+}(aq) + 6Fe^{3+}(aq) + 7H_2O(l)$. $Cr_2O_7^{2-}$ is the oxidizing agent and Fe^{2+} is the reducing agent. (b) $2VO_2^+(aq) + Zn(s) + 4H^+(aq) \longrightarrow 2VO^{2+}(aq) + Zn^{2+}(aq) + 2H_2O(l)$. VO_2^+ is the oxidizing agent and Zn is the reducing agent. (c) $VO_2^+(aq) + Zn(s) + 4H^+(aq) \longrightarrow V^{3+}(aq) + Zn^{2+}(aq) + 2H_2O(l)$. VO_2^+ is the oxidizing agent and Zn is the reducing agent. (d) $2VO_2^+(aq) + 3Zn(s) + 8H^+(aq) \longrightarrow 2V^{2+}(aq) + 3Zn^{2+}(aq) + 4H_2O(l)$. VO_2^+ is the oxidizing agent and Zn is the reducing agent. **16.47** (a) $48NO_3^-(aq) + S_8(s) + 32H^+(aq) \longrightarrow 48NO_2(g) + 8SO_4^{2-}(aq) + 16H_2O(l)$. NO_3^- is the oxidizing agent and S_8 is the reducing agent. (b) $2MnO_4^-(aq) + 5SO_2(g) + 2H_2O(l) \longrightarrow 2Mn^{2+}(aq) + 5SO_4^{2-}(aq) + 4H^+(aq)$. MnO_4^- is the oxidizing agent and SO_2 is the reducing agent. (c) $Cr_2O_7^{2-}(aq) + 3HNO_2(aq) + 5H^+(aq) \longrightarrow 2Cr^{3+}(aq) + 3NO_3^-(aq) + 4H_2O(l)$. $Cr_2O_7^{2-}$ is the oxidizing agent and HNO_2 is the reducing agent. (d) $2MnO_4^-(aq) + 5HNO_2(aq) + H^+(aq) \longrightarrow 2Mn^{2+}(aq) + 5NO_3^-(aq) + 3H_2O(l)$. MnO_4^- is the oxidizing agent and HNO_2 is the reducing agent. **16.49** (a) $8Al(s) + 3NO_3^-(aq) + 5OH^-(aq) + 18H_2O(l) \longrightarrow 8Al(OH)_4^-(aq) + 3NH_3(aq)$. NO_3^- is the oxidizing agent and Al is the reducing agent. (b) $S^{2-}(aq) + 4I_2(aq) + 8OH^-(aq) \longrightarrow SO_4^{2-}(aq) + 8I^-(aq) + 4H_2O(l)$. I_2 is the oxidizing agent and S^{2-} is the reducing agent. (c) $Mn(OH)_2(s) + H_2O_2(aq) \longrightarrow MnO_2(s) + 2H_2O(l)$. H_2O_2 is the oxidizing agent and $Mn(OH)_2$ is the reducing agent. (d) $Cl_2(g) + IO_3^-(aq) + 2OH^-(aq) \longrightarrow 2Cl^-(aq) + IO_4^-(aq) + H_2O(l)$. Cl_2 is the oxidizing agent and IO_3^- is the reducing agent. **16.51** Sn is the anode and Cu is the cathode; $Sn(s) + Cu^{2+}(aq) \longrightarrow Sn^{2+}(aq) + Cu(s)$. **16.53** Cr is the anode and Pb is the cathode; $Cr(s) + Pb^{2+}(aq) \longrightarrow Cr^{2+}(aq) + Pb(s)$ **16.55** $AlBr_3$ is an ionic compound 3consisting of Al^{3+} and Br^- ions. $Al^{3+} + 3e^- \longrightarrow Al$ and $2Br^- \longrightarrow Br_2 + 2e^-$. **16.57** $2Cr(OH)_3(s) + 3H_2O_2(aq) + 4OH^-(aq) \longrightarrow 2CrO_4^{2-}(aq) + 8H_2O(l)$ **16.59** $4Fe(OH)_2(s) + O_2(g) + 2H_2O(l) \longrightarrow 4Fe(OH)_3(s)$ **16.61** $Cl_2(aq) + 2OH^-(aq) \longrightarrow Cl^-(aq) + ClO^-(aq) + H_2O(l)$.

Chapter 17

17.27 ${}^{87}_{37}Rb \longrightarrow {}^{87}_{38}Sr + {}^{0}_{-1}e$ **17.29** ${}^{232}_{90}Th \longrightarrow {}^{228}_{88}Ra + {}^{4}_{2}He$ **17.31** ${}^{210}_{84}Po \longrightarrow {}^{206}_{82}Pb + {}^{4}_{2}He$ **17.33** ${}^{18}_{9}F \longrightarrow {}^{18}_{8}O + {}^{0}_{1}e$ **17.35** ${}^{4}_{2}He$, an alpha particle. **17.37** ${}^{239}_{94}Pu$ **17.39** After three half-lives nuclide, there will be $1/2 \times 1/2 \times 1/2 = 1/8$ of the original sample remaining. **17.41** 1×10^{-4} g **17.43** 5730 years old; 3700 B.C. **17.45** 11,400 years old **17.47** ${}^{209}_{83}Bi + {}^{4}_{2}He \longrightarrow {}^{211}_{85}At + 2\,{}^{1}_{0}n$ **17.49** ${}^{238}_{92}U + {}^{12}_{6}C \longrightarrow {}^{246}_{98}Cf + 4\,{}^{1}_{0}n$

Chapter 18

18.33 (a) tetrahedral (b) trigonal planar (c) linear

18.35

```
(a)   H H H H                    H
      | | | |                    |
    H-C-C-C-C-H                H-C-H
      | | | |              H     |     H
      H H H H              |     |     |
                         H-C-----C-----C-H
                           |     |     |
                           H     H     H

(b)   H H H            H   H H           O-H
      | | |            |   | |         H |  H
    H-C-C-C-O-H      H-C-O-C-C-H       | |  |
      | | |            |   | |       H-C-C-C-H
      H H H            H   H H         | | |
                                       H H H
```

18.37

```
(a)   H Cl          Cl Cl
      | |           |  |
    H-C-C-Cl      H-C--C-H
      | |           |  |
      H H           H  H

(b)   H H H          H Cl H
      | | |          | |  |
    H-C-C-C-Cl     H-C-C--C-H
      | | |          | |  |
      H H H          H H  H
```

18.39

```
(a)          H
             |
           H-C-H
       H     |     H
       |     |     |
     H-C-----C-----C-O-H
       |     |     |
       H     H     H

(b)                 H
                    |
                  H-C-H
         H          |        H H
         |          |        | |
       H-C-O--------C--------C-C-H
         |          |        | |
         H          H        H H
```

18.41 C_8H_{18}; $CH_3CH_2CH_2CH_2CH_2CH_2CH_2CH_3$

```
    H H H H H H H H
    | | | | | | | |
  H-C-C-C-C-C-C-C-C-H
    | | | | | | | |
    H H H H H H H H
```

18.43 (a) 3-methylhexane (b) 2,2-dimethylbutane (c) 3-ethylhexane

18.45

```
(a)   CH3                 (b)   CH3
      |                         |
    CH3CHCH2CH3               CH3CHCHCH2CH3
                                   |
                                   CH2CH3
```

18.47 $C_7H_{16} + 11O_2 \longrightarrow 7CO_2 + 8H_2O$

18.49 $CH_3CH_2CH_2CH_2CH_3 + Cl_2 \longrightarrow CH_3CH_2CH_2CH_2CH_2Cl + HCl$

18.51 (a) 4-methyl-1-pentene (b) 3-methylbutyne

18.53

```
                                   Br
                                   |
CH3CH=CH2 + Br2 ⟶ CH3CHCH2Br
```

18.55

```
F F     F F     F F         F F F F F F
| |     | |     | |         | | | | | |
C=C  +  C=C  +  C=C   ⟶   -C-C-C-C-C-C-
| |     | |     | |         | | | | | |
F F     F F     F F         F F F F F F
```

18.57

```
         CH3                          CH3
         |                            |
  H     C      H               H     C      H
    \ //  \   /                  \  /  \\  /
     C     C                      C      C
     |     ||        ⟷            ||     |
     C     C                      C      C
    / \\  /  \                   /  \  //  \
  H     C      H               H     C      H
        |                            |
        H                            H
```

18.59

```
(a)  CH2CH3 on a benzene ring, with a second CH2CH3 meta to it
(b)  CH3 on a benzene ring, with Br ortho to it
(c)  Br on a benzene ring, with Br and Br at the two meta positions
```

18.61 (a) *m*-dibromobenzene (b) 1,2,3-tribromobenzene

18.63 (a) tertiary (b) primary (c) secondary

18.65 3-methyl-2-pentanol **18.67** (a) $CH_3CH_2CH_2CH_2OH + 6O_2 \longrightarrow 4CO_2 + 5H_2O$ (b) $CH_3CH_2CH_2CH_2OH + (O) \longrightarrow CH_3CH_2CH_2COOH + H_2O$ (c) $CH_3CHOHCH_2CH_3 + (O) \longrightarrow CH_3COCH_2CH_3 + H_2O$ (d) no reaction **18.69** (a) 2-methylbutanal (b) 3-methyl-2-butanone

18.71

```
(a)        CH3 O                (b)   O  CH3
           |   ||                     ||  |
   CH3CH2CHCH—C—H                 CH3C—C—CH2CH3
           |                              |
           CH3                            CH3
```

18.73 2,2-dimethylpropanoic acid **18.75**

```
   CH3  O
   |    ||
CH3CH—C—OH
```

18.77 $CH_3CH_2CH_2\overset{O}{\overset{\|}{C}}-OH + CH_3\underset{}{\overset{OH}{\overset{|}{C}}}HCH_3 \xrightarrow{H^+} CH_3CH_2CH_2\overset{O}{\overset{\|}{C}}-O-\underset{CH_3}{\underset{|}{\overset{CH_3}{\overset{|}{C}}}}H + H_2O$

18.79 $CH_3CH_2CH_2CH_2\overset{H}{\overset{|}{N}}-H + H_3O^+ \longrightarrow \left[CH_3CH_2CH_2CH_2\underset{H}{\underset{|}{\overset{H}{\overset{|}{N}}}}-H\right]^+ + H_2O$

18.81 $H-\overset{H}{\overset{|}{N}}-H + HO-\overset{O}{\overset{\|}{C}}CH_2CH_3 \xrightarrow{\Delta} CH_3CH_2\overset{O}{\overset{\|}{C}}-NH_2 + H_2O$

18.83 (a) $H-\underset{H}{\underset{|}{\overset{H}{\overset{|}{C}}}}-\underset{H}{\underset{|}{\overset{H}{\overset{|}{C}}}}-OH$ (b) $H-\underset{H}{\underset{|}{\overset{H}{\overset{|}{C}}}}-\overset{O}{\overset{\|}{C}}-H$

(c) $H-\underset{H}{\underset{|}{\overset{H}{\overset{|}{C}}}}-\overset{O}{\overset{\|}{C}}-OH$ (d) $H_2C=CH_2$ (each C bonded to two H)

18.85 (a) $C_6H_5-\overset{O}{\overset{\|}{C}}-H$ — aldehyde (b) $CH_3CH_2\underset{CH_3}{\underset{|}{C}}H-\overset{O}{\overset{\|}{C}}-O-H$ — carboxylic acid

(c) benzene ring bearing NH_2 (amine) and $-\overset{}{C}(=O)-O-H$ (carboxylic acid)

18.87 (a) 2,3-dimethyl-2-butene (b) 3,5-dimethyloctane (c) 1,3,5-triethylbenzene

18.89

(a) $CH_3\underset{CH_3}{\underset{|}{\overset{CH_3}{\overset{|}{C}}}}CH_2\overset{CH_2CH_3}{\overset{|}{C}}HCH_2CH_2CH_2CH_3$ (b) $CH_3\overset{CH_3}{\overset{|}{C}}=\overset{CH_3}{\overset{|}{C}}CH_2CH_2CH_2CH_3$

(c) benzene ring with three CH_2CH_3 groups on adjacent carbons (1,2,3-triethylbenzene)

18.91 $CH_3CH_2CH_2CH_2CH_2Cl$

$CH_3CH_2CH_2\underset{Cl}{\underset{|}{C}}HCH_3$

$CH_3CH_2\underset{Cl}{\underset{|}{C}}HCH_2CH_3$

18.93 $CH_3CH=CHCH_3 + Br_2 \longrightarrow CH_3\underset{Br}{\underset{|}{C}}H\underset{Br}{\underset{|}{C}}HCH_3$

18.95 $3CH_3CH_2\underset{CH_3}{\underset{|}{\overset{CH_3}{\overset{|}{C}}}}CH_2OH + 2Cr_2O_7^{2-} + 16H^+ \longrightarrow$

$3CH_3CH_2\underset{CH_3}{\underset{|}{\overset{CH_3}{\overset{|}{C}}}}-\overset{O}{\overset{\|}{C}}-OH + 4Cr^{3+} + 11H_2O$

18.97 (a) ${}^4CH_3{}^3\underset{CH_3}{\underset{|}{\overset{CH_3}{\overset{|}{C}}}}{}^2CH={}^1CH_2$ 3,3-dimethyl-1-butene

(b) ${}^5CH_3{}^4\overset{CH_3}{\overset{|}{C}}H{}^3CH_2{}^2\overset{O}{\overset{\|}{C}}{}^1CH_3$ 4-methyl-2-pentanone

18.99 $HO-\overset{O}{\overset{\|}{C}}-(CH_2)_2-\overset{O}{\overset{\|}{C}}-OH + H-\underset{H}{\underset{|}{N}}-(CH_2)_4-\underset{H}{\underset{|}{N}}-H \longrightarrow$

$-\overset{O}{\overset{\|}{C}}-(CH_2)_2-\overset{O}{\overset{\|}{C}}-\underset{H}{\underset{|}{N}}-(CH_2)_4-\underset{H}{\underset{|}{N}}-\overset{O}{\overset{\|}{C}}-(CH_2)_2-\overset{O}{\overset{\|}{C}}-\underset{H}{\underset{|}{N}}-(CH_2)_4-\underset{H}{\underset{|}{N}}-$

Chapter 19

19.35 (a) alanine $CH_3\underset{NH_2}{\underset{|}{C}}HCOOH$ (b) β-amino acid $\underset{NH_2}{\underset{|}{C}}H_2CH_2COOH$

19.37 $H-\underset{H}{\underset{|}{N}}-\underset{CH_3}{\underset{|}{\overset{H}{\overset{|}{C}}}}-\overset{O}{\overset{\|}{C}}-OH + H-\underset{H}{\underset{|}{N}}-\underset{CH_2C_6H_5}{\underset{|}{\overset{H}{\overset{|}{C}}}}-\overset{O}{\overset{\|}{C}}-OH \longrightarrow$

$H-\underset{H}{\underset{|}{N}}-\underset{CH_3}{\underset{|}{\overset{H}{\overset{|}{C}}}}-\overset{O}{\overset{\|}{C}}-\underset{H}{\underset{|}{N}}-\underset{CH_2C_6H_5}{\underset{|}{\overset{H}{\overset{|}{C}}}}-\overset{O}{\overset{\|}{C}}-OH + H_2O$

```
     H  O               H  O
     |  ||              |  ||
H—N—C—C—OH + H—N—C—C—OH ⟶
  |  |                |  |
  H  |                H  CH3
  H—C—(C6H5)
     |
     H

           H  O     H  O
           |  ||    |  ||
      H—N—C—C—N—C—C—OH + H2O
        |  |     |  |
        H  |     H  CH3
        H—C—(C6H5)
           |
           H
```

19.39 Glycine (Gly), alanine (Ala), methionine (Met), and leucine (Leu)

Gly−Ala−Met−Leu	Ala−Gly−Met−Leu
Gly−Ala−Leu−Met	Ala−Gly−Leu−Met
Gly−Met−Ala−Leu	Ala−Leu−Gly−Met
Gly−Met−Leu−Ala	Ala−Leu−Met−Gly
Gly−Leu−Ala−Met	Ala−Met−Leu−Gly
Gly−Leu−Met−Ala	Ala−Met−Gly−Leu
Met−Gly−Ala−Leu	Leu−Met−Gly−Ala
Met−Gly−Leu−Ala	Leu−Met−Ala−Gly
Met−Ala−Gly−Leu	Leu−Gly−Met−Ala
Met−Ala−Leu−Gly	Leu−Gly−Ala−Met
Met−Leu−Ala−Gly	Leu−Ala−Gly−Met
Met−Leu−Gly−Ala	Leu−Ala−Met−Gly

The amine groups are on the left and the carboxylic acid groups are on the right.

19.41 aldoses: 5-carbon 6-carbon

```
   5-carbon          6-carbon
      O                 O
      ||                ||
      C—H               C—H
      |                 |
   H—C—OH            H—C—OH
      |                 |
   H—C—OH            H—C—OH
      |                 |
   H—C—OH            H—C—OH
      |                 |
   H—C—OH            H—C—OH
      |                 |
      H              H—C—OH
                        |
                        H
```

19.43 $C_{12}H_{22}O_{11} + H_2O \xrightarrow{H^+} C_6H_{12}O_6 + C_6H_{12}O_6$

Lactose Glucose Galactose

19.45 T−A−C **19.47** One is: GCAGGAUUU.
19.49 CGTCCTAAA

19.51

```
   H        O
   |        ||
H—C—O—C—(CH2)10CH3
   |        O
   |        ||
H—C—O—C—(CH2)16CH3
   |        O
   |        ||
H—C—O—C—(CH2)7CH=CH(CH2)7CH3
   |
   H
```

19.53

```
                                      H      O
                                      |      ||
                                   H—C—O—C—(CH2)10CH3
                                      |      O
                                      |      ||
                                   H—C—O—C—(CH2)16CH3
      CH3                 O           |
      |                   ||          |
CH3—N+—CH2CH2—O—P—O—C—H
      |                   |           |
      CH3                 O−          H
```

19.55 Ala−Gly−Val, Ala−Val−Gly, Gly−Ala−Val, Gly−Val−Ala, Val−Gly−Ala, Val−Ala−Gly

19.57 The disulfide bond forms between the cysteine units

```
Glu—Gln—Cys—Thr
         |    |
         S   His
         |    |
         S   Ser
         |    |
His—Leu—Cys—Gly
```

19.59 Asn−Glu−Val

Appendix A **1** (a) 1.15×10^{10} (b) 3.170×10^6 (c) 3.114×10^6 (d) 4.5×10^8 (e) 3.7×10^8
2 (a) 9.7×10^1 (b) 5.0×10^{-10} (c) 5.0×10^8 (d) 1.4×10^{12} (e) 5.0×10^{-3} **3** (a) $\frac{V}{T}$ (b) $\frac{PV}{nR}$
(c) $\frac{V_2}{V_1} \times T_1$ (d) $\frac{\text{Mass}}{\text{Density}}$
(e) 9/5°C + 32 **4** volume is 1 L, pressure is 4 atm

Appendix B **1** (a) 5.88×10^{-3} (b) 2.7×10^7 (c) 4×10^9 (d) 2.0×10^2 (e) 3.3×10^8 (f) 2.5 (g) -1.0×10^1

Answers to Practice Exams

Chapter 1 **1.** a **2.** c **3.** e **4.** d **5.** a **6.** b **7.** a **8.** b **9.** c **10.** e

Chapter 2 **1.** c **2.** a **3.** c **4.** c **5.** d **6.** b **7.** c **8.** a **9.** e **10.** e **11.** b **12.** a **13.** a **14.** d **15.** a

Chapter 3 **1.** b **2.** a **3.** b **4.** a **5.** d **6.** a **7.** d **8.** d **9.** c **10.** c **11.** b **12.** b **13.** a **14.** d **15.** e

Chapter 4 **1.** c **2.** b **3.** c **4.** a **5.** b **6.** e **7.** c **8.** d **9.** a **10.** b **11.** e **12.** c **13.** d **14.** b **15.** a

Chapter 5 **1.** c **2.** c **3.** e **4.** b **5.** d **6.** b **7.** d **8.** a **9.** d **10.** c **11.** b **12.** e **13.** a **14.** e **15.** c **16.** c **17.** d **18.** d **19.** c **20.** d

Chapter 6 **1.** b **2.** d **3.** a **4.** c **5.** b **6.** c **7.** a **8.** d **9.** c **10.** a **11.** c **12.** a **13.** e **14.** c **15.** e

Chapter 7 **1.** c **2.** d **3.** a **4.** d **5.** d **6.** e **7.** b **8.** d **9.** e **10.** b **11.** d **12.** c **13.** d **14.** d **15.** d

Chapter 8 **1.** e **2.** a **3.** c **4.** e **5.** e **6.** a **7.** e **8.** d **9.** b **10.** c **11.** a **12.** a **13.** c **14.** c **15.** b

Chapter 9 **1.** e **2.** b **3.** d **4.** b **5.** c **6.** b **7.** a **8.** e **9.** e **10.** d

Chapter 10 **1.** a **2.** d **3.** e **4.** e **5.** b **6.** c **7.** e **8.** d **9.** a **10.** c **11.** d **12.** a **13.** e **14.** b **15.** b

Chapter 11 **1.** d **2.** c **3.** d **4.** c **5.** a **6.** d **7.** c **8.** b **9.** c **10.** a **11.** a **12.** d **13.** d **14.** a **15.** d

Chapter 12 **1.** a **2.** c **3.** e **4.** e **5.** a **6.** a **7.** a **8.** e **9.** e **10.** e **11.** a **12.** b **13.** b **14.** a **15.** c

Chapter 13 **1.** d **2.** c **3.** b **4.** d **5.** a **6.** b **7.** a **8.** e **9.** b **10.** e **11.** b **12.** b **13.** a **14.** c **15.** d

Chapter 14 **1.** a **2.** e **3.** a **4.** c **5.** d **6.** a **7.** c **8.** c **9.** b **10.** b **11.** c **12.** c **13.** b **14.** a **15.** a

Chapter 15 **1.** b **2.** c **3.** c **4.** a **5.** c **6.** a **7.** a **8.** b **9.** c **10.** b **11.** a **12.** c **13.** d **14.** b **15.** d

Chapter 16 **1.** c **2.** c **3.** c **4.** a **5.** d **6.** c **7.** c **8.** a **9.** b **10.** a **11.** e **12.** a **13.** a **14.** d **15.** b

Chapter 17 **1.** a **2.** b **3.** b **4.** d **5.** b **6.** e **7.** b **8.** d **9.** d **10.** d **11.** e **12.** b **13.** b **14.** e **15.** c

Chapter 18 **1.** c **2.** b **3.** b **4.** e **5.** d **6.** b **7.** c **8.** e **9.** d **10.** a

Chapter 19 **1.** a **2.** b **3.** d **4.** b **5.** c **6.** e **7.** b **8.** b

Glossary

The number given in parentheses at the end of a definition indicates the section (or page) where the term was introduced. In a few cases, several sections are indicated.

acid A compound that produces hydrogen ions (H^+) when it is dissolved in water. In the Brønsted–Lowry theory, an acid is a proton donor. (p. 138, 6.9, 15.1, 15.3)

acid–base indicator A weak acid whose solution will change color within a small pH range and indicate the pH of the solution by the color. (15.8)

acid strength A measure of an acid's ease of donating a proton to a base, such as water. (15.4)

activation energy The minimum energy of a collision that leads to a reaction. (14.1)

activity series A list of metallic elements (plus hydrogen) ordered by their relative activities in single-replacement reactions to form a given ion in aqueous solution. (6.5)

actual yield The mass of a product that is actually recovered. (8.6)

addition polymer A polymer formed by linking many monomer molecules through addition reactions. (18.5)

addition reaction A reaction in which parts of a reactant molecule are added to each carbon atom of a carbon–carbon multiple bond; a carbon–carbon double bond becomes a single bond and a carbon–carbon triple bond becomes a double bond. (18.4)

alcohol A compound ROH that contains an —OH group bonded to a hydrocarbon group. (18.7)

aldehyde A compound RCHO that contains a carbonyl group bonded to a hydrocarbon group R and to a hydrogen atom. (18.8)

alkane A hydrocarbon containing only single bonds and having the general molecular formula C_nH_{2n+2}. (18.3)

alkene A hydrocarbon containing a carbon–carbon double bond and having the general formula C_nH_{2n}. (18.4)

alkyl group A group of atoms obtained by removing one hydrogen atom from an alkane. (18.3)

alkyne A hydrocarbon containing a carbon–carbon triple bond and having the general formula C_nH_{2n-2}. (18.4)

alpha emission Emission of a ^4_2He nucleus (called an alpha particle) from a decaying nucleus. (17.2)

alpha particle A ^4_2He nucleus. (17.2)

amide A compound derived from ammonia or an amine and a carboxylic acid by elimination of water. (18.11)

amide bond See *peptide bond.*

amine A compound that we can regard as structurally derived from ammonia, NH_3, through the replacement of one or more hydrogen atoms by hydrocarbon groups. (18.10)

amino acid An organic molecule that contains an amine group and a carboxyl group. (19.3)

amorphous solid A solid that lacks order and does not have a well-defined arrangement of structural units. (12.6)

amphoteric substance A substance that can behave as both an acid and a base. (15.5)

anion A negatively charged ion. (4.8)

anode The electrode at which oxidation occurs. (16.5)

aromatic hydrocarbon A hydrocarbon that has a structure based on the benzene ring. (18.6)

atmosphere (atm) A unit of pressure equal to exactly 760 mmHg. (11.2)

atom An extremely small, chemically indivisible particle. (4.1)

atomic mass unit (amu) Exactly one-twelfth the mass of a carbon-12 isotope. (4.3)

atomic number The number of protons in the nucleus. (4.2)

atomic solid A solid containing atoms from a single nonmetallic element. (12.6)

atomic theory An explanation of the structure and chemical reactions of matter in terms of atoms. (4.1)

atomic weight The weighted average mass (expressed in atomic mass units) of an atom of the naturally occurring element. (4.3)

Avogadro's law The law stating that equal volumes of any two gases at the same temperature and pressure contain equal moles of molecules. (11.7)

Avogadro's number The number of atoms in exactly 12 g of carbon-12. It equals 6.022×10^{23}. (7.2)

balanced chemical equation A chemical equation in which the number of atoms of each element is the same on both sides of the arrow. (6.3)

barometer A device used to measure atmospheric pressure. (11.2)

base A compound that produces hydroxide ions (OH^-) when it is dissolved in water. In the Brønsted–Lowry theory, a base is a proton acceptor. (6.9, 15.1, 15.3)

battery A single voltaic cell or a series of voltaic cells. (16.7)

beta emission Emission of a high-speed electron (called a beta particle) from a decaying nucleus. (17.2)

beta particle An electron ejected from a nucleus. (17.2)

binary acid An acid solution that forms when you dissolve a binary molecular compound of hydrogen and another nonmetallic element in water. (5.6)

binary ionic compound A compound composed of ions from only two elements. (5.1)

binary molecular compound A molecular compound composed of only two elements. (p. 134)

boiling point The temperature at which the vapor pressure of a liquid equals the atmospheric pressure. (12.3)

boiling-point elevation (ΔT_b) The colligative property of a solution equal to the boiling point of the solution minus the boiling point of the pure solvent. (13.10)

bond length The normal distance between nuclei whose atoms form a bond in a molecule. (10.6)

bonding pair An electron pair that is shared between two atoms. (10.3)

Boyle's law The law stating that the volume of a fixed amount of a gas at a given temperature is inversely proportional to the applied pressure. (11.4)

Brønsted–Lowry acid A proton (H^+) donor. (15.3)

Brønsted–Lowry base A proton (H^+) acceptor. (15.3)

buffer solution A solution that can resist changes in pH when limited amounts of acid or base are added to it. (15.9)

calorie (cal) An energy unit defined as exactly 4.184 J; originally defined as the quantity of energy needed to raise the temperature of 1 g of water by 1°C. (3.6)

carbohydrate A substance that is either a polyhydroxy aldehyde or a polyhydroxy ketone; a substance that yields such compounds if it hydrolyzes. (p. 630)

carbonyl compound A compound containing the carbonyl, or C=O, group. (18.8)

carboxylic acid A compound containing the carboxyl group, —COOH. (18.9)

catalyst A substance that causes a chemical reaction to speed up, even though it is not consumed by the reaction. A catalyst increases the rate of a chemical reaction without being consumed by it. (6.2, 14.2)

cathode The electrode at which reduction occurs. (16.5)

cation A positively charged ion. (4.8)

Celsius scale The temperature scale for general use in much of the world and for scientific use worldwide. On this scale, the freezing point of water is 0°C (zero degrees Celsius), and the boiling point of water at normal barometric pressure is 100°C. (2.6)

chain reaction A self-sustaining series of nuclear fissions caused by the absorption of neutrons from previous nuclear fissions. (17.8)

Charles's law The law stating that the volume of a fixed amount of gas is directly proportional to its temperature as long as the pressure is kept constant. (11.5)

chemical change A change in which one or more kinds of matter are transformed into one or more new kinds of matter. (3.2)

chemical equation A symbolic way of expressing a chemical reaction. (6.2)

chemical equilibrium A dynamic state in which the rates of the forward and the reverse reactions have become equal. (14.3)

chemical property A characteristic of a material involving its chemical change. (3.2)

chemical reaction A rearrangement of the atoms present in reacting substances to give new chemical combinations in the substances formed by the reaction. (4.1) See *chemical change.*

chemistry The science concerned with describing and explaining the different forms of matter and the chemical reactions (and accompanying energy changes) of matter. (1.1)

coefficient The number in front of the formula in a chemical equation. (6.2)

colligative property A property of a solution that depends only on the number of solute particles (molecules and ions) in a given quantity of solution and not on the particular characteristics of the solute particles. (p. 420)

collision theory The theory that reacting molecules must come so close that they collide, and the energy of the collision must be greater than a certain minimum value. (14.1)

combination reaction A reaction in which two substances chemically combine to form a third. (6.4)

combined gas law A combination of Boyle's law and Charles's law. (11.6)

combustion reaction A reaction of a substance with either pure oxygen or oxygen in the air that causes the rapid release of heat and the appearance of a flame. (6.10)

complementary base pairing The pairing (through hydrogen bonding) of certain bases in DNA (or RNA) with other bases. (19.9)

compound A type of matter composed of two or more elements chemically combined. Compounds consist of atoms of two or more elements chemically combined in fixed proportions. (3.4, 4.1)

concentrated solution A solution whose concentration is relatively high. (p. 406)

condensation The change of a gas to a liquid. (12.1)

condensation polymer A polymer formed from monomer molecules by condensation reactions. (18.11)

condensed structural formula A structural formula that uses established abbreviations for various groups of atoms. (18.2)

conjugate acid–base pair Two substances (one an acid and the other a base) in an acid–base reaction that differ by the gain or loss of a proton. (15.3)

conversion factor A factor equal to 1 that converts a quantity in one unit to the same quantity in another unit. (2.7)

cosmic rays Radiation that originates from nuclear reactions in the sun and other stars. (17.3)

covalent bond A chemical bond formed by the sharing of a pair of electrons between two atoms. (10.3)

covalent network solid A solid containing atoms held in large networks or chains by covalent bonds. (12.6)

critical mass The minimum mass that will allow fission to become a chain reaction. (17.8)

crystalline solid A solid that is composed of one or more crystals with each crystal having a well-defined, ordered arrangement of structural units. (12.6)

curie The unit describing the amount of radioactivity; 37 billion disintegrations per second. (17.5)

Dalton's law of partial pressures The law stating that the total pressure exerted by a gaseous mixture is the sum of the partial pressures of the components of the mixture. (11.9)

decomposition reaction A reaction in which a single compound breaks up into two or more other substances. (6.4)

denaturation The loss of a protein's three-dimensional shape through the unfolding and uncoiling of the protein as the weak forces that hold the protein in its normal three-dimensional shape (such as hydrogen bonding) are broken. (19.5)

density The mass of a substance per unit volume. (2.8)

deoxyribonucleic acid (DNA) A polymer of deoxyribonucleotides; a polynucleotide. (19.9)

deposition The change of a gas to a solid. (12.1)

dilute solution A solution whose concentration is relatively low. (p. 406)

dimensional analysis A general problem-solving method that uses the units of quantities to help one decide how to set up the problem. (2.7)

dipole–dipole force An attractive intermolecular force resulting from the interaction of the positive end of one molecule with the negative end of another. (12.5)

diprotic acid An acid capable of transferring two protons per acid molecule to a base. (15.4)

dispersion forces The weak attractive forces between temporarily polarized atoms (or molecules) caused by the varying positions of the electrons during their motion about nuclei. (12.5)

double bond A covalent bond formed by the sharing of two pairs of electrons between two atoms. (10.3)

double-replacement reaction A type of reaction in which two compounds exchange parts to form two new compounds. (6.6)

electricity The flow of electrons through a conductor. (p. 524)

electrochemical cell An apparatus in which a chemical reaction either generates or uses an electric current. (p. 525)

electrochemistry The study of the conversion of stored chemical energy into electrical energy and vice versa. (p. 524)

electrolytic cell An electrochemical cell in which an external electric current drives a nonspontaneous oxidation–reduction reaction. (16.6)

electron A very light, negatively charged subatomic particle. (4.2)

electron configuration A particular distribution of electrons among the different subshells of an atom. (9.4)

electron-dot formula A formula that uses pairs of dots to represent covalent bonds, as well as dots to represent the electrons on individual atoms; a *Lewis formula* (10.3)

electron-dot symbol A symbol of an atom or ion in which the valence-shell electrons are represented by dots placed around the letter symbol of the element. Also known as a *Lewis symbol.* (10.2)

electron shell A set of orbitals of approximately the same size and energy. (9.3)

electronegativity A measure of the ability of an atom in a covalent bond to draw bonding electrons to itself. (10.4)

element A substance that cannot be decomposed by any chemical reaction into simpler substances; a type of matter composed of only one kind of atom, which always has certain specific properties. (3.4, 4.1)

empirical formula The chemical formula with the smallest possible whole-number subscripts. (7.8)

endothermic reaction A reaction that absorbs heat. (14.8)

energy The potential or capacity to move matter. (3.6)

energy level One of the allowed energy values that an electron can have. (9.2)

equilibrium constant The value obtained from the equilibrium expression when equilibrium concentrations are substituted. (15.4)

equivalent of an acid The amount of acid that yields one mole of hydrogen ion, H^+. (13.8)

equivalent of a base The amount of base that yields one mole of hydroxide ion, OH^-. (13.8)

ester A compound formed from a carboxylic acid, RCOOH, and an alcohol, R′OH, with the general formula RCOOR′. (18.9)

exothermic reaction A reaction that liberates heat. (14.8)

experiment The observation of some natural phenomenon under controlled circumstances. (1.3)

Fahrenheit scale The temperature scale in common use in the United States in which the freezing point of water is 32°F and the boiling point of water at normal barometric pressure is 212°F. (2.6)

family See *group*.

formula unit The group of atoms or ions explicitly symbolized in the formula of a substance. (4.8)

formula weight The sum of the atomic weights of all the atoms in a formula unit of a substance. (7.1)

free radical A molecule with an unpaired valence electron. (17.7)

freezing The change of a liquid to a solid. (12.1)

freezing point The temperature at which the liquid state of a substance freezes, or changes to the solid state. (12.1)

freezing-point depression (ΔT_f) The colligative property of a solution equal to the freezing point of the pure solvent minus the freezing point of the solution. (13.10)

functional group A reactive portion of a molecule that undergoes predictable reactions. (p. 591)

fusion See *melting*.

gamma emission Emission of pure energy (a photon) with a single wavelength from an excited nucleus. (17.2)

gas The form of matter having indefinite shape and volume (taking the shape and volume of its container). (3.1, 12.1)

group The elements in any one column of the periodic table. Also called a *family* of elements. (4.4)

half-cell That portion of an electrochemical cell in which a half-reaction takes place. (16.5)

half-life The time required for half of any amount of a radioactive nuclide to decay. (17.4)

heat of fusion See *molar heat of fusion.*

heat of vaporization See *molar heat of vaporization.*

heterogeneous equilibrium An equilibrium in which more than one state of matter is involved. (14.7)

heterogeneous mixture A mixture that consists of physically distinct parts with different properties. (3.3)

homogeneous equilibrium An equilibrium in which all substances are in a single state of matter. (14.7)

homogeneous mixture A mixture that is uniform in its properties throughout. Also called a *solution.* (3.3)

Hund's rule The rule stating that electrons do not pair up in an orbital unless all orbitals in a subshell already contain one electron. (9.4)

hydrocarbon A compound containing only carbon and hydrogen. (p. 571)

hydrogen bond A dipole–dipole attractive force between two polar molecules containing a hydrogen atom covalently bonded to an atom of nitrogen, oxygen, or fluorine. (12.5)

hydronium ion An ion whose formula is H_3O^+. (15.3)

hypothesis A tentative explanation of a law or regularity of nature. (1.3)

ideal gas A gas that obeys a simple mathematical relationship between its pressure, volume, temperature, and amount. (11.3)

ideal gas law A combination of the laws of Boyle, Charles, and Avogadro. (11.8)

immiscible liquids Two liquids that do not mix together or dissolve in one another in significant amounts, but rather tend to separate into two distinct layers. (13.1)

insoluble A term applied to compounds that do not dissolve appreciably in a liquid. (6.7)

ion An electrically charged particle obtained from an atom or chemically bonded group of atoms by adding or removing electrons. (4.5)

ion-product constant for water (K_w) The equilibrium value of the ion product $[H_3O^+][OH^-]$. (15.5)

ionic bond The strong attractive force that exists between a positive ion and a negative ion in an ionic compound. (10.1)

ionic solid A solid consisting of ions. (12.6)

ionization energy The energy required to remove an electron from an atom. (9.9)

isomers Compounds that have the same molecular formula but different structural formulas. (18.2)

isotopes Atoms whose nuclei have the same atomic number but different mass numbers. (4.2)

joule (J) The SI unit of energy. (3.6)

kelvin (K) The SI base unit of temperature. (2.6)

Kelvin scale An absolute temperature scale on which the lowest temperature is zero. (2.6)

ketone A compound RCOR′ that contains a carbonyl group bonded to two hydrocarbon groups. (18.8)

kilogram The SI base unit of mass. (2.3)

kinetic energy The energy associated with an object by virtue of its motion. (3.6)

kinetic molecular theory of gases The theory that provides an explanation for the behavior of gases in terms of molecules. (11.3)

law A simple generalization from experiment. (1.3)

law of conservation of energy The law stating that energy may be converted from one form to another, but the total quantity of energy remains constant (the total energy is conserved). (3.8)

law of conservation of mass The law stating that the total mass remains constant during a chemical reaction. (3.5)

law of constant composition See *law of definite proportions.*

law of definite proportions The law stating that a pure compound, whatever its source, always contains definite or constant proportions of the elements by mass. (3.4)

law of mass action The law stating each reaction has an equilibrium constant with its own characteristic value at a given temperature. (14.4)

Le Chatelier's principle The principle that when a reaction mixture at equilibrium is disturbed by a change in volume, temperature, or the concentration of one of the components, the system will shift its equilibrium composition in a way that tends to counteract the change in an attempt to reestablish the equilibrium. (14.8)

Lewis formula See *electron-dot formula.*

Lewis symbol See *electron-dot symbol.*

limiting reactant The reactant that is used up when a reaction goes to completion even though other reactants are not consumed completely. (p. 233)

lipid A biological substance belonging to one of several structurally different classes of substances that are soluble in organic solvents. (p. 642)

liquid The form of matter having indefinite shape (taking the shape of its container), but having a definite volume. (3.1, 12.1)

liter The derived SI unit of volume equal to a cubic decimeter. (2.3)

logarithm of a number The power to which 10 must be raised to equal that number. (15.7)

lone pair An electron pair that is on one atom and that is not involved in bonding. Also called a *nonbonding pair.* (10.3)

main group An A group of elements in the periodic table. Also called a *representative group.* (4.4)

manometer A device that measures a gas's pressure in a container. (11.2)

mass The quantity of matter an object contains. (2.3)

mass number The total number of protons and neutrons in a nucleus. (4.2)

mass percent of solute A unit that expresses the concentration of solute as a percentage of the mass of solution. (13.7)

mass percentage The grams of an element in 100 grams of a compound. (7.6)

material A particular kind of matter. (3.2)

matter The material things around you. (1.1)

melting The change of a solid to a liquid. Also called *fusion.* (12.1)

melting point The temperature at which a solid melts. (12.1)

metal A material that has a characteristic luster or shine and that is a relatively good conductor of heat and electricity. (4.4)

metallic solid A solid containing atoms from a metallic element. (12.6)

metalloid An element that has properties intermediate between those of a metal and a nonmetal. (4.4)

meter (m) The SI base unit of length. (2.3)

metric system A system of units based on the decimal number system. (2.3)

millimeter of mercury (mmHg) A unit of pressure equal to that exerted by a column of mercury that is exactly 1 mm high. Also called a *torr.* (11.2)

miscible liquids Two liquids that mix completely in one another to form a solution, regardless of their proportions. (13.1)

mixture A material that can be separated by physical processes into two or more substances. (3.3)

molality A concentration unit equal to the moles of solute dissolved in a kilogram of solvent. (13.9)

molar heat of fusion The energy required to melt (fuse) one mole of a solid. (12.2)

molar heat of vaporization The energy required to vaporize one mole of a liquid. (12.2)

molar mass The mass of one mole. (7.3)

molarity A concentration unit equal to the moles of solute dissolved in a liter of solution. (13.8)

mole The quantity of a substance that contains Avogadro's number (6.022×10^{23}) of atoms, molecules, or formula units. (7.2)

molecular equation An equation in which each substance is written as if it were a molecular substance, even though it may actually exist as ions in solution. (15.2)

molecular formula A notation that uses atomic symbols with numeric subscripts to convey the exact number of atoms of the different elements that are in a molecule of a substance; a multiple of an empirical formula. (4.7, 7.9)

molecular solid A solid consisting of molecules. (12.6)

molecular weight The sum of the atomic weights of all the atoms in the formula of a molecule. (7.1)

molecule A specific group of atoms that are chemically bonded—that is, tightly connected by attractive forces—and electrically neutral. (4.5)

monatomic ion An ion consisting of only one atom. (4.8)

monomer A compound used to prepare a polymer; the monomer gives rise to the polymer's repeating unit. (18.5)

monoprotic acid An acid that is capable of transferring only one proton per acid molecule to a base. (15.4)

monosaccharide A carbohydrate that cannot be broken down by hydrolysis into simpler carbohydrates. (19.6)

net ionic equation An equation that shows only the ions that take part in a reaction. (15.2)

neutralization reaction The reaction that occurs between an acid and a base with the formation of an ionic compound and usually water. (6.9)

neutron An electrically neutral subatomic particle found in a nucleus. (4.2)

noble gases The Group VIIIA elements. (9.5)

nonbonding pair See *lone pair.*

nonmetal An element that does not exhibit the properties of a metal. (4.4)

normal boiling point The temperature at which a liquid boils (at normal atmospheric pressure at sea level); the constant temperature at 1 atm pressure at which a liquid changes into a gas. (12.1)

normal freezing point The constant temperature at 1 atm pressure at which a liquid changes into a solid. (12.1)

normal melting point The constant temperature at 1 atm pressure at which a solid changes into a liquid. (12.1)

normality The number of equivalents of acid or base solution in one liter of solution. (13.8)

nuclear equation A symbolic representation of a nuclear reaction. (17.2)

nuclear fission A process in which a heavy nucleus splits to form two lighter and more stable nuclei. (p. 557)

nuclear fusion A process in which light nuclei combine to give heavier, more stable nuclei. (p. 557)

nucleons Neutrons and protons. (17.1)

nucleotide A molecule consisting of a sugar, which is either ribose or 2-deoxyribose, to which is attached a phosphate group and a nitrogen-containing base. (19.8)

nucleus A positively charged atomic core that takes up very little space in an atom but that has most of its mass. (4.2)

nuclide The nucleus of a specific isotope. (17.1)

octet rule The rule stating that atoms tend to lose or gain electrons when bonding to give eight electrons in their valence shells (two electrons for hydrogen). (10.2)

oligosaccharide An oligomer, or short polymer, containing from two to ten monosaccharide (simple sugar) units. (19.7)

orbital The wave function for an electron in an atom. (9.3)

organic chemistry The chemistry of carbon compounds. (p. 568)

osmosis The process of solvent flow through a semipermeable membrane in order to equalize the concentrations of solutes on the two sides of the membrane. (13.11)

osmotic pressure The pressure that must be exerted on a solution to stop osmosis. (13.11)

oxidation The loss of electrons. (16.1)

oxidation number The charge on an atom or monatomic ion, or the charge that an atom in a substance would have if the shared pair of electrons belonged to the more electronegative atom in a bond. (16.3)

oxidation–reduction reaction A reaction in which electrons are transferred from one reactant to another. (p. 503)

oxidizing agent A substance that causes the oxidation of another substance in an oxidation–reduction reaction. (16.1)

oxyacid A molecular substance containing hydrogen, oxygen, and another element that when added to water yields hydrogen ion (H^+) and the corresponding oxyanion. (5.7)

oxyanion A negatively charged polyatomic ion that contains an atom of some element plus one or more oxygen atoms. (5.4)

partial pressure The pressure exerted by each component of a mixture of gases. (11.9)

particle accelerator A device used to accelerate electrons, protons, and nuclei (such as alpha particles) to very high speeds. (17.3)

pascal (Pa) The SI unit of pressure. (11.2)

Pauli exclusion principle The principle stating that an orbital can hold no more than two electrons. (9.4)

peptide (amide) bond The bond linking the carbon atom and the nitrogen atom in the amide group of a protein molecule. (19.4)

percentage abundance The percentage of an isotope in a naturally occurring sample of an element. (4.2)

percentage composition The mass percentages of each element in a compound. (7.6)

percentage yield The actual yield (from experiment) expressed as a percentage of the theoretical yield (calculated). (8.6)

period The elements in any one horizontal row of the periodic table. (4.4)

periodic law The law stating that when the elements are arranged by atomic number, their chemical and physical properties vary periodically. (p. 275)

periodic table A table of the elements ordered by increasing atomic number into rows and columns so that the elements in any one column have similar or regularly changing properties. (4.4)

pH The negative of the logarithm of the hydronium-ion concentration. (15.7)

phospholipid A lipid compound that contains a phosphate group. (19.12)

photon A particle of light consisting of an extremely small packet of energy. (9.1)

physical change A change in the form of matter but not in its chemical identity. (3.2)

physical property A characteristic of a material that can be observed without changing its chemical identity. (3.2)

polar bond See *polar covalent bond.*

polar covalent bond A covalent bond in which the bonding electrons spend more time near one atom than the other. (10.4)

polar molecule A covalent molecule in which the centers of partial positive charge and partial negative charge are separated. (10.4)

polyatomic ion An ion consisting of two or more atoms chemically bonded but having an excess or a deficiency of electrons so that the entire unit has an electric charge. (4.8)

polymer A very large molecule consisting of many repeating units of low molecular weight. (18.5)

polysaccharide A polymer containing many monosaccharide (simple sugar) units. (19.7)

positron A particle with the same mass as an electron, but with a positive rather than a negative charge. (17.2)

positron emission Emission of a positron from a decaying nucleus. (17.2)

potential energy The energy an object has because of its position in a gravitational field, magnetic field, or similar environment that affects the object. (3.6)

precipitate A solid product that forms when two solutions are mixed. (6.7)

precipitation reaction A reaction in which a precipitate forms. (6.7)

pressure The force exerted on a unit area. (11.2)

primary structure The order, or sequence, of the amino acid units in a protein or peptide molecule. (19.4)

product A chemical substance that results from a chemical reaction. (p. 149)

protein A biological polymer whose monomer units are amino acids linked by peptide (amide) bonds. (19.4)

protein biosynthesis The building of protein molecules in a cell. (19.10)

proton A positively charged subatomic particle found in a nucleus. (4.2)

radioactive dating A technique for determining the age of certain old objects that relies on the known decay rate of radioactive nuclides in the object. (17.4)

radioactive decay The spontaneous disintegration of a nucleus. (17.2)

radioactive decay series A sequence of decay steps that continues until a stable nucleus is reached. (17.2)

radioactive tracer A radioactive isotope that is added to a chemical or biological system to trace the path of a nonradioactive isotope that is normally used by the system. (17.6)

radioactivity Radiation in the form of particles or energy coming from the nucleus of an atom undergoing spontaneous disintegration. (p. 540)

reactant A chemical substance involved at the start of a chemical reaction. (p. 149)

reducing agent A substance that causes the reduction of another substance in an oxidation–reduction reaction. (16.1)

reduction The gain of electrons. (16.1)

representative group See *main group.*

ribonucleic acid (RNA) A polymer of ribonucleotides. (19.10)

rounding The procedure of dropping nonsignificant digits in a calculation result and perhaps adjusting the last remaining digit upward. (2.5)

salt The ionic compound formed in a neutralization reaction. It contains the cation from the base and the anion from the acid. (6.9, 15.2)

saponification The base-catalyzed hydrolysis of an ester, especially a fat. (19.11)

saturated solution A solution that contains the maximum amount of solute in a solvent at a given temperature. (13.4)

scientific method The general process whereby scientific knowledge grows. (1.3)

scientific notation The representation of a number in the form $A \times 10^n$, where A is a number with a single nonzero digit to the left of the decimal point, and n is a whole number. (2.2)

self-ionization A process in which two identical molecules react to give ions. (15.5)

SI base units Those SI units from which all others are derived. (2.3)

SI prefix A prefix used in SI to indicate a power of ten. (2.3)

significant figures Those digits in a measured number (or in the result of calculations with measured numbers) that include all certain digits plus a final one that is somewhat uncertain. (2.4)

single bond A covalent bond formed by the sharing of a single pair of electrons between two atoms. (10.3)

single-replacement reaction A reaction in which one element reacts by replacing another element in a compound. (6.5)

solid The form of matter having a definite shape and volume. (3.1, 12.1)

solubility The maximum amount of substance that dissolves in a given volume of solvent at a specified temperature. (13.4)

soluble A term applied to compounds that dissolve in a liquid. (6.7)

solute The substance in a solution that is dissolved by a solvent; or if it is not clear which substance does the dissolving, the substance in the solution that is in smaller amount. (13.1)

solution A homogeneous mixture of substances. (3.3, 13.1) See *homogeneous mixture.*

solvent The substance in a solution that dissolves another substance; or if it is not clear which substance does the dissolving, it is the substance in the solution that is in greater amount. (13.1)

specific heat The quantity of heat required to raise the temperature of 1 g of a substance by 1°C. (3.7)

spectator ions Ions that do not take part in an ionic reaction. (15.2)

spontaneous process A physical or chemical change that occurs by itself. (p. 524)

standard temperature and pressure (STP) Arbitrarily chosen to be exactly 0°C and 1 atm pressure. (11.7)

states of matter The three forms of matter—solid, liquid, and gas. (3.1)

stoichiometric amounts The amounts of reactants that will be entirely consumed if the reaction goes to completion. You obtain the amounts by doing a stoichiometric calculation. (8.4)

stoichiometry The use of a balanced chemical equation to calculate the quantities of the reactants and products of a reaction. (p. 219)

strong acid An acid that is completely dissociated in water. (15.4)

strong base A base that is totally ionized in water. (15.4)

structural formula A chemical formula that shows what atoms are bonded to one another. (4.7)

sublimation The change of a solid to a gas. (12.1)

subshell A subset of orbitals of an electron shell, all of which have the same energy and similar shape. (9.3)

substance A material that cannot be separated into different materials by any physical process. (3.3)

substitution reaction A reaction in which one atom (or atom group) substitutes for another atom (or group) in a molecule. (18.3)

supersaturated solution A solution that contains more of a solute than is contained in the saturated solution. (13.4)

symbol A one- or two-letter abbreviation for an element. (3.4)

theoretical yield The maximum mass of a product that can be obtained from given amounts of reactants. (8.6)

theory A tested explanation of some body of natural phenomena. (1.3)

titration A procedure for determining the amount of one substance by determining the volume of a solution of known molarity of another substance that reacts completely with the first substance. (13.8)

torr See *millimeter of mercury.*

total ionic equation An equation that shows all ions in solution. (15.2)

transition-metal group A B group of elements in the periodic table. (4.4)

transmutation The change of one element into another. (17.3)

transuranium elements Elements with atomic numbers greater than that of uranium ($Z = 92$). (17.3)

triacylglycerol An ester of glycerol (a trihydroxy alcohol) and three fatty acids (long-chain carboxylic acids). (19.11)

triple bond A covalent bond formed by the sharing of three pairs of electrons between two atoms. (10.3)

unsaturated solution A solution that contains less solute in a given volume of solvent than is contained in the saturated solution (13.4)

valence electrons Outer electrons; those electrons corresponding to the largest shell quantum number, n. (9.7)

valence-shell electron-pair repulsion (VSEPR) model A model for predicting the shapes of molecules and ions, in which the valence-shell electron pairs are arranged about each atom in such a way as to keep electron pairs as far away from one another as possible. (10.7)

vapor pressure The pressure exerted by a liquid's vapor at equilibrium. (12.3)

vaporization The change of a liquid to a gas. (12.1)

volt An SI unit describing the energy per unit charge. (16.5)

voltaic (galvanic) cell A cell in which a spontaneous chemical reaction generates an electric current. (16.5)

volume The amount of space an object occupies. (2.3)

wavelength The distance between two peaks (or the distance between two troughs) of a wave. (9.1)

weak acid An acid that is not completely dissociated in water. (15.4)

weak base A base that is not totally ionized in water. (15.4)

Photo Credits

All photos by Sean Brady unless otherwise noted.

Chapter 1 opener M.P.L. Fogden/Bruce Coleman, Inc. **Figure 1.1(A)** Charlie Ott/Photo Researchers, Inc. **Figure 1.1(B)** Ken & Donna Dannen/Photo Researchers, Inc. **Figure 1.4** James Scherer **Figure 1.5** Paul Shambroom/Photo Researchers, Inc. **Figure 1.6** Northrup Grumman **Figure 1.7** James Scherer **Figure 1.10** James Scherer **Figure 1.11** Image generated with a NanoScope Scanning Probe Microscope, manufactured by Digital Instruments, Santa Barbara, CA **Figure 2.1** E.R. Degginger **Figure 2.7** James Scherer **Figure 2.12** Permission from Jeffrey L. Gage (photographer) and Michael L. Pollock, Ph.D., Center for Exercise Science, University of Florida. **Chapter 3 opener** Joanna McCarthy/The Image Bank **Figure 3.1** NASA **Figure 3.9** Bruce Iverson **Figure 3.13** James Scherer **Figure 3.15** Barth Falkenberg/Stock, Boston **Figure 3.17** Brownie Harris/The Stock Market **Figure 3.18** National Air and Space Museum/Smithsonian Institutio **Chapter 4 opener** Joseph Van Os/The Image Bank **Figure 4.1** Image generated with a NanoScope Scanning Probe Microscope, manufactured by Digital Instruments, Santa Barbara, CA **Figure 4.8** Edgar Fahs Smith Collection, Van Pelt-Dietrich Library, University of Pennsylvania **Figure 4.11** Newdoll Productions and Rick Sayre, Oakland, CA **Figure 4.17** James Scherer **Figure 4.19** IBM Corporation, Research Division, Almaden Research Center **Figure 5.2** R. Weldon, Gemological Institute of America **Figure 5.4** Dallas & John Heaton/Stock, Boston **Figure 5.4 inset** Ward's Natural Sciences **Figure 5.7** Newdoll Production and Rick Sayre, Oakland, CA **Figure 5.8** Courtesy, Schwarz Pharmaceuticals **Figure 5.10** Thomas Eisner and Daniel Aneshansley, Cornell University **Chapter 6 opener** James Scherer **Figure 6.1(A)** American Gas Association **Figure 6.2** James Scherer **Figure 6.5** Newdoll Productions and Rick Sayre, Oakland, CA **Figure 6.7** Newdoll Productions and Rick Sayre, Oakland, CA **Figure 6.8** James Scherer **Figure 6.10** James Scherer **Figure 6.11** James Scherer **Figure 6.13** R. Oberlander/Stock, Boston **Figure 6.16** Peter Turnley/Black Star **Figure 6.18** James Scherer **Chapter 7 opener** E.R. Degginger **Figure 7.3** Mark Boulton/Photo Researchers, Inc. **Figure 7.3 inset** Yoav Levy/Phototake **Figure 7.4** E.R. Degginger **Figure 7.8** Peter Bosted/Photo 20-20 **Figure 7.13** James Scherer **Chapter 8 opener** James Scherer **Figure 8.2** Photo courtesy of The M.W. Kellogg Company **Figure 8.4** Keith Kent/Science Photo Library/Photo Researchers, Inc. **Figure 8.6** Diane Schiumo/Fundamental Photographs **Figure 8.7** Science Photo Library/Photo Researchers, Inc. **Figure 8.11** Richard Megna/Fundamental Photographs **Figure 8.12(A)** Susan Leavines/Science Source/Photo Researchers, Inc. **Figure 8.13** James Scherer **Figure 8.15** Charlton Photos, Inc. **Figure 9.1** Richard Megna/Fundamental Photographs **Figure 9.10** Derek Berwin/The Image Bank **Figure 9.17** IBM Research Division, Almaden Research Center **Figure 9.18** Kerem Su/Tony Stone Images **Chapter 10 opener** Newdoll Productions and Rick Sayre, Oakland, CA **Figure 10.1** Newdoll Productions and Rick Sayre, Oakland, CA **Figure 10.7** Craig Hammell/The Stock Market **Figure 10.8** Newdoll Productions and Rick Sayre, Oakland, CA **Figure 10.10** Newdoll Productions and Rick Sayre, Oakland, CA **Figure 10.11** Newdoll Productions and Rick Sayre, Oakland, CA **Figure 10.12** Newdoll Productions and Rick Sayre, Oakland, CA **Figure 10.13** Newdoll Productions and Rick Sayre, Oakland, CA **Figure 10.14** Newdoll Productions and Rick Sayre, Oakland, CA **Figure 10.15(A)** From *The Double Helix* by James D. Watson, Atheneum Press, NY/Photo Researchers, Inc. **Figure 10.15(B)** Newdoll Productions and Rick Sayre, Oakland, CA **Figure 10.16(A)** Newdoll Productions and Rick Sayre, Oakland, CA **Figure 10.16(B)** R.M. Stroud et al. (Created by Julie Newdoll at the U.C.S.F. Computer Graphics Lab using Midas Plus) **Chapter 11 opener** Joe Azzara/The Image Bank **Figure 11.1** NASA **Figure 11.2** David R. Frazier/Photo Researchers, Inc. **Figure 11.9** Jim Goodwin/Photo Researchers, Inc. **Figure 11.13** The Granger Collection **Figure 11.14** NOAA **Chapter 12 opener** Eric Meola/The Image Bank **Figure 12.1(A)** W. Perry Conway/Tom Stack & Associates **Figure 12.1(B)** Jan Halaska/Photo Researchers, Inc. **Figure 12.1(C)** David Hiser/Tony Stone Images **Figure 12.2** Jan Halaska/Photo Researchers, Inc. **Figure 12.7(A)** H. Eisenbeiss/Photo Researchers, Inc. **Figure 12.7(B)** Rod Planck/Photo Researchers, Inc. **Figure 12.14** Courtesy of Diamond Promotion Service **Figure 12.15** Didier Givois/Agence Vandystadt/Photo Researchers, Inc. **Figure 12.17** Robert Frerck/Tony Stone Images **Figure 13.1** Eric Lessing/Art Resource **Chapter 14 opener** Ken O'Donoghue **Figure 14.1(left)** Ted Kerasote/Photo Researchers, Inc. **Figure 14.1(right)** Courtesy, Alpine Ascents International **Figure 14.7** Delphi Automotive Systems **Figure 14.8(A)** Billy E. Barnes/PhotoEdit **Figure 14.8(B)** Tony Dietrich/Tony Stone Images **Figure 14.14** Ken O'Donoghue **Figure 14.15** James Scherer **Figure 15.3** Ken O'Donoghue **Figure 15.10(A)** NYC Photo Archive/Fundamental Photographs **Figure 15.10(B)** Kristen Brochman/Fundamental Photographs **Chapter 16 opener** Richard Megna/Fundamental Photographs **Figure 16.1** Reuters/Bettmann **Figure 16.3** NASA **Figure 16.5** Will & Deni McIntyre/Photo Researchers, Inc. **Figure 16.7** Dennis MacDonald/PhotoEdit **Figure 16.7 inset** Breathalyzer © National Draeger, Pittsburgh, PA **Figure 16.12** Corbis-Bettmann **Chapter 17 opener** John Shaw/Tom Stack & Associates **Figure 17.1(A)** UPI/Bettmann **Figure 17.1(B)** Charles Gupton/Tony Stone Images **Figure 17.1(C)** Skeletal image courtesy DuPont Pharma Radiopharmaceuticals **Figure 17.3** Richard Megna/Fundamental Photographs **Figure 17.5** David Parker/Photo Researchers, Inc. **Figure 17.7** UPI/Bettmann **Figure 17.8** Courtesy of BICRON **Figure 17.10(A)** Hank Morgan/Science Source/Photo Researchers, Inc. **Figure 17.10(B)** NIH/Science Source/Photo Researchers, Inc. **Figure 17.14** Earl Roberge/Photo Researchers, Inc. **Figure 17.15** NASA **Figure 17.16(A)** Culver Pictures **Figure 17.16(B)** UPI/Bettmann **Chapter 18 opener** Harris Calorific Division, The Lincoln Electric Company **Figure 18.2** Newdoll Productions and Rick Sayre, Oakland, CA **Figure 18.4** Newdoll Productions and Rick Sayre, Oakland, CA **Figure 18.5** Newdoll Productions and Rick Sayre, Oakland, CA **Figure 18.6** James Scherer **Figure 18.7** Courtesy, American Petroleum Institute **Figure 18.8** James Scherer **Figure 18.9** Harris Calorific Division, The Lincoln Electric Company **Figure 18.12** Gary Milburn/Tom Stack & Associates **Figure 18.15** Tom Pantages **Figure 18.16** James Scherer **Figure 18.17** James Scherer **Chapter 19 opener** Newdoll Productions and Rick Sayre, Oakland, CA **Figure 19.7(A)** Arthur Beck/Science Source/Photo Researchers, Inc. **Figure 19.7(B)** Harry Rogers/National Audubon Society/Photo Researchers, Inc. **Figure 19.7(C)** Newdoll Productions and Rick Sayre, Oakland, CA **Figure 19.20** J. Koivula/Science Source/Photo Researchers, Inc.

Index

Some Common Cations and Anions

Cation Symbol	Name	Anion Symbol	Name
Li^+	Lithium ion	F^-	Fluoride ion
Na^+	Sodium ion	Cl^-	Chloride ion
K^+	Potassium ion	Br^-	Bromide ion
Rb^+	Rubidium ion	I^-	Iodide ion
Cs^+	Cesium ion	O^{2-}	Oxide ion
Mg^{2+}	Magnesium ion	S^{2-}	Sulfide ion
Ca^{2+}	Calcium ion	N^{3-}	Nitride ion
Ba^{2+}	Barium ion	P^{3-}	Phosphide ion
Al^{3+}	Aluminium ion		

Stock and Classical Names for Some Cations

Cation	Name According to	
Symbol	Stock System	Classical System
Fe^{2+}	Iron(II) ion	Ferrous ion
Fe^{3+}	Iron(III) ion	Ferric ion
Co^{2+}	Cobalt(II) ion	Cobaltous ion
Co^{3+}	Cobalt(III) ion	Cobaltic ion
Cu^+	Copper(I) ion	Cuprous ion
Cu^{2+}	Copper(II) ion	Cupric ion
Sn^{2+}	Tin(II) ion	Stannous ion
Sn^{4+}	Tin(IV) ion	Stannic ion
Pb^{2+}	Lead(II) ion	Plumbous ion
Pb^{4+}	Lead(IV) ion	Plumbic ion

Names of Some Common Polyatomic Ions

Ion Formula	Name	Ion Formula	Name
NH_4^+	Ammonium ion	HCO_3^-	Hydrogen carbonate ion†
OH^-	Hydroxide ion	PO_4^{3-}	Phosphate ion
CN^-	Cyanide ion	HPO_4^{2-}	Hydrogen phosphate ion
MnO_4^-	Permanganate ion	$H_2PO_4^-$	Dihydrogen phosphate ion
NO_2^-	Nitrite ion	ClO^-	Hypochlorite ion
NO_3^-	Nitrate ion	ClO_2^-	Chlorite ion
SO_3^{2-}	Sulfite ion	ClO_3^-	Chlorate ion
SO_4^{2-}	Sulfate ion	ClO_4^-	Perchlorate ion
HSO_4^-	Hydrogen sulfate ion*		
CO_3^{2-}	Carbonate ion		

* Sometimes called bisulfate ion

† Sometimes called bicarbonate ion